TABLE C-1 *Cumulative probabilities and percentiles of the standard normal distribution* (Continued)

(a)
Cumulative Probabilities

Entry is area *a* under the standard normal curve from $-\infty$ to $z(a)$

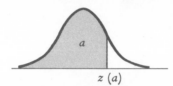

$z\,(a)$

z	.00	.01	.02	.03	.04	.05	.06	.07	.08	.09
2.5	.9938	.9940	.9941	.9943	.9945	.9946	.9948	.9949	.9951	.9952
2.6	.9953	.9955	.9956	.9957	.9959	.9960	.9961	.9962	.9963	.9964
2.7	.9965	.9966	.9967	.9968	.9969	.9970	.9971	.9972	.9973	.9974
2.8	.9974	.9975	.9976	.9977	.9977	.9978	.9979	.9979	.9980	.9981
2.9	.9981	.9982	.9982	.9983	.9984	.9984	.9985	.9985	.9986	.9986
3.0	.9987	.9987	.9987	.9988	.9988	.9989	.9989	.9989	.9990	.9990
3.1	.9990	.9991	.9991	.9991	.9992	.9992	.9992	.9992	.9993	.9993
3.2	.9993	.9993	.9994	.9994	.9994	.9994	.9994	.9995	.9995	.9995
3.3	.9995	.9995	.9995	.9996	.9996	.9996	.9996	.9996	.9996	.9997
3.4	.9997	.9997	.9997	.9997	.9997	.9997	.9997	.9997	.9997	.9998

(b)
Selected Percentiles

Entry is $z(a)$ where $P[Z \leq z(a)] = a$

a:	.10	.05	.025	.02	.01	.005	.001
z(a):	−1.282	−1.645	−1.960	−2.054	−2.326	−2.576	−3.090
a:	.90	.95	.975	.98	.99	.995	.999
z(a):	1.282	1.645	1.960	2.054	2.326	2.576	3.090

EXAMPLE: $P(Z \leq 1.96) = .9750$ so $z(.9750) = 1.96$.
TEXT REFERENCE: Use of this table is discussed on pp. 169–173.

Applied Statistics

SECOND EDITION

John Neter University of Georgia

William Wasserman Syracuse University

G. A. Whitmore McGill University

ALLYN AND BACON, INC.

Boston London Sydney Toronto

Library of Congress Cataloging in Publication Data

Neter, John.
  Applied statistics.

  Includes bibliographical references and index.
  1. Social sciences—Statistical methods. 2. Management—Statistical methods. 3. Economics—Statistical methods. 4. Statistics. I. Wasserman, William.
II. Whitmore, G. A. III. Title.
HA29.N437 1982   519.5   81-22758
ISBN 0-205-07613-0   AACR2

Printed in the United States of America.
10 9 8 7 6 5 4 3 2   86 85 84 83 82

To Dottie, Cathy, and Lonnie

Contents

UNIT FIVE   Linear Statistical Models   449

Preface

Applied Statistics, Second Edition, is written for students taking basic statistics courses in management, economics, and other social sciences, as well as for people already engaged in these fields who desire an introduction to statistical methods and their application. Our aim has been to offer a balanced presentation of fundamental statistical concepts and methods, along with practical advice on their effective application to real-world problems. Conceptual rigor is not sacrificed, however. The conceptual foundation of each subject is developed carefully up to that level needed for prudent and beneficial use of statistical methods in practice.

The second edition differs from the first in a number of important ways. We have sought to clarify the main track of presentation by placing major topics that are not essential to the basic development of statistical ideas in *Optional Topic* sections at the ends of chapters. In this way, they can be omitted without loss of continuity by instructors who do not have time to consider them but can be chosen selectively for coverage by other instructors where a topic fits the curricular requirements. Several important topics have been added or expanded, including exploratory data analysis, P-values, use of standardized test statistics for testing, use of the jackknife method for nonparametric estimation, coefficients of partial correlation, and simultaneous confidence interval estimation in the analysis of variance. To emphasize the reference nature of the materials on the χ^2, t, and F distributions, they have been placed in Appendix B. Also, the materials on the sampling distribution of $\bar{p}$ and inferences for a population proportion have been combined into one chapter.

In addition, we have made many revisions to improve the clarity of the presentation. For instance, the chapters on probability and random variables have been largely rewritten. The problems at the end of each chapter have been revised and expanded and grouped into three categories to facilitate their assignment by instructors. The *Problems* category includes basic problems and questions, while *Exercises* contain conceptual and theoretical questions and *Studies* contain comprehensive and more advanced problems.

As in the first edition of this Text, explanation of important principles and concepts always precedes discussion of detailed statistical procedures. Each new technique is illustrated by one or more examples drawn from real life. In addition, many case applications are presented, so that statistical concepts and methods can be understood in the context of their actual use. These case applications demonstrate the usefulness of statistical methods and statistically rigorous thinking in management, economics, and other social sciences.

The topical coverage of the text is broad, but it is in no sense a miscellany of statistical tools. Topics have been selected on the basis of two criteria: (1) their importance in real applications of statistics, and (2) their contribution to the development and understanding of material presented subsequently in the book. Units, chapters, and sections have been organized and sequenced in a way that always keeps the main track of the subject clear to the reader. Technical notes and secondary observations are presented in *Comments* sections. Major topics of interest that are not essential to the main development of statistical ideas are presented in *Optional Topic* sections and may be included in the basic statistics course at the instructor's discretion to meet particular course objectives. These sections always appear at the ends of chapters and can be omitted without loss of continuity. Extensive use is made of figures and tables. Important definitions and formulas are set out in a distinctive manner to aid in learning and to facilitate ready reference.

Use of this book requires knowledge of college-entrance algebra but not calculus. We believe that fundamental statistical ideas can be introduced with little mathematics, and that this approach in an introductory text leads to a fuller appreciation of the uses of statistical methods than a more mathematical approach. Mathematical demonstrations are included where they make a significant contribution to the reader's understanding of the subject. All such demonstrations are self-contained and illustrated by numerical examples. Occasionally, a mathematical demonstration is presented which requires calculus. These sections are marked "calculus needed" and may be omitted without loss of continuity. Those readers who will continue their study of statistics and, therefore, must become more familiar with the mathematical basis of the subject will find that this book provides a strong foundation upon which they can build.

We assist the reader with the mathematical and computational aspects of the subject in a number of ways. Notation is used only where needed, and then a uniform and straightforward notational system is employed throughout the Text. A mathematical review is included in Appendix A for readers desiring a brief summary of: (1) summation notation, (2) rules for exponents and logarithms, (3) set notation, operations, and rules, and (4) the basics of permutations and combinations. In preparing the text and problems, we have assumed that readers will use pocket calculators to assist with computational work. Because computers and statistical computer packages play a major role in real applications of statistics, we have sought to familiarize readers with ways in which they assist in statistical analyses (such as regression) and in data handling (such as tabulations by SPSS). Computer printout is presented in several units of the Text to illustrate its form and its usefulness in various practical contexts. We also discuss some output commonly presented by computer packages to facilitate analysis, such as P-values.

A large number of basic problems and questions are presented in *Problems* sections at the ends of chapters, while conceptual questions are presented in *Exercises* sections, and major, comprehensive problems in *Studies* sections. The problems, exercises, and studies each follow the sequence of topics in the text. Numerical answers for selected problems (identified by asterisks) are given at the end of the book to facilitate immediate checking by the reader. The problems, exercises, and studies have been designed to assist the reader's understanding of concepts and to enable him or her to obtain experience in applying statistical techniques in practical situations and in interpreting results of statistical investigations. Many different types of applications are employed in order to expose the reader to the rich and varied settings in which statistics is applied in real life. Intermediate computational results are presented in many problems in the form in which they might be available from computer or calculator output. A number of the problems and studies draw on the data set in Appendix D.

The book is divided into seven units, as shown in Figure I:

Unit One, Data (Chapters 1–3). However sophisticated a statistical procedure, its successful application depends on data. This unit is concerned with the acquisition, classification, and summarization of data. Appropriately, the unit emphasizes large data sets, computerized data handling, and exploratory data analysis. Since a large portion of day-to-day dealings with statistics for many persons is concerned with raw data and its examination by descriptive (as opposed to inferential) means, the chapters in this unit are designed to prepare the reader for these common encounters. The important role of computer data-processing packages, data banks, and retrieval systems is also recognized in this unit, and the groundwork is laid for data handling in subsequent chapters.

Unit Two, Probability (Chapters 4–7 and Appendix B). Probability theory is the foundation of statistical inference, so this unit comes next in the book. The chapters of this unit contain an integrated presentation of basic probability concepts, random variables, and common probability distributions utilized in applications. The last topic is covered in two chapters, the first presenting common discrete distributions and the second, common continuous distributions. Appendix B covers the χ^2, t, and F distributions and explains their relationships to one another and to the standard normal distribution. These three distributions are presented in the appendix as a handy reference so that they can be accessed readily from any place in the main Text where they might be encountered for the first time in a course. Extensive probability tables, prepared in a consistent fashion, are provided in Appendix C to support this and later units.

Unit Three, Estimation and Testing—I (Chapters 8–14). This unit opens with two chapters discussing sampling and the sampling distribution of $\overline{X}$. In contrast to conventional theoretical explanations, an empirical demonstration of the sampling distribution of $\overline{X}$ and the central limit theorem is presented first, using simulated, repeated sampling from an actual population. Then the key theorems are stated and explained. This approach allows the reader to anticipate the theoretical results on the basis of empirical experience, and it should lead to greater understanding of these conceptually important topics. Next, estimation and testing for a population mean are presented, followed by a discussion of the sampling distribution of $\overline{p}$ and inference procedures for a population proportion. Inference procedures in comparative studies involving two population means and proportions and inferences for population variances are discussed in the next-to-last chapter of this unit. The final

FIGURE I *Structure of **Applied Statistics**. Chart shows chapter interdependencies—arrows point to prerequisite chapters.*

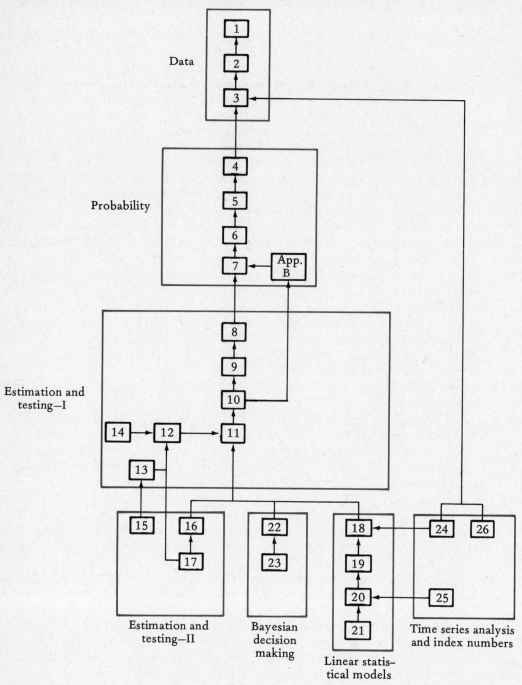

FIGURE I STRUCTURE OF *APPLIED STATISTICS* (Continued)

Unit One   Data

1. Data Acquisition
2. Data Classification and Exploratory Analysis
3. Data Summarization

Unit Two   Probability

4. Basic Probability Concepts
5. Random Variables
6. Common Discrete Probability Distributions
7. Common Continuous Probability Distributions

Unit Three   Estimation and Testing—I

8. Statistical Sampling
9. Sampling Distribution of $\bar{X}$
10. Estimation of Population Mean
11. Tests for Population Mean
12. Inferences for Population Proportion
13. Comparisons of Two Populations and Other Inferences
14. Sampling Procedures

Unit Four   Estimation and Testing—II

15. Nonparametric Procedures
16. Goodness of Fit
17. Multinomial Populations

Unit Five   Linear Statistical Models

18. Simple Linear Regression
19. Inferences in Simple Linear Regression
20. Multiple Regression
21. Analysis of Variance

Unit Six   Bayesian Decision Making

22. Bayesian Decision Making—I
23. Bayesian Decision Making—II

Unit Seven   Time Series Analysis and Index Numbers

24. Time Series—I
25. Time Series—II
26. Price and Quantity Indexes

Appendices

A. Mathematical Review
B. Chi-Square, *t*, and *F* Distributions
C. Tables
D. Data Set

chapter considers sampling of finite populations and various sampling procedures other than simple random sampling.

Unit Four, Estimation and Testing—II (Chapters 15–17). In the first chapter of this second unit on estimation and testing, nonparametric procedures which have found extensive use are discussed. The approach is not one of presenting a grab bag of statistical tools—rather, procedures are presented which match those discussed in the previous unit, but which do not depend on large samples or restrictive distributional assumptions. Goodness of fit procedures and techniques for multinomial populations are taken up in the next two chapters. Separation of these topics into two chapters emphasizes the separate classes of problems involved and is a departure from the usual textbook coverage.

Unit Five, Linear Statistical Models (Chapters 18–21). This unit meets a pressing need for a thorough treatment of regression and analysis of variance to prepare the reader for the extensive use of these techniques in management, economics, and social sciences in general. The topics of regression and analysis of variance are handled in a unified fashion in these chapters. Computer-assisted analysis is stressed, as is the facility to interpret standard printout from a computer regression program. Also stressed are practical aspects in using linear statistical models, such as study of the aptness of the model employed.

Unit Six, Bayesian Decision Making (Chapters 22 and 23). This unit presents a discussion of Bayesian decision analysis, showing how the decision-theoretic aspects of a statistical problem can be handled formally. The relation between statistical decision making and statistical testing is made clear by our initial emphasis on normal-form analysis and later presentation of extensive-form analysis.

Unit Seven, Time Series Analysis and Index Numbers (Chapters 24–26). A compact treatment of the classical time series model is presented in the first chapter of this unit. The second chapter is devoted to regression time series models. The emphasis again is on concepts and procedures needed for effective application. In this same spirit a third chapter is devoted to price and quantity indexes, which play an important role today for many administrators, economists, and social scientists.

This book can be used for a wide variety of one- or two-quarter or one- or two-semester courses. Figure I shows the logical interdependencies of the units and their chapters, with the arrows indicating prerequisite chapters. As can be seen, various chapter sequences may be used, depending on the length of the course and the emphasis desired. Here are four examples:

1. A course covering descriptive statistics and the basics of statistical inference might include data (Unit One), probability (Unit Two), inferences for population mean and proportion (Chapters 8–12), simple linear regression (Chapters 18 and 19), and the descriptive portions of time series analysis and index numbers (Chapters 24 and 26).
2. A course on inferential statistics emphasizing linear statistical models might include data (Unit One), probability (Unit Two), inferences for population mean (Chapters 8–11), linear statistical models (Unit Five), and regression models for time series analysis (Chapter 25).

3. A course covering both parametric and nonparametric inferences might include data (Unit One), probability (Unit Two), parametric inferences (Unit Three), nonparametric inferences (Unit Four), and as much of linear statistical models (Unit Five) as time permits.
4. A course emphasizing decision theory might include data (Unit One), probability (Unit Two), the major elements of statistical inference (Chapters 8–13), Bayesian decision making (Unit Six), and other topics as time permits.

We are greatly indebted to many individuals and organizations who have helped us in the preparation of this book. Our sincere thanks go to all who have provided us with case materials and illustrations that demonstrate the usefulness of statistical methods in management, economics, and other social sciences. Many persons, including colleagues and reviewers, have made helpful suggestions on the manuscript and otherwise assisted us, for which we are most grateful. We particularly wish to thank for their help Professors Robert F. Berner of the State University of New York–Buffalo, R. V. Erickson of Michigan State University, Edgar Hickman of the University of South Carolina, Oswald Honkalehto of Colgate University, H. K. Hsieh of the University of Massachusetts at Amherst, Allan Humphrey of the University of Rhode Island, Raj Jaganathan of the University of Iowa, William Meeker of Iowa State University, Robert Norland, Jr., of Bowling Green State University, Thomas Pray of the State University of New York–Geneseo, Thomas Rothrock of the University of Florida, Barbara Rufle of Syracuse University, J. Michael Ryan of the University of South Carolina, Kenneth C. Schneider of St. Cloud State University, Randolph Shen of the University of Rhode Island, Erland Sorensen of Northeastern University, Stephen Vardeman of Purdue University, Dean Wichern of the University of Wisconsin at Madison, and Morty Yalovsky of McGill University.

We are also indebted to our respective universities for providing an environment in which the undertaking could be brought to fruition. A number of people were most helpful in working with us in preparing and checking the manuscript. We wish to express our appreciation to Rebecca G. Baggett, Stephen A. Floyd, Donna Fredrickson, Jeannine Hall, and Gayle Zalmanovitch for their valuable assistance. Allyn and Bacon, Inc., our publishers, helped us in many ways. Of course, we thank our families for their patience and encouragement while we completed this book.

Introduction

What is Statistics?

Almost everyone uses statistics and is affected by applications of statistics. The word *statistics* refers in common usage to numerical data. Vital statistics, for example, are numerical data on births, deaths, marriages, divorces, and communicable diseases; business and economic statistics are numerical data on employment, production, prices, and sales; social statistics are numerical data on housing, delinquency and crime, education, and social security and welfare.

Statistics has an additional meaning that is more specialized. In this second sense, *statistics* refers to the methodology for the collection, presentation, and analysis of data, and for the uses of such data. Unless data are accurate, properly presented, and correctly analyzed, they may be dangerously misleading. Since we all are "consumers" of statistics, it is important for all of us, not only professional statisticians, to acquire some knowledge of statistical methodology.

The word *statistician* also has several meanings. It can refer to: (1) a person who performs routine operations with statistical data; or (2) an analyst who is highly trained in statistical methodology and uses this methodology in the collection and interpretation of data; or finally, (3) an applied mathematician who utilizes advanced mathematics in the development of new statistical methods. Statisticians are needed in all these capacities in order to make statistical data most useful.

The Expanding Role of Statistics

Statistical data have been used for many centuries by governments as an aid in administration. In antiquity, statistics were compiled to ascertain the number of citizens liable for military service and taxation. After the Middle Ages, governments in Western Europe were interested in vital statistics because of the widespread fear of devastating epidemics and the belief that population size could affect political and military power. As a result, data were compiled from registrations of christenings, marriages, and burials. In the sixteenth through eighteenth centuries, when mercantilistic aspirations set nation-states in search of economic power for political purposes, data began to be collected on such economic subjects as foreign trade, manufacturing, and food supply.

Today, data are collected, classified, stored, and retrieved in diverse and comprehensive information systems that supply individuals and organizations with the statistical intelligence required to carry out their activities. The expansion in the collection, transmission, storage, and retrieval of statistical data, facilitated by computers, has been accompanied by the rapid development of statistical methodology and data analysis.

Statistical concepts have exercised a profound influence in almost every field of human activity and have been incorporated into the basic principles of such sciences as physics, genetics, meteorology, and economics. Statistical methods have been used to improve agricultural products, to design space equipment, to plan traffic control, to forecast epidemics, and to attain better management in business and in government. Students of the natural and social sciences study statistics to become better scientists; students of economics study statistics to become better economists; students of administration study statistics to become more effective administrators.

Uses of Statistics

Some knowledge of statistics is essential today for people pursuing careers in almost every area of industry, government, public service, or the professions. Not only are more comprehensive networks of data available to serve as a basis for drawing valid conclusions and making decisions, but the purpose in assembling the data has shifted from record keeping to evaluation and action, based on timely information. Until recently, statistics were collected primarily as a record of past events. Although such statistics were analyzed to gain insights into current problems, the emphasis was essentially upon the past. At the present time, the collection of numerical information for the record still takes place, but because of the needs for improved planning and control, data-collection systems and data repositories have been designed to provide data that are as up-to-date as possible. Statistical analysis, in turn, has become chiefly concerned with the present and the future, rather than the past.

The increasing uses of statistics are part of the trend toward basing evaluations and making decisions on the most objective and scientific foundations possible. Modern organizations are becoming more dependent on statistical data to obtain factual information about their internal operations and their social, economic, and ecological surroundings. Statistical data are concise, specific, capable of being analyzed objectively with powerful formal procedures, and well suited for making comparisons. Hence, they are especially useful in such key organizational functions as choosing among alternatives, setting goals, evaluating performance, measuring progress, and locating weaknesses.

UNIT ONE

—◆—

Data

1
Data Acquisition

1.1 DATA SETS

Our dictionary defines data as follows:

(1.1) *Data* are facts or figures from which conclusions may be drawn (Ref. 1.1, p. 340; *see reference at end of chapter*).

Data are the raw material of statistics. Statistical analysis cannot proceed until the data of interest in an investigation are assembled and organized in a useful manner. We shall refer to the data collected for a particular study as a *data set*.

Data sets are found all around us. The financial section of our daily paper contains price data for securities and commodities; an economic report shows inflation rates for different countries; a newspaper article contains data on achievement test scores for all schools in a city.

☐ **Example**

Figure 1.1 shows a data set for a study of physical-fitness profiles of the six police officers in a township. Information about age, sex, and various physical and fitness characteristics of each officer was obtained in this study. ☐

Characteristics of Data Sets

A data set represents a collection of facts and figures. We now consider the key characteristics of data sets.

Element. A data set provides data about a collection of elements and contains, for each element, information about one or more characteristics of interest. In Figure 1.1, an element of the data set is a particular police officer.

Variable. A variable is a characteristic of interest about an element. In Figure 1.1, one characteristic of interest is age of police officer. This characteristic takes on different values for different officers, hence is called a variable. Age is a quantita-

FIGURE 1.1   *Data set for study of physical fitness profiles of police officers in township*

Officer	Age	Sex	Systolic Blood Pressure	Variable ↓ Diastolic Blood Pressure	Tricep Skinfold Thickness[1] (cm)	Number of Situps	Fitness Rank	
Anders	32	M	120	80	1.50	100	1	
Colm	28	F	118	75	1.96	35	3	←Observation on element
Greene	46	M	138	90	1.79	45	4	
Keene	23	F	121	75	2.30	29	5	
Osman	36	M	141	95	3.05	18	6	
Waldorn	22	M	123	75	1.91	75	2	

[1]Index of body fat

Element   Observation on variable

tive variable because its outcomes are numerical in nature. On the other hand, the variable sex in Figure 1.1 is a qualitative variable because its outcomes are nonnumerical. The following are formal definitions:

(1.2)   A characteristic that can take on different possible outcomes is called a *variable*. When the outcomes are expressed numerically, the variable is said to be *quantitative*. When the outcomes refer to nonnumerical qualities or attributes, the variable is said to be *qualitative*.

Of the seven variables in the data set in Figure 1.1, six are quantitative and one (sex) is qualitative.

Observation.   The information on all variables for one element in the data set is called an *observation*. Thus, the information on the seven variables for Officer Colm constitutes an observation. In Figure 1.1, each row of data represents an observation, so there are six observations in the data set.

The information for an element of the data set about a single variable is also called an observation, when this variable is the only one considered at the moment. The information about a single variable for an element of the data set is also called an *outcome*. Thus, age 32 of Officer Anders is an observation on, or an outcome of, the variable age for this officer.

Comments

1. Sometimes a quantitative variable is converted into a qualitative one by grouping the possible numerical outcomes into nonnumerical classes. This is done to facilitate reporting, interpreting, or analyzing the data. Thus, with eggs for retail sale, the quantitative variable weight (measured in grams) is converted into a qualitative variable by partitioning all possible weights into categories such as extra large and large. At other times the reverse is done to convert a qualitative variable into a quantitative one by assigning a numerical value or code to each nonnumerical category. For instance, the

variable sex in Figure 1.1 might be quantified by using the number 0 for male and the number 1 for female.

2. Data sets may be distinguished on the basis of the number of variables they contain. A *univariate* data set contains one variable, a *bivariate* data set two variables, and a *multivariate* data set three or more variables. Thus, the data set in Figure 1.1 is multi-variate.

Types of Data

Statistical data are obtained in a variety of ways. For instance, the data on blood pressure in Figure 1.1 were obtained by employing a blood-pressure-measuring instrument, and these data are called *measurement data*. The data on number of situps were obtained by counting and are called *count data*. The data on fitness were obtained by ranking the officers, with the most fit officer assigned rank 1 and the least fit officer, rank 6. These data are called *rank data*. Finally, the data on sex involve a classification process where classes or categories (male, female) are set up and each data-set element is assigned to the appropriate category. These data are called *classification data*.

1.2 DATA SOURCES

Statistics is not only concerned with organizing and analyzing data once they are assembled, but also with the sources of data and how data are collected for study. The first stage of any investigation involves a specification or definition of the problem to be studied. From this specification comes an identified need for particular types of data to clarify the problem. At this point, the question of where to obtain the necessary data is posed.

Internal and External Sources

Some data sets may be partially or totally available from an *internal source*, such as records of operating and accounting data, which most organizations maintain routinely. Indeed, many organizations store such data in records that are grouped into computerized data files for efficient entry and retrieval of information. These data files, together with the associated computer programs, constitute the organization's internal data base. Design and maintenance of the data base have become an important managerial function in many firms and other organizations.

Other data are obtainable from an *external source*. Such external data may be published in a reference book or a statistical periodical, or the data may be in a computerized form such as cards, tapes, or on-line access to an external data bank. The external-source organization may be a government agency, a trade association, or a private specialized service company.

To illustrate the use of these sources of data, consider the case of a market researcher who wishes to study the geographic distribution of customers who use a particular company product. The researcher might find much of the required data in the internal accounting records of the business. On the other hand, a store-location analyst for a retail-outlet chain who is concerned with crime patterns in different cities might find the necessary data in external sources such as government reports on criminal and judicial statistics or computerized police records.

Cautions in the Use of Data Sources

When using any data source, the user should be thoroughly acquainted with the nature and limitations of the data. Limitations may include imperfect or improper methods of data collection, recording, and classification, as well as errors of omission or commission when data are transferred from one record to another. The user also needs to determine whether the definitions employed in compiling the data are appropriate for the purpose. It is also important to check whether changes in concepts, definitions, and data-collection methods have occurred over the time period of interest and, if so, to determine the effect of these changes on the data. The user should have access to pertinent information and explanatory comments about the manner in which the data were compiled, the data's accuracy, and how the data should be interpreted. The need for caution in the use of data applies to all data sources, whether they be internal or external, manual or computerized.

1.3 EXPERIMENTAL AND NONEXPERIMENTAL STUDIES

When the data needed for an investigation are not available in existing sources, one must consider some method for obtaining them directly. Two major methods of data collection are the experimental study and the nonexperimental or observational study. We now illustrate the distinction between these two types of studies.

☐ Illustration

An insurance company hired a large number of new programmers when it moved its headquarters to another city. The director of training for the company set up a voluntary training program that could be taken by the new programmers. Later, the director compared data on the job progress of the two groups of new programmers, i.e., those who took the training program and those who did not. He found that the programmers with the training, on the whole, had made more progress on the job than the others.

The director could not, however, assess the contribution of the training program to job progress since it is likely that the programmers who volunteered for the training were ones who were better motivated and more achievement-oriented. Such individuals might have made more job progress than the others even without the training program.

The fact that the training program was voluntary so that no control was exercised over any factors which might affect progress on the job makes this a *nonexperimental* or *observational study.*

Some years later, the director of training had the opportunity to assess the effect of the training program more definitively. The company expanded its data-processing staff substantially and hired 50 new programmers. Twenty-five of these were selected at random and assigned to the training program, and the other 25 programmers were immediately assigned to operations. Follow-up studies showed that the job progress was about the same for both groups of programmers.

The second study is an *experimental study* because control was exercised over the factor under study (training program) and randomization was employed to balance out all other factors which might affect job progress. Randomization here tended to balance highly motivated programmers in both groups, and similarly tended to balance programmers by age, experience, and any other factors which might affect job progress. Thus, if any difference in job progress had been observed between the two groups, it could more convincingly be attributed to the training program because of the experimental nature of the study than in the previous use of the training program where programmers volunteered for training. □

We now consider experimental and nonexperimental studies in more detail.

Experimental Studies

The use of experiments for collecting data in the fields of business, economics, and the social sciences has expanded in recent years.

 (1.3) An *experimental study* is a study where the factors under consideration are controlled so as to obtain information about their influence on the variable of interest.

In statistically designed experiments the control over the factors under study is accompanied by randomization to balance out the influence of all extraneous factors which might affect the variable of interest.

□ Examples

1. Thirty stores were randomly assigned one of three store displays to study the effect of the display on store sales of a product.
2. Four hundred consumers were randomly assigned one of four brands of all-purpose flour to study consumer reactions to each brand in routine home use over a period of one month. □

Nonexperimental Studies

In spite of the increasing use of experimental studies, researchers and analysts in business, economics, and the social sciences must often still rely on nonexperimental or observational studies.

(1.4) A *nonexperimental* or *observational study* is a study where no special controls are exercised in the collection of the data over any of the factors influencing the variable of interest.

□ Examples

1. A survey of 1,000 residents of a metropolitan area was conducted to obtain information about frequency of attending concerts and plays (the variable of interest) and about possible explanatory factors, including income, age, and education of the resident.
2. A study was undertaken by a large corporation of five plants to obtain information about plant productivity (the variable of interest) and about possible explanatory factors, including type and age of machinery used in the plant, education and age characteristics of the production employees in the plant, and type of wage incentive program in the plant. □

Choice between Experimental and Nonexperimental Studies

Both experimental and nonexperimental studies can be extremely useful for studying the effects of one or more factors on the variable of interest. However, as noted earlier, experimental studies provide stronger evidence of these effects than nonexperimental studies. Experiments are especially advantageous in investigating cause-and-effect patterns, such as determining whether or not a high cholesterol level tends to cause atherosclerosis.

Despite the advantages of experimental studies, much of statistical analysis in business, economics, and the social sciences is based on nonexperimental studies. One reason is that most available data, such as internal data on company operations and external data on the economy and consumer behavior, are nonexperimental data. Another reason is that it is often not feasible or may not be desirable to exercise the experimental controls required in experimental studies. For example, an economist interested in the effect of family size on the proportion of income saved cannot select a group of newlyweds and tell each couple to have a family of certain size. Observational studies of existing families, on the other hand, can provide useful information on family size, income, and expenditures, from which the relation between size of family and proportion of income saved can be studied.

1.4   DATA ACQUISITION

A variety of procedures for acquiring data are employed in experimental and non-experimental studies. Three commonly used ones are observation, personal interview, and self-enumeration.

Data-Acquisition Procedures

Observation.   Observation entails direct examination and recording of an on-going activity.

Examples

1. In a study of family decision making, a researcher observed and recorded interactions between husband and wife as they decided on the brand of color television to buy.
2. In an engineering study, data about the internal temperature of a kiln were obtained by reading an instrument inserted in the kiln.
3. In a marketing study, an analyst monitored customer flow in a department store by means of closed-circuit television.

The observation procedure has certain advantages:

1. The directness of the procedure avoids problems such as incomplete or distorted recall.
2. Data can be gathered more or less continuously over an extended time period.

Limitations of the method include the following:

1. The observer (or the instrument) must be free of bias and must accurately record the events of interest. Human observers usually require thorough training so that they will record precisely what they observe and so that different observers will record the same events in the same manner.
2. Individuals who are under observation and aware of this fact may alter their behavior and, as a result, observations of their behavior may be biased.

Personal Interview.   In a personal interview, an interviewer asks questions that are printed on a questionnaire and records the respondent's answers in designated spaces on the questionnaire form.

Examples

1. A household member was interviewed at home about purchases of toothpastes and mouthwashes and about family characteristics, such as income and family size.
2. A household member was interviewed over the telephone about television viewing, including viewing at the moment of call, the station viewed, and number of persons viewing.
3. A company's financial executive was interviewed at the office by a representative of a trade association about the firm's plans for capital expenditures in the coming year.

Both the advantages and limitations of securing data through personal interviews arise from the direct contact between the respondent and the interviewer. Advantages include the following:

1. Persons will tend to respond when they are approached directly; hence, the personal-interview procedure usually yields a high proportion of usable returns from those persons who are contacted.
2. The direct contact generally enables the interviewer to clear up misinterpretations of questions by the respondent, to observe the respondent's reactions to particular questions, and to collect relevant supplementary information.

There are several limitations of the personal-interview method:

1. The interviewer may not follow directions for selecting respondents. For instance, if a member of the family other than the one designated is interviewed, a bias may be introduced into the results.
2. The interviewer may influence the respondent by the manner in which the questions are asked or by other actions. A slight inadvertent gesture of surprise at an answer, for example, can exert subtle, undetected pressures on the respondent.
3. The interviewer may make errors in recording the respondent's answers.

Self-Enumeration.   With self-enumeration, the respondent is provided with a questionnaire to complete which often also contains necessary instructions.

☐ Examples

1. A student who graduated from high school recently received a self-enumeration questionnaire through the mail, a page of which is shown in Figure 1.2, to provide information about educational activities since graduation.
2. A new magazine subscriber received a questionnaire through the mail to provide information about age, type of job held, income, and amounts of money spent last year on
   specified recreational activities.
3. A person completed a certificate of registration for a motor vehicle, supplying information on make, model, and year of car.
4. A purchaser of a toaster filled out the warranty card, giving information on family
   characteristics and on the primary method by which attention was directed to this
   appliance (e.g., word of mouth, television commercial). ☐

Both the advantages and limitations of the self-enumeration procedure arise
from the elimination of interviewers. The type of interviewer errors discussed earlier are thus avoided. On the other hand, the absence of interviewers creates two
serious problems:

1. When a questionnaire is sent to a household or an organization, there is no
   control over which person answers the questions.
2. The absence of interviewers can lead to low response rates. A low response rate
   can be a source of serious bias in survey results, because the persons who *do*
   answer the questionnaires are often not representative of the entire group contacted. The user of data collected by a self-enumeration procedure should
   therefore know the rate of nonresponse as one factor affecting the magnitude of
   the potential bias.

Where there is a material rate of nonresponse, some follow-up of nonrespondents will be valuable. In most well-conducted mail surveys, some or all of
the nonrespondents are contacted as a routine procedure. Nonrespondents may be
contacted by means of "reminder" letters, telephone calls, or special personal interviews.

Questionnaire Design

Since questionnaires are a major vehicle in data collection, we now examine briefly
some important considerations in designing and using a questionnaire so that
accurate data are obtained.

Types of Questions.   The two basic types of questions employed in questionnaires
are multiple-choice and free-answer or open-end questions.

The *multiple-choice question* presents the respondent with a choice from
among two or more prespecified answers. In Figure 1.2, Questions 7a and 7b are of
this type.

The alternatives in a multiple-choice question should be clear-cut; mutually
exclusive; and when dealing with an issue, more or less evenly distributed on both
sides of the issue. Ideally, the choices should cover all answers likely to arise in

FIGURE 1.2 *Example of part of a self-enumeration questionnaire*

3. Since you left high school, have you taken any courses in a technical or trade school, hospital school, beauty school, business school, or other vocational school? (*DO NOT include regular college courses.*)	⑪⑤ 1 ☐ Yes — *Continue with question 4* 2 ☐ No — *Go directly to question 8*
4. When did you first begin classes at a vocational, technical, trade, or business school?	Month ⎮ Year ⑪⑥ _____ ⎮ 19 _____
5. What was your field of study when you began taking these classes (*for example, beautician, auto mechanics, accounting, etc.*)?	_____ OR ⑪⑦ 0 ☐ No specific field of study ☐☐☐ **CENSUS USE ONLY**
6. What is the full name and address of the school you attended? *If you attended more than one school, enter the one you attended longest.*	Name _____ Address (*Number and street*) _____ City State ZIP code ⑪⑧ ☐☐☐☐☐ **CENSUS USE ONLY**
7a. At the school you entered in question 6, were you enrolled in a program leading to a degree or certificate?	⑪⑨ 1 ☐ Yes — *Continue with question 7b* 2 ☐ No — *Go directly to question 8*
7b. How many years of continuous full-time study does it usually take to complete the requirements for the degree or certificate for which you were enrolled? (*Count only years after high school.*)	⑫⓪ 1 ☐ Less than ½ year 2 ☐ ½ to 1 year 3 ☐ 2 years (13–24 months) 4 ☐ 3 years (25–36 months) 5 ☐ 4 years (37–48 months) 6 ☐ 5 years or more
7c. Did you complete that program?	☐ YES→ **When did you complete that program?** ⑫① Month ⎮ Year ⎮ 19 } *Go directly to question 8* ☐ NO → **When did you last attend classes at the school you entered in question 6?** ⑫② Month ⎮ Year ⎮ 19 } *Continue with question 7d*
7d. What are the reasons that you did not complete the program? (*Mark (X) all that apply*)	⑫③ 1 ☐ Financial reasons * 2 ☐ Health reasons 3 ☐ Marriage or pregnancy 4 ☐ Family or household responsibilities 5 ☐ Other personal or family reasons 6 ☐ Academic problems ⑫④ 7 ☐ Took a job * 8 ☐ Entered military service 9 ☐ Wanted to leave school 0 ☐ Other – *Specify*_____
8. Since you left high school, have you been enrolled in a college or university?	⑫⑤ 1 ☐ Yes — *Continue with question 9 on page 4* 2 ☐ No — *Go directly to question 17a on page 5*

SOURCE: U.S. Bureau of the Census.

response to the question. If too many alternatives are presented, however, they may not be clear-cut enough for the respondent to make a meaningful selection. One disadvantage of multiple-choice questions is that they tend to suggest an answer to the respondent in terms of the alternatives stated when actually the question would have been answered differently if no choices had been suggested.

This danger is avoided by the other type of question, the *free-answer* or *open-end question*, which elicits answers in the respondent's own words. The following question is of the free-answer type:

> In your opinion, what things are good or bad about the Blank Company as a place to work?

In requesting answers in the respondent's own words, the free-answer question yields a multitude of replies that must be classified into a limited number of categories before statistical analysis can be undertaken. This classification becomes a difficult task when thousands of replies to a free-answer question have been received, since the wording of each reply must be interpreted carefully. Hence, free-answer questions are usually employed only in small-scale studies or in the exploratory stages of large-scale ones to determine the patterns into which the responses to a question tend to fall. The information obtained through the use of exploratory free-answer questions in the early stages of a large-scale study is then used to construct the multiple-choice questions utilized later for the bulk of the data collection.

Multiple-choice and free-answer formats sometimes are combined in constructing a question. An example of this is the use of a category titled "other" among the multiple choices, which allows the respondent to provide an answer other than the ones given. When the respondent chooses the "other" category, usually he or she is asked to provide additional details. Question 7d in Figure 1.2 is of this type.

Order of Questions.   A questionnaire consists of a battery of questions arranged in a certain order. The first questions should establish rapport with the respondent, and all early questions should be simple. Where several topics are involved, it is usually best to complete one topic before going to another rather than to jump back and forth. The order of questions often affects the answers given by the respondent, because questions draw the respondent's attention to a particular complex of thoughts or feelings in the context of which later questions are answered. In market-research surveys, for example, questions mentioning a specific product or firm tend to bias answers to questions that follow; consequently, such identifying questions should be placed toward the end whenever possible.

Directness of Approach.   Many respondents are prone to rationalize or to exaggerate their replies when they are questioned directly about their motives, accomplishments, or other subjects involving their prestige or self-esteem. In order to avoid biased data, an indirect approach is often taken in framing "prestige questions." For example, instead of asking, "Did you complete high school?" interviewers might be instructed to ask, "What grade did you complete when you left school?" In the latter question, an attempt is made to avoid implications adverse to a respondent who did not graduate from high school.

Need for Clarity in Questions. A question must have approximately the same meaning for all respondents if the data obtained from the replies are to be meaningful. Vaguely defined terms should be avoided in framing questions, since such terms will be interpreted differently by different persons. In addition, questions should be stated simply. If long and involved questions are asked, there will usually be some respondents who do not interpret them correctly.

Need to Avoid Bias in Question Framing. The replies of respondents are frequently influenced in a one-sided manner by the way in which questions are phrased. One type of question that tends to produce biased replies is a *leading question*—that is, one suggesting a particular answer. The question, "Wouldn't you agree that the extra quality obtained in this brand is well worth the few extra pennies in cost?" so obviously suggests a particular answer that few persons would take the replies seriously. In many instances, though, the suggestive quality of a question is less easily recognized.

Value of Pretesting. Once a questionnaire has been designed, it is often pretested in a preliminary small-scale study. Pretesting, usually involving between a few dozen to several hundred respondents, brings out unforeseen difficulties, such as the layout of the questionnaire, arrangement and wording of the questions, and even the clarity of instructions to interviewers when interviewing is used. The difficulties can then be corrected before the full-scale study is begun.

1.5   ERRORS IN DATA

Data sets often contain errors. These may arise in the data-acquisition stage or in other ways. It is important for users of data to be on guard against possible errors. In this section we discuss errors in data and show examples of how errors arise.

Errors Arising in Data Acquisition

Errors in data can be introduced by defects in the acquisition procedure.

(1.5)   An *error in data acquisition* is any discrepancy between the actual result obtained and the correct result that would be provided by an ideal procedure.

☐   Examples

1. Some Canadians completing a travel questionnaire overlooked trips to the United States in answering a question on the number of foreign trips taken in the past year. These errors led to incorrect counts of total number of foreign trips.
2. In answering a question relating to occupation, some respondents gave an incorrect category for occupation, possibly because the definitions of the various categories were misunderstood. For instance, some of the respondents who are "retired" mistakenly recorded themselves as "unemployed." These are classification errors, which led to incorrect counts.

3. A meteorologist, developing data on a weather system, measured the wind velocity inaccurately and thereby recorded an incorrect measurement.
4. An industrial scale showed incorrect weights for loaded trucks because it was not calibrated properly, and thereby provided incorrect measurements. □

As the examples show, errors in acquiring data may be caused by imperfect recall by respondents, inaccurate measurements by an instrument, misinterpretations of a question, misunderstandings of a definition—the list is long. In acquiring data, it is important to select a procedure that will keep errors at tolerable levels. Thus, a personal interview may be preferred over a mailed self-enumeration questionnaire in a study involving many technical definitions, since the interviewer can help respondents avoid misunderstandings. Of course, not all errors in data are of equal importance. The procedure for data acquisition must be chosen to control the important errors so that the data set obtained will be useful.

Other Errors in Data

Errors in data acquisition are not the only kind of errors found in data sets. Other errors may also be present, such as when a data set is incomplete or contains typographical errors.

□ Examples

1. In a survey of families residing in a community to obtain information about labor skills, families away on vacation at the time were not included. As a result, the data set is incomplete.
2. A data set on financial operating characteristics of small businesses in a region contained some incorrect codings of the survey results, some errors because of incorrect key punching of computer punch cards, and some typographical errors made when the data were set in type. □

Cited Reference

1.1 *Funk & Wagnalls Standard College Dictionary*. Canadian ed. Toronto: Fitzhenry and Whiteside, 1976.

PROBLEMS

1.1 Consider the following data set relating to candidates for promotion to product manager, compiled by a firm's senior management committee:

Name	Age	Sex	Number of Training Programs Attended	Committee's Order of Preference
Bixby, Milton	46	M	2	2
Morris, Wm.	31	M	0	3
Parker, Kim	42	F	1	1

a. What constitute the elements of this data set? How many elements are there?

    b. Identify the variables in this data set. For each variable, indicate whether the data were obtained by measurement, count, classification, or ranking, and whether the variable is quantitative or qualitative.

    c. What constitutes the observation for Milton Bixby?

1.2 Refer to the data set in Appendix D for crude-oil producers.

    a. What constitute the elements of this data set? How many elements are there?

    b. Is the data set univariate or multivariate?

    c. Identify the variables in this data set. For each variable, indicate whether it is quantitative or qualitative.

    d. What constitutes the observation for firm 17?

1.3 A multinational corporation operates in 27 countries. Would the *Monthly Bulletin of Statistics* published by the United Nations be an internal or external source of data for this corporation? Explain.

1.4 A computer tape contains production data for the various divisions of a firm. Is this tape an internal or external source for the firm? Explain.

1.5 A company has been selling one of its popular beverage products in bottles in supermarkets, and in cans in other outlets. An analyst is examining sales data for the beverage to see which type of container is preferred by customers.

    a. Are these data nonexperimental or experimental? Why?

    b. Does the distinction in **a** matter to the analyst? Explain.

1.6 Thirty sales trainees were grouped into 15 pairs, each pair being similar in age, sales experience, and educational background. One trainee in each pair was selected at random to attend a sales training program based on role playing; the other trainee was assigned to a sales training program based on case studies. At the end of the programs, the trainees were tested for sales skills and knowledge. The test scores are now being evaluated to determine whether one type of program produces better scores than the other.

    a. Is this study experimental or nonexperimental? Why?

    b. An observer asks why the trainees in each pair could not simply have decided between themselves who would attend each program. Answer the observer's question.

1.7 For each of the following, state which method of data acquisition (observation, personal interview, or self-enumeration) is most appropriate for obtaining the information and justify your choice. Consider cost, need to obtain respondent's cooperation, need for consistent interpretation of definitions, and any other factors you believe are relevant.

    a. Quarterly data from employers on the number of employees who worked at any time during the quarter.

    b. Data from persons in a national register of scientists concerning field of specialization, nature of present employment, and academic training.

    c. Data from technicians of a large research organization on hazardous aspects of their jobs.

    d. Data on order in which shoppers purchase items in supermarkets.

    e. Data on blood-cholesterol levels of Canadians.

1.8 Identify any major faults in each of the following questions and reword the question to eliminate the faults. Indicate any assumptions you made when rewording the question.

    a. How many calls for telephone directory assistance did you make in the past 24 months?

    b. Does the name Rolls-Royce or some other name come to mind first when the term *chauffeur-driven car* is mentioned?

    c. Is your family income low, average, or high compared to incomes of other families living on this block?

    d. Have you or anyone in your family recently suffered from an elevation of bodily temperature?

1.9 Consider the following alternative questions for ascertaining consumer purchasing plans: (1) Do you plan to buy a new small car within the next year? (2) Do you plan to buy a new small car within the next year, and how certain are you of this plan on a scale from 1 (highly uncertain) to 10 (highly certain)?

    a. Which question do you think will yield more useful information? Explain.

    b. Would the clarity of the question be improved if "within the next year" were replaced by "within the next 12 months"? Explain.

1.10 An instructor divided a class on opinion research at random into two groups. One group was asked the following question: "Should university officials take primary responsibility for reducing rowdyism at basketball games?" The second group was asked the same question, except that "student leaders" was substituted for "university officials." About 80 percent of the students in each group answered affirmatively.

    a. What fault in question design is illustrated by the instructor's experiment?

    b. Design a better question to obtain student opinions on who should take primary responsibility to reduce rowdyism at basketball games. Explain how your approach avoids the fault identified in **a**.

1.11 Explain whether each of the following entails an error arising from defects in the data-acquisition procedure or some other type of error:

    a. A school child, when asked to indicate the number of brothers or sisters at home who have not yet started school, overlooked a new baby brother and wrote "1" instead of "2" on the questionnaire.

    b. In a data set on new births in area hospitals, incorrect keypunching caused one baby's weight to be shown as 15 pounds, 8 ounces when the actual weight entered on the hospital form was 8 pounds, 15 ounces.

    c. In filling out a questionnaire sent to owner-operated restaurants, an owner misinterpreted the definition of "employee" in answering a question on number of full-time employees and counted himself as one of the employees.

1.12 In a mail survey of 2,000 practicing physicians conducted by a pharmaceutical firm, a series of questions related to prescriptions for an antidepressant drug code-named ADV. The proportion of responding physicians who stated they had prescribed the drug at least once in the previous month was .40.

    a. Suppose that 95 percent of the questionnaires had been completed and returned. By how much might the proportion .40 be changed if every questionnaire had been completed and returned?

    b. Answer **a** on the supposition that the response rate was 20 percent.

    c. What are the implications of your answers to **a** and **b** for interpreting data based on low response rates?

    d. What factors might cause a physician to not respond in this type of mail survey? Might any of these factors be related to whether or not the physician prescribes this particular drug? Does it matter, for the purposes of the survey, whether or not such a relation exists? Discuss.

STUDIES

1.13 For each of the following, ascertain a library source where data for the most recent period can be obtained:
a. U.S. annual birthrate.
b. U.S. monthly unemployment rate.
c. U.S. monthly Consumer Price Index for All Urban Consumers.

1.14 Refer to Problem 1.13. Answer it for data pertaining to Canada.

1.15 For each published data source assigned from the following list, state who publishes the source, how often it is published, and give three specific items of information from the most recent issue of that source:

U.S. Statistical Abstract　　*Moody's Industrial Manual*
Survey of Current Business　　*U.S. Census of Housing*
Federal Reserve Bulletin　　*Canada Year Book*
Monthly Labor Review　　*Bank of Canada Review*
Construction Review　　*U.N. Statistical Yearbook*
Business Conditions Digest　　*World Health Statistics Annual*
Life Insurance Fact Book　　*International Financial Statistics*

1.16 A study indicated that high-school students who took a voluntary driver training course tended later to have better driving records than students who did not take the course. However, the study team noted that this difference could be due simply to differences in personal qualities of the students in the two groups. Among other things, relatively more women than men took the course, and young women tend to have better driving records than young men in any case. The team also noted that students taking the course might have been better motivated to be good drivers than students not taking the course.
a. Why was this study not an experimental study?
b. Develop a design for an experimental study in a high school to evaluate the impact of the driver training course. Your design should control the effects of sex and motivation of students.

1.17 A professional association composed of economists and statisticians surveyed its members to study how they liked the association's journal. A high proportion of the responses indicated an unfavorable opinion. Subsequently, it was learned that the response rate was much higher for economists than for statisticians. Construct a numerical example to show how the differential response rates could have biased the survey results.

1.18 It is frequently stated that errors in response are not too serious because they will tend to balance out. One researcher conducted a survey of 1,643 library users in a community to study the nature of response errors. He compared responses as to the number of books borrowed from the library in the preceding three months with the library's computerized records. The results were as follows:

Respondent Statement	Percent of Respondents	Average Number of Books Borrowed	
		Library Record	Respondent's Recollection
Accurate	23	6.1	6.1
Overstated	43	2.8	5.6
Understated	34	6.4	5.4

a. By what percent did overstaters overstate their borrowings? Did understaters understate their borrowings to the same extent?

b. What is the overall relative bias in the reported average number of books borrowed? Did the response biases largely balance out?

c. What are some factors that might account for the biases in response here?

2

Data Classification and Exploratory Analysis

Data sets are often so large that the data must be organized and reduced to manageable proportions before any study of them can begin. The human mind is simply not capable of assimilating and interpreting hundreds of facts and figures in raw form. In this chapter we present a number of common statistical methods for initially summarizing data sets and performing exploratory analyses. These methods facilitate subsequent detailed analyses and contribute to an understanding of the problem at hand.

2.1 QUALITATIVE DISTRIBUTIONS

Construction of Qualitative Distributions

☐ Illustration

An analyst for a retailer wished to study the occupational profile of the company's credit-card customers. Her intention was to compare this profile with the occupational profile of the total labor force as reported in a recent government statistical publication. The analyst expected that the comparison would provide insights useful for credit policy, advertising, and billing.

    The analyst had available on computer tape occupational data obtained from a recent survey of the company's customers. Altogether, information was available for 131,845 customers who were in the labor force. The analyst concluded that a simple computer listing of the customers' occupations would be of little use since she could not digest such a large collection of raw data nor effectively compare these data with the occupational profile of the total labor force. Instead, she decided to classify all occupations into the seven categories utilized for the government labor-force study. These seven categories are shown in Table 2.1. She wrote a computer program that reads each customer's occupation from the tape (carpenter, dietitian, civil engineer, etc.), assigns it to the appropriate occupational category, and counts the number of customers falling into each category. The results are shown in Table 2.1 (column 1). Finally, so that she could effectively compare the customer profile with that of the total labor force, she computed the percent of customers who are in each occupational category (column 2 of Table 2.1). As an illustration of the calculations, $100(38,835/131,845) = 29.5$ percent of customers have managerial

TABLE 2.1   *Example of a qualitative distribution: Distribution of customers by major occupational categories*

Occupational Category	(1) Number of Customers	(2) Percent of Customers	(3) Percent of Total Labor Force
Managerial	38,835	29.5	9.2
Professional and technical	31,262	23.7	9.9
Service and recreation	14,011	10.6	10.9
Clerical and sales	12,797	9.7	20.7
Craftsmen and production workers	11,090	8.4	24.2
Laborers and unskilled workers	1,577	1.2	5.0
Others	22,273	16.9	20.1
Total	131,845	100.0 (131,845)	100.0

occupations. To facilitate the comparison, the analyst added in column 3 the occupational profile of the total labor force.

Comparing the occupational profiles in columns 2 and 3 of Table 2.1, the analyst discovered, among other things, that over half of the customers (53.2 percent) fall into the managerial and professional and technical categories whereas these categories comprise only 19.1 percent of the total labor force. This one fact alone has significant implications for the company.   □

The analyst used classification and counting to simplify her data set. She established a set of classes or categories and counted the number of elements (customers) that belong in each one. The data-classification scheme shown in Table 2.1 is called a qualitative distribution because the classes refer to a characteristic that is nonnumerical (occupation).

(2.1)   A *qualitative distribution* is the classification of the elements of a data set by a nonnumerical characteristic.

System of Classification

Any system of classification used for a qualitative distribution must meet certain formal requirements.

(2.2)   The classes in any system of classification must be *mutually exclusive* and *exhaustive*.

In other words, every element in the data set must fall into one and only one class of the system. The classification system used in Table 2.1 satisfies this requirement.

Whenever data are classified, some information is lost. The amount lost depends on the system of classification employed. For example, the classification in Table 2.1 has resulted in loss of exact information about customers' occupation types. We know, for instance, that 31,262 customers have professional and technical occupations but the qualitative distribution does not tell us how many of these are architects or economists.

The determination of the classes to be utilized in constructing a qualitative distribution therefore requires care. The objective is to select a system of classification that loses as little essential information as possible, while still achieving an effective summarization of the data. For example, the class "defective" would be too coarse for recording inspection results for a production process if it is important to distinguish between defects caused by faulty material and those caused by poor workmanship. Often, several systems of classification need be tried before a final selection is made.

Graphic Presentation of Qualitative Distributions

Often, it is useful to study a qualitative distribution by displaying it pictorially. The display usually takes the form of a *bar chart*, such as the one shown in Figure 2.1a for the qualitative distribution of customers' occupations given in Table 2.1. Each bar corresponds to one class of the qualitative distribution. The length of each bar corresponds to the number (or percent) of customers in the class.

Several aspects of the arrangement of the bars should be noted: (1) The bars differ only in length, not in width. (2) A space has been left between each bar to make it easier to identify each bar by its label. (3) The bars have been ranked by order of magnitude to facilitate analysis. The order may be a decreasing one, as here, or it may be an increasing one. If an "others" or "miscellaneous" category is present, it is usually shown as the lowest bar regardless of its magnitude, because this type of category most often is a collection of relatively unimportant classes.

When a pictorial comparison of two or more qualitative distributions is desired, it is often possible to combine their bar charts. This is done in Figure 2.1b, where the occupational profiles of the retail-store customers and the total labor force are compared. Note that both distributions are expressed in percent terms to provide a meaningful comparison. Modifications can be made in the bar charts in Figure 2.1 to show additional detail or to highlight particular features of the data. For example, the bars in Figure 2.1a could be segmented to show the numbers of male and female customers in each occupational category. Computerized systems for statistical analysis enable users to readily generate various types of bar charts for display and analysis of qualitative distributions.

2.2 FREQUENCY DISTRIBUTIONS

Often, the characteristic of interest that is classified in a distribution is quantitative (i.e., the variable has numerical outcomes). The observations for a quantitative variable can be listed in increasing or decreasing order of magnitude. Such an ordering, called an *array*, greatly facilitates inspection of the data. For example, we can easily ascertain the smallest and largest observations and can study whether there are concentrations or clusters of particular values.

FIGURE 2.1 *Examples of bar charts of qualitative distributions*

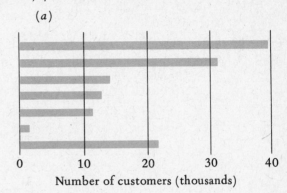

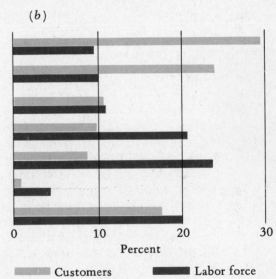

The observations can also be classified into a distribution, with or without first making an array.

☐ Example

Table 2.2 shows a frequency distribution of the balances of a bank's 30,794 savings accounts. The classes in this case are defined in terms of the amount of the balance, and each class is expressed as an interval. Thus, the first class contains all accounts that have balances under $5,000. We readily see from Table 2.2 that there are 10,196 such accounts.

The frequency distribution in Table 2.2 was prepared to assist management in revising the service-charge schedule for savings accounts transactions. It shows, for example, that over 17 percent of the savings accounts had balances of $10,000 or more. Two years ago, when a similar study was conducted, the corresponding figure was only 6 percent. ☐

TABLE 2.2 *Example of a frequency distribution: Distribution of savings accounts by amount of balance*

Amount of Balance (*dollars*)	Number of Accounts	Percent of Accounts
0–under 5,000	10,196	33.1
5,000–under 10,000	15,335	49.8
10,000–under 15,000	1,812	5.9
15,000–under 20,000	1,798	5.8
20,000–under 25,000	1,653	5.4
Total	30,794	100.0
		(30,794)

The classification in Table 2.2 is called a frequency distribution.

(2.3) A *frequency distribution* is the classification of the elements of a data set by a quantitative (i.e., numerical) characteristic.

Frequency and Percent Frequency

The number of elements in a given class of a frequency distribution is called the *frequency* of that class. When the number of elements in each class is expressed as a percent of the total number of elements in the data set, the resulting figure for a class is called the *percent frequency* of that class, and the resulting distribution is called a *percent frequency distribution*. When a percent frequency distribution is given for a data set, it is customary to show the total number of elements in the data set. This has been done in Table 2.2, where the total number of accounts (30,794) has been shown in parentheses under the 100.0 percent total. Showing the total number of elements for a percent distribution is also customary for qualitative distributions, as illustrated in Table 2.1.

2.3 CONSTRUCTION OF FREQUENCY DISTRIBUTIONS

The general principles of classification discussed for qualitative distributions apply equally to frequency distributions. Thus, the classes of a frequency distribution must be mutually exclusive and exhaustive. Also, the number of classes utilized should be small enough to provide an effective summary, yet large enough to avoid losing essential information.

In general, the construction of a satisfactory frequency distribution requires experimentation with alternative distributions since there are no rules available that fit all situations. Nevertheless, there exist some general considerations that are helpful for constructing a frequency distribution. We take these up now.

Number of Classes

The larger the number of classes in a frequency distribution, the more detail is shown. If the number of classes is too large, though, the classification loses its effectiveness for summarizing the data. Too few classes, on the other hand, condense the information so much as to leave little insight into the pattern of the distribution. The best number of classes in a frequency distribution often needs to be determined by experimentation. Usually, an effective number of classes is somewhere between 4 and 20.

☐   **Example**

In a study of the capability of untrained subjects to discriminate between different food preparations, 40 subjects were each given two food samples and asked for a rating of each sample. Unknown to the subjects, both food samples were identical in every respect. Each subject tasted both food samples and gave each sample a rating between 0 and 10. The difference between each subject's ratings for the first and second samples was taken as a measure of preference for the first sample over the second. Tables 2.3a, 2.3b, and 2.3c show the subjects' rating differences classified into 3, 7, and 21 classes, respectively.

TABLE 2.3   *Subjects' rating differences of two food samples: Three frequency distributions with different numbers of classes*

(a) Three classes		(b) Seven classes		(c) Twenty-one classes	
Rating Difference	Number of Subjects	Rating Difference	Number of Subjects	Rating Difference	Number of Subjects
−10 to − 4	13	−10 to − 8	1	−10	0
− 3 to + 3	19	− 7 to − 5	7	− 9	1
+ 4 to +10	8	− 4 to − 2	12	− 8	0
Total	40	− 1 to + 1	4	− 7	2
		+ 2 to + 4	11	− 6	3
		+ 5 to + 7	3	− 5	2
		+ 8 to +10	2	− 4	5
		Total	40	− 3	3
				− 2	4
				− 1	2
				0	0
				+ 1	2
				+ 2	3
				+ 3	5
				+ 4	3
				+ 5	1
				+ 6	2
				+ 7	0
				+ 8	1
				+ 9	0
				+10	1
				Total	40

At one extreme, the frequency distribution in Table 2.3c has so many classes that it is difficult to discern a meaningful pattern in the rating differences. At the other extreme, the frequency distribution in Table 2.3a has so few classes that any pattern present is almost lost. Finally, in Table 2.3b we have a frequency distribution that reveals the pattern of variation in the rating differences in a meaningful way. We note first that the rating differences are approximately symmetrically distributed about zero. This is not unexpected because both food samples were identical. An even more interesting observation is that the frequency of the class -1 to $+1$ is much lower than the frequencies of the classes immediately above and below it. It would appear from this observation that the subjects tended to give different ratings to the two food samples, even though the two food samples were identical. What appears to have happened is that the subjects believed the two samples must be different (for what other reason would the study be undertaken?) and consequently gave different ratings to them. These important characteristics of the rating differences, which are readily discernible from Table 2.3b, are not readily apparent from the other two frequency distributions.   □

The very clear statistical lesson provided by this example is that consideration of alternative numbers of classes is helpful for finding that frequency distribution which best yields insights into the data set.

Width of Classes

The choice of the class width or class interval is related to the determination of the number of classes. It is generally best if all the classes have the same width. If the classes are not equally wide, one often cannot tell readily whether differences in class frequencies are due mainly to differences in the concentration of items or to differences in the class widths.

Sometimes one must use unequal class intervals. Consider the construction of a frequency distribution of annual salaries of employees in a company. Assume that the annual salaries range from $5,000 to $120,000 and that about six classes are to be used in the distribution. If equal class intervals were used, each would have a width of about $20,000, the first class perhaps being $0–under $20,000. This grouping provides equal class intervals but destroys the usefulness of the classification. Most of the employees earn less than $20,000 and hence would be included in the first class. Thus, no information would be provided about the distribution of salaries of the majority of the employees except that they earned less than $20,000. Subsequent classes would not reveal much information either, since relatively few employees earn over $20,000 in this company, yet five of the six classes would be devoted to a classification of their salaries.

In cases of this nature, unequal class intervals are generally used. For instance, equal class intervals of, say, $5,000 in width might be used for the range wherein most of the salaries fall, after which the interval might increase to, say, $20,000. In fact, when all but a few of the salaries have been classified, an open-end class might be used to account for the remainder. An *open-end class* is one that has only one limit, either upper or lower. Thus, if only three officials in the company earned over $60,000, the last class might be $60,000 and more. In this way, only one class is needed to account for these top three salaries. Open-end classes may be necessary at the upper end of the distribution, as just illustrated, sometimes at the lower end of the distribution, and occasionally at both ends.

Class Limits

Still another problem is the choice of class limits. Calculations from a frequency distribution often use the midpoint of each class to represent all the items in the class. The *midpoint* of a class is the value halfway between the two class limits. It is usually suggested as good statistical practice that the class limits be chosen so that the midpoint of each class is approximately equal to the arithmetic average of the values falling in that class. Often, this can be accomplished without special concern about the class limits. If, however, the values tend to be bunched at certain periodic points throughout the range of the data—for example, rents are often multiples of $5 or $10—one may have to experiment in deciding upon class limits so that each midpoint will approximately equal the arithmetic average of the values in that class.

In stating class limits, one should be careful to be unambiguous. For instance, the limits $300–$400, $400–$500 are not clear because one cannot be sure in which class $400 is included. Stating limits as $300–$399, $400–$499 is clear when the data are expressed in dollars. When they are, the midpoint of the first class is $(300 + 399)/2 = \$349.50$, or for most practical purposes, $350.

The limits $300–under $400, $400–under $500 are clear. However, without additional information it may not be possible to determine the midpoints accurately. If no additional information is provided, we shall follow the convention of considering the midpoint of a class to be the arithmetic average of the two limits. Thus, the midpoint of the class $300–under $400 would be taken to be $(300 + 400)/2 = \$350$.

In some frequency distributions, each class corresponds to a single value, as shown in Table 2.3c earlier. As another example, in a frequency distribution of number of automobiles owned by a family, the classes might be the numbers 0, 1, 2, etc. In these cases, the individual values correspond to the class midpoints.

2.4   GRAPHIC PRESENTATION OF FREQUENCY DISTRIBUTIONS

Graphic displays of frequency distributions often are useful for presentation and analysis. Two common methods of graphic display are the histogram and the frequency polygon.

(2.4)   A *histogram* is a bar graph of a frequency distribution. A *frequency polygon* is a line graph of a frequency distribution.

Construction When Class Intervals Equal

Figure 2.2 illustrates both graphic methods when the frequency distribution has equal class intervals. Figure 2.2a presents a percent frequency distribution of ages of 27,303 farm operators classified in equal, 10-year intervals. The age data were obtained from a recent survey of farm operators conducted by a farm-equipment manufacturer. Figures 2.2b and 2.2c show a histogram and frequency polygon, respectively, of this frequency distribution. In the histogram, adjoining bars are drawn with a width spanning the class interval and a height corresponding to the frequency (or percent frequency) of that class. In the frequency polygon, each class frequency (or percent frequency) is plotted corresponding to the midpoint of that class. The plotted points are then connected by a series of straight lines. Both the histogram and frequency polygon show readily that the ages of the farm operators tend to be concentrated in the range 35 to 75 years.

Note that in both graphic presentations, the magnitudes of the class frequencies are conveyed visually by areas—in the case of histograms by the areas of the bars, and for frequency polygons by the areas under the line graph. Thus, when the class widths are equal, as they are in Figure 2.2, and the heights of the histogram bars or polygon points are equal to the class frequencies, the areas are automatically proportional to these frequencies.

FIGURE 2.2 *Illustration of graphic presentation of a frequency distribution with equal class intervals: Age distribution of farm operators*

(a)
Frequency distribution

Age (years)	Percent of Farm Operators
15–under 25	2
25–under 35	10
35–under 45	19
45–under 55	27
55–under 65	25
65–under 75	16
75–under 85	1
Total	100 (27,303)

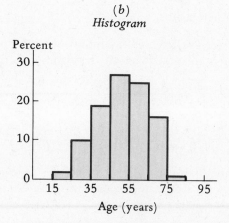

(b)
Histogram

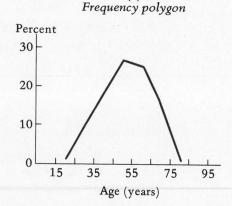

(c)
Frequency polygon

Construction When Class Intervals Unequal

When the class intervals are not equal, the heights of the histogram bars or the polygon points must be adjusted so as to make the areas proportional to the frequencies (or percent frequencies) of the classes. This is done by choosing one class width as the unit width and expressing all frequencies (or percent frequencies) in terms of this unit width. The procedure is illustrated in Figure 2.3. Figure 2.3a

---

FIGURE 2.3 *Illustration of graphic presentation of a frequency distribution with unequal class intervals: Income distribution of taxpayers*

(*a*)

Total Income (*dollars*)	Percent of Taxpayers	Number of $1,000 Widths	Percent of Taxpayers per $1,000 Width
0–under 1,000	.6	1	.60
1,000–under 2,000	10.2	1	10.20
2,000–under 3,000	13.4	1	13.40
3,000–under 5,000	29.3	2	14.65
5,000–under 10,000	36.4	5	7.28
10,000–under 20,000	10.1	10	1.01
Total	100.0		
	(1,049,916)		

(*b*)

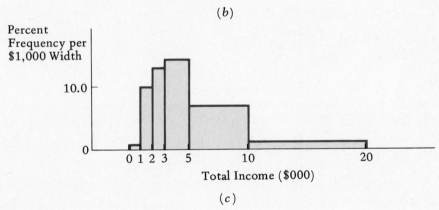

(*c*)

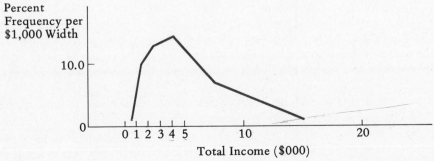

contains the percent frequency distribution of incomes of taxpayers in a state who have incomes under $20,000. The distribution is of interest to a legislative committee reviewing income-tax exemptions for lower- to middle-income taxpayers of the state. The interval of $1,000 has been chosen as the unit width, and all percent frequencies have been expressed in terms of this unit width. Thus, the percent frequency for the class 3,000–under 5,000 in terms of the unit width is $29.3/2 = 14.65$ because that class has a width of 2 thousand-dollar units. The procedure then is to plot the percent frequencies per unit width in the usual fashion. This has been done in Figures 2.3b and 2.3c.

Comparison of Frequency Polygons

Two or more frequency polygons can be compared readily if: (1) they have the same class intervals, and (2) they have the same total frequency (or are both expressed in percent form). Figure 2.4 provides an illustration. It shows the age distribution of farm operators (taken from Figure 2.2c) as well as that of the U.S. nonfarm civilian labor force (as reported in a recent economic study), each expressed in percent frequencies. It is easy to compare the two distributions from Figure 2.4. For example, we readily see that the farm operators are relatively much older than the nonfarm civilian labor force.

In contrast to frequency polygons, it is quite difficult to obtain an effective comparison of two or more frequency distributions by superimposing their histograms.

2.5   CUMULATIVE FREQUENCY DISTRIBUTIONS

When one wishes to know what proportion of the elements of a data set have values above or below certain levels, the cumulative form of the frequency distribution is a very effective device for summarizing this information.

(2.5)   The cumulative form of a frequency distribution is called a *cumulative frequency distribution*. A graph of a cumulative frequency distribution is called an *ogive*.

Figure 2.5a repeats the frequency distribution of the ages of farm operators, Figure 2.5b shows the cumulative distribution, and Figure 2.5c contains the ogive. The cumulative distribution in Figure 2.5b is called a *less-than cumulative frequency distribution* because, for each value of the quantitative characteristic (age, in this example), the number (or percent) of elements having less than that value is recorded. For instance, the cumulative age distribution indicates that 2 percent of farm operators are less than 25 years of age, and 12 percent are less than 35 years of age.

In graphing the cumulative distribution, the cumulative number (or percent) of elements is plotted against the corresponding value of the variable. Thus, the points plotted begin with (15, 0) and (25, 2) and end with (85, 100). The points are then connected by straight lines to permit interpolation. Thus, one can estimate

FIGURE 2.4 *Age distributions of farm operators and U.S. nonfarm civilian labor force*

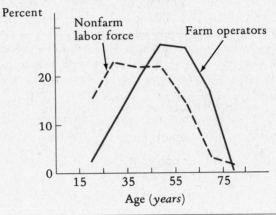

FIGURE 2.5 *Less-than cumulative age distribution of farm operators*

(a) Frequency distribution		(b) Cumulative frequency distribution	
Age (years)	Percent of Farm Operators	Less Than This Age (years)	Cumulative Percent of Farm Operators
15–under 25	2	15	0
25–under 35	10	25	2
35–under 45	19	35	12
45–under 55	27	45	31
55–under 65	25	55	58
65–under 75	16	65	83
75–under 85	1	75	99
Total	100	85	100

(c)
Ogive

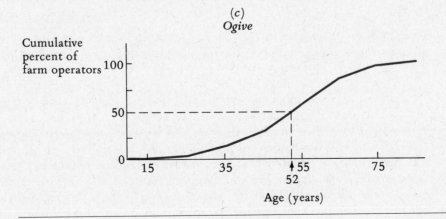

from Figure 2.5c (see the dashed line in the figure) that 50 percent of farm operators are less than 52 years of age.

Comments

1. When straight lines are used to connect the plotted points of the ogive, it is assumed that the observations within each class are uniformly distributed throughout the class. Since this condition is usually not met exactly, the resulting graph is only an approximation to the exact distribution of the values. The straight-line approximation is sufficiently good in many cases.

2. A situation where it is *not* appropriate to use straight lines to connect the known points of an ogive occurs when the discrete nature of the data set must be recognized. In this case, the ogive should be drawn as a step function. For example, Figure 2.6 shows the cumulative frequency distribution and the ogive of family size for families belonging to a community recreation center. Since family size must be a whole number, the ogive has a "step" at each whole number. Note that the cumulative frequency distribution in this case shows the percent of families with size equal to or less than the given number.

3. An ogive will not reach the 100 percent level if the frequency distribution has an open-end class at the upper end. The reason is that the maximum value of the elements is not known in this case. A similar situation arises with the 0 percent level when there is an open-end class at the lower end of the distribution.

4. A cumulative distribution and ogive can be constructed so they indicate the percent (or number) of items having magnitudes "more than" any specified value. The resulting distribution is called a *more-than cumulative frequency distribution*. The method of construction is analogous to that for the less-than cumulative distribution and ogive. For example, the more-than cumulative distribution of the ages of farm operators, based on the data in Figure 2.5a, is:

This Age or More (*years*)	Cumulative Percent of Farm Operators
15	100
25	98
35	88
45	69
etc.	etc.

5. In plotting an ogive, no special procedures are required when the class intervals are unequal. (Recall that unequal class intervals entail special procedures for constructing a histogram or frequency polygon.)

6. Ogives of different frequency distributions can be compared readily on the same graph if the total frequencies are the same or if percent frequencies are used. Class intervals of the respective distributions need not be the same.

2.6   MULTIVARIATE DATA CLASSIFICATION

When data sets are multivariate and it is desired to classify the elements in the set simultaneously in terms of two or more characteristics, the classification process is called *cross-classification* or *cross-tabulation*. The principles discussed earlier for classification systems for univariate data sets apply equally to multivariate data sets.

FIGURE 2.6 *Example of ogive for discrete data: Cumulative distribution of family size.* The points for the cumulative distribution are plotted and the ogive is drawn as a step function.

(a)
Cumulative frequency distribution

This Size or Less	Cumulative Percent of Families
1	0
2	33
3	57
4	76
5	87
6	100

(b)
Ogive

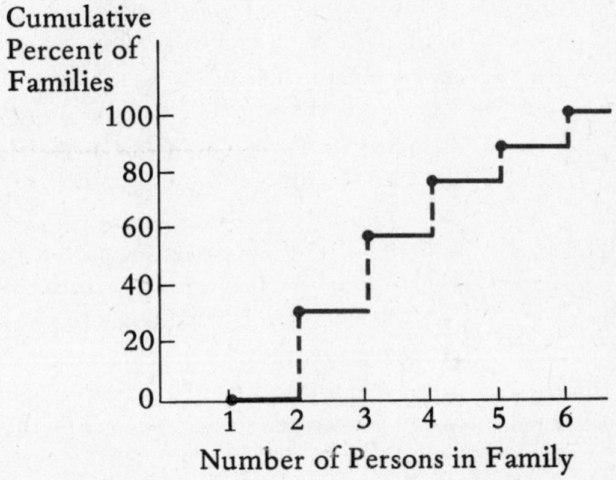

Bivariate Cross-Classification

We shall illustrate cross-classification for bivariate data sets by two examples.

☐ Examples

1. A European government tourist bureau compiled a list of 322 escorted tours of 8 days or less, which are offered by various private agencies. There are two characteristics of interest for each tour in the data set—its duration in days and its cost in U.S. dollars. To summarize the data set with respect to these two characteristics, the bureau set up the classification system shown in Table 2.4. Each tour from the data set is counted in that cell of the table which corresponds to its duration in days and its cost category.

   The data-classification scheme in Table 2.4 is called a *bivariate frequency distribution* because it is the result of cross-classifying two quantitative variables, duration and cost. Note that when the frequencies in the table are summed by row and column, one obtains univariate frequency distributions of the two variables considered separately at the margins of Table 2.4.

TABLE 2.4 *Escorted tours cross-classified by cost and duration*

Cost (U.S. $)	Duration (days)						Total
	3	4	5	6	7	8	
0– 300	53	43	9	–	–	–	105
301– 600	20	45	37	12	26	1	141
601– 900	4	9	11	6	17	12	59
901–1,200	–	–	1	1	5	10	17
Total	77	97	58	19	48	23	322

The bivariate frequency distribution in Table 2.4 has summarized the data set in a way that brings a number of interesting facts to light. One is that the cost of an escorted tour tends to vary directly with its duration—a very reasonable relationship. Another is that an escorted tour of 3 days, costing up to $300, is the most commonly offered type of tour.

2. In Figure 2.7, a bivariate classification is shown for 5,722 psychiatric patients cross-classified by: (1) their diagnosis, and (2) whether or not they were admitted to a psychiatric in-patient facility within the past year. In this cross-classification, four values are given in each cell of the distribution. The topmost value represents the actual count or frequency for the cell, i.e., the number of patients in the cell. Thus, the values in the first position constitute the bivariate distribution of the data set.

The second value in each cell is the percent of patients in that row who are in the cell. The third value is the percent of patients in that column who are in the cell. The fourth value is the percent of all 5,722 patients who are in the cell. Thus, the values in the fourth position constitute the bivariate percent distribution of the data set.

From Figure 2.7 we can note several facts which might be of interest to a health-care administrator. For instance, over one-third of all admissions to in-patient facilities were schizophrenics (37.4 percent). Also, proportionately fewer patients with neuroses were admitted to in-patient facilities (7.2 percent) than for any other diagnostic class. ☐

Multivariate Cross-Classification

Classification systems where more than two characteristics are studied simultaneously may be set up along the same lines as those in Table 2.4 and Figure 2.7.

Computer-Assisted Classification

Classifying and cross-classifying large data sets used to require enormous amounts of tedious clerical work. Today, large data sets are readily handled automatically in computerized information-processing and data-sorting systems. The data may be recorded on punch cards, magnetic tapes, or magnetic disks. High-speed printers and visual-display screens are two of the common devices used for presenting the results of computerized data classification.

A number of computer software packages are available which are designed specially to facilitate data classification. The user of these packages has considerable flexibility with respect to the variables to be classified, the specifications of the classification systems to be employed, and the selection of reporting formats for presenting the results. Because of the high speeds and large memories of computers, these packages allow the user to handle large data sets, to demand extensive classifications, and to investigate the data from many vantage points.

Many of these computer packages allow the user to experiment with different classification schemes in a real-time interactive manner. Thus, the user may request that the data be classified according to a trial scheme. The data set is then displayed in that format immediately. The user then studies the classified data. If some aspect looks interesting, the user can request a reclassification of the data set according to a new scheme which will bring that interesting feature into clearer

FIGURE 2.7 *Example of computer-assisted classification: Psychiatric patients cross-classified by diagnosis and in-patient admission status*

```
                     ADMIT
            COUNT  I
            ROW PCT INO          YES           ROW
            COL PCT I                          TOTAL
            TOT PCT I      0   I     1   I
DIAGNOSE    --------I--------I--------I                      Frequency in cell
         1  I    1192   I    306   I   1498
SCHIZOPHRENIA  I  79.6  I   20.4   I   26.2
            I    24.3   I   37.4   I                         Percent of row
            I    20.8   I    5.3   I                         Percent of column
         -I--------I--------I                                Percent of total
         2  I    1126   I    214   I   1340                  Frequency of row
DEPRESSION  I    84.0   I   16.0   I   23.4                  Percent of total
            I    23.0   I   26.2   I
            I    19.7   I    3.7   I
         -I--------I--------I
         3  I     323   I     58   I    381
DISEASES OF AGE  I  84.8  I   15.2   I    6.7
            I     6.6   I    7.1   I
            I     5.6   I    1.0   I
         -I--------I--------I
         4  I     149   I     38   I    187
OTHER PSYCHOSES  I  79.7  I   20.3   I    3.3
            I     3.0   I    4.6   I
            I     2.6   I    0.7   I
         -I--------I--------I
         5  I     595   i     46   I    641
NEUROSES    I    92.8   I    7.2   I   11.2
            I    12.1   I    5.6   I
            I    10.4   I    0.8   I
         -I--------I--------I
         6  I     713   I     56   I    769
CHARACTER DISORD I  92.7  I    7.3   I   13.4
            I    14.5   I    6.8   I
            I    12.5   I    1.0   I
         -I--------I--------I
         7  I     259   I     41   I    300
ADDICTIONS  I    86.3   I   13.7   I    5.2
            I     5.3   I    5.0   I
            I     4.5   I    0.7   I
         -I--------I--------I
         8  I     201   I     17   I    218
RETARDATION I    92.2   I    7.8   I    3.8
            I     4.1   I    2.1   I
            I     3.5   I    0.3   I
         -I--------I--------I
         9  I     346   I     42   I    388
OTHER       I    89.2   I   10.8   I    6.8
            I     7.1   I    5.1   I
            I     6.0   I    0.7   I
         -I--------I--------I
   COLUMN      4904        818       5722          Total
    TOTAL      85.7       14.3      100.0
```

Frequency of column   Percent of total

focus. The desired reclassification of the data is carried out immediately and again displayed for the user. The user is able to classify and reclassify the data set in this manner until satisfied that it has been fully analyzed. The bivariate frequency distribution in Figure 2.7 illustrates output from one such computer package, SPSS (Statistical Package for the Social Sciences).

2.7   EXPLORATORY DATA ANALYSIS

We have seen that frequency distributions are an effective means of initially summarizing a data set so that basic patterns emerge. We shall now discuss two other analytical techniques which supplement frequency distribution analysis to further assist in the exploratory examination of the underlying structure of a data set. These techniques are *stem-and-leaf displays* and *residual plots*, both of which are frequently used in exploratory data analysis.

Stem-and-Leaf Displays

A stem-and-leaf display provides information about the individual observations on a variable in a data set so that underlying patterns can be seen. It combines the features of an array and a histogram.

◻   **Example**

An analyst, commissioned to study characteristics of concert subscribers in a large metropolitan area, pretested the questionnaire on a sample of 70 subscribers. Figure 2.8a shows a stem-and-leaf display for the responses to the question on subscriber's age. The digits to the left of the vertical line are the "stems" and the digits to the right are the "leaves." The digit 2 constitutes the first stem. All age responses from 20 to 29 inclusive are recorded on this stem. Only the second digit in each age response is written because the first digit is given by the stem. We see that the leaves on the first stem consist of the digits 0, 5, 5, 5, 9. This indicates that five respondents reported ages in the interval 20–29; one reported 20, three reported 25, and one reported 29.

The analyst, viewing the display as a histogram, concluded that it fitted the usual age pattern for concert subscribers. In turning to the specific responses, however, he noted a disproportionately large number of 0s and 5s among the second digits (leaves) in the age responses. This was unanticipated since he had expected each final digit from 0 to 9 to be roughly equally represented. We shall further consider this finding in the next section.   ◻

Note how the stem-and-leaf display provides a histogram of the observations in a data set and also shows special concentrations of observations.

Residual Plots

Residual plots are useful for highlighting major departures in the observed values of a data set from anticipated patterns. A *residual* is the difference between an observed value and the corresponding anticipated value. Often, the anticipated values are based on theoretical models or previous experience.

FIGURE 2.8 *Stem-and-leaf display and residual plot for concert subscribers example*

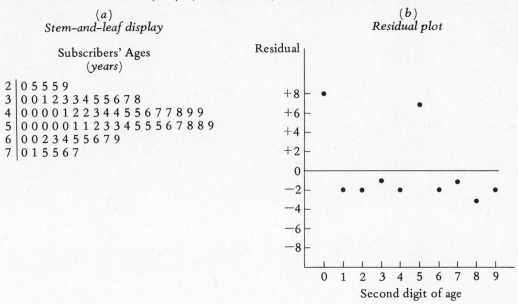

(a)
Stem–and–leaf display

Subscribers' Ages
(*years*)

```
2 | 0 5 5 5 9
3 | 0 0 1 2 3 3 4 5 5 6 7 8
4 | 0 0 0 0 1 2 2 3 4 4 5 5 6 7 7 8 9 9
5 | 0 0 0 0 0 1 1 2 3 3 4 5 5 5 6 7 8 8 9
6 | 0 0 2 3 4 5 5 6 7 9
7 | 0 1 5 5 6 7
```

(b)
Residual plot

☐ **Examples**

1. Figure 2.8b shows a residual plot for the concert subscribers example where the analyst anticipated that the final digits in age would occur with equal frequency. Since the data set contains 70 readings and 10 digits are involved, each digit should appear about 7 times in the leaves under the expectation of equal representation. The digit 0 appears 15 times, so the residual for this digit is $15 - 7 = 8$. This residual is represented by the leftmost dot in the residual plot, corresponding to 0 on the horizontal scale and 8 on the vertical scale. The digit 1 appears 5 times, hence the residual is $5 - 7 = -2$ and the dot is plotted at 1 on the horizontal scale and -2 on the vertical scale. The residuals for the other digits are calculated and plotted in the same manner. Note how readily the residual plot shows the excessive number of 0 and 5 final digits and the correspondingly lower-than-anticipated frequencies for the other digits.

   The analyst concluded that the most likely explanation for this behavior of the residuals is end-digit preferences by some respondents who tend to round their reported ages to multiples of 5 or 10. The analyst therefore decided to ask for year of birth on the questionnaire to be used in the large-scale study and then calculate a subscriber's age from the year of birth while editing the questionnaire.

2. (Adapted from Ref. 2.1, pp. 230–31.) A company which manufactures small motors learned that some of the motors produced recently were breaking down. One possible cause was a rod in the motor being too loose in the bearing within which it rotates. Other possible causes included defects in the bearing. A consultant who was asked to locate the cause of the breakdowns learned that the diameter of each rod is inspected by the manufacturing department and that any rod with a diameter less than the lower tolerance limit of 1.000 centimeter is to be scrapped. A histogram of the diameters of rods recently inspected, based on records kept by the inspectors, is shown in Figure 2.9a.

FIGURE 2.9 *Histogram and residual plot for motor rod example*

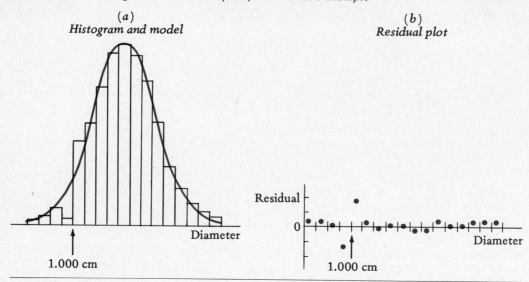

Also shown there is a smooth curve that the consultant drew as the expected model for the data. This model reflects the pattern of variability found by the consultant in similar situations in other companies.

Note that the histogram and the model are in close accord except in the interval around the lower tolerance limit, which is marked by an arrow in the figure. The residuals for each class were obtained and are graphed in Figure 2.9b. The consultant was interested in the size of the large negative residual just below the tolerance limit and the large positive residual in the next-higher class. Some systematic factor appeared to be operating here which tended to create large residuals in opposite directions in the two adjacent classes. Upon further investigation, the consultant found that the inspectors were under pressure to keep the scrappage rate low and had been misclassifying borderline-defective rods by recording their diameters as borderline-acceptable. This misclassification led to too few rods recorded just below 1.000 centimeter and too many rods recorded just above the tolerance limit. Further investigation showed that the misclassified rods were in most cases the cause of the motors breaking down. □

These two examples illustrate how effectively residual plots can point to major differences between the observed values in a data set and anticipated patterns based on theory or related experience. Residual plots often also provide useful clues for investigating the causes of the departures from anticipations. We shall discuss the use of residual plots for some other situations in later chapters.

Cited Reference

2.1 Deming, W. Edwards. "Making Things Work." Pages 229–36 in *Statistics: A Guide to the Unknown.* Edited by J. W. Tanur. San Francisco: Holden-Day, 1972.

PROBLEMS

2.1 A bus company in a large northeastern city has categorized complaints from riders for last May and June as follows:

	Number of Complaints	
Type of Complaint	May	June
Buses too hot	14	31
Fare too high	11	12
Buses sometimes don't stop at bus stops	17	16
Bus stops have no shelters	10	21
Bus travel too expensive	12	13
Buses should be air conditioned	14	41
Drivers are rude	13	11
Bus schedules make transfers difficult	14	14
Route maps not available	25	21
Total	130	180

  a. For each month, classify the complaints data using the complaint categories comfort, service, cost.
  b. Would other persons necessarily assign the complaints to the same complaint classes in **a** as you have? What is the implication of this for interpreting qualitative distributions for free-answer data classified by others?
  c. Construct a bar chart patterned after Figure 2.1b to compare the distributions for the two months. Briefly report your findings.

2.2 Refer to Problem 2.1
  a. For each month, classify the complaints data using the complaint categories bus schedules, bus drivers, equipment, other.
  b. Construct a bar chart patterned after Figure 2.1b to compare the distributions for the two months. Briefly report your findings.

*2.3 Refer to Appendix D.
  a. Construct a qualitative distribution of companies classified by industry.
  b. Construct a bar chart of your distribution.
  c. Which industry has the greatest representation among companies in the data set? If the three electronic industrial categories were combined to form a single class, would your answer change? What is the implication of this for the construction and interpretation of qualitative distributions?

2.4 Refer to Figure 1.2 (p. 10).
  a. Do the response classes for Question 7b satisfy the formal requirements of a system of classification as defined in (2.2)? Explain.
  b. Do the choices in Question 7d constitute a system of classification as defined in (2.2)? Explain.

*2.5. Twenty-five merchants donated the following dollar amounts in a fund-raising campaign to beautify a nearby park: 70, 90, 25, 50, 30, 100, 75, 50, 125, 100, 150, 15, 80, 100, 50, 25, 50, 65, 60, 100, 50, 100, 75, 90, 80.
  a. Array these observations by increasing order of magnitude.
  b. (1) What are the smallest and largest donations? (2) What percent of the donations are for $100 or more? (3) What is the most frequently contributed dollar amount?

2.6 Refer to Problem 2.5. Answer both parts for the following dollar donations: 150, 200, 100, 175, 140, 150, 225, 125, 140, 165, 180, 150, 175, 145, 200, 125, 145, 225, 100, 150, 200, 125, 175, 150, 160.

*2.7 Refer to Problem 2.5.
  a. Construct a frequency distribution for the donation data using six classes with equal class widths and midpoints which are multiples of $25.
  b. Plot the frequency distribution as a histogram.

2.8 Refer to Problem 2.6. For the donation data given there:
  a. Construct a frequency distribution using six classes with equal class widths and midpoints which are multiples of $25.
  b. Plot the frequency distribution as a frequency polygon.

2.9 For each of the following systems of classification, indicate whether or not it meets the formal requirements (2.2):
  a. Classes $0–$3.00, $3.00–$10.00, $10.00 or more, for hourly wage rates.
  b. Classes 3 or less, 4 or more, for number of children in families.
  c. Classes 0°–under 90°, 100°–under 190°, 200° and over, for the temperature of ovens.

2.10 Refer to the response classes for Question 7b in Figure 1.2 (p. 10). (1) Do the classes have equal widths? (2) Is there an open-end class? (3) What is the midpoint of the second class? Of the third class? Explain any assumptions you made in your answers.

2.11 A frequency distribution of monthly rents in middle-income housing is to be constructed. The rents tend to be in multiples of $25 and range from $275 to $475. Class intervals of $25 are to be employed. What class limits would be most appropriate? Why?

*2.12 The distribution of response times for alarms answered last year by the volunteer fire company in a township follows.

Response Time (minutes)	Number of Alarms
2.5–under 7.5	5
7.5–under 12.5	18
12.5–under 17.5	15
17.5–under 22.5	9
22.5–under 27.5	3
Total	50

  a. Convert the frequency distribution into a percent distribution and plot it as a frequency polygon and as a histogram, in separate graphs.
  b. Which of the two graphs would be more effective for comparing the distribution of response times with that for a fire department in a nearby township? Why?

2.13 Average weekly wages and salaries (in dollars) in 60 urban areas are:

209	255	186	195	210	191	227	175	175	207
201	182	225	198	202	221	179	246	209	235
192	183	210	189	227	203	201	187	231	186
204	210	198	221	200	197	210	172	231	237
169	182	208	205	199	205	178	218	253	209
225	189	200	250	176	165	215	245	215	185

a. Construct a frequency distribution of these data using equal class intervals. Justify your choice of classes.

b. Present your distribution as a frequency polygon and as a histogram, in separate graphs. Analyze your graphs and briefly describe your findings.

c. Suppose you had constructed a distribution of the individual weekly wages and salaries of the wage earners in the 60 urban areas. How would you expect this distribution to compare with that in **a**? Explain. What generalization does this suggest?

2.14 The age distribution of horses competing in the Winston County Trials during a recent year follows.

Age (years at last birthday)	Number of Horses	Age (years at last birthday)	Number of Horses
5	26	10	11
6	60	11	8
7	51	12	3
8	44	13	0
9	23	14	1
		Total	227

a. What is the midpoint of the age class 5?

b. Convert this frequency distribution to a percent distribution and graph it as a frequency polygon and as a histogram, in separate graphs.

c. Does it appear from the graphs that horses under 5 years of age were not permitted to compete? Horses over 14 years of age? Explain.

d. Suppose the frequencies for the age classes 5 and 6 had been reversed to give 60 horses aged 5 and 26 horses aged 6. Would this new pattern for the distribution indicate that an unusual factor is affecting the age data? Explain.

2.15 The age distribution (to nearest birthday) of employees who were absent more than 10 days last year follows.

Age Class	Percent of Employees	Age Class	Percent of Employees
15–19	7	40–49	7
20–24	18	50–64	4
25–29	30	65–79	1
30–39	33	Total	100

a. Would you agree that employees in the 30–39 age class had the worst absentee record of all the employees, and that the employees in the 25–29 age class were a close second? Explain.

b. Construct a histogram of the age distribution, using five years as the unit width. Does your graph support your answer in **a**? Explain.

2.16 Refer to Appendix D.

a. Construct frequency distributions of 1980 and 1981 net assets of the textile apparel manufacturers. Use identical classes for both distributions. Do not use open-end classes, but use unequal class widths if appropriate.

b. Plot the two distributions on one graph using frequency polygons. How do the two distributions compare? What unit width did you use in constructing your graphs?

2.17 Refer to Problem 2.16. Answer both parts for the textile products companies.

*2.18 Refer to Problem 2.12.
a. Construct a less-than cumulative percent distribution of the response times. Use classes less than 2.5, less than 7.5, etc.
b. Plot the ogive of the cumulative percent distribution in **a**. Reading from the ogive, what percent of calls had a response time of less than 20 minutes?

2.19 The distribution of a company's stock according to the number of shares held (including fractional shares) follows.

Number of Shares Held	Percent of Stockholders
0–under 25	41
25–under 50	25
50–under 100	19
100–under 500	12
500 and over	3
Total	100

a. Construct a less-than cumulative percent distribution. Use classes less than 0, less than 25, etc.
b. Plot the ogive of the cumulative distribution in **a**. Reading from the ogive: (1) What percent of the stockholders held fewer than 75 shares each? (2) The 50 percent of the stockholders with the most shares each held what number of shares or more?

2.20 Refer to Problem 2.19. Answer both parts, but construct a more-than cumulative distribution. Use classes 0 or more, 25 or more, etc.

2.21 Refer to Problem 2.14.
a. Construct a less-than cumulative percent distribution. Use classes less than 5, less than 6, etc.
b. Plot the step-function ogive patterned after Figure 2.6 for your cumulative percent distribution in **a**. Did you treat the age variable here as continuous or discrete? Explain.

*2.22 Refer to Figure 2.7.
a. (1) What percent of all the patients were admitted? (2) What percent of patients diagnosed as schizophrenic were admitted? (3) What percent of patients admitted were diagnosed as schizophrenic? (4) Which diagnostic class contributed least to the number of patients admitted?
b. Present the univariate frequency distribution of patients by in-patient admission status.

2.23 Refer to Table 2.4.
a. What is an element of this data set? How many elements are there?
b. Identify the variables in this data set. Present the univariate frequency distribution for each variable.
c. If you had *only* the univariate frequency distributions in **b**, could you construct Table 2.4? Explain.

2.24 Refer to Table 2.4.
a. (1) How many tours cost $300 or less? (2) How many tours have a duration of three days? (3) How many tours cost $300 or less *and* have a duration of three days?
b. What is the second most commonly offered type of tour?

*2.25 Refer to Figure 1.1 (p. 3).
   a. Set up a cross-classified table for age and sex of officers. Use age classes under 30, 30–39, 40 and over.
   b. Is there a relationship between age and sex for the officers in the township? Explain.

2.26 Refer to Figure 1.1 (p. 3).
   a. Set up a cross-classified table for systolic blood pressure and age of officers. Use blood-pressure classes under 123, 123–127, 128 and over; use age classes under 25, 25–34, 35 and over.
   b. Is there a relationship between blood pressure and age for the officers in the township? Explain.

2.27 The coded observations that follow pertain to errors made in handling "emergencies" in training sessions on a flight simulator. The observations are cross-classified by phase of flight—takeoff (T), cruising (C), or landing (L)—and by cause of error—misinterpretation of instrument readings (M), or other cause (O). For instance, the coded observation TM involves a takeoff error caused by misinterpretation.

TM	TO	LM	LO	CO	TM	CM	LO	TM	CO
LO	CM	LO	TM	LO	CM	TM	CO	LO	LM
TM	TO	TO	LM	TM	LO	LO	TM	TM	LO
LO	CO	LO	LM	TM	LM	TM	LM	TM	CM
TM	TM	TO	LO	TO					

   a. Construct the bivariate distribution for these observations.
   b. (1) What percent of the M-type errors occurrred in the landing phase? (2) What percent of the landing-phase errors were of the O type?
   c. Is there a relationship between cause of error and phase of flight for these observations? Explain.

*2.28 Refer to Problem 2.13.
   a. Make a stem-and-leaf display for the data on average weekly wages and salaries. Use as stems 16, 17, etc. What class intervals in a frequency distribution would correspond to your stems here?
   b. A social and economic goals committee has suggested a target of $195 or higher for the average weekly wage and salary in each of the 60 areas. Use your stem-and-leaf display to count the number of areas currently below this target. Could you have made this determination equally well from the corresponding frequency distribution? Explain.

2.29 The following data were collected on average daily hospital costs paid by a medical insurance program in 1980 in the 50 U.S. states, Puerto Rico, and the District of Columbia. (The figures are rounded to the nearest dollar for convenience.)

157	215	227	292	218	205	242	208	284	209
174	213	167	212	172	154	157	145	176	186
219	268	218	165	135	171	152	152	241	168
182	206	226	149	141	187	182	218	189	235
153	130	155	176	209	158	167	231	149	178
169	149								

   a. Make a stem-and-leaf display for these data. Use as stems 13, 14, etc. Some of the stems contain no leaves. Did you include such stems in your display or did you omit them? Justify your procedure.
   b. Which cost observations appear to you to be unusually small or large?

*2.30 Eight packers in a mail-order department fill orders for shipment to customers. Packers 1–4 work full-time during the day, and packers 5–8 work full-time at night. Orders are assigned to packers at random, and all packers fill orders at about the same speed. Of the orders filled by the department during a recent four-month period, 60 proved to be incorrectly filled. The packers' identification numbers for these orders are as follows (number 1 refers to packer 1, etc.):

2	7	4	1	7	1	3	4	4	7
1	5	7	3	8	7	6	3	7	2
3	8	3	8	3	7	3	5	3	4
7	3	1	7	3	8	7	8	5	5
8	6	7	2	3	2	6	3	1	7
5	3	6	2	7	1	7	3	8	4

a. What are the anticipated numbers of incorrectly filled orders for each packer if all packers are equally accurate and time of the shift does not affect the accuracy of filling orders? Explain your reasoning.

b. Construct a residual plot to assess the comparative accuracy of the packers during this four-month period. What does your plot show?

2.31 Technicians in a health survey were instructed to take and record blood-pressure readings for each subject to the nearest millimeter. The distribution of end digits in 500 blood-pressure readings expressed to the nearest millimeter follows.

End digit:	0	1	2	3	4	5	6	7	8	9
Number of readings:	60	0	80	0	112	40	88	0	120	0

a. What are the expected frequencies for the end digits if all were equally likely?

b. Construct a residual plot to assess whether each end digit actually was equally likely. What does your plot show? Discuss.

2.32 A new product was expected to show a 10 percent sales increase each quarter for the first three years, starting with a sales volume of 40 thousand units. Actual and expected sales during this period were as follows (in thousands):

Quarter:	1	2	3	4	5	6	7	8	9	10	11	12
Actual:	37.9	46.2	51.1	50.4	55.7	67.0	74.4	74.8	84.0	96.5	105.7	111.4
Expected:	40.0	44.0	48.4	53.2	58.6	64.4	70.9	77.9	85.7	94.3	103.7	114.1

Obtain the residuals and plot them against time. What does your plot show? Discuss.

STUDIES

2.33 Refer to Problem 2.29. An analyst wishes to investigate whether the frequency distribution of costs has one or two peaks and whether any of the areas has relatively extreme costs. To determine the most effective number of classes, construct frequency distributions with 3, 5, 7, 9, and 11 classes, each using equal class intervals. Plot the histograms for these frequency distributions on different graphs. Which frequency distribution appears to you to be most effective here? Explain.

2.34 The distributions of number of days of bed care for recent burn cases in two treatment centers follow.

Number of Days	Center A	Center B
Under 5	174	52
5–under 10	225	120
10–under 15	372	210
15–under 25	159	36
25–under 50	180	80
Total	1,110	498

    a. Plot frequency polygons of these distributions on the same graph for ready comparison. Use 0 as the lower limit of the first class and employ a unit width of 5 days.

    b. Compare the frequency polygons plotted in **a** and state your findings.

    c. Plot less-than ogives of the two distributions on the same graph for ready comparison. Explain how your findings in **b** are indicated by the ogives.

    d. Reading from your ogives: (1) What percent of the cases entailed a bed stay of less than 8 days in center A? In center B? (2) The longest 25 percent of the bed stays in center A entailed a stay of how many days or more? What is the comparable figure for center B?

2.35 In a large training experiment for a new welding technology, trainees were classified by previous welding experience—no previous experience (A_1) or previous experience (A_2)—and half of each group was randomly assigned to a concentrated program (B_1) while the other half was assigned to an extended program (B_2). The following four tables show different possible outcome patterns for the percents of trainees in each experience-training group who later used the new technology successfully:

Pattern 1	B_1	B_2		Pattern 2	B_1	B_2		Pattern 3	B_1	B_2		Pattern 4	B_1	B_2
A_1	60%	60%		A_1	60%	90%		A_1	60%	60%		A_1	60%	80%
A_2	60%	60%		A_2	60%	90%		A_2	90%	90%		A_2	70%	90%

Interpret each of the four possible outcome patterns as to the effects of previous experience and type of training on the likelihood of successful use of the new technology.

2.36 Refer to Appendix D.

    a. Construct a bivariate frequency distribution of 1980 and 1981 net sales for textile apparel manufacturers. Consider such points as: (1) number of classes to be used for each variable, (2) use of equal or unequal class widths, and (3) possible use of open-end classes.

    b. Is there a relationship between 1980 and 1981 net sales for these manufacturers? Discuss.

2.37 Refer to Problem 2.36. Answer both parts for the textile products companies.

2.38 A city police official has released the following data on the outcomes of charges for serious crimes by type of crime, for last year and five years ago. The data are annual data, and exclude pending cases.

Outcome of Charge	Five Years Ago		Last Year	
	Violent Crimes	Property Crimes	Violent Crimes	Property Crimes
Adults—guilty	2,499	10,402	3,152	15,704
Adults—acquitted or dismissed	1,427	3,754	4,461	10,543
Referred to juvenile court	1,159	14,970	1,925	13,461
Total	5,085	29,126	9,538	39,708

Construct appropriate tables and graphs to answer the following questions, and discuss your findings:

a. For adults only, has the proportion of charges that led to acquittal or were dismissed changed between five years ago and last year, for each type of crime?

b. Combining both types of crimes, what changes have taken place in the relative occurrence of each outcome of charge between five years ago and last year?

c. Are your findings in b largely the result of changes in violent crimes, in property crimes, or in both types of crimes?

3
Data Summarization

The preceding chapter has shown how the elements of a data set can be classified to facilitate reporting, interpretation, and analysis. In this chapter, some measures that summarize important characteristics of univariate data sets are presented. These measures fall into three broad categories: measures of position, dispersion, and skewness.

3.1 MEASURES OF POSITION

Measures of position for a data set describe where the values are concentrated and are useful in many situations.

☐ Illustration

In 1975, a government agency was established to monitor and control pollution of an inland waterway. Since then, the agency has recorded, among other things, occurrences of oil spills in the waterway. An agency analyst now wishes to compare the time intervals between oil spills after the agency was established with comparable data available for a limited period prior to the agency's establishment. The intention is to determine whether or not the agency's controls have lengthened the period between oil spills, which, of course, would imply that the frequency of oil spills has been reduced.

The data sets for the two periods are given in Table 3.1a; the values shown represent time intervals in days between oil spills. The data sets are classified into percent frequency distributions in Table 3.1b to facilitate comparison.

A comparison of the two frequency distributions indicates that the intervals between oil spills have tended to be longer after the creation of the agency. This is evident because: (1) the proportion of cases where the time interval is under 15 days has decreased from 37.5 percent to 30.3 percent, and (2) the proportion of cases where the time interval is long has increased. A simple summary measure of this lengthening in the intervals between oil spills is desired. ☐

A summary measure of the location of a frequency distribution is a number which provides information about where the distribution is located. A comparison

TABLE 3.1 *Time intervals between oil spills in an inland waterway*

(a)
Number of days between oil spills

Before Pollution Controls			After Pollution Controls							
17	16	4	37	38	87	6	20	11	22	3
3	1	7	3	11	4	39	17	23	37	21
5	37	15	21	10	12	20	15	2	20	15
31	2	22	3	18	45	36	9	15	51	24
20	36	6	31	23	49	33	5	36	161	12
64	16	45	1	2	4	33	31	31	5	163
31	11	1	13	22	23	1	5	35	32	2
2	24	38	0	6	18	32	16	25	131	31
14	41	18	1	27	25	33	0	27	34	20
48	19	1	17	19	29	31	21	17	74	1
21	73	20	19	32	32	17	38	1	21	135
28	19	5	12	27	56	24	34	120	12	18
3	30	8	25	15	39	7	22	64	27	46
33			12	21	7	36	2			

(b)
Percent frequency distributions

Time Interval (days)	Percent of Cases	
	Before Controls	After Controls
0–under 15	37.5	30.3
15–under 30	32.5	35.8
30–under 45	20.0	22.0
45–under 60	5.0	4.6
60–under 75	5.0	1.8
75–under 90		.9
90–under 105		
105–under 120		
120–under 135		1.8
135–under 150		.9
150–under 165		1.8
Total	100.0	100.0[1]
	(40)	(109)

[1]Percents do not add to 100 because of rounding.

of such a measure for two frequency distributions can indicate in direct fashion a shift in the frequency distribution, such as the one noted in the illustration.

We now discuss several useful measures of position.

Mean

One measure of position is the arithmetic average, or mean.

(3.1) The *mean* $\bar{X}$ (read "X bar") of a set of values $X_1, X_2, \ldots, X_n$ is the sum of the values divided by the number of items; i.e.:

$$\bar{X} = \frac{\sum_{i=1}^{n} X_i}{n}$$

(Appendix A, Section A.1 contains a review of the summation symbol Σ.)

☐ **Example**

For the data set before pollution controls in Table 3.1a, there are 40 time intervals whose sum is 835. Therefore, the mean time interval between oil spills in the period before the agency was created is $\bar{X} = 835/40 = 20.9$ days. By a similar calculation, it is found that the mean of the 109 intervals after pollution controls is 27.6 days. Thus, the two means summarize the lengthening of the time intervals between oil spills after pollution controls quickly and easily, much more so than the comparison of the two frequency distributions.

☐

Properties of Mean. The mean has a number of unique properties that should be considered when deciding on an appropriate measure of position.

1. The sum of the values of a set of items is equal to the mean multiplied by the number of items. In other words:

(3.2)
$$\sum_{i=1}^{n} X_i = n\bar{X}$$

This result is obtained by multiplying both sides of (3.1) by n.

☐ **Example**

A restaurant owner is informed that the mean expenditure per customer was $5.800 for the past week when 650 customers visited the restaurant. Total receipts for the week therefore were $650(5.800) = \$3,770$. ☐

2. The sum of the deviations of the X_i values from their mean $\bar{X}$ is zero. In other words:

(3.3)
$$\sum_{i=1}^{n} (X_i - \bar{X}) = 0$$

☐ **Example**

Consider the set of three values 2, 6, and 13. Their mean is $\bar{X} = 7$, and we have:

$$(2 - 7) + (6 - 7) + (13 - 7) = -5 - 1 + 6 = 0$$

Figure 3.1 illustrates the deviations $X_i - \bar{X}$, and shows that the X_i values deviate from the mean $\bar{X}$ in a balanced fashion. Thus, $\bar{X}$ is in this sense centered among the X_i values.

☐

FIGURE 3.1 *Illustration of balancing of deviations around mean*

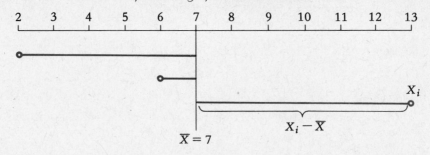

To derive (3.3), we utilize (3.2) as follows:

$$\sum_{i=1}^{n} (X_i - \bar{X}) = \sum_{i=1}^{n} X_i - \sum_{i=1}^{n} \bar{X} = \sum_{i=1}^{n} X_i - n\bar{X} = 0$$

3. In the next section, we shall be concerned with the dispersion or variability of values in a data set. One measure of dispersion we shall consider depends on an expression of the form:

$$\sum_{i=1}^{n} (X_i - A)^2 \qquad \text{where } A \text{ is a fixed value}$$

Note that this expression represents the sum of the squared deviations of the X_i values from the value A. An important property of the mean is that the sum of these squared deviations is a minimum when $A = \bar{X}$. In other words:

(3.4) $$\sum_{i=1}^{n} (X_i - A)^2 \text{ is a minimum when } A = \bar{X}$$

☐ Example

Consider again the three values 2, 6, and 13. For $A = \bar{X} = 7$, we have:

$$\Sigma(X_i - 7)^2 = (2 - 7)^2 + (6 - 7)^2 + (13 - 7)^2 = 62$$

while for any other value of A this sum is larger. For $A = 0$, for instance:

$$\Sigma(X_i - 0)^2 = (2 - 0)^2 + (6 - 0)^2 + (13 - 0)^2 = 209 \qquad ☐$$

4. When a set of items is divided into two or more subsets, the overall mean is a weighted average of the means of the subsets:

(3.5) $$\bar{X} = \frac{f_1\bar{X}_1 + f_2\bar{X}_2 + \cdots + f_k\bar{X}_k}{n}$$

where: k is the number of subsets

$\overline{X}_i$ is the mean of the ith subset

f_i is the number of items in the ith subset

$n = \sum_{i=1}^{k} f_i$ is the total number of items

☐ Example

A trade union has 1,500 male members whose mean age is 41 years and 300 female members whose mean age is 25 years. The mean age of the entire trade union membership therefore is:

$$\overline{X} = \frac{1{,}500(41) + 300(25)}{1{,}500 + 300} = 38.3 \text{ years}$$

☐

Calculation of Mean from Frequency Distribution. When data are presented only in terms of a frequency distribution, an approximation of the mean $\overline{X}$ can be obtained as follows:

(3.6)
$$\overline{X} \simeq \frac{\sum_{i=1}^{k} f_i M_i}{n}$$

where: k is the number of classes in the frequency distribution

M_i is the midpoint of the ith class

f_i is the frequency of the ith class

$$n = \sum_{i=1}^{k} f_i$$

The symbol $\simeq$ is used throughout the text to denote approximate equality.

☐ Example

In Table 3.2 the mean of the frequency distribution of time intervals between oil spills before pollution controls is calculated. This distribution was presented in percent form in Table 3.1b. Note that the calculated mean $\overline{X} = 23.6$ days is simply a weighted average of the midpoints M_i with the class frequencies f_i used as weights. ☐

Comments

1. Formula (3.6) for computing $\overline{X}$ from a frequency distribution assumes that the midpoint of each class is approximately equal to the mean value of the items included in that class. This assumption is widely used and is often, but not always, appropriate. In Table 3.2, for instance, the approximation of $\overline{X}$ computed from the frequency distribution (23.6 days) differs substantially from the exact mean of the 40 time intervals, namely 20.9 days.

In general, the more detailed the frequency distribution (i.e., the smaller the class widths), the better is the approximation of $\overline{X}$ obtained from (3.6). If the midpoints of the classes in the frequency distribution coincide exactly with the means of the items in

TABLE 3.2 *Calculation of mean from frequency distribution—Oil spills example*

Time Interval (*days*)	Number of Cases f_i	Class Midpoint M_i	$f_i M_i$
0–under 15	15	7.5	112.5
15–under 30	13	22.5	292.5
30–under 45	8	37.5	300.0
45–under 60	2	52.5	105.0
60–under 75	2	67.5	135.0
Total	40		945.0

$$\bar{X} = \frac{945.0}{40} = 23.625$$

the respective classes, then (3.6) becomes identical to (3.5) and the weighted average of the midpoints gives the exact mean of the data set.

Whenever there is a choice, the mean should be calculated from the original data set rather than from a frequency distribution to avoid approximation errors. Ready access to computers makes this calculation routine, even for large data sets.

2. Unequal class widths in a frequency distribution do not affect the computation of the mean by formula (3.6). On the other hand, if the distribution has an open-end class, it is not possible to calculate the mean by (3.6) without additional information because an open-end class has no defined midpoint.

3. The mean of a percent frequency distribution can be computed by rewriting formula (3.6) as follows:

(3.6a)
$$\bar{X} \simeq \sum_{i=1}^{k} \left(\frac{f_i}{n}\right) M_i$$

Here, f_i/n is the percent frequency of the *i*th class expressed in decimal form.

Trimmed Mean. A major disadvantage of the arithmetic mean is that it is strongly influenced by extreme values in a data set. Consider the following array of 12 values:

$$1, 1, 2, 4, 5, 7, 8, 9, 12, 12, 13, 130$$

for which the arithmetic mean is $\bar{X} = 17$. Note that 11 of the 12 values in the data set fall below the mean. The mean is so large here because of the influence of the extreme 12th value, 130.

To lessen the effect of extreme values on the mean, a modified mean called the trimmed mean is employed at times.

(3.7). A *50 percent trimmed mean* is the mean of the central 50 percent of the values of the data set arranged by order of magnitude.

Thus, to calculate a 50 percent trimmed mean, we simply eliminate 25 percent of the values at each end of the array and calculate the mean of the remaining middle items.

☐ **Example**

Obtain the 50 percent trimmed mean for the data set given earlier, repeated here for convenience:

$$1; 1, 2, 4, 5, 7, 8, 9, 12, 12, 13, 130$$

Fifty percent of this data set represents six items, so we eliminate the first three and last three values in the array. The 50 percent trimmed mean is:

$$\frac{4 + 5 + 7 + 8 + 9 + 12}{6} = 7.5$$

☐

Comment

A trimmed mean can be based on percentages other than the central 50 percent. For example, when in a data set consisting of 50 values the five smallest and five largest values are trimmed, the resulting trimmed mean is based on the central 80 percent of items.

Median

In frequency distributions that are not symmetrical, such as the oil-spill distributions, the mean tends to be located somewhat away from the concentration of items. Thus, the mean time interval between oil spills for the period before pollution controls, namely 20.9 days, is only exceeded in 16 of the 40 cases. Similarly, the mean of 27.6 days for the period after pollution controls is only exceeded in 38 of the 109 cases. A measure of position which is more central in the distribution, in the sense that it divides the items ordered by magnitude into two equal parts, is the median.

(3.8) The *median* of a set of items is the value of the middle item when all items are arranged by order of magnitude. Expressed symbolically, the median Md is the value of the item with rank $(n + 1)/2$ when the n items in the data set are ranked from 1 to n by order of magnitude.

☐ **Examples**

1. Changes in annual sales between 1979 and 1980 for five convenience stores operated by a single owner were as follows (data in thousands of dollars): 14, 17, −13, 41, 12. By arranging the items in ascending order and giving them ranks, we obtain:

Value:	−13	12	14	17	41
Rank:	1	2	3	4	5

Rank $(n + 1)/2 = (5 + 1)/2 = 3$ corresponds to the value 14. Hence, $Md = 14$. Note that the median divides the set of items equally, with two items being to one side and two to the other of the median.

2. When the data set contains an even number of items, rank $(n + 1)/2$ is not an integer, and the median is taken to be the average of the middle two values, i.e., the average of the values with ranks $n/2$ and $(n/2) + 1$. Suppose the owner operated six stores, with the following changes in annual sales: 14, 17, −13, 41, 12, 66. By arranging the values in ascending order and giving them ranks, we obtain:

Value:	−13	12	14	17	41	66
Rank:	1	2	3	4	5	6

Rank $n/2 = 6/2 = 3$ and rank $(n/2) + 1 = 4$ correspond to values 14 and 17, respectively. Hence, $Md = (14 + 17)/2 = 15.5$. Again, we find that the median divides the set of items equally, three being to one side and three being to the other side of the median.

3. For the before-pollution-controls data set in Table 3.1, it can be determined that rank $(40 + 1)/2 = 20.5$ corresponds to value $(18 + 19)/2 = 18.5$, while for the after-pollution-controls data set rank $(109 + 1)/2 = 55$ corresponds to value 21. Thus, the medians of the two data sets are 18.5 and 21 days, respectively. Consequently, comparing the two medians shows, just as did the comparison of the means, that the time intervals between oil spills have tended to be longer after the pollution controls were instituted.

□

Properties of Median

1. As definition (3.8) indicates, the median Md of a set of items is in the middle of the values ordered by magnitude. For large data sets where the values do not tend to repeat themselves extensively, the median can be interpreted to indicate that about one-half of the items are smaller than the median and about one-half are larger. Thus, if the median family income in a community is $15,000, this indicates that about one-half the families have incomes under $15,000 and about one-half have incomes over $15,000.

   When the values in a data set repeat themselves frequently, this interpretation may not be proper. For example, suppose 20 percent of families in a community consist of two persons, 40 percent consist of three persons, and the other 40 percent consist of four persons. The data set appears as follows:

$$2 \quad 2 \quad 2 \quad \ldots \quad 2 \quad 3 \quad 3 \quad 3 \quad \ldots \quad 3 \quad 4 \quad 4 \quad 4 \quad \ldots \quad 4$$

   The median value is $Md = 3$, but it does not indicate here that about one-half the family sizes are less than 3. In fact, only 20 percent of family sizes are smaller.

2. Consider the sum of absolute deviations around an arbitrary fixed value A for the values in a data set, i.e., consider the expression:

$$\sum_{i=1}^{n} |X_i - A|$$

   An important property of the median is that this sum is a minimum when A is set equal to the median, i.e., when $A = Md$.

(3.9)
$$\sum_{i=1}^{n} |X_i - A| \text{ is a minimum when } A = Md$$

□ Example

Consider the values 14, 17, −13, 41, 12. For $A = Md = 14$, we have:

$$\Sigma |X_i - 14| = |14 - 14| + |17 - 14| + |-13 - 14| + |41 - 14| + |12 - 14| = 59$$

For any other value of A, this sum would be larger. For instance, if A were set equal to the mean of these values ($\bar{X} = 14.2$), the sum would equal:

$$\Sigma |X_i - 14.2| = |14 - 14.2| + |17 - 14.2| + |-13 - 14.2|$$
$$+ |41 - 14.2| + |12 - 14.2| = 59.2 \quad \square$$

3. The median is not affected by any extreme items in a data set. For this reason, it is often preferred to the mean as a measure of position for frequency distributions containing a few extreme values. Income distributions are a case in point, often containing some extremely large values. The median in this situation tends to be more indicative of typical income levels than the mean.

4. The median can be viewed as a special case of a trimmed mean, where all items other than the middle one or two are trimmed away or excluded. Thus, a 50 percent trimmed mean is a compromise between the mean, which is based on all items in the data set, and the median, which excludes all but the middle one or two items.

Calculation of Median from Frequency Distribution.   When data are presented only as a frequency distribution, an approximation of the median Md can be obtained as follows:

(3.10)
$$Md \simeq L + \left(\frac{n_1}{n_2}\right) I$$

where: L is the lower limit of the median class (the class containing the middle item)

n_1 is the number of items that must be covered in the median class in order to reach the middle item

n_2 is the frequency in the median class

I is the width of the median class

□   Example

The calculation of Md for the frequency distribution of time intervals between oil spills before pollution controls is illustrated in Table 3.3. Note that $n/2$ is used to locate the middle of the distribution (rather than using the rank $(n + 1)/2$ as we did for locating the middle item in the original data set). In Table 3.3, the middle of the distribution corresponds to a cumulative frequency of $n/2 = 40/2 = 20$, which lies in the second class. □

Comment

Formula (3.10) for computing Md from a frequency distribution assumes that the items in the median class are spread evenly throughout that class. This assumption is adequate in many cases, but not always. For our example, the approximation is only fair, the median for the original data set being 18.5 while the approximate value calculated from the frequency distribution is 20.8. If there is a choice, the median should be calculated from the original data set to avoid approximation errors.

TABLE 3.3 *Calculation of median from frequency distribution—Oil spills example*

Time Interval (days)	Number of Cases	Cumulative Number of Cases	
0–under 15	15	Class	15
15–under 30	13	containing 20→28	
30–under 45	8		36
45–under 60	2		38
60–under 75	2		40
Total	40		

$$Md = 15 + \frac{5}{13}(15) = 20.77$$

since: $\text{middle} = \frac{n}{2} = \frac{40}{2} = 20$

$L = 15$
$n_1 = 5$
$n_2 = 13$
$I = 30 - 15 = 15$

Mode

The mode is still another measure of position that may be used in investigating a data set. Consider, for instance, the age distribution of farm operators shown previously in Figure 2.2. Suppose interest centers on the most common age of farm operators. We see from the frequency distribution that ages in the class 45–under 55 are the most common. The class 45–under 55 is called the modal age class.

(3.11) The *modal class* in a frequency distribution with equal class intervals is the class with the largest frequency. If the frequency polygon of a distribution has only a single peak, it is said to be *unimodal*. If the frequency polygon has two peaks, it is said to be *bimodal*.

☐ Example

Consideration of whether a frequency polygon is unimodal or bimodal can sometimes provide insights into the nature of the underlying data set. This is illustrated in Figure 3.2. A metal-alloy manufacturer was concerned with customer complaints about the lack of uniformity in the melting points of one of the firm's alloy filaments. Sixty filaments were selected from the production process and their melting points determined. The resulting frequency polygon is shown in Figure 3.2a. Note its bimodal nature. Closer examination of the data revealed that 26 of the filaments were produced by the first shift and the other 34 by the second shift. The two frequency polygons for the melting points of the filaments for each shift separately are shown in Figure 3.2b. It is now clear that the bimodal nature of the original frequency polygon is the result of a difference in the locations of the distributions for the two shifts. An investigation revealed that the second shift was using the wrong alloy-mixture specification. Correction of the specification resulted in production of filaments with more uniform melting points. ☐

FIGURE 3.2 *A bimodal frequency polygon*

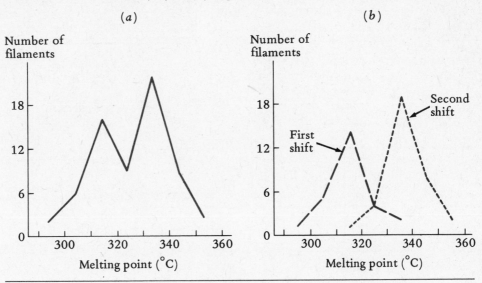

Percentiles

The median, as we saw, divides a set of items ordered by magnitude into two equal parts and may therefore be considered the 50th percentile. Other percentiles, such as the 25th and 75th percentiles, often are useful for indicating the positions of the noncentral values in a data set. For small data sets, an easy way to find percentiles is from a step-function ogive such as the one illustrated in Figure 2.6.

Example

The owner of the five convenience stores mentioned earlier wanted to compare the changes in annual sales for these stores with industry data which showed that 25 percent of convenience stores had annual increases of sales of $10 thousand or less, and 75 percent of stores had annual sales increases of $15 thousand or less. The changes in annual sales for the five stores are shown in array form in Figure 3.3a. To obtain comparable percentiles for the five stores, a step-function ogive has been plotted in Figure 3.3b. Reading from this graph, we see that the 25th percentile for the five stores is $12 thousand, and the 75th percentile is $17 thousand. Thus, the five convenience stores performed comparably to the industry. Note also from Figure 3.3b that the median, or 50th percentile, is $14 thousand, as we determined previously.

The correspondences employed in Figure 3.3b can also be expressed by the line diagram of Figure 3.3c. Since there are five items in the data set, the cumulative percent scale ranging from 0 to 100 is divided into five equal intervals and the corresponding value from the array is associated with each interval. For example, the smallest value (−13) is placed in the percent interval from 0 to 20. The value associated with any percentile can then be read from the line diagram. Thus, the 25th percentile is $12 thousand because the value 12

FIGURE 3.3   *Determination of percentiles from step-function ogive*

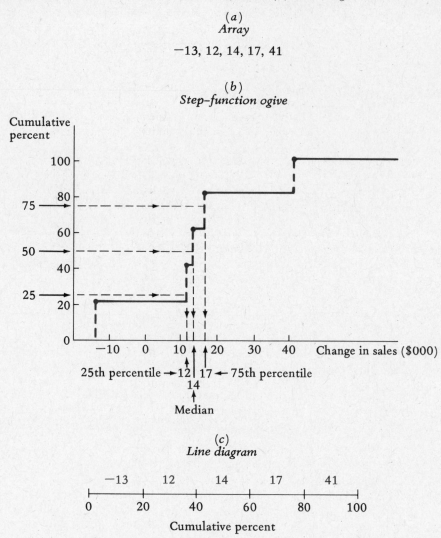

(a)
Array

−13, 12, 14, 17, 41

(b)
Step-function ogive

(c)
Line diagram

is in the interval in which the cumulative percent 25 lies. Similarly, the 50th percentile is $14 thousand and the 75th percentile is $17 thousand.

Occasionally, the desired percentile falls on the boundary between two intervals in the line diagram in Figure 3.3c, or equivalently falls at the top of a step in the ogive in Figure 3.3b. In that case, the percentile is not uniquely defined. We shall follow the convention employed for the median of taking the mean of the two adjacent values. For example, the 40th percentile of the data set in Figure 3.3a is $(12 + 14)/2 = \$13$ thousand.   □

Meaning of Percentiles. In large data sets where values do not tend to repeat themselves extensively, the 25th percentile indicates that about 25 percent of the items are less than this value and about 75 percent are more. Other percentiles are interpreted correspondingly. When the values in a data set do tend to repeat themselves, this interpretation is no longer appropriate, as was illustrated for the median.

Comments

1. A single value can correspond to more than one percentile. For instance, the value 12 in Figure 3.3 is not only the 25th percentile, but it is also every percentile between 20 and 40.
2. Many percentiles are known by other names. Percentiles that are multiples of 25 are called *quartiles*. Thus, the 25th percentile is also called the first quartile, the 50th percentile the second quartile (also the median), and the 75th percentile the third quartile. Similarly, percentiles that are multiples of 10 are called *deciles*. For instance, the 70th percentile is also called the 7th decile.
3. If data are only available in the form of a frequency distribution, an approximation of the *p*th percentile can be obtained in the same manner as for the median, using the formula:

(3.12)
$$p\text{th percentile} \simeq L + \left(\frac{n_1}{n_2}\right)I$$

where: L is the lower limit of the *p*th percentile class (the class containing the *p*th percentile)
n_1 is the number of items which must be covered in the *p*th percentile class in order to reach the *p*th percentile
n_2 is the frequency of the *p*th percentile class
I is the width of the *p*th percentile class

As for the median, the formula assumes the items are spread evenly throughout the *p*th percentile class.

For instance, the 80th percentile for the before-controls oil-spill distribution in Table 3.3 corresponds to cumulative frequency $.80(40) = 32$. To reach this cumulative frequency, 4 items must be covered in the interval 30–under 45. This interval contains 8 items altogether, so:

$$80\text{th percentile} \simeq 30.0 + \frac{4}{8}(15) = 37.5$$

Thus, the estimate of the 80th percentile based on the frequency distribution is 37.5 days.
4. Plotting two ogives together makes it very easy to compare the percentiles of two different distributions. For instance, Figure 3.4 presents the ogives for the two percent frequency distributions of time intervals between oil spills presented in Table 3.1b. Note that the ogive for the period after pollution controls lies entirely to the right of the ogive for the earlier period. It follows, therefore, that all percentiles for the period after pollution controls are larger than the corresponding ones for the period before pollution controls. This is convincing evidence that the time intervals between oil spills have lengthened since the institution of pollution controls.

FIGURE 3.4 *Comparison of percentiles of two percent frequency distributions*

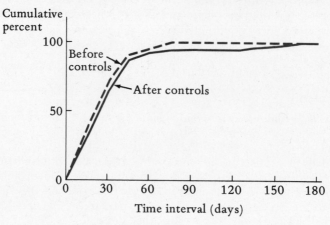

3.2 MEASURES OF DISPERSION

In summarizing data sets, the variability or dispersion of the values is often an important feature of interest. We now consider the major summary measures of dispersion.

Range

Suppose a data set consists of the amount of active ingredient in each of 300 tablets included in a shipment from a pharmaceutical manufacturer. One factor that might be important in assessing the quality of these tablets is the maximum extent of variation in the amount of active ingredient per tablet.

(3.13) The *range* is the difference between the largest and smallest values in a set of items.

☐ Examples

1. The largest value in the data set for the 300 tablets is 1.38 grams and the smallest value is .96 gram. The range then is $1.38 - .96 = .42$ gram, indicating a relatively large extent of variability in the amounts of active ingredient.
2. In our pollution-controls example, it can be ascertained that the range for the before-pollution-controls data set is $73 - 1 = 72$ days while that for the after-pollution-controls data set is $163 - 0 = 163$ days. Thus, the after-pollution-controls data set has a larger spread. ☐

Comments

1. A limitation of the range as a measure of variation of a set of items is that it depends only on the extreme items and does not consider the others. Thus, a data set might contain items quite close to each other with the exception of one extreme value.

Despite the concentration of almost all the items, the range would be large since it is based only on the extreme values. Another limitation of the range as a measure of variability is that it depends on the number of items in the set. The larger the number of items, the larger the range tends to be.

2. When data are classified in a frequency distribution, the range of the original items usually cannot be determined exactly. For instance, from the frequency distribution in Table 3.3, it can only be seen that the range of the 40 time intervals is not larger than $75 - 0 = 75$ days. Actually, from the original data in Table 3.1a, the range is $73 - 1 = 72$ days, as noted in Example 2.

3. Sometimes, the two extreme values in a set of items are presented instead of the difference between the two. Thus, stock-market summaries in newspapers often provide the high and low prices for the day. This form of presentation has the advantage of not only providing information about the variability of a stock's price during the day since the range can be easily calculated, but it also provides information about the location of the stock's price distribution. Another advantage is that the daily highs and lows can be studied to determine the price range for a longer period, such as a week.

Interquartile Range

Because the range depends only on the two extreme items, a modified range is sometimes used which reflects the variability of the middle 50 percent of the items in a data set. This modified range is called the interquartile range.

(3.14) The *interquartile range* is the difference between the third and first quartiles of a set of items.

☐ Examples

1. An economist, studying the variation in family incomes in a community, found that the middle 50 percent of families have incomes which vary from $12,400 to $19,100, a range of $6,700. Here, the first quartile is $12,400 and the third quartile is $19,100. The range of $6,700 is the interquartile range, reflecting the extent of variation among the middle 50 percent of family incomes.

2. In our pollution-controls example, the first and third quartiles and the interquartile ranges for the two data sets are:

	Before Controls	After Controls
Third quartile	31	33
First quartile	5.5	11.5
Interquartile range	25.5	21.5

The upward shift of the quartiles reflects the lengthening of the durations between oil spills after pollution controls were instituted. The decline in the interquartile range indicates that the middle 50 percent of items in the after-pollution-controls data set are somewhat more concentrated than those in the before-pollution-controls data set.

☐

Variance

The most commonly used measure of dispersion in statistical analysis is called the variance. It is a measure that takes into account all the values in a set of items.

☐ Illustration

Earlier, we considered the situation where customer complaints about increased variability in the melting points of alloy filaments led the manufacturer to investigate the cause of the problem. One of the customers had selected a sample of 15 filaments from a shipment received in April and another sample of 15 filaments from a shipment received the following October. The data on the melting points for these two samples are given in Figure 3.5.

To compare the variability of melting points for the two data sets, the customer considered the deviations of values in each data set around their respective means. These two sets of deviations are shown graphically in Figure 3.5. The means of the two data sets are 330.8 and 339.3, respectively. A comparison of the two sets of deviations in Figure 3.5 shows readily that there was greater variability around the center of the distribution (as measured by the mean) for the October filaments than for the April filaments.

The variance is a measure which provides quantified information about the variability in a data set. The first step in calculating the variance is to square the deviations around the mean. Then, to obtain an overall measure of variability per item, the average of the squared deviations is taken. This mean squared deviation is called the variance. For reasons to be explained later, we divide by $n - 1$ rather than by n in obtaining the mean of the squared deviations. The calculation of the variance for each of the two data sets

FIGURE 3.5 *Deviations around mean for two data sets*

(a) Melting points (°C) of filaments from April shipment				(b) Melting points (°C) of filaments from October shipment		
330	334	321		302	365	343
358	318	337		348	318	317
373	325	328		374	378	385
346	295	348		279	294	304
343	288	318		364	357	362

$\overline{X} = 330.8$ $\overline{X} = 339.3$

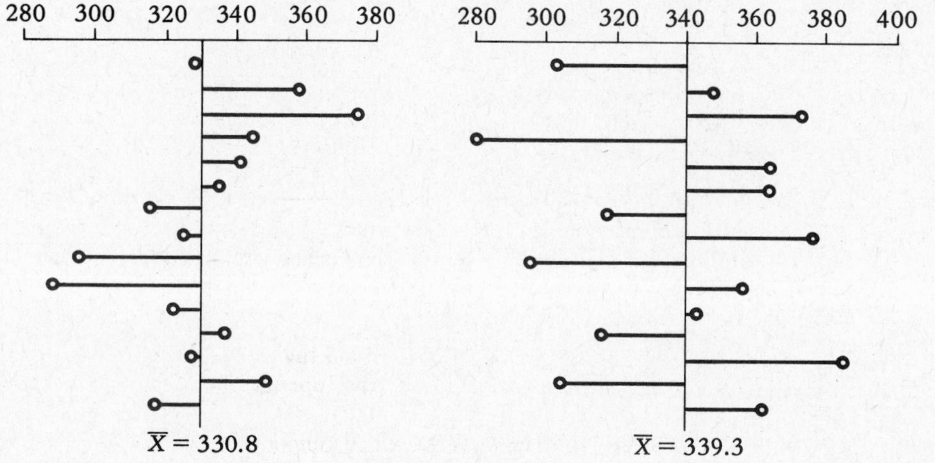

proceeds as follows:

April shipment: $\dfrac{(330 - 330.8)^2 + \cdots + (318 - 330.8)^2}{15 - 1} = 487.5$

October shipment: $\dfrac{(302 - 339.3)^2 + \cdots + (362 - 339.3)^2}{15 - 1} = 1{,}161.1$

The variance for the October filaments is more than twice that for the April filaments. It was this observation which prompted the customer to complain about the increased variability of the melting points of the alloy filaments.

We now define the variance formally.

(3.15) The *variance* s^2 of a set of values $X_1, X_2, \ldots, X_n$ is defined:

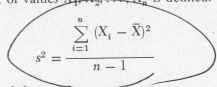

$$s^2 = \frac{\displaystyle\sum_{i=1}^{n} (X_i - \overline{X})^2}{n - 1}$$

where: $\overline{X}$ is the mean of the n values

An equivalent formula for calculating the variance s^2 is:

(3.15a)

$$s^2 = \frac{\displaystyle\sum_{i=1}^{n} X_i^2 - \frac{\left(\displaystyle\sum_{i=1}^{n} X_i\right)^2}{n}}{n - 1}$$

Example

Consider the data set 5, 17, 12, 10, whose mean is $\overline{X} = 11$. We can calculate s^2 by (3.15):

$$s^2 = \frac{(5 - 11)^2 + (17 - 11)^2 + (12 - 11)^2 + (10 - 11)^2}{4 - 1}$$

$$= \frac{(-6)^2 + 6^2 + 1^2 + (-1)^2}{3} = 24.67$$

or by (3.15a):

$$s^2 = \frac{5^2 + 17^2 + 12^2 + 10^2 - \dfrac{(5 + 17 + 12 + 10)^2}{4}}{4 - 1} = 24.67$$

Frequently, the latter formula is easier for manual calculations.

Comments

1. The wider the dispersion of the values around their mean, the greater the variance. If there is no dispersion of the values—i.e., if all are equal and hence all are equal to the mean—then $s^2 = 0$.
2. Many calculators contain a preprogrammed function to produce the variance. Some calculator models employ a different definition of the variance than the one in (3.15),

using n as the divisor instead of $n - 1$. Unless the data set is small, either n or $n - 1$ in the denominator gives essentially the same result. To determine which divisor is being used in a calculator, the user's manual should be consulted or a small sample problem run. A variance with n in the denominator can be converted to one with $n - 1$ by multiplying by $n/(n - 1)$.

Calculation of Variance from Frequency Distribution. When the data are presented in a frequency distribution, only an approximation to the variance can be obtained.

(3.16)
$$s^2 \simeq \frac{\sum_{i=1}^{k} f_i(M_i - \bar{X})^2}{n - 1}$$

where: k is the number of classes in the frequency distribution

f_i is the frequency of the ith class

M_i is the midpoint of the ith class

$\bar{X}$ is the mean of the frequency distribution as approximated by (3.6)

$$n = \sum_{i=1}^{k} f_i$$

☐ **Example**

The calculation of the variance from a frequency distribution is illustrated in Table 3.4, which contains the frequency distribution of the time intervals between oil spills before pollution controls from Table 3.2. Note that the approximate value of $\bar{X}$ as computed from the frequency distribution in Table 3.2 is used in the calculation of s^2. ☐

Comments

1. Formula (3.16) for computing s^2 from a frequency distribution assumes that each item in a class has a value equal to the midpoint of that class. This assumption may appear

TABLE 3.4 *Calculation of variance from frequency distribution—Oil spills example*

Time Interval (*days*)	Number of Cases f_i	Class Midpoint M_i	Deviation $M_i - \bar{X}$	Deviation Squared $(M_i - \bar{X})^2$	$f_i(M_i - \bar{X})^2$
0–under 15	15	7.5	−16.125	260.0156	3,900.234
15–under 30	13	22.5	−1.125	1.2656	16.453
30–under 45	8	37.5	13.875	192.5156	1,540.125
45–under 60	2	52.5	28.875	833.7656	1,667.531
60–under 75	2	67.5	43.875	1,925.0156	3,850.031
Total	40				10,974.374

$$\bar{X} \simeq 23.625 \qquad \text{(From Table 3.2)}$$

$$s^2 \simeq \frac{10,974.374}{40 - 1} = 281.39$$

somewhat crude, but frequently it yields results close to the variance that would have been obtained from the original data. If there is a choice, the variance should be calculated from the original data to avoid approximation errors.

2. Since the mean cannot be computed from an open-end frequency distribution without further information, neither can the variance. Also like for the mean, the calculation of the variance is not affected by unequal class intervals.

3. A formula algebraically equivalent to (3.16) that is often easier to use for manual calculation is:

(3.16a)
$$s^2 \simeq \frac{\sum\limits_{i=1}^{k} f_i M_i^2 - \dfrac{\left(\sum\limits_{i=1}^{k} f_i M_i\right)^2}{n}}{n-1}$$

Standard Deviation

The variance s^2 is expressed in units that are the square of the units of measure of the characteristic under study. For instance, the variance of melting points in our earlier example is expressed in Celsius degrees squared. Often, it is desirable to return to the original units of measure. This is done by taking the positive square root of the variance. The resulting value is called the standard deviation and is also used as a measure of dispersion.

(3.17) The positive square root of the variance is called the *standard deviation* and is denoted by s:

$$s = \sqrt{s^2}$$

As with the variance, the greater the dispersion of the values around their mean, the larger the standard deviation. If there is no dispersion, i.e., if all values are equal, the standard deviation is zero.

☐ Example

For the earlier filament melting points example, the standard deviations are $s = \sqrt{487.5} = 22.1°C$ for the April filaments and $s = \sqrt{1,161.1} = 34.1°C$ for the October filaments. The standard deviations again show that there was greater variability of the filament melting points in the October sample than in the April sample. ☐

Chebyshev Inequality. One of the uses of the standard deviation is found in the Chebyshev inequality.

(3.18) *Chebyshev Inequality.* In any data set, no matter what the nature of the pattern of variation, the proportion of items falling beyond k standard deviations from the mean is at most $1/k^2$.

☐ Example

A data set consists of the waiting times of customers at the checkout counter of a food store during the past week. The mean waiting time was 4.0 minutes and the standard deviation was .9 minute. From the Chebyshev inequality we know that for, say, $k = 2$, at most $1/k^2 = 1/2^2 = .25$ or 25 percent of the waiting times differed by more than 2 standard deviations from the mean. Hence, at most 25 percent of the customers waited less than $4.0 - 2(.9) = 2.2$ minutes or more than $4.0 + 2(.9) = 5.8$ minutes. The Chebyshev inequality does not, however, provide any information about how these customers are divided between very short and very long waits. ☐

Coefficient of Variation

The standard deviation is a measure of the *absolute variability* in a set of items. For a number of problems, however, the *relative variability* is a more significant measure. The most commonly used measure of relative variability is the coefficient of variation.

(3.19) The *coefficient of variation*, denoted by C, is the ratio of the standard deviation to the mean expressed as a percent:

$$C = 100\left(\frac{s}{\overline{X}}\right)$$

☐ Examples

1. A sales analyst was studying the variability of her company's sales revenues over time periods of different lengths. She determined that weekly sales revenues during the past year had a mean value of $1.36 million and a standard deviation of $.28 million, while sales revenues for four-week intervals during the past year had a mean value of $5.44 million and a standard deviation of $.50 million. The analyst felt that a direct comparison of the two standard deviations would not be appropriate because the absolute variability in revenues is affected by the fact that sales revenues for four-week periods, on the average, are four times as large as weekly sales revenues. Instead, the analyst computed the coefficient of variation for each case as follows:

Weekly Sales Revenues	Four-weekly Sales Revenues
$C = 100\left(\frac{.28}{1.36}\right) = 20.6\%$	$C = 100\left(\frac{.50}{5.44}\right) = 9.2\%$

Based on these results, the analyst concluded that sales revenues for four-week periods ($C = 9.2\%$) were relatively less variable than weekly sales revenues ($C = 20.6\%$) during the past year. In particular, she noted that measuring revenues over a period four times as long cut the relative variability of revenues approximately in half.

2. A traffic analyst for an air carrier needed to compare the variability in size and weight of containerized air-cargo shipments. The analyst found that the shipments had a mean size of 12.6 cubic feet and standard deviation of 2.1 cubic feet, while the corresponding measures for weight were 64.0 pounds and 18.3 pounds. The standard deviations cannot be compared here because size is measured in cubic feet and weight in pounds. The coefficients of variation, in contrast, are dimensionless because the units of measurement cancel out when the standard deviation is divided by the mean. The coefficients of variation here are:

Size	Weight
$C = 100\left(\frac{2.1}{12.6}\right) = 16.7\%$	$C = 100\left(\frac{18.3}{64.0}\right) = 28.6\%$

Hence, the relative variability of the container sizes is smaller than that of the container weights.   □

Comment

The coefficient of variation is usually employed only when the values are all positive, such as for sales, size, and weight data. When the values can be both positive and negative, as for profit-and-loss data, the coefficient of variation is not an appropriate measure of relative dispersion.

3.3   MEASURES OF SKEWNESS

Measures of position are concerned with the location around which items are concentrated, whereas measures of dispersion consider the extent to which the items vary. Measures of skewness are still another kind of measure, summarizing the extent to which the items are symmetrically distributed.

(3.20) A *skewed* frequency distribution is one that is not symmetrical.

Nature of Skewness

One way of studying the skewness of a frequency distribution is to compare the values of the mode, median, and mean. The mode is the position on the scale that has the greatest concentration of items. For the median, we know that generally about half the items will lie below it and about half above it. In contrast, the mean tends to be pulled in the direction of the extreme values because of the disproportionate influence of very large or very small values on it. The three frequency polygons in Figure 3.6 show the comparative positions of the mean, median, and mode in unimodal frequency distributions with differing degrees and directions of asymmetry.

Figure 3.6a is an example of a symmetrical unimodal frequency distribution. The values of the mean, median, and mode in any such distribution are identical.

Figures 3.6b and 3.6c show, respectively, an example of a unimodal frequency distribution skewed to the left, or skewed negatively, and one skewed to the right, or skewed positively. These figures indicate the relationship typically existing between the mean, median, and mode in unimodal frequency distributions that are moderately skewed: the mean is furthest out toward the tail of the distribution, and the median is between the mean and the mode.

Pearson's Coefficient of Skewness

Pearson's coefficient of skewness is a measure which quantifies the extent of asymmetry in a set of items in terms of the difference between the mean and median relative to the standard deviation.

(3.21) *Pearson's coefficient of skewness S is defined:*

$$S = \frac{3(\overline{X} - Md)}{s}$$

FIGURE 3.6 *Examples of symmetrical and skewed unimodal frequency distributions.* In skewed unimodal distributions, the mean is typically furthest out toward the tail and the median falls between the mean and the mode.

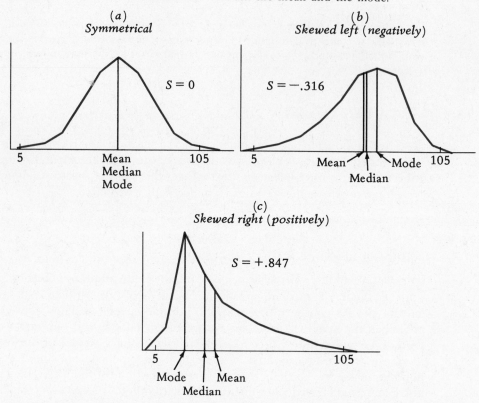

The coefficient may be computed from a data set directly or from a frequency distribution.

If the frequency distribution is symmetrical, the mean and median are identical and the coefficient S will be zero. If the distribution is skewed to the left, as in Figure 3.6b, the mean will be less than the median and the coefficient S will be negative. If the distribution is skewed to the right, as in Figure 3.6c, the mean will be greater than the median and the coefficient S will be positive.

☐ Example

Pearson's coefficient of skewness has been computed for each of the frequency distributions in Figure 3.6. For the right-skewed frequency distribution in Figure 3.6c, for instance, $\bar{X} = 37.00$, $Md = 31.19$, and $s = 20.57$ (calculations not shown), so:

$$S = \frac{3(37.00 - 31.19)}{20.57} = +.847$$

☐

3.4 OPTIONAL TOPIC—GEOMETRIC AND HARMONIC MEANS

Geometric Mean

When a data set contains ratios, a geometric mean may be a useful measure of position. Suppose that beginning-of-year prices of a product in four consecutive years were $30.00, $33.00, $33.66, and $41.74, respectively. Consider the ratio of the price at the beginning of one year to the price at the beginning of the previous year. The first price ratio is $33.00/30.00 = 1.10$, and the other two are 1.02 and 1.24, respectively. If a single measure is desired to describe the typical value of the three price ratios, the geometric mean may be appropriate.

(3.22) The *geometric mean* G of a set of values $X_1, X_2, \ldots, X_n$ is the antilogarithm of the arithmetic mean of the logarithms of the values, i.e.:

$$\log G = \frac{\log X_1 + \log X_2 + \cdots + \log X_n}{n}$$

The logarithms in (3.22) are to the base 10, but can be to any base. Many calculators are preprogrammed to produce logarithms. (Appendix A, Section A.2 contains a review of logarithms.)

☐ **Example**

The geometric mean price ratio in our example, where $X_1 = 1.10$, $X_2 = 1.02$, and $X_3 = 1.24$, is:

$$\log G = \frac{\log 1.10 + \log 1.02 + \log 1.24}{3} = \frac{.04139 + .00860 + .09342}{3} = .04780$$
$$G = \text{antilog } .04780 = 1.11635$$

The geometric mean here is an average ratio showing that the annual price increase, on the average, was 11.635 percent. It is a property of the geometric mean that the price in the final period can be obtained from the price in the initial period and the average year-to-year price ratio, as follows:

$$30.00(1.11635)(1.11635)(1.11635) = \$41.74$$ ☐

Comments

1. By taking the antilogarithm of both sides of formula (3.22), it can be seen that an equivalent formula for the geometric mean is:

(3.22a) $$G = (X_1 \cdot X_2 \cdot \cdots \cdot X_n)^{1/n}$$

2. The geometric mean is undefined for a data set that contains zero or negative values.

Harmonic Mean

When a data set contains values which represent rates of change, the harmonic mean may be a useful measure of position. For instance, in a study of traffic flow, it was noted that a vehicle traveled at 50 miles per hour over one 10-mile stretch of highway and at 30 miles per hour over a second 10-mile stretch. Suppose it is desired to obtain a measure of the average of the two speeds, reflecting the average speed traveled over the entire 20-mile stretch. In this case, the harmonic mean may be appropriate.

(3.23) The *harmonic mean* H of a set of values $X_1, X_2, \ldots, X_n$ is the reciprocal of the arithmetic mean of the reciprocals of the values, i.e.:

$$\frac{1}{H} = \frac{\dfrac{1}{X_1} + \dfrac{1}{X_2} + \cdots + \dfrac{1}{X_n}}{n}$$

☐   **Example**

In our example, where $X_1 = 50$ and $X_2 = 30$, the harmonic mean speed is:

$$\frac{1}{H} = \frac{\dfrac{1}{50} + \dfrac{1}{30}}{2} = .0267 \quad \text{so} \quad H = \frac{1}{.0267} = 37.5 \text{ miles per hour}$$

The harmonic mean is an appropriate measure here because the vehicle took .533 hour to cover the entire 20 miles of highway and this represents an average speed of $20/.533 = 37.5$ miles per hour.   ☐

Comments

1. By taking reciprocals of both sides of the equation in (3.23), it can be seen that an equivalent formula for the harmonic mean is:

(3.23a)

$$H = \frac{n}{\dfrac{1}{X_1} + \dfrac{1}{X_2} + \cdots + \dfrac{1}{X_n}}$$

2. The harmonic mean is undefined for a data set which contains zero or negative values.

PROBLEMS

3.1 Using Figure 2.4 (p. 29) as an illustration, briefly explain what is meant by the *position* or *location* of a frequency distribution.

*3.2 The numbers of restaurants in 12 hotels in a city are: 0, 0, 1, 1, 2, 3, 3, 3, 4, 4, 5, 7. Let X_i denote the number of restaurants in the ith hotel. Calculate ΣX_i and $\bar{X}$.

3.3 The numbers of admissions to the emergency ward of a hospital during the evening shift for the past three weeks were as follows: 6, 0, 3, 1, 5, 7, 4, 2, 1, 0, 2, 3, 2, 9, 0, 3, 5, 3, 1, 4, 2. Calculate the mean number of admissions per evening shift. Is the mean number of admissions necessarily a whole number?

3.4 The numbers of defectives in eight recent lots of 100 items each were as follows: 3, 1, 0, 0, 2, 21, 4, 1. Calculate the mean number of defectives per lot. Use $\bar{X}$ to obtain the total number of defectives in the eight lots, and verify your result from the original data.

3.5 In a shipment of 500 frozen turkeys, the mean weight is guaranteed to be at least 12.5 pounds. What can be said about the total weight of these turkeys?

3.6 Refer to Problem 3.3. Verify that $\Sigma (X_i - \bar{X}) = 0$ for these data.

*3.7 A police department has 125 officers whose mean length of service is 8.5 years and 23 civilian employees whose mean length of service is 6.1 years. Calculate the mean length of service of all 148 members of the department.

3.8 The mean age of the 25 women employees in a department is 28.3 years and the mean age of the 41 male employees is 39.2 years. Calculate the mean age of all employees in the department.

*3.9 Refer to Problem 2.12. Calculate the mean of the distribution of response times.

3.10 The frequency distribution of the amounts of bonus pay earned during the last year for 65 salespersons follows.

Amount of Bonus Pay ($)	Number of Salespersons
0–under 100	4
100–under 200	17
200–under 400	27
400–under 600	11
600–under 1,000	5
1,000–under 2,000	1
Total	65

Calculate the mean amount of bonus pay per employee. What is the approximate total amount of bonuses paid to all salespersons? Why is this an approximation?

3.11 The percent frequency distribution of department A employees of the Calder Company according to their straight-time hourly earnings at the end of last year follows.

Hourly Earnings ($)	Percent of Employees
8.10–under 8.30	6
8.30–under 8.50	14
8.50–under 8.70	44
8.70–under 8.90	31
8.90–under 9.10	5
Total	100
Number of employees	(2,100)

Calculate the mean hourly earnings. What assumption did you make in your calculation?

*3.12 Refer to Problem 3.2.
   a. Calculate the 50 percent trimmed mean for the data set.
   b. Compare the values of the trimmed mean and the arithmetic mean. Is the difference between them relatively small here? Will this always be the case? Discuss.

3.13 Refer to Problem 3.4. Calculate the 50 percent trimmed mean for the data set. Is this trimmed mean close to $\bar{X}$ here? Explain.

3.14 A discussant suggests that it is better to multiply the number of items in a data set by the 50 percent trimmed mean than by $\bar{X}$ to obtain the sum of the values in a data set when the data set contains extreme values "because a trimmed mean tends to be more representative than $\bar{X}$ in such cases." Do you agree with this recommendation? Discuss.

*3.15 Refer to Problem 3.2. Obtain the median number of restaurants per hotel.

3.16 Refer to Problem 3.3.
  a. (1) Array the observations in ascending order. (2) Obtain the median number of admissions per evening shift.
  b. Is the median equal to an actual value in the data set here? Must this be the case for every data set? Explain.

3.17 Refer to Problem 3.4. Obtain the median number of defectives per lot. Why does the median differ substantially from the mean, $\bar{X} = 4$, here?

3.18 Refer to Problem 3.2. Calculate the sum of the absolute deviations of the values in the data set around their median and also around their mean. Are the two sums consistent with theorem (3.9)? Explain.

*3.19 Refer to Problem 2.12. Calculate the median of the distribution of response times.

3.20 Refer to Problem 3.10. Calculate the median of the distribution of bonus amounts. Why is the median less than the mean here?

3.21 Refer to Problem 3.11.
  a. To calculate the median hourly earnings here, is it necessary to first convert the percent frequencies to actual frequencies? Explain.
  b. Calculate the median hourly earnings. Explain how the median is interpreted here. What assumption did you make in calculating the median?

3.22 Refer to Problem 3.11. The corresponding distribution of hourly earnings of department B employees follows.

Hourly Earnings ($)	Percent of Employees
7.50–under 7.70	11
7.70–under 7.90	13
7.90–under 8.10	20
8.10–under 8.30	52
8.30–under 8.50	4
Total	100
Number of employees	(850)

  a. Calculate the mean and median hourly earnings of department B employees.
  b. The mean and median hourly earnings of department A employees are $8.63 and $8.64, respectively. Would it be better to use the two means, the two medians, or the two modal-class midpoints to compare the locations of the two hourly earnings distributions? Discuss.

3.23 Refer to Problem 3.11. Is the distribution of employees' hourly earnings unimodal or bimodal? Discuss.

3.24 A commentator, learning that the modal number of children in U.S. families is zero, stated: "The birthrate has fallen so low that the majority of U.S. families have no children at all!" Comment.

3.25 A service worker stated: "The Internal Revenue Service has calculated that my average tip is $1.50. I have to admit they are correct. But I'll tell you, most of my tips are below the average!" What type of average must the Internal Revenue Service have used here? Would some other average be more appropriate for their purposes? Discuss.

3.26 A promotional brochure reported: "The average family in our condominium development operates 1.79 automobiles." What type of average must have been used? Would some other type of average have been more meaningful here? Discuss.

3.27 An analyst stated: "When a data set has one or several extreme outcomes that might have arisen from errors in data acquisition, I use the median or a trimmed mean in preference to the arithmetic mean." Why might the analyst take this approach?

3.28 A chain of high-fidelity equipment stores sells four stereo systems. Data on sales during the past six months follow.

Price of System ($)	Number of Systems Sold
250	68
400	110
650	91
1,000	56
Total	325

In calculating the mean or median price for this frequency distribution, why will the results be exactly the same as when these measures are calculated from the actual 325 observations?

*3.29 Rates of return (in percent) earned on 10 utility stocks during the past year were: 15.7, 21.5, 22.7, 29.2, 21.8, 20.5, 24.2, 20.8, 28.4, 20.9.
   a. Array these observations and plot their step-function ogive, as in Figure 3.3b.
   b. Obtain the 25th and 75th percentiles from the ogive.
   c. Draw the line diagram corresponding to the step-function ogive, as in Figure 3.3c. Show how the 25th and 75th percentiles are obtained from this diagram.

3.30 Refer to Problem 3.2.
   a. Plot the step-function ogive and the corresponding line diagram for this data set, as in Figures 3.3b and 3.3c. Did any difficulties arise in making these plots because some of the values repeat themselves in this data set? Discuss.
   b. Obtain the 20th, 50th, and 80th percentiles, first from your ogive and then from the line diagram. Label your ogive to show these percentiles.
   c. Does the existence of repeated values in the data set need to be taken into account in interpreting any of the percentiles in b? Discuss.

*3.31 Refer to Problem 2.12.
   a. Calculate the first and third quartiles of the distribution of response times.
   b. Find the 90th percentile of the distribution of response times. Interpret its meaning here.

3.32 The G. Stokes Company, as part of a profit-improvement program, studied customer order sizes. The distribution of order sizes for the past fiscal year follows.

Order Size ($)	Number of Orders	Order Size ($)	Number of Orders
5–under 25	1,852	100–under 250	354
25–under 50	1,120	250–under 500	471
50–under 100	693	500–under 1,000	530
		Total	5,020

   a. Calculate the 40th and 80th percentiles of the distribution of order sizes. Interpret the meaning of these measures.
   b. What percent of orders were for less than $200? What percent of the total dollar volume do these orders represent?

3.33 Swimming team A has nine divers whose scores had a range of 2.3 in a recent competition. Swimming team B has five divers whose scores had a range of 1.6 in the same competition. A sports reporter commented that team B has more uniform diving talent because its scores had a smaller range. Do you agree with the reporter? Discuss.

*3.34 Refer to Problem 3.2. Obtain the range and the interquartile range for the data set.

3.35 Refer to Problem 3.4. Obtain the range and the interquartile range for the data set. Can the interquartile range ever be larger than the range? Explain.

*3.36 Refer to Problems 2.12 and 3.31. Calculate the interquartile range for the distribution of response times. In what units is this measure expressed? Is the interquartile range substantially smaller than the range here? Explain.

3.37 Refer to Problems 3.11 and 3.22.
   a. For each distribution of hourly earnings, obtain the first quartile, third quartile, and interquartile range.
   b. Compare the corresponding quartiles and interquartile ranges for the two distributions and interpret your findings.
   c. An observer states that the interquartile range can be viewed as a trimmed range. Do you agree? Explain.

*3.38 Refer to Problem 3.2. Calculate the variance and the standard deviation for the data set.

3.39 Refer to Problem 3.4. Calculate the variance and the standard deviation for the data set. Can the standard deviation ever be larger than the variance? Can either ever be negative? Explain.

3.40 Refer to Problem 3.3.
   a. Calculate the range, variance, and standard deviation for the data set. In what units is each of these measures expressed?
   b. If observations for six additional weeks were added to the data set, would all three measures in a be likely to increase? Discuss.

*3.41 Refer to Problem 2.12. The mean of the frequency distribution of response times is 13.7 minutes. Calculate the variance and the standard deviation of the distribution.

3.42 Refer to Problem 3.10. The mean of the frequency distribution of bonus amounts is $336.2. Calculate the variance and standard deviation of the distribution. In what units are these measures expressed? The standard deviation of the distribution of bonus amounts two years ago was $151.7. What change in the distribution of bonus amounts appears to have taken place?

3.43 Refer to Problem 2.14. Calculate the variance and standard deviation of the distribution of the ages of the horses. Use as class midpoints 5.5, 6.5, etc. If horses of all ages were allowed to compete in the horse trials, would you expect the standard deviation of the age frequency distribution to change? Explain.

3.44 A student entered the values -1 and 1 in a pocket calculator, pressed the key for the standard deviation, and obtained 1.414 in the display. Does this calculator use n or $n - 1$ in the divisor of the formula for the standard deviation?

*3.45 A national door-to-door selling organization released the following information on earnings of 3,850 college students employed last summer in a special selling campaign: maximum earnings $2,830.50, minimum earnings $263.40, mean $1,730.50, median $1,564.40, standard deviation $215.25.
   a. In applying the Chebyshev inequality, what can be said about the proportion of students earning less than $1,300.00 or more than $2,161.00?
   b. What can be said about the proportion of students earning between $869.50 and $2,591.50?

3.46 A research chemist reported in an article that in 48 repeated measurements on the yield of a particular chemical reaction, the mean yield was 103.31 grams and the standard deviation was .25 gram. What is the minimum number of measurements which were within ±.5 gram of the mean? Within ±1.0 gram?

*3.47 Refer to Problem 2.12. The mean and standard deviation of the frequency distribution are 13.7 minutes and 5.33 minutes, respectively. Calculate the coefficient of variation. Is this coefficient expressed in minutes, as are the mean and standard deviation? Explain.

3.48 Refer to Problem 3.45. Information is also available on the hours worked by the 3,850 students during the selling campaign: mean 346 hours, median 325 hours, standard deviation 20 hours. Was there greater variability in dollar earnings or in hours worked? Be specific, and explain your choice of the measure of variability utilized.

3.49 A detergent used in hospitals comes in package sizes of 5 and 25 kilograms. The exact weight varies somewhat from one package to the next, for each package size. An inspector for a buying consortium wishes to compare the variability in package weights for the two package sizes. Which measure do you recommend be used? Why?

3.50 A speaker stated: "A quick way to measure the variability in a data set is to note the difference between the mean and the median. If this difference is small, there is relatively little variability; the larger the difference, the greater the variability." Comment.

*3.51 Some characteristics of family income distributions (in dollars) in two regions follow.

Region	Mean	Median	Standard Deviation
A	16,450	15,200	2,900
B	16,980	15,500	2,500

Calculate Pearson's coefficient of skewness for each region. For which region is the income distribution more skewed?

3.52 Refer to Problem 3.10. Calculate Pearson's coefficient of skewness for the distribution of bonus amounts. In what units is the coefficient expressed?

3.53 Refer to Problems 3.45 and 3.48.
   a. Calculate Pearson's coefficient of skewness for earnings and for hours worked. Interpret your results.
   b. Why can the coefficients in a be compared even though earnings and hours worked are in different units?

*3.54 The charge per day for a semiprivate room in a hospital was as follows in four consecutive years:

Year:	1	2	3	4
Charge ($):	50	60	84	105

   a. Obtain the ratio of the charge in one year to that in the preceding year for years 2, 3, and 4.
   b. Obtain the geometric mean of the three ratios in a. Show how the charge in year 4 can be obtained from knowledge of the charge in year 1 and the geometric mean.

3.55 Refer to Problem 3.54. Do both parts, but use the following data on charges for a private room:

Year:	1	2	3	4
Charge ($):	90	110	150	165

*3.56 In flying a distance of 1,200 miles between two airports a cargo aircraft encountered strong headwinds and made the trip at a ground speed of 300 miles per hour. The 1,200-mile return flight was made at a ground speed of 400 miles per hour. Calculate the harmonic mean of the ground speeds.

3.57 The accident record for a plant, showing number of manhours worked per accident during a four-year period in which 4 million manhours were worked in each year, follows.

Year	Manhours per Accident	Year	Manhours per Accident
1	12,535	3	11,691
2	10,810	4	14,735

    a. Calculate the harmonic mean number of manhours per accident for the four-year period.

    b. Is the harmonic mean an appropriate measure of position here? (*Hint*: How many accidents occurred during the four-year period, in which 16 million manhours were worked?)

    c. Would your answer in **b** be affected if the annual number of manhours worked was not constant during the four-year period? Explain.

EXERCISES

3.58 Derive (3.15a) from (3.15).

3.59 Derive (3.16a) from (3.16).

3.60 Using (3.16), derive the appropriate formula for calculating the variance from a percent frequency distribution.

3.61 Refer to Problem 3.56. Show that a weighted arithmetic mean of the two ground speeds, using the number of hours flown at each speed as weights, is equivalent to the harmonic mean.

3.62 Prove for any set of values $X_1, X_2, \ldots, X_n$ that $\Sigma (X_i - A)^2$ is smallest when $A = \bar{X}$. [*Hint*: Write the sum as $\Sigma (X_i - \bar{X} + \bar{X} - A)^2$.]

STUDIES

3.63 The price movements of common stocks traded on an exchange such as the New York Stock Exchange are of importance to government, business, academicians, and the public. Effective summarization of the mass of price data generated each trading day requires careful selection of appropriate statistical measures.

    a. Describe the principal statistical measures used by newspapers to report the general trend of common stock prices on a daily basis and to report the price movements of any single stock on a daily, weekly, monthly, and yearly basis.

    b. How appropriate are the measures used? What are their advantages and disadvantages compared to other measures that might be used for summarizing the price data?

3.64 Refer to Appendix D. Consider the data on 1980 net assets of textile apparel manufacturers.

a. Calculate the mean, median, interquartile range, and standard deviation for these observations.

b. Construct a frequency distribution using five classes of equal widths, and calculate the same measures from the frequency distribution.

c. Compare the results obtained in **a** and **b**. How large is the approximation error in each case resulting from working with the frequency distribution?

d. Would the approximation errors tend to have been smaller if 10 classes had been used in the frequency distribution? Explain.

3.65 In a sample survey about family health insurance, the following data on family income, health-insurance coverage, and age of family head were obtained:

Family Income ($)	Percent of All Families	Percent of Families with Health Insurance	Mean Age of Family Head
Under 10,000	18	24	28
10,000–under 15,000	43	55	39
15,000–under 25,000	27	83	51
25,000 and over	12	91	57

a. What percent of all families had health insurance?

b. What is the mean age of the family head for all families?

c. How is it possible that the mean age can be calculated in **b**, but the mean income of all families cannot because of the open-end class?

d. How does the median income of families with health insurance compare to the median income of all families? Make specific reference to the data to support your answer.

3.66 A management-labor team initiated a training program to increase productivity and thereby raise the average earnings of piecework employees in a garment plant by $8.00 per week without any change in the piecework rate. "Before" and "after" data on hourly earnings follow for the 1,668 employees involved in the program.

Hourly Earnings ($)	Number of Employees	
	Before	After
4.50–under 5.00	146	148
5.00–under 5.50	395	400
5.50–under 6.00	735	731
6.00–under 6.50	332	23
6.50–under 7.00	43	146
7.00–under 7.50	17	220
Total	1,668	1,668

a. If you used the mean hourly earnings as the measure to assess the impact of the training program, how would you evaluate the success of the program?

b. Would you reach the same conclusion if the median were used to measure the impact of the program? Discuss fully.

c. The management-labor team had two objectives in running the program: (1) to improve the output of employees, and (2) to improve the economic welfare of employees. Evaluate the extent to which these goals were achieved by comparing the means, medians, and ogives of the two distributions. State your findings.

UNIT TWO

---◆---

Probability

Review

4
Basic Probability Concepts

The collection and summarization of data, discussed in the preceding chapters, are important first steps in using data for analysis and decision making. We now take up probability theory. This theory is important in its own right, being used by managers and analysts in assessing odds, in constructing probability models, and in selecting courses of action where risk must be taken into account. In addition, the theory of probability provides the logical foundation of statistical inference, which is the central topic in later units of this book.

We begin with basic probability concepts.

4.1 RANDOM TRIALS, SAMPLE SPACES, AND EVENTS

Random Trial

Activities that have uncertain or chance outcomes are common.

☐ **Examples**

1. An auditor selects a voucher and examines it. Whether or not the voucher contains an error is uncertain in advance of the examination. If it does contain an error, the type of error is uncertain in advance, there being a number of different types of errors that may occur.
2. A farmer plants a crop and eventually harvests it. The crop yield is uncertain in advance because of the chance influences of weather and other natural factors.
3. A laboratory centrifuge experiences operational breakdowns from time to time. The cause of the next breakdown is uncertain, there being several possible causes, including different kinds of mechanical and electrical failures. ☐

Each of the foregoing examples involves an activity referred to as a random trial.

(4.1) A *random trial* is an activity having two or more different possible outcomes, with uncertainty in advance as to which outcome prevails.

In Example 1, for instance, the random trial is the auditor's examination of a voucher. The possible outcomes are no error, or one or more errors of particular types. In Example 2, the random trial is the planting and harvesting of a crop and the possible outcomes are different crop yields. In each example, the outcome is uncertain until the trial has been conducted.

Sample Space

The outcomes of a random trial usually can be defined in different ways, depending on the purpose of the study.

☐ **Example**

A department store is studying its delivery service. The delivery of a customer's order is the random trial. If interest lies in whether or not the delivery is made on the day the customer's order is placed, the outcomes of interest are the two shown in Figure 4.1a. Here, day 1 signifies delivery on the same day, and delayed delivery signifies delivery after day 1. One of these two outcomes will result from the random trial.

If the purpose of the study is to investigate in detail the nature of delivery delays, interest might be in knowing whether delayed delivery occurred on the second day, third day, or after the third day. This more detailed array of outcomes is shown in Figure 4.1b.   ☐

We refer to the given set of outcomes of interest for a random trial as the sample space.

(4.2)   The different possible outcomes of a random trial are called the *basic outcomes* of the trial. The set of all basic outcomes for a random trial is called the *sample space* of the trial.

Figures 4.1a and 4.1b contain two different sample spaces for the delivery-service random trial. The first has two basic outcomes; the second is more detailed and has four basic outcomes.

FIGURE 4.1   *Two univariate sample spaces for random trial delivery of customer's order*

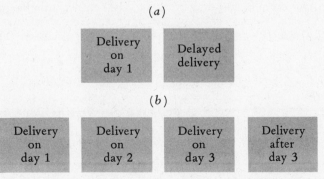

Comments

1. The basic outcomes of a sample space are mutually exclusive and exhaustive—that is, the outcome of a random trial will always be one and only one of the basic outcomes in its sample space. In Figure 4.1b, for example, the timing of delivery of a customer's order must be one and only one of the four basic outcomes shown.
2. A sample space and its basic outcomes are equivalent in structure to a system of classification and the classes of the system, respectively. This correspondence carries over to other characteristics as well, such as the dimension of a sample space, discussed next.

Univariate, Bivariate, and Multivariate Sample Spaces. The sample spaces in Figures 4.1a and 4.1b are called *univariate sample spaces* because the basic outcomes refer to a single characteristic—in this case, the timing of delivery. For many random trials, each basic outcome refers to two, or more than two, characteristics. The corresponding sample spaces are then called *bivariate* and *multivariate sample spaces,* respectively.

☐ Example

In the study of delivery service, the timing of the delivery may not be the only characteristic of interest. There may also be concern about whether or not the correct order is delivered to the customer. The resulting bivariate sample space can be displayed in a tabular arrangement, such as in Figure 4.2a. Note that the sample space here contains eight basic outcomes and that these have been arrayed in the same manner as in a bivariate cross-classification system. The univariate sample spaces for the two characteristics considered separately, status of order and timing of delivery, are found at the margins of the bivariate sample space.   ☐

A *tree diagram* is a useful graphic device for visualizing bivariate or multivariate sample spaces. It is especially helpful when there are more than two characteristics, since multivariate sample spaces are difficult to portray in a table.

☐ Example

Figure 4.2b shows a tree diagram for the bivariate sample space in Figure 4.2a. Note that the first branching (on the left) shows the two possible outcomes for the status of the order. Each of these outcomes has a second branching corresponding to the four possible outcomes for the timing of delivery. Thus, the tree ends on the right with eight branches corresponding to the eight basic outcomes of the bivariate sample space. Of course, the tree diagram could have been drawn with the first branching for the timing of delivery and the second for status of the order and it would still represent the same sample space.
                                                                                                ☐

Event

Interest frequently centers on subsets of basic outcomes of the sample space.

☐ Examples

1. The sample space of a student's grade in a course consists of the basic outcomes corresponding to the letter grades A, B, C, D, and F. A subset of basic outcomes of interest consists of those corresponding to a pass in the course (grades A, B, C, and D).

FIGURE 4.2 *Bivariate sample space for random trial delivery of customer's order. A tree diagram representation is shown in part b*

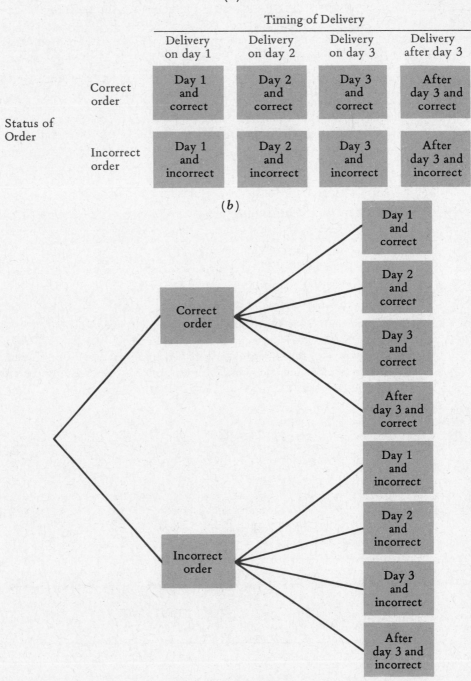

2. In the delivery-service example, the department store is concerned with poor delivery service. The store defines poor service as the subset of basic outcomes which involve either delivery of an incorrect order, or delivery after two days, or both. □

Figure 4.3 shows the six basic outcomes constituting poor delivery service in Example 2. Such a subset of basic outcomes is called an event of the sample space. We say that the poor-service event occurs if any one of these six outcomes is realized in the random trial.

(4.3) An *event* is a subset of basic outcomes of a sample space. An event is said to *occur* if one of its basic outcomes is realized in the random trial.

Many events can be defined for the same sample space. In Example 1, another event of interest might be failure in the course. This event consists of a single basic outcome, grade F. In Example 2, another event of interest might be delivery of incorrect order (whatever the day of delivery).

Set Notation and Terminology

To facilitate subsequent discussion, it is useful to introduce some mathematical notation for representing basic outcomes and events of a sample space. (For a review of mathematical set notation and operations, the reader should refer to Appendix A, Section A.3.)

Sample Space and Basic Outcomes.   The sample space of a random trial will be denoted by S, and the basic outcomes contained in it will be denoted by o_1, $o_2, \ldots, o_k$, where o_i represents the ith basic outcome and k the number of basic outcomes in S. We view the sample space S as the set containing all the basic outcomes and represent it as $S = \{o_1, o_2, \ldots, o_k\}$.

FIGURE 4.3   *Poor-service event for random trial delivery of customer's order*

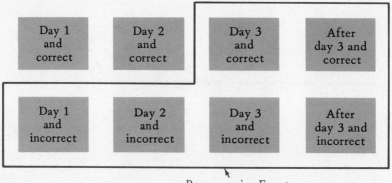

Poor-service Event

FIGURE 4.4   *Illustrations of various types of events for delivery service example*

(*a*)
Sample space, S

| | | Delivery | | | |
		On Day 1	On Day 2	On Day 3	After Day 3
Order	Correct	o_1	o_2	o_3	o_4
	Incorrect	o_5	o_6	o_7	o_8

(*b*)
Event, E

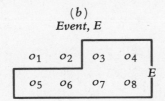

(*c*)
Complementary event, E*

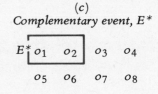

(*d*)
Event intersection, F ∩ G

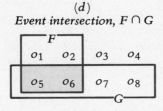

(*e*)
Event union, F ∪ G

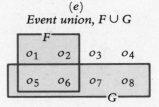

(*f*)
Mutually exclusive events, H and J

o_1	o_2	o_3	o_4
o_5	o_6	o_7	o_8

☐   **Example**

Figure 4.4a shows the sample space S consisting of the eight basic outcomes for the delivery-service example.   ☐

Events.   Symbols such as E and F, or E_1 and E_2, will be used to denote events.

☐ Example

The poor-service event described in Figure 4.3 may be denoted by E. Since this event E consists of the basic outcomes o_3, o_4, o_5, o_6, o_7, and o_8 in Figure 4.4a, we will write $E = \{o_3, o_4, o_5, o_6, o_7, o_8\}$. Event E is shown in Figure 4.4b. ☐

Complementary Event. Occasionally, interest centers on all the basic outcomes of a sample space that are *not* contained in some specific event.

(4.4) The set of all basic outcomes not contained in an event E is called the *complementary event to E* and is denoted by E^*.

☐ Example

Figure 4.4c shows the basic outcomes in the complement to the poor-service event E defined in Figure 4.4b. In this case, complementary event E^* consists of all basic outcomes that constitute good delivery service; i.e., $E^* = \{o_1, o_2\}$. ☐

Event Intersection

(4.5) The set of basic outcomes that belong to both of the events E_1 and E_2 of a sample space is called the *intersection of E_1 and E_2* and is denoted by:

$$E_1 \cap E_2 \text{ or, equivalently, } E_2 \cap E_1$$

☐ Example

In Figure 4.4d, two events are shown: $F = \{o_1, o_2, o_5, o_6\}$ and $G = \{o_5, o_6, o_7, o_8\}$. Here, F represents all basic outcomes in which delivery occurs within two days, and G represents all basic outcomes in which the order is incorrect. The intersection $F \cap G$ is shaded in Figure 4.4d. It is an event consisting of the two basic outcomes that F and G have in common—i.e., $F \cap G = \{o_5, o_6\}$. Thus, this intersection represents basic outcomes with both delivery within two days *and* an incorrect order. ☐

If the intersection of several events, say E_1, E_2, . . . , E_n, is of interest, it is written symbolically as $E_1 \cap E_2 \cap \cdots \cap E_n$ and it represents the set of all basic outcomes that the n events have in common.

Event Union

(4.6) The set of basic outcomes that belong to at least one of the events E_1 and E_2 of a sample space is called the *union of E_1 and E_2* and is denoted by:

$$E_1 \cup E_2 \text{ or, equivalently, } E_2 \cup E_1$$

☐ Example

In Figure 4.4e, the union of events F and G defined previously is shown by the shaded region. It is an event consisting of the six basic outcomes found in one or the other of events F and G or in both—i.e., $F \cup G = \{o_1, o_2, o_5, o_6, o_7, o_8\}$. Thus, this union represents the set of basic outcomes with delivery within two days *or* an incorrect order *or both*. ☐

If the union of several events, say E_1, E_2, . . . , E_n, is of interest, it is written symbolically as $E_1 \cup E_2 \cup \cdots \cup E_n$ and it represents the set of all basic outcomes each of which is found in at least one of the n events.

Mutually Exclusive Events

(4.7) If events E_1 and E_2 of a sample space have no basic outcomes in common, they are said to be *mutually exclusive events*. In this case, $E_1 \cap E_2 = \varnothing$, where $\varnothing$ denotes the empty set.

☐ Example

Events H and J in Figure 4.4f are mutually exclusive. Here, H represents all basic outcomes in which delivery occurs on day 1 and J represents all basic outcomes in which the correct order is delivered on day 3 or later. ☐

Any event E and its complement E^* are mutually exclusive. Thus, in our delivery-service example, the good- and poor-service events E and E^* are mutually exclusive. Similarly, in our course-grade example, the events pass and fail are mutually exclusive.

4.2 PROBABILITY AND ITS POSTULATES

Probability Measure

We have seen that the outcome of a random trial will be one and only one of the basic outcomes in its sample space. However, in advance of the random trial, it is not known with certainty which particular basic outcome will be realized. In probability theory we assign to each basic outcome o_i a value called its *probability*, which is a measure of how likely it is that o_i will be the realized outcome of the random trial. We shall denote the probability of o_i by $P(o_i)$. In like manner, we shall let $P(E)$ denote the probability that event E will occur, that is, that one of the basic outcomes of event E will be the realized outcome of the random trial. We temporarily postpone discussion about the meaning of the probability measure and how probability values are obtained.

☐ Examples

1. If the probability of outcome o_1 in Figure 4.4a is .57, we write $P(o_1) = .57$. Thus, the probability that the delivery is correct and on day 1 is .57.
2. If the probability of the event E in Figure 4.4b is .25, we write $P(E) = .25$. Thus, the probability that the outcome is either o_3, o_4, o_5, o_6, o_7, or o_8—one of the poor-service outcomes—is .25. ☐

Probability Postulates

Mathematicians have developed a formal structure whereby certain properties of probability are defined, and from which numerous important consequences flow. One set of postulates on which this formal structure can be built follows.

(4.8)   *Postulate 1.* $0 \leq P(o_i) \leq 1$ for any basic outcome o_i of the sample space S.

Postulate 2.   For any event E of a sample space S:

$$P(E) = \sum_E P(o_i)$$

where the summation is over the basic outcomes contained in E

Postulate 3. $P(S) = 1$ and $P(\emptyset) = 0$.

The first postulate states that the probability value of each basic outcome in a sample space is a number between 0 and 1. If the occurrence of basic outcome o_i is impossible, then $P(o_i) = 0$. If basic outcome o_i is certain to occur, then $P(o_i) = 1$. The closer the probability measure is to 0 the less likely it is that o_i will occur, and the closer $P(o_i)$ is to 1 the more likely it is that o_i will occur.

The second postulate states that the probability of any event equals the sum of the probability values of the basic outcomes that constitute the event. For instance, if $E = \{o_1, o_5, o_7\}$, then $P(E) = P(o_1) + P(o_5) + P(o_7)$.

The third postulate states that the probability value associated with the entire sample space is 1 and the probability associated with the empty set $\emptyset$ is 0.

Some Consequences of the Postulates.   A number of important consequences follow directly from the probability postulates. Several of these are given now. Others are presented later as formal probability theorems.

1. Since the sample space S contains all the basic outcomes, i.e., $S = \{o_1, o_2, \ldots, o_k\}$, it follows from postulate 2 that $P(S) = P(o_1) + P(o_2) + \cdots + P(o_k)$. Furthermore, from postulate 3, we have $P(S) = 1$. Therefore, the sum of the probabilities of all the basic outcomes is 1:

(4.9)

$$\sum_{i=1}^{k} P(o_i) = 1$$

2. If all basic outcomes contained in an event E_1 are also contained in an event E_2, so E_1 is a subset of E_2, it follows from postulate 2 and from the fact that probability values for basic outcomes are nonnegative (postulate 1) that $P(E_1)$ cannot exceed $P(E_2)$:

(4.10) $P(E_1) \leq P(E_2)$ when event E_1 is a subset of event E_2

For instance, in our delivery-service example, E denotes poor delivery service and G denotes an incorrect order. It can be seen from Figure 4.4 that G is a subset of E. Hence, $P(G) \leq P(E)$.

3. Since every event E is a subset of the sample space S on which it is defined, it follows from (4.10) that $P(E) \leq P(S)$. Furthermore, postulates 1 and 2 imply that the probability of an event E cannot be negative, and postulate 3 states that $P(S) = 1$. Hence:

(4.11)
$$0 \leq P(E) \leq 1 \text{ for any event } E$$

In other words, every event has a probability value between 0 and 1.

4. If E_1 and E_2 are any two mutually exclusive events of a sample space, then by definition they have no basic outcomes in common; i.e., $E_1 \cap E_2$ is the empty set $\varnothing$. Thus, by postulate 3, $P(E_1 \cap E_2) = 0$.

(4.12) If E_1 and E_2 are mutually exclusive events of a sample space:

$$P(E_1 \cap E_2) = 0$$

Meaning of Probability

It has been seen that probability is a measure which, for any basic outcome or event, is a number between 0 and 1. For instance, if event E denotes the occurrence "It rains in Seattle on August 17," the probability value assigned to this event might be $P(E) = .3$. How is this probability value interpreted, and how does one determine such a value in practice? We now present two interpretations of probability, which will allow the reader to view its meaning from two useful points of view. We shall follow this with a brief discussion of how probability values are assessed.

Objective Interpretation.   In the *objective* interpretation of probability, the probability value of an event is equated with the relative frequency of occurrence of the event in the long run under constant causal conditions. Let us again consider the event E "It rains in Seattle on August 17." With the objective interpretation, the probability statement $P(E) = .3$ is understood to mean that in an indefinitely large number of August 17s in Seattle, with basic climatic conditions the same as at the present time, there will be rain on about 30 percent of them.

The objective interpretation of probability applies only to repeatable events and not to unique events. For instance, from this point of view, one would not talk about the probability that John William Jones, aged 40, will die during the next year; he either will or won't. John William Jones' 41st year is not a repeatable event. But it does make sense, according to this interpretation of probability, to say that the probability of a male of age 40 dying during the next year is .002. This is interpreted to mean that among many males of age 40, about .2 percent of them will die during their 41st year. Thus, probability is interpreted from the objective point of view with reference to a large number of random trials under the same causal conditions, and not with reference to a specific random trial.

Subjective Interpretation.   Since many events of interest are not repeatable under constant causal conditions, there are many situations where the objective interpretation of probability cannot be applied. The *subjective* interpretation of probability (also called the *personal* interpretation) relates probability to degree of personal belief. It is not restricted to repeatable events but applies also to unique events. Thus, under this approach one would be willing to consider the probability that John William Jones, aged 40, will die this coming year, or the probability that the marketing of a new breakfast cereal will be successful. The subjective interpretation even allows one to assign a probability number to an event for a random trial that has occurred in the past. For instance, a police detective investigating the burglary of a store last week might assign a probability of $P(A) = .9$ to suspect A having committed this burglary. For all of these events, it would not be possible to interpret probabilities as relative frequencies in the long run since these events are inherently nonrepeatable.

Personal probability is intimately related to the person making the probability evaluation, including subjective feelings and the information available to him or her at the time. Thus, two insurance examiners who have identical information about John William Jones might still arrive at different personal probabilities that he will die during his 41st year. Indeed, each insurance examiner might revise his or her probability assessment upon more reflection. Even with reference to a repeatable event, two individuals might subjectively assess the probabilities differently, depending on the information available to each and how this information is evaluated. On the other hand, with the objective probability interpretation, the probabilities of repeatable events are considered to be determined by the causal conditions that are operating and therefore do not depend on personal factors.

Comment

Often, probability values are expressed in terms of odds. We say that the *odds in favor* of an event E are a to b if:

$$P(E) = \frac{a}{a + b}$$

For instance, if the odds in favor of candidate A winning an election are 3 to 2, then the probability of candidate A winning is $3/(3 + 2) = .6$. If the probability of dying in the next year is .002, then the odds of death can be expressed as .002 to .998 or 1 to 499.

Assessment of Probability Values

The probability postulates provide conditions that a probability measure for any random trial must satisfy, but by themselves do not provide probability numbers. We now consider one method for obtaining probability values consistent with the objective interpretation of probability and then take up two methods consistent with the subjective interpretation.

Observed Relative Frequency. The probability of an event is equated by the objective interpretation to the relative frequency of occurrence of the event in a long series of trials conducted under the same conditions. Hence, the observed relative frequency of an event in a large number of trials under approximately constant causal conditions may be used as an estimate of the probability of occurrence. For instance, if a production process has produced 30 defective items in the last 5,000 produced and conditions for this production run have been stable and are likely to remain unchanged, then $30/5,000 = .006$ is an estimate of the probability that the next item produced will be defective.

The use of observed relative frequency is not without difficulties. It requires repeated experience with the random process and constant causal conditions for the repeated trials. It is also assumed that the observed relative frequency has not departed substantially from the "true" long-run relative frequency as the result of chance influences. Statistical procedures (covered in subsequent chapters) exist for determining the precision with which an observed relative frequency approximates the underlying probability value. Nevertheless, the application of any of these statistical procedures involves assumptions, so it is impossible to avoid judgmental factors completely even with the use of observed relative frequencies.

Direct Questioning.   If a subjective probability for an event is required, the most straightforward way to obtain it is to ask a direct question about it. There are various ways in which a probability assessment question can be phrased. We illustrate three questions for a product manager about the likelihood that the sales of a new product will exceed the sales target.

1. "What is the probability that this new product will exceed its sales target?"
2. "How many chances out of 10 does this new product have of exceeding its sales target?"
3. "What are the odds in favor of this new product exceeding its sales target?"

Like observed relative frequency, the direct-question approach to probability assessment has some inherent difficulties. The respondent may find it difficult to express probabilistic beliefs in quantitative terms, or the answer may vary depending on when the question is asked and how much thought is given to it. There is also the problem of who should be the source of the probability assessment. Should the company president ask the sales manager for a probability assessment about the sales of a new product or should the marketing research director be asked? If the president asks both and different probability assessments are furnished, how are they to be combined?

Another difficulty is the tendency of some respondents to provide biased probability assessments. For instance, the director of a computer center has assigned a probability of .3 each week to the possibility of a computer breakdown the following week. Yet there were computer breakdowns in only 10 percent of the weeks during the past two years, suggesting that the director tends to overestimate the probability of a computer breakdown.

A final difficulty is that the probability assessments provided by respondents may not satisfy the probability postulates. For instance, a respondent might state that the probability of rain tomorrow is .6, yet subsequently indicate that the probability of rain or fog tomorrow is .5, violating (4.10).

Indirect Methods.   A host of methods exist for eliciting subjective probabilities by indirect means. We shall illustrate one of these for ascertaining the sales manager's assessment that the sales of a new product will exceed the sales target. The first question is:

Would you accept the gamble of receiving $5 if sales exceed the target and paying $5 if they do not?

If the answer is yes, one may impute that the subjective probability of sales exceeding the target is at least .5. A second question then might be:

> Would you accept the gamble of receiving $4 if sales exceed the target and paying $5 if they do not?

If the answer again is yes, one may impute that the manager views the odds as at least 5 to 4 in favor of sales exceeding the target. Hence, the subjective probability is at least $5/(5 + 4) = 5/9$ that sales will exceed the target.

By considering additional gambles with still other payoffs, the sales manager's subjective probability of sales exceeding the target can be narrowed to a small interval.

4.3 PROBABILITY DISTRIBUTIONS

Definition

(4.13) An assignment of probabilities to each of the basic outcomes in a sample space is called a *probability distribution* for the sample space.

☐ Example

Consider the sample space in Figure 4.1b relating to our example on the timing of delivery of customers' orders. Probabilities have been assigned in Table 4.1a to each of the four basic outcomes in this sample space, based on relative frequency in the recent past. Notice that none of the probability values is less than 0, and that the sum of all the probabilities equals 1. Postulates 1 and 3 in (4.8) state that all probability distributions must have these two properties. ☐

Univariate, Bivariate, and Multivariate Probability Distributions

The probability distribution in Table 4.1a is called a *univariate probability distribution* because it is based on a univariate sample space. Probability distributions for bivariate or multivariate sample spaces can be constructed as well. These are referred to as *bivariate* and *multivariate probability distributions*, respectively.

☐ Example

Table 4.1b contains a bivariate probability distribution for the bivariate sample space given earlier in Figure 4.2a. The two variables forming the bivariate sample space are status of order and timing of delivery. ☐

Joint, Marginal, and Conditional Probability Distributions

Joint Probabilities. As discussed earlier, in bivariate and multivariate sample spaces each basic outcome refers to two or more characteristics. Thus, in Table 4.1b, the basic outcome of delivery of a correct order on day 1 refers to the two characteristics: (1) correct order (A_1), and (2) delivery on day 1 (B_1). Each of these

TABLE 4.1 *Univariate and bivariate probability distributions*

(a)

Basic Outcome	Delivery	Probability
B_1	On day 1	.60
B_2	On day 2	.20
B_3	On day 3	.10
B_4	After day 3	.10
	Total	1.00

(b)

Status of Order	Timing of Delivery				
	Delivery on Day 1 B_1	Delivery on Day 2 B_2	Delivery on Day 3 B_3	Delivery after Day 3 B_4	Total
A_1: Correct order	.57	.18	.08	.07	.90
A_2: Incorrect order	.03	.02	.02	.03	.10
Total	.60	.20	.10	.10	1.00

(c)

	B_1	B_2	B_3	B_4	Total
A_1:	$P(A_1 \cap B_1)$	$P(A_1 \cap B_2)$	$P(A_1 \cap B_3)$	$P(A_1 \cap B_4)$	$P(A_1)$
A_2:	$P(A_2 \cap B_1)$	$P(A_2 \cap B_2)$	$P(A_2 \cap B_3)$	$P(A_2 \cap B_4)$	$P(A_2)$
Total	$P(B_1)$	$P(B_2)$	$P(B_3)$	$P(B_4)$	1.00

characteristics can, in our new terminology, be called an event of the bivariate sample space. Thus, the basic outcome of delivery of a correct order on day 1 can be considered a *joint outcome* of the two events A_1 and B_1.

Examples

1. The basic outcome of delivery of a correct order on day 1 is denoted symbolically by $A_1 \cap B_1$, since it is the joint outcome of events A_1 and B_1. The probability of this joint outcome is denoted by $P(A_1 \cap B_1)$ in Table 4.1c. From Table 4.1b we see that this probability is $P(A_1 \cap B_1) = .57$.
2. The basic outcome of an incorrect order delivered on day 3 is the joint outcome of the event incorrect order (A_2) and the event delivery on day 3 (B_3). This joint outcome is denoted by $A_2 \cap B_3$ and the corresponding probability according to Table 4.1b is $P(A_2 \cap B_3) = .02$.

The probability of a joint outcome is called a *joint probability*. Thus, $P(A_1 \cap B_1) = .57$ is a joint probability.

A bivariate or multivariate probability distribution, such as that in Table 4.1b, is called a *joint probability distribution* when it is desired to emphasize the fact that the basic outcomes are determined jointly by two or more characteristics. Table 4.1c shows the joint probability distribution for our example in symbolic form.

Marginal Probabilities. With a bivariate probability distribution, one is also frequently interested in the probability distributions of the individual variables considered separately. Thus, management may wish to know the probability that an order is incorrect, irrespective of the timing of delivery, or to know the probability that an order is delivered on day 1, irrespective of whether or not it is correct. The univariate probability distribution for each of the individual variables can be obtained by summing the joint probabilities across the columns or rows, as the case may be.

☐ Examples

1. The univariate probability distribution in Table 4.1a corresponding to the variable timing of delivery may be obtained by summing each column in Table 4.1b. The resulting probabilities are shown in the bottom row of Table 4.1b and are boxed for emphasis. For example, we have $P(B_1) = .60$.
2. The univariate probability distribution for the variable status of order is obtained by summing each row in Table 4.1b. The resulting probabilities are shown in the rightmost column and are boxed for emphasis. For example, we have $P(A_2) = .10$. ☐

The rationale for summing will be explained for $P(B_1)$, the probability of delivery on day 1. Since event B_1, if it occurs, must occur jointly with correct order (A_1) or incorrect order (A_2) but not both, it follows that the joint outcomes $A_1 \cap B_1$ and $A_2 \cap B_1$ are mutually exclusive and exhaustive of the possible outcomes leading to the occurrence of B_1. Therefore:

$$P(B_1) = P(A_1 \cap B_1) + P(A_2 \cap B_1) = .57 + .03 = .60$$

Since the probabilities obtained by summing across either one of the classifications are shown in the margins of Table 4.1b, they are often called *marginal probabilities,* and the resulting probability distributions are then called *marginal probability distributions.*

☐ Example

Table 4.1a contains the marginal probability distribution for the timing of delivery shown in the horizontal box of Table 4.1b. The marginal probabilities for the second variable, status of order, are shown in the vertical box in Table 4.1b. ☐

In general, the marginal probabilities are found from a bivariate probability distribution as follows:

(4.14)

$$P(A_i) = \sum_j P(A_i \cap B_j)$$

$$P(B_j) = \sum_i P(A_i \cap B_j)$$

where the summations are over all events B_j and A_i, respectively. Table 4.1c illustrates the appropriate summations to be performed for obtaining the marginal probabilities.

Conditional Probabilities. Often, it is desired to know the probability of one event occurring, given that a second event occurs. For instance, referring to Table 4.1b, we may wish to know the probability that an order is incorrect (A_2), given that the order is delivered on the third day (B_3). We shall denote this probability by $P(A_2|B_3)$. The vertical rule in the notation is read as "given." In other words, $P(A_2|B_3)$ means "probability of A_2 occurring, given B_3 occurs." We refer to this type of probability as a conditional probability.

(4.15) If E_1 and E_2 are any two events of a sample space and $P(E_2)$ is not equal to zero, the *conditional probability of E_1, given E_2*, is denoted by $P(E_1|E_2)$ and defined:

$$P(E_1|E_2) = \frac{P(E_1 \cap E_2)}{P(E_2)}$$

☐   **Examples**

1. Table 4.2 shows, for job placement at a college, the joint probability distribution for the two variables: (1) location of a graduate's first job (C_i), and (2) length of time the graduate remains in this job (D_j). We wish to obtain $P(D_1|C_1)$, the conditional probability that a graduate remains in the first job for less than two years, given that the job is in the private sector. In accordance with the relative frequency interpretation of the probability $P(C_1) = .60$ in Table 4.2, we know that of every 100 placements in first jobs, on the average 60 are in the private sector. Further, we can state on the basis of the probability $P(C_1 \cap D_1) = .45$ that of every 100 first-job placements, on the average 45 are in the private sector *and* have a duration of less than two years. Thus, of each 60 first-job placements in the private sector, on the average 45 entail a stay of less than two years. The proportion $45/60 = .75$ is the conditional probability that a graduate remains in the first job for less than two years, given that the job is in the private sector.
   Formula (4.15) provides this same result formally, as follows:

$$P(D_1|C_1) = \frac{P(C_1 \cap D_1)}{P(C_1)} = \frac{.45}{.60} = .75$$

2. For our delivery-service example in Table 4.1b, we wish to obtain $P(A_2|B_3)$, i.e., the conditional probability of A_2 occurring, given B_3 occurs. We can reason that, on the

TABLE 4.2 *Joint probability distribution for job-placement example*

Job Location C_i	Job Duration D_j		Total
	Less Than Two Years D_1	Two Years or Longer D_2	
C_1: Private sector	.45	.15	.60
C_2: Public sector	.35	.05	.40
Total	.80	.20	1.00

average, out of every 100 deliveries, 10 are made on day 3. Out of these 10 deliveries, 2 involve incorrect orders. Therefore, the desired probability is $P(A_2|B_3) = 2/10 = .20$.

Formula (4.15) gives this result formally. Since $P(B_3) = .10$ and $P(A_2 \cap B_3) = .02$, we obtain:

$$P(A_2|B_3) = \frac{P(A_2 \cap B_3)}{P(B_3)} = \frac{.02}{.10} = .20$$

3. For the delivery-service example in Table 4.1b, we wish to find the conditional probability of delivery on day 1, given that the order is correct. We obtain by (4.15):

$$P(B_1|A_1) = \frac{P(A_1 \cap B_1)}{P(A_1)} = \frac{.57}{.90} = .63$$ □

Conditional Probability Distributions.   A conditional probability distribution is the set of conditional probabilities conditioned on a given event.

□   Examples

1. We wish to obtain for the job-placement example the conditional probability distribution for duration of first job (D_j), given that the job is in the private sector (C_1). We require the set of conditional probabilities $P(D_j|C_1)$. From previous work, we know that $P(D_1|C_1) = .75$. We calculate from Table 4.2 that $P(D_2|C_1) = P(D_2 \cap C_1)/P(C_1) = .15/.60 = .25$. Hence, the conditional probability distribution for duration of first job, given that the job is in the private sector, is:

Duration	Conditional Probability, Given C_1
D_1: Less than two years	.75
D_2: Two years or more	.25
Total	1.00

Note that the conditional probability distribution has the attributes of any probability distribution: it involves outcome probabilities between 0 and 1 and these probabilities sum to 1.

2. For our delivery-service example in Table 4.1b, we wish to obtain two conditional probability distributions: (1) for status of order, given that delivery is on day 3, and (2) for timing of delivery, given that the order is correct. Table 4.3 shows these two conditional distributions and how they were obtained.   □

4.4   BASIC PROBABILITY THEOREMS

Several useful theorems follow from the probability postulates and concepts presented in the last section.

Addition Theorem

The addition theorem is useful in a variety of situations for obtaining the probability that either one of two events, or both, occur.

TABLE 4.3 *Two conditional probability distributions for delivery-service example*

(a)

Status of Order	Conditional Probability, Given B_3
A_1: Correct order	$P(A_1 \mid B_3) = .08/.10 = .80$
A_2: Incorrect order	$P(A_2 \mid B_3) = .02/.10 = \underline{.20}$
	Total 1.00

(b)

Timing of Delivery	Conditional Probability, Given A_1
B_1: On day 1	$P(B_1 \mid A_1) = .57/.90 = .63$
B_2: On day 2	$P(B_2 \mid A_1) = .18/.90 = .20$
B_3: On day 3	$P(B_3 \mid A_1) = .08/.90 = .09$
B_4: After day 3	$P(B_4 \mid A_1) = .07/.90 = \underline{.08}$
	Total 1.00

□ **Examples**

1. In our delivery-service example in Table 4.1b, what is the probability that the delivery is correct, or the delivery is on day 1, or both; i.e., what is $P(A_1 \cup B_1)$?
2. In the job-placement example in Table 4.2, what is the probability that the first job is in the private sector, or the graduate remains in the job for less than two years, or both; i.e., what is $P(C_1 \cup D_1)$? □

(4.16) *Addition Theorem.* For any two events E_1 and E_2 of a sample space:

$$P(E_1 \cup E_2) = P(E_1) + P(E_2) - P(E_1 \cap E_2)$$

□ **Examples**

1. Refer to Table 4.1b. We wish to find $P(A_1 \cup B_1)$. We have $P(A_1) = .90$, $P(B_1) = .60$, and $P(A_1 \cap B_1) = .57$. Hence, by (4.16):

$$P(A_1 \cup B_1) = .90 + .60 - .57 = .93$$

2. Refer to Table 4.2. We wish to obtain $P(C_1 \cup D_1)$. We have $P(C_1) = .60$, $P(D_1) = .80$, and $P(C_1 \cap D_1) = .45$. Thus, by (4.16):

$$P(C_1 \cup D_1) = .60 + .80 - .45 = .95$$

3. Refer to the sample space in Figure 4.5a. The six basic outcomes of the sample space are indicated by circles, and the probability of each outcome is shown in its circle. Two events are defined on this sample space. Note that E contains four basic outcomes, F contains three, and two outcomes are common to both events. For these two events, we have:

$$P(E) = .34 + .15 + .10 + .07 \qquad = .66$$

$$P(F) = \qquad\qquad .10 + .07 + .22 = .39$$

$$P(E \cap F) = \qquad\qquad .10 + .07 \qquad = .17$$

FIGURE 4.5 *Illustrations of addition and complementation theorems*

(a)

(b)

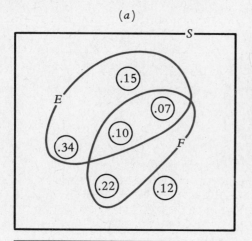

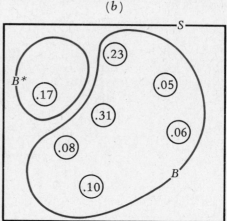

Substituting these values into addition theorem (4.16), we obtain:

$$P(E \cup F) = .66 + .39 - .17 = .88$$

By adding the probabilities for the basic outcomes in the event $E \cup F$ directly from Figure 4.5a, we obtain:

$$P(E \cup F) = .34 + .15 + .10 + .07 + .22 = .88$$

which is the result provided by the addition theorem. ☐

From the third example, it is clear why $P(E \cap F)$ must be subtracted from $P(E) + P(F)$ to obtain $P(E \cup F)$. Note that when $P(E)$ and $P(F)$ are added, the probabilities of the two basic outcomes these events have in common are included twice. Since $E \cap F$ is the set of basic outcomes the two events have in common, $P(E \cap F)$ is subtracted to correct for the double counting.

Mutually Exclusive Events. The addition theorem takes on a special form when events E_1 and E_2 are mutually exclusive. From (4.12), we know that $P(E_1 \cap E_2) = 0$ in this case. Therefore, the addition theorem simplifies to:

(4.17) For any two mutually exclusive events E_1 and E_2:

$$P(E_1 \cup E_2) = P(E_1) + P(E_2)$$

☐ Example

In Table 4.1b, the events delivery on day 1 (H) and delivery of correct order after day 2 (J) are mutually exclusive. Since $P(H) = .57 + .03 = .60$ and $P(J) = .08 + .07 = .15$, we have by (4.17) that $P(H \cup J) = .60 + .15 = .75$. ☐

Formula (4.17) may be extended to n mutually exclusive events E_1, $E_2, \ldots, E_n$ of a sample space:

(4.18) $$P(E_1 \cup E_2 \cup \cdots \cup E_n) = P(E_1) + P(E_2) + \cdots + P(E_n)$$

Complementation Theorem

Sometimes it is more difficult to find the probability of an event than to find the probability of its complement. One can then obtain the probability of the event by the complementation theorem.

(4.19) *Complementation Theorem.* For any event E of a sample space:

$$P(E) = 1 - P(E^*)$$

where: E^* is the complementary event to E

☐ Examples

1. For our delivery-service example in Table 4.1b, we wish to find the probability that service is not perfect, i.e., that the correct order is not delivered on day 1. This event E consists of all outcomes other than $A_1 \cap B_1$. Hence, $E^* = A_1 \cap B_1$. We have $P(E^*) = .57$ and by the complementation theorem $P(E) = 1 - .57 = .43$.
2. In Figure 4.5b, $P(B) = 1 - P(B^*) = 1 - .17 = .83$. ☐

Multiplication Theorem

By rearranging the definition of conditional probability (4.15), we obtain the multiplication theorem, which is frequently helpful in finding the joint probability of two events.

☐ Examples

1. Sally McQue has applied to two law schools. What is the probability that she will be accepted by both schools?
2. At a police road checkpoint each vehicle is checked for having a current safety inspection sticker and the driver is checked for having a current driver's license. What is the probability that a vehicle will have a current inspection sticker and the driver a current driver's license? ☐

(4.20) *Multiplication Theorem.* For any two events E_1 and E_2 of a sample space:

$$P(E_1 \cap E_2) = P(E_1)P(E_2|E_1) = P(E_2)P(E_1|E_2)$$

☐ Examples

1. In our law school admissions example, let H and B denote acceptance by schools 1 and 2, respectively. The probability that Sally McQue is accepted by law school 1 is known to be $P(H) = .6$. Given acceptance by law school 1, the conditional probability of acceptance by law school 2 is $P(B|H) = .9$. We wish to find the probability of acceptance by both schools. Using (4.20), we obtain:

$$P(H \cap B) = P(H)P(B|H) = .6(.9) = .54$$

2. For the traffic-inspection example, let events S and D denote current safety inspection sticker and current driver's license, respectively. It is known that 90 percent of vehicles have current safety inspection stickers and that 95 percent of drivers of vehicles with current inspection stickers have current driver's licenses. We wish to find the probabil-

ity that both driver and vehicle will be found to have current documents. We have $P(S) = .90$ and $P(D|S) = .95$. Hence, we obtain by (4.20):

$$P(S \cap D) = P(S)P(D|S) = .90(.95) = .855$$

3. A computer manufacturer receives memory chips, which are inspected 100 percent before entering assembly operations. Let D denote a chip is defective and D^* it is not defective. Also, let A denote that a chip is approved for assembly by the inspector, and A^* that it is not approved. From past experience, it is known that $P(D) = .10$. Also, it is known that the probability of an inspector passing a chip, given it is defective, is $P(A|D) = .005$ while the corresponding probability, given the chip is acceptable, is $P(A|D^*) = .999$. The tree diagram in Figure 4.6, called a *probability tree*, summarizes the situation. We use multiplication theorem (4.20) to find the joint probability that a chip is defective and is approved for assembly:

$$P(D \cap A) = P(D)P(A|D) = .10(.005) = .0005$$

Similarly, the joint probability that a chip is acceptable and is approved for assembly is:

$$P(D^* \cap A) = P(D^*)P(A|D^*) = .90(.999) = .8991$$

The other joint probabilities in Figure 4.6 have been obtained in similar fashion.

FIGURE 4.6   *Probability tree for inspection example*

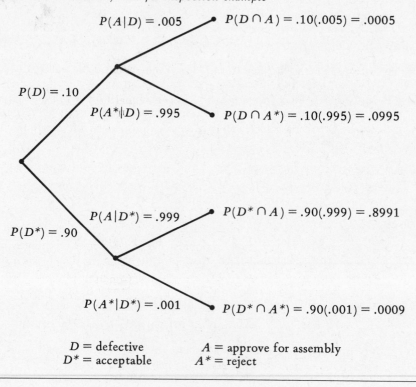

$$P(A|D) = .005 \qquad P(D \cap A) = .10(.005) = .0005$$

$$P(D) = .10$$

$$P(A^*|D) = .995 \qquad P(D \cap A^*) = .10(.995) = .0995$$

$$P(A|D^*) = .999 \qquad P(D^* \cap A) = .90(.999) = .8991$$

$$P(D^*) = .90$$

$$P(A^*|D^*) = .001 \qquad P(D^* \cap A^*) = .90(.001) = .0009$$

D = defective A = approve for assembly
D^* = acceptable A^* = reject

Note that since a chip is either acceptable or defective, we can find the probability that a chip is approved for assembly as follows:

$$P(A) = P(D \cap A) + P(D^* \cap A) = .0005 + .8991 = .8996 \qquad \square$$

The multiplication theorem can be extended to more than two events. We have for three events:

(4.21)
$$P(E_1 \cap E_2 \cap E_3) = P(E_1) P(E_2|E_1) P(E_3|E_1 \cap E_2)$$

The three events in (4.21) can be ordered in any fashion desired.

4.5   STATISTICAL INDEPENDENCE

The concept of statistical independence, and its contrast with statistical dependence, is the subject of the following examples.

☐   Examples

1. Table 4.4 contains, for our earlier inspection example, the probability distributions of inspector's action for a chip inspected, conditional on whether or not the chip is defective. Note that the probability of the chip being approved for assembly varies, depending on whether or not it is defective. This suggests that inspector's action is related to the quality of the chip. One would hope, of course, that such a relation exists so that it is both highly likely that good chips are approved and that defective ones are rejected.

2. A hospital administrator is studying the relationship between length of stay in hospital and whether or not the patient has hospital insurance, for patients with a certain illness. The joint probability distribution is shown in Table 4.5a. The conditional probabilities for length of stay, given insurance status, are shown in Table 4.5b. The conditional probabilities were obtained in the usual manner. For example, $P(A_1|B_1) = .42/.70 = .60$. Note that for both the insured and uninsured, the conditional probabilities of any given length of stay are the same, and are equal to the marginal probability of that class; i.e., $P(A_i|B_j) = P(A_i)$ for all i and j. This indicates that insurance status of the patient is unrelated to length of stay. A similar result is found if we look at conditional probabilities of insurance status, given the length of stay of the patient.   ☐

Before considering statistical independence of two or more variables, we must first take up statistical independence of events.

TABLE 4.4 *Conditional probability distributions for inspector's action, given the quality of the chip.* The differences between the two conditional probability distributions indicate that quality of chip and inspector's action are not independent.

Action		Conditional on:	
		Chip Defective D	Chip Acceptable D*
Approve (A)		.005	.999
Reject (A*)		.995	.001
	Total	1.000	1.000

TABLE 4.5 *Joint and conditional probability distributions for length of hospital stay and insurance status.* The identity of the conditional probability distributions indicates that length of stay and insurance status are independent.

(a)
Joint probability distribution

| Length of Stay (days) | Insurance Status | | Total |
	Insured B_1	Uninsured B_2	
A_1: Less than 5	.42	.18	.60
A_2: 5–10	.21	.09	.30
A_3: Over 10	.07	.03	.10
Total	.70	.30	1.00

(b)
Conditional probability of length of stay, given insurance status

| Length of Stay (days) | Conditional on: | | Total |
	Insured B_1	Uninsured B_2	
A_1: Less than 5	.60	.60	.60
A_2: 5–10	.30	.30	.30
A_3: Over 10	.10	.10	.10
Total	1.00	1.00	1.00

Independent Events

We say that two events are unrelated statistically if the probability that one event occurs is unaffected by whether or not the other event occurs. When two events are unrelated in this sense, they are said to be statistically independent. If two events are not statistically independent, they are related statistically and are said to be statistically dependent.

Symbolically, the statistical independence of two events E_1 and E_2 means that $P(E_2|E_1) = P(E_2)$ or, equivalently, that $P(E_1|E_2) = P(E_1)$. Substituting either of these relations into multiplication theorem (4.20) yields the formal definition of statistical independence:

(4.22) Two events E_1 and E_2 of a sample space are said to be *statistically independent* if:

$$P(E_1 \cap E_2) = P(E_1)P(E_2)$$

Thus, two events are statistically independent if their joint probability equals the product of their marginal probabilities.

☐ Examples

1. For the probability distribution in Table 4.5a, A_1 and B_1 are statistically independent events since $P(A_1 \cap B_1) = .42$ and $P(A_1)P(B_1) = .60(.70) = .42$. In other words, a

length of stay of less than five days is statistically independent of a patient being insured.

2. For the inspection example in Figure 4.6, the events D and A are statistically dependent because $P(D \cap A) = .0005$ while $P(D)P(A) = .10(.8996) \doteq .08996$. [Recall that $P(A) = .8996$ was computed earlier.] □

Formula (4.22) can be extended to n independent events $E_1, E_2, \ldots, E_n$:

(4.23) $$P(E_1 \cap E_2 \cap \cdots \cap E_n) = P(E_1)P(E_2) \cdots P(E_n)$$

□   **Example**

In a new security system being tested at a military base, visitors wishing to enter a secure area must pass three security checks in succession—a voice-pattern check, a fingerprint check, and a handwriting check. Studies indicate a probability of .02 that an unauthorized visitor will pass any one check. Assuming that the three checks are statistically independent, we wish to find the probability that an unauthorized visitor passes all three checks.

Let V, F, and H denote the events that an unauthorized person passes the voice, fingerprint, and handwriting checks, respectively. We have $P(V) = P(F) = P(H) = .02$. Hence, by (4.23):

$$P(V \cap F \cap H) = P(V)P(F)P(H) = .02(.02)(.02) = .000008$$

Of course, a materially different probability might be obtained if independence does not hold for the three checks. □

Comment

Mutually exclusive events are generally statistically dependent. For mutually exclusive events E_1 and E_2, we have from (4.12) that $P(E_1 \cap E_2) = 0$. Hence, whenever both $P(E_1) > 0$ and $P(E_2) > 0$, condition (4.22) for statistical independence cannot be met and E_1 and E_2 are statistically dependent. Intuitively, the statistical dependence is reasonable since if one event occurs we know that the other has not occurred because they are mutually exclusive.

Independent Variables

The concept of two independent events extends to the concept of two statistically independent variables or characteristics. We denote the classes of the two variables by A_1, A_2, etc., and B_1, B_2, etc., respectively.

(4.24) Two variables A and B are said to be *statistically independent* if:

$$P(A_i \cap B_j) = P(A_i)P(B_j) \quad \text{for all } A_i \text{ and } B_j$$

Note that *all* joint probabilities must equal the product of their corresponding marginal probabilities for the two variables to be statistically independent.

□   **Example**

In Table 4.5a, it is seen that the two variables (length of stay and insurance status) are statistically independent since the joint probability of any two events A_i and B_j is equal to the product of the respective marginal probabilities. For example, $P(A_1 \cap B_2) = .18$, and $P(A_1)P(B_2) = .60(.30) = .18$. □

Comments

1. Statistical independence of two variables implies, for our hospital-stay example, that the conditional distribution of length of stay for insured patients is the same as the conditional distribution of length of stay for uninsured patients, and that both conditional distributions are equal to the marginal distribution of length of stay.

2. Statistical dependence is present in many bivariate probability distributions. In the joint probability distribution in Table 4.1b, for instance, status of the order and timing of delivery are statistically dependent variables. This can be seen quickly by multiplying corresponding pairs of marginal probabilities. For instance, $P(A_2)P(B_1) = .10(.60) = .06$, which does not agree with $P(A_2 \cap B_1) = .03$. To be sure, some of the joint probabilities in Table 4.1b are equal to products of corresponding marginal probabilities. However, since not *all* are equal, there is statistical dependence between the two variables. The nature of the statistical relationship can be seen readily when we obtain the conditional probability distributions of day of delivery, for status of the order:

Status of Order	Day of Delivery				Total
	1	2	3	After 3	
Correct	.63	.20	.09	.08	1.00
Incorrect	.30	.20	.20	.30	1.00

Note that incorrect orders are less likely to be delivered on day 1 than correct orders; the opposite is the case for day 3 or thereafter.

4.6 INFINITE SAMPLE SPACES

To this point, the discussion of sample spaces, probability postulates, and probability distributions has assumed implicitly that a random trial has only a finite number of basic outcomes, namely $\{o_1, o_2, \ldots, o_k\}$. This has been a convenient assumption for the presentation of basic concepts. There are, however, many settings in which an infinity of possible outcomes is conceivable. For example, the weight of the wheat crop in North America next year can be any of an infinity of values in an interval; thus, the sample space has an infinity of possible outcomes. With only a few qualifications, the concepts for finite sample spaces carry over directly to infinite sample spaces.

4.7 OPTIONAL TOPIC—BAYES' THEOREM

Bayes' theorem represents a special application of conditional probabilities where information on certain marginal and conditional probabilities is used to obtain other conditional probabilities.

☐ Examples

1. Martin Wilson has submitted an application for a trainee position. It is known that the company hires 4 percent of applicants. We shall denote by A_1 the event applicant

hired, so $P(A_1) = .04$. We shall denote by A_2 the event applicant not hired, so $P(A_2) = 1 - .04 = .96$ because A_1 and A_2 are complementary events. Only some of the applicants are called for a formal interview. We shall denote the event called for interview by B_1, and the event not called for interview by B_2. It is known that among all applicants hired, 98 percent receive interviews (the other 2 percent being hired on the basis of written credentials alone). Hence, we know that $P(B_1|A_1) = .98$. Furthermore, it is known that among all applicants not hired, only 1 percent are interviewed; hence, $P(B_1|A_2) = .01$. Martin Wilson has been called for an interview and wishes to know the probability that he will be hired; i.e., he wishes to determine $P(A_1|B_1)$.

The marginal probabilities $P(A_1)$ and $P(A_2)$ constitute initial information about the chances of being hired and are called *prior probabilities*. The desired conditional probability $P(A_1|B_1)$ is called a *posterior probability* because it is based on additional information, namely that Martin Wilson has been called for an interview.

2. A new frozen-food product can either turn out to be a failure in the market (A_1), a marginal success (A_2), or a major success (A_3). It is known from past experience with similar products that the prior probabilities are $P(A_1) = .80$, $P(A_2) = .15$, and $P(A_3) = .05$. The product has been introduced in a test market to see whether it would achieve a small market share (B_1) or a large market share (B_2) in the test market. It was found that the frozen-food product obtained a large market share (B_2) in the test market. It is known from past experience that the conditional probabilities of a large share in the test market (B_2), given the ultimate success of the product (A_i), are as follows:

$$P(B_2|A_1) = .02 \qquad P(B_2|A_2) = .24 \qquad P(B_2|A_3) = .98$$

The marketing manager now wishes to know the probabilities of ultimate success of the product, given the test-market results. In other words, the posterior probabilities $P(A_1|B_2)$, $P(A_2|B_2)$, and $P(A_3|B_2)$ are desired. □

We denote the outcomes of the variable of interest (job hiring, ultimate market success) by A_1, A_2, etc. The outcomes of the additional information variable (call for interview, test-market results) are denoted by B_1, B_2, etc. The prior probabilities $P(A_1)$, $P(A_2)$, etc., are known, as are the conditional probabilities $P(B_j|A_1)$, $P(B_j|A_2)$, etc., for the known outcome B_j of the additional information variable. The conditional probability $P(A_i|B_j)$ of outcome A_i occurring, given that B_j occurs, is then given by *Bayes' theorem*, as follows:

(4.25) *Bayes' Theorem:*
$$P(A_i|B_j) = \frac{P(A_i)P(B_j|A_i)}{\sum_i P(A_i)P(B_j|A_i)}$$

□ Examples

1. In our example of Martin Wilson's job application, we know that $P(A_1) = .04$, $P(A_2) = .96$, $P(B_1|A_1) = .98$, and $P(B_1|A_2) = .01$. Substituting into Bayes' theorem (4.25), we obtain:

$$P(A_1|B_1) = \frac{P(A_1)P(B_1|A_1)}{P(A_1)P(B_1|A_1) + P(A_2)P(B_1|A_2)} = \frac{.04(.98)}{.04(.98) + .96(.01)} = .803$$

Thus, the posterior probability of Martin Wilson being hired, given that he has been called for an interview, is $P(A_1|B_1) = .803$. Note how much greater is this posterior probability of being hired than is the prior probability, $P(A_1) = .04$, as a result of Wilson having been called for an interview.

Since A_2 is the complementary event of A_1, the posterior probability that Martin will not be hired, given that he has been called for an interview, is:

$$P(A_2|B_1) = 1 - P(A_1|B_1) = 1 - .803 = .197$$

2. In our test-market example, the desired posterior probabilities can best be obtained by means of systematic calculations, such as those shown in Table 4.6. Column 2 contains the prior probabilities $P(A_i)$ of ultimate success for the new product. Column 3 contains the known conditional probabilities $P(B_2|A_i)$ of large test-market share, given the ultimate success of the product. Column 4 contains the products of columns 2 and 3, representing the joint probabilities $P(A_i \cap B_2)$. The sum of this column is the denominator of Bayes' theorem (4.25). The ratio of each entry in column 4 to the sum of column 4 is the posterior probability of each ultimate success outcome, given large test-market share. Thus, the probability of major ultimate success, given large test-market share, is $P(A_3|B_2) = .0490/.1010 = .485$.

The probabilities in column 5 constitute the posterior probability distribution of marketing success, given large test-market share. Note how this posterior probability distribution differs from the prior probability distribution in column 2. As a result of the test-market findings, it is much more likely now that the new product will be either a marginal success or a major success. ☐

Comment

Bayes' theorem is a compact formula for finding posterior probabilities, but actually we could have obtained the same results by using our basic probability theorems. We illustrate this for Martin Wilson's job-application example where we wished to find $P(A_1|B_1)$. We start with the following information about the joint probability distribution of job hiring (A_i) and call for interview (B_j):

	B_1	B_2	Total
A_1			.04
A_2			.96
Total	—	—	1.00

$$P(B_1|A_1) = .98 \qquad P(B_1|A_2) = .01$$

TABLE 4.6 *Calculation of conditional probabilities using Bayes' theorem for test-market example*

(1) Ultimate Market Success A_i	(2) Prior Probability $P(A_i)$	(3) $P(B_2|A_i)$	(4) Joint Probability $P(A_i \cap B_2)$ (2) × (3)	(5) Posterior Probability $P(A_i|B_2)$ (4) ÷ .1010
A_1: Failure	.80	.02	.0160	.158
A_2: Marginal success	.15	.24	.0360	.356
A_3: Major success	.05	.98	.0490	.485
Total	1.00		.1010	1.00

We use this information to complete the joint probability distribution. Multiplication theorem (4.20) enables us to obtain:

$$P(A_1 \cap B_1) = P(A_1)P(B_1|A_1) = .04(.98) = .0392$$

$$P(A_2 \cap B_1) = P(A_2)P(B_1|A_2) = .96(.01) = .0096$$

The remaining joint probabilities can be obtained by subtraction:

$$P(A_1 \cap B_2) = .04 - .0392 = .0008$$

$$P(A_2 \cap B_2) = .96 - .0096 = .9504$$

We have therefore obtained the following joint probability distribution:

	B_1	B_2	Total
A_1	.0392	.0008	.04
A_2	.0096	.9504	.96
Total	.0488	.9512	1.00

Note that $P(B_1) = .0488$ and $P(B_2) = .9512$.

To find $P(A_1|B_1)$, we use definition (4.15) of conditional probability:

$$P(A_1|B_1) = \frac{P(A_1 \cap B_1)}{P(B_1)}$$

We readily find the needed probabilities from the joint probability distribution and obtain:

$$P(A_1|B_1) = \frac{.0392}{.0488} = .803$$

PROBLEMS

4.1 For each of the following outcomes, explain whether or not a random trial is involved. Indicate any assumptions you make.
   a. The air pressure in an automobile tire 30 days after the tire was last checked.
   b. The speed of a baseball pitched to a batter.
   c. The amount of light furnished by a light fixture that is turned off.

4.2 A food product is inspected at the processing plant by an inspector and is given a quality grade of A, B, C, D, or E. Major food stores sell only the grade A product. Grade B, C, and D products are sold only through discount outlets. Grade E products are not fit for human consumption and are sold to processors of animal foods. Develop sample spaces that: (1) describe different *quality* outcomes for inspected food product, (2) distinguish between different channels of *market distribution,* and (3) relate to whether or not the food product is *fit for human consumption.* Are the sample spaces univariate or bivariate? Explain.

*4.3 Refer to Problem 4.2. Let grade A be denoted by o_1, grade B by o_2, etc.
   a. Let T_1 denote the event that the product is sold through discount stores. What basic outcomes constitute T_1? What basic outcomes constitute T_1^*?
   b. Let T_2 denote the event that the product is fit for human consumption. What basic outcomes constitute T_2? Are T_1 and T_2 mutually exclusive here? Complementary? Explain.

4.4 Two reviewers for a publishing house independently screen manuscripts that arrive unsolicited in the mail. Each reviewer gives a grade of good, fair, or poor to a manuscript.

    a. Develop a sample space to describe the possible outcomes of the joint review of a manuscript. Is this sample space univariate or bivariate?

    b. Label the basic outcomes in your sample space o_1, o_2, etc., following the pattern in Figure 4.4a. Let E_1 denote the event that at least one reviewer assigns a grade of good. What basic outcomes constitute E_1? What basic outcomes constitute E_1^*?

    c. Let E_2 denote the event that the reviewers give different grades. What basic outcomes constitute E_2?

    d. Are E_1 and E_2 mutually exclusive here? Complementary? Explain.

4.5 Refer to Figure 4.2b.

    a. Construct the corresponding tree diagram where the first branching is for timing of delivery.

    b. Is the validity of the tree diagram affected by which variable is involved in the first branching? Explain.

4.6 Refer to Problem 4.4. Construct a tree diagram representing the sample space. Which variable did you associate with the first branching? Why?

*4.7 Refer to Figure 4.4. List the basic outcomes for each of the following sets: (1) $S \cap H$, (2) H^*, (3) $E \cup H$, (4) $E \cap H$, (5) $H \cup J$, (6) $J \cap H$.

4.8 Refer to Problem 4.4. List the basic outcomes for each of the following sets: (1) $E_1 \cap E_2$, (2) $E_1 \cup E_2$, (3) $E_1^* \cap E_2$, (4) $E_1^* \cup E_2^*$.

4.9 Consider the following probabilities: (1) probability that a certain type of automobile fuel pump will not fail during the first 1,000 kilometers of operation; (2) probability that the price of Sabine, Inc., common stock will fall more than $5 per share during the next five trading days; (3) probability that a shipment of tires from the regular supplier will contain no defectives; (4) probability that leukemia contracted by John Doe was caused by exposure to hazardous chemical agents during a period of employment overseas 10 years ago. For each of these probabilities, indicate whether it is amenable to an objective interpretation, a subjective interpretation, or both. Also, comment on how one might make the probability assessment for each, and on the factors that should be considered in making the assessment.

4.10 An engineer in a construction company states: "I'm willing to bet $3 against $1 that we'll get the contract." A cost analyst replies: "I'll take you up on that!" What does this exchange imply about the engineer's probability assessment of getting the contract? About the cost analyst's assessment?

4.11 A young accounting professor has sent an article to a professional journal and states: "There is 1 chance in 5 that this article will be accepted for publication."

    a. Define the sample space here and give the professor's probability distribution.

    b. Since the submission of this article to the journal is not a repeatable event, what interpretation can be given to the probabilities in the probability distribution?

    c. A colleague states: "I'll bet $5 against your $5 that the article will be accepted." Does there appear to be a discrepancy between the probability assessments of the two individuals? Explain.

    d. The professor's statement is based on the fact that this journal accepts 20 percent of the articles submitted. Would you now conclude that the professor's probability assessment is more credible than the colleague's? Discuss.

4.12 For each of the following statements, give your probability assessment that it is correct and explain whether the probability is objective or subjective. Do not consult any outside sources in arriving at your assessments. (1) Ottawa, Canada, lies farther east than Washington, D.C. (2) Cleopatra's first child was male. (3) A thumbtack will land point up if tossed to the floor.

*4.13 A device for checking internal welds in metal kegs is designed to signal when the weld is defective. The probability distribution for the status of the weld and the response of the device follows.

		Response of Device		
Status of Weld		B_1 Signals	B_2 Does Not Signal	Total
A_1: Defective		.2	.0	.2
A_2: Not defective		.1	.7	.8
	Total	.3	.7	1.0

a. Give the symbolic notation for the probability of each of the following events: (1) weld is defective; (2) weld is defective and device signals; (3) device signals, given that weld is defective; (4) weld is defective, given that device signals.
b. Determine each of the probabilities in **a**.

*4.14 Refer to Problem 4.13.
a. Explain what each of the following probabilities means here: (1) $P(A_1 \cap B_2)$, (2) $P(B_1|A_2)$, (3) $P(B_2)$.
b. Obtain the marginal probability distribution for response of device.
c. Obtain the conditional probability distribution for response of device, given that the weld is not defective.

4.15 The probability distribution for directions of traffic movements through an intersection in late afternoon follows.

		Vehicle Going to:				
Vehicle Coming from:		B_1 North	B_2 South	B_3 East	B_4 West	Total
A_1: North		.00	.28	.06	.07	.41
A_2: South		.12	.00	.03	.03	.18
A_3: East		.08	.04	.00	.05	.17
A_4: West		.07	.06	.11	.00	.24
	Total	.27	.38	.20	.15	1.00

a. Give the symbolic notation for the probability of each of the following events: (1) Vehicle goes north. (2) Vehicle comes from the west and goes south. (3) Vehicle goes north, given that it comes from the west.
b. Explain what each of the following probabilities means here: (1) $P(A_3)$, (2) $P(A_3 \cap B_4)$, (3) $P(A_3 \cup B_4)$.
c. Determine each of the probabilities in **a** and **b**.

4.16 Refer to Problem 4.15.
a. Obtain the marginal probability distribution for the direction from which a vehicle enters the intersection.
b. Obtain the conditional probability distribution for the direction from which a vehicle enters the intersection, given that it leaves to the south.

4.17 Refer to Problem 4.15. Do all parts of this problem, employing the following probability distribution for traffic movements:

Vehicle Coming from:	Vehicle Going to:				Total
	B_1 North	B_2 South	B_3 East	B_4 West	
A_1: North	.00	.15	.13	.06	.34
A_2: South	.10	.00	.07	.05	.22
A_3: East	.08	.05	.00	.14	.27
A_4: West	.05	.03	.09	.00	.17
Total	.23	.23	.29	.25	1.00

Also obtain: (1) the marginal probability distribution for the direction in which the vehicle leaves the intersection, and (2) the conditional probability distribution for the direction from which a vehicle enters the intersection, given that it leaves to the north.

*4.18 Refer to Problem 4.13. Find the following probabilities by the multiplication, complementation, or addition theorems and verify your answers by obtaining the probabilities directly from the joint probability distribution: (1) $P(A_1 \cup B_1)$, (2) $P(A_2 \cap B_1)$, (3) $P(A_2^*)$.

4.19 Refer to Problem 4.15. Find the following probabilities by the multiplication, complementation, or addition theorems and verify your answers by obtaining the probabilities directly from the joint probability distribution: (1) $P(A_2 \cup B_1)$, (2) $P(A_2 \cap B_4)$, (3) $P(A_1^*)$.

4.20 Answer Problem 4.19 employing the probability distribution for traffic movements given in Problem 4.17.

*4.21 Refer to Problem 4.13.
  a. Are the two variables statistically independent? Explain.
  b. Display the joint probability distribution in the form of a probability tree, as in Figure 4.6. Place variable A in the first branching.

4.22 Refer to Problem 4.15.
  a. Are the variables A and B statistically independent? If so, what is the significance of this fact? If not, describe the nature of the relationship between them.
  b. Present the joint probability distribution in the form of a probability tree, as in Figure 4.6. Place variable A in the first branching.

4.23 Answer both parts of Problem 4.22 for the bivariate probability distribution given in Problem 4.17.

*4.24 Given:

A_1: Family owns home B_1: Family income is under \$15,000

A_2: Family rents home B_2: Family income is \$15,000–under \$25,000

B_3: Family income is \$25,000 or more

and that:

$$P(A_2) = .52 \qquad P(A_1|B_1) = .20$$
$$P(B_1) = .50 \qquad P(A_1|B_3) = .80$$
$$P(B_3) = .10$$

    a.   Find $P(A_1 \cap B_3)$, $P(A_1 \cup B_3)$, and $P(B_3|A_1)$. Explain what each probability means here.

    b.   Develop the joint probability distribution of variables A and B.

    c.   Are home ownership and family income statistically independent variables? If so, what is the significance of this fact? If not, describe the nature of the relationship between them.

4.25   Refer to Problem 4.24. Answer all parts using the following probabilities: $P(A_2) = .60$, $P(B_1) = .50$, $P(B_3) = .10$, $P(A_1|B_1) = .40$, $P(A_1|B_3) = .40$.

4.26   Several leading law schools have developed an information exchange under which applicants may apply to at most two of the schools for admission. Applicant Blanque wishes to study at one of three schools (A, B, C), all of which are participants in the information exchange program. Based on experiences of similar graduates from his college, Blanque makes the following probability assessments, where $P(A)$ denotes the probability of his being accepted by law school A and the other probabilities are interpreted analogously: $P(A) = .60$, $P(B) = .80$, $P(C) = .50$, $P(B|A) = .90$, $P(C|A) = .20$, $P(C|B) = .50$.

    a.   To which two of the three schools should Blanque apply to have the greatest probability of being accepted by one or the other (or both)? Show your calculations.

    b.   Do schools A and B appear to have similar or dissimilar criteria for admission? Schools A and C? Discuss.

4.27   An early version of an intercontinental ballistic missile was deemed to have a probability of .11 of destroying an enemy missile silo if fired at it. The probability of destroying the silo if two such missiles were fired was stated to be .21. Are the outcomes of the two firings taken to be statistically independent? Explain.

*4.28   A salesperson is evaluated according to whether the week's sales exceed the sales quota (A), meet the sales quota (B), or fall below the quota (C). The probabilities for an experienced salesperson are: $P(A) = .3$, $P(B) = .6$, $P(C) = .1$. Consider the performances of an experienced salesperson in two successive weeks, and assume these are statistically independent.

    a.   Obtain the joint probability distribution. What is the probability that the salesperson will exceed the quota in both weeks?

    b.   If the performances in two successive weeks were not independent, could you obtain the joint probability distribution with no additional information? Explain.

4.29   In handling a customer's order, the order department will fill it incorrectly with probability .04. Also, the order will be delivered to the wrong address with probability .02. Assume that the two variables filling of order and delivery of order are statistically independent.

    a.   Obtain the bivariate probability distribution. What is the probability that a customer's order is filled incorrectly and delivered to the wrong address? That it is filled correctly and delivered to the right address?

    b.   If the two variables were statistically dependent, then what would be the maximum probability that a customer's order is filled incorrectly and delivered to the wrong address?

4.30   A speculator specializing in commodities A and B has made the following subjective probability assessments about the prices of these commodities during the next two days: (1) The probability is .90 that the price of A will rise. (2) The probability is .80

that the price of B will rise. (3) If the price of A rises, the conditional probability is .65 that the price of B will rise.

a. Are these subjective probability assessments mutually consistent? Explain.

b. Would the assessments be consistent if the probability in (2) were .50 instead of .80?

*4.31 Consider the Text security system example on p. 101. Suppose: (1) the probability is .02 that an unauthorized visitor will pass the voice pattern check; (2) the probability an unauthorized visitor will pass the fingerprint check if he or she passes the voice pattern check is .50; and (3) the probability an unauthorized visitor will pass the handwriting check if he or she passes both the other two checks is .90.

a. Use (4.21) to obtain the probability that an unauthorized visitor will pass all three checks.

b. How does the probability in **a** compare with the corresponding probability in the Text example? What are the implications of this for the security system?

4.32 A consultant reports to a city executive that family use of the golf course and swimming pool operated by the city are statistically independent events. An observer states that this report must be incorrect because she personally knows of families that use both facilities. Comment.

4.33 Suppose the whole numbers 1–10 comprise the basic outcomes in a sample space, and each basic outcome has probability .1 of occurring. Consider the following events: E_1, the number is odd; E_2, the number is 6 or less.

a. Are E_1 and E_2 statistically independent? Explain.

b. Are E_1 and E_2 mutually exclusive? Explain.

*4.34 Refer to Problem 4.13.

a. Obtain the following probabilities: $P(A_1)$, $P(A_2)$, $P(B_1|A_1)$, $P(B_1|A_2)$. Using these values, calculate $P(A_1|B_1)$ and $P(A_2|B_1)$ by means of Bayes' theorem (4.25).

b. Which probabilities in **a** are the prior probabilities? Which are the posterior probabilities? Interpret the prior and posterior probabilities for this application.

c. Verify the probability value $P(A_1|B_1)$ obtained in **a** by using (4.15) directly.

4.35 Sixty percent of the graduates of a driver training school pass the official driver's test on the first attempt and the other 40 percent fail. The school gives a pretest to graduates before they take the official test. Of the graduates who pass the official test on the first attempt, 80 percent passed the pretest. Of the graduates who fail the official test on the first attempt, 10 percent passed the pretest. Let A_1 denote that a graduate passes the official test on the first attempt and A_2 that the graduate fails, and let B_1 denote that the graduate passed the pretest.

a. Give the values of the following probabilities: $P(A_1)$, $P(A_2)$, $P(B_1|A_1)$, $P(B_1|A_2)$.

b. A graduate has passed the pretest. Use Bayes' theorem to obtain the posterior probabilities $P(A_1|B_1)$ and $P(A_2|B_1)$ and interpret them. Has the information provided by the pretest led to a substantial modification of the prior probabilities? Discuss.

4.36 Refer to Problem 4.35. Answer both parts but assume that 50 percent of the graduates pass the official test on the first attempt and 50 percent fail.

EXERCISES

4.37 For each of the following probability statements, state whether it is always true, always false, or neither, for all pairs of events E and F of a sample space. Justify each answer.

    a. $P(E) < P(E \cap F)$

    b. $P(E) > P(E \cup F)$

    c. $P(E \cup F) \leq P(E) + P(F)$

    d. $P(E) + P(F) = 1$

4.38 For each of the following probability statements, state whether it is always true, always false, or neither, for all pairs of events E and F of a sample space. Justify each answer.

    a. Both $P(E|F) < P(E)$ and $P(E|F^*) < P(E)$

    b. $P(E^*|F) = 1 - P(E|F)$

    c. $P(E^* \cap F^*) = 1 - P(E \cup F)$

4.39 Prove extension (4.21) of the multiplication theorem (4.20).

4.40 Addition theorem (4.16) extends to more than two events. For any three events E_1, E_2, and E_3 of a sample space, use (4.16) to show that:

$$P(E_1 \cup E_2 \cup E_3) = P(E_1) + P(E_2) + P(E_3) - P(E_1 \cap E_2) - P(E_1 \cap E_3)$$
$$- P(E_2 \cap E_3) + P(E_1 \cap E_2 \cap E_3)$$

[*Hint:* Denote $E_2 \cup E_3$ by F initially and expand $P(E_1 \cup F)$.]

4.41 The probability that a household has: (1) a color television is .55, (2) one automobile is .56, (3) two or more automobiles is .22, (4) a color television and one automobile is .28, (5) a color television and two or more automobiles is .21. Use the extension in Problem 4.40 to find the probability that a household has either a color television or at least one automobile.

4.42 Two students will be selected at random from the top three students in a political science class to attend a mock political convention. The first student selected will be the voting delegate and the second one selected will be the alternate. Before the selection begins, one of the three students states that she has 1 chance in 3 of being the voting delegate and 1 chance in 2 of being the alternate. Do you agree? (*Hint:* Before the selection begins, is the probability of being chosen second a marginal probability, a joint probability, or a conditional probability?)

STUDIES

4.43 A large construction firm has a supervisor responsible for preparing project bids. In an internal report on each bid, he states his subjective probability that the bid will be successful. A tabulation of his record on the last 205 bids is on p. 112.

    a. Calculate for each probability of successful bid category the expected number of successful bids if the supervisor's probability assessments are accurate. Obtain the residuals and prepare a residual plot.

    b. Overall, how does the total number of bids expected to be successful compare with the actual number (87)? Based on the residuals plotted in **a**, is there any range of probability values where the supervisor's assessment performance is particularly poor?

    c. Since the firm has won 87 of the last 205 bids, it has been suggested that the probability $87/205 = .42$ be used as the probability of winning each bid rather than relying on the supervisor's probability assessment. Evaluate this proposal.

Supervisor's Subjective Probability of Successful Bid	Number of Bids	Number of Successful Bids
.10	3	0
.20	10	2
.30	29	9
.40	48	18
.50	37	18
.60	35	22
.70	21	10
.80	11	5
.90	7	2
1.00	4	1
Total	205	87

4.44 Consider the event: "The next U.S. president will hold office for less than four years."

a. Use the indirect method described in the Text to elicit the subjective probability for this event from 10 persons not in your class. Record your results.

b. Are the 10 probability assessments highly variable? Calculate a suitable measure and interpret it.

4.45 The probability distribution of the week of settlement of strikes in a particular industry follows. (Week 1 is the week when the strike begins, etc.)

Week of settlement:	1	2	3	4	5 or later
Probability:	.63	.23	.09	.03	.02

a. A strike has just begun. What is the probability that it will be settled in week 1? In week 5 or later?

b. If a strike is just entering week 2, what is the probability that it will be settled this week?

c. Calculate the probability that a strike is *not* settled in week 1 and *is* settled in week 2.

d. For each of weeks 2, 3, and 4, obtain the probability that a strike will be settled in that week, given that it has not been settled before that week. Does it appear to become more or less difficult to settle strikes as the duration of the strike increases?

4.46 On the basis of a physical examination and symptoms, a physician assesses the probabilities that the patient has no tumor, a benign tumor, or a malignant tumor as .70, .20, and .10, respectively. A thermographic test is subsequently given to the patient. This test gives a negative result with probability .90 if there is no tumor, with probability .80 if there is a benign tumor, and with probability .20 if there is a malignant tumor.

a. What is the probability that a thermographic test will give a negative result for this patient?

b. Obtain the posterior probability distribution for this patient when the test result is negative. Interpret this probability distribution. What is the most likely state for the patient?

c. Obtain the posterior probability distribution for this patient when the test result is positive. How does this probability distribution differ from the one in b? Discuss.

5
Random Variables

In Chapter 4, we considered probability distributions in general. In this chapter, we are concerned with probability distributions for sample spaces relating to quantitative characteristics. Special interest exists in these distributions because of the importance and pervasiveness of quantitative characteristics in practical problems and the relative ease with which they lend themselves to statistical analysis.

5.1   BASIC CONCEPTS

Definition of Random Variable

When the sample space of a random trial relates to a quantitative characteristic, a number or value can be associated with each basic outcome. Consider a random trial involving automobile tires that are molded in pairs. The quantitative characteristic of interest might be the number of defective tires in a pair. Thus, the basic outcomes in the sample space are o_1: zero defective tires, o_2: one defective tire, and o_3: two defective tires. If X is used to denote the number of defective tires in the pair, clearly X can assume one and only one of the values 0, 1, and 2 corresponding to basic outcomes o_1, o_2, and o_3, respectively, in a random trial. Which value X will assume, however, is uncertain. For this reason, X is called a random variable.

(5.1)   A *random variable* is a variable whose numerical value is determined by the outcome of a random trial.

Initially, it is useful to distinguish notationally between a random variable and the possible values it can assume. We shall use an uppercase letter, such as X, to denote the random variable and the corresponding lowercase letter, x in this case, to denote a particular value assumed by the random variable.

☐   Examples

1. In the tire-molding example, random variable X denotes the number of defective tires in a pair, a number which is uncertain before the random trial occurs. The notation $X = x$ designates that the actual trial outcome is x, where x is either the value 0, 1, or 2 here.

2. Let random variable X denote the number of football games a team will win in a nine-game regular season. The notation $X = x$ designates that the team actually wins x games, where x is either 0, 1, . . . , or 9 here. ☐

Discrete Random Variable.   A random variable that takes on distinct values only is called a *discrete* random variable. For example, the number of defective tires in a pair is a discrete random variable because it may assume one of three distinct values: 0, 1, 2. Similarly, the number of games the team wins is a discrete random variable that may assume one of ten distinct values: 0, 1, . . . , 9.

Some discrete random variables are treated as if they can assume one of an infinity of possible distinct values. For instance, one may treat the number of traffic violations committed in a large city during a one-month period as having possible outcomes 0, 1, 2, . . . , ad infinitum. While actually there is a realistic upper limit to the number of violations, it is often useful to model the number of possible outcomes as potentially infinite.

Continuous Random Variable.   A random variable that may take on any value on a continuum is called a *continuous* random variable. For example, the temperature in a warehouse may be any value on the temperature continuum between, say, $-40°C$ and $45°C$. Of course, any measurement can be made only to a finite number of significant digits and so, strictly speaking, measured temperature is a discrete random variable. However, it is often conceptually advantageous to view measured temperature as being inherently continuous. Additional examples of measured variables that are often treated as continuous are family income, I.Q., and height of person.

We consider discrete random variables first, and then take up continuous random variables in Section 5.8.

5.2   DISCRETE RANDOM VARIABLES

Probability Distribution

Let X be a discrete random variable and let x_i, $i = 1, 2, . . . , k$, denote the k distinct values that X may assume. Since each x_i corresponds to a basic outcome of the sample space of the random trial, a probability distribution for the sample space will associate a probability value with each x_i. The probability that random variable X will assume value x_i will be denoted by $P(X = x_i)$. For instance, $P(X = 0)$ denotes the probability that X will assume the value 0.

(5.2)   The *probability distribution* for a discrete random variable X associates with each of the distinct outcomes x_i $(i = 1, 2, . . . , k)$ a probability $P(X = x_i)$.

☐   Example

In our tire-molding example, the probabilities assigned to the basic outcomes are $P(o_1) = .75$, $P(o_2) = .10$, and $P(o_3) = .15$. The probability distribution for X then is:

x:	0	1	2
$P(X = x)$:	.75	.10	.15

Figure 5.1a presents this probability distribution in graphic form and Figure 5.1b shows an alternative, tabular form. □

Comments

1. The probability distribution for a discrete random variable X has the usual properties of a probability distribution:
   a. $0 \leq P(X = x_i) \leq 1 \qquad i = 1, 2, \ldots, k$

   b. $\displaystyle\sum_{i=1}^{k} P(X = x_i) = 1$

2. The probability distribution for a discrete random variable is also called *probability mass function* or simply *probability function*.

Notation. For convenience, we will often abbreviate the notation $P(X = x_i)$ to $P(x_i)$. $P(0)$ then denotes $P(X = 0)$, the probability that the value of random variable X is 0; $P(1)$ denotes $P(X = 1)$, the probability that the value of X is 1; etc.

Cumulative Probability

In some applications, we are interested in the probability that a random variable X takes a value less than or equal to some specified value x. Such a probability is called a *cumulative probability* and is denoted by $P(X \leq x)$.

Example

In our tire-molding example, a production trial is profitable if the number of defective tires X is one or less. Thus, a trial is profitable if the event $X \leq 1$ occurs. This event

---

FIGURE 5.1 *Graphic and tabular representations of a probability distribution for a discrete random variable*

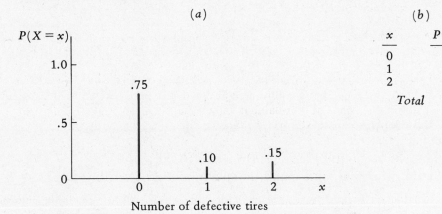

x	$P(X = x)$
0	.75
1	.10
2	.15
Total	1.00

happens if either 0 or 1 defective tire occurs in the trial. Using the abbreviated notation $P(0)$ and $P(1)$ to denote the probabilities $P(X = 0)$ and $P(X = 1)$, respectively, we see from Figure 5.1b that:

$$P(X \leq 1) = P(0) + P(1) = .75 + .10 = .85$$

Thus, the probability is .85 that a trial will be profitable. □

Cumulative Probability Distribution

Just as a probability distribution gives probabilities $P(X = x_i)$ for all outcomes x_i, a cumulative probability distribution gives cumulative probabilities $P(X \leq x)$ for various values of x:

(5.3)   The *cumulative probability distribution* for a discrete random variable X provides the cumulative probabilities $P(X \leq x)$ for all values x.

□   Example

A health systems analyst obtained the probability distribution tabulated in Figure 5.2a for the annual number of visits by families to a clinic. The cumulative probability distribution for this random variable is developed in Figure 5.2b. The cumulative probabilities are $P(X \leq 0) = P(0) = .37$, $P(X \leq 1) = P(0) + P(1) = .37 + .40 = .77$, etc. We thus see that the probability is .97 that a family makes three or fewer visits during a year.

A graph of the cumulative probability distribution for clinic visits is shown in Figure 5.2c. The graph is a step function because the random variable is discrete. The step function is constructed in exactly the same way as a step-function ogive for a cumulative frequency distribution. Each step of the graph occurs at an outcome x_i and has a size equal to $P(X = x_i)$. For instance, the step size at $x = 2$ is $P(X = 2) = .15$. Figure 5.2c shows how one can graphically determine any cumulative probability by reading the height of the step function. We see that $P(X \leq 1) = .77$. □

Comments

1. The definition of the cumulative probability distribution in (5.3) associates cumulative probability values with all values of x, not just those constituting the basic outcomes of the random variable. For the clinic visits example, we note in Figure 5.2c that $P(X \leq 2.6) = .92$, even though 2.6 visits cannot occur. $P(X \leq 2.6)$ denotes the probability that 2.6 or fewer visits take place, and this happens when 0, 1, or 2 visits occur.
2. Two important properties of any cumulative probability distribution are:
   a. $P(X \leq x)$ is always a value between 0 and 1.
   b. The cumulative distribution never decreases as x increases.
   These two properties are illustrated in Figure 5.2c, where we see that the cumulative distribution starts at 0 on the left and ends at 1 on the right.
3. The cumulative probability distribution is also called *cumulative probability function*.

Joint, Marginal, and Conditional Probability Distributions

In Chapter 4, joint, marginal, and conditional probability distributions were defined for bivariate and multivariate sample spaces. When the sample spaces relate

FIGURE 5.2 *Cumulative probability distribution of family clinic visits*

(a) Probability distribution		(b) Cumulative probability distribution	
Number of Visits x	$P(X = x)$	x	$P(X \leq x)$
0	.37	0	.37
1	.40	1	.77
2	.15	2	.92
3	.05	3	.97
4	.03	4	1.00
Total	1.00		

(c)
Graph of cumulative probability distribution

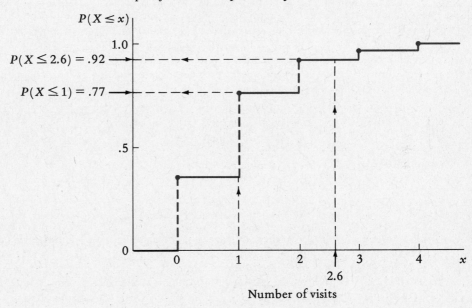

to discrete random variables, the joint, marginal, and conditional probability distributions play the same roles as before. We review the main ideas by means of an example.

☐ Example

A *bivariate probability distribution* for two discrete random variables is illustrated in Table 5.1. The two variables are the number of trucks from a delivery fleet of three trucks that will be in a repair depot on two consecutive nights. X denotes the number of trucks in

TABLE 5.1 *Example of bivariate probability distribution for discrete random variables*

		Number of Trucks in Depot during Second Night y				
		0	1	2	3	Total
Number of Trucks	0	.58	.06	.01	.00	.65
in Depot during	1	.06	.10	.03	.01	.20
First Night	2	.01	.03	.05	.01	.10
x	3	.00	.01	.01	.03	.05
	Total	.65	.20	.10	.05	1.00

the repair depot during the first night, and Y denotes the number during the second night. Thus, the possible outcomes for X are 0, 1, 2, 3, and the same holds for Y.

The notation for joint probabilities for two discrete random variables follows that used in Chapter 4 for events. The joint probability that x_i trucks are in the depot the first night and y_j are in the depot the second night is written $P(X = x_i \cap Y = y_j)$. We see from Table 5.1 that $P(X = 0 \cap Y = 0) = .58$ and $P(X = 1 \cap Y = 2) = .03$.

The two *marginal probability distributions* are found in the margins of Table 5.1, that for X in the rightmost column and that for Y in the bottom row. The marginal probabilities are obtained as usual, in accordance with (4.14). Thus, we have:

$$P(X = 0) = P(X = 0 \cap Y = 0) + P(X = 0 \cap Y = 1) + P(X = 0 \cap Y = 2)$$
$$+ P(X = 0 \cap Y = 3)$$
$$= .58 + .06 + .01 + .00 = .65$$

Similarly:

$$P(Y = 2) = P(X = 0 \cap Y = 2) + P(X = 1 \cap Y = 2) + P(X = 2 \cap Y = 2)$$
$$+ P(X = 3 \cap Y = 2)$$
$$= .01 + .03 + .05 + .01 = .10$$

Thus, the probability of 0 trucks in the depot the first night is .65 and that of 2 trucks in the depot the second night is .10.

Conditional probability distributions are obtained analogously to (4.15). Thus, the conditional probability distribution for the number of trucks in the depot during the second night, given that one truck is in the depot during the first night, is:

Number of Trucks in Depot during Second Night y	Conditional Probability $P(Y = y\|X = 1)$
0	.06/.20 = .30
1	.10/.20 = .50
2	.03/.20 = .15
3	.01/.20 = .05
	Total 1.00

Note, for instance, that $P(Y = 0|X = 1) = P(X = 1 \cap Y = 0)/P(X = 1)$ according to definition (4.15). Hence, we obtain $P(Y = 0|X = 1) = .06/.20 = .30$. ☐

5.3 STATISTICAL INDEPENDENCE OF RANDOM VARIABLES

We can apply the definition of statistical independence given in (4.24) to random variables:

(5.4) Two random variables X and Y are said to be statistically independent if:

$$P(X = x_i \cap Y = y_j) = P(X = x_i)P(Y = y_j) \qquad \text{for all } x_i \text{ and } y_j$$

☐ Example

In Table 5.1, the two random variables are not independent. For example, $P(X = 0 \cap Y = 1) = .06$, while $P(X = 0)P(Y = 1) = .65(.20) = .13$. An examination of Table 5.1 indicates the nature of the relation, namely, that the number of trucks in the depot on two successive nights is likely to be the same. ☐

5.4 EXPECTATION

Expected Value of Random Variable

Often, we are interested in the average value of a random variable in many trials. This value is known as the expected value of a random variable:

(5.5) The *expected value* of a discrete random variable X is denoted by $E\{X\}$ and defined:

$$E\{X\} = \sum_{i=1}^{k} x_i P(x_i)$$

where: $P(x_i)$ denotes $P(X = x_i)$

$E\{\ \}$ (read "expectation of") is called the *expectation operator*

☐ Examples

1. The probability distribution of X, the annual number of family clinic visits, is given in Figure 5.2 and is:

x:	0	1	2	3	4
$P(x)$:	.37	.40	.15	.05	.03

The expected value of X is obtained by using (5.5):

$$E\{X\} = 0(.37) + 1(.40) + 2(.15) + 3(.05) + 4(.03) = .97 \text{ visit}$$

2. A piece of jewelry worth $10,000 has probability .003 of being stolen or lost during a year. Let X denote the amount of loss during a year. The probability distribution of X is:

Loss, x:	0	10,000
$P(x)$:	.997	.003

The expected loss during a year is, using (5.5):

$$E\{X\} = 0(.997) + 10,000(.003) = \$30 \qquad \square$$

Meaning of $E\{X\}$. The expected value of a random variable may be likened to a measure of position for the probability distribution. Indeed, $E\{X\}$ is simply a weighted mean of the possible outcomes, with the probability values used as weights. It is for this reason that $E\{X\}$ is often called the *mean* of the probability distribution of X, or simply the mean of X.

There is another way of understanding the meaning of $E\{X\}$. If, for instance, the random trial associated with the annual number of family clinic visits is repeated independently many times, the relative frequency interpretation of probability suggests that about 37 percent of families will have no visits in a year, 40 percent will have one visit, etc. The mean outcome over many independent trials would be about $E\{X\} = .97$ visit per family.

Observe that $E\{X\}$, like any mean, may be a number that does not correspond to any of the possible outcomes. In the clinic visits example, we found $E\{X\} = .97$ visit, yet all the outcomes of X are whole numbers.

Expected Value of Linear Function of Random Variable

Often, interest centers on a linear function of a random variable X, i.e., $Y = a + bX$.

☐ Examples

1. X denotes the number of local calls made from a pay telephone in a day. Y denotes the daily revenue from the local calls. Each call costs \$.15, so we have $Y = .15X$. Here, $a = 0$ and $b = \$.15$. If $X = 3$, for instance, $Y = \$.45$.
2. X denotes the quantity of output of a plant in a day and Y the total cost of this output. It is known that $Y = 200 + 4X$, where $a = \$200$ is the fixed setup cost and $b = \$4$ is the cost per unit produced. If $X = 40$, for instance, $Y = \$360$. ☐

To find the expected value of Y, one can obtain the probability distribution of Y from that of X and then use (5.5) to find $E\{Y\}$. In Example 1, for instance, the probability distribution of the number of phone calls X is known to be:

x:	0	1	2	3
$P(x)$:	.2	.4	.3	.1

Since each call costs \$.15, the probability distribution of the total revenue from these calls (Y) must correspond to that of the number of calls (X), as follows:

y:	0	.15	.30	.45
$P(y)$:	.2	.4	.3	.1

Using (5.5) to obtain $E\{Y\}$, we find:

$$E\{Y\} = 0(.2) + .15(.4) + .30(.3) + .45(.1) = \$.195$$

When $E\{X\}$ is already known, there is no need to find the probability distribution of Y first. The following theorem can be used to find $E\{Y\}$ from $E\{X\}$ directly when $Y = a + bX$:

(5.6)
$$E\{a + bX\} = a + bE\{X\}$$

This says that $E\{Y\}$ is the same linear function of $E\{X\}$ as Y is of X. There are three important special cases of (5.6):

(5.6a)
$$E\{a\} = a$$

(5.6b)
$$E\{bX\} = bE\{X\}$$

(5.6c)
$$E\{a + X\} = a + E\{X\}$$

☐ Examples

1. In our telephone calls example, $E\{X\} = 1.3$ (calculations not shown). Hence, using theorem (5.6b), we obtain for $Y = .15X$:

$$E\{Y\} = .15(1.3) = .195$$

This, of course, is the same result as when $E\{Y\}$ was calculated from the probability distribution of Y.

2. In our plant output example, it is known that $E\{X\} = 50$. Hence, using theorem (5.6), we obtain for $Y = 200 + 4X$:

$$E\{Y\} = 200 + 4(50) = 400$$ ☐

5.5   VARIANCE

Variance of Random Variable

Since the outcomes of a random variable are probabilistic, it is useful to have a measure of the dispersion or variability of the outcomes. A key measure is the variance of a random variable.

(5.7)   The *variance* of a discrete random variable X is denoted by $\sigma^2\{X\}$ and defined:

$$\sigma^2\{X\} = \sum_{i=1}^{k} (x_i - E\{X\})^2 P(x_i)$$

where: $P(x_i)$ denotes $P(X = x_i)$

$\sigma^2\{\ \ \}$ (read "variance of") is called the *variance operator*

☐ Examples

1. In our family clinic visits example, the probability distribution of X, the annual number of family visits, is:

x:	0	1	2	3	4
P(x):	.37	.40	.15	.05	.03

We found earlier that $E\{X\} = .97$. Hence, the variance of X according to (5.7) is:

$$\sigma^2\{X\} = (0 - .97)^2(.37) + (1 - .97)^2(.40) + (2 - .97)^2(.15) + (3 - .97)^2(.05)$$
$$+ (4 - .97)^2(.03) = .9891$$

2. In our jewelry loss example, the probability distribution of the dollar loss from theft is:

Loss, x:	0	10,000
$P(x)$:	.997	.003

We found previously that $E\{X\} = 30$. Hence:

$$\sigma^2\{X\} = (0 - 30)^2(.997) + (10,000 - 30)^2(.003) = 299,100 \qquad \square$$

Meaning of $\sigma^2\{X\}$. The variance is a weighted average of squared deviations, the deviations being the outcomes of X around their expected value or mean $E\{X\}$, and the weights being the respective probabilities of occurrence. Therefore, $\sigma^2\{X\}$ measures the extent to which the outcomes of X depart from their expected value in the same way that the variance of a data set measures the dispersion of values in the set about their mean.

Comment

Since the variance of a random variable X is a weighted average of the squared deviations, $(X - E\{X\})^2$, it may be defined equivalently as an expected value:

(5.8) $$\sigma^2\{X\} = E\{(X - E\{X\})^2\}$$

An algebraically identical expression is:

(5.8a) $$\sigma^2\{X\} = E\{X^2\} - (E\{X\})^2$$

Standard Deviation

Note that $\sigma^2\{X\}$ is expressed in the squared units of X. When we take its positive square root, we return to the original units of measure and obtain the standard deviation of X.

(5.9) The positive square root of the variance of X is called the *standard deviation* of X and is denoted by $\sigma\{X\}$:

$$\sigma\{X\} = \sqrt{\sigma^2\{X\}}$$

$\sigma\{\ \}$ (read "standard deviation of") is called the *standard deviation operator*

□ **Examples**

1. In the clinic visits example, we found that the variance is $\sigma^2\{X\} = .9891$. Hence, the standard deviation of X is $\sigma\{X\} = \sqrt{.9891} = .9945$ clinic visit per family.
2. In the jewelry loss example, we found the variance to be $\sigma^2\{X\} = 299,100$. Hence, $\sigma\{X\} = \sqrt{299,100} = \$546.90$. We see that the standard deviation is much larger than the mean, $E\{X\} = \$30$. □

Chebyshev Inequality. The Chebyshev inequality provides information about the extent of variation for any random variable:

(5.10) *Chebyshev Inequality.* For any random variable X, with expected value $E\{X\}$ and standard deviation $\sigma\{X\}$, the probability of outcomes beyond k standard deviations from $E\{X\}$ is at most $1/k^2$. Symbolically:

$$P(|X - E\{X\}| > k\sigma\{X\}) \le \frac{1}{k^2}$$

☐ Example

Consider the number of trucks arriving at a warehouse in a day. For this random variable X, it is known that $E\{X\} = 18$ and $\sigma\{X\} = 2$. The Chebyshev inequality then tells us that for, say, $k = 3$, the probability is at most $1/3^2 = 1/9$ that on any one day fewer than $18 - 3(2) = 12$ trucks or more than $18 + 3(2) = 24$ trucks will arrive. ☐

Variance of Linear Function of Random Variable

Frequently, one requires the variance of a linear function of a random variable X, i.e., $Y = a + bX$. When the variance of X is known, one can obtain the variance of Y directly from the following theorem:

(5.11)
$$\sigma^2\{a + bX\} = b^2\sigma^2\{X\}$$

Two important special cases of (5.11) are:

(5.11a)
$$\sigma^2\{a + X\} = \sigma^2\{X\}$$

(5.11b)
$$\sigma^2\{bX\} = b^2\sigma^2\{X\}$$

☐ Examples

1. In our telephone calls example, the variance of X is $\sigma^2\{X\} = .81$ (calculations not shown). Hence, using theorem (5.11b), we find for $Y = .15X$:

$$\sigma^2\{Y\} = (.15)^2(.81) = .0182$$

The standard deviation of Y is $\sigma\{Y\} = \sqrt{.0182} = \$.135$.

2. In our plant output example, it is known that $\sigma^2\{X\} = 400$. Hence, using (5.11), we obtain for $Y = 200 + 4X$:

$$\sigma^2\{Y\} = (4)^2(400) = 6{,}400$$

so $\sigma\{Y\} = \sqrt{6{,}400} = \$80$. Note that the constant setup cost $a = \$200$ had no effect on the variance of Y. ☐

Comments

1. Example 2 illustrates that the constant a does not affect the variance of $a + bX$. This is intuitively reasonable, since adding a constant to a variable shifts the position of the distribution but does not affect the variability of the distribution.

2. The standard deviation of $Y = a + bX$ is related to $\sigma\{X\}$ as follows:

(5.11c)
$$\sigma\{Y\} = |b|\sigma\{X\}$$

In Example 1, for instance, $\sigma\{X\} = \sqrt{.81} = .90$ and $Y = .15X$. Hence,

$$\sigma\{Y\} = |.15|(.90) = \$.135.$$

5.6   STANDARDIZED RANDOM VARIABLE

In later work, we frequently use a standardized random variable:

(5.12) If X is a random variable with expected value $E\{X\}$ and standard deviation $\sigma\{X\}$, then:

$$Y = \frac{X - E\{X\}}{\sigma\{X\}}$$

is known as the *standardized* form of random variable X.

The standardized form of a random variable is called a *standardized random variable,* because each standardized random variable has expected value 0 and variance 1. Thus, for Y defined in (5.12), we have $E\{Y\} = 0$ and $\sigma^2\{Y\} = 1$.

Comment

The proof that $E\{Y\} = 0$ and $\sigma^2\{Y\} = 1$ for a standardized random variable utilizes (5.6) and (5.11). To show $E\{Y\} = 0$, for instance, we proceed as follows:

$$E\{Y\} = E\left\{\frac{1}{\sigma\{X\}}(X - E\{X\})\right\} = \frac{1}{\sigma\{X\}}E\{X - E\{X\}\} \qquad \text{by (5.6b)}$$

$$= \frac{1}{\sigma\{X\}}(E\{X\} - E\{X\}) = 0 \qquad \text{by (5.6c)}$$

5.7   SUMS AND DIFFERENCES OF INDEPENDENT RANDOM VARIABLES

Sum and Difference of Two Independent Random Variables

The sum and difference of two independent random variables are frequently encountered.

☐   Examples

1. X denotes the bonus amount received by salesperson A and Y the bonus amount received by salesperson B. Then $T = X + Y$ represents the total bonus amount received by the two salespeople.
2. X denotes the number of responses to a classified advertisement in city 1 and Y the number of responses to the advertisement in city 2. Then $R = X + Y$ represents the total number of responses to the advertisement in the two cities.
3. X denotes quarterly sales revenues and Y denotes direct costs. Then $P = X - Y$ represents quarterly gross profit.   ☐

The expected value and variance of a sum or difference of two independent random variables are often required. One could obtain these characteristics by first finding the probability distribution of the sum or difference. For instance, the probability distribution of X, the bonus amount received by salesperson A, is:

Bonus, x:	0	500
$P(x)$:	.6	.4

Y, the bonus amount received by salesperson B, follows the same probability distribution as X. Further, X and Y are independent random variables, so the joint probability distribution is:

		y	
		0	500
x	0	.36	.24
	500	.24	.16

These joint probabilities are obtained from the independence definition (5.4). For instance, we have $P(X = 0 \cap Y = 0) = P(X = 0)P(Y = 0) = .6(.6) = .36$.

The probability distribution of $T = X + Y$, the total bonus amount, is obtained by recognizing that $T = 0$ if neither salesperson receives a bonus. Hence, $P(T = 0) = .36$. Similarly, $P(T = 1,000) = .16$ since both salespeople must receive a bonus if the total amount is to be \$1,000. Finally, $P(T = 500) = .24 + .24 = .48$ since the total bonus amount is \$500 when either salesperson, but not the other, receives the bonus. Thus, the probability distribution of T is:

Total bonus, t:	0	500	1,000
$P(t)$:	.36	.48	.16

From this probability distribution, $E\{T\}$ and $\sigma^2\{T\}$ can be found in the usual fashion. They are $E\{T\} = 400$ and $\sigma^2\{T\} = 120,000$ (calculations not shown).

When the expectations and variances of each random variable are already known, one need not obtain the probability distribution of the sum since the expected value and variance of the sum can be obtained directly from the following theorem:

(5.13) If X and Y are two independent random variables, then $X + Y$ has:

(5.13a) *Expected Value:* $\quad E\{X + Y\} = E\{X\} + E\{Y\}$

(5.13b) *Variance:* $\quad \sigma^2\{X + Y\} = \sigma^2\{X\} + \sigma^2\{Y\}$

Note that the expected value of the sum of two independent random variables is simply the sum of the expected values of each of the two random variables, and similarly for the variance.

The expected value and variance of a difference are presented next:

(5.14) If X and Y are two independent random variables, then $X - Y$ has:

(5.14a) *Expected Value:* $\quad E\{X - Y\} = E\{X\} - E\{Y\}$

(5.14b) *Variance:* $\quad \sigma^2\{X - Y\} = \sigma^2\{X\} + \sigma^2\{Y\}$

Note that the expected value of the difference of two independent random variables is simply the difference of the expected values of the two random variables.

Note also that the variance of the difference is the same as the variance of the sum when the two random variables are independent.

☐ **Examples**

1. In our salesperson bonus example, $E\{X\} = E\{Y\} = 200$ and $\sigma^2\{X\} = \sigma^2\{Y\} = 60,000$ (calculations not shown). Hence, for $T = X + Y$, we obtain using (5.13):

$$E\{T\} = 200 + 200 = 400 \qquad \sigma^2\{T\} = 60,000 + 60,000 = 120,000$$

These are the same results, of course, as those obtained from the probability distribution of T. The standard deviation of T is $\sigma\{T\} = \sqrt{120,000} = 346.4$.

2. In our advertisement response example, it is known that $E\{X\} = 40$, $E\{Y\} = 70$, $\sigma^2\{X\} = 15$, $\sigma^2\{Y\} = 10$, and that X and Y are independent. For $R = X + Y$, the total number of responses, we obtain using (5.13):

$$E\{R\} = 40 + 70 = 110 \qquad \sigma^2\{R\} = 15 + 10 = 25$$

The standard deviation of R is $\sigma\{R\} = \sqrt{25} = 5$.

3. In our gross profit example, it is known that $E\{X\} = 100,000$, $E\{Y\} = 70,000$, $\sigma^2\{X\} = 8,000,000$, $\sigma^2\{Y\} = 4,000,000$, and X and Y are independent. For $P = X - Y$, the gross profit, we obtain using (5.14):

$$E\{P\} = 100,000 - 70,000 = 30,000$$
$$\sigma^2\{P\} = 8,000,000 + 4,000,000 = 12,000,000$$

The standard deviation of P is $\sigma\{P\} = \sqrt{12,000,000} = 3,464$. ☐

Sum of More Than Two Independent Random Variables

In statistical analysis, we are frequently concerned with the sum of more than two random variables. For example, suppose $X_1, X_2, \ldots, X_s$ represent the amounts an insurance company must pay out for fire losses during the year on s policies. Then $T = X_1 + X_2 + \cdots + X_s$ represents the total disbursement for fire losses during the year on these policies.

The expected value and variance of a sum of independent random variables are given in the next theorem:

(5.15) If $T = X_1 + X_2 + \cdots + X_s$ is the sum of s independent random variables, then T has:

(5.15a) *Expected Value:* $\qquad E\{T\} = \sum_{i=1}^{s} E\{X_i\}$

(5.15b) *Variance:* $\qquad \sigma^2\{T\} = \sum_{i=1}^{s} \sigma^2\{X_i\}$

☐ **Example**

A carton contains 100 cans of peas. The weight X_i of the peas in the ith can ($i = 1, 2, \ldots,$ 100) has expected value $E\{X_i\} = 8$ ounces and variance $\sigma^2\{X_i\} = .017$. In other words, all cans have the same mean and variance. In addition, the weights X_i are independent random variables. Hence, if $T = X_1 + X_2 + \cdots + X_{100}$ denotes the total weight of the peas in the carton, we obtain by (5.15):

$$E\{T\} = 100(8) = 800 \text{ ounces} \qquad \sigma^2\{T\} = 100(.017) = 1.7$$

The standard deviation of T is $\sigma\{T\} = \sqrt{1.7} = 1.3$ ounces.   □

Effect of Statistical Dependence

The influence of statistical dependence on the variance formulas for sums and differences of random variables is taken up in detail in Section 5.9. We point out here, however, that formulas (5.13a), (5.14a), and (5.15a) for expected values of sums and differences are valid even if the random variables are statistically dependent. In contrast, formulas (5.13b), (5.14b), and (5.15b) for the corresponding variances generally will not be valid if the random variables are not independent.

5.8   CONTINUOUS RANDOM VARIABLES

Probability Density Function

As noted at the beginning of the chapter, a continuous random variable assumes values on a continuum. Therefore, it is not meaningful to speak of a probability value being associated with each possible point on the continuum. Instead, we associate probability values with intervals on the continuum. To illustrate this point, consider a continuous random variable X which represents the yield of a crop in tons per acre. Suppose the yield can be any value between 0 and 1 ton. Figure 5.3 shows a mathematical function which has the property that, for any interval, the area between the axis and the curve of the function corresponds to the probability that X will have a value in this interval. For example, the area for the interval .5–.7 has been shaded and equals .229. Therefore, the probability that the yield will lie somewhere in this interval is .229. Symbolically, we state: $P(.5 \leq X \leq .7) = .229$. The probability corresponding to any other interval is

FIGURE 5.3   *Example of probability density function:* $f(x) = 12x(1 - x)^2$ *for* $0 \leq x \leq 1$. Area under the curve represents probability.

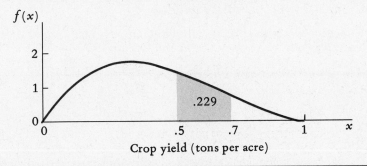

found in a similar manner, namely, by determining the area under the curve corresponding to that interval.

The mathematical function of a curve of the type shown in Figure 5.3 is called a probability density function:

(5.16) The *probability density function* of a continuous random variable X is a mathematical function for which the area under the curve corresponding to any interval is equal to the probability that X will take on a value in the interval. The probability density function is denoted by $f(x)$. The value $f(x)$ is called the *probability density* at x.

☐ **Example**

The probability density function for the curve shown in Figure 5.3 is:

(5.17)
$$f(x) = \begin{cases} 12x(1 - x)^2 & 0 \leq x \leq 1 \\ 0 & \text{elsewhere} \end{cases}$$

For instance, the probability density at $x = .5$ (and therefore the height of the curve at $x = .5$) is:

$$f(.5) = 12(.5)(1 - .5)^2 = 1.5$$ ☐

Comments

1. A probability density function can never lie below the x-axis because the probability that X will lie in any interval cannot be negative. Thus, $f(x)$ is always nonnegative. Further, since X must assume some value on the axis under the curve, the total area under a probability density function must be 1.

2. The probability that a continuous random variable will take on any particular value on the continuum is zero. This is because a single point corresponds to an interval of zero width, and hence there is zero area under the probability density function. As a result of this fact, in any probability statement of the type $P(a \leq X \leq b)$ for a *continuous* random variable, it is immaterial whether the endpoints of the interval are included or not, i.e:

$$P(a \leq X \leq b) = P(a < X \leq b) = P(a \leq X < b) = P(a < X < b)$$

3. The probability density function is often simply called *probability function, density function,* or *probability distribution* of the random variable.

Cumulative Probability Function

Sometimes we wish to work with cumulative probabilities for continuous random variables:

(5.18) The *cumulative probability function* of a continuous random variable is denoted by $F(x)$ and is defined:

$$F(x) = P(X \leq x)$$

where: $-\infty < x < +\infty$

In other words, the cumulative probability function $F(x)$ indicates the probability that the outcome of X in a random trial will be less than or equal to any specified value x. Thus, $F(x)$ corresponds to the area under the probability density function to the left of x. Two probability density functions and their cumulative probability functions are shown in Figure 5.4. Note that $F(x)$ is a smooth continuous curve when the random variable is continuous, rather than a step function as for discrete random variables. However, the properties of the cumulative probability function for discrete random variables apply also for continuous random variables, i.e., $F(x)$ always has values between 0 and 1 and never decreases as x increases.

Calculational Procedures with Continuous Probability Functions (Calculus Needed)

Determining Probabilities. Integral calculus provides a convenient means for determining probabilities for continuous random variables. To find an area under a probability function $f(x)$ corresponding to the interval from a to b, we determine the value of the definite integral:

(5.19)
$$P(a \leq X \leq b) = \int_a^b f(x)\, dx$$

FIGURE 5.4 *Examples of probability density functions and their cumulative probability functions*

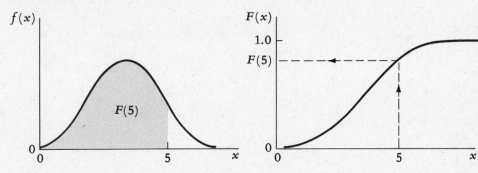

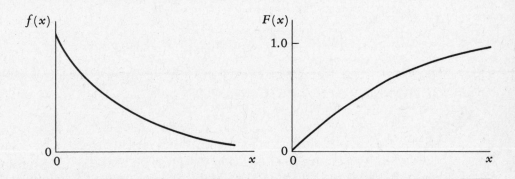

For example, for the probability function in Figure 5.3 we wish to find $P(.5 \leq X < .7)$, the area illustrated in the figure. We obtain:

$$P(.5 \leq X \leq .7) = \int_{.5}^{.7} 12x(1 - x)^2 \, dx = .229$$

Cumulative Probability Function. In calculus notation, the cumulative probability function $F(x)$ for a continuous random variable is defined:

(5.20)
$$F(x) = \int_{-\infty}^{x} f(u) \, du$$

Here, u denotes the variable of integration.

For example, the cumulative probability function for probability function (5.17) is:

$$F(x) = \int_{0}^{x} 12u(1 - u)^2 \, du = 3x^4 - 8x^3 + 6x^2 \qquad 0 \leq x \leq 1$$

We see, for instance, that $F(.5) = 3(.5)^4 - 8(.5)^3 + 6(.5)^2 = .688$.

Expected Value. The expected value of a continuous random variable is defined:

(5.21)
$$E\{X\} = \int_{-\infty}^{+\infty} xf(x) \, dx$$

Note that (5.21) is analogous to (5.5); integration has simply replaced summation in the definition.

To illustrate (5.21), the expected value for probability function (5.17) is:

$$E\{X\} = \int_{0}^{1} x[12x(1 - x)^2] \, dx = .4$$

Variance. The variance of a continuous random variable is defined:

(5.22)
$$\sigma^2\{X\} = \int_{-\infty}^{+\infty} (x - E\{X\})^2 f(x) \, dx$$

Note that (5.22) is analogous to (5.7). Integration has simply replaced summation.

To illustrate (5.22), the variance for probability function (5.17) is (recall that $E\{X\} = .4$):

$$\sigma^2\{X\} = \int_{0}^{1} (x - .4)^2[12x(1 - x)^2] \, dx = .04$$

5.9 OPTIONAL TOPIC—COVARIANCE AND CORRELATION

In later chapters, we shall be concerned with the extent to which two random variables are associated with each other. Two measures of association for a pair of random variables are their *covariance* and their *coefficient of correlation*.

Covariation

We begin with the concept of concurrent variation or covariation.

☐   **Examples**

1. In a midwestern city, the expected midday temperature on July 4 is 25°C and the expected relative humidity is 45 percent. Last July 4, the midday temperature was 30°C and the relative humidity was 70 percent. The covariation of temperature and humidity on that day is the product of the deviations from the expected values, i.e., $(30 - 25)(70 - 45) = 125$. Covariation can be positive, as here, or negative, or zero. Positive covariation here indicates that both temperature and humidity exceeded their respective expectations.
2. The expected yield of stock A is 5 percent and that of stock B is 10 percent. The actual yields last month were 8 percent each. Hence, the covariation is $(8 - 5)(8 - 10) = -6$. Here, the covariation is negative, indicating that one stock performed above its expectation and the other below its expectation.   ☐

As has been seen in the preceding examples, covariation for the outcomes x_i and y_j is defined:

(5.23) $$\text{Covariation} = (x_i - E\{X\})(y_j - E\{Y\})$$

Covariance

The measure covariance is simply the average covariation over repeated trials:

(5.24) The *covariance* of two discrete random variables X and Y is denoted by $\sigma\{X, Y\}$ and defined:

$$\sigma\{X, Y\} = \sum_i \sum_j (x_i - E\{X\})(y_j - E\{Y\})P(x_i, y_j)$$

where: $P(x_i, y_j)$ denotes $P(X = x_i \cap Y = y_j)$

$\sigma\{\ ,\ \}$ (read "covariance of") is called the *covariance operator*

Readers who are unfamiliar with double summation notation should refer to Appendix A, Section A.1.

☐   **Examples**

1. Random variables X and Y have the joint probability distribution:

		y	
		5	10
x	10	.30	.20
	30	.10	.40

Note that the two most likely outcomes are $X = 10$, $Y = 5$ and $X = 30$, $Y = 10$, indicating that random variables X and Y tend to vary in the same direction, or positively. We find that $E\{X\} = 20$ and $E\{Y\} = 8$ from the marginal distributions of the random variables (calculations not shown).

TABLE 5.2 *Calculation of covariance for two examples*

(a)
Example 1: $E\{X\} = 20$, $E\{Y\} = 8$

(1)		(2)	(3)	(4)	(5)	(6)
						Weighted
			Deviation	Deviation	Covariation	Covariation
x	y	$P(x, y)$	$x - E\{X\}$	$y - E\{Y\}$	(3) × (4)	(5) × (2)
10	5	.30	$10 - 20 = -10$	$5 - 8 = -3$	$(-10)(-3) = 30$	9.0
10	10	.20	$10 - 20 = -10$	$10 - 8 = 2$	$(-10)(2) = -20$	−4.0
30	5	.10	$30 - 20 = 10$	$5 - 8 = -3$	$(10)(-3) = -30$	−3.0
30	10	.40	$30 - 20 = 10$	$10 - 8 = 2$	$(10)(2) = 20$	8.0
Total		1.00				$\sigma\{X, Y\} = +10.0$

(b)
Example 2: $E\{W\} = 10.5$, $E\{Z\} = 3$

(1)		(2)	(3)	(4)	(5)	(6)
						Weighted
			Deviation	Deviation	Covariation	Covariation
w	z	$P(w, z)$	$w - E\{W\}$	$z - E\{Z\}$	(3) × (4)	(5) × (2)
5	2	.10	−5.5	−1	5.5	.55
5	4	.20	−5.5	1	−5.5	−1.10
10	2	.10	− .5	−1	.5	.05
10	4	.20	− .5	1	− .5	− .10
15	2	.30	4.5	−1	−4.5	−1.35
15	4	.10	4.5	1	4.5	.45
Total		1.00				$\sigma\{W, Z\} = -1.50$

The calculation of the covariance of X and Y is best carried out in a systematic fashion, as illustrated in Table 5.2a. Columns 1 and 2 contain the joint probability distribution of X and Y in a columnar form, where $P(x, y)$ in column 2 denotes $P(X = x \cap Y = y)$. Columns 3 and 4 contain the deviations $x_i - E\{X\}$ and $y_j - E\{Y\}$, respectively. Column 5 contains the covariations $(x_i - E\{X\})(y_j - E\{Y\})$. Finally, column 6 contains the covariations weighted by their probabilities $P(x_i, y_j)$. Thus, for the random outcome $X = 10$, $Y = 5$, the deviations are $(10 - 20) = -10$ and $(5 - 8) = -3$ and the covariation is $(-10)(-3) = 30$. This covariation is weighted by the probability $P(X = 10 \cap Y = 5) = P(10, 5) = .30$ to yield the weighted covariation $30(.30) = 9.0$, as shown in column 6.

The sum of column 6 is the covariance. We see that $\sigma\{X, Y\} = +10.0$. Note that the covariance is positive because the positive covariations in column 5 are the most likely ones to occur.

2. The joint probability distribution of W and Z is:

		z	
		2	4
	5	.10	.20
w	10	.10	.20
	15	.30	.10

We find $E\{W\} = 10.5$ and $E\{Z\} = 3$ from the marginal distributions of the random variables (calculations not shown). Table 5.2b contains the calculation of the covariance. We see that $\sigma\{W, Z\} = -1.50$. The covariance is negative because the negative covariations in column 5 are the most likely ones to occur. ☐

Interpretation of Covariance

The magnitude of the covariance measure $\sigma\{X, Y\}$ depends on the units of X and Y and generally will change when the units of X and Y are changed. Hence, the main information provided by the covariance measure about the association between X and Y is whether $\sigma\{X, Y\}$ is positive, negative, or zero.

In Table 5.3, we have presented three bivariate probability distributions exhibiting different amounts and direction of covariance. In all three distributions, $E\{X\} = 1$ and $E\{Y\} = 3$. Since covariance is expected covariation, the interpretation of covariance parallels that of covariation for an individual joint outcome. When $\sigma\{X, Y\}$ is positive, negative, and zero, we say that X and Y exhibit *positive*, *negative*, and *zero covariance*, respectively. Thus, Table 5.3a contains a bivar-

TABLE 5.3 *Three bivariate probability distributions with varying amounts of covariance and correlation*

		$y = 1$	$y = 3$	$y = 5$	Total
		(a)			
$x = 0$		.00	.05	.25	.30
$x = 1$		.05	.30	.05	.40
$x = 2$		.25	.05	.00	.30
	Total	.30	.40	.30	1.00

$$\sigma\{X, Y\} = -1.0 \qquad \rho_{XY} = -.833$$

		$y = 1$	$y = 3$	$y = 5$	Total
		(b)			
$x = 0$		.30	.00	.00	.30
$x = 1$		.00	.40	.00	.40
$x = 2$		.00	.00	.30	.30
	Total	.30	.40	.30	1.00

$$\sigma\{X, Y\} = +1.2 \qquad \rho_{XY} = +1.0$$

		$y = 1$	$y = 3$	$y = 5$	Total
		(c)			
$x = 0$		.10	.10	.10	.30
$x = 1$		.10	.20	.10	.40
$x = 2$		.10	.10	.10	.30
	Total	.30	.40	.30	1.00

$$\sigma\{X, Y\} = 0 \qquad \rho_{XY} = 0$$

iate probability distribution exhibiting negative covariance. Note, in this case, that X and Y tend to vary inversely. If X has a value greater than its mean 1, Y is likely to have a value less than its mean 3, and if X has a value less than 1, Y is likely to have a value greater than 3. The bivariate probability distribution in Table 5.3b exhibits positive covariance. Here, X and Y tend to vary in the same direction. Finally, Table 5.3c shows a bivariate probability distribution exhibiting zero covariance. Here, the covariations between X and Y tend to balance out.

An important property of the covariance measure is:

(5.25)
$$\sigma\{X, Y\} = 0 \text{ when } X \text{ and } Y \text{ are independent}$$

Thus, the covariance between X and Y in our bonus amount example on p. 125 must be zero because X and Y are independent variables.

Comments

1. Although statistical independence implies zero covariance, the converse is *not* always true. It is possible that $\sigma\{X, Y\} = 0$, yet X and Y are statistically dependent. This is illustrated by Table 5.3c where $\sigma\{X, Y\} = 0$, yet X and Y are statistically dependent. Note, for example, that $P(X = 2 \cap Y = 1) = .10$, while the product of the marginal probabilities is $.30(.30) = .09$. This shows that two random variables may be associated in a fashion which their covariance does not reflect.

2. The covariance can be defined formally as expected covariation:

(5.26)
$$\sigma\{X, Y\} = E\{(X - E\{X\})(Y - E\{Y\})\}$$

An algebraically identical expression is:

(5.26a)
$$\sigma\{X, Y\} = E\{XY\} - E\{X\}E\{Y\}$$

Coefficient of Correlation

As we noted earlier, the main information provided by the covariance measure about the association between X and Y is whether the covariance is positive, negative, or zero. The magnitude of $\sigma\{X, Y\}$ depends on the units of X and Y and thus usually cannot be directly compared for different pairs of variables.

The coefficient of correlation is a measure which adjusts the covariance so that the resulting measure is unit-free:

(5.27) The *coefficient of correlation* of two random variables X and Y is denoted by ρ_{XY} (Greek rho) and defined:

$$\rho_{XY} = \frac{\sigma\{X, Y\}}{\sigma\{X\}\sigma\{Y\}}$$

where: $\sigma\{X\}$ is the standard deviation of X

$\sigma\{Y\}$ is the standard deviation of Y

$\sigma\{X, Y\}$ is the covariance of X and Y

☐ **Examples**

1. For the example in Table 5.2a, we found that $\sigma\{X, Y\} = +10.0$. Further, $\sigma\{X\} = 10$ and $\sigma\{Y\} = 2.449$ (calculations not shown). Hence, by (5.27):

$$\rho_{XY} = \frac{+10.0}{10(2.449)} = +.41$$

Note that ρ_{XY} is positive, just like $\sigma\{X, Y\}$. This must be the case because the standard deviations in the denominator of ρ_{XY} in (5.27) are always positive.

2. In Table 5.3, we have shown the coefficients of correlation for each of the three bivariate distributions. The coefficient of correlation in Table 5.3a is negative, as is the covariance, and the coefficient of correlation in Table 5.3b is positive, as is the covariance. ☐

Interpretation of Coefficient of Correlation

It can be shown that ρ_{XY} takes on values between -1 and $+1$:

(5.28)
$$-1 \leq \rho_{XY} \leq +1$$

Coefficients of $+1$ and -1 indicate perfect positive and negative linear associations, respectively. Note that there is a perfect positive linear association between X and Y in Table 5.3b since for any outcome X, only one outcome for Y can arise, namely, that where $Y = 1 + 2X$. Thus, if $X = 2$, the only possible Y outcome is $Y = 1 + 2(2) = 5$.

A coefficient of correlation of 0 indicates that there is no linear association between X and Y, as is illustrated in Table 5.3c. We say that X and Y are uncorrelated then. The closer ρ_{XY} is to -1 or to $+1$, the stronger is the degree of linear association between the two random variables.

Variances of Sum and Difference of Two Dependent Random Variables

One use of the covariance is found in the variances of the sum and difference of two dependent random variables. We noted earlier that the expected values for the sum and difference of two random variables in (5.13a) and (5.14a) apply whether the random variables are independent or dependent. The variances given earlier, however, do not apply to dependent variables in general.

If two random variables are dependent, the variances of the sum and difference also include the covariance term:

(5.29)
$$\sigma^2\{X + Y\} = \sigma^2\{X\} + \sigma^2\{Y\} + 2\sigma\{X, Y\}$$

(5.30)
$$\sigma^2\{X - Y\} = \sigma^2\{X\} + \sigma^2\{Y\} - 2\sigma\{X, Y\}$$

☐ **Examples**

1. X denotes output by machine 1 and Y output by machine 2. Suppose $E\{X\} = 200$, $E\{Y\} = 100, \sigma^2\{X\} = 400, \sigma^2\{Y\} = 80$, and $\sigma\{X, Y\} = -10$. If $T = X + Y$ denotes the total output, we then have:

$$E\{T\} = 200 + 100 = 300 \qquad \text{by (5.13a)}$$

$$\sigma^2\{T\} = 400 + 80 + 2(-10) = 460 \qquad \text{by (5.29)}$$

The standard deviation of T is $\sigma\{T\} = \sqrt{460} = 21.4$. Note that if the covariance were positive, $\sigma^2\{T\}$ would be larger. For instance, if $\sigma\{X, Y\} = +10$, $\sigma^2\{T\} = 400 + 80 + 2(10) = 500$.

2. In Example 1, suppose interest is in the difference D in output, where $D = X - Y$. We then have:

$$E\{D\} = 200 - 100 = 100 \quad \text{by (5.14a)}$$

$$\sigma^2\{D\} = 400 + 80 - 2(-10) = 500 \qquad \text{by (5.30)} \qquad \square$$

Comment

When random variables X and Y are statistically independent, we know from (5.25) that $\sigma\{X, Y\} = 0$. Hence, (5.29) and (5.30) in that case reduce to (5.13b) and (5.14b) for independent variables.

PROBLEMS

5.1 In each of the following situations, indicate whether the random variable is discrete or continuous and describe its sample space (i.e., its set of possible outcomes):
  a. The number of passengers on a scheduled flight having a capacity of 220 passengers.
  b. The number of employees of a firm with 300 employees who are absent because of sickness on a given day.
  c. The length of time a machine is idle during an eight-hour working day.

*5.2 The probability distribution of X, the number of positions held previously by a job applicant, follows.

x:	0	1	2	3	4
$P(X = x)$:	.60	.20	.10	.05	.05

  a. Interpret each of the following expressions: (1) $P(0)$, (2) $P(2)$, (3) $P(X \leq 1)$, (4) $P(1 \leq X \leq 3)$.
  b. Obtain each of the probabilities in **a**.

5.3 The probability distribution of X, the number of passengers on a daily helicopter shuttle run from airport A to airport B, follows.

x:	1	2	3	4
$P(X = x)$:	.1	.3	.2	.4

  a. Interpret each of the following expressions: (1) $P(3)$, (2) $P(4)$, (3) $P(X \leq 2)$, (4) $P(2 \leq X \leq 4)$.
  b. Obtain each of the probabilities in **a**.

5.4 The probability distribution of X, the length of a long-distance telephone call in minutes, follows.

x:	5	10	15	20
$P(x)$:	.3	.5	.1	.1

a. Interpret each of the following expressions: (1) $P(10)$, (2) $P(X \leq 15)$, (3) $P(X \geq 10)$, (4) $P(5 \leq X \leq 10)$. .

b. Obtain each of the probabilities in **a**.

*5.5 Refer to Problem 5.2.

a. Construct a graph of the probability distribution.

b. Construct a graph of the cumulative probability distribution. From your graph, find $P(X \leq 3)$. Explain how this probability is interpreted.

5.6 Refer to Problem 5.3.

a. Construct a graph of the probability distribution.

b. Construct a graph of the cumulative probability distribution. From your graph, find: (1) $P(X \leq 2)$, (2) $P(X \leq 3.5)$. Explain how the latter probability is interpreted.

5.7 Refer to Problem 5.4.

a. Construct a graph of the probability distribution.

b. Construct a graph of the cumulative probability distribution. From your graph find: (1) $P(X \leq 5)$, (2) $P(X \leq 25)$. Explain how the latter probability is interpreted.

*5.8 The bivariate probability distribution for number of computer stoppages in a day (X) and number of computer operators sick for the day (Y) follows.

| | | Number of Sick Operators | | |
| | | y | | |
		0	1	2
Number of	0	.40	.15	.02
Stoppages	1	.30	.05	.01
x	2	.04	.03	0

From inspection of the bivariate probability distribution, find and interpret each of the following probabilities: (1) $P(X = 1 \cap Y = 0)$, (2) $P(X = 2 \cap Y = 1)$, (3) $P(X \leq 1)$, (4) $P(Y \geq 2)$, (5) $P(X = 1|Y = 0)$.

5.9 Refer to Problem 5.3. Let X_1 be the number of passengers on day 1 and X_2 the number of passengers on day 2. Assume that X_1 and X_2 are independent, each with the probability distribution in Problem 5.3.

a. Construct the bivariate probability distribution for X_1 and X_2.

b. From inspection of the bivariate probability distribution, find and interpret each of the following probabilities: (1) $P(X_1 = 2 \cap X_2 = 1)$, (2) $P(X_1 = 4 \cap X_2 = 4)$, (3) $P(X_2 = 4|X_1 = 2)$, (4) $P(X_1 = 2|X_2 = 4)$, (5) $P(X_2 = 4)$.

*5.10 Refer to Problem 5.8.

a. Obtain the marginal probability distribution of number of computer stoppages.

b. Obtain the conditional probability distribution of number of computer stoppages when no operator is sick. How does this conditional distribution compare with the marginal distribution obtained in **a**? Does this imply that X and Y are dependent? Explain.

5.11 Refer to Problems 5.3 and 5.9.

a. From the joint probability distribution, obtain the conditional probability distribution of number of passengers on day 2, given that there are two passengers on day 1.

    b. Why must the conditional distribution in **a** be the same as the marginal distribution of number of passengers on day 2? Explain.

*5.12 Refer to Problem 5.2. Calculate $E\{X\}$. Interpret its meaning here in relative frequency terms.

5.13 Refer to Problem 5.3. Calculate $E\{X\}$ and interpret this measure. Should one expect the measure to be a whole number here?

5.14 Refer to Problem 5.4. Calculate the mean of the probability distribution and interpret this measure. What symbol is used in the Text to denote this quantity?

5.15 The probability that a $35 million North Sea oil-drilling platform will be totally lost at sea next year is .009, and the probability that a total loss will not occur is .991. How much should the platform's owners expect to pay for complete insurance against this contingency in the next year (excluding administrative and other costs and profits of the insurance company)? Explain.

*5.16 Refer to Problem 5.3. The fare for the trip is $28. Let Y denote the total fare revenue for the run.
    a. Express Y in terms of X.
    b. Obtain $E\{Y\}$ by (5.6). (*Hint:* $E\{X\} = 2.9$.)
    c. Verify your result in **b** by obtaining the probability distribution of Y and calculating $E\{Y\}$ by (5.5).

5.17 The probability distribution of X, the number of persons in a restaurant party, follows.

x:	1	2	3	4	5	6
$P(x)$:	.05	.15	.25	.40	.10	.05

The restaurant serves a buffet dinner costing $7.50 per person. Let Y denote the total cost of the buffet dinner for a party.
    a. Express Y in terms of X.
    b. Obtain $E\{X\}$, and then obtain $E\{Y\}$ by (5.6).
    c. Verify your result in **b** by obtaining the probability distribution of Y and calculating $E\{Y\}$ by (5.5).

5.18 Refer to Problem 5.4. Let Y denote the cost of a long-distance telephone call in dollars. The cost per minute is $.80 and the initial connection charge is $.60.
    a. Express Y as a function of X. Calculate $E\{Y\}$, using (5.6). Interpret the meaning of $E\{Y\}$ here. (*Hint:* $E\{X\} = 10.0$.)
    b. Which would increase the expected cost of the telephone call by more, an increase in the initial connection charge to $1.00 or an increase in the cost per minute to $.85?

*5.19 Refer to Problem 5.2. Calculate the variance and standard deviation of the probability distribution. In what units is the standard deviation expressed? (*Hint:* $E\{X\} = .75$.)

5.20 Refer to Problem 5.3. Calculate $\sigma^2\{X\}$ and $\sigma\{X\}$. In what units are $\sigma^2\{X\}$ and $\sigma\{X\}$ expressed? Interpret $\sigma\{X\}$ in relative frequency terms.

5.21 Refer to Problem 5.4. Calculate the variance and standard deviation of the probability distribution. Is the standard deviation here large relative to the mean? Discuss.

*5.22 A bank's study of midmonth balances of personal checking accounts shows that they follow a probability distribution with mean $300 and standard deviation $40.

    a. From the Chebyshev inequality, determine the minimum probability that the midmonth balance of an account is between $180 and $420.

    b. The probability distribution of midmonth balances is skewed to the right, and large negative balances do not occur. Explain if either of these two facts affects the validity of the probability bound computed in **a**.

    c. Consider the probability distribution of the end-of-month balances for these same accounts. Would the probability bound in **a** apply to the end-of-month balances? Explain.

5.23 Refer to Problem 5.22. Answer all parts using a mean of $300 and a standard deviation of $60.

*5.24 Refer to Problems 5.3 and 5.16.

    a. Calculate $\sigma^2\{Y\}$ and $\sigma\{Y\}$, using (5.11). In what units is $\sigma\{Y\}$ expressed? (*Hint:* $\sigma\{X\} = 1.044$.)

    b. Verify your result in **a** by calculating $\sigma^2\{Y\}$ from the probability distribution of Y, using (5.7).

5.25 Refer to Problem 5.17.

    a. Obtain the standard deviation of the probability distribution of Y, using (5.11). In what units is the standard deviation expressed? (*Hint:* $\sigma\{X\} = 1.1619$.)

    b. Verify your result by calculating $\sigma\{Y\}$ directly from the probability distribution of Y, using (5.7).

5.26 Refer to Problems 5.4 and 5.18.

    a. Obtain the standard deviation of the probability distribution of Y, using (5.11). Interpret this measure in relative frequency terms.

    b. Is the standard deviation affected by the initial connection charge? By the cost per minute? Explain.

    c. Compare the coefficients of variation (ratio of standard deviation to mean in percent) for the probability distributions of X and Y. Which distribution is relatively more variable?

*5.27 Refer to Problem 5.2. Here, $E\{X\} = .75$ and $\sigma\{X\} = 1.135$.

    a. Obtain the probability distribution of the standardized random variable for X, defined in (5.12). In what units is the standardized variable expressed?

    b. Calculate the mean and standard deviation of the standardized random variable from the probability distribution in **a**. Are the results consistent with the definition of a standardized random variable? Explain.

5.28 Refer to Problem 5.4.

    a. Obtain the probability distribution of the standardized random variable for X, defined in (5.12). In what units is the standardized random variable expressed?

    b. Calculate the mean and standard deviation of the standardized random variable from the probability distribution in **a**. Must these be 0 and 1, respectively? Explain.

*5.29 Refer to Problems 5.3 and 5.9. Here, $E\{X_1\} = E\{X_2\} = 2.9$, $\sigma^2\{X_1\} = \sigma^2\{X_2\} = 1.09$, and X_1 and X_2 are independent random variables. Let $T = X_1 + X_2$ denote the total number of passengers on the two days' runs.

    a. Calculate $E\{T\}$ and $\sigma^2\{T\}$, using (5.13). Does either of these results depend on the fact that X_1 and X_2 are statistically independent?

    b. Obtain the probability distribution of T. Verify your results in **a** by direct calculations from this probability distribution.

    c. What does the difference $D = X_2 - X_1$ represent? Find $E\{D\}$ and $\sigma\{D\}$, using (5.14). Interpret each of these measures in terms of relative frequency.

5.30 The probability distributions of the number of goals scored by the home team (X) and the visiting team (Y) in a sports match follow.

x:	0	1	2
$P(x)$:	.4	.3	.3

y:	0	1	2
$P(y)$:	.5	.3	.2

X and Y are independent random variables. Let $T = X + Y$ denote the total number of goals scored in the match.
  a. Calculate $E\{T\}$ and $\sigma\{T\}$, using (5.13). Interpret each measure in terms of relative frequency.
  b. Obtain the probability distribution of T. Verify your results in **a** by direct calculations from this probability distribution.

5.31 Refer to Problem 5.30.
  a. What does the difference $D = X - Y$ represent here? Calculate $E\{D\}$ and $\sigma\{D\}$.
  b. Obtain the probability distribution of D. What is the probability of a tie game? Of the home team winning?

*5.32 Refer to Problem 5.3. Here, $E\{X\} = 2.9$ and $\sigma^2\{X\} = 1.09$. Let T denote the total number of passengers on five statistically independent runs. Obtain $E\{T\}$ and $\sigma^2\{T\}$, using (5.15). Is the standard deviation of T five times as great as the standard deviation of X? Discuss.

5.33 The daily output X in a bottling plant is a random variable with $E\{X\} = 12,000$ bottles and $\sigma\{X\} = 500$ bottles. Assume that the outputs on successive days are statistically independent.
  a. What is the expected total output for 10 successive days? For 30 successive days?
  b. Obtain the standard deviation of total output for 10 days. Also obtain the standard deviation of total output for 30 days. How do these two standard deviations compare relative to their respective means?

*5.34 Refer to Figure 5.3. For this density function, $P(X \leq .5) = .688$.
  a. Find: (1) $P(0 \leq X \leq .7)$, (2) $P(X \geq .7)$. Interpret the probability in (2).
  b. If you were asked to find $P(X > .7)$ rather than $P(X \geq .7)$, why would the answer not change? Explain.

5.35 A highway contains a 200-kilometer stretch through a desolate area. A breakdown in this stretch is equally likely to occur at any point. X, the distance into the stretch at which the breakdown occurs, is treated as a continuous random variable whose probability density function is $f(x) = 1/200$, for $0 \leq x \leq 200$.
  a. Graph the probability density function for X. Does the area under your density function equal 1?
  b. Use the geometric properties of the probability density function to find the following probabilities: (1) $P(X \leq 100)$, (2) $P(X \leq 50)$, (3) $P(40 \leq X \leq 80)$.
  c. Obtain the cumulative probability function for X and graph it. [*Hint:* $P(X \leq 0) = 0$ and $P(X \leq 200) = 1$.]
  d. From your graph of the cumulative probability function in **c**, find the values of: (1) $F(80)$, (2) $F(150)$. Interpret the meaning of the latter value.

5.36 A melon's shelf life in days (X) is treated as a continuous random variable. The density function of X is the right triangular probability function:

$$f(x) = \begin{cases} .005(20 - x) & 0 \le x \le 20 \\ 0 & \text{elsewhere} \end{cases}$$

a. Graph this probability function. Is the area under the density function equal to 1?
b. Use the geometric properties of $f(x)$ to find $P(X \le 10)$.

*5.37 The bivariate probability distribution of the price received (P) and quantity sold (Q) of a product follows.

		q (cartons)	
		30	40
p	6	.2	.4
($ per carton)	8	.3	.1

a. Calculate $E\{P\}$ and $E\{Q\}$.
b. Obtain the covariation of P and Q when $p = 6$ and $q = 40$. What information is provided by the sign of the measure of covariation?
c. Calculate $\sigma\{P, Q\}$. Does its value indicate that P and Q are statistically related? Explain.

5.38 The bivariate probability distribution of the number of bubble flaws (X) and solid particle flaws (Y) in hand-blown decorative bottles follows.

		Particle Flaws y	
		0	1
Bubble Flaws	0	.2	.1
x	1	.1	.2
	2	.1	.3

a. Calculate $E\{X\}$, $E\{Y\}$, and $\sigma\{X, Y\}$.
b. Are the two random variables statistically independent? Discuss.

*5.39 Refer to Problem 5.37.
a. Calculate the coefficient of correlation between price received and quantity sold. What range of values can this coefficient take on?
b. Using your result in a, describe the direction and strength of the linear association between price and quantity.

5.40 Refer to Problem 5.38.
a. Calculate ρ_{XY}. In what units is this measure expressed?
b. What range of values can ρ_{XY} take on? Using your result in a, describe the direction and strength of the linear association between X and Y.

*5.41 Refer to Table 5.1. Here, $E\{X\} = E\{Y\} = .55$, $\sigma^2\{X\} = \sigma^2\{Y\} = .7475$, and $\sigma\{X, Y\} = .5675$.
a. Calculate $E\{X + Y\}$, $E\{X - Y\}$, $\sigma\{X + Y\}$, and $\sigma\{X - Y\}$. Interpret each of these numbers.
b. What would be the values calculated in a if X and Y had the marginal probability distributions shown in Table 5.1 but were statistically independent?

5.42 Refer to Problem 5.38. Let T denote the total number of flaws in a bottle. Find $E\{T\}$ and $\sigma\{T\}$. Interpret each of these measures.

EXERCISES

5.43 Refer to Problem 5.8. Find the following probabilities: (1) $P(X = 2|Y \leq 1)$, (2) $P(X \leq 1|Y \geq 1)$, (3) $P(X = Y|Y \leq 1)$.

5.44 Verify the following identities: (5.6), (5.8a), (5.11).

5.45 Let Y be a standardized random variable.
  a. Prove that $\sigma\{Y\} = 1$.
  b. From the Chebyshev inequality, determine the maximum probability that the absolute value of Y exceeds 2.

5.46 A student, citing theorem (5.13b), states that $\sigma\{X + Y\} = \sigma\{X\} + \sigma\{Y\}$. Comment.

5.47 (Calculus needed.) Refer to Problem 5.35.
  a. Obtain $E\{X\}$ and $\sigma\{X\}$.
  b. Obtain the cumulative probability function.
  c. Obtain the following, using the cumulative probability function: (1) $F(10)$, (2) $F(16)$, (3) $P(10 \leq X \leq 16)$.

5.48 (Calculus needed.) Refer to Problem 5.36. Obtain: (1) $E\{X\}$, (2) $\sigma^2\{X\}$.

5.49 (Calculus needed.) The probability density function of a random variable X follows.

$$f(x) = \begin{cases} \dfrac{4}{3} - x^2 & 0 \leq x \leq 1 \\ 0 & \text{elsewhere} \end{cases}$$

  a. Graph the probability density function.
  b. Find the mean and variance of this probability distribution.
  c. Obtain the cumulative probability function and graph it.

5.50 Verify the following identities: (5.25), (5.26a).

STUDIES

5.51 A potato farmer has a contract to sell the forthcoming harvest to a food cannery. The entire harvest will receive a grade of A, B, or C. If the harvest is graded A, the farmer will receive \$4.20 per bushel. For grades B and C, the prices per bushel will be \$3.80 and \$3.20, respectively. The probabilities for the grade of the harvest, if the harvesting is done at the normal time, are $P(A) = .35$, $P(B) = .45$, and $P(C) = .20$. The yield of the harvest at the normal time is expected to be 60,000 bushels. By harvesting earlier, the probabilities would be changed to $P(A) = .40$, $P(B) = .60$, and $P(C) = 0$, but the yield would be reduced to 57,000 bushels. Should the farmer harvest earlier or at the normal time if he wishes to maximize expected revenue?

5.52 A national chain operates 400 similar restaurants. The probability distribution of loss by fire in any one of these restaurants (X, in dollars) in a year follows.

x:	0	100,000
$P(x)$:	.997	.003

Assume that the losses in the different restaurants ($X_1, X_2, \ldots, X_{400}$) are statistically independent. Let $T = \Sigma X_i$ denote the total loss by fire for the whole chain in a year.

a. Calculate $E\{X_i\}$ and $\sigma\{X_i\}$ for any one restaurant. Then find $E\{T\}$ and $\sigma\{T\}$. Is random variable T relatively more predictable than the random variable X_i for any one restaurant? Comment.

b. Use the Chebyshev inequality to obtain an upper bound on the probability that total loss by fire in a year for the chain will differ from the expected amount by more than \$380,000. Is this a bound on the probability $P(T \le 500,000)$ here? Explain.

c. The chain self-insures its 400 restaurants against fire loss because the expected loss by fire for the chain is substantially less than the cost of fire insurance for a year. If a middle-income person would own and operate a single one of these restaurants, why might he or she not be willing to self-insure against fire loss?

5.53 A truck driver's emergency kit includes four highway flares. The burning times of the flares are independent random variables, with mean of 15.0 minutes and standard deviation of 1.50 minutes for each flare. Let T denote the total burning time of the four flares when they are burned consecutively. From the Chebyshev inequality, determine the minimum probability that the flares will last between 54.0 and 66.0 minutes when burned consecutively.

5.54 A small investor is considering a \$200 investment in growth stocks for one year. Let X denote the gain in buying and holding for one year one \$100 share of stock 1, and let Y be defined similarly for one \$100 share of stock 2. Assume that $E\{X\} = 50$, $E\{Y\} = 50$, $\sigma^2\{X\} = 64$, $\sigma^2\{Y\} = 81$, and $\sigma\{X, Y\} = -60$. Consider the following options: (1) buy two shares of stock 1; (2) buy two shares of stock 2; (3) buy one share each of stocks 1 and 2.

a. Is any one of these options better than the others in terms of expected gain?

b. Calculate $\sigma^2\{2X\}$, $\sigma^2\{2Y\}$, and $\sigma^2\{X + Y\}$, and explain what each quantity represents. The investor states that option 3 "is less risky" than options 1 and 2. How is she interpreting risk here?

c. Repeat the calculations in b for $\sigma\{X, Y\} = +60$. Which option now has the smallest variance? What does this finding suggest about the covariance conditions which will "spread the risk" when diversifying with a number of different securities?

5.55 Refer to Problem 5.37.

a. Suppose the dollar cost of producing quantity Q is given by $C = 100 + 3Q$. Obtain $E\{C\}$ and $\sigma^2\{C\}$.

b. Denote the revenue from the sale of the product by R, so $R = PQ$. Find $E\{R\}$ and interpret its meaning. [*Hint:* Use (5.26a).]

c. The profit from the sale of the product is $R - C$. Find $E\{R - C\}$.

6
Common Discrete Probability Distributions

There exists an infinite variety of probability distributions. It has been found, however, that a limited number of types or *families* of probability distributions are used in a wide range of applications. In this chapter, we study several important families of distributions for discrete random variables. In the next chapter, we extend our study to distributions for continuous random variables.

6.1 BINOMIAL PROBABILITY DISTRIBUTIONS

Bernoulli Random Trial

A *Bernoulli random trial* is a random trial that has two basic outcomes of a qualitative nature. To quantify these outcomes, one outcome is assigned the value 0 and the other the value 1. Which outcome is assigned the value 0 and which the value 1 is arbitrary. The random variable X associated with a Bernoulli random trial is called a *Bernoulli random variable*.

☐ Examples

1. The calculation of a payroll check may be correct or incorrect. We define the Bernoulli random variable for this trial so that $X = 0$ corresponds to a correctly calculated check and $X = 1$ to an incorrectly calculated one.
2. A consumer either recalls a television commercial ($X = 1$) or does not recall ($X = 0$). X is a Bernoulli random variable.
3. Screening of a credit card application gives either the rating $X = 1$ (accept) or $X = 0$ (reject). X is a Bernoulli random variable.   ☐

Bernoulli Process

Sequence of Bernoulli Trials. In statistical analysis, one is seldom interested in a single Bernoulli trial. More commonly, a sequence of Bernoulli trials is under consideration.

◻ Examples

1. In a process for manufacturing spoons, each spoon may either be defective or not. This process may be viewed as a sequence of Bernoulli trials in which the ith spoon has an associated random variable X_i, with $X_i = 1$ if the ith spoon is defective and $X_i = 0$ if it is not defective. Interest may center on how many defective spoons are produced in the most recent production run of 1,000 spoons.

2. Thirty plant specimens are treated for a particular fungus. At the end of the test period, each specimen is examined to see if it is fungus-free. This may be viewed as a sequence of Bernoulli trials in which the ith specimen has an associated random variable X_i, with $X_i = 1$ if the ith specimen is found to be fungus-free and $X_i = 0$ if fungus is present. Although no time sequence is involved in this case, one may still view the 30 specimen outcomes as a sequence of 30 Bernoulli trials. ◻

Bernoulli Process Postulates. Suppose $X_1, X_2, \ldots, X_n$ are n Bernoulli random variables associated with a sequence of n trials. If these n trials meet the following conditions, we shall call the sequence of trials a *Bernoulli process*:

Postulate 1. The Bernoulli random trials in the sequence are statistically independent, i.e.; the Bernoulli random variables $X_1, X_2, \ldots, X_n$ are statistically independent.

Postulate 2. The probabilities that $X_i = 1$ and $X_i = 0$ are the same for all Bernoulli trials in the sequence; i.e., that $P(X_i = 1) = p$ and $P(X_i = 0) = 1 - p$ for $i = 1, 2, \ldots, n$, where p is the common probability that the trial outcome is 1.

◻ Example

For the manufacturing process cited earlier, the first postulate requires that whether one spoon is defective or not is statistically independent of the outcomes for the other spoons in the production run. The second postulate requires that the probability of a spoon being defective is the same for all spoons produced in the run. Of course, if the probability of a defective spoon is p, since there are only two possible outcomes, the probability of a spoon not being defective is $1 - p$. ◻

The two postulates for a Bernoulli process describe two conditions that are frequently assumed for a sequence of random variables. The conditions are those of statistical independence and stationarity:

(6.1) Let $X_1, X_2, \ldots, X_n$ be a sequence of random variables associated with a random process. The process will be said to be *independent and stationary* if the random variables are statistically independent of one another and if each has the same probability distribution. Under these conditions, the random variables will be said to be *independent and identically distributed*.

Binomial Random Variable

In a sequence of Bernoulli trials, interest often focuses on how many trials result in the outcome that has been assigned the value 1. In our earlier plant specimens example, the number of specimens among the 30 which are fungus-free at the end of the test period may be the characteristic of interest. Similarly for the manufacturing process example, interest may be in the number of defective spoons in the production run of 1,000 spoons.

In the latter example, $X_i = 1$ if the ith spoon is defective and $X_i = 0$ otherwise. Hence, the sum $X_1 + X_2 + \cdots + X_{1,000}$ contains as many 1s as there are defective spoons among the 1,000, and 0s otherwise. Thus, $X_1 + X_2 + \cdots + X_{1,000}$ equals the number of defective spoons in the run of 1,000 spoons. Similarly for our plant specimens example, $X_1 + X_2 + \cdots + X_{30}$ equals the number of fungus-free specimens among the 30 in the study, since $X_i = 1$ if the ith specimen is free of fungus and $X_i = 0$ otherwise.

(6.2) Let the sum of n independent and identically distributed Bernoulli random variables be denoted by X:

$$X = X_1 + X_2 + \cdots + X_n$$

X is called a *binomial random variable*.

Binomial Probability Function

As we have seen in Chapter 5, the probability distribution for a discrete random variable X associates a probability $P(X = x)$ with each basic outcome x. Often, the probability function $P(X = x) = P(x)$ can be expressed by a mathematical formula so that the probability $P(x)$ can be calculated for each outcome x from the formula. The probability function for a binomial random variable is of this type:

(6.3) The *binomial probability function* is:

$$P(x) = \binom{n}{x} p^x (1 - p)^{n-x}$$

[handwritten: $X = $ Success]

where: $P(x) = P(X = x)$
$x = 0, 1, \ldots, n$
$0 < p < 1$

[handwritten: P^X]

Here, $\binom{n}{x}$ is a *binomial coefficient*, which is defined as:

(6.4)
$$\binom{n}{x} = \frac{n!}{x!(n - x)!}$$

where $a! = a(a - 1) \cdots (2)(1)$, and $0! = 1$. For example, we have:

$$\binom{5}{2} = \frac{5!}{2!3!} = \frac{5(4)(3)(2)(1)}{2(1)(3)(2)(1)} = 10$$

(A review of permutations and combinations can be found in Appendix A, Section A.4.)

The binomial probability distribution is a discrete probability distribution since X can take only one of the $n + 1$ values $0, 1, \ldots, n$. Once values are assigned to

p and n, we have identified one binomial probability distribution from the family of all such probability distributions; p and n are said to be the *parameters* of the binomial probability distribution.

☐ **Example**

Consider the plant specimens example presented earlier, but let us suppose that only three plant specimens are under study, i.e., $n = 3$. Assume that the postulates for a Bernoulli process are satisfied by the test situation and that $p = .2$ is the probability of any individual specimen being free of fungus at the end of the test period. We compute the binomial probability distribution for the number X of fungus-free specimens in the group of $n = 3$, using (6.3):

x		$P(x)$
0	$\binom{3}{0}(.2)^0(.8)^3 =$	.5120
1	$\binom{3}{1}(.2)^1(.8)^2 =$	.3840
2	$\binom{3}{2}(.2)^2(.8)^1 =$	.0960
3	$\binom{3}{3}(.2)^3(.8)^0 =$	.0080
	Total	1.0000

We see, for instance, that the probability of all three specimens being fungus-free is $P(3) = .0080$. We also note that the probability of one or fewer specimens being fungus-free is $P(X \le 1) = P(0) + P(1) = .5120 + .3840 = .8960$. ☐

Derivation. To derive the binomial probability function (6.3), consider again the plant specimens example in which $n = 3$ and $p = .2$. The probability tree diagram in Figure 6.1 is helpful in this connection. It shows all possible outcome sequences for the three trials. Suppose we wish to find $P(X = 1) = P(1)$. Figure 6.1 shows that there are three sequences yielding exactly one fungus-free specimen, namely, S_4, S_6, and S_7.

Let us obtain the probability that sequence S_4 occurs. Since the random variables X_1, X_2, and X_3 are independent, and since $P(X_i = 1) = .2$ and $P(X_i = 0) = .8$ for each, it follows from a generalization of definition (5.4) for independent random variables that:

$$P(S_4) = P(X_1 = 1)P(X_2 = 0)P(X_3 = 0) = .2(.8)(.8) = (.2)^1(.8)^2$$

Similarly, we find:

$$P(S_6) = (.2)^1(.8)^2$$
$$P(S_7) = (.2)^1(.8)^2$$

Finally, since S_4, S_6, and S_7 are mutually exclusive events, we have by (4.18):

$$P(X = 1) = P(1) = P(S_4) + P(S_6) + P(S_7) = 3(.2)^1(.8)^2 = .3840$$

This is the same result noted earlier.

FIGURE 6.1 *Illustration of derivation of binomial probability function for $n = 3$ and $p = .2$*

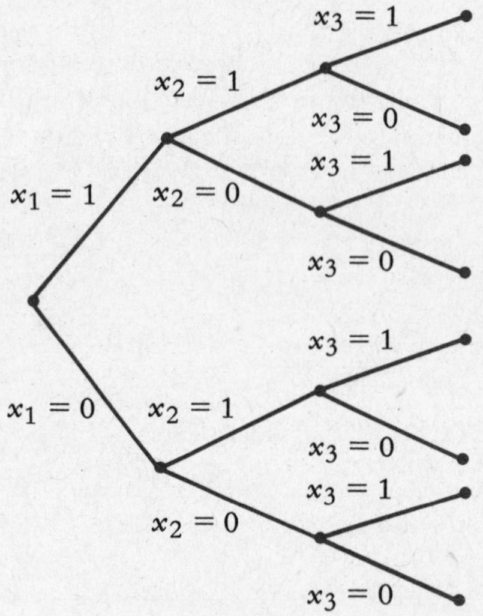

	Sequence	$x = x_1 + x_2 + x_3$	Probability of Sequence
$x_3 = 1$	S_1	3	$(.2)^3(.8)^0$
$x_3 = 0$	S_2	2	$(.2)^2(.8)^1$
$x_3 = 1$	S_3	2	$(.2)^2(.8)^1$
$x_3 = 0$	S_4	1	$(.2)^1(.8)^2$
$x_3 = 1$	S_5	2	$(.2)^2(.8)^1$
$x_3 = 0$	S_6	1	$(.2)^1(.8)^2$
$x_3 = 1$	S_7	1	$(.2)^1(.8)^2$
$x_3 = 0$	S_8	0	$(.2)^0(.8)^3$

Figure 6.1 shows the probabilities for all eight possible sample sequences. When these are combined appropriately, the binomial probabilities obtained directly from the probability function (6.3) are produced. For instance:

$$P(2) = P(S_2) + P(S_3) + P(S_5) = 3(.2)^2(.8)^1 = .0960$$

We generalize the derivation now:

1. When there are n independent Bernoulli trials, for each of which $P(X_i = 1) = p$, the probability of obtaining x 1s and $n - x$ 0s in a *specific sequence* or *permutation* is:

$$p^x(1 - p)^{n-x}$$

2. The number of sequences or distinct permutations of x 1s and $n - x$ 0s is given by the binomial coefficient $\binom{n}{x}$.

3. Combining these results, we see that $P(x)$ is simply the sum of $\binom{n}{x}$ sequence probabilities, each of which is $p^x(1 - p)^{n-x}$:

$$P(x) = \binom{n}{x}p^x(1 - p)^{n-x}$$

Probability Table

Binomial probabilities can be readily computed with many types of pocket calculators. In addition, they can be obtained from tables of binomial probabilities. Table C-5 (see Appendix C) gives binomial probabilities for selected values of p and selected values of n up to 20. For large n, approximation methods may be used. We shall take these up in Chapter 12.

☐ Examples

1. In a psychological experiment, $n = 20$ children are each asked independently to solve a puzzle in a fixed amount of time. The characteristic of interest is whether or not the subject succeeds in completing the task. Suppose the experimental conditions make it reasonable to assume that the number who succeed is a binomial random variable with $p = .4$. We wish to find the probability of exactly 6 successes. We see from Table C-5 that it is $P(6) = .1244$. This value is located in the column marked $p = .40$ at the top and in the row corresponding to $n = 20$ and $x = 6$, as labeled on the left-hand side of the table.

   If $p = .7$, we would then find $P(6) = .0002$ in the probability table. This time the value is located in the column marked $p = .70$ at the bottom and in the row corresponding to $n = 20$ and $x = 6$, as labeled on the right-hand side of the table. (Note, therefore, that the (n, x) values on the left-hand side correspond to p values of .5 or less, and those on the right-hand side correspond to p values of .5 or more.)

   To obtain cumulative binomial probabilities from Table C-5, we simply sum the appropriate individual probabilities. For example, if $P(X \leq 2)$ is desired for the case in which $p = .40$ and $n = 20$, we find that $P(X \leq 2) = P(0) + P(1) + P(2) = .0000 + .0005 + .0031 = .0036$.

2. In an insurance company, $n = 10$ claims adjusters are asked, as part of a continuing training program, to consider independently a hypothetical claim. Interest is in the number of adjusters who handle the claim correctly. Suppose it is reasonable to assume that the number who handle the claim correctly is a binomial random variable with $p = .9$. It is desired to find $P(X \leq 8)$. We use the complementation theorem (4.19) to recognize that $P(X \leq 8) = 1 - P(X \geq 9)$. From Table C-5, we find $P(X \geq 9) = P(9) + P(10) = .3874 + .3487 = .7361$. Hence, $P(X \leq 8) = 1 - .7361 = .2639$. ☐

Characteristics of Binomial Probability Distributions

Mean and Variance. The mean and variance of a binomial probability distribution are:

(6.5) $$E\{X\} = np$$

(6.6) $$\sigma^2\{X\} = np(1 - p)$$

☐ Example

The number of substandard bindings produced in a production run of $n = 3,000$ books is a binomial random variable X, with probability $p = .02$ of a substandard binding. The expected number of substandard bindings in the production run is then $E\{X\} = 3,000(.02) = 60$. The variance of the number of substandard bindings in the production run is $\sigma^2\{X\} = 3,000(.02)(.98) = 58.8$. Thus, $\sigma\{X\} = \sqrt{58.8} = 7.67$ bindings is the standard deviation of the number of substandard bindings in the run. □

Comment

The mean and variance of the binomial random variable can be readily derived. To show that $E\{X\} = np$, for instance, we first find $E\{X_i\}$ for each Bernoulli random variable. Using (5.5), we obtain $E\{X_i\} = 1(p) + 0(1 - p) = p$. Since $X = X_1 + X_2 + \cdots + X_n$, we have by (5.15a) that $E\{X\} = np$.

Distribution Shape.   The binomial probability distribution can take on a variety of shapes, as illustrated by Figure 6.2, which shows three binomial probability distributions. The distribution is skewed right if $p < .5$, is skewed left if $p > .5$, and is symmetrical if $p = .5$. The closer p is to 0 or 1, for a given n, the more pronounced is the skewness.

Sum of Binomial Random Variables

In applications where interest is in the sum of two or more independent binomial random variables, the following theorem can be helpful:

(6.7)   If V and W are two independent binomial random variables with common probability parameter p and based on n_1 and n_2 trials, respectively, the sum $V + W$ is

---

FIGURE 6.2   *Three binomial probability distributions.* The binomial distribution is skewed right when $p < .5$, skewed left when $p > .5$, and symmetrical when $p = .5$.

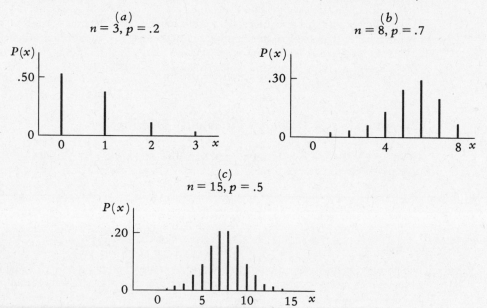

---

also a binomial random variable with parameters p and n, where $n = n_1 + n_2$, the combined number of trials.

☐   **Example**

Two chess-playing computer programs will compete against one another in two tournaments, consisting of $n_1 = 5$ and $n_2 = 7$ games, respectively. The probability that program A wins in any given game is .6 (the probability that program A draws or loses is .4). The numbers of games to be won by program A in the two tournaments are independent binomial random variables. What is the probability that program A wins seven or more games in both tournaments?

By (6.7), the total number of wins by program A in both tournaments is a binomial random variable with parameters $p = .6$ and $n = 5 + 7 = 12$. In consulting Table C-5 for $n = 12$ and $p = .6$, we find that $P(X \geq 7) = P(7) + P(8) + \cdots + P(12) = .2270 + .2128 + \cdots + .0022 = .6652.$ ☐

6.2   POISSON PROBABILITY DISTRIBUTIONS

The Poisson probability distribution applies to many random phenomena occurring in a period of time.

☐   **Examples**

1. The number of machines in a plant that break down during a day (X) is a Poisson random variable. Management wishes to know how likely it is that there are no breakdowns in a day.
2. The number of units of an item sold from stock during a week (X) is a Poisson random variable. The inventory controller wishes to know the probability that more than five units will be demanded in a week.
3. The number of persons arriving at a bank teller per quarter-hour (X) is a Poisson random variable. A management consultant needs to know for designing a staffing plan the probability that more than 10 persons will arrive in a 15-minute period.   ☐

The Poisson probability distribution also applies to other types of random phenomena such as the number of typographical errors on a page.

Poisson Probability Function

A Poisson random variable is a discrete variable which can take on any integer value from 0 on up indefinitely.

(6.8)   The *Poisson probability function* is:

$$P(x) = \frac{\lambda^x \exp(-\lambda)}{x!}$$

where: $P(x) = P(X = x)$

$x = 0, 1, \ldots, \infty$

$0 < \lambda < \infty$

[handwritten margin notes:]
X Event
λ average of past event
e = Nat lg

Here and elsewhere in the Text, $\exp(a)$ represents e^a, where $e = 2.71828...$ is the base of natural logarithms. Thus, $\exp(-\lambda)$ represents $e^{-\lambda}$. (A review of exponentiation and natural logarithms can be found in Appendix A, Section A.2.)

Even though the Poisson random variable may assume any indefinitely large integer value while real-world phenomena are bounded, the Poisson random variable is often a reasonable model, as we shall see.

The Poisson probability distribution has only one parameter, λ (Greek lambda). Each different value of λ corresponds to a different member of the family of Poisson probability distributions. Observe in (6.8) that λ may be any positive number.

☐ Example

The number of crimes occurring in a city district in the one-hour period between 1 A.M. and 2 A.M. is a Poisson random variable with $\lambda = .2$. We obtain the Poisson probability distribution in part, using (6.8):

x	$P(x)$	
0	$(.2)^0\exp(-.2)/0! = .8187$	[Note that $\exp(-.2) = .8187$]
1	$(.2)^1\exp(-.2)/1! = .1637$	
2	$(.2)^2\exp(-.2)/2! = .0164$	
3	$(.2)^3\exp(-.2)/3! = .0011$	
4	$(.2)^4\exp(-.2)/4! = .0001$	

We see, for instance, that the probability of no crime occurring during this hour is $P(0) = .8187$. Further, the probability of two or fewer crimes during the hour is $P(X \leq 2) = P(0) + P(1) + P(2) = .8187 + .1637 + .0164 = .9988$.

For outcomes of five crimes or more, the probabilities of occurrence are so small that we have not shown calculations. Here is the reason why the Poisson random variable is often a useful model for bounded phenomena; namely, because the probabilities of large values are negligible. ☐

Probability Table

Poisson probabilities can be computed readily from many pocket calculators. Extensive Poisson probability tables also exist. Table C–6 contains Poisson probabilities for selected values of λ up to 20.

☐ Examples

1. Daily demand for a certain replacement part of a cassette recorder model is Poisson-distributed with $\lambda = .9$. We see from Table C–6 that the probability of no demand for the part on a particular day is $P(0) = .4066$. This value is found in the column headed $\lambda = .9$ and the row marked $x = 0$. For cumulative probabilities, we need only add the appropriate individual probabilities. For example, we see the probability that no more than 4 replacement parts are demanded is $P(X \leq 4) = P(0) + P(1) + \cdots + P(4) = .4066 + .3659 + \cdots + .0111 = .9977$.

2. The number of accidents in an office building during a four-week period is a Poisson random variable with $\lambda = 2$. Interest is in the probability that there are one or fewer accidents in a four-week period. We find from the table that $P(X \leq 1) = P(0) + P(1) = .1353 + .2707 = .4060$. ☐

Characteristics of Poisson Probability Distributions

Mean and Variance.   The mean and variance of a Poisson probability distribution are:

(6.9) $$E\{X\} = \lambda$$

(6.10) $$\sigma^2\{X\} = \lambda$$

Thus, the mean and variance of a Poisson distribution are equal.

☐ **Example**

For the crime example, $\lambda = .2$, so $E\{X\} = .2$ crime and $\sigma^2\{X\} = .2$. The standard deviation therefore is $\sigma\{X\} = \sqrt{\lambda} = \sqrt{.2} = .447$ crime. ☐

Distribution Shape.   Figure 6.3 contains three Poisson probability distributions with parameter values $\lambda = .3$, 2.0, and 5.0, respectively. All Poisson probability distributions are skewed to the right, although they become more symmetrical as λ becomes larger.

FIGURE 6.3   *Three Poisson probability distributions.* Right-skewness decreases as λ increases.

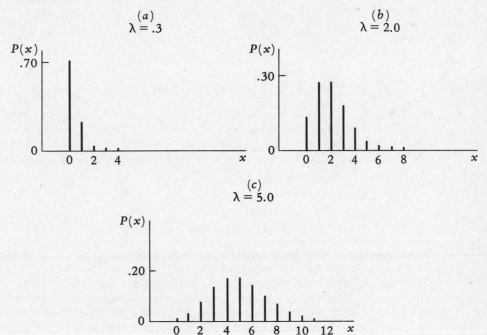

Poisson Process

To obtain an understanding of the type of random phenomena described by the Poisson distribution, it is useful to examine the formal conditions or postulates that lead to this distribution.

A Poisson distribution is concerned with the number of occurrences of some event in a fixed length of time, fixed amount of space, etc. For the sake of simplifying the exposition in setting out the postulates that follow, we shall consider phenomena or processes generating occurrences over time, such as number of phone calls received at a telephone switchboard in a 5-minute period. With minor alterations in the discussion, the postulates would apply to occurrences distributed in space, such as blemishes occurring in a roll of vinyl wallpaper or forest fires occurring in a wilderness region.

Poisson Process Postulates. Consider occurrences that happen randomly on the time continuum. Figure 6.4 illustrates three such occurrences. Imagine this continuum to be divided into many small nonoverlapping intervals of size Δt, as shown in Figure 6.4. The time between t and $t + \Delta t$ denotes a typical interval.

Postulate 1. The numbers of occurrences in nonoverlapping time intervals are statistically independent.

Postulate 2. The number of occurrences in a time interval has the same probability distribution for all time intervals.

Postulate 3. The probability of one occurrence in any time interval $(t, t + \Delta t)$ is approximately proportional to the size of the interval Δt. Specifically, if λ is the constant of proportionality (by Postulate 2, it must be the same for all time intervals), the probability of one occurrence is approximately $\lambda \Delta t$.

Postulate 4. The probability of two or more occurrences in any time interval $(t, t + \Delta t)$ is negligibly small, relative to the probability of one occurrence in the interval.

When these four postulates hold, the number of occurrences in a unit time interval follows a Poisson probability distribution with parameter λ. Incidentally, the conditions specified in Postulates 1 and 2, respectively, are those of statistical independence and stationarity. We encountered these same conditions in our discussion of the Bernoulli process.

☐ Example

Consider the number of phone calls received at a switchboard in a 5-minute period. Divide this time period into a large number of nonoverlapping intervals of size Δt—for instance, into 3,000 intervals of size $\Delta t = .1$ second. Postulate 1 requires that the number of phone calls received during each of these intervals be statistically independent of one another. Postulate 2 requires that the number of phone calls have the same probability distribution in each interval. This postulate would be violated, for instance, if some of the intervals correspond to busy periods and others to periods of slack telephone activity. Postulate 3 requires that the probability of receiving one phone call in any interval of size Δt is

FIGURE 6.4 *Illustration of a Poisson process*

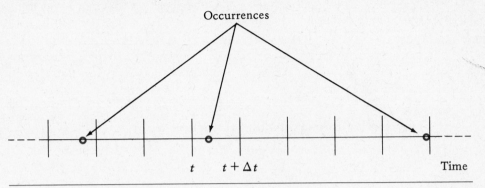

approximately proportional to Δt; for instance, the probability for $\Delta t = .1$ second must be about twice as large as the probability for $\Delta t = .05$ second. Finally, Postulate 4 requires that the probability of receiving two or more phone calls in any interval is very small. If all four of these postulates are satisfied by the telephone calling process, then the number of calls received in a time period of fixed length will be a Poisson random variable. □

Sum of Poisson Random Variables

In applications where interest is in the sum of two or more independent Poisson random variables, the following theorem is helpful:

(6.11) If V and W are two independent Poisson random variables with parameters λ_1 and λ_2, respectively, the sum $V + W$ is also a Poisson random variable, with parameter $\lambda = \lambda_1 + \lambda_2$.

□ Example

Broadcasting interruptions at a television station arise from mechanical breakdowns and human errors. The numbers of interruptions per 1,000 broadcasting hours from these sources are independent Poisson random variables with parameters $\lambda_1 = .4$ and $\lambda_2 = 1.1$, respectively. By theorem (6.11), we know that the total number of interruptions from both sources is a Poisson random variable with parameter $\lambda = .4 + 1.1 = 1.5$. Thus, the probability of no interruptions from either source in the next 1,000 hours of broadcasting is $P(0) = .2231$ (refer to Table C-6, column $\lambda = 1.5$, row $x = 0$). □

6.3 DISCRETE UNIFORM PROBABILITY DISTRIBUTIONS

A random variable that can take on integer values within a given interval with equal probabilities is useful for statistical sampling. This random variable is called a *discrete uniform random variable*.

☐ Example

X, the final digit in today's volume of stock transactions, can assume any of the 10 digits from 0 to 9, each with probability .10, i.e.:

x:	0	1	2	3	4	5	6	7	8	9
P(x):	.10	.10	.10	.10	.10	.10	.10	.10	.10	.10

☐

Discrete Uniform Probability Function

(6.12) The *discrete uniform probability function* is:

$$P(x) = \frac{1}{s}$$

where: $P(x) = P(X = x)$

$x = a, a + 1, \ldots, a + (s - 1)$; a and s are integers, with $s > 0$

The discrete uniform probability distribution has two parameters, a and s. Parameter a denotes the smallest outcome, and s denotes the number of distinct outcomes. For instance, when the possible outcomes are the integers from 0 to 9, $a = 0$ and $s = 10$. A graph of this probability distribution is given in Figure 6.5.

Characteristics of Discrete Uniform Probability Distributions

The mean and variance of a discrete uniform probability distribution are:

(6.13)
$$E\{X\} = a + \frac{s - 1}{2}$$

(6.14)
$$\sigma^2\{X\} = \frac{s^2 - 1}{12}$$

FIGURE 6.5 *Example of a discrete uniform probability distribution: a = 0, s = 10*

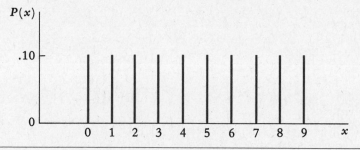

☐ Example

For the discrete uniform probability distribution presented in Figure 6.5, $a = 0$ and $s = 10$, so:

$$E\{X\} = 0 + \frac{10 - 1}{2} = 4.5 \qquad \sigma^2\{X\} = \frac{10^2 - 1}{12} = 8.25$$

Thus, the standard deviation of the distribution is $\sigma\{X\} = \sqrt{8.25} = 2.87$. ☐

6.4 OPTIONAL TOPIC—HYPERGEOMETRIC PROBABILITY DISTRIBUTIONS

The hypergeometric probability distribution is a discrete distribution which finds applications in statistical sampling. Unlike the binomial probability distribution which is applicable when independent Bernoulli trials are made, the hypergeometric distribution applies when the trials are dependent.

☐ Examples

1. A state senate consists of $N = 50$ senators, of whom $C = 10$ are women. What is the probability that a committee of $n = 5$ senators formed at random contains $X = 0$ women?

   The binomial probability distribution is not applicable here since the trials are not independent. The probability that the first senator assigned to the committee is female is 10/50. However, the probability that the second committee member selected at random is female depends on the first selection; it is 10/49 if the first committee assignment is a male senator and 9/49 if the first assignment is a female senator. Thus, the probabilities of selecting a female (or male) senator vary throughout the selection process, depending on the previous selections.

2. Of the $N = 15$ generators owned by a utility company, $C = 2$ have a defective safety valve. If $n = 4$ different generators are selected at random for inspection, what is the probability that exactly $X = 1$ of them has a defective safety valve? ☐

Hypergeometric Probability Function

A hypergeometric random variable X has possible outcomes $0, 1, \ldots,$ up to n or C, whichever is smaller (the smallest possible outcome is $n - N + C$ when $n > N - C$):

(6.15) The *hypergeometric probability function* is:

$$P(x) = \frac{\binom{C}{x}\binom{N - C}{n - x}}{\binom{N}{n}}$$

The combinations symbol $\binom{a}{b}$ is defined in (6.4). (A review of combinations can be found in Appendix A, Section A.4.)

☐ **Examples**

1. In our senate committee example, we have $N = 50$, $C = 10$, and $n = 5$, and we wish to find $P(X = 0)$. Substituting $x = 0$ into (6.15), we obtain:

$$P(0) = \frac{\binom{10}{0}\binom{40}{5}}{\binom{50}{5}} = \frac{\frac{10!}{0!10!}\frac{40!}{5!35!}}{\frac{50!}{5!45!}} = .311$$

Thus, the probability is $P(0) = .311$ that a committee of five senators formed at random will not contain any women.

2. In our utility generator example, we have $N = 15$, $C = 2$, and $n = 4$, and wish to find $P(X = 1)$. Using (6.15), we obtain for $x = 1$:

$$P(1) = \frac{\binom{2}{1}\binom{13}{3}}{\binom{15}{4}} = \frac{\frac{2!}{1!1!}\frac{13!}{3!10!}}{\frac{15!}{4!11!}} = .419$$

Thus, the probability is $P(1) = .419$ that among four generators selected at random, exactly one will be found to have a defective safety valve. ☐

Characteristics of Hypergeometric Probability Distributions

We shall denote the proportion C/N by p. In Example 1, $p = 10/50 = .20$ is the proportion of senators who are women. In Example 2, $p = 2/15 = .133$ is the proportion of reactors with a defective safety valve.

Using this notation, we can express the mean and variance of a hypergeometric probability distribution as follows:

(6.16)
$$E\{X\} = np$$

(6.17)
$$\sigma^2\{X\} = \left(\frac{N - n}{N - 1}\right)np(1 - p)$$

where: $p = C/N$

Note the similarity to the mean and variance of a binomial random variable in (6.5) and (6.6), except for the term $(N - n)/(N - 1)$ in the variance expression in (6.17).

☐ **Examples**

1. In our senate committee example, $p = .20$, $n = 5$, and $N = 50$. Hence:

$$E\{X\} = 5(.20) = 1.0 \qquad \sigma^2\{X\} = \left(\frac{50 - 5}{50 - 1}\right)(5)(.20)(.80) = .735$$

Thus, the hypergeometric distribution has mean 1.0 and standard deviation $\sigma\{X\} = \sqrt{.735} = .86$.

2. The mean and variance of the hypergeometric distribution for our utility generator example are:

$$E\{X\} = 4(.133) = .53 \qquad \sigma^2\{X\} = \left(\frac{15 - 4}{15 - 1}\right)(4)(.133)(.867) = .363$$

☐

Comment

When N is large relative to n, the hypergeometric probability function (6.15) can be approximated well by the binomial probability function (6.3).

PROBLEMS

6.1 For each of the following outcomes, indicate whether or not the random trial is a Bernoulli trial: (1) the inspection classification (accept, rework, or scrap) for a component produced on an assembly line, (2) the sex of a chick in a brood bred for egg production, (3) the weight of a package of hamburger in a meat counter of a supermarket.

6.2 There are 10 participants in a six-mile race for persons over the age of 50. Let $X_i = 1$ if the ith participant finishes the race and $X_i = 0$ otherwise.
   a. What is represented by the random variable $X = X_1 + X_2 + \cdots + X_{10}$?
   b. Under what conditions is X a binomial random variable?

*6.3 The number of inaccurate gauges in a group of four is a binomial random variable X with $n = 4$ and $p = .25$.
   a. Describe the underlying Bernoulli trial. What is the sample space of X (i.e., what are the different values that X can take on)?
   b. Obtain $P(0)$, $P(2)$, and $P(X \leq 2)$, using (6.3). Confirm your answers using Table C-5.

6.4 The number of trucks in the fleet of a hauling firm which will require major transmission work next year is a binomial random variable X with $n = 8$ and $p = .2$.
   a. Describe the underlying Bernoulli trial. What different values can X take on (i.e., what is its sample space)?
   b. Obtain $P(0)$, $P(1)$, and $P(X \leq 3)$, using (6.3). Confirm your answers using Table C-5.

*6.5 Use Table C-5 to obtain the following binomial probabilities: (1) $P(1)$ for $p = .09$, $n = 10$; (2) $P(3)$ for $p = .35$, $n = 6$; (3) $P(8)$ for $p = .95$, $n = 9$; (4) $P(X \geq 4)$ for $p = .7$, $n = 6$.

6.6 Use Table C-5 to obtain the following binomial probabilities: (1) $P(2)$ for $p = .15$, $n = 12$; (2) $P(4)$ for $p = .25$, $n = 8$; (3) $P(3)$ for $p = .90$, $n = 7$; (4) $P(X \leq 2)$ for $p = .6$, $n = 10$.

*6.7 Refer to Problem 6.3.
   a. Graph the probability distribution of X. Is the distribution symmetrical or skewed?
   b. Find $E\{X\}$, $\sigma^2\{X\}$, and $\sigma\{X\}$. In what units is $\sigma\{X\}$ expressed?

6.8 Refer to Problem 6.4.
   a. Graph the probability distribution of X. Is the distribution skewed? If so, in what direction?
   b. Find the mean and standard deviation of the probability distribution of X. In what units are these expressed?

*6.9 The probability is .01 that a watch will require repairs if it is dropped. Seven watches have just been dropped from a tray by a jewelry store clerk. Assume that the number of watches that require repair (X) is a binomial random variable.
   a. What is the probability that none of the seven watches requires repair? That at most two watches require repair?

b. Obtain the probability distribution of X and graph it. Is the distribution highly skewed?

6.10 The probability is .06 that a patient will cancel a dental appointment. Consider a group of 12 patients scheduled for appointments this morning, and let X denote the number of cancellations in this group. Assume that X is a binomial random variable.
a. What is the underlying Bernoulli trial here?
b. What is the probability that exactly 2 out of the 12 appointments will be cancelled? That 2 or fewer will?
c. What is the expected number of cancellations? What is the standard deviation of the number of cancellations?

6.11 Refer to Problem 6.10. The postulates for a Bernoulli process have been assumed to hold in this case. Explain if these are reasonable in this situation, considering each postulate in turn.

6.12 A firm has just hired four management trainees for its Houston office and five for its Dallas office. Let X and Y denote, respectively, the number of trainees who remain in their original offices for at least three years. Assume that X and Y are independent binomial random variables, each with $p = .7$.
a. What does $T = X + Y$ represent here? What are the parameters of the binomial distribution for T?
b. Find $E\{T\}$ and $\sigma\{T\}$.
c. What is the probability that all nine trainees will remain in their original offices for at least three years?
d. Would the probability distribution of T differ if six trainees had been hired by the Houston office and three by the Dallas office?

*6.13 The number of calls (X) to a police dispatcher between 8:00 P.M. and 8:30 P.M. on Fridays is Poisson-distributed with $\lambda = 3.5$.
a. Using (6.8), find the probability of no calls during this period. Of three calls. Confirm your answers using Table C–6.
b. What is the expected number of calls received by the dispatcher during this period? What is the variance of the number of calls?
c. Graph the probability distribution of X, using Table C–6 to obtain the needed probabilities. Is the distribution skewed? If so, in which direction?

6.14 The number of pinholes (X) in 1,000 feet of plastic coating for cable is Poisson-distributed with $\lambda = .9$ pinhole per 1,000 feet.
a. Using (6.8), find $P(0)$ and $P(1)$. Confirm your answers using Table C–6.
b. Obtain $E\{X\}$ and $\sigma\{X\}$. In what units are these expressed?
c. Graph the probability distribution of X, using Table C–6 to obtain the needed probabilities. Is the distribution skewed? If so, in which direction?

6.15 Refer to Problem 6.13.
a. Obtain the cumulative probability distribution for X.
b. Plot the step-function ogive for the cumulative probability distribution of X. From your graph, find the 90th percentile. What is its meaning here?

*6.16 A new word-processing unit is fully warranted during its first year. The number of warranty service calls (X) for a unit during its first year is Poisson-distributed with $\lambda = 4.0$.
a. What is the probability of exactly four warranty service calls for a unit during the first year? Of four or more warranty service calls?
b. Graph the probability distribution of X. Is the distribution skewed? If so, in which direction?

    c.   What are the mean and standard deviation of the probability distribution of X? In what units are these measures expressed?

6.17   The number of coating blemishes (X) in 10-square-meter rolls of customized wallpaper is a Poisson random variable with $\lambda = .3$.

    a.   Find: (1) $P(X = 2)$, (2) $P(X \leq 1)$, (3) $P(1 \leq X \leq 3)$.

    b.   Find $E\{X\}$ and $\sigma\{X\}$. In what units are these measures expressed?

    c.   Plot the probability distribution of X on a graph. Is the distribution skewed? If so, in which direction?

6.18   Refer to Problem 6.17. Y, the number of printing blemishes in 10-square-meter rolls of customized wallpaper, is also Poisson-distributed but with $\lambda = .1$. Y is independent of X, the number of coating blemishes.

    a.   What does the random variable $T = X + Y$ represent here?

    b.   What are the mean and standard deviation of the probability distribution of T?

    c.   What is the most probable total number of blemishes in a roll? If rolls with a total of more than one blemish are scrapped, what is the probability a roll will be scrapped?

6.19   Refer to Problem 6.16. For each of the following situations, explain which postulate of a Poisson process is violated: (1) Typically, the warranty calls for a unit are bunched toward the end of the warranty year. (2) When the production process has a serious undetected quality failure, at least 100 units are produced and sold which will require extensive warranty work before corrective action on the process is taken. (3) About half of the units have no warranty calls during the warranty year, while the other half have six or more calls.

*6.20   The number of students (X) who participate on any day in discussion in Professor Bird's class is a discrete uniform random variable with $a = 0$ and $s = 14$.

    a.   Obtain each of the following probabilities: (1) $P(0)$, (2) $P(X > 10)$, (3) $P(1 \leq X \leq 5)$.

    b.   What is the expected number of students who participate in discussion on any day? What is the standard deviation of X?

    c.   Graph the probability distribution of X.

6.21   The calculation of a home insurance premium involves seven consecutive computational steps. An auditor is reviewing a large number of premium calculations that contain an error. Let X denote the step at which the error is made. Assume X has a discrete uniform probability distribution.

    a.   Graph the probability distribution of X.

    b.   On the average, how many steps must the auditor check in each premium calculation to find the error?

    c.   Obtain $\sigma\{X\}$. In what units is this measure expressed?

6.22   Refer to Problem 6.21.

    a.   If locations of the errors for different premium calculations are statistically independent, what is the probability that two premium calculations have errors that occur at step 4?

    b.   Would you expect an error to occur with equal probability at each step in actuality? Explain.

*6.23   An accounting department has six tenured and four untenured faculty members. A committee of $n = 3$ members is to be selected at random. Let X denote the number of committee members who are tenured.

    a.   What is the probability that : (1) exactly one of the committee members is tenured, (2) the number of tenured members on the committee exceeds the number of untenured members?

    b. Obtain the expected number of committee members who are tenured. Also obtain the standard deviation of X.

6.24 Refer to Problem 6.23. Answer all parts but assume the department contains five tenured and three untenured faculty members.

EXERCISES

6.25 Let X_i $(i = 1, 2, \ldots, n)$ be independent and identically distributed Bernoulli random variables with $P(X_i = 1) = p$.
    a. Prove that $\sigma^2\{X_i\} = p(1 - p)$.
    b. Use the result in **a** to verify (6.6).

6.26 Let X be a binomial random variable with $p = .3$. Plot the probability distributions of X for $n = 5$ and $n = 20$ on two graphs, one below the other. What effect does the increase in n have on the shape of the distribution?

6.27 For binomial random variable X with $n = 20$, obtain $\sigma\{X\}$ for $p = 0, .1, \ldots, .9$, 1.0. Plot these $\sigma\{X\}$ values on a graph as a function of p. What effect does the magnitude of p have on the variability of the probability distribution?

6.28 Each use of an electrical switch is an independent Bernoulli trial in which the switch either functions ($X = 0$) or fails ($X = 1$). Consider the associated Bernoulli process and let Y be a random variable representing the number of trials up to and including the trial in which the switch fails for the first time. Show that the probability function of Y is:

$$P(Y = y) = p(1 - p)^{y-1} \qquad y = 1, 2, \ldots$$

6.29 Prove (6.9). [*Hint:* $\exp(\lambda) = \sum_{x=0}^{\infty} \lambda^x/x!$.]

STUDIES

6.30 A glass company has received an order for two large lenses that must be specially cast. The probability that a given lens will prove to be acceptable (i.e., free from flaws) when the glass has cooled is .6. The number of acceptable lenses (X) in a casting run of n lenses is a binomial random variable. Management wishes to know how many lenses to cast in the run to have a probability of at least .95 that a minimum of two lenses are acceptable. Find the smallest size of the casting run that will provide this assurance.

6.31 Ten persons in a taste-test experiment will be served a portion of sausage prepared according to the current recipe (A) and another portion prepared from a recipe designed to give the same taste as the current recipe but having a longer shelf life (B). The order of the two servings will be randomized. Each person must independently identify which of the two servings is the preferred one. Let X denote the number of persons of the 10 who prefer serving B.
    a. Describe the underlying Bernoulli trial here. Does X necessarily follow a binomial distribution here?

b. Suppose sausage B has exactly the same taste as sausage A, so that a person will prefer A or B with equal probability. What is the probability, then, that the proportion X/n of persons in the study who prefer B is not less than .4 or greater than .6? How much larger would this probability be if 20 persons were included in the testing group? What are the implications of these results?

6.32 The number of typographical errors (X) on a page is a Poisson random variable with $\lambda = .8$.

a. Obtain the expected number of errors on a page and the coefficient of variation of the probability distribution of X.

b. If the numbers of errors on different pages are statistically independent, what is the expected total number of errors on 400 pages? What is the coefficient of variation of the probability distribution for the total number of errors on 400 pages? [*Hint:* Theorem (6.11) generalizes directly to more than two independent Poisson random variables.]

c. Based on your results in **a** and **b**, what is the effect on the relative variability of the probability distribution of the total number of errors as the number of pages increases? Discuss.

6.33 The daily number of disabling breakdowns in cabs of a taxi fleet (X) follows a Poisson distribution with $\lambda = .7$. The loss of revenue to the company on the day of the breakdown of a taxi is $100 if a standby cab is not available and $0 if a standby cab is available. Disabled cabs can be repaired by the next day. Standby cabs (without drivers) can be leased for $35 per day. Assume that the probability of a breakdown in a standby cab is negligibly small.

a. Let C denote the daily revenue loss from breakdowns plus the cost of leasing, and let m denote the number of standby cabs leased for the day. Express C as a function of X and m.

b. Evaluate $E\{C\}$ for $m = 0, 1, 2,$ and 3.

c. How many standby cabs should the company have on hand at the beginning of a day to minimize the expected value of C?

7

Common Continuous Probability Distributions

In this chapter, we take up several important families of continuous probability distributions. Recall from Chapter 5 that the probability that a continuous random variable X takes on a value in a specified interval is found by determining the corresponding area under its probability density function $f(x)$.

7.1 CONTINUOUS UNIFORM (RECTANGULAR) PROBABILITY DISTRIBUTIONS

The discrete uniform probability distribution discussed in the last chapter has a continuous analog known as the *continuous uniform* or *rectangular probability distribution*. This distribution has uniform probability density over an interval. The rectangular probability distribution is useful in a variety of situations, including as an approximation of the discrete uniform probability distribution.

Density Function

(7.1) The *continuous uniform probability density function* is:

$$f(x) = \frac{1}{b - a}$$

where: $a \leq x \leq b$

□ Example

An analyst has concluded that a continuous uniform probability distribution is a good approximation for the distribution of incomes of taxpayers in the $5,000–$7,500 income bracket. The density function here is:

$$f(x) = \frac{1}{2,500} \qquad 5,000 \leq x \leq 7,500$$

FIGURE 7.1   *Example of uniform probability distribution: a -- 5,000, b = 7,500*

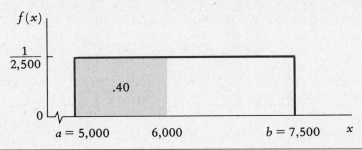

In this example, $b = 7,500$ and $a = 5,000$. A graph of this probability distribution is shown in Figure 7.1.   □

Note that the continuous uniform probability distribution has two parameters, a and b. The smallest value the random variable can assume is a and the largest is b. As the graph in Figure 7.1 illustrates, the use of the term *rectangular* in the name of this distribution is quite descriptive.

Characteristics of Uniform Probability Distributions

The mean and variance of a uniform probability distribution are:

(7.2)
$$E\{X\} = \frac{b + a}{2}$$

(7.3)
$$\sigma^2\{X\} = \frac{(b - a)^2}{12}$$

□   Example

The continuous uniform probability distribution in Figure 7.1 has mean and variance:

$$E\{X\} = \frac{7,500 + 5,000}{2} = 6,250 \qquad \sigma^2\{X\} = \frac{(7,500 - 5,000)^2}{12} = 520,833$$

The standard deviation is $\sigma\{X\} = \sqrt{520,833} = 721.7$.   □

Determining Probabilities and Percentiles

Any desired probability or percentile for a uniform probability distribution may be found by using its cumulative probability function:

(7.4)
$$F(x) = P(X \leq x) = \frac{x - a}{b - a}$$

where: $a \leq x \leq b$

□ Example

We wish to find for our taxpayers' incomes example the probability that a taxpayer's income in the $5,000–$7,500 income bracket is $6,000 or less. We find:

$$F(6,000) = P(X \leq 6,000) = \frac{6,000 - 5,000}{7,500 - 5,000} = .40$$

This probability is shown as the shaded area in Figure 7.1. Note that the 40th percentile of incomes of taxpayers in this bracket is $6,000.　□

7.2　NORMAL PROBABILITY DISTRIBUTIONS

The family of normal probability distributions is one of the most important in statistics. Many types of random trials that arise in applications involve the *normal random variable*.

□ Examples

1. The temperature X at noon on August 15 in a southeastern city is a normal random variable. A weather forecaster wishes to know the probability that the noon temperature exceeds 35°C.
2. The weight X of a metal ingot produced in a smelter is a normal random variable. An industrial engineer needs to know the probability that the weight of an ingot will be between 500 and 540 pounds.
3. The height X of women who are 20-29 years old is a normal random variable. A clothing designer wishes to know the proportion of women in this age group who are less than five feet tall.　□

In addition to the uses of the normal distribution to model many phenomena, it will be seen in later chapters that the normal distribution is also frequently utilized in drawing inferences from data.

Density Function

The normal random variable is a continuous random variable that may take on any value between $-\infty$ and $+\infty$. While real-world phenomena are bounded in magnitude, nevertheless the normal random variable, which is not bounded, is often a good model for bounded phenomena—as we shall explain.

(7.5)　The *normal probability density function* is:

$$f(x) = \frac{1}{\sqrt{2\pi}\sigma} \exp\left[-\frac{1}{2}\left(\frac{x-\mu}{\sigma}\right)^2\right]$$

where:　$-\infty < x < \infty$
$-\infty < \mu < \infty$
$\sigma > 0$
$\pi = 3.14159\ldots$

The notation exp(a) represents e^a, where $e = 2.71828\ldots$ is the base of natural logarithms. (A review of exponentiation and natural logarithms can be found in Appendix A, Section A.2.)

Characteristics of Normal Probability Distributions

Distribution Shape. The normal probability distribution has two parameters, μ (Greek mu) and σ (Greek sigma), with σ positive. Each different pair of (μ, σ) values corresponds to a different probability distribution of the family of normal distributions. Every normal distribution is bell-shaped and symmetrical, as illustrated in Figure 7.2. Each normal probability distribution is centered at μ, which is the mean of the distribution and determines its position on the x-axis. The parameter σ is the standard deviation of the probability distribution and determines the spread of the distribution—the larger the value of σ, the more spread out is the distribution. The examples in Figure 7.2 illustrate these facts. The distributions in Figures 7.2a and 7.2b have the same standard deviation but different means, while the distributions in Figures 7.2b and 7.2c have the same mean but different standard deviations.

Figure 7.2 also shows that almost all of the probability in a normal probability

FIGURE 7.2 *Three normal probability distributions.* The mean μ determines the location and the standard deviation σ the variability of the distribution

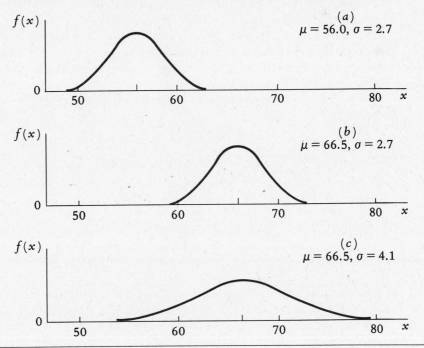

distribution is located in a limited range about its mean, so the probability of an observation falling outside this range is practically zero. It is for this reason that the normal distribution can be used to model bounded phenomena, such as heights of persons and tensile strengths of iron bars.

Mean and Variance. As the foregoing discussion has indicated, the mean and variance of a normal probability distribution are:

(7.6)
$$E\{X\} = \mu$$

(7.7)
$$\sigma^2\{X\} = \sigma^2$$

Note that we have used the same Greek symbol for the variance operator, $\sigma^2\{X\}$, as for the variance of a normal distribution, σ^2. The context and notation will make it clear whether the symbol refers to the variance operator or the variance parameter.

$N(\mu, \sigma^2)$ Notation. The normal distribution is used frequently throughout the Text, so we shall adopt the compact notation $N(\mu, \sigma^2)$ to denote a normal distribution with mean μ and variance σ^2. Thus, when we state that X is distributed as $N(100, 20)$, we mean that X is a normal random variable with mean $\mu = 100$ and variance $\sigma^2 = 20$.

Standard Normal Probability Distribution

The *standard normal probability distribution* is a particular member of the family of normal distributions, namely, that normal distribution which has a mean of 0 and a standard deviation of 1; i.e., $\mu = 0$ and $\sigma = 1$. The normal random variable corresponding to the standard normal distribution is called the *standard normal variable*, and is denoted by Z. Thus, in terms of our compact notation, Z is distributed as $N(0, 1)$.

The significance of the standard normal distribution results from an important theorem:

(7.8) Any linear function of a normal random variable is also a normal random variable.

The foregoing theorem permits any normal distribution to be transformed into the standard normal distribution, and hence the probability table for the standard normal distribution may be used for all normal distributions.

Any normal random variable can be transformed into a standard normal variable by utilizing the standardized form of a random variable as given in (5.12):

(7.9)
$$Z = \frac{X - \mu}{\sigma}$$

We know from our earlier discussion that for any standardized variable Z, we have $E\{Z\} = 0$ and $\sigma^2\{Z\} = 1$. Theorem (7.8) tells us that Z is normally distributed when X is a normal random variable. Thus, Z here is a standard normal random variable.

Determining Probabilities and Percentiles for Standard Normal Distribution

Standard Normal Probability Table. As we mentioned earlier, probabilities for any normal random variable can be obtained from a table for the standard normal variable Z. We now explain Table C-1, the probability table for the standard normal variable, and then consider how this table is used to obtain probabilities for any normal random variable.

In accordance with our earlier notation, we employ the lowercase letter z to denote a particular outcome of the standard normal random variable Z. The row and column labels in Table C-1 specify different z outcomes, the row label providing the first decimal place value, and the column label the second decimal place value. Each cell entry for a given value z is the cumulative probability $P(Z \leq z)$. This cumulative probability is represented by the shaded area labeled a in the figure at the top of Table C-1. The percentile corresponding to cumulative area a is denoted by $z(a)$.

We can use Table C-1 in two ways: (1) to determine the area a that corresponds to a specified value of z, and (2) to find the percentile $z(a)$ that corresponds to a specified cumulative probability a. We begin by explaining the first use.

Probabilities. Figure 7.3 contains several examples of finding probabilities for the standard normal variable Z.

☐ Examples

1. We wish to find $P(Z \leq .45)$; see Figure 7.3a. In entering Table C-1 at the row labeled .4 and the column labeled .05, we find for $z = .45$ the cumulative area $a = .6736$. Hence, $P(Z \leq .45) = .6736$.

    Note, therefore, that $z = .45$ is the 67.36th percentile of the standard normal distribution.

2. We wish to find $P(Z \geq 1.00)$; see Figure 7.3b. Table C-1 shows areas to the left of any z value. To find areas to the right, we use the complementation theorem (4.19):

$$P(Z \geq 1.00) = 1 - P(Z < 1.00)$$

    Remember that $P(Z < 1.00) = P(Z \leq 1.00)$ because Z is a continuous random variable. From Table C-1, we find that $P(Z < 1.00) = .8413$. Therefore, $P(Z \geq 1.00) = 1 - .8413 = .1587$.

3. We wish to find $P(Z \leq -1.00)$; see Figure 7.3c. Table C-1 does not show probabilities for negative values of z. The reason is the symmetry of the standard normal distribution about its mean 0. From this symmetry, we have:

$$P(Z \leq -1.00) = P(Z \geq 1.00)$$

    Using the result from the previous example, we conclude that $P(Z \leq -1.00) = .1587$.

4. We wish to find $P(-1.00 \leq Z \leq 1.00)$; see Figure 7.3d. We know from Example 2 that $P(Z \leq 1.00) = .8413$. Further, we know from Example 3 that $P(Z < -1.00) = .1587$. Hence:

$$P(-1.00 \leq Z \leq 1.00) = P(Z \leq 1.00) - P(Z < -1.00)$$
$$= .8413 - .1587 = .6826$$

☐

FIGURE 7.3 *Determining probabilities for the standard normal distribution*

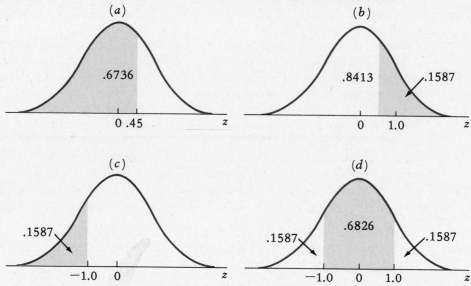

Percentiles. We stated earlier that $z(a)$ denotes the value on the z scale to the left of which the cumulative probability is a, as illustrated in Figure 7.4a. Thus, $z(a)$ is the $100a$ percentile of the standard normal distribution:

(7.10)
$$P[Z \leq z(a)] = a$$

☐ Examples

1. We wish to find the 67.36th percentile, i.e., $z(.6736)$. We found earlier in Table C–1 that for $z = .45$, the cumulative area to the left is .6736. Hence, $z(.6736) = .45$, so .45 is the 67.36th percentile.

2. We wish to find $z(.96)$, the 96th percentile. The cell entry nearest .96 in Table C–1 is $a = .9599$. The corresponding z value is 1.75 because entry .9599 is found in the row labeled 1.7 and the column labeled .05. Thus, $z(.96) = 1.75$. This percentile is illustrated in Figure 7.4b.

3. We wish to find $z(.95)$, the 95th percentile. The cell entries nearest .95 in Table C–1 are $a = .9495$ and $a = .9505$. The corresponding z values are 1.64 and 1.65, respectively. In interpolating linearly, we obtain 1.645 for the 95th percentile; i.e., $z(.95) = 1.645$.

☐

In the preceding examples we found percentiles above the 50th. Because of the symmetry of the standard normal distribution about its mean 0, percentiles below 50 are related to those above 50 in the following fashion:

(7.11)
$$z(a) = -z(1 - a)$$

☐ Example

We wish to find the 4th percentile, i.e., $z(.04)$. By theorem (7.11), we know that

FIGURE 7.4 *Determining percentiles for the standard normal distribution*

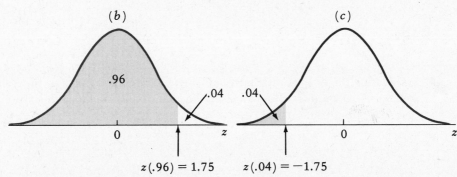

$z(.04) = -z(.96)$. In Example 2, we found $z(.96) = 1.75$; hence, $z(.04) = -1.75$, as illustrated in Figure 7.4c. □

Some percentiles of the standard normal distribution are used so frequently that they are shown separately in part b of Table C-1. These percentiles may be obtained from the cumulative probabilities in part a of Table C-1 in the manner just illustrated, although not to the accuracy shown in part b of Table C-1.

Examples

1. We wish to find $z(.98)$. From the table of selected percentiles, we obtain $z(.98) = 2.054$.
2. We wish to find $z(.001)$. From the table of selected percentiles, we obtain $z(.001) = -3.090$. □

Determining Probabilities and Percentiles for Any Normal Distribution

The standard normal probability table may be used to calculate probabilities for any normal random variable X by transforming it into the standard normal variable Z, using the standardizing transformation (7.9). We illustrate the procedure by

FIGURE 7.5 *Relationship between normal distribution with* $\mu = 520$ *and* $\sigma = 11$
and standard normal distribution

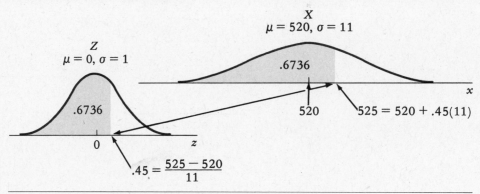

considering the normal distribution in Figure 7.5, which refers to the weight X of
a metal ingot produced in a smelter. The mean and standard deviation of X are
$\mu = 520$ pounds and $\sigma = 11$ pounds, respectively.

☐   Examples

1. We wish to find $P(X \leq 525)$. Figure 7.5 illustrates the relation between the area to the
   left of 525 and the corresponding area in the standard normal distribution. The value z
   for the standard normal distribution corresponding to $x = 525$ is obtained by the standardizing transformation (7.9):

$$z = \frac{x - \mu}{\sigma} = \frac{525 - 520}{11} = .45$$

   This indicates that 525 is .45 standard deviation above the mean.
       Consequently, $P(X \leq 525) = P(Z \leq .45)$, as shown in Figure 7.5. We can find this
   probability directly from Table C-1. Actually, we found this probability earlier in
   Figure 7.3a, where it equals .6736. Hence, the probability that an ingot will weigh 525
   pounds or less is $P(X \leq 525) = .6736$.

2. We wish to find $P(509 \leq X \leq 531)$. We calculate the z values corresponding to the
   specified values of x as follows:

$$z = \frac{x - \mu}{\sigma} = \frac{531 - 520}{11} = +1.00 \qquad z = \frac{x - \mu}{\sigma} = \frac{509 - 520}{11} = -1.00$$

   Hence, $P(509 \leq X \leq 531) = P(-1.00 \leq Z \leq +1.00)$. Figure 7.6a illustrates the correspondence between the probability distributions of X and Z by the alignment of the
   two scales. The desired probability equals .6826, and was found earlier in Figure 7.3d.
   Hence, the probability that an ingot weighs between 509 and 531 pounds is
   $P(509 \leq X \leq 531) = .6826$.

3. We wish to find the central 98 percent probability limits of ingot weight. We see from
   Figure 7.6b that we must find the 1st and 99th percentiles of ingot weight. From part b
   of Table C-1, we find that the 99th percentile of the standard normal distribution is

FIGURE 7.6 *Determining probabilities for a normal probability distribution with* $\mu = 520$ *and* $\sigma = 11$

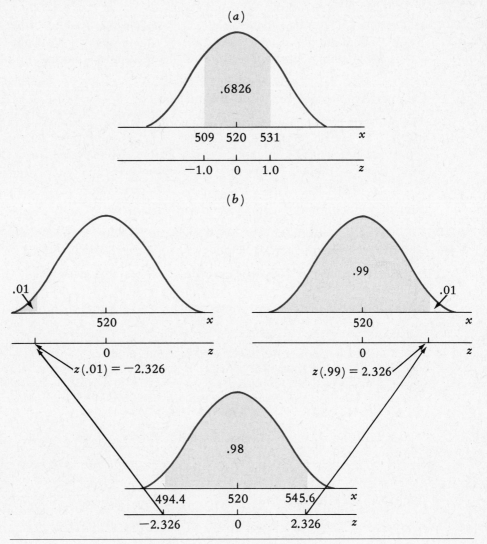

$z(.99) = 2.326$ and that the 1st percentile is $z(.01) = -2.326$. The weight limits are therefore located 2.326 standard deviations on either side of the mean; i.e., at the weights:

$$520 - 2.326(11) = 494.4 \text{ pounds}$$

$$520 + 2.326(11) = 545.6 \text{ pounds}$$

In other words, $P(494.4 \leq X \leq 545.6) = .98$.

Three Important Normal Probabilities. Three sets of central probability limits for the normal distribution are used so frequently that it is worthwhile recording them. They are:

1. $\mu \pm 1\sigma$ contains about 68.3 percent of the area under a normal distribution.
2. $\mu \pm 2\sigma$ contains about 95.4 percent of the area under a normal distribution.
3. $\mu \pm 3\sigma$ contains about 99.7 percent of the area under a normal distribution.

Cumulative Normal Probability Function

The cumulative probability function of a normal random variable is illustrated in Figure 7.7. Note the S-shaped appearance of the cumulative probability curve. The function is drawn by obtaining a series of cumulative probabilities of the form $P(X \leq x)$ from Table C–1 and then plotting these.

Sum of Independent Normal Random Variables

When we work with the sum of two or more independent normal random variables, the following theorem is helpful:

(7.12) If V and W are two independent normal random variables with means μ_1 and μ_2 and variances σ_1^2 and σ_2^2, respectively, the sum $V + W$ is also a normal random variable, with mean $\mu = \mu_1 + \mu_2$ and variance $\sigma^2 = \sigma_1^2 + \sigma_2^2$.

□ **Example**

The material cost for a construction project is a normal random variable V with mean $\mu_1 = \$60$ million and standard deviation $\sigma_1 = \$4$ million. The labor cost for this same project is an independent normal random variable W with mean $\mu_2 = \$20$ million and standard deviation $\sigma_2 = \$3$ million. What is the probability that the total cost of material and labor for this project will not exceed $85 million?

FIGURE 7.7 *Cumulative probability function for normal distribution*

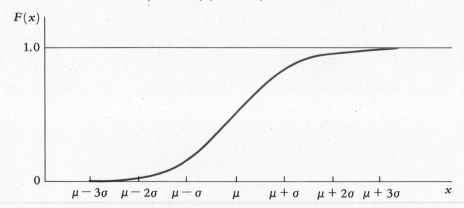

By (7.12), the total cost is a normal random variable with mean $\mu = 60 + 20 = \$80$ million and variance $\sigma^2 = (4)^2 + (3)^2 = 25$; hence, the standard deviation is $\sigma = \sqrt{25} = \$5$ million. We employ the standardizing transformation (7.9) and obtain $z = (85 - 80)/5 = 1.00$. From Table C–1, we see that $P(Z \leq 1.00) = .8413$. Hence, the probability is .8413 that the total cost of material and labor will not exceed $85 million. ☐

7.3 EXPONENTIAL PROBABILITY DISTRIBUTIONS

Exponential probability distributions are an important family of probability distributions useful in describing duration phenomena.

☐ Examples

1. A police department has been criticized about the length of time required by police cars to answer calls. The mayor wishes to know what percent of calls are answered within 15 minutes. Here X, the number of minutes required to answer a call, has an exponential distribution. The desired probability is $P(X \leq 15)$.
2. The length of a long-distance telephone call (X) has an exponential distribution. An analyst with a telephone company would like to know the proportion of calls completed within three minutes for the purpose of rate setting. She therefore requires $P(X \leq 3)$. ☐

Density Function

The *exponential random variable* is a continuous random variable that may take on any positive value:

(7.13) The *exponential probability density function* is:

$$f(x) = \lambda \exp(-\lambda x)$$

where: $0 < x < \infty$
$\lambda > 0$

The notation $\exp(a)$ represents e^a, where $e = 2.71828\ldots$ is the base of natural logarithms. Thus, $\exp(-\lambda x)$ represents $e^{-\lambda x}$. (A review of exponentiation and natural logarithms can be found in Appendix A, Section A.2.)

λ is the only parameter of the exponential probability distribution. It is always positive. Each different value of λ yields a different probability distribution of the exponential family.

☐ Example

The number of minutes (X) required by a police patrol car to respond to a call is an exponential random variable with $\lambda = .2$. The corresponding exponential density function is:

$$f(x) = .2 \exp(-.2x)$$ ☐

FIGURE 7.8 *Two exponential probability distributions.* The larger is λ, the smaller are the mean and variability of the distribution.

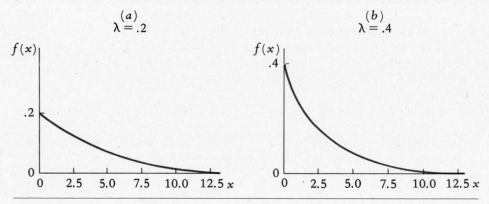

Characteristics of Exponential Probability Distributions

The mean and variance of an exponential probability distribution are:

(7.14)
$$E\{X\} = \frac{1}{\lambda}$$

(7.15)
$$\sigma^2\{X\} = \frac{1}{\lambda^2}$$

Note that the standard deviation for an exponential distribution is $\sigma\{X\} = 1/\lambda$ and is equal to the mean.

☐ Example

In our police car example, the mean time required by a patrol car to respond to a call is $E\{X\} = 1/.2 = 5$ minutes. The standard deviation of the time required is the same, i.e., $\sigma\{X\} = 5$ minutes, and hence the variance is $\sigma^2\{X\} = (5)^2 = 25$. ☐

Figure 7.8 contains two exponential probability distributions, corresponding to parameter values $\lambda = .2$ and $\lambda = .4$, respectively. Note that the exponential distribution is markedly right-skewed, and that the probability density decreases steadily as x increases. Finally, note that the larger the value of λ, the less spread out is the distribution and the closer is the mean to the origin.

Determining Probabilities and Percentiles

Any desired probability or percentile for an exponential random variable may be found by using its cumulative probability function:

(7.16)
$$F(x) = P(X \leq x) = 1 - \exp(-\lambda x)$$

where: $0 < x < \infty$

The exponential cumulative probability function for $\lambda = .2$ is shown in Figure 7.9.

FIGURE 7.9   *Cumulative probability function for exponential distribution with* $\lambda = .2$

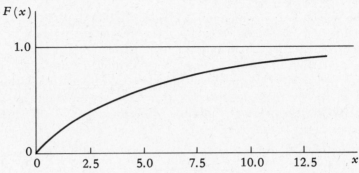

☐   Example

For our police car example, we wish to find the probability that a car's response time is within 15 minutes. This probability equals:

$$P(X \le 15) = 1 - \exp[-.2(15)] = .9502$$

In words, the probability is .95 that a police car's response time is within 15 minutes.   ☐

Exponential cumulative probabilities are also tabulated. Table C–7 presents cumulative exponential probabilities for different values of λx.

☐   Examples

1. For our police car example, we shall obtain the probability that a patrol car's response time will be less than or equal to 15 minutes. Since $x = 15$ and $\lambda = .2$ in this case, we enter Table C–7 at $\lambda x = .2(15) = 3$. The cumulative probability is found to be .9502, so $P(X \le 15) = .9502$.
2. The waiting time (in minutes) at a ticket counter is an exponential variable with $\lambda = .5$. We wish to find the probability that a customer has to wait more than 5 minutes. Here, $\lambda x = .5(5) = 2.5$. We obtain:

$$P(X > 5) = 1 - P(X \le 5) = 1 - .9179 = .0821$$

3. In Example 2, we also wish to find the 95th percentile of the exponential distribution with $\lambda = .5$. Referring to Table C–7, we find that the cell entry .9502 corresponds to $\lambda x = 3.00$. Since $\lambda = .5$, we have $.5x = 3.00$ or $x = 6$ minutes. Hence, the 95th percentile waiting time is 6 minutes.   ☐

Comment

(Calculus needed.) The exponential cumulative probability function is readily derived from its probability function (7.13) as follows:

(7.17)
$$F(x) = \int_0^x \lambda \exp(-\lambda u)\, du = 1 - \exp(-\lambda x)$$

7.4 OPTIONAL TOPIC—RELATION BETWEEN POISSON AND EXPONENTIAL DISTRIBUTIONS

Poisson and exponential probability distributions both pertain to occurrences generated by a Poisson process and are closely related. Figure 7.10 shows a time scale in which occurrences are indicated by small circles. The time scale is marked off in unit intervals. In the first interval there are three occurrences, in the second there is one, and so on. The lengths of time between consecutive occurrences are denoted by x_1, x_2, etc., and are shown by braces. Statistical theory gives us the following relationship between the number of occurrences and the times between occurrences in a Poisson process:

(7.18) If occurrences are generated by a Poisson process (i.e., the numbers of occurrences in the different unit intervals are independent and have the same Poisson distribution), then the lengths of time between successive occurrences are statistically independent random variables from an exponential distribution. Moreover, the Poisson process and exponential distribution will have the same parameter value λ.

The last sentence of theorem (7.18) implies that if the mean number of occurrences per time interval is λ (the Poisson parameter), the mean length of time between successive occurrences is $1/\lambda$.

☐ Example

The numbers of accidents in a plant each month are statistically independent outcomes from a Poisson distribution with mean $E\{X\} = \lambda = 1.5$. The lengths of time between accidents are then statistically independent outcomes from an exponential distribution with mean $E\{X\} = 1/\lambda = 1/1.5 = .67$ month. ☐

FIGURE 7.10 *Relation between Poisson and exponential probability distributions.* The numbers of occurrences in the unit intervals are independent Poisson variables with parameter λ. The times between successive occurrences are independent exponential variables with parameter λ.

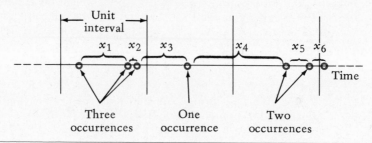

PROBLEMS

*7.1 A team preparing a bid on an excavation project assesses that the lowest competitive bid (X) is a continuous uniform random variable with $a =$ $250 thousand and $b =$ $300 thousand.
  a. State the density function for X and plot it on a graph.
  b. Find the mean and variance of the probability distribution of X.
  c. What is the probability that the lowest competitive bid is between $250 thousand and $270 thousand? Between $275 thousand and $300 thousand?

7.2 A service station is located at one end of a four-kilometer tunnel. Any vehicle breakdown in the tunnel that requires servicing is equally likely to be at any point in the tunnel. Let X denote the position of a disabled vehicle (measured in kilometers from the service station).
  a. Graph the probability distribution of X.
  b. What is the probability that a service call to the tunnel will involve a vehicle located in the first two kilometers? Located between the first and third kilometers?
  c. Obtain $E\{X\}$ and $\sigma\{X\}$. In what units are these measures expressed?

*7.3 a. Identify the probability distributions denoted by: (1) $N(200, 100)$, (2) $N(0, 100)$, (3) $N(5, 1)$.
  b. For each of the normal random variables in a, obtain the expression (7.9) for the standardized normal variable Z.

7.4 a. Identify the probability distributions denoted by: (1) $N(40, 400)$, (2) $N(0, 400)$, (3) $N(-3, 1)$.
  b. For each of the normal random variables in a, obtain the expression (7.9) for the standardized normal variable Z.

*7.5 Consider the standard normal variable Z and Table C-1.
  a. Find the areas a that correspond to the z values: (1) 0, (2) .42, (3) 1.06, (4) 2.50.
  b. Find the following probabilities: (1) $P(Z \geq 0)$, (2) $P(Z \geq 1.50)$, (3) $P(Z < -1.25)$.
  c. Find the following percentiles from Table C-1a: (1) $z(.5000)$, (2) $z(.9066)$, (3) $z(.0934)$.
  d. Find the following percentiles from Table C-1b: (1) $z(.99)$, (2) $z(.05)$, (3) $z(.975)$.

7.6 Consider the standard normal variable Z and Table C-1.
  a. Find the areas a that correspond to the z values: (1) .57, (2) 1.73, (3) 2.46.
  b. Find the following probabilities: (1) $P(Z \leq 0)$, (2) $P(Z \geq 1.41)$, (3) $P(Z \leq -2.40)$.
  c. Find the following percentiles from Table C-1a: (1) $z(.6179)$, (2) $z(.8159)$, (3) $z(.1841)$.
  d. Find the following percentiles from Table C-1b: (1) $z(.95)$, (2) $z(.025)$, (3) $z(.995)$.

*7.7 The weekly total long-distance telephone charge (X) for a travel agency is normally distributed with $\mu =$ $2,800 and $\sigma =$ $150.
  a. For a randomly selected week, what is the probability the total charge will be: (1) less than $2,950, (2) more than $3,000, (3) between $2,560 and $3,040?
  b. Find the 75th percentile of the probability distribution and interpret its meaning here.

c. Within what interval centered about μ will the weekly total charge fall with probability .50?

7.8 Refer to Problem 7.7. Answer all parts using $\mu = \$2,700$ and $\sigma = \$125$.

7.9 The duration (X) of a flight between two cities is normally distributed with mean $\mu = 3.6$ hours and standard deviation $\sigma = .15$ hour.
   a. Find the following probabilities: (1) $P(X \le 4.0)$, (2) $P(3.2 \le X \le 4.0)$, (3) $P(X > 3.7)$.
   b. Find the 80th percentile of the probability distribution. Interpret its meaning here.
   c. During 90 percent of the flights, within what interval centered about μ will the flight duration fall?

7.10 The temperature (X) at noon on August 15 in a southeastern city is normally distributed with $\mu = 30.1\,°C$ and $\sigma = 2.3\,°C$. What is the probability the noon temperature will exceed $35.0\,°C$? What is the 95th percentile of the probability distribution?

7.11 In restoring the exterior of a historic mansion, the required number of carpentry manhours (X) is assessed to be $N(265, 144)$. The required number of painting manhours (Y) is assessed to be $N(208, 25)$. X and Y are assumed to be independent random variables.
   a. What does $T = X + Y$ represent here? Describe the probability distribution of T.
   b. The labor cost of either type of work is $\$9$ per hour. A sum of $\$4,230$ has been budgeted for the total labor cost of carpentry and painting. What is the probability that the total labor cost will not exceed the budget?

*7.12 The time required to check out a customer at a supermarket (X) follows an exponential distribution with mean of 200 seconds.
   a. What is the value of the parameter λ here?
   b. Obtain each of the following probabilities: (1) $P(X \le 100)$, (2) $P(X \le 250)$, (3) $P(X > 400)$.
   c. Obtain $\sigma\{X\}$.
   d. Sketch the cumulative probability function. How can you tell from it that the probability distribution is skewed?

7.13 Refer to Problem 7.12. Answer all parts assuming that the mean is 250 seconds.

7.14 The waiting time $(X$, in minutes) to place a call on an office long-distance telephone system with limited capacity has an exponential distribution with $\lambda = .8$.
   a. What is the probability that a call is placed with a wait under 1.0 minute? With a wait between .6 and 2.0 minutes?
   b. Obtain $E\{X\}$ and $\sigma\{X\}$.
   c. Find the 25th and 75th percentiles of the probability distribution. What is the interquartile range and how is this range interpreted here?
   d. Sketch the cumulative probability function. How can you tell from your graph that the probability distribution is skewed?

*7.15 The number of calls (X) to an emergency ambulance service in a large city is generated by a Poisson process with $\lambda = 3$ calls per hour.
   a. What is the nature of the probability distribution for the lengths of time between successive calls to the service?
   b. Find the mean length of time between calls to the service. What is the probability that less than .25 hour will elapse between two consecutive calls?

7.16 In a canal serving ocean-going ships the time durations (in years) between collisions

(X_1, X_2, etc.) are independent exponential random variables with parameter $\lambda = 1$.

a. Let Y denote the number of collisions in a year. What is the nature of the probability distribution of Y?

b. Obtain $P(X > 1)$ and $P(Y = 0)$. Why must these be the same?

EXERCISES

7.17 (Calculus needed.) Derive (7.2) and (7.3).

7.18 X is a normal random variable with mean $\mu = \$500$.
a. Find the probability $P(490 \leq X \leq 510)$ for $\sigma = \$2$, $\$4$, and $\$10$. What is the implication of your results? Discuss.
b. Find the following conditional probabilities where X is $N(500, 25)$:
   (1) $P(X \geq 510|X \geq 500)$, (2) $P(495 \leq X \leq 505|490 \leq X \leq 510)$.

7.19 Use (7.5) to calculate the densities of the normal distribution $N(3, 25)$ at $x = -7$, -2, 0, 3, 6, 8, and 13. Sketch the normal density function on a graph.

7.20 Refer to the after-pollution-controls data set on time intervals between oil spills in Table 3.1a, p. 46. An analyst wished to study the nature of the probability distribution of these time intervals and began by calculating the mean and standard deviation of the observations to determine whether they were more or less equal. Why would this be of interest to the analyst?

7.21 (Calculus needed.) Derive (7.14). (*Hint:* Use integration by parts.)

7.22 (Calculus needed.) Derive (7.16) by carrying out the integration shown in (7.17).

7.23 X is an exponential random variable with parameter $\lambda = .004$.
a. Find the following probabilities: (1) $P(X > 100)$, (2) $P(X > 200|X > 100)$, (3) $P(X > 400|X > 300)$.
b. Find the following probabilities: (1) $P(X > 400)$, (2) $P(X > 600|X > 200)$, (3) $P(X > 900|X > 500)$.
c. What property of an exponential probability distribution is illustrated by your results in **a** and **b**? Explain.

7.24 Show that for any exponential random variable X, the following relation holds: $P(X > x + x_0|X > x_0) = P(X > x)$ for any positive x and x_0.

STUDIES

7.25 Weights of canned hams are normally distributed with a mean of 4.15 kilograms and a standard deviation of .12 kilogram. The label weight is stated as 4.00 kilograms. Which one of the following two modifications in the canning process would lead to a greater reduction in the proportion of hams below the label weight? (1) increase the process mean weight to 4.20 kilograms while keeping the standard deviation unchanged, (2) decrease the process standard deviation to .10 kilogram while keeping the mean unchanged.

7.26 A safety light with two batteries is designed so that the second battery begins to operate when the first battery fails. The safety light fails when the second battery has failed. The lifetime of each battery is normally distributed with a mean of 200 hours and a variance of 16. Moreover, the lifetimes of the batteries are independent. Let T denote the lifetime of the safety light.

a. Specify the nature of the probability distribution of T.

b. Show that this light satisfies the current company specification that the probability is at least .95 of the light operating a minimum of 375 hours before failing.

c. A new version of the light with three batteries is being designed to have a probability of at least .95 of operating for a minimum of 585 hours before failing. Will three batteries with independent lifetimes each distributed as $N(200, 16)$ satisfy this specification?

7.27 A machine fills "100 lb." bags with white sand. The actual weight of the sand when the machine operates at its standard speed of 100 bags per hour follows a normal distribution with a standard deviation of 1.5 pounds. The mean of the distribution depends on the setting of the machine.

a. At what mean weight should the machine be set so that only five percent of the bags are underweight, i.e., contain less than 100 pounds of sand?

b. If the machine operates at low speed, the standard deviation will be reduced to 1.0 pound but only 50 bags will be filled in an hour. The sand costs the owner of the machine $.08 per pound. An employee who operates the machine is paid $5.00 per hour. At what speed—standard or low—should the machine be operated so that the expected cost to the owner of filling 100 bags will be minimized? Assume that with either setting, only five percent of the bags are to be underweight.

7.28 A large display will contain many colored light bulbs. Because the labor cost of replacing the bulbs individually when they burn out will be expensive, the operators plan to replace all the bulbs simultaneously. Two replacement schedules are under consideration: (1) Replace when 15 percent of the bulbs have burned out. (2) Replace when 30 percent of the bulbs have burned out.

a. For bulbs of make A, the lifetimes in hours are independent and exponentially distributed with $\lambda = .0004$. Approximately how many operating hours will elapse before the bulbs are replaced under schedule 1? Under schedule 2?

b. For bulbs of make B, the lifetimes are independent and normally distributed with $\mu = 1,100$ hours and $\sigma = 250$ hours. Approximately how many operating hours will elapse before the bulbs are replaced under schedule 1? Under schedule 2?

c. Which make of bulb do you recommend that the operators use under schedule 1? Under schedule 2? Why?

7.29 Refer to the frequency distribution in Table 3.1b, p. 46, of the number of days between oil spills after pollution controls were instituted. An analyst believes that these time intervals follow an exponential distribution. The mean of the 109 observations is 27.6 days. Using this mean value and ignoring the rounded nature of the data, find the exponential probabilities for the frequency classes in Table 3.1b; i.e., find $P(0 \leq X < 15)$, $P(15 \leq X < 30)$, etc. Express these probabilities in percent form and compare them with the actual percent frequencies. Obtain the residuals and plot them. Does the analyst's belief appear to be justified? Discuss.

UNIT THREE

Estimation and Testing—I

8
Statistical Sampling

We are now ready to begin applying the principles of probability to the analysis of statistical data. Data sets are utilized in all statistical investigations. We now need to distinguish between those data sets that are considered to be populations and others that are considered to be samples.

In any statistical investigation, there is a collection of elements of interest, called the population. For instance, a U.S. congressional assistant was studying Medicaid abuses during the previous year to help in drafting new legislation. The data set consisting of all Medicaid claims in the previous year was the population of interest in this study.

Often, it is not possible to consider all elements in the population, and a subset, or sample, is selected to provide information about the population. For instance, the congressional staff assistant could not examine all Medicaid claims in the year because the number was much too large. Instead, she selected a probability sample of 1,000 claims. This data set was then utilized to make inferences about the characteristics of Medicaid abuses in all claims for the year.

The use of probability principles to draw sound inferences from a sample about the entire population is the subject of *statistical inference*. In this chapter, we consider some basic concepts of sampling. We define a population and a sample, discuss why sampling is so widely used, and explain what a simple random sample is and how such a sample can be selected.

8.1 POPULATIONS

(8.1) A *population* or *universe* is the total set of elements of interest for a given problem.

☐ Examples

1. The manager of an automobile-leasing agency is interested in the fuel economy of the passenger cars in the company's fleet. Here, the population consists of all passenger cars in the fleet. The elements of the population are the individual cars.

2. A candidate for the city council is concerned about the attitudes of voters toward a

metropolitan transit agency. The population here consists of all eligible voters in the city. The elements of the population are the individual voters.

3. A quality assurance manager wishes information about the current quality level of the firm's process for manufacturing computer memory chips. Here, the population consists of all the chips that could be produced by the current production process operating under the same conditions. The elements of the population are the individual chips. □

Examples 1 and 2 are illustrations of finite populations, while Example 3 illustrates an infinite population. We consider each type of population in turn.

Finite Population

A *finite population* consists of a finite number of elements, such as the set of passenger cars owned by a leasing agency or the set of eligible voters in a city. Many of the populations of interest in business, economics, and the social sciences are finite, although they are often large. The set of all persons living in the United States is an example of a large finite population.

Mean and Variance.   When the variable of interest for the finite population is quantitative, we are often interested in the mean and variance or standard deviation of the variable. In Example 1, for instance, we may be interested in the mean fuel economy per car in the company's fleet, as well as the variability of fuel economies between cars.

The mean, variance, and standard deviation for a data set representing a finite population are defined identically (with one small exception) as for a data set representing a sample. However, we shall use some new notation to distinguish the summary measures for population data sets from those for sample data sets, which were considered in Chapter 3.

The number of elements in a finite population will be denoted by N to distinguish it from the number of elements in a sample, denoted by n. Greek letters will be used to represent summary measures for populations. We shall use μ (Greek mu) to denote the population mean, σ (Greek sigma) for the population standard deviation, and σ^2 (read "sigma squared") for the population variance. The corresponding measures for a sample, $\bar{X}$, s, and s^2, were defined in (3.1), (3.17), and (3.15), respectively.

The definitions of a population mean, variance, and standard deviation follow.

(8.2) Let $X_1, X_2, \ldots, X_N$ represent the values of the N population elements.

(8.2a) *Population Mean:*
$$\mu = \frac{\sum_{i=1}^{N} X_i}{N}$$

(8.2b) *Population Variance:*
$$\sigma^2 = \frac{\sum_{i=1}^{N} (X_i - \mu)^2}{N}$$

(8.2c) *Population Standard Deviation:* $\qquad \sigma = \sqrt{\sigma^2}$

Note that the definition of a population mean μ is the same as that for a mean $\bar{X}$ of a sample; see the definition of $\bar{X}$ in (3.1). Also, a population variance σ^2 is defined almost the same as a sample variance s^2 in (3.15) except that the denominator is N and not $N - 1$. When N is large, the difference in the two definitions is negligible. We shall explain the reason for this definitional difference subsequently. For the moment, it suffices to remember to use divisor N for the variance when the data set refers to a population and $n - 1$ when it refers to a sample.

☐ **Example**

A population consists of the $N = 5$ Ph.D. students currently enrolled in an economics department. Their grade-point scores are:

$$X_1 = 3.74 \qquad X_2 = 3.89 \qquad X_3 = 4.00 \qquad X_4 = 3.68 \qquad X_5 = 3.69$$

The population mean is:

$$\mu = \frac{3.74 + 3.89 + 4.00 + 3.68 + 3.69}{5} = 3.80$$

The population variance is:

$$\sigma^2 = \frac{(3.74 - 3.80)^2 + (3.89 - 3.80)^2 + \cdots + (3.69 - 3.80)^2}{5} = .01564$$

Thus, the population standard deviation is $\sigma = \sqrt{.01564} = .125$. ☐

Comment

A finite population may be related to a probability distribution in the following way. Consider selecting one household from a population of N households such that each household in the population has equal probability of being chosen. The probability of any household being chosen must then be $1/N$. Let X be the random variable denoting the household income observed for the household selected at random. The random variable X can then take on the values $X_1, X_2, \ldots, X_N$ (the household incomes in the population) with probability $1/N$ each. Hence, by the definition of the expected value of a discrete random variable in (5.5), we have:

$$E\{X\} = \sum_{i=1}^{N} X_i \left(\frac{1}{N}\right) = \frac{\sum_{i=1}^{N} X_i}{N} = \mu$$

Similarly, by the definition of the variance of a discrete random variable in (5.7), we have:

$$\sigma^2\{X\} = \sum_{i=1}^{N} (X_i - \mu)^2 \left(\frac{1}{N}\right) = \sigma^2$$

Thus, the population mean μ and variance σ^2 for a finite population correspond respectively to the expected value and variance of a random variable—namely, the random variable associated with the equal-probability selection of one population element.

Infinite Population

An *infinite population* consists of an indefinitely large number of elements. In general, an infinite population refers to a *process* and its elements consist of all the outcomes of the process if it were to operate indefinitely under the same conditions. Thus, the process of manufacturing memory chips is represented by an infinite population whose elements are the chips that would be produced by the process if it were to operate indefinitely under the same conditions.

The characteristic of interest in an infinite population is described by a probability distribution. Consider the infinite population pertaining to the process of molding tires in pairs. Let X, the number of defective tires in a pair, be the characteristic of interest. This characteristic of the infinite population might then be described by the following probability distribution:

x:	0	1	2
$P(x)$:	.75	.10	.15

The probability $P(0) = P(X = 0) = .75$ indicates that 75 percent of the trials in the infinite population lead to zero defective tires in the pair. Similarly, $P(1) = .10$ indicates that 10 percent of the trials in the infinite population lead to one defective tire in the pair.

Mean and Variance. The population mean and variance for an infinite population correspond to the expected value and variance of the associated random variable:

(8.3) Let X represent the random variable associated with the infinite population of interest.

(8.3a) *Population Mean:* $\mu = E\{X\}$

(8.3b) *Population Variance:* $\sigma^2 = \sigma^2\{X\}$

(8.3c) *Population Standard Deviation:* $\sigma = \sqrt{\sigma^2} = \sigma\{X\}$

☐ Examples

1. Consider the infinite population pertaining to the process of molding tires in pairs. The random variable X is the number of defective tires in the pair. The probability distribution for X is that previously given. The population mean then is, using definition (5.5) for the expected value of a discrete random variable:

$$\mu = E\{X\} = 0(.75) + 1(.10) + 2(.15) = .40$$

The population variance is found using definition (5.7) for the variance of a discrete random variable:

$$\sigma^2 = \sigma^2\{X\} = (0 - .40)^2(.75) + (1 - .40)^2(.10) + (2 - .40)^2(.15) = .54$$

Hence, the population standard deviation is $\sigma = \sigma\{X\} = \sqrt{.54} = .73$ tire.

2. The infinite population of interest pertains to the weight (X) of metal ingots produced in a smelter. Suppose the probability distribution of X is N(520, 121). Then the popula-

tion of interest is a normal distribution, and the population mean and variance are, respectively, $\mu = 520$ pounds and $\sigma^2 = 121$.

3. The infinite population of interest pertains to the response time (X, in minutes) of a police patrol car to a call. Suppose the probability distribution of X is exponential, with $\lambda = .2$. Then the population of interest is an exponential distribution with population mean $\mu = 1/.2 = 5$ minutes by (7.14) and population variance $\sigma^2 = 1/(.2)^2 = 25$ by (7.15). □

8.2   CENSUSES AND SAMPLES

Information about a finite population can be obtained either by a census or by a sample.

Census

(8.4)   A *census* of a finite population is a study that includes every element of the population.

In some cases, it is feasible to conduct a census of the population.

□   Examples

1. An organization with 50 employees is interested in employees' preferences for a new pension plan. The population here is small, and it is easy to reach every employee; hence, a census is appropriate.
2. A state employment office urgently needs information about the total number of hours devoted last week to processing new unemployment compensation claims by all of the offices in the state. Data are needed for the entire state as well as for the five planning regions into which the state is divided. There are 23 such offices, and since they can be contacted readily, a census is practicable. A census is also necessary for obtaining reliable data for each of the five planning regions. □

Comment

An attempted census need not necessarily be successful and include every population element. Thus, the U.S. and Canadian population censuses are known to miss significant numbers of persons even though great efforts are made to include everyone.

Sample

When finite populations are large, a census is often not feasible because it would be too costly and time-consuming. Consequently, many studies of large finite populations utilize sampling.

(8.5)   A *sample* is a part of the population under study selected so that inferences can be drawn from it about the population.

□   Examples

1. An economist, as part of a study of Canadian fiscal policy, considers data on savings

plans by households based on a sample of 1,000 households from the population of Canada.

2. A state analyst uses data on the unemployment rate for the state and on the incidence of unemployment for different groups in the population, based on a sample of households from all households in the state, for setting state priorities for public works.

3. An inspector in an automobile assembly plant uses data from a sample of automobile tires selected from a shipment of tires to decide whether or not the shipment should be accepted.   □

Reasons for Sampling

There are a number of reasons why sampling is utilized so often for finite populations:

1. _Cost_ Typically, a sample can provide reliable and useful information at much lower cost than a census. For example, the cost of a census of the population of a city to obtain information for an educational television station about the viewing habits of inhabitants would be extremely high. A sample of the population can provide data with sufficient reliability at a fraction of this cost.

2. _Timeliness_ A sample usually provides more timely information than a census, because fewer data have to be collected and processed. This is particularly important when information is needed quickly, such as information about how much of a proposed tax reduction might be saved by families.

3. _Accuracy_ A sample often will provide information as accurate as, or even more accurate than, a census, because errors typically can be controlled more effectively in a small undertaking than in a large one. Thus, use of sampling often permits better training and supervision of a smaller work force than would be possible with a census.

4. _Detailed information_ Frequently, more time can be spent in probing each respondent's attitudes and motivations and in getting more detailed information with a sample than with a census. This point is related to cost, because the cost per respondent of detailed probing or of getting more detailed information can be high.

5. _Destructive testing_ When a test involves the destruction of an item, sampling must be used. Thus, a manufacturer who wishes to study the life of batteries in a production lot by subjecting batteries to a life test must use a sample if there are to be any batteries left to sell.

When a population is infinite, information about it can be obtained only from a sample. Thus, observations for obtaining information about a process always constitute a sample, no matter how large the number of observations.

Probability and Judgment Samples

We distinguish between two types of samples, probability samples and judgment samples, based on the manner in which population elements are selected for the sample. The following examples illustrate the distinction.

☐   Examples

1. A sample of 36 members was selected from the 900 members of a professional association by giving each member probability 1/900 of being chosen as the first sample element, probability 1/899 of being chosen as the second sample element (if not already chosen before), etc. This sample is a probability sample because the probabilities of selection are known for all population elements.
2. The director of a television show which was previewed in an auditorium selected 40 persons she judged to be representative of the audience of 435 for detailed interviews about their impressions of the show. The selection of persons here was not carried out by means of a probability mechanism. The sample is a judgment sample because the director judged it to be a "representative" sample. ☐

Probability Sample

(8.6)   A *probability sample* is one where the selection of elements from the population is made according to known probabilities.

The selection of population elements by known probabilities allows no discretion as to which particular elements in the population enter the sample. Probability selection has two major advantages:

1. The sample data can be evaluated by statistical methods to provide information about the margin of error due to sampling in the results.
2. Biases are avoided that could enter if judgment were used to select the population elements for the sample.

Throughout much of the remainder of this Text, we shall be concerned with probability samples and their statistical evaluation.

Judgment Sample

(8.7)   A *judgment sample* is one where judgment is used to select "representative" elements from the population or to infer that a sample is "representative" of the population.

Although the use of probability sampling has expanded greatly in recent years, judgment samples still are widely used. In contrast to probability sampling, where no discretion is allowed as to which population elements enter the sample, in judgment sampling expert opinion is used to select "representative" elements for the sample or to determine whether the sample is "representative" of the population.

Judgment samples may provide useful results. Statistical methods cannot, however, be utilized to evaluate the sample results for purposes of assessing the margin of error due to sampling. Also, some judgment samples may be very poor because the judgment is not sound and the sample turns out to be highly unrepresentative.

A common type of judgment sample used in surveys is a *quota sample*. For instance, in a household survey interviewers are given "quotas" to provide a cross section of the population under study with respect to certain characteristics such as age, sex, income, and residence. The actual selection of persons is left to the interviewers. While interviewers are supposed to use good judgment in selecting persons who meet the quotas, they often choose persons readily available such as homemakers at home during the day. Also, an interviewer can simply substitute another person from the same quota for a person who is not immediately available for interview. In these and other ways, biases can enter the survey. These biases do not enter a probability sample that is properly carried out.

Comments

1. Some samples which are designed as probability samples turn out to be judgment samples upon completion. For example, a probability sample of new-car purchasers was carefully selected. Unfortunately, only 50 percent of the purchasers in the sample replied to the questionnaire about the performance of their car. The market researcher decided, however, that the purchasers whose responses were obtained are typical of all new-car purchasers. Hence, the sample actually obtained should be viewed as a judgment sample because the market researcher made a key judgment about the "representativeness" of the responses obtained.

2. Some studies are based on data which fulfill neither the definition of a probability sample nor that of a judgment sample. This occurs when a study is based on population elements that happen to be conveniently at hand, such as the students enrolled in a marketing course who were used in a study of the effectiveness of different advertisements. These students were not selected as representative from the relevant population nor were they selected according to known probabilities. Such a study group is called a *chunk* or a *convenience sample*. A chunk can be useful for limited purposes but it cannot be relied on for credible inferences about the population.

8.3   SIMPLE RANDOM SAMPLING FROM A FINITE POPULATION

There are many types of probability samples, the most basic being a simple random sample.

Definition

(8.8)   A *simple random sample* from a finite population is a sample selected such that each possible sample combination has equal probability of being chosen.

☐   Example

Consider our earlier population consisting of five Ph.D. students, and let us denote these by A, B, C, D, and E. If a sample of two students is to be selected from this population, there are altogether 10 possible sample combinations, as follows:

A, B	A, D	B, C	B, E	C, E
A, C	A, E	B, D	C, D	D, E

To obtain a simple random sample of two students from the population of five, each of the 10 sample combinations must have probability .10 of being chosen as the actual sample.   ☐

Comments

1. For brevity, a simple random sample is often called a *random sample*.
2. Definition (8.8) refers to simple random sampling *without replacement*, where a population element can enter the sample only once. Simple random sampling *with replacement*, where the same population element may enter the sample more than once, is rarely used with finite populations.
3. It can be shown that definition (8.8) implies that each population element has an equal probability of being selected. However, simple random sampling requires more than this. Equal selection probability for each population element is a necessary but not a sufficient condition for a sample to be a simple random one. For example, suppose that a random sample of 50 oranges is to be selected from a crate of 100 oranges. We could divide the 100 oranges into two groups consisting of the 50 largest and the 50 smallest oranges, respectively, and then select one of the two groups with equal probability. This procedure gives each orange in the population an equal (and known) probability of .5 of entering the sample, and hence qualifies as probability sampling. Yet it does not qualify as simple random sampling because every combination of 50 oranges does not have equal probability of being selected. For example, a mixture of small and large oranges has no chance of being selected.

Selection of Simple Random Sample

A practical method of meeting the requirement that each possible sample combination have equal selection probability involves selecting sample elements one at a time:

(8.9) The requirements of selecting a simple random sample of n elements without replacement from a finite population of N elements are met by the following procedure:

1. Select the first sample element by giving each of the N population elements equal probability—i.e., probability $1/N$—of being chosen.
2. Select the second sample element by giving each of the remaining $N - 1$ population elements equal probability—i.e., probability $1/(N - 1)$—of being chosen.
3. Repeat this process until all n sample elements have been selected.

It can be shown that this procedure gives all possible sample combinations of size n equal probability of being chosen.

Frame. Selecting a simple random sample requires a frame:

(8.10) A *frame* is a listing of all the elements of the population.

Thus, a computer listing of the students at a university is a frame for sampling the population of students.

Sometimes, the frame available for sampling is not a perfect one for the population of interest. Consider the case where a random sample of lawyers is to be selected from the population of all lawyers in a county. The frame for sampling is a listing of all lawyers who are members of the county bar association as of last month. This listing does not include lawyers who joined the association since last month nor any lawyers in the county who are not members of the county bar association. In cases like this, we distinguish between the *target population* and the *sampled population*. The target population is the population of interest (all lawyers in the county), while the sampled population consists of the elements in the frame (all lawyers in the county who belong to the county bar association as of last month). When the frame available for sampling differs materially from the target population, the sample results may have only limited relevance. Hence, great efforts are made in practice to obtain a frame that coincides closely with the target population.

Use of Table of Random Digits. A table of random digits, such as the one in Table C–8, can readily be used to select a simple random sample by the procedure in (8.9). A table of random digits contains outcomes of independent random trials from the discrete uniform probability distribution in Figure 6.5, which has possible outcomes 0, 1, . . . , 9 with equal probability. Thus, for each position in the table, every digit from 0 to 9 has equal probability of appearing in that position, and the outcomes for the various positions in the table are independent.

Tables of random digits are generated by computer and are usually tested carefully to assure close adherence to the required properties of equal probability and independence. These properties permit the user to form random numbers from 00 to 99, or from 000 to 999, or still larger numbers. Consider forming pairs of random digits. Since every digit has equal probability of appearing in a position and all positions are filled independently, pairs of positions are filled with numbers from 00 to 99 with equal probability and independently. It is this capability of forming random numbers of any size, which are equally likely and independent, that enables us to use a table of random digits for selecting a simple random sample.

☐ Example

Consider again the frame of lawyers who are members of the county bar association as of last month. Altogether, there are 950 lawyers in this frame. We assign numbers to the lawyers, for convenience from 001 to 950. Since the frame contains 950 elements, we shall select three-digit numbers. Table 8.1 contains an extract of Table C–8, consisting of the

TABLE 8.1 *Use of table of random digits*

Row	Columns 1–5	Lawyer
1	13284 ⟶	132
2	21224 ⟶	212
3	99052 ⟶	Disregarded
4	00199 ⟶	001
5	60578 ⟶	605
6	91240 ⟶	912
7	97458 ⟶	Disregarded
8	35249 ⟶	352
9	38980 ⟶	389
10	10750 ⟶	107

first 10 rows of columns 1–5. Suppose we use the first three digits in each row as the three-digit number, and read downward. The procedure, then, is as follows:

1. The first number is 132. Lawyer 132 is therefore the first element for the sample. A number over 950 would have been disregarded. Thus, each of the 950 lawyers has equal probability of being the first sample element.
2. The second number is 212. Lawyer 212 is therefore the second element for the sample. A number over 950 would have been disregarded. Also, number 132 would have been disregarded since that lawyer is already in the sample. Thus, each of the remaining 949 lawyers has equal probability of being the second sample element.
3. This procedure is repeated until the required number of lawyers for the sample has been selected.

Table 8.1 shows the disposition for the first 10 random numbers.   □

Comments

1. Numbers may be chosen from a table of random digits in any manner so long as the procedure is systematic and determined in advance. Thus, three-digit numbers in our example might have been obtained by reading across in a row, or by taking the first digit in each of three columns, or by reading upward in a column.
2. Sometimes, the elements in a frame are prenumbered, such as invoices which have serial numbers or students who have been assigned I.D. numbers. These numbers can be used for sampling identification provided that there are no duplicate assignments of the same number to several population elements. Often, these number assignments have gaps, such as when an invoice is voided or when a student has graduated. Such gaps cause no problem; when a selected random number refers to a blank, it is simply disregarded and the next random number is selected.
3. At times, numbers need not be assigned explicitly to the elements in the frame. Suppose a sample of employees is to be selected. The frame consists of 8,500 employee file folders organized in 85 files of 100 folders each. The employee designated by a random number can then be identified by counting. For example, random number 017 refers to the employee whose folder is the 17th one in the first file. Similarly, random number 317 refers to the employee whose folder is the 17th one in the fourth file.
4. Random numbers often are produced directly by a computer routine using a random-number generator. This is easier than use of a table of random digits when the sample size is large, especially since many computer packages order the random numbers for the user. In some computer applications, the random numbers generated by the computer are used internally to identify those elements of a computerized data bank which are selected for the sample, and only the descriptions of the sample elements are printed out. Some hand-held calculators also can generate random numbers.

8.4   SIMPLE RANDOM SAMPLING FROM AN INFINITE POPULATION

When the population is infinite, the only available method for obtaining information about it involves sampling. However, one cannot use a probability mechanism for selecting the sample. The process associated with the infinite population simply furnishes sample observations. Thus, the process of molding tires in pairs provides sample observations on the number of defective tires in a pair day after day, without any probability selection mechanism.

Observations generated by a process are random variables prior to the random trial, and constitute a simple random sample from an infinite population if two conditions are met:

(8.11) The n random variables $X_1, X_2, \ldots, X_n$ generated by a process constitute a *simple random sample from an infinite population* if:

1. They come from the same probability distribution.
2. They are statistically independent.

Thus, the n random variables must be independent and identically distributed. Note that these are the conditions of an independent and stationary process as described in (6.1).

Statistical tests are available for examining whether a set of observations generated by a process appears to meet the requirements of simple random sampling. One such test is discussed in Chapter 15.

Comment

A sample from an infinite population may, for some other purpose, be considered a finite population. Thus, the memory chips produced during a week represent a sample from the infinite population associated with the production process. When, however, these chips are sent to a computer manufacturer who wishes to sample them to determine whether this shipment is of acceptable quality, the chips in the shipment constitute a finite population for this purpose.

8.5   SAMPLE STATISTICS

Once a sample has been selected and observations on the sample elements have been made, the observations constitute a data set and summary measures may be computed in the usual way as explained in Chapter 3. For convenience, we repeat here the definitions of the mean, variance, and standard deviation for a sample:

(8.12) Let $X_1, X_2, \ldots, X_n$ denote the n sample observations.

(8.12a) *Sample Mean:*
$$\bar{X} = \frac{\sum_{i=1}^{n} X_i}{n}$$

(8.12b) *Sample Variance:*
$$s^2 = \frac{\sum_{i=1}^{n} (X_i - \bar{X})^2}{n - 1}$$

(8.12c) *Sample Standard Deviation:* $s = \sqrt{s^2}$

Sample measures are usually called *sample statistics*, or *statistics* for short. Thus, $\bar{X}$ and s are examples of statistics. In contrast, population measures are called *population parameters*, or *parameters* for short. For example, μ and σ denote population parameters.

Intuition tells us that the sample statistic $\overline{X}$ provides information about the population mean μ and that the sample statistic s provides information about the population standard deviation σ. In succeeding chapters, we examine the nature of the information provided by these and other sample statistics about the population measures of interest.

PROBLEMS

8.1 Explain in each of the following cases whether the sampled population is finite or infinite:
    a. Tax returns for last year are sampled to estimate the mean deduction for interest expenses.
    b. A company sends a large number of letters by mail to estimate the mean number of days required for delivery.
    c. Outgoing orders for books are sampled prior to mailing to estimate the proportion of orders filled incorrectly by the current process.

*8.2 A population consists of $N = 7$ employees. Their ages are 27, 34, 47, 44, 37, 29, and 41. Find the population mean and standard deviation. Are these measures parameters or statistics? Explain.

8.3 A population consists of $N = 5$ plants. The floor spaces in these plants (in thousand square meters) are 10.1, 6.8, 12.4, 7.9, and 8.0. Find the population mean and standard deviation. In what units are these measures expressed?

*8.4 The number of photocopying machines in a large office which are out of order at any one time is denoted by X. The probability distribution of X follows.

x:	0	1	2	3	4
$P(x)$:	.80	.11	.05	.03	.01

    a. Describe the infinite population associated with this probability distribution.
    b. Find the population mean and variance.

8.5 An office building contains three elevators. The probability distribution of the number of elevators waiting on the ground floor at any time during business hours (X) is:

x:	0	1	2	3
$P(x)$:	.40	.30	.20	.10

    a. Describe the infinite population associated with this probability distribution.
    b. Find the population mean and standard deviation. In what units are these measures expressed?

8.6 The following numbers of wings were found in the five packages labeled "one cut-up frying chicken" in the display case in a supermarket on Tuesday morning: 1, 2, 0, 3, 1. Under what circumstances might this data set be viewed as a population? As a sample?

8.7 For each of the following, discuss whether a sample or a census would be preferable. Indicate any assumptions you make.
    a. A survey of Canadian households to obtain information on the proportion that used a microwave oven at least once during the preceding week.

    b. An examination of three turbines to be installed in a dam to obtain information about possible damage in shipment.

    c. An examination of a large shipment of light bulbs to obtain information on the proportion that burn out prematurely in normal use.

    d. An examination of dwellings in a large city to obtain information about the extent of present home insulation and the cost of providing additional insulation to bring substandard dwellings up to minimum standards.

8.8 For each of the following situations, identify the population of interest and discuss whether a census or a sample would be preferable to obtain the desired information:

    a. A regional association of 25 universities wishes to obtain current data on enrollments in the member institutions.

    b. A company with 3,000 employees wishes to study their attitudes toward the company.

    c. An appliance manufacturer wishes to study consumers' color preferences for refrigerators.

8.9 Respond to each of the following arguments:

    a. "Samples are to be preferred to censuses because samples cost less."

    b. "Censuses are to be preferred to samples because censuses are more accurate."

8.10 An information specialist in O.K. Financial Services, Inc., selected six branch offices in which to test a new computerized decision-support system, because these offices represent the variety of informational requirements found in the company's branch offices throughout the country.

    a. Do these six branch offices represent a probability sample, a judgment sample, or a convenience sample? Explain.

    b. Would your answer in **a** change if the offices were selected because they were within easy driving distance of the firm's home office? Discuss.

8.11 The editors of a campus newspaper have selected 200 students from a computer printout of the 9,730 students currently attending the university to investigate the extent of cheating on campus. A procedure was used that gives each student a probability of 1/9,730 of being selected. Each selected student is to be interviewed to obtain information on the topic of the study.

    a. Do the 200 selected students constitute a probability sample or a judgment sample? Explain.

    b. Is it possible that the completed study will not be based on the kind of sample indicated in your answer in **a**? Explain.

8.12 The retail establishments in a county in existence in March are to be sampled, using a directory published in October of the preceding year.

    a. What is the frame here?

    b. Describe several ways in which the sampled population might differ from the target population.

8.13 The adult population of a city is to be sampled next week. Some persons will be out of town during that time. What is the target population here? The sampled population?

8.14 Refer to Problem 8.12. The retail establishment directory contains 100 pages. Fifty establishments are listed to a page, except on the last page, where 31 establishments are listed.

    a. Explain how you would use a table of random digits to select a simple random sample of 100 establishments without replacement from the directory.

    b. If a table of random digits were used to select one establishment at random from each page of the directory, would the resultant sample be a probability sample? A simple random sample? Explain.

8.15 Explain how you would select a simple random sample without replacement by means of a table of random digits in each of the following cases:
a. Fifty school buses from all the school buses registered in a state.
b. One hundred vinyl 8-inch floor tiles from a warehouse floor laid with 150 rows of such tiles, 70 tiles to the row.
c. One hundred points in time (designated in minutes) during working hours next week (9:00 A.M.–5:00 P.M. Monday through Friday).

*8.16 A simple random sample of two stores is to be selected without replacement from a population of four stores (A, B, C, and D).
a. List the different possible sample combinations. How many are there?
b. What probability of selection must each sample combination in **a** have?
c. Is it necessary to list all the possible sample combinations in order to select the simple random sample? Explain.

8.17 A simple random sample of three cars is to be selected without replacement from a population of six cars (A, B, C, D, E, and F).
a. List the different possible sample combinations. How many are there?
b. What probability of selection must each sample combination in **a** have?
c. Is it necessary to list all the possible sample combinations in order to select the simple random sample? Explain.

8.18 A student, explaining how he constructed a table of random digits on his own, stated that he never placed the same digit in adjoining positions in the table because such an arrangement would not be random. Comment.

8.19 A city planning commission, using an up-to-date listing of all addresses in the city, will select a simple random sample of 300 addresses for a survey of households. The number of households which currently live at each address may be zero, one, two, or more than two.
a. Suppose that all households living at the 300 addresses will be included in the survey. Does each household in the city have an equal chance of being included? Will the households at the 300 addresses constitute a simple random sample of households in the city? Explain.
b. Alternatively, suppose that when a selected address contains more than one household, one of those households is chosen at random for inclusion in the survey and the other households at that address are not included. Will the households included in the survey constitute a simple random sample of all households in the city? Explain.

EXERCISES

8.20 If a two-digit number is selected from a table of random digits, what is the probability that it is 12? That it is even? That the two digits are the same?

8.21 In a table of random digits, what is the probability that three successive digits are 2 each? That three successive digits are the same?

8.22 Refer to Problem 8.2. An employee is to be picked at random and the age (X) noted.
a. What values can X take on? State the probability distribution of X.

  b. Find the mean and standard deviation of the probability distribution of X. What is the correspondence between these values and the population mean and standard deviation calculated in Problem 8.2?

8.23 A simple random sample of 200 vouchers is to be selected from a population of 3,000 vouchers serially numbered from 0001 to 3000. The analyst plans to select four-digit random numbers. If the first digit is 0, 1, or 2, it will be treated as 0; if it is 3, 4, or 5, it will be treated as 1; if it is 6, 7, or 8, it will be treated as 2; and if it is 9, the four-digit number will be discarded. The other three digits in the number will remain unchanged. Thus, 1245 becomes 0245; 7391 becomes 2391; and 9218 is discarded. The transformed number 0000 will correspond to voucher 3000. Is this a valid way of obtaining a simple random sample in this case? Explain.

8.24 Show for a population of $N = 4$ elements that a random sample of $n = 2$ elements selected according to (8.9) yields equal probabilities for each possible sample combination. Draw a probability tree representing selection process (8.9) here, and identify the outcomes constituting a given sample combination.

8.25 Explain why simple random sampling with replacement from a finite population is equivalent to simple random sampling from an infinite population, referring to definition (8.11).

STUDIES

8.26 The following 10 hospitals have agreed to participate in a joint research program. The research procedures are to be tested initially in a simple random sample of 3 of these 10 hospitals.

Carmel	Maple Hill	St. Francis
Central Valley	Mercy	Winchester General
Community	Northmount	
Franklin General	Southridge	

  a. Use Table C–8 to select a simple random sample of 3 hospitals without replacement.
  b. To study how a different random sample of $n = 3$ hospitals might be composed, select another simple random sample of 3 hospitals without replacement from the 10 hospitals in the population, using random numbers independent of those used in **a**. How much overlap is there between the hospitals in the samples in **a** and **b**? What is the probability of no overlap in the second sample, given the results for your first sample? What is the probability of no overlap in advance of selecting either of the two samples?
  c. Continue selecting independent random samples of $n = 3$ hospitals from the population of 10 hospitals until you have selected 50 random samples of $n = 3$. For each hospital in the population, determine the number of samples in which it was selected. What is the expected number for each hospital? Are the actual frequencies close to the expected frequencies? Discuss.

8.27 Refer to Appendix D. The population of textile apparel manufacturing firms is to be sampled to obtain information about 1981 net sales. You are asked to take a simple random sample of 10 companies without replacement.
  a. How many different possible sample combinations of 10 companies can be selected from this population? [*Hint*: Use (A.29) in Appendix A.]

    b. Select a random sample of 10 companies, using the procedure in (8.9).

    c. Calculate the sample mean and sample standard deviation.

    d. The population mean is $\mu = \$153.43$ million and the population standard deviation is $\sigma = \$256.00$ million. Given the large variability in the population (the coefficient of variation is $256.00/153.43 = 1.67$ or 167 percent), do you expect the sample mean and standard deviation based on a sample of 10 firms to be close to the population parameters? How close are your sample statistics in c to the corresponding population parameters?

9
Sampling
Distribution of $\overline{X}$

We noted in the last chapter that samples are selected to provide information about population characteristics. Often the characteristic of interest is the population mean μ.

☐ Examples

1. The unemployment rate in a western state is 7.8 percent. The governor needs to know whether the unemployed tend to be unemployed for a short time while shifting to another job or whether they tend to be unemployed for a long time, unable to find employment. The mean duration of unemployment (μ) of the unemployed in the state during the week of January 12 is to be estimated from a random sample of unemployed persons.
2. An environmental control agency requires information about the mean pollution emission per car (μ) for a particular make of passenger car. A random sample of cars from the production line for this make of automobile will be selected and tested.
3. An auditor needs to estimate the population mean audit amount per account (μ) in a population of N accounts from a random sample of accounts. The total audit amount for all accounts in the populaton is given by $N\mu$, which can be estimated once the estimate of μ is in hand. ☐

These examples illustrate the variety of problems where interest exists in the population mean μ. The sample mean $\overline{X}$ provides information about the population mean μ. Intuition tells us, however, that if we were to take a number of simple random samples of given size from a population and calculate the sample mean $\overline{X}$ for each, we would obtain different values of $\overline{X}$ for the several samples. The reason is that different samples usually will include different population elements. While in practice we select only a single sample, knowledge about the behavior of the sample statistic $\overline{X}$ in repeated samples from a population will enable us to assess and control the margin of error due to sampling for any given sample.

In this chapter, we first study, by means of an experiment, the behavior of the sample mean $\overline{X}$ as it varies from one sample to another. Then we present relevant results from statistical theory and show how the experiment illustrates the theoretical results.

9.1 STUDY OF BEHAVIOR OF $\bar{X}$ BY EXPERIMENTATION

Population

An auditor conducted an experiment involving the 8,042 freight accounts receivable of a freight company. He wished to demonstrate the use of sampling to estimate the mean audit account balance in the population—i.e., the mean audit amount receivable per account for the 8,042 accounts. The audit amount may differ from the book amount for various reasons, including use of improper tariff, incorrect calculations, and dispute over damages. For purposes of this experiment, each of the 8,042 accounts receivable was initially audited. Ordinarily, the auditor would only audit a sample of these accounts and base his conclusions on the evaluation of the sample results.

Table 9.1 contains a frequency distribution of the audit amounts of the 8,042 accounts in the population, as well as the population mean and standard deviation (calculated from the actual 8,042 observations). Figure 9.1 shows the frequency distribution graphically. It is readily apparent that the distribution is substantially skewed to the right, which is typical of many populations in business and economics.

Simple Random Sampling with $n = 3$

We begin our study of the behavior of the sample mean $\bar{X}$ by starting with a very small sample size, $n = 3$. A simple random sample of three accounts was selected from the population, and the audit amounts shown in the first row of Table 9.2 were obtained. The sample mean for this sample is:

$$\bar{X} = \frac{\Sigma X_i}{n} = \frac{30.96 + 38.20 + 22.45}{3} = 30.537$$

TABLE 9.1 *Frequency distribution of audit amounts of 8,042 accounts receivable*

Audit Amount (dollars)	Number of Accounts	
0–under 10	1,008	
10–under 20	2,856	
20–under 30	1,926	
30–under 40	780	
40–under 50	326	
50–under 60	258	Population mean: $\mu = 30.303$
60–under 70	160	Population standard deviation: $\sigma = 30.334$
70–under 80	148	
80–under 90	99	
90–under 100	116	
100–under 150	244	
150–under 200	121	
Total	8,042	

FIGURE 9.1   *Frequency distribution of audit amounts of 8,042 accounts receivable.*
The population is highly skewed to the right.

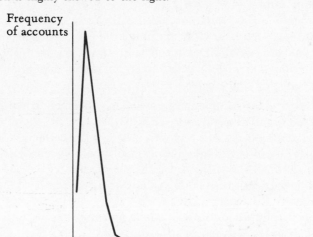

Additional random samples of size $n = 3$ were selected, always from the full population of 8,042 accounts. Table 9.2 contains the results for five of these samples.

It is clear from Table 9.2 that all five sample means differ from one another, and that none of them is equal to the population mean $\mu = 30.303$. In fact, some of the sample means differ substantially from the population mean. The departure of the sample mean from the population mean is called the sampling error:

(9.1)   *Sampling error* is the difference between the result obtained from a sample and that which would be obtained from a census of the population elements in the frame, conducted under the same procedures as the sample.

Sampling error is distinguished from errors in data as discussed in Chapter 1 in that sampling error is due to sampling. In contrast, data errors—such as those due to

TABLE 9.2   *Results for first five random samples of $n = 3$*

	Observation			
Sample	1	2	3	$\bar{X}$
1	30.96	38.20	22.45	30.537
2	18.91	6.75	15.45	13.703
3	10.60	14.08	9.15	11.277
4	51.82	20.76	50.79	41.123
5	23.05	31.20	25.15	26.467

faulty recall by a respondent or calculational errors by an auditor—can be present whether a census or a sample is conducted. Such errors often are called *non-sampling errors*.

Altogether, 600 samples of size $n = 3$ were selected for this experiment. Table 9.3 contains a frequency distribution of the 600 sample means, as well as the mean and standard deviation of the 600 $\bar{X}$ values. Figure 9.2a shows the frequency distribution of the 600 $\bar{X}$ values in graphic form. Three important results emerge from Table 9.3 and Figure 9.2a:

1. While the 600 sample means are widely divergent, their mean of 30.68 is close to the population mean $\mu = 30.303$.
2. The standard deviation of the 600 $\bar{X}$ values, which is 17.60, shows that the variability of the sample means is substantially smaller than the variability of the audit account balances in the population ($\sigma = 30.334$). In fact, the standard deviation of the 600 $\bar{X}$'s is only about six-tenths as great as the population standard deviation.
3. The distribution of the 600 sample means is skewed to the right, like the population, but not as much.

Simple Random Sampling with $n = 10$

Next in the experiment, 600 random samples of size $n = 10$ were selected from the population of 8,042 accounts receivable. Table 9.4 presents the frequency distribution of the 600 $\bar{X}$ values, as well as the mean and standard deviation of the 600 sample means. Figure 9.2b shows the frequency distribution of the 600 $\bar{X}$ values graphically. Four important results are evident:

1. The 600 sample means for sample size $n = 10$ have a mean of 30.23, which is close to the population mean $\mu = 30.303$. This was also true for sample size $n = 3$.

TABLE 9.3   *Frequency distribution of 600 $\bar{X}$ values: $n = 3$*

Sample Mean	Number of Samples	
Under 5.80	1	
5.80–under 12.80	18	
12.80–under 19.80	188	
19.80–under 26.80	132	
26.80–under 33.80	77	
33.80–under 40.80	52	
40.80–under 47.80	30	Mean of 600 $\bar{X}$'s: 30.68
47.80–under 54.80	29	Standard deviation of 600 $\bar{X}$'s: 17.60
54.80–under 61.80	20	
61.80–under 68.80	26	
68.80–under 75.80	12	
75.80–under 82.80	9	
82.80 and more	6	
Total	600	

FIGURE 9.2 *Frequency distributions of 600 $\bar{X}$ values for samples of n = 3 and n = 10 accounts receivable.* As the sample size was increased, the sampling distribution of $\bar{X}$ has remained centered at μ and has become less variable and less skewed.

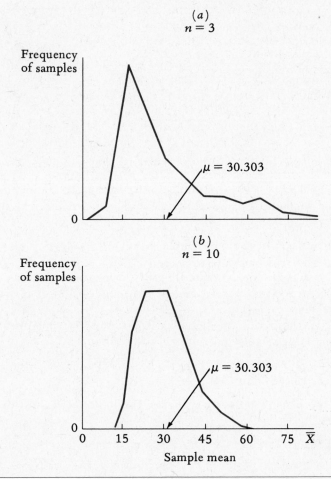

2. The standard deviation of the 600 $\bar{X}$ values, which is 9.13, is substantially smaller than the population standard deviation $\sigma = 30.334$. In fact, it is only about three-tenths as large.

3. The 600 $\bar{X}$ values for sample size $n = 10$ are less variable than the 600 $\bar{X}$ values for sample size $n = 3$. The respective standard deviations are 9.13 and 17.60.

4. The distribution of the 600 $\bar{X}$ values for sample size $n = 10$ is only moderately skewed to the right, in contrast to the marked positive skewness for the population.

TABLE 9.4 *Frequency distribution of 600 $\overline{X}$ values: n = 10*

Sample Mean	Number of Samples
12.80–under 19.80	71
19.80–under 26.80	170
26.80–under 33.80	173
33.80–under 40.80	108
40.80–under 47.80	51
47.80–under 54.80	21
54.80–under 61.80	5
61.80–under 68.80	1
Total	600

Mean of 600 $\overline{X}$'s: 30.23
Standard deviation of 600 $\overline{X}$'s: 9.13

Simple Random Sampling with *n* = 100

The final part of the experiment was to select 600 random samples of size $n = 100$. Table 9.5 and Figure 9.3 contain the relevant results. The following conclusions are of interest:

1. The mean of the 600 sample means, 30.31, again is close to the population mean $\mu = 30.303$.
2. The standard deviation of the 600 $\overline{X}$ values is smaller than that for both sample sizes $n = 3$ and $n = 10$, and is only about one-tenth as great as the population standard deviation.
3. The distribution of the 600 $\overline{X}$ values is quite symmetrical and appears, when smoothed, similar to a normal distribution.

Empirical Conclusions

On the basis of these experiments, the following conjectures appear to be warranted:

1. The distribution of $\overline{X}$ values for simple random sampling is centered around the population mean, regardless of sample size.

TABLE 9.5 *Frequency distribution of 600 $\overline{X}$ values: n = 100*

Sample Mean	Number of Samples
22.80–under 25.80	39
25.80–under 28.80	157
28.80–under 31.80	213
31.80–under 34.80	148
34.80–under 37.80	37
37.80–under 40.80	5
40.80–under 43.80	1
Total	600

Mean of 600 $\overline{X}$'s: 30.31
Standard deviation of 600 $\overline{X}$'s: 3.05

FIGURE 9.3   *Frequency distribution of 600 $\bar{X}$ values for samples of n = 100 accounts receivable.* As the sample size has become large, the sampling distribution of $\bar{X}$ has approached a normal distribution.

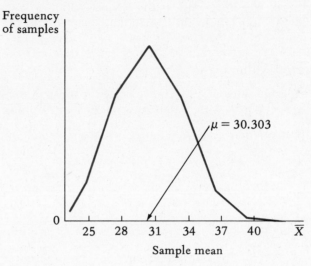

2. The standard deviation of the $\bar{X}$ values decreases with increasing sample size—i.e., the distribution of $\bar{X}$ values becomes more concentrated around the population mean as the sample size gets larger.
3. The distribution of the $\bar{X}$'s becomes more symmetrical as the sample size gets large and is approximately normal for large sample sizes.

9.2   THEORY ABOUT BEHAVIOR OF $\bar{X}$

We now turn to theoretical results about the behavior of the sample mean $\bar{X}$ illustrated by the previous experiment. These results are applicable for infinite populations and for finite populations whenever the sample size n is a small portion of the population size N. This latter condition applies to our experimental study where $N = 8,042$ and n varied from 3 to 100.

Sampling Distribution of $\bar{X}$

When a simple random sample of given size is selected from a population, the actual sample obtained is the result of a random trial, i.e., the random selection process. Hence, the observed sample mean $\bar{X}$ is also the outcome of a random trial. In advance of sampling, therefore, the sample mean $\bar{X}$ is a random variable and has an associated probability distribution:

(9.2)   The probability distribution associated with the random variable $\bar{X}$ is called the *sampling distribution of $\bar{X}$* or the *sampling distribution of the mean.*

In our experiment the 600 $\bar{X}$ values for a given sample size n represent 600 observations from the sampling distribution of $\bar{X}$ for that sample size.

There is a different sampling distribution of $\bar{X}$ for each population and random sample size. We now consider the characteristics of sampling distributions of $\bar{X}$ and how these relate to the population sampled and the sample size.

Expected Value of $\bar{X}$

The expected value of $\bar{X}$, or equivalently the mean of the sampling distribution of $\bar{X}$, is always equal to the population mean μ with simple random sampling:

(9.3)   For all populations and sample sizes, when simple random sampling is employed:

$$E\{\bar{X}\} = \mu$$

The results of our experiment certainly are consistent with theorem (9.3):

Sample Size	Mean of 600 $\bar{X}$'s
3	30.68
10	30.23
100	30.31

Population mean $\mu = 30.303$

Of course, we would not expect the mean of 600 trials to equal μ exactly, since there is always some random variation present in a limited number of trials.

Comment

Theorem (9.3) can be proven readily when a simple random sample is selected from an infinite population. In that case, the expected value of any sample observation is $E\{X_i\} = \mu$, and we have:

$$E\{\bar{X}\} = E\left\{\frac{X_1 + X_2 + \cdots + X_n}{n}\right\} = \frac{1}{n}E\{X_1 + X_2 + \cdots + X_n\} \qquad \text{by (5.6b)}$$

$$= \frac{1}{n}(E\{X_1\} + E\{X_2\} + \cdots + E\{X_n\}) = \frac{n\mu}{n} = \mu \qquad \text{by (5.15a)}$$

Variance of $\bar{X}$

The variance of $\bar{X}$, or equivalently the variance of the sampling distribution of $\bar{X}$, is denoted by $\sigma^2\{\bar{X}\}$. It is often also called the *variance of the mean*. Similarly, the standard deviation of $\bar{X}$ is denoted by $\sigma\{\bar{X}\}$ and is often called the *standard deviation of the mean* or the *standard error of the mean*.

The variance $\sigma^2\{\bar{X}\}$ and the standard deviation $\sigma\{\bar{X}\}$ measure the variability of the possible sample means $\bar{X}$ which can be obtained with simple random sampling from a population. It turns out that these measures depend on the population variance σ^2 and sample size n in a simple way:

(9.4)   When simple random sampling is employed:

$$\sigma^2\{\bar{X}\} = \frac{\sigma^2}{n} \qquad \sigma\{\bar{X}\} = \frac{\sigma}{\sqrt{n}}$$

The results of our experiment certainly are consistent with theorem (9.4):

Sample Size n	Standard Deviation of 600 $\bar{X}$'s	$\frac{\sigma}{\sqrt{n}}$
3	17.6	17.5
10	9.1	9.6
100	3.0	3.0

Population standard deviation $\sigma = 30.334$

Again, we must anticipate some variation from theoretical expectations in the results of a limited number of trials.

Effect of Sample Size.   Theorem (9.4) indicates that the standard deviation of $\bar{X}$ decreases proportionately with the square root of the sample size. Thus, the sampling distribution of $\bar{X}$ will be more concentrated the larger the sample size. As we shall see, this is the counterpart to our intuition that larger samples lead to more precise results.

However, since the standard deviation of $\bar{X}$ decreases proportionately with the *square root* of the sample size, it becomes increasingly difficult to reduce $\sigma\{\bar{X}\}$ by increasing n. Thus, if the standard deviation of $\bar{X}$ based on sample size $n = 100$ is to be cut in half, the sample size must be quadrupled to $n = 400$. In turn, if that standard deviation is to be cut in half, the sample size must be quadrupled to $n = 1,600$.

Effect of Population Variability.   Theorem (9.4) indicates that, for any given sample size, the greater the population variability, the larger the variability of the sampling distribution of $\bar{X}$. Thus, for any given n, $\bar{X}$ will tend to vary more about the population mean μ when the population is highly variable than when it is concentrated.

Comment

Theorem (9.4) can be readily proven for infinite populations. Simple random sampling from infinite populations requires independent and identically distributed observations, the variance of each being $\sigma^2\{X_i\} = \sigma^2$. Hence, we obtain:

$$\sigma^2\{\bar{X}\} = \sigma^2\left\{\frac{X_1 + X_2 + \cdots + X_n}{n}\right\} = \frac{1}{n^2}\sigma^2\{X_1 + X_2 + \cdots + X_n\} \qquad \text{by (5.11b)}$$

$$= \frac{1}{n^2}\left(\sigma^2\{X_1\} + \sigma^2\{X_2\} + \cdots + \sigma^2\{X_n\}\right) = \frac{n\sigma^2}{n^2} = \frac{\sigma^2}{n} \qquad \text{by (5.15b)}$$

Central Limit Theorem

One of the most important theorems in statistics is the central limit theorem:

(9.5) *Central Limit Theorem.* For almost all populations, the sampling distribution of $\bar{X}$ is approximately normal when the simple random sample size is sufficiently large.

This theorem is important because from it we know the nature of the sampling distribution of $\bar{X}$ (approximately normal) when the random sample size is reasonably large for almost any population.

Our experimental results are consistent with this theorem. While the distribution of the 600 $\bar{X}$'s is skewed for $n = 3$, it is less skewed for $n = 10$ and is almost symmetrical for $n = 100$. Further, the appearance of the distribution of the $\bar{X}$ values in Figure 9.3 for $n = 100$ suggests normality. To study this aspect more formally, recall that the population mean and standard deviation are $\mu = 30.303$ and $\sigma = 30.334$. Hence, we know that the sampling distribution of $\bar{X}$ for $n = 100$ has the following characteristics according to (9.3) and (9.4):

$$E\{\bar{X}\} = \mu = 30.303 \qquad \sigma\{\bar{X}\} = \frac{\sigma}{\sqrt{n}} = \frac{30.334}{\sqrt{100}} = 3.033$$

If the sampling distribution of $\bar{X}$ for $n = 100$ is approximately normal, we can determine from Table C-1 the proportion of $\bar{X}$ values expected to fall in any interval around the population mean. Here are some results for our experiment with $n = 100$.

	Experimental Results		Theoretical Expectations
Interval	Number of $\bar{X}$'s in Interval	Proportion of $\bar{X}$'s in Interval	Normal Probability Based on Central Limit Theorem
30.30 ± 3.50	451	.752	.750
30.30 ± 5.50	559	.932	.930
30.30 ± 7.50	594	.990	.986
30.30 ± 9.50	599	.998	.998

The close accord between the experimental results and the theoretical expectations based on the central limit theorem strongly supports the applicability of the central limit theorem here.

Comments

1. What is a sufficiently large sample size for the central limit theorem to apply depends on the nature of the population and on the degree of approximation to the normal distribution required. In general, for skewed populations such as the one in our experiment, a larger random sample size is required for the sampling distribution of $\bar{X}$ to be approximately normal than for a population that is fairly symmetrical.
2. The central limit theorem applies whenever the population standard deviation σ is defined. This is the case in almost all problems encountered in practice.
3. The theoretical expectations for our experiment were calculated by using the standard deviation of the sampling distribution of $\bar{X}$ for $n = 100$, namely, $\sigma\{\bar{X}\} = 3.033$. For

instance, in the first interval the deviation 3.50 represents $3.50/3.033 = 1.15$ standard deviations. We find from Table C–1 that $P(-1.15 \leq Z \leq 1.15) = .750$.

Sampling Finite Populations

As we noted earlier, theorems (9.3) and (9.4) for the mean and variance of the sampling distribution of $\bar{X}$ and central limit theorem (9.5) apply to the sampling of infinite and finite populations, provided in the latter case that the sample size n is not large relative to the finite population size N. In practice, the sample size is usually small relative to the population size when finite populations are sampled. Thus, in our audit sampling experiment, the sampling fraction for $n = 100$ was only $n/N = 100/8,042 = .012$. Similarly, when a sample of $n = 1,500$ households is selected from the population of a state containing $N = 1,000,000$ households, the sampling fraction is only $n/N = 1,500/1,000,000 = .0015$. Even when a sample of $n = 75$ retail outlets is selected from the $N = 1,800$ retail outlets of a chain, the sampling fraction is only .042.

Throughout this Text, unless otherwise noted, we shall assume that finite population sizes are large relative to the sample sizes. In Chapter 14, we shall take up the special case where this condition does not hold.

9.3 PROBABILITY STATEMENTS ABOUT BEHAVIOR OF $\bar{X}$ USING NORMAL APPROXIMATION

Central limit theorem (9.5) can be utilized to make probability statements about the behavior of $\bar{X}$ for reasonably large sample sizes. Use of the standard normal distribution in Table C–1 in this case requires the following standardized variable:

$$(9.6) \qquad Z = \frac{\bar{X} - E\{\bar{X}\}}{\sigma\{\bar{X}\}} = \frac{\bar{X} - \mu}{\sigma/\sqrt{n}}$$

This corresponds to the earlier definition (7.9) of a standard normal variable, except that the variable standardized now is $\bar{X}$ instead of X.

We illustrate use of the central limit theorem to make probability statements about the behavior of $\bar{X}$ with three examples.

☐ Examples

1. Suppose that the auditor in our earlier illustration was actually planning to use a simple random sample of 250 accounts receivable. Invoking the central limit theorem, he would expect that the sampling distribution of $\bar{X}$ is approximately normal for that sample size. Theorem (9.3) indicates that the mean of the sampling distribution of $\bar{X}$ is equal to the population mean $\mu = 30.303$, and theorem (9.4) states that the standard deviation of the sampling distribution of $\bar{X}$ is:

$$\sigma\{\bar{X}\} = \frac{\sigma}{\sqrt{n}} = \frac{30.334}{\sqrt{250}} = 1.92$$

The auditor would like to know the probability that the sample mean will be within $4 of the population mean audit amount, i.e., between 26.30 and 34.30. Figure 9.4a shows the sampling distribution of $\bar{X}$ as approximated by the normal distribution. The shaded area represents the desired probability. We calculate the z values for the specified $\bar{X}$ values by (9.6) in the usual fashion:

$$z = \frac{26.30 - 30.30}{1.92} = -2.08 \qquad z = \frac{34.30 - 30.30}{1.92} = 2.08$$

We find that $P(26.30 \leq \bar{X} \leq 34.30) = P(-2.08 \leq Z \leq 2.08) = .96.$

Thus, almost always the auditor will obtain a sample mean that is within $4 of the population mean. In ordinary practice, of course, the auditor takes only one random sample of 250 accounts, on the basis of which he estimates the population mean audit amount. Since the probability is so high that the sample mean is within $4 of the population mean for a sample of 250 accounts, the auditor has by this knowledge a measure of the margin of error due to sampling for the particular sample value of $\bar{X}$ which he will obtain.

It is in this way that the sampling distribution of $\bar{X}$ provides information about the margin of error due to sampling for the one sample result that is actually obtained—namely, by indicating the probability that the sample mean $\bar{X}$ will be within any given proximity of the population mean μ.

FIGURE 9.4   *Sampling distributions of $\bar{X}$ for n = 100 and n = 250*

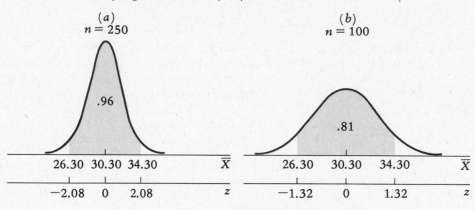

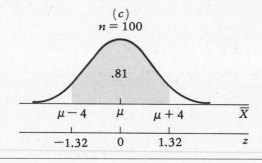

2. To study the effect of sample size, let us find the same probability for a random sample of 100 accounts. In that case, the sampling distribution of $\bar{X}$ would be approximately normal according to the central limit theorem. The mean of the sampling distribution of $\bar{X}$ would still be $\mu = 30.303$, because it is unaffected by sample size. The standard deviation of the sampling distribution of $\bar{X}$ would, however, be larger:

$$\sigma\{\bar{X}\} = \frac{30.334}{\sqrt{100}} = 3.03$$

Figure 9.4b shows the effect of the smaller sample size on the sampling distribution of $\bar{X}$. It is more variable than the distribution for the larger sample size shown in Figure 9.4a. Hence, the probability of $\bar{X}$ falling in a given interval around the population mean μ is smaller for $n = 100$ than for $n = 250$. Specifically, let us find the probability that $\bar{X}$ falls within \$4 of μ:

$$z = \frac{26.30 - 30.30}{3.03} = -1.32 \qquad z = \frac{34.30 - 30.30}{3.03} = 1.32$$

Hence, $P(26.30 \leq \bar{X} \leq 34.30) = P(-1.32 \leq Z \leq 1.32) = .81$. Recall that this probability for $n = 250$ is .96. The larger probability for the larger sample size corresponds to our intuition that the sample mean is more precise for larger samples.

3. Suppose that the auditor does not know the population mean μ but wishes to know the probability that the sample mean $\bar{X}$ based on $n = 100$ will fall within \$4 of μ. From Example 2, we know that $\sigma\{\bar{X}\} = 3.03$ for this sample size. Figure 9.4c portrays the situation. The z values are the same as for Example 2:

$$z = \frac{(\mu - 4) - \mu}{3.03} = -1.32 \qquad z = \frac{(\mu + 4) - \mu}{3.03} = 1.32$$

Hence, the probability that $\bar{X}$ based on a random sample of size $n = 100$ will fall within \$4 of the population mean μ is .81. ☐

Comment

In actual practice, the population mean and standard deviation, which are required for obtaining the mean and standard deviation of the sampling distribution of $\bar{X}$, are usually unknown. We shall explain in the next chapter how the sample mean $\bar{X}$ and the sample standard deviation s can be used to estimate the population mean μ and the standard deviation σ, respectively, and how these estimates in turn enable us to estimate the mean and standard deviation of the sampling distribution of $\bar{X}$.

9.4 EXACT SAMPLING DISTRIBUTION OF $\bar{X}$

The central limit theorem tells us the approximate nature of the sampling distribution of $\bar{X}$ when the sample size is reasonably large. In certain cases, the exact sampling distribution of $\bar{X}$ can be ascertained.

Normal Population

When the population sampled is normal, the following theorem provides us with the exact sampling distribution of $\bar{X}$.

(9.7) When the population sampled is a normal probability distribution, the sampling distribution of $\bar{X}$ is exactly normal for *any* random sample size.

We shall utilize this theorem later when we consider sampling of normal populations.

Other populations for which statistical theory permits us to derive the exact sampling distribution of $\bar{X}$ include exponential and Poisson populations.

Construction by Enumeration

For simple discrete populations, the exact sampling distribution of $\bar{X}$ can be derived by enumeration of all possible samples. This procedure is tedious, however, and is rarely used. We present a very simple example to illustrate the basic principles.

☐ Example

The number of microcomputers sold in week 1 is denoted by X_1 and the number sold in week 2 by X_2. X_1 and X_2 are independent random variables with the same probability distribution:

x:	2	4
$P(x)$:	.9	.1

For this probability distribution, $E\{X\} = 2.2$ and $\sigma\{X\} = .6$ (calculations not shown). We wish to obtain the exact sampling distribution of $\bar{X}$, the mean number of microcomputers sold in the two weeks.

We list all possible sample outcomes and proceed as follows:

Sample outcome $\begin{cases} X_1: \\ X_2: \end{cases}$	2 2	2 4	4 2	4 4
Sample mean $\bar{X}$:	2	3	3	4
Probability:	.81	.09	.09	.01

The probabilities are obtained by theorem (5.4) because X_1 and X_2 are independent random variables. Thus, $P(X_1 = 2 \cap X_2 = 2) = P(X_1 = 2)P(X_2 = 2) = .9(.9) = .81$.

We can now obtain the exact sampling distribution of $\bar{X}$ for $n = 2$:

$\bar{X}$:	2	3	4
$P(\bar{X})$:	.81	.18	.01

Note that $P(\bar{X} = 3) = .09 + .09 = .18$ because two sample outcomes yield this value of $\bar{X}$.

According to theorems (9.3) and (9.4), respectively, this sampling distribution of $\bar{X}$ has mean $E\{\bar{X}\} = \mu = 2.2$ and standard deviation $\sigma\{\bar{X}\} = \sigma/\sqrt{n} = .6/\sqrt{2} = .42$. We confirm these results by direct calculation of the mean and standard deviation from the preceding sampling distribution, using (5.5) and (5.7):

$$E\{\bar{X}\} = 2(.81) + 3(.18) + 4(.01) = 2.2$$

$$\sigma^2\{\bar{X}\} = (2 - 2.2)^2(.81) + (3 - 2.2)^2(.18) + (4 - 2.2)^2(.01) = .18$$

Hence, $\sigma\{\bar{X}\} = \sqrt{.18} = .42$, as anticipated.

PROBLEMS

*9.1 Bricks are produced with mean weight $\mu = 1.74$ kilograms and standard deviation $\sigma = .03$ kilogram.
   a. What is the mean of the sampling distribution of $\bar{X}$ when $n = 10$? When $n = 50$?
   b. What is the standard deviation of the sampling distribution of $\bar{X}$ when $n = 10$? When $n = 50$?

9.2 Scholastic aptitude scores of college-bound high school seniors have a mean of $\mu = 450$ and a standard deviation of $\sigma = 75$. What are the mean and standard deviation of the sampling distribution of $\bar{X}$ when $n = 100$? When $n = 40$?

9.3 The mean time from germination to harvest for a certain variety of bean under laboratory conditions is $\mu = 61$ days and the standard deviation is $\sigma = 3$ days.
   a. What is the mean of the sampling distribution of $\bar{X}$ for a random sample of $n = 8$ seedlings? For $n = 20$ seedlings?
   b. What is the standard deviation of the sampling distribution of $\bar{X}$ when $n = 8$? When $n = 20$?

9.4 Refer to Problem 9.1.
   a. For which sample size ($n = 10$ or $n = 50$) is it more likely that the sample mean is within $\pm .01$ kilogram of μ? Why?
   b. If a random sample of 10 bricks were selected from the process, and independently another random sample of 50 bricks were selected, would the sample mean based on the larger sample necessarily be closer to $\mu = 1.74$ kilograms? Discuss.

9.5 A manager in an executive development program stated: "Sampling errors are inherent in any sampling process, just as measurement errors are inherent in any measurement process." Comment.

9.6 A student, on studying sampling, was puzzled about how there can be a sampling distribution of $\bar{X}$ when in practice only a single sample is selected. Explain to this student the concept of a sampling distribution and its significance.

9.7 Refer to Figure 9.2. The frequency polygons for $n = 3$ and $n = 10$ are located in almost identical positions on the $\bar{X}$-scale, but they differ in variability. What theorems are illustrated by this comparison?

9.8 When the sample size is large, is the sample mean $\bar{X}$ more likely to fall above or below the population mean? Discuss.

*9.9 A sample of $n = 3,000$ persons is to be taken from a population of 1,000,000 persons in the labor force. What is the sampling fraction here? Is it large or small?

9.10 A sample of $n = 200$ parts is to be taken from a lot of 50,000 parts. What is the sampling fraction here? Is it large or small?

*9.11 Refer to Table 9.1. A random sample of 200 accounts is to be selected from this population of accounts receivable.
   a. Obtain the mean and standard deviation of the sampling distribution of $\bar{X}$. What is the approximate functional form of the sampling distribution?
   b. What is the probability that $\bar{X}$ will be within $4 of the population mean—i.e., within the interval $\mu \pm \$4$?
   c. Within what interval centered around μ will $\bar{X}$ fall 99 percent of the time? What is the corresponding interval when $n = 100$? Is the latter interval two times as wide as the former? Comment.

9.12 A large population of music stores is to be sampled to estimate the mean loss per store due to theft of phonograph records during the past year. The population mean and standard deviation are $\mu = \$380$ and $\sigma = \$210$.
   a. For a random sample of 250 stores, what is the probability that the sample mean will be within $\pm\$15$ of the population mean?
   b. Would the probability in a be increased by 20 percent if the sample size were increased by 20 percent, to $n = 300$? Be specific.
   c. Would the probability in a be affected substantially if the population standard deviation were $\sigma = \$300$, for $n = 250$?
   d. Would the probability in a be affected substantially if $\mu = \$600$, for $n = 250$ and $\sigma = \$210$?

9.13 An automatic machine cuts ribbons of hot steel into bars of specified length. The variability in the lengths of the bars is $\sigma = .07$ meter. For a random sample of 60 bars, what is the probability that the sample mean $\bar{X}$ will not differ from the process mean μ by more than $\pm.01$ meter? Assume that the central limit theorem applies.

9.14 For each of the following cases, state the functional form of the sampling distribution of $\bar{X}$ and indicate whether it is exact or approximate:
   a. The diameters of shafts produced by a robot-operated machine are uniformly distributed. A random sample of 225 shafts is selected and $\bar{X}$ is determined.
   b. A population of 18,000 employees is sampled to estimate the mean distance employees have to travel from home to place of work. A random sample of $n = 250$ employees is selected and $\bar{X}$ is determined.
   c. The weights of flagstones follow a normal distribution. A random sample of $n = 5$ flagstones is selected and $\bar{X}$ is determined.

*9.15 The weight of solvent in a steel drum is normally distributed, with mean $\mu = 69$ kilograms and standard deviation $\sigma = .9$ kilogram.
   a. For a random sample of three drums, what is the functional form of the sampling distribution of $\bar{X}$? What is the probability that the sample mean will not exceed 70 kilograms?
   b. For a random sample of six drums, what is the interval centered around μ within which $\bar{X}$ will fall 99 percent of the time?
   c. If the sample size were nine drums, what would be the interval in b? What is the effect of increased sample size?

9.16 Refer to Problem 9.1. The weights of bricks are normally distributed.
   a. For a random sample of 16 bricks, what is the probability that $\bar{X}$ will be within $\pm.01$ kilogram of the population mean?
   b. For a random sample of 16 bricks, what is the interval centered around μ within which $\bar{X}$ will fall 95 percent of the time?
   c. If the population standard deviation were $\sigma = .07$ kilogram, what would be the interval in b? What is the effect of increased population variability?

*9.17 The probability distribution of X, the number of computer breakdowns per week, is:

x:	0	1	2
$P(x)$:	.70	.20	.10

   a. Obtain the mean and standard deviation of this infinite population.
   b. Obtain the exact sampling distribution of $\bar{X}$ for a random sample of $n = 2$ weeks.

c. Calculate the mean and standard deviation of the sampling distribution of $\bar{X}$ in **b** and verify that these agree with theorems (9.3) and (9.4).

9.18 Refer to Problem 9.17. Answer all parts using the probability distribution:

x:	0	1	2
$P(x)$:	.50	.40	.10

EXERCISES

9.19 Cartons of powdered milk have mean weight $\mu = 1.13$ kilograms and standard deviation $\sigma = .03$ kilogram. An inspector, measuring the net weights for a random sample of five cartons on the regular scale employed for quality control testing, obtained $\bar{X} = 1.26$ kilograms. When this sample was weighed on a high-precision scale just calibrated, the inspector obtained $\bar{X} = 1.20$ kilograms. What are the sampling and nonsampling errors in the estimate $\bar{X} = 1.26$ kilograms based on the regular scale? Explain.

9.20 The standard deviation of a population is $\sigma = 10$.
a. Calculate $\sigma\{\bar{X}\}$ for $n = 2, 5, 10, 20, 50,$ and 100.
b. Plot the $\sigma\{\bar{X}\}$ values obtained in **a** against n on a graph. Describe the effect of increasing the sample size on the variability of the sampling distribution of $\bar{X}$.

9.21 A population is normal with $\mu = 200$ and $\sigma = 20$.
a. Calculate $P(195 \leq \bar{X} \leq 205)$ for $n = 4, 10, 20, 50,$ and 100.
b. Plot the probabilities obtained in **a** against n on a graph. Describe the effect of increasing the sample size on the probability that $\bar{X}$ will be within ± 5 of the population mean μ.
c. Calculate $P(195 \leq \bar{X} \leq 205)$ for $n = 20$ if $\sigma = 2, 4, 12, 16, 20,$ and 30. Plot the probabilities against σ on a graph and describe the effect of a larger population standard deviation on the probability that $\bar{X}$ will be within ± 5 of the population mean μ.

STUDIES

9.22 Consider the discrete uniform probability distribution (6.12) with possible outcomes $X = 0, 1, 2, 3,$ and 4.
a. Obtain the mean and standard deviation of this infinite population.
b. Using Table C-8, draw 50 independent random samples, each of size $n = 2$, from this infinite population and obtain $\bar{X}$ for each sample.
c. Construct a frequency distribution of your 50 sample means and plot it as a frequency polygon. Does the theoretical shape of the sampling distribution of $\bar{X}$ appear to be symmetrical here? To be normal?
d. Obtain the exact sampling distribution of $\bar{X}$ by enumeration. Calculate $E\{\bar{X}\}$ and $\sigma\{\bar{X}\}$ from the exact sampling distribution and verify that these agree with theorems (9.3) and (9.4).

9.23 (Computer needed.)
a. Use a uniform random-number generator to generate 100 random samples of $n = 2$ each from the continuous uniform probability distribution (7.1) with

$a = 0$ and $b = 100$. Make a frequency distribution of the 100 $\bar{X}$ values and plot it as a frequency polygon.

b. Repeat **a** for $n = 5$ and $n = 10$.

c. For which sample sizes does the sampling distribution of $\bar{X}$ appear to be approximately normal?

9.24 (Computer needed.) Refer to Appendix D. Consider the population of all firms.

a. Determine the population mean and standard deviation for 1981 net income of all firms.

b. Using the computer, select 300 independent random samples, each of $n = 15$ companies. Use sampling with replacement—i.e., when a company is selected, do not remove it from the population. Calculate the sample mean $\bar{X}$ for each sample.

c. Using the computer, select 300 independent random samples, each of $n = 50$ companies. Use sampling with replacement—i.e., when a company is selected, do not remove it from the population. Calculate the sample mean $\bar{X}$ for each sample.

d. Use your simulation results to demonstrate that the sampling distribution of $\bar{X}$ approaches the normal distribution as the sample size increases, and that its mean and standard deviation are given by (9.3) and (9.4), respectively.

10
Estimation of
Population Mean

Statistical estimation of population characteristics from sample data is needed in a wide variety of circumstances.

☐ Examples

1. A market researcher wishes to estimate the mean shift in attitude induced by a television advertisement, based on a sample survey of television viewers.
2. In the development of inventory policy for a wholesaler, an operations analyst needs to estimate the mean levels of demand for stocked items, based on sample time periods.
3. A government energy agency requires an estimate of the total amount of heating oil burned in residential dwellings in a region during last winter, based on a sample of dwellings using heating oil. ☐

In this chapter, we build on earlier theoretical foundations to explain how a population mean (or population total) is estimated from simple random sample data. This is a common type of statistical estimation problem. In later chapters, we shall consider the estimation of other population characteristics.

10.1   POINT ESTIMATION

We examine the main features of point estimation by means of an illustration first and then summarize these features more formally.

☐ Illustration

A company requires an estimate of the mean length of service (μ) for the population of 3,580 employees, for current collective bargaining negotiations. Let us consider how to estimate the unknown parameter μ from a simple random sample of employees. One obvious way to estimate μ is to use the sample mean $\bar{X}$. Thus, the personnel office might select a simple random sample of 50 employees, determine the length of service for each, and calculate the sample mean $\bar{X}$ of the 50 lengths of service. If the sample mean is $\bar{X} = 6.3$ years, then 6.3 years would be the estimate of μ. ☐

The estimate 6.3 years in our example is called a point estimate.

(10.1) When a population characteristic is estimated by a single number, the number is called a *point estimate* of the characteristic.

Features of Point Estimation

Our illustration has shown the main features of point estimation with simple random sampling. These are the following:

1. There is an unknown population characteristic or parameter, denoted by θ (Greek theta), which is to be estimated.
2. To obtain a point estimate of θ, a simple random sample of n observations, X_1, $X_2, \ldots, X_n$, is selected from the population. Some statistic S, which is a function of the sample values, is used to estimate θ.
3. Prior to the actual selection of the sample, $X_1, X_2, \ldots, X_n$ are random variables. Hence, S is a random variable. The probability distribution of S is also called the sampling distribution of S.

With reference to our illustration, the population parameter θ is μ. The sample statistic S is $\overline{X}$, and from the definition of $\overline{X}$, we know it is the following function of $X_1, X_2, \ldots, X_n$:

$$\overline{X} = \frac{X_1 + X_2 + \cdots + X_n}{n}$$

Prior to the selection of the sample, $\overline{X}$ is a random variable whose probability distribution is called the sampling distribution of $\overline{X}$. We discussed this distribution in Chapter 9.

Since a sample statistic prior to sample selection is a random variable but after sample selection is simply a number, statisticians use the terms estimator and estimate to distinguish between these two situations:

(10.2) An *estimator* is a random variable used to estimate a population characteristic. An actual numerical value obtained for an estimator is called an *estimate*.

In our illustration, $\overline{X}$ is used as a point estimator of μ. For the particular sample selected in the study, $\overline{X} = 6.3$ years. The number 6.3 is an estimate of μ.

Alternative Point Estimators

It is possible to estimate a population mean μ by many point estimators. One possibility is $\overline{X}$. Another is the sample *midrange*:

(10.3)
$$\frac{X_S + X_L}{2}$$

which is the mean of the smallest (X_S) and largest (X_L) observations in the sample. Another possibility is the sample median Md, and there are still many other estimators of μ. How does one choose between these different possible point estimators? Basically, one wishes to choose a point estimator with desirable properties. We now consider some important properties of good point estimators.

10.2 PROPERTIES OF POINT ESTIMATORS

The statistical properties of point estimators which we now take up are related to the general criterion that a good estimator should provide reasonable assurance that the estimate obtained will be "close" to the population parameter being estimated.

Unbiasedness

It is desirable that the sampling distribution of an estimator be located near the parameter to be estimated, rather than far from it. This is the motivation of the property of unbiasedness:

(10.4) An estimator S is *unbiased* if the mean of its sampling distribution is equal to the population characteristic θ to be estimated; i.e., it is unbiased if:

$$E\{S\} = \theta$$

If estimator S is biased, the amount of its *bias* is:

$$\text{Bias} = E\{S\} - \theta$$

Figure 10.1 shows two estimators, S_1 being unbiased and S_2 having substantial bias. Clearly, S_2 will tend to give estimates far from θ, while estimates obtained from S_1 will tend to be nearer θ.

☐ Examples

1. In Chapter 9, it was shown that $E\{\bar{X}\} = \mu$. Hence, $\bar{X}$ is an unbiased estimator of μ.
2. The sample median Md is a biased estimator of μ when the population being sampled

---

FIGURE 10.1 *Sampling distributions of biased and unbiased estimators.* The estimator S_2 is biased because $E\{S_2\} \neq \theta$. The magnitude of the bias is $E\{S_2\} - \theta$.

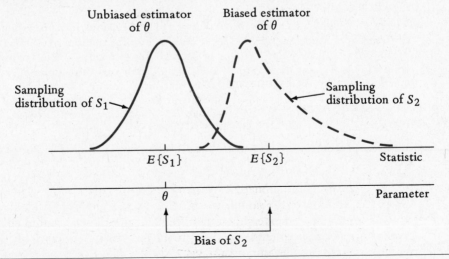

is skewed; i.e., $E\{Md\} \neq \mu$ then. For example, median income in a random sample of families will, on the average, understate mean family income in the population when the income distribution is right-skewed.

3. The sample variance s^2, defined in (3.15), is an unbiased estimator of the population variance σ^2 for infinite populations; i.e., $E\{s^2\} = \sigma^2$. Indeed, the reason for the denominator $n - 1$ in the sample variance is to obtain an unbiased estimator of σ^2.

4. The sample standard deviation s, defined in (3.17), is a biased estimator of the population standard deviation σ; i.e., $E\{s\} \neq \sigma$. However, the bias is small for reasonably large samples. □

Comments

1. Unbiasedness in point estimators refers to the tendency of sampling errors to cancel out over all possible samples. For any one sample, of course, the sample estimate will usually differ from the population parameter.

2. A biased estimator may still be a desirable estimator when the bias is not too large, if it has other good properties.

Efficiency

If two estimators have no bias, it is reasonable to prefer the estimator that has the smaller variability (i.e., the tighter sampling distribution) for the given sample size because its outcomes will tend to lie closer to the population characteristic. This observation leads to the property of relative efficiency:

(10.5) The *efficiency* of an unbiased estimator is measured by the variance of its sampling distribution. If two estimators based on the same sample size are both unbiased, the one with the smaller variance is said to have greater *relative efficiency* than the other; i.e., if:

$$\sigma^2\{S_1\} < \sigma^2\{S_2\} \quad \text{and} \quad E\{S_1\} = E\{S_2\} = \theta$$

then S_1 is relatively more efficient than S_2 in estimating θ.

Example

The mean shelf life of a new breakfast cereal is to be estimated. It can be assumed that the distribution of shelf lives of packages is normal, and the question is whether to use the mean life or median life of a sample of packages for estimating μ. For random sampling from a normal population, both $\overline{X}$ and Md are unbiased estimators of μ. We saw in Chapter 9 that $\sigma^2\{\overline{X}\} = \sigma^2/n$. From statistical theory, it can be shown that $\sigma^2\{Md\} \simeq 1.57(\sigma^2/n)$ for random sampling from a normal population when n is large. Thus, $\sigma^2\{\overline{X}\} < \sigma^2\{Md\}$ for given sample size n, and consequently $\overline{X}$ is relatively more efficient than Md as an estimator of μ here. Figure 10.2 illustrates the situation. Note that $\overline{X}$ has a greater probability than Md of falling within any specified interval about μ. □

Consistency

It is also desirable that an estimate tend to lie nearer the population characteristic as the sample size becomes larger. This is the basis of the property of consistency:

FIGURE 10.2   *Two unbiased estimators when sampling a normal population that differ in relative efficiency.* Both estimators are unbiased, but the sampling distribution of $\bar{X}$ is tighter than that of Md.

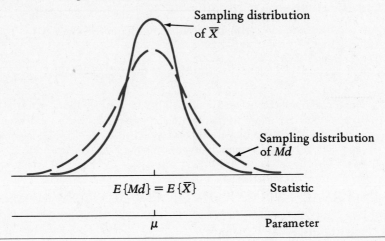

(10.6)   An estimator is a *consistent* estimator of a population characteristic θ if, the larger the sample size, the more likely it is that the estimate will be close to θ.

☐   Example

$\bar{X}$ is a consistent estimator of μ. Figure 10.3 illustrates, for sampling from a normal population, how the sampling distribution of $\bar{X}$ tightens around μ as the sample size increases.   ☐

FIGURE 10.3 $\bar{X}$ *as a consistent estimator of* μ. The sampling distributions of $\bar{X}$ for increasing sample sizes become more concentrated around μ.

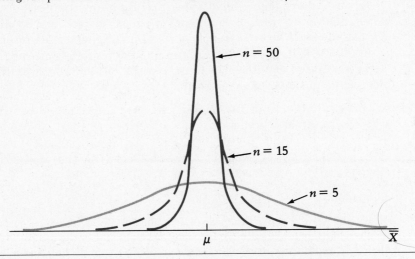

Comments

1. A consistent estimator may be biased. Consistency assures, however, that the bias becomes smaller as the sample size becomes larger.
2. A formal definition of consistency is:

(10.6a) S is a *consistent* estimator of population characteristic θ if for any small positive value ϵ:

$$\lim_{n \to \infty} P(|S - \theta| < \epsilon) = 1$$

The definition states that the probability is nearly 1 that a consistent estimator will have an outcome close to the parameter value (i.e., within $\pm\epsilon$) when n is sufficiently large. The only difficulty with this criterion is that the sample size may have to be very large before the assurance of closeness of the estimate takes effect.

10.3 INTERVAL ESTIMATION OF POPULATION MEAN

Need for Interval Estimate

Any point estimate has the limitation that it does not provide information about the *precision* of the estimate—i.e., about the magnitude of the error due to sampling. Often, such information is essential for proper interpretation of the sample result. For instance, a television audience rating agency reported that 24 percent of the sample households watched a particular show during the past four weeks. The sponsor of the program is assessing whether or not to continue sponsorship. If the margin of sampling error of the estimate were ±20 percent points, so that the proportion of households in the population watching the program could be as high as 44 percent or as low as 4 percent, the estimate would not be useful to the sponsor because of the wide margin of error due to sampling. On the other hand, if the margin of sampling error were ±2 percent points, the estimate would be quite useful since the sponsor would know that the proportion of households in the population viewing the program is somewhere between 22 and 26 percent. A point estimate by itself does not supply this information about its precision.

In this section, we consider statistical procedures for estimating the population mean in terms of an interval, where the width of the interval provides an indication of the precision of the estimate:

(10.7) An *interval estimate* of the population mean μ consists of two bounds within which μ is estimated to lie:

$$L \leq \mu \leq U$$

where L is the *lower bound* and U is the *upper bound*.

Illustration

In Chapter 9, we described the case of an auditor sampling the 8,042 accounts receivable of a freight company to estimate the mean audit amount of the receivables in the population. The auditor has selected a simple random sample of 100 accounts to be used in estimating the population mean μ. A fragment of the sample data is presented in Table 10.1. The sample mean is $\overline{X} = 33.19$, and we know that this estimator has desirable properties such as unbiasedness and consistency. Still, we also know that almost surely $\overline{X} = 33.19$ is not equal to μ. To develop an interval estimate that reflects the margin of error due to sampling, we need to estimate the variability of the sampling distribution of $\overline{X}$. ☐

Estimated Standard Deviation of $\overline{X}$

Recall from Chapter 9 that the variability of the sampling distribution of $\overline{X}$ indicates how likely it is that the sample mean $\overline{X}$ is close to the population mean μ. The smaller the variability, the greater the probability that $\overline{X}$ will fall within any specified interval around μ.

Even though we take only one sample from the population, we can estimate the variability of the sampling distribution of $\overline{X}$. The reason is that $\sigma^2\{\overline{X}\}$, the variance of $\overline{X}$, is a simple function of the population variance σ^2, as shown by (9.4), namely, $\sigma^2\{\overline{X}\} = \sigma^2/n$.

An estimate of $\sigma^2\{\overline{X}\}$ is obtained by the following reasoning. We noted earlier that the sample variance s^2 is an unbiased estimator of the population variance when the population is infinite. Hence, an unbiased estimator of $\sigma^2\{\overline{X}\} = \sigma^2/n$ is s^2/n. We denote this estimator by $s^2\{\overline{X}\}$ and the estimated standard deviation of $\overline{X}$ by $s\{\overline{X}\}$. Thus, $s^2\{\overline{X}\} = s^2/n$ and $s\{\overline{X}\} = s/\sqrt{n}$. This latter standard deviation is often called the *estimated standard error of the mean.*

(10.8) Estimators of the variance and standard deviation of the sampling distribution of $\overline{X}$ are, respectively:

$$s^2\{\overline{X}\} = \frac{s^2}{n} \qquad s\{\overline{X}\} = \frac{s}{\sqrt{n}}$$

These estimators are also applicable when a finite population is sampled as long as the sample size n is not large relative to the population size N, i.e., as long as the sampling fraction n/N is small.

TABLE 10.1 *Sample results for random sample of 100 accounts receivable*

i	X_i	$X_i - \overline{X}$	$(X_i - \overline{X})^2$
1	80.29	47.10	2,218.41
2	6.97	−26.22	687.49
3	4.55	−28.64	820.25
⋮	⋮	⋮	⋮
99	51.51	18.32	335.62
100	10.30	−22.89	523.95
Total	3,318.73	0	117,674.67

$$n = 100 \qquad \overline{X} = \frac{3,318.73}{100} = 33.19 \qquad s^2 = \frac{117,674.67}{100 - 1} = 1,188.63 \qquad s = 34.48$$

☐   **Example**

We wish to estimate the variability of the sampling distribution of $\bar{X}$ from the sample results in Table 10.1; we have $s^2 = 1{,}188.63$ and $n = 100$. By (10.8), we obtain:

$$s^2\{\bar{X}\} = \frac{1{,}188.63}{100} = 11.886 \qquad s\{\bar{X}\} = \sqrt{11.886} = 3.448$$

Since the sample size is large $(n = 100)$, the sampling distribution of $\bar{X}$ is approximately normal by central limit theorem (9.5). Consequently, most of the time the sample mean $\bar{X}$ should be within, say, two standard deviations of the population mean. Thus, $s\{\bar{X}\} = 3.45$ implies that most of the time the sample mean $\bar{X}$ should be within $2(3.45) = 6.90$ of μ.

Incidentally, we know that $\sigma = 30.334$ here and $\sigma\{\bar{X}\} = 3.03$ (see Chapter 9), so our sample estimate $s\{\bar{X}\} = 3.45$ is reasonably close.   ☐

Behavior of Interval Estimate

We just found for our example that $s\{\bar{X}\} = 3.45$ and concluded that most of the time the sample mean $\bar{X}$ should be within $2(3.45) = \$6.90$ of the population mean. If that is so, we might reason that starting with $\bar{X}$ and adding and subtracting $\$6.90$ should give us limits that are likely to include μ. Here, the limits would be:

$$L = \bar{X} - 2s\{\bar{X}\} = 33.19 - 2(3.45) = 26.29$$
$$U = \bar{X} + 2s\{\bar{X}\} = 33.19 + 2(3.45) = 40.09$$

and we would therefore state:

$$26.29 \leq \mu \leq 40.09$$

Thus, we would estimate that the mean audit amount per account in the population is somewhere between $\$26.29$ and $\$40.09$.

Since we happen to know $\mu = \$30.303$ in this case, we see that this particular interval estimate is correct. But what is the behavior of this interval estimate in general?

Note that our approach involves the following steps:

1. Select a random sample of 100 accounts.
2. Obtain $\bar{X}$ and s for that sample.
3. Estimate $\sigma\{\bar{X}\}$ by $s\{\bar{X}\}$.
4. Calculate the interval:

(10.9) $$\bar{X} - 2s\{\bar{X}\} \leq \mu \leq \bar{X} + 2s\{\bar{X}\}$$

The term $2s\{\bar{X}\}$ represents our calculated margin of error due to sampling.

We shall now study the behavior of interval estimate (10.9) by experimentation. We have selected 600 independent samples of 100 accounts from the population of 8,042 accounts receivable. The auditor's sample in Table 10.1 is the first of the 600. In the second sample, we obtained $\bar{X} = 31.89$ and $s = 34.94$. If the auditor had selected this second sample, we would have calculated

$s\{\bar{X}\} = 34.94/\sqrt{100} = 3.494$ and $2(3.494) = 6.988$,   giving   limits   of 31.89 ± 6.988. Hence, the interval is $24.90 \leq \mu \leq 38.88$. Again, this interval contains $\mu = 30.303$. Figure 10.4 shows the interval estimates for a few of the 600 samples, including the two already mentioned. Note that all but one of the interval estimates displayed contain $\mu = 30.303$. The exception is sample 39, which provides an interval estimate lying entirely above μ. In all, 559 of the 600 intervals, or 93.2 percent, contain μ.

This proportion of correct intervals is a measure of the confidence we can have in the interval estimation procedure. Here, it is quite likely that the interval estimate obtained for the one sample actually selected will contain μ. Clearly, the proportion of correct intervals is a function of the multiple of $s\{\bar{X}\}$. In our study, we used a multiple of 2. In general, as we shall now see, we can use any multiple and, indeed, will pick that multiple which yields a specified probability of a correct interval.

FIGURE 10.4   *Interval estimates of the form* $\bar{X} \pm 2s\{\bar{X}\}$ *for several of 600 samples. The intervals differ in location because* $\bar{X}$ *varies from sample to sample and they differ in width because* s *varies from sample to sample.*

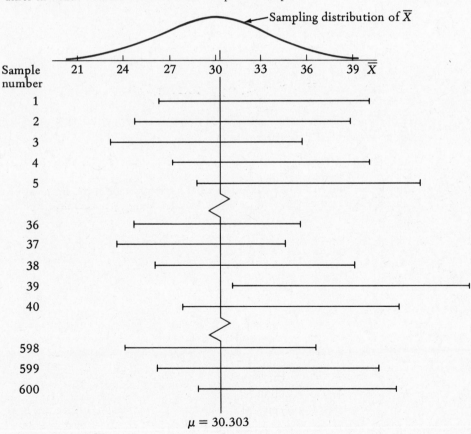

Confidence Interval for μ—Large Sample

We now consider the construction of an interval estimate $L \leq \mu \leq U$ for any specified probability of a correct interval, when the sample size is large.

(10.10) The probability that a correct interval estimate is obtained is called the *confidence coefficient* and is denoted by $1 - \alpha$. The interval:

$$L \leq \mu \leq U$$

is called a *confidence interval*.

L and U are called the *lower* and *upper confidence limits*, respectively. The numerical confidence coefficient (e.g., .95) is often expressed as a percent (e.g., 95 percent). A confidence interval that has an associated confidence coefficient of $1 - \alpha$ is frequently called a $100(1 - \alpha)$ percent confidence interval. For instance, a 95 percent confidence interval is a confidence interval with a confidence coefficient of .95.

(10.11) The confidence limits L and U for the population mean μ with approximate confidence coefficient $1 - \alpha$, when the sample size is reasonably large, are:

$$\bar{X} \pm zs\{ \bar{X} \}$$

where:　$z = z(1 - \alpha/2)$
　　　　$s\{\bar{X}\}$ is given by (10.8)

The $100(1 - \alpha)$ percent confidence interval for μ is:

$$\bar{X} - zs\{ \bar{X} \} \leq \mu \leq \bar{X} + zs\{\bar{X}\}$$

Recall from Chapter 7 that $z(1 - \alpha/2)$ denotes the $100(1 - \alpha/2)$ percentile of the standard normal distribution.

☐　Examples

1. Refer to Table 10.1. We wish to construct a confidence interval for μ with confidence coefficient $1 - \alpha = .954$. Hence, $\alpha = .046$ and $1 - \alpha/2 = .977$. We find from Table C-1 that $z(.977) = 2.0$. We previously obtained:

$$\bar{X} = 33.19 \qquad s\{\bar{X}\} = 3.448$$

Hence:

$$L = 33.19 - 2(3.448) = 26.29 \qquad U = 33.19 + 2(3.448) = 40.09$$

and:

$$26.29 \leq \mu \leq 40.09$$

This is the confidence interval we obtained before. Now, we know that it has an approximate confidence coefficient of 95.4 percent. Recall that the proportion of correct intervals in the 600 trials was close to 95.4 percent, namely, 93.2 percent.

2. Refer to Table 10.1 again. Suppose we wish to use a confidence coefficient of $1 - \alpha = .90$. Hence, we require $z(.95) = 1.645$. Our confidence limits will then be $33.19 \pm 1.645(3.448)$, and the confidence interval is:

$$27.52 \leq \mu \leq 38.86$$

3. In a random sample of $n = 40$ bricks from a production process, the mean weight was $\bar{X} = 3.724$ pounds and the standard deviation was $s = .071$ pound. We wish to estimate the process mean μ with a 99 percent confidence interval. We require:

$$s\{\bar{X}\} = \frac{.071}{\sqrt{40}} = .011 \qquad z(.995) = 2.576$$

and obtain the confidence limits $3.724 \pm 2.576(.011)$. The confidence interval therefore is:

$$3.70 \leq \mu \leq 3.75$$

We thus conclude, with 99 percent confidence, that the process mean weight of bricks is between 3.70 and 3.75 pounds. ☐

Development of Confidence Interval

We now show the formal basis of confidence interval (10.11). First, we need a formal definition of a confidence interval:

(10.12) The interval $L \leq \mu \leq U$ is a $1 - \alpha$ confidence interval for the population mean μ if prior to sampling:

$$P(L \leq \mu \leq U) = 1 - \alpha$$

This definition simply states that a confidence interval with confidence coefficient $1 - \alpha$ is an interval estimate such that the probability is $1 - \alpha$ that the calculated limits include μ for any random trial (i.e., for any random sample of size n). In other words, in many random samples of n from a population, $100(1 - \alpha)$ percent of the interval estimates will include μ and therefore will be correct.

The validity of confidence interval (10.11) depends on an extension of central limit theorem (9.5):

(10.13) For almost any population, $(\bar{X} - \mu)/s\{\bar{X}\}$ follows approximately a standard normal distribution when the random sample size is sufficiently large.

For populations that are not highly skewed, sample sizes need not be too large for theorem (10.13) to apply. For highly skewed populations, however, the sample size may need to be quite large.

Assuming n is large enough that $(\bar{X} - \mu)/s\{\bar{X}\}$ is distributed approximately as $N(0, 1)$, we can state:

$$P\left[z(\alpha/2) \leq \frac{\bar{X} - \mu}{s\{\bar{X}\}} \leq z(1 - \alpha/2)\right] \simeq 1 - \alpha$$

Figure 10.5 illustrates why this probability is $1 - \alpha$. In each tail, the area is $\alpha/2$, so the central area is $1 - \alpha$. But by (7.11), $z(\alpha/2) = -z(1 - \alpha/2)$. Hence:

$$P\left[-z(1 - \alpha/2) \leq \frac{\bar{X} - \mu}{s\{\bar{X}\}} \leq z(1 - \alpha/2)\right] \simeq 1 - \alpha$$

Rearranging the inequalities, we obtain:

$$P[\bar{X} - z(1 - \alpha/2)s\{\bar{X}\} \leq \mu \leq \bar{X} + z(1 - \alpha/2)s\{\bar{X}\}] \simeq 1 - \alpha$$

FIGURE 10.5 *Two-sided confidence interval for μ—large sample.* Before the sample is taken, the probability is $1 - \alpha$ that the quantity $(\bar{X} - \mu)/s\{\bar{X}\}$ will fall in the shaded interval. The interval estimate $\bar{X} \pm z(1 - \alpha/2)s\{\bar{X}\}$ will be correct (i.e., will contain μ) if $(\bar{X} - \mu)/s\{\bar{X}\}$ does fall in the shaded interval.

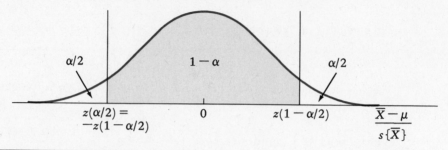

But this expression fits the definition of a confidence interval in (10.12) where the lower and upper confidence limits are:

$$L = \bar{X} - z(1 - \alpha/2)s\{\bar{X}\} \qquad U = \bar{X} + z(1 - \alpha/2)s\{\bar{X}\}$$

These are, of course, the limits given in (10.11).

Figure 10.5 summarizes the reasoning that underlies the confidence interval for μ in (10.11). In effect, the risk α of an incorrect confidence interval is divided equally in the two tails of the standard normal distribution.

Comment

We can now see why confidence interval (10.11) has a confidence coefficient of *approximately* $1 - \alpha$. The reason is that $(\bar{X} - \mu)/s\{\bar{X}\}$, in general, is only approximately normally distributed for large n.

Confidence Coefficient

Meaning of Confidence Coefficient. From the definition of a confidence interval in (10.12), we know that, *in advance* of selecting the sample, the probability is $1 - \alpha$ that the confidence interval we obtain will contain the population mean μ. The particular confidence interval obtained, however, may or may not contain μ. Thus, any one confidence interval result will be either correct or incorrect, and we do not know for certain which is the case. We act as if the result is correct when $1 - \alpha$ is sufficiently large because the procedure gives us assurance that in $100(1 - \alpha)$ percent of the cases, the confidence interval result will be correct.

Selecting the Confidence Coefficient. In the best of all worlds we would like the confidence interval to be very precise (i.e., very narrow) and would like to be very confident that it contains μ. Unfortunately, for any fixed sample size, the confidence coefficient can only be increased by increasing the width of the confidence interval. Figure 10.6 illustrates the relationship graphically. It shows how the

FIGURE 10.6 *Relation between confidence coefficient and confidence interval width*

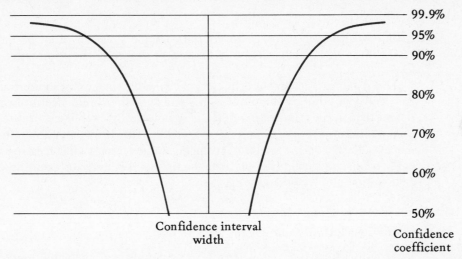

width of confidence interval (10.11) varies with the confidence coefficient for any given sample result. Note how rapidly the confidence interval widens as the confidence coefficient gets near 100 percent.

The choice of $1 - \alpha$ will vary from case to case, depending on how much risk of obtaining an incorrect interval can be taken. Confidence coefficients of 90, 95, and 99 percent are often used in practice.

Comment

Confidence intervals provide an evaluation of the magnitude of the sampling error only. They do not take into account biases in the data, such as systematic understatements of income by respondents.

10.4 PLANNING OF SAMPLE SIZE

Need for Planning

The user of an interval estimate often would like to specify in advance the precision required in an estimate and the confidence level it should have.

Examples

1. A market manager specifies that it is necessary to know the mean income of households living in the vicinity of a shopping center within ±$500 with a 95 percent confidence coefficient. The number of households to be included in the sample needs to be determined.
2. In a merger negotiation, the mean number of days of sick leave accrued by the employees of one of the corporations must be estimated. Lawyers on both sides agree that it

will be sufficient to estimate this mean within ± 2 days with a confidence coefficient of 99 percent. The number of employee files to be included in the sample needs to be determined. ☐

Based on such specifications, appropriate sample sizes can often be determined that will approximately yield the desired precision at the specified confidence level.

This planning approach is often more reasonable than using an arbitrary sample size. With an arbitrary sample size, the width of the confidence interval may be tighter than needed, indicating that the sample size was too large, or it may be too wide, indicating that the sample size was inadequate. If too large a sample is taken, money will have been wasted in getting greater precision than necessary. If, on the other hand, too small a sample is taken, it may be costly or even impossible to enlarge the sample subsequently.

Planning Method

We consider here the case where: (1) the sample size ultimately determined is reasonably large, and (2) the population is either infinite or, if finite, is large relative to the resulting sample size.

☐ Illustration

A farming region contains over 40,000 farms. In the early spring, it is planned to survey a simple random sample of farm operators to determine the acreage each operator intends to sow in summer wheat. The survey forms part of a larger study designed to estimate total agricultural production in the next twelve months. It is desired to have the wheat acreage survey produce an interval estimate of the mean wheat acreage per farm within ± 5 acres with a confidence level of 95 percent. ☐

As shown by the illustration, two specifications must be made for determining the needed sample size. One is the desired *margin of sampling error*, denoted by $\pm h$, which indicates how close the confidence limits must be around the point estimate $\overline{X}$. In the farming illustration, $h = 5$ acres. Thus, it is desired that the confidence limits $\overline{X} \pm h$ should be $\overline{X} \pm 5$. The symbol h denotes the desired *half-width* of the confidence interval, since the total width of the confidence interval is desired to be $2h$. The second specification is the desired confidence coefficient $1 - \alpha$. In the farming illustration, it is specified that $1 - \alpha = .95$.

From (10.11), we know that the half-width h of the confidence interval is $h = z\sigma\{\overline{X}\}$, where $z = z(1 - \alpha/2)$. Note that here we use the notation for the true standard deviation of the mean, $\sigma\{\overline{X}\}$, rather than that for the estimated standard deviation, $s\{\overline{X}\}$, because we are in the planning phase.

We have by (9.4):

$$\sigma\{\overline{X}\} = \frac{\sigma}{\sqrt{n}}$$

Hence:

$$h = z\sigma\{\bar{X}\} = z\frac{\sigma}{\sqrt{n}}$$

Squaring both sides and solving for n, we obtain:

$$n = \frac{z^2\sigma^2}{h^2}$$

To use this formula, we need a planning value for the population standard deviation σ, since σ is usually unknown. Information about σ is often available from past experience with the same or a similar problem, or can be obtained from a pilot study. Approximate information about the order of magnitude of σ is usually adequate for planning purposes.

(10.14) The needed sample size, based on the desired half-width h of the confidence interval and the specified confidence coefficient $1 - \alpha$, for a given planning value of σ is:

$$n = \frac{z^2\sigma^2}{h^2}$$

where: $z = z(1 - \alpha/2)$

This planning procedure is applicable when the resulting sample size n is large and the population is either infinite or, if finite, is large relative to the resulting sample size.

Example

In our farming illustration, $h = 5$ acres and $1 - \alpha = .95$. Hence, $\alpha = .05$ and $1 - \alpha/2 = .975$. Thus, $z(.975)$ is required. We find from Table C–1 that $z(.975) = 1.960$. Ample recent historical data about farm wheat acreage in the region are available and consultation with agricultural experts indicates that the variability of wheat acreages among farms in the coming year is not likely to differ sharply from that exhibited in the recent past. Based on the historical data, a planning value of $\sigma = 85$ acres seems reasonable. Substituting this and the other values into (10.14) gives the following needed sample size for the survey:

$$n = \frac{(1.960)^2(85)^2}{(5)^2} = 1{,}110$$

If the planning value of σ is at all reliable, this sample size should produce a 95 percent confidence interval with a half-width of about 5 acres. Note that formula (10.14) is appropriate here because the resultant n is large while n/N is small.

Once the sample size has been determined and the actual sample selected, the construction of a confidence interval proceeds in the usual manner. We use, of course, the sample standard deviation s and not the planning value for σ in constructing the interval.

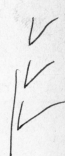

☐ **Example**

Suppose a random sample of 1,110 farms was selected in our farming illustration and yielded $\bar{X} = 75.6$ acres and $s = 80.1$ acres. We calculate $s\{\bar{X}\} = 80.1/\sqrt{1,110} = 2.404$ by (10.8). Hence, the 95 percent confidence limits for the mean wheat acreage per farm are, by (10.11), $75.6 \pm 1.960(2.404)$. The confidence interval therefore is:

$$70.9 \leq \mu \leq 80.3$$

Observe that the attained half-width is 4.7 acres. This is slightly smaller than the 5-acre value specified in the planning stage. The extra precision resulted because the sample standard deviation of 80.1 acres is a little smaller than the planning value of 85 acres. ☐

10.5 CONFIDENCE INTERVAL FOR μ—SMALL SAMPLE

When the sample size is small, $(\bar{X} - \mu)/s\{\bar{X}\}$ no longer follows the standard normal distribution and the construction of the confidence interval for μ depends on the nature of the underlying population.

Normal Population

When the population is normal, the following theorem helps us to construct confidence intervals for μ:

(10.15) For a simple random sample of size n from a normal population:

$$\frac{\bar{X} - \mu}{s\{\bar{X}\}} = t(n - 1)$$

In other words, $(\bar{X} - \mu)/s\{\bar{X}\}$ has a t distribution with $n - 1$ degrees of freedom when the sampled population is normal. (If a review of the t distribution is needed, the reader should turn to Appendix B, Section B.2 before proceeding in this section.)

In view of theorem (10.15), the confidence limits for μ when n is small and the population is normal will be of exactly the same form as those for large samples except that an appropriate percentile of the t distribution is used instead of a percentile from the standard normal distribution.

(10.16) The confidence limits for μ with confidence coefficient $1 - \alpha$, when the population is normal or the departure is not too marked, are:

$$\bar{X} \pm ts\{\bar{X}\}$$

where: $t = t(1 - \alpha/2; n - 1)$
$s\{\bar{X}\}$ is given by (10.8)

☐ **Example**

A sample of five cans of standard-grade tomatoes in puree was taken at random from a production line immediately after filling. The solid content of each can was weighed to serve as a basis for estimating the process mean drained weight per can (μ). Past experience indicates that the distribution of drained weights is normal. It is desired to construct a 99 percent confidence interval for the process mean μ.

The following are the sample results:

i:	1	2	3	4	5
X_i:	23.0	23.5	23.5	25.0	24.5

$$n = 5 \qquad \bar{X} = 23.9 \qquad s = .822$$

Hence, $s\{\bar{X}\} = s/\sqrt{n} = .822/\sqrt{5} = .37$. The desired confidence coefficient is $1 - \alpha = .99$, so $\alpha = .01$ and $1 - \alpha/2 = .995$. The degrees of freedom for the t distribution are $n - 1 = 5 - 1 = 4$. From Table C-3, we find that $t(.995; 4) = 4.604$. Using (10.16), the confidence limits are $23.9 \pm 4.604(.37)$. This gives the confidence interval:

$$22.20 \leq \mu \leq 25.60$$

It can now be stated, with 99 percent confidence, that the mean drained weight for the current process is somewhere between 22.20 and 25.60 ounces. □

Robustness of t Distribution

Confidence interval (10.16) is applicable even if the population is not exactly normal. Statistical theory shows that as long as the population does not depart too markedly from normality (e.g., it is not highly skewed) and n is not exceedingly small, $(\bar{X} - \mu)/s\{\bar{X}\}$ is distributed approximately as $t(n - 1)$. Hence, confidence interval (10.16) may still be used in these cases, although the confidence coefficient now is only approximately $1 - \alpha$. This property of the t distribution, i.e., that it applies approximately for many other populations besides normal ones, is called *robustness*.

Comments

1. As the degrees of freedom $n - 1$ increase, we know from (B.6) in Appendix B that the t distribution becomes more concentrated. Hence, the t multiple in (10.16) becomes smaller for larger n. In fact, for large n we use the standard normal distribution to approximate the t multiple.
2. Occasionally, the population standard deviation σ is known and need not be estimated by the sample standard deviation s. In this case, $\sigma\{\bar{X}\}$ is also known exactly. If the population is normal, then $(\bar{X} - \mu)/\sigma\{\bar{X}\}$ follows a standard normal distribution by (7.9). Consequently, the appropriate confidence limits in this case are $\bar{X} \pm z\sigma\{\bar{X}\}$, where $z = z(1 - \alpha/2)$, irrespective of sample size.
3. The confidence limits (10.16) for sampling a normal population are obtained by using theorem (10.15). It follows from this theorem that:

$$P\left[-t(1 - \alpha/2; n - 1) \leq \frac{\bar{X} - \mu}{s\{\bar{X}\}} \leq t(1 - \alpha/2; n - 1)\right] = 1 - \alpha$$

when sampling from a normal population. Recall that $t(1 - \alpha/2; n - 1)$ is the $100(1 - \alpha/2)$ percentile of the t distribution with $n - 1$ degrees of freedom. Recall also, that because of the symmetry of the t distribution about its mean 0,

$t(\alpha/2; n - 1) = -t(1 - \alpha/2; n - 1)$. The preceding probability is analogous to the one illustrated in Figure 10.5 for the case of large n.

Rearranging the expression inside the brackets, as we did in the large-sample case, gives:

$$P[\bar{X} - t(1 - \alpha/2; n - 1)s\{\bar{X}\} \leq \mu \leq \bar{X} + t(1 - \alpha/2; n - 1)s\{\bar{X}\}] = 1 - \alpha$$

Thus, we have obtained $1 - \alpha$ confidence limits for μ.

10.6 CONFIDENCE INTERVAL FOR POPULATION TOTAL

A confidence interval for the population total of a finite population is often desired. For example, in the study of the population of 8,042 accounts receivable, the auditor was actually interested in the total audit amount of these accounts, and his concern with the mean audit amount (μ) was merely a step toward obtaining a confidence interval for the population total.

Let τ (Greek tau) denote the population total:

(10.17)
$$\tau = \sum_{i=1}^{N} X_i$$

In the auditing example, τ denotes the total of all audit amounts for the $N = 8,042$ accounts in the population. Since, by definition (8.2a) of the population mean, we have $N\mu = \Sigma X_i$, it follows that:

(10.18)
$$\tau = N\mu$$

Consequently, confidence limits for τ can be obtained by multiplying the confidence limits for μ by N.

☐ Example

Refer to Table 10.1. We wish to obtain a 95.4 percent confidence interval for the total audit amount of the accounts in the population. Earlier, we found the 95.4 percent confidence interval for μ:

$$26.29 \leq \mu \leq 40.09$$

Since $N = 8,042$, the 95.4 percent confidence limits for the total audit amount of the company's accounts receivable are:

$$26.29(8,042) = 211,424 \qquad 40.09(8,042) = 322,404$$

so the 95.4 percent confidence interval is:

$$211,424 \leq \tau \leq 322,404$$

We conclude, with 95.4 percent confidence, that the total audit amount of the 8,042 accounts is between \$211.4 thousand and \$322.4 thousand. ☐

10.7 OPTIONAL TOPIC—ONE-SIDED CONFIDENCE INTERVAL FOR μ

Thus far, our confidence intervals have had both upper and lower confidence limits. Sometimes, however, one-sided confidence intervals are useful. For example, a bank officer contemplating making a loan to a firm on its inventory would be interested in an upper limit for the total value of obsolete items in the inventory. Again, a purchasing agent would be interested in a lower limit for the mean life of electronic components in a shipment just received.

One-sided confidence intervals are constructed in a similar fashion to two-sided intervals. The only difference is that the risk α is placed all in one side. This may be seen by examining Figure 10.4 again. If we were only interested in an upper confidence limit here, note that all upper limits shown are correct, including that for sample 39, where the lower limit was too high. Thus, when constructing a one-sided confidence limit, we incur risks of incorrect limits in only one side of Figure 10.5. Hence, for one-sided confidence limits with confidence coefficient $1 - \alpha$, the risk α of an incorrect limit is placed entirely in one side of the distribution, and $100(1 - \alpha)$ percentiles are used in constructing the confidence limits.

We present the one-sided confidence limits for μ for large sample sizes:

(10.19) One-sided confidence limits for μ with approximate confidence coefficient $1 - \alpha$, when the sample size is reasonably large, are:

Lower Confidence Limit: $L = \bar{X} - z(1 - \alpha)s\{\bar{X}\}$

Upper Confidence Limit: $U = \bar{X} + z(1 - \alpha)s\{\bar{X}\}$

☐ Example

Refer to Table 10.1. We wish to obtain a lower 95 percent confidence limit for the mean audit amount per account receivable in the population. We know from before that $\bar{X} = 33.19$ and $s\{\bar{X}\} = 3.448$. For $1 - \alpha = .95$, we require $z(.95) = 1.645$. Hence, the lower confidence limit is $33.19 - 1.645(3.448) = 27.52$. Thus, with 95 percent confidence, it can be asserted that $\mu \geq 27.52$. In terms of the total audit amount of the company's accounts receivable, the auditor can conclude, with 95 percent confidence, that $\tau \geq 8,042(27.52) = 221,316$ or \$221.3 thousand.   ☐

Comments

1. Note that the two-sided $1 - \alpha$ confidence limits for μ in (10.11) can be viewed as the respective one-sided lower and upper confidence limits in (10.19), each with confidence coefficient $1 - \alpha/2$.
2. One-sided confidence limits for small sample sizes when the population is normal or the departure is not too marked are the same as in (10.19), except that $z(1 - \alpha)$ is replaced by $t(1 - \alpha; n - 1)$.

10.8 OPTIONAL TOPIC—MAXIMUM LIKELIHOOD METHOD OF FINDING GOOD POINT ESTIMATORS

Good point estimators for particular circumstances are not always intuitively obvious. We now discuss one important method of finding point estimators with desirable properties, namely, the *maximum likelihood method*.

☐ **Illustration**

The weekly demand for a programmable pocket calculator at a retailer follows a Poisson distribution with unknown parameter λ. Three weeks' observations are available; they are $X_1 = 2$, $X_2 = 4$, and $X_3 = 3$. We assume that these observations are independent.

To find the maximum likelihood estimate of λ here, we first need to find an expression for the probability of obtaining the three sample observations. Recall that the Poisson probability function is given by (6.8), here denoted by $P(X; \lambda)$ to show explicitly that the probability of outcome X depends on the value of the parameter λ:

$$P(X; \lambda) = \frac{\lambda^X \exp(-\lambda)}{X!}$$

We shall denote the joint probability of obtaining the observations X_1, X_2, X_3 by $P(X_1, X_2, X_3; \lambda)$. Because of the independence of the observations, we have:

$$P(X_1, X_2, X_3; \lambda) = \frac{\lambda^{X_1} \exp(-\lambda)}{X_1!} \cdot \frac{\lambda^{X_2} \exp(-\lambda)}{X_2!} \cdot \frac{\lambda^{X_3} \exp(-\lambda)}{X_3!}$$

$$= \frac{\lambda^{X_1 + X_2 + X_3} \exp(-3\lambda)}{X_1! X_2! X_3!}$$

For our specific outcomes, we have:

$$P(2, 4, 3; \lambda) = \frac{\lambda^{2+4+3} \exp(-3\lambda)}{2! 4! 3!}$$

Since the joint probability of the observed sample outcomes is just a function of λ, we shall denote it by $L(\lambda)$ and call it the likelihood function.

The method of maximum likelihood now requires that value of λ which maximizes the likelihood function for the observed sample outcomes; this value is the maximum likelihood estimate of λ. It is instructional to proceed by trial and error to find the appropriate λ value. If, for instance, $\lambda = 2$, we have:

$$L(2) = \frac{2^{2+4+3} \exp(-6)}{2! 4! 3!} = .0044$$

If $\lambda = 3$, we obtain:

$$L(3) = \frac{3^{2+4+3} \exp(-9)}{2! 4! 3!} = .0084$$

These and other likelihood values are plotted in Figure 10.7. It appears from this figure that $L(\lambda)$ is maximized for $\lambda = 3$, so this is our maximum likelihood estimate. Note also that $\bar{X} = 3$. ☐

FIGURE 10.7   *Poisson likelihood function for calculator demand example.* The maximum likelihood estimate is that value of λ which maximizes the likelihood function.

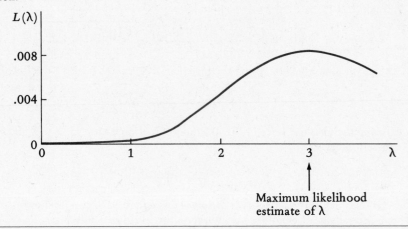

Maximum likelihood
estimate of λ

In lieu of a trial-and-error approach, one can find the value of λ which maximizes $L(\lambda)$ analytically. We demonstrate in the next example that this value is always $\overline{X}$ when sampling a Poisson population, so the sample mean $\overline{X}$ is the maximum likelihood estimator of λ here.

General Approach

We now summarize the essence of the maximum likelihood method when the random variable is discrete:

1. We start with the probability function of a sample observation X_i, i.e., $P(X_i; \theta)$, where θ is the unknown parameter.
2. The joint probability function for the sample, because of the independence of the sample observations $X_1, X_2, \ldots, X_n$, is:

$$P(X_1, X_2, \ldots, X_n; \theta) = P(X_1; \theta) P(X_2; \theta) \cdots P(X_n; \theta)$$

When this is viewed as a function of θ, for given $X_1, X_2, \ldots, X_n$, it is called the *likelihood function $L(\theta)$.*
3. We maximize the likelihood function with respect to θ. This value of θ is a function of the sample observations and is the maximum likelihood estimator of θ.

(10.20) The maximum likelihood method for estimating a parameter θ selects as a point estimator that value of θ which maximizes the likelihood function:

$$L(\theta) = P(X_1; \theta) P(X_2; \theta) \cdots P(X_n; \theta)$$

☐ **Example**

(Calculus needed.) When a random sample of size n is selected from a Poisson distribution, $P(X; \lambda)$ is given by (6.8). Substituting into the likelihood function in (10.20), we obtain:

$$L(\lambda) = \left[\frac{\lambda^{X_1}\exp(-\lambda)}{X_1!}\right]\left[\frac{\lambda^{X_2}\exp(-\lambda)}{X_2!}\right] \cdots \left[\frac{\lambda^{X_n}\exp(-\lambda)}{X_n!}\right]$$

Algebraic simplification, using the fact that $\Sigma X_i = n\bar{X}$, leads to the likelihood function:

$$L(\lambda) = \frac{\lambda^{n\bar{X}}\exp(-n\lambda)}{X_1!X_2! \cdots X_n!}$$

Taking the first derivative of the likelihood function with respect to λ and setting it equal to zero gives:

$$\frac{dL}{d\lambda} = n\left(\frac{\bar{X}}{\lambda} - 1\right)\frac{\lambda^{n\bar{X}}\exp(-n\lambda)}{X_1!X_2! \cdots X_n!} = 0$$

It is seen that setting λ equal to $\bar{X}$ solves this equation. It can be shown that this solution corresponds to an absolute maximum of the likelihood function. Hence, $\bar{X}$ is the maximum likelihood estimator of λ for a random sample from a Poisson distribution. ☐

Properties of Maximum Likelihood Estimators

Under quite general conditions, maximum likelihood estimators are consistent estimators. Also, for reasonably large samples, they are usually approximately normally distributed and more efficient than other estimators. This last property makes the maximum likelihood estimators particularly useful in statistical analysis.

Comment

Another method which is widely used for finding point estimators with desirable properties is the method of *least squares*. We shall discuss this method in Chapter 18.

PROBLEMS

10.1 The mean damage per crate in a random sample of 50 crates from a shipment damaged in a train derailment was $350. Distinguish between the estimator and the estimate here.

10.2 A sample of four tires of a new type is subjected to forced-life testing. The first tire in the sample to wear out has X miles on it. Is X an unbiased estimator of the mean life of this type of tire? Explain.

10.3 As noted in the Text, $\bar{X}$ is relatively more efficient than Md as an estimator of μ in a normal population. Does this mean: (1) that $\bar{X}$ will lie closer to μ than Md will in every random sample from a normal population? (2) that a smaller sample size is needed to estimate μ if $\bar{X}$ is used as the estimator than if Md is used? Explain.

10.4 A travel agents' association commissioned a survey of the winter travel plans and preferences of a random sample of retired persons. One item of particular interest was the mean anticipated expenditure per retired person for travel during the coming winter. The survey was conducted during the third week in September and the response rate was 70 percent. Interviewers, when commenting on reasons for nonresponses, often noted that the nonrespondent was away on a trip. Explain what is the population mean μ here which $\bar{X}$, the mean anticipated expenditure per respondent in the sample survey, is estimating. Is μ here the quantity the travel agents' association wishes to estimate? Might a larger sample size have helped? Comment.

*10.5 Amounts of postage were recorded for a random sample of $n = 400$ parcels handled by the postal system on a particular day. The sample mean and standard deviation were $\bar{X} = \$3.47$ and $s = \$2.50$. It is desired to obtain an interval estimate of μ, the mean amount of postage on all parcels handled by the postal system that day.

a. Estimate $\sigma\{\bar{X}\}$, the standard deviation of the sampling distribution of $\bar{X}$. What notation is used for this estimate?

b. Calculate the confidence limits $\bar{X} \pm 2s\{\bar{X}\}$. What confidence coefficient is associated with these limits?

c. Obtain a 90 percent confidence interval for μ.

d. If 100 independent random samples of 400 parcels each had been selected and if a 90 percent confidence interval for μ had been calculated for each of the 100 samples, what proportion of the 100 confidence intervals would be expected to contain μ?

10.6 An apparel and shoe trade association has a large number of member firms. As part of a study of the impact of new minimum wage legislation on association members, a random sample of 225 member firms was selected to estimate μ, the mean number of hourly-paid employees in member firms. A computer analysis of the sample results showed that $\bar{X} = 8.31$ and $s = 4.80$, where X denotes the number of hourly-paid employees in a firm.

a. Construct a 99 percent confidence interval for μ. Interpret your confidence interval.

b. Explain why the confidence coefficient of your interval estimate is only approximately 99 percent.

c. If you had constructed a 95 percent confidence interval, would it have been wider or narrower than the one in a? Which interval would involve the greater risk of an incorrect estimate?

10.7 An airline researcher studied reservation records for a random sample of 100 days in order to estimate μ, the mean number of persons who fail to keep their reservations (no-shows) on the daily 4 P.M. commuter flight to New York City. The records revealed the following:

Number of no-shows:	0	1	2	3	4	5	6
Number of days:	20	37	23	15	4	0	1

a. Verify that $\bar{X} = 1.500$ and $s = 1.185$. Calculate $s\{\bar{X}\}$. What does $s\{\bar{X}\}$ estimate here?

b. Construct a 95 percent confidence interval for μ. Explain the meaning of your confidence interval.

c. Would a different random sample of 100 days have provided the same interval estimate as the one in b? Explain.

10.8 As part of a profitability analysis, a refrigeration company wished to estimate μ, the mean number of manhours expended by company personnel per service call last year. A random sample of 144 service calls yielded $\bar{X} = 1.34$ manhours and $s = 1.32$ manhours, where X denotes the manhours expended in a service call.

a. Calculate the confidence limits $\bar{X} \pm 1.0s\{\bar{X}\}$. How confident can one be that these limits cover μ?

b. A person, commenting on the analysis, said: "Since the confidence limits always include the mean $\bar{X}$, the confidence interval corresponding to $\bar{X} \pm s\{\bar{X}\}$ should always be correct." Do you agree? Comment.

c. Obtain a 90 percent confidence interval for the population mean and interpret it.

10.9 A research report stated that the mean return on invested capital earned by furniture manufacturers is between 6.1 and 10.8 percent per year, with a confidence coefficient of 95 percent. One person interpreted this as meaning that 95 percent of furniture manufacturers had investment returns between 6.1 and 10.8 percent. Another person interpreted this in the sense that if many random samples were taken, 95 percent of them would have sample means between 6.1 and 10.8 percent. Why are these interpretations incorrect?

*10.10 A sample survey of kindergarten children in a state is being planned to estimate, among other things, the mean number of older siblings of such children. It is desired to estimate this mean within $\pm.08$, with a 90 percent confidence coefficient. A reasonable planning value for σ is .6. What sample size is needed to estimate the mean number of older siblings?

10.11 A nationwide survey of practicing physicians is to be undertaken to estimate μ, the mean number of prescriptions written per day. The desired margin of sampling error is $\pm.75$, with a 99 percent confidence coefficient. A pilot study revealed that a reasonable planning value for the population standard deviation is 5.

a. How many physicians should be contacted in the survey to estimate μ?

b. If the desired confidence coefficient were lowered to 95 percent, would the required sample size be reduced substantially?

10.12 Refer to Problem 10.7. Sometime after this study, the airline instituted a new reservations system that discourages customers from holding simultaneous reservations on different flights to the same destination. It is now desired to reestimate the mean number of no-shows for the daily 4 P.M. flight to New York City. A 95 percent confidence interval, with a half-width of .25, is wanted. The standard deviation of the original sample ($s = 1.185$) is to be taken as the planning value for σ.

a. How many days should be included in the sample to prepare the estimate?

b. If the desired confidence coefficient were increased to 99 percent (other factors being unchanged), how many days would have to be included in the sample? Is the resultant change in sample size substantial relative to the increase in the confidence coefficient?

c. What is the likely effect on the confidence interval if the planning value for σ is too large? Why can you not discuss the effect on the confidence interval for any one sample?

*10.13 In a study of the effects of nutrition on work efficiency, a random sample of 16 workers were asked to follow a rigid dietary regimen. In one part of this study, the blood-sugar level (X) of each sample worker was measured two hours after breakfast. The results (in milligrams of sugar per 100 cubic centimeters of blood) were $\bar{X} = 112.8$ and $s = 9.6$. Assume that X is normally distributed. Construct a 95 percent confidence interval for the mean blood-sugar level under this regimen and interpret your interval estimate.

10.14 An examination of the records for a random sample of nine motor vehicles in a large fleet shows the following operating costs (in cents per mile): 28.3, 26.4, 27.0, 22.5, 23.5, 29.1, 26.8, 26.7, 30.9. Let μ denote the mean operating cost in cents per mile for vehicles in the fleet. Construct a 90 percent confidence interval for μ. Assume that operating costs are nearly normally distributed. Interpret your interval estimate.

10.15 A scientist is studying the effect of communication modes on the problem-solving capability of teams. For one mode, a random sample of 16 teams completed a specific task in an average of 25.9 minutes with a standard deviation of 3.6 minutes. Construct a 95 percent confidence interval for the mean time required to complete the task. Assume that completion times are normally distributed. Interpret your interval estimate.

*10.16 Refer to Problem 10.5. Two million parcels were handled by the postal system on the particular day. Obtain a 90 percent confidence interval for the total amount of postage on parcels handled that day. Interpret your confidence interval.

10.17 Refer to Problem 10.6. The association has 9,100 member firms. Construct a 99 percent confidence interval for the total number of hourly-paid employees of all member firms. Interpret your interval estimate.

10.18 Refer to Problem 10.14. All of the vehicles in the fleet travel about the same number of miles during a month. Construct a 90 percent confidence interval for the expected total cost of operating the fleet in a month when a total of 5 million vehicle-miles is accumulated. Interpret your confidence interval.

*10.19 Refer to Problem 10.5.
   a. Construct a lower 95 percent confidence interval for μ. Explain why the limit of this lower 95 percent confidence interval is the same as the lower limit of the two-sided 90 percent confidence interval calculated in Problem 10.5c.
   b. Can the postal authorities confidently claim that the total amount of postage for the two million parcels handled that day exceeded $6 million? Explain.

10.20 Refer to Problem 10.6.
   a. Obtain upper 99 percent confidence intervals for: (1) the mean number of hourly-paid employees per firm, (2) the total number of hourly-paid employees in the 9,100 member firms.
   b. The trade association is contemplating developing a mailing list containing all hourly-paid employees and needs an indication of the minimum number of hourly-paid employees that will be contained in the list. Obtain the appropriate 99 percent confidence limit.

10.21 Refer to Problem 10.13.
   a. Construct a lower 95 percent confidence interval for the mean blood-sugar level under this regimen.
   b. Why can one not conclude that 95 percent of workers' blood-sugar levels two hours after eating the regimen breakfast will exceed the confidence limit in a?

*10.22 The number of annual visits to a dentist by a child is a Poisson random variable with unknown parameter value λ. In a random sample of two children, the numbers of visits made to the dentist last year were $X_1 = 0$ and $X_2 = 3$.
   a. State the likelihood function $L(\lambda)$ for this sample outcome.
   b. Evaluate $L(\lambda)$ at $\lambda = 0, 1, 1.5, 2,$ and 3. Sketch a graph of $L(\lambda)$, as in Figure 10.7. Does the sample mean $\bar{X}$ appear to be the maximum likelihood estimator of λ, as the theory in the Text states?

10.23 Refer to Problem 10.22. Answer all parts for the sample outcomes $X_1 = 1$ and $X_2 = 2$.

EXERCISES

10.24 If a point estimator is biased but consistent, does this imply that the bias must approach zero as the sample size gets large? Explain.

10.25 Consider the following probability distribution for an infinite population:

x:	0	2
$P(x)$:	.5	.5

   a. Verify that $\sigma = 1$ for this population.
   b. Construct the sampling distributions of s^2 and s for a simple random sample of $n = 2$ from this population.
   c. As noted in the Text, s^2 is an unbiased estimator of σ^2 for an infinite population, but s is a biased estimator of σ. Verify this for the sampling distributions of s^2 and s constructed in **b**. What is the amount of bias in s here?

10.26 Given that $E\{s^2\} = \sigma^2$, prove that $s^2\{\bar{X}\}$ is an unbiased estimator of $\sigma^2\{\bar{X}\}$ for a random sample from an infinite population.

10.27 With simple random sampling, what distribution does each of the following statistics follow if: (1) the population is normal and n is small, (2) the population is normal and n is large, (3) the population is not normal and n is small, (4) the population is not normal and n is large?
   a. $\bar{X}$
   b. $(\bar{X} - \mu)/\sigma\{\bar{X}\}$
   c. $(\bar{X} - \mu)/s\{\bar{X}\}$

10.28 Show that the interval estimate $\mu \geq L$, where $L = \bar{X} - t(1 - \alpha; n - 1)s\{\bar{X}\}$, is a lower $1 - \alpha$ confidence interval for μ based on a random sample of size n from a normal population. [*Hint:* Show that $P(L \leq \mu) = 1 - \alpha$ prior to sampling.]

10.29 The time until relapse for a patient suffering from a certain chronic disease has an exponential probability distribution. In a random sample of two patients, the numbers of days until relapse were $X_1 = 60$ and $X_2 = 100$.
   a. State the likelihood function $L(\lambda)$ for this sample outcome and evaluate it at $\lambda = .010, .0125, .015$, and .020.
   b. Sketch a graph of the likelihood function of λ for this sample outcome. Confirm by inspection that $1/80$ is the maximum likelihood estimate of λ.
   c. Suppose the relapse times of a random sample of n patients were observed. Verify that the likelihood function of λ in this case is $L(\lambda) = \lambda^n \exp(-\lambda n \bar{X})$, where $\bar{X}$ denotes the sample mean relapse time.

10.30 (Calculus needed.) Refer to Problem 10.29c. Show that $1/\bar{X}$ is the maximum likelihood estimator of λ in this case.

STUDIES

10.31 The 10 consecutive sets of five numbers in column 1 of Table C-9 are equivalent to 10 independent random samples of size $n = 5$ from $N(0, 1)$. The sample means and standard deviations for these 10 samples follow.

Sample	$\overline{X}$	s	Sample	$\overline{X}$	s
1	.8852	.6117	6	.3080	1.2288
2	.2532	1.2749	7	.3896	.5858
3	−.2960	1.0780	8	−.3112	1.0929
4	−.3822	1.1650	9	−.1750	.7936
5	−.8342	.9706	10	.2522	.6594

a. For each sample, calculate an 80 percent confidence interval for μ of the form $\overline{X} \pm z\sigma\{\overline{X}\}$. (*Hint:* See Comment 2, p. 235.) How many of the 10 intervals are correct; i.e., how many contain $\mu = 0$?

b. Redo **a** using intervals of the form $\overline{X} \pm ts\{\overline{X}\}$.

c. Explain why in both **a** and **b**, 8 out of 10 intervals are expected to contain $\mu = 0$. How do the actual numbers compare with the expected numbers in the two cases?

10.32 An aluminum company is experimenting with a new design for electrolytic cells in smelter potrooms. A major design objective is to maximize a cell's expected service life. Thirty cells of the new design were started and operated under similar conditions, and failed at the following ages (in days):

634	976	1,184	1,355	1,585	1,820
751	1,022	1,252	1,468	1,608	1,943
812	1,120	1,295	1,477	1,683	1,980
855	1,158	1,310	1,502	1,698	1,992
947	1,168	1,341	1,535	1,711	2,192

Two items of concern to management are: (1) the mean service life of this design, and (2) the comparative performance of this design with the standard industry design, which is known to have a mean service life of 1,300 days. Management does not want to conclude that the new design is superior to the standard one unless the evidence is fairly strong.

a. Assuming that the distribution of service life of the cells is normal, calculate an appropriate 95 percent confidence interval for the mean service life of cells of the new design. Justify your choice of one- or two-sided confidence interval. Should management conclude that the new design is superior to the standard one with respect to mean service life? Comment.

b. It has been suggested that the logarithms of service life are more normally distributed than the original observations. Take logarithms (to base 10) of the service-life data and calculate the same type of confidence interval as in **a** for the mean log-service-life of cells of the new design. Cells of the standard design are known to have a mean log-service-life of 3.095 (to base 10). Can management claim with confidence that the mean log-service-life is greater for cells of the new design than for cells of the standard design?

c. Graph histograms of the original data and the log-data. Does the distribution of log-service-lives appear to be more normal than the distribution of service lives, as suggested? Comment.

d. A management objective is to obtain a large total service life for the cells. Is the mean service life or the mean log-service-life the more relevant measure here? Explain.

11
Tests for Population Mean

Often, it is desired to test on the basis of sample data whether the population mean differs from a specified standard or historical value.

Examples

1. A manufacturing concern wishes to test whether or not the mean service life of a modified machine component exceeds the mean service life of the original component, based on a sample of 36 modified components.
2. A financial analyst wishes to test whether or not mutual funds currently offer investors a lower average return than the historical return in earlier decades, based on a sample of 15 mutual funds.
3. A psychologist concerned with job performance wishes to determine whether or not the average span of concentration on a task when the employee works alone is the same as the current span where several employees work in the same room. □

In this chapter, we discuss statistical tests for the population mean based on simple random samples. This type of test is common in practice and forms the basis for other statistical tests discussed subsequently.

11.1 STATISTICAL TESTS

Specifying the Alternative Conclusions

In this chapter, we are concerned with *testing* or *drawing a conclusion* about the value of the population mean μ based on simple random sample data. Let us look more closely at Example 1.

□ **Illustration**

An ultrasonic equipment manufacturer employs a component in one of its machines that must withstand considerable stress from vibration while the machine is operating. An all-metal component has been available for some years from a supplier. Extensive historical experience has shown that this component has a mean service life of 1,100 hours. The research division of the supplier has just developed a modified component constructed from plastic and metal, bonded in a special way. The manufacturer wishes to know whether or not the mean service life of the modified component (μ) exceeds the mean service life of 1,100 hours for the original component. Thus, the manufacturer wishes to choose between two alternative conclusions: (1) The mean service life of the modified component does not exceed 1,100 hours. (2) The mean service life exceeds 1,100 hours. Symbolically, these alternatives may be stated as follows:

$$H_0: \mu \leq 1,100$$

$$H_1: \mu > 1,100$$

Here, H_0 and H_1 are symbols denoting the alternative conclusions. To conclude H_0 in this case is to conclude that the mean service life of the modified component (μ) does not exceed 1,100 hours. To conclude H_1 is to conclude the opposite of H_0, namely, that μ exceeds 1,100 hours. ☐

Three Types of Alternatives.   In our illustration, the manufacturer was concerned whether or not the mean service life of the new component exceeds 1,100 hours, and the alternative conclusions took the form:

$$H_0: \mu \leq 1,100$$

$$H_1: \mu > 1,100$$

A different type of alternatives is considered by an industrial engineer who is concerned whether or not the mean hourly output per machine is at least 400 units. Here, the alternatives take the form:

$$H_0: \mu \geq 400$$

$$H_1: \mu < 400$$

To conclude H_0 here is to conclude that the mean hourly output per machine (μ) is at least 400 units. To conclude H_1 is to conclude the opposite, namely, that μ is less than 400 units.

There is a third form of alternatives that often arises. Consider a work operation which, according to a collective bargaining agreement, requires 25 seconds to perform. Both management and the union wish to know whether this standard is still appropriate under current operating conditions. The alternative conclusions here are:

$$H_0: \mu = 25$$

$$H_1: \mu \neq 25$$

To conclude H_0 is to conclude that the current mean time for the operation (μ) is 25 seconds. To conclude H_1 is to conclude the opposite, namely, that μ is not 25 seconds currently. In the latter case, further analysis would be required to determine whether μ is below or above 25 seconds.

We now summarize the three forms of alternative conclusions concerning μ:

(11.1a) *One-sided (Upper-tail) Alternatives:*

$$H_0: \mu \leq \mu_I$$
$$H_1: \mu > \mu_I$$

(11.1b) *One-sided (Lower-tail) Alternatives:*

$$H_0: \mu \geq \mu_I$$
$$H_1: \mu < \mu_I$$

(11.1c) *Two-sided Alternatives:*

$$H_0: \mu = \mu_I$$
$$H_1: \mu \neq \mu_I$$

In these pairs of alternatives, μ_I denotes the special value of the population mean that is the standard or historical value against which the mean of the sampled population is to be compared. In our three examples, the values of μ_I were, respectively, 1,100 hours, 400 units, and 25 seconds.

☐ Examples

1. In Example 2 on p. 246, the historical rate of return in earlier decades was 6.3 percent. This level is denoted by μ_I. For testing whether or not mutual funds currently offer a lower mean return, the two alternatives are:

$$H_0: \mu \geq 6.3$$
$$H_1: \mu < 6.3$$

2. In Example 3 on p. 246, the current mean span of concentration is $\mu_I = 3.2$ minutes. For testing whether or not the mean span is at the same level when the employees work alone, the two alternatives are:

$$H_0: \mu = 3.2$$
$$H_1: \mu \neq 3.2$$ ☐

Comments

1. In statistical testing, the alternative conclusions H_0 and H_1 are often called *hypotheses* since each conclusion may be thought of as a hypothesis about the true value of the parameter. For instance, $H_0: \mu \leq 1,100$ represents the hypothesis that μ is 1,100 or less.
2. The names of the three forms of alternatives in (11.1) are associated with conclusion H_1. Thus, in (11.1c), H_1 includes values of μ on both sides of μ_I; hence, this form is called two-sided alternatives. Similarly, in (11.1a), H_1 includes values of μ that are larger than μ_I; hence, this form is called one-sided upper-tail alternatives. The reason for the term *upper-tail* will become apparent shortly.
3. In all cases, we shall follow the convention of labeling the alternative containing the equality sign as H_0. Sometimes H_0 is then referred to as the *null hypothesis* of the test.

Testing by Means of Confidence Interval

Statistical testing is closely connected with interval estimation. To see this, let us return to our illustration in Chapter 10 where an auditor sampled the 8,042 accounts receivable of a freight company in order to estimate the mean audit amount per account (μ). Based on a simple random sample of 100 accounts, the following 95 percent confidence interval for μ was constructed:

$$26.29 \leq \mu \leq 40.09$$

Suppose now that the accounting records of the company show the mean amount of accounts receivable to be \$47.86. The auditor would like to test whether or not the sample results support the book figure of \$47.86. Since any material overstatement or understatement would represent a serious matter for the auditor, he considers the following two-sided alternatives for testing:

$$H_0: \mu = 47.86$$

$$H_1: \mu \neq 47.86$$

How can the auditor now use the confidence interval for μ, the mean audit amount, to decide which of the two alternatives is the appropriate conclusion?

Returning to the two-sided 95 percent confidence interval, we note the auditor is 95 percent confident that \$26.29 $\leq \mu \leq$ \$40.09. But, if that is the case, then the auditor is confident that μ does not equal \$47.86. Therefore, the auditor would decide H_1 is the correct conclusion—i.e., that the mean audit amount differs from the book amount of \$47.86.

As has been seen from this example, confidence intervals that are appropriately constructed can be used for choosing one of the alternative conclusions H_0 and H_1. We shall return to this topic later. Now we shall present new concepts and terminology that are important for understanding statistical tests.

Errors of Inference

Two Types of Errors. Whenever sample data are used to choose between the two alternative conclusions, there are risks of choosing the incorrect conclusion because of sampling. Such errors in choosing the incorrect conclusion are called *inferential errors,* because they entail drawing an incorrect inference or conclusion from the sample about the value of the population parameter. Inferential errors are of two types, which we label Type I and Type II errors according to the following definition:

(11.2) A *Type I error* is made if conclusion H_1 is selected as being correct when, in fact, H_0 is the correct conclusion.

A *Type II error* is made if conclusion H_0 is selected as being correct when, in fact, H_1 is the correct conclusion.

TABLE 11.1 *Illustration of the two types of errors of inference*

Conclusion Reached	True State	
	$\mu \leq 1,100$ (H_0 is correct)	$\mu > 1,100$ (H_1 is correct)
H_0: ($\mu \leq 1,100$)	Correct conclusion	Type II error
H_1: ($\mu > 1,100$)	Type I error	Correct conclusion

Example

In our example concerning the mean service life of the modified machine component, the two alternative conclusions were:

$$H_0: \mu \leq 1,100$$

$$H_1: \mu > 1,100$$

Table 11.1 illustrates the two types of errors that are possible here. A Type I error corresponds to concluding that $\mu > 1,100$ (i.e., concluding H_1) when, in fact, $\mu \leq 1,100$ (H_0 is correct). On the other hand, a Type II error corresponds to concluding that $\mu \leq 1,100$ (i.e., concluding H_0) when, in fact, $\mu > 1,100$ (H_1 is correct). ☐

Need to Control Risks of Making Errors. In adopting a procedure for choosing between H_0 and H_1, it is important that the risks of inferential errors of both types be at tolerable levels. Some risks must be accepted, of course, whenever conclusions are based on sample data, but the magnitudes of the risks can be controlled.

Clearly, the amounts of risk that can be accepted depend on the seriousness of the Type I and Type II errors. When a chemical processor is testing the mean toxicity level in a batch of product, for instance, the more serious of the two types of errors might be to conclude that the toxicity of the batch is within safety specifications when, in fact, it is not. Hence, this risk must be kept low. On the other hand, the risk of concluding the batch has excess toxicity when, in fact, it does not may be less serious. In other instances, the seriousness of the two types of errors is more nearly balanced. Thus, the test procedure should provide for varying degrees of control over the risks of making Type I and Type II errors.

Indeed, the seriousness of a Type I error or of a Type II error generally will vary with the value of the parameter—μ in our case. For example, in our machine component illustration, a conclusion that H_0 is true ($\mu \leq 1,100$) when, in fact, μ is 1,500 hours represents a serious Type II error because of the substantial superiority of the modified component over the present one. On the other hand, concluding that $\mu \leq 1,100$ when, in fact, μ equals 1,101 hours is not serious even though technically a Type II error is committed.

Statistical decision making enables the decision maker to specify the tolerable risks of Type I and Type II errors, whereupon an appropriate test procedure can be determined to assure no greater risks. We shall explain the control of risks of incorrect decisions in the following sections.

11.2 STATISTICAL DECISION RULE

Control of the risks of incorrect decisions requires the use of a statistical decision rule. We explain the nature of a statistical decision rule with an illustration.

☐   Illustration

Let us return again to the machine component case. Recall that the alternative conclusions are:

$$H_0: \mu \leq 1,100$$
$$H_1: \mu > 1,100$$

The choice will be based on a sample of 36 modified components prepared as a pilot batch by the supplier. Each of these will be subjected to a forced-life test to determine its service life in hours.   ☐

Nature of Statistical Decision Rule

If the mean service life $(\bar{X})$ of the 36 sample components is much smaller than 1,100 hours, it would be sensible to conclude that $\mu \leq 1,100$ (i.e., conclude H_0). Similarly, if $\bar{X}$ is much larger than 1,100 hours, it would be sensible to conclude that $\mu > 1,100$ (i.e., conclude H_1). This follows from the fact that $\bar{X}$ is expected to lie somewhere in the neighborhood of μ. This common-sense argument leads us to consider the following type of rule for choosing between H_0 and H_1:

$$\text{If } \bar{X} \leq A, \text{ conclude } H_0 \ (\mu \leq 1,100)$$
$$\text{If } \bar{X} > A, \text{ conclude } H_1 \ (\mu > 1,100)$$

Here, A is some number in the neighborhood of 1,100. Observe that this rule decides which conclusion should be selected for each possible value of the sample mean $\bar{X}$. For this reason, it is called a statistical decision rule.

(11.3) A *statistical decision rule* specifies, for each possible sample outcome, which alternative should be selected. The value A in the decision rule is called the *action limit* of the decision rule.

In our illustration, suppose management arbitrarily specified $A = 1,125$ hours. The decision rule then is:

$$\text{If } \bar{X} \leq 1,125, \text{ conclude } H_0 \ (\mu \leq 1,100)$$
$$\text{If } \bar{X} > 1,125, \text{ conclude } H_1 \ (\mu > 1,100)$$

The action limit here is $A = 1,125$ hours. The decision rule states that if $\bar{X}$ does not exceed 1,125, conclusion H_0 is reached; if $\bar{X}$ does exceed 1,125, conclusion H_1 is reached.

11.3 EVALUATING RISKS INHERENT IN DECISION RULE

Inherent in each statistical decision rule are risks of making incorrect decisions. We shall now consider how to evaluate these risks. We continue with our machine component illustration where the alternatives are:

(11.4)
$$H_0: \mu \leq 1,100$$
$$H_1: \mu > 1,100$$

the sample size is $n = 36$ components, and the statistical decision rule is:

(11.5)
$$\text{If } \bar{X} \leq 1,125, \text{ conclude } H_0 \ (\mu \leq 1,100)$$
$$\text{If } \bar{X} > 1,125, \text{ conclude } H_1 \ (\mu > 1,100)$$

The procedure to be described for evaluating the risks assumes that the sample size is reasonably large so that $\bar{X}$ is approximately normally distributed. Here, $n = 36$ is large enough since the lifetimes of the machine components are distributed fairly symmetrically. We assume that the standard deviation of the distribution of lifetimes is known to be $\sigma = 300$ hours.

Power Curve

To study the risks of making Type I and Type II errors with a given decision rule, it is necessary to determine the probability that the rule will lead to the selection of each of the alternative conclusions H_0 and H_1. We shall consider the probability of concluding H_1 first. Since decision rule (11.5) states that H_1 is concluded when $\bar{X}$ exceeds 1,125, the probability of concluding H_1 clearly must depend on the value of μ. Hence, we denote the probability of concluding H_1 by $P(H_1; \mu)$. For decision rule (11.5), this probability is:

$$P(H_1; \mu) = P(\bar{X} > 1,125; \mu)$$

Computing $P(H_1; \mu)$. Since we do not know the true value of μ, we need to evaluate $P(H_1; \mu)$ at different possible values of μ. We shall begin with $\mu = 1,050$. Here, interest is in the probability:

$$P(H_1; 1,050) = P(\bar{X} > 1,125; \mu = 1,050)$$

Figure 11.1a illustrates the calculation of this probability. We know from Chapter 9 that the sampling distribution of $\bar{X}$ is approximately normal for reasonably large n, with mean $E\{\bar{X}\} = \mu$ and standard deviation $\sigma\{\bar{X}\} = \sigma/\sqrt{n}$. Since we are evaluating $P(H_1; \mu)$ at $\mu = 1,050$, the distribution is centered at $E\{\bar{X}\} = 1,050$. Further, since $n = 36$ and $\sigma = 300$, the distribution has standard deviation $\sigma\{\bar{X}\} = 300/\sqrt{36} = 50$. The probability that $\bar{X} > 1,125$, i.e., that the decision rule leads to H_1, is shaded in Figure 11.1a. The z value is calculated as usual:

$$z = \frac{1,125 - 1,050}{50} = 1.50$$

FIGURE 11.1 *Computation of* $P(H_1; \mu)$ *for several possible values of* μ

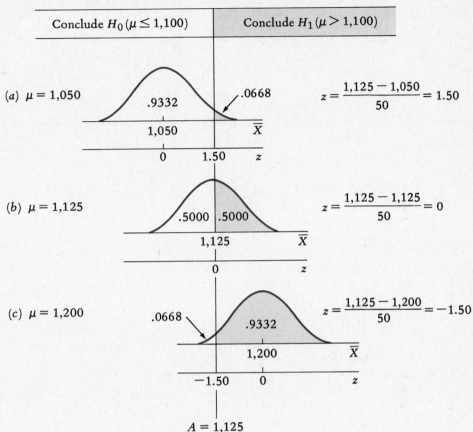

and we find:

$$P(\overline{X} > 1,125; \mu = 1,050) = P(Z > 1.50) = .0668$$

In other words, the probability of concluding H_1 when $\mu = 1,050$ is .0668 with decision rule (11.5) because $\overline{X}$ must fall more than 1.50 standard deviations above its mean to exceed the action limit of 1,125.

Values of $P(H_1; \mu)$ need also be computed for other possible values of μ. Figures 11.1b and 11.1c illustrate the calculations for $\mu = 1,125$ and $\mu = 1,200$. The following probabilities of concluding H_1 are obtained:

$$P(H_1; 1,125) = P(Z > 0) = .5000$$

$$P(H_1; 1,200) = P(Z > -1.50) = .9332$$

Figure 11.2a shows a graph of $P(H_1; \mu)$ for values of μ ranging from 1,000 to 1,250. Note how the probability of concluding H_1 increases as μ increases.

FIGURE 11.2 *Power curve and error curve for decision rule (11.5). The α risk portion of the error curve is given by $P(H_1; \mu)$ for $\mu \leq 1,100$; the β risk portion is given by $P(H_0; \mu) = 1 - P(H_1; \mu)$ for $\mu > 1,100$.*

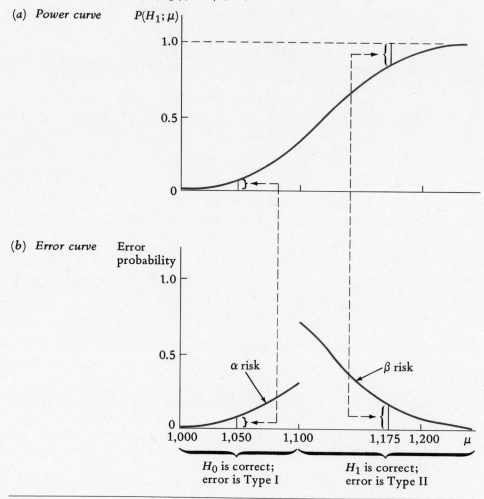

(a) *Power curve*

(b) *Error curve*

Power. The probability $P(H_1; \mu)$ is called the *power* of the decision rule at parameter value μ. For instance, since $P(H_1; 1,050) = .0668$ for decision rule (11.5), .0668 is the power of the rule at $\mu = 1,050$. The graph of $P(H_1; \mu)$ for different values of μ is called the power curve of the decision rule.

The concept of a power curve for a decision rule applies also to tests for parameters other than the population mean. A general definition is now given:

(11.6) The *power curve* of a statistical decision rule is a graph of the probability of concluding H_1 for different possible values of the parameter of interest.

Computing $P(H_0; \mu)$. The probability that decision rule (11.5) will lead to conclusion H_0 is denoted by $P(H_0; \mu)$. Decision rule (11.5) states that H_0 is to be concluded if $\overline{X}$ does not exceed 1,125. Hence:

$$P(H_0; \mu) = P(\overline{X} \le 1{,}125; \mu)$$

Since decision rule (11.5) leads either to H_0 or to H_1, $P(H_0; \mu)$ and $P(H_1; \mu)$ are complementary probabilities for each value of μ. Therefore, from the complementation theorem (4.19), we obtain the following important relationship:

(11.7) $$P(H_0; \mu) + P(H_1; \mu) = 1$$

Formula (11.7) provides a simple way of computing $P(H_0; \mu)$ for a decision rule from its power $P(H_1; \mu)$. For instance, we know that $P(H_1; 1{,}050) = .0668$ for decision rule (11.5); hence:

$$P(H_0; 1{,}050) = 1 - P(H_1; 1{,}050) = 1 - .0668 = .9332$$

Note that this probability corresponds to the unshaded area in Figure 11.1a.

Other probabilities $P(H_0; \mu)$ may be obtained in similar fashion once the power $P(H_1; \mu)$ has been calculated.

Comment

A plot of the probabilities $P(H_0; \mu)$ against μ is called the *operating characteristic curve* of the statistical decision rule. This curve is frequently employed in quality control applications and is illustrated in Chapter 14.

Error Curve

Computing Error Probabilities. Once we have the probabilities $P(H_1; \mu)$ for different values of μ, we can immediately derive the probability that the decision rule will lead to an erroneous conclusion for each value of μ. For the alternatives in (11.4), we will be in error if we conclude H_1 when $\mu \le 1{,}100$ (a Type I error) or if we conclude H_0 when $\mu > 1{,}100$ (a Type II error). Thus, $P(H_1; \mu)$ is the error probability for $\mu \le 1{,}100$ and $P(H_0; \mu) = 1 - P(H_1; \mu)$ is the error probability for $\mu > 1{,}100$.

Selecting these probability values from the power curve in Figure 11.2a, we have plotted the error probabilities in Figure 11.2b for μ values ranging from 1,000 to 1,250. For instance, since $P(H_1; 1{,}050) = .0668$ and H_0 is the correct conclusion when $\mu = 1{,}050$, it follows that the error probability at $\mu = 1{,}050$ is .0668. This is a Type I error probability and is shown in Figure 11.2b. Similarly, for $\mu = 1{,}175$, we have $P(H_1; 1{,}175) = .8413$, H_1 is the correct conclusion when $\mu = 1{,}175$, and the probability of making an error in this case is $P(H_0; 1{,}175) = 1 - P(H_1; 1{,}175) = 1 - .8413 = .1587$. This is a Type II error probability and is also shown in Figure 11.2b.

Note from Figure 11.2b that decision rule (11.5), based on $n = 36$ components, involves some substantial inferential risks. For example, if the mean life for the modified component is $\mu = 1{,}175$ and management really would like to recognize this superiority of the modified component, there exists a risk of .1587 of failing to detect this improvement.

In our discussion of risks of making Type I and Type II errors, we shall employ the following terms and notation:

(11.8) The probability of a Type I error will be denoted by α (Greek alpha) and will be called an α *risk.*

(11.9) The probability of a Type II error will be denoted by β (Greek beta) and will be called a β *risk.*

Thus, we shall say that the β risk at $\mu = 1,175$ is .1587 for decision rule (11.5), and that the α risk at $\mu = 1,050$ is .0668.

11.4   EFFECTS OF ACTION LIMIT AND SAMPLE SIZE ON RISKS

We have seen how to evaluate the risks of making incorrect decisions that are inherent in a statistical decision rule. Now, we wish to examine the particular effects of the action limit and the sample size on these risks.

Effect of Action Limit

The error curve plotted in Figure 11.2b summarizes the ability of decision rule (11.5) to control inferential errors. This curve is reproduced in Figure 11.3a and is labeled for the action limit in rule (11.5)—namely, $A = 1,125$. If we change the action limit, the new decision rule will have a different error curve. For instance, Figure 11.3a also shows the error curve for a decision rule with $A = 1,150$. Since the calculations of these error probabilities are similar to those for the decision rule with $A = 1,125$, we show only the final error curve. Note how the increase in the action limit from 1,125 to 1,150 has decreased the α risk (the Type I error probabilities) but increased the β risk (the Type II error probabilities). This example illustrates an important statistical principle:

(11.10) For a given random sample size, one type of error probability can be reduced only at the expense of increasing the other type.

This principle is evident from Figure 11.1. Note how a shift of the action limit to the right will decrease the shaded area, or $P(H_1; \mu)$, and increase the unshaded area, or $P(H_0; \mu)$, for all values of μ. Since Figure 11.2 indicates that a Type I error is made when $\mu \leq 1,100$ and H_1 is concluded, the decrease in $P(H_1; \mu)$ implies a lower α risk. Similarly, a Type II error is made when $\mu > 1,100$ and H_0 is concluded, so the increase in $P(H_0; \mu)$ implies a larger β risk. The opposite occurs when the action limit is shifted to the left.

FIGURE 11.3 *Effects of change in action limit and sample size on error probabilities*

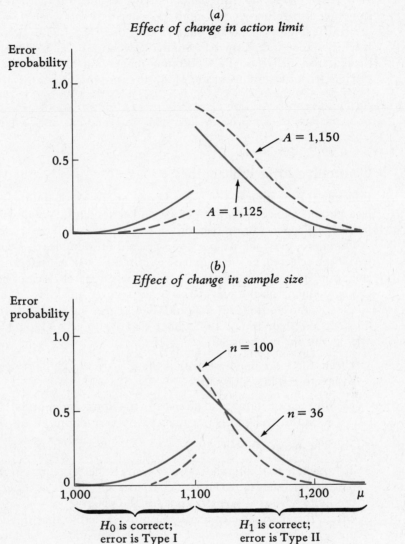

(a)
Effect of change in action limit

(b)
Effect of change in sample size

Effect of Sample Size

The error curve is also altered by changing the sample size. For instance, Figure 11.3b shows the error curves for decision rule (11.5) with $A = 1,125$ when based on sample size $n = 36$ (the case we have just considered) and when based on sample size $n = 100$. The error probabilities for $n = 100$ are calculated in the same way as before, except $\sigma\{\bar{X}\} = \sigma/\sqrt{n}$ now equals $\sigma\{\bar{X}\} = 300/\sqrt{100} = 30$ hours.

Note how the increase in sample size has decreased both α and β risks almost everywhere (except in the interval between $\mu_I = 1,100$ and the action limit $A = 1,125$). To see how this happens, refer again to Figure 11.1. For $n = 100$, the sampling distribution of $\overline{X}$ is more concentrated than when $n = 36$. Hence, when $\mu = 1,050$ (Figure 11.1a) the shaded area, or $P(H_1; 1,050)$, becomes smaller and the α risk therefore is smaller. Similarly, when $\mu = 1,200$ (Figure 11.1c) the unshaded area, or $P(H_0; 1,200)$, becomes smaller and the β risk therefore is smaller.

Here, then, is an illustration of another important statistical principle:

(11.11) A simultaneous reduction in both types of error probabilities in a given test situation can only be achieved by increasing the random sample size.

Controlling Error Probabilities

Principle (11.11) indicates that both types of error probabilities can be controlled by a suitable choice of the sample size and action limit. We discuss how to control both the α and β risks in Section 11.6.

Often, however, the sample size is given. For example, the sample may already have been selected, or the sample size is fixed by cost considerations. In this situation, principle (11.10) indicates that only one type of risk can be controlled and the consequent magnitude of the other type of risk must simply be accepted.

Clearly, the controlled risk should then be the one with the more serious consequences. We shall assume throughout that the more important risk is the α risk. We do this for two reasons:

1. Often, the more important risk is the α risk naturally. For example, a scientist wishes to test the alternatives:

   H_0: The mean number of fillings, extractions, and untreated dental cavities found in children is not reduced by fluoridation of drinking water

   H_1: The mean number is reduced

   To conclude that fluoridated water leads to better dental hygiene for children (i.e., to conclude H_1) when, in fact, it does not (i.e., H_0 is correct) may represent the more serious inferential error for the scientist.

2. Often, the alternative conclusions can be stated so that the α risk is the more important risk. In our machine component illustration, we stated that the manufacturer was most concerned about using the modified component when, in fact, it represented no improvement. Hence, for the alternatives:

$$H_0: \mu \leq 1,100$$

$$H_1: \mu > 1,100$$

the α risk is precisely the risk of concluding that the modified component is an improvement when, in fact, it is not.

Suppose, however, that the cost of shifting to the modified component and, if need be, back to the original component is small, but the payoff from a substantially longer life of the modified component, say 1,200 hours or more, is very substantial. If the sample size is fixed, we would then use the alternatives:

$$H_0\!: \mu \geq 1,\!200$$

$$H_1\!: \mu < 1,\!200$$

With this set of alternatives, the α risk is the risk of failing to note that the modified component is a substantial improvement when, in fact, it is. Again, then, the α risk is the more serious risk.

◻ Examples

1. In Example 2 on p. 246, the historical rate of return in earlier decades was 6.3 percent. If it is more important that the analyst not err by concluding that the current mean rate of return for mutual funds is below 6.3 percent when it is 6.3 percent or greater, then the two alternatives would be:

$$H_0\!: \mu \geq 6.3$$

$$H_1\!: \mu < 6.3$$

If, however, it is more important that the analyst find out when the current mean rate of return has declined to 5.5 percent or less, the alternatives would be:

$$H_0\!: \mu \leq 5.5$$

$$H_1\!: \mu > 5.5$$

2. A supplier guarantees that the mean weight per bar of dehydrated food is at least 100 grams. The purchaser, however, is more concerned about accepting a shipment whose mean weight per bar is 90 grams or less. The two alternatives to be considered by the purchaser therefore would be:

$$H_0\!: \mu \leq 90$$

$$H_1\!: \mu > 90$$ ◻

11.5 CONSTRUCTION OF DECISION RULE TO CONTROL α RISK

We explain now how to control the α risk when the given random sample size is reasonably large. We return to our machine component illustration. Recall that the alternatives are:

$$H_0\!: \mu \leq \mu_I = 1,\!100$$

$$H_1\!: \mu > \mu_I = 1,\!100$$

and that a sample of $n = 36$ components will be selected. Because the cost of an unnecessary switchover to the modified component is so great, management wishes to control the α risk at $\mu_I = 1,\!100$ hours. Specifically, when $\mu = 1,\!100$, management wants to control the risk of making a Type I error (concluding H_1 and making an undesired switch) at $\alpha = .01$.

Determination of Action Limit

Figure 11.4a shows the type of decision rule required here. In Figure 11.4b, the sampling distribution of $\bar{X}$ is shown when $\mu = \mu_I = 1{,}100$, the level of μ for which the α risk is specified. Note that this sampling distribution is approximately normal because the sample size is reasonably large here. As is shown in Figure 11.4c, we wish to locate the action limit so that the shaded area is .01. The shaded area represents:

$$P(\bar{X} > A; \mu = 1{,}100) = P(H_1; 1{,}100) = \alpha$$

Since the shaded area should be .01, the action limit A must be $z(.99) = 2.326$ standard deviations to the right of $\mu_I = 1{,}100$. Recalling that $\sigma\{\bar{X}\}$ denotes the standard deviation of the sampling distribution of $\bar{X}$, it is clear from Figure 11.4c that the desired action limit is:

$$A = 1{,}100 + 2.326\sigma\{\bar{X}\}$$

FIGURE 11.4   *Construction of decision rule to control α risk—Large sample, one-sided upper-tail test for μ*

(a) *Appropriate type of decision rule*

| Conclude H_0 ($\mu \leq \mu_I = 1{,}100$) | Conclude H_1 ($\mu > \mu_I = 1{,}100$) |

A $\bar{X}$

Large values of $\bar{X}$ imply H_1 is correct.

(b) *Sampling distribution of $\bar{X}$ located at $\mu_I = 1{,}100$*

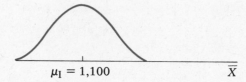

$\mu_I = 1{,}100$ $\bar{X}$

Distribution of $\bar{X}$ is centered at μ_I since α risk is controlled there.

(c) *Placement of action limit A*

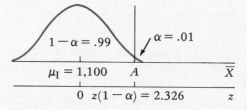

$1 - \alpha = .99$ $\alpha = .01$

$\mu_I = 1{,}100$ A $\bar{X}$

0 $z(1 - \alpha) = 2.326$ z

A is placed $z(1 - \alpha)$ deviations above μ_I so upper–tail area equals α.

(d) *Calculation of action limit A*

$$A = \mu_I + z(1 - \alpha)s\{\bar{X}\}$$
$$= 1{,}100 + 2.326(37.0) = 1{,}186$$

$\sigma\{\bar{X}\}$ estimated by $s\{\bar{X}\}$.

(e) *Decision rule*

If $\bar{X} \leq A = 1{,}186$, conclude H_0
If $\bar{X} > A = 1{,}186$, conclude H_1

Conclude appropriate alternative based on observed $\bar{X}$.

Since we usually do not know $\sigma\{\bar{X}\}$, we estimate it by $s\{\bar{X}\}$. Theorem (10.13) tells us that $(\bar{X} - \mu)/s\{\bar{X}\}$ generally has an approximate standard normal distribution when n is large. Hence, use of $s\{\bar{X}\}$ in place of $\sigma\{\bar{X}\}$ does not materially affect the z multiple, and the action limit becomes:

$$A = 1{,}100 + 2.326s\{\bar{X}\}$$

We now summarize the determination of the decision rule:

(11.12) When the alternatives are:

$$H_0: \mu \leq \mu_{\mathrm{I}}$$

$$H_1: \mu > \mu_{\mathrm{I}}$$

and the sample size is reasonably large, the appropriate decision rule to control the α risk at μ_{I} is:

If $\bar{X} \leq A$, conclude H_0

If $\bar{X} > A$, conclude H_1

where: $A = \mu_{\mathrm{I}} + z(1 - \alpha)s\{\bar{X}\}$

$s\{\bar{X}\}$ is given by (10.8)

☐ Example

In our machine component illustration, the alternatives are:

$$H_0: \mu \leq 1{,}100$$

$$H_1: \mu > 1{,}100$$

The α risk is to be controlled at .01 for $\mu_{\mathrm{I}} = 1{,}100$. Hence, $z(.99) = 2.326$. A sample of $n = 36$ components yielded $\bar{X} = 1{,}121$ and $s = 222$ hours. The estimated standard deviation $s\{\bar{X}\}$ is given by (10.8):

$$s\{\bar{X}\} = \frac{s}{\sqrt{n}} = \frac{222}{\sqrt{36}} = 37.0$$

Hence, $A = 1{,}100 + 2.326(37.0) = 1{,}186$ hours and the decision rule is:

If $\bar{X} \leq 1{,}186$, conclude H_0 ($\mu \leq 1{,}100$)

If $\bar{X} > 1{,}186$, conclude H_1 ($\mu > 1{,}100$)

Figures 11.4d and 11.4e show the calculation of A and the final form of the decision rule, respectively. Inasmuch as the sample mean of the 36 observations is $\bar{X} = 1{,}121$, the decision rule leads to conclusion H_0—that the mean service life of the modified component is 1,100 hours or less. This means that the sample evidence is not strong enough to warrant shifting to the modified component. ☐

Comments

1. The reason why the test for the alternatives in (11.12) is called an upper-tail test is apparent from Figure 11.4c. Conclusion H_1 is reached when the sample mean falls in the upper tail of the sampling distribution of $\bar{X}$.

2. When stating the alternatives in (11.12), the notation μ_I is used because this is the level of μ for which the Type I error is controlled.
3. Note from Figure 11.2b that the α risk is a maximum at $\mu_I = 1,100$. Hence, we know that if the mean life of the modified component is less than $\mu_I = 1,100$, the α risk of concluding H_1 in our example is less than .01.
4. The specified level of α is often called the *level of significance*.
5. Because the sampling distribution of $(\bar{X} - \mu)/s\{\bar{X}\}$ is only *approximately* normal for large n, the actual α risk when $\mu = \mu_I$ is usually only approximately the specified level of α.
6. The choice between H_0 and H_1 in (11.12) can also be made from a one-sided confidence interval. In our machine component illustration, we would set up a 99 percent one-sided *lower* confidence interval, corresponding to the specified risk of $\alpha = .01$ of making a Type I error when $\mu = \mu_I$. We have $\bar{X} = 1,121$, $s\{\bar{X}\} = 37$, and $z(.99) = 2.326$. Hence, by (10.19) we obtain $L = 1,121 - 2.326(37) = 1,034.9$ and the lower confidence interval is:

$$\mu \geq 1,034.9$$

If μ_I falls within the confidence interval, we conclude H_0; otherwise we conclude H_1. Since $\mu_I = 1,100$ falls within the confidence interval, we conclude H_0, just as we did by the decision rule.

11.6 CONTROL OF BOTH TYPES OF RISKS THROUGH DETERMINATION OF SAMPLE SIZE

When the sample size is not predetermined and can be set at whatever level is required, both the α and β risks can be controlled. To see how this can be done, let us again consider our machine component illustration but assume this time that the sample size has not yet been determined. Figure 11.5 illustrates the necessary steps to find the needed sample size n. The procedure assumes that the resulting sample size is reasonably large.

Figure 11.5a shows the appropriate type of decision rule, which is the same as before. Figure 11.5b shows the sampling distribution of $\bar{X}$ when $\mu = \mu_I = 1,100$, where the α risk is to be controlled at .01. The control of the α risk is the same as before.

Figure 11.5c pertains to the control of the β risk. The manufacturer and the supplier have agreed that if the modified component has a substantially longer mean service life, the test must provide some assurance that conclusion H_1 will be reached. Specifically, it was agreed that if the mean service life of the modified component is 1,200 hours—i.e., 100 hours above the mean service life of the original component (1,100)—the probability that the decision rule will lead erroneously to conclusion H_0 ($\mu \leq 1,100$) should only be .05.

Figure 11.5c shows the sampling distribution of $\bar{X}$ when $\mu = \mu_{II} = 1,200$, the level at which the risk of a Type II error is to be controlled. The shaded area represents:

$$P(\bar{X} \leq A; \mu_{II} = 1,200) = P(H_0; 1,200) = \beta$$

FIGURE 11.5 *Construction of decision rule to control α and β risks—Large sample, one-sided upper-tail test for μ*

(a) *Appropriate type of decision rule*

| Conclude H_0 ($\mu \le \mu_{\mathrm{I}} = 1{,}100$) | Conclude H_1 ($\mu > \mu_{\mathrm{I}} = 1{,}100$) | Large values of $\overline{X}$ imply H_1 is correct. |

$$A \qquad \qquad \overline{X}$$

(b) *Sampling distribution of $\overline{X}$ located at $\mu_{\mathrm{I}} = 1{,}100$*

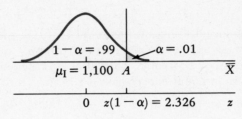

Distribution of $\overline{X}$ is centered at μ_{I} since α risk is controlled there.

(c) *Sampling distribution of $\overline{X}$ located at $\mu_{\mathrm{II}} = 1{,}200$*

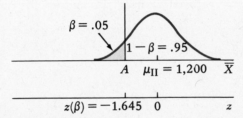

Distribution of $\overline{X}$ is centered at μ_{II} since β risk is controlled there. Note that μ_{II} is above μ_{I}.

(d) *Simultaneous specifications of action limit*

$$A = \mu_{\mathrm{I}} + z(1 - \alpha)\frac{\sigma}{\sqrt{n}}$$

$$= 1{,}100 + 2.326\frac{\sigma}{\sqrt{n}}$$

$$A = \mu_{\mathrm{II}} + z(\beta)\frac{\sigma}{\sqrt{n}}$$

$$= 1{,}200 - 1.645\frac{\sigma}{\sqrt{n}}$$

Simultaneous determination of action limit A and sample size n, where $\sigma\{\overline{X}\} = \sigma/\sqrt{n}$.

(e) *Calculation of sample size n*

$$n = \left[\frac{z(\beta) - z(1 - \alpha)}{\mu_{\mathrm{II}} - \mu_{\mathrm{I}}}\right]^2 \sigma^2$$

$$= \left(\frac{-1.645 - 2.326}{1{,}200 - 1{,}100}\right)^2 (300)^2 = 142$$

Planning value for σ is used. Resultant value of n must be large.

(f) *Calculation of action limit A*

$$A = \mu_{\mathrm{I}} + z(1 - \alpha)\frac{s}{\sqrt{n}}$$

$$= 1{,}100 + 2.326\frac{278}{\sqrt{142}}$$

$$= 1{,}154$$

$\sigma\{\overline{X}\} = \sigma/\sqrt{n}$ estimated by $s\{\overline{X}\} = s/\sqrt{n}$.

(g) *Decision rule*

If $\overline{X} \le A = 1{,}154$, conclude H_0
If $\overline{X} > A = 1{,}154$, conclude H_1

Conclude appropriate alternative based on observed $\overline{X}$.

Observe that we use μ_{II} to denote the level of μ at which the Type II error is controlled. The shaded area in Figure 11.5c is .05, the specified level of β.

Figure 11.5d shows the simultaneous specifications of the action limit for the two types of risks. From the error specification in Figure 11.5b, it is seen that the action limit must lie $z(1 - \alpha) = z(.99) = 2.326$ standard deviations to the right of $\mu_I = 1,100$ in order to leave only a probability $\alpha = .01$ that $\bar{X}$ will exceed A when $\mu = \mu_I = 1,100$. Expressing this as an equation, we have:

$$(11.13) \qquad A = \mu_I + z(1 - \alpha)\sigma\{\bar{X}\} = 1,100 + 2.326\sigma\{\bar{X}\}$$

From the error specification in Figure 11.5c, it is seen that A must lie at $z(\beta) = z(.05) = -1.645$ standard deviations from $\mu_{II} = 1,200$ in order to leave only a probability $\beta = .05$ that $\bar{X}$ will fall at A or below when $\mu = \mu_{II} = 1,200$. Expressing this as an equation, we have:

$$(11.14) \qquad A = \mu_{II} + z(\beta)\sigma\{\bar{X}\} = 1,200 - 1.645\sigma\{\bar{X}\}$$

Substituting $\sigma/\sqrt{n}$ for $\sigma\{\bar{X}\}$ in (11.13) and (11.14) yields the following pair of simultaneous equations in A and n:

$$(11.15) \qquad A = \mu_I + z(1 - \alpha)\frac{\sigma}{\sqrt{n}} \qquad \text{(From } \mu_I \text{ risk specification)}$$

$$A = \mu_{II} + z(\beta)\frac{\sigma}{\sqrt{n}} \qquad \text{(From } \mu_{II} \text{ risk specification)}$$

When the equations in (11.15) are solved simultaneously, we obtain a formula for n:

$$n = \left[\frac{z(\beta) - z(1 - \alpha)}{\mu_{II} - \mu_I}\right]^2 \sigma^2$$

This formula involves the population standard deviation σ. Since σ is generally unknown, we need to use a planning value of σ for finding n. Use of a planning value of σ is also necessary, as we have seen, for planning the sample size for estimation purposes:

(11.16) For the one-sided upper-tail test (11.12), the necessary sample size to control both the α and β risks, for a given planning value of σ, is:

$$n = \left[\frac{z(\beta) - z(1 - \alpha)}{\mu_{II} - \mu_I}\right]^2 \sigma^2$$

Once the sample size has been determined, the sample is chosen. The action limit A is then determined in the same way as for the case where the sample size is given, using the sample standard deviation s as an estimate of σ.

☐ Example

In our machine components illustration, the risk specifications were:

$$\mu_I = 1,100 \qquad \alpha = .01$$
$$\mu_{II} = 1,200 \qquad \beta = .05$$

Hence, $z(1 - \alpha) = z(.99) = 2.326$ and $z(\beta) = z(.05) = -1.645$. Suppose that it is agreed by the manufacturer and the supplier that a reasonable planning value of the standard deviation is $\sigma = 300$ hours. We now substitute into (11.16) to obtain:

$$n = \left(\frac{-1.645 - 2.326}{1,200 - 1,100}\right)^2 (300)^2 = 142$$

Hence, a random sample of 142 modified components should provide approximately the desired control over the Type I and Type II error probabilities if the planning value of σ is reasonably good.

Suppose that a random sample of $n = 142$ modified components was given forced-life tests with the results $\bar{X} = 1,113$ and $s = 278$ hours. Thus, $s\{\bar{X}\} = s/\sqrt{n} = 278/\sqrt{142} = 23.3$ and the action limit is, by (11.12):

$$A = 1,100 + 2.326(23.3) = 1,154 \text{ hours}$$

The decision rule therefore is:

$$\text{If } \bar{X} \leq 1,154, \text{ conclude } H_0 \ (\mu \leq 1,100)$$

$$\text{If } \bar{X} > 1,154, \text{ conclude } H_1 \ (\mu > 1,100)$$

Since $\bar{X} = 1,113$, the decision rule leads to conclusion H_0—that the mean service life of the modified component is 1,100 hours or less.

Figures 11.5e, 11.5f, and 11.5g summarize the calculation of the sample size n and the action limit A, and the final form of the decision rule. □

Comments

1. We repeat that the procedure just described for controlling the risks of both types of errors is appropriate if the resulting sample size is reasonably large. Modifications in the procedure are needed if this sample size is small.
2. As Figure 11.2b shows, the β risk declines as μ increases. For our example, this risk is controlled at $\mu_{II} = 1,200$ at the level $\beta = .05$, and hence is smaller than .05 for $\mu > 1,200$. For μ values between $\mu_I = 1,100$ and $\mu_{II} = 1,200$, the probability of a Type II error can be very large. But as a practical matter, β risks tend to be unimportant throughout most of this interval because μ is close to μ_I and a Type II error is relatively less costly for μ values in this range.
3. The sample size is determined from (11.16) using a planning value of σ. Since this planning value is only judgmental, the action limit (which controls the Type I error probability) is based on the sample standard deviation s. Consequently, it is possible that the Type II error probability will depart from the specified level. However, if the planning value of σ is reasonably accurate, the departure from the specified Type II error probability likely will be small. For instance, in our example with $n = 142$, $A = 1,154$, and $s\{\bar{X}\} = 23.3$, we estimate the Type II error probability at $\mu_{II} = 1,200$ to be:

$$P(H_0; 1,200) = P(\bar{X} \leq 1,154; \mu_{II} = 1,200) \simeq .0244$$

This probability value is approximate mainly because we do not know the actual value of σ. Note that the approximate risk is fairly close to the specified level of .05.

11.7 ONE-SIDED LOWER-TAIL TESTS

Construction of Decision Rule to Control α Risk

To this point, we have considered statistical tests involving one-sided upper-tail alternatives. When we turn to one-sided lower-tail alternatives:

$$H_0: \mu \geq \mu_I$$

$$H_1: \mu < \mu_I$$

the decision rules are constructed in a fashion corresponding to that for the upper-tail case:

(11.17) When the alternatives are:

$$H_0: \mu \geq \mu_I$$

$$H_1: \mu < \mu_I$$

and the sample size is reasonably large, the appropriate decision rule to control the α risk at μ_I is:

$$\text{If } \bar{X} \geq A, \text{ conclude } H_0$$

$$\text{If } \bar{X} < A, \text{ conclude } H_1$$

where: $A = \mu_I + z(\alpha)s\{\bar{X}\}$
$s\{\bar{X}\}$ is given by (10.8)

Figure 11.6 summarizes the procedure for controlling the α risk when n is given.

☐ Example

A grocery store is concerned whether or not shoppers tend to purchase fewer items at a time than in the past because of rising prices. In the past, the mean number of items purchased per customer was 12.3. The alternatives are:

$$H_0: \mu \geq 12.3$$

$$H_1: \mu < 12.3$$

Management would like to control the α risk at .025 when $\mu = \mu_I = 12.3$.

A random sample of $n = 400$ customers is selected and the following results obtained:

$$n = 400 \qquad \bar{X} = 11.9 \qquad s = 7.2$$

We require $z(.025) = -1.960$ and $s\{\bar{X}\} = 7.2/\sqrt{400} = .360$. Hence, by (11.17) we obtain $A = 12.3 - 1.960(.360) = 11.6$ and the decision rule is:

$$\text{If } \bar{X} \geq 11.6, \text{ conclude } H_0 \ (\mu \geq 12.3)$$

$$\text{If } \bar{X} < 11.6, \text{ conclude } H_1 \ (\mu < 12.3)$$

Since $\bar{X} = 11.9$, we conclude H_0—that there is no decline in the mean number of items purchased per customer. ☐

FIGURE 11.6 *Construction of decision rule to control α risk—Large sample, one-sided lower-tail test for μ*

(a) *Appropriate*
type of
decision
rule

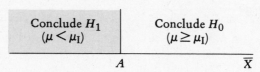

Small values of $\overline{X}$ imply H_1 is correct.

(b) *Sampling*
distribution
of $\overline{X}$
located
at μ_I

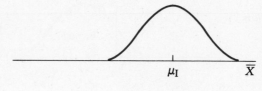

Distribution of $\overline{X}$ is centered at μ_I since α risk is controlled there.

(c) *Placement*
of action
limit A

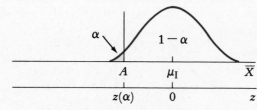

A is placed $z(\alpha)$ deviations below μ_I so lower–tail area equals α.

(d) *Calculation*
of action
limit A

$$A = \mu_I + z(\alpha)s\{\overline{X}\}$$

$\sigma\{\overline{X}\}$ estimated by $s\{\overline{X}\}$.

(e) *Decision*
rule

If $\overline{X} \geq A$, conclude H_0
If $\overline{X} < A$, conclude H_1

Conclude appropriate alternative based on observed $\overline{X}$.

Control of Both Types of Risks through Sample Size Determination

When the sample size is not predetermined, both the α and β risks can be controlled when conducting a lower-tail test, as in the case of an upper-tail test. Figure 11.7 summarizes the procedure. The resulting formula for the sample size n is analogous to formula (11.16):

(11.18) For the one-sided lower-tail test (11.17), the necessary sample size to control both the α and β risks, for a given planning value of σ, is:

$$n = \left[\frac{z(1-\beta) - z(\alpha)}{\mu_I - \mu_{II}} \right]^2 \sigma^2$$

FIGURE 11.7 *Construction of decision rule to control α and β risks—Large sample, one-sided lower-tail test for μ*

(a) *Appropriate type of decision rule*

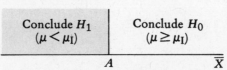

Small values of $\overline{X}$ imply H_1 is correct.

(b) *Sampling distribution of $\overline{X}$ located at μ_I*

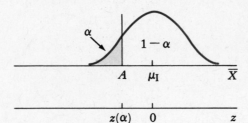

Distribution of $\overline{X}$ is centered at μ_I since α risk is controlled there.

(c) *Sampling distribution of $\overline{X}$ located at μ_{II}*

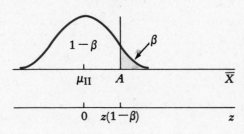

Distribution of $\overline{X}$ is centered at μ_{II} since β risk is controlled there. Note that μ_{II} is below μ_I.

(d) *Simultaneous specifications of action limit*

$$A = \mu_I + z(\alpha)\frac{\sigma}{\sqrt{n}}$$

$$A = \mu_{II} + z(1-\beta)\frac{\sigma}{\sqrt{n}}$$

Simultaneous determination of action limit A and sample size n, where $\sigma\{\overline{X}\} = \sigma/\sqrt{n}$.

(e) *Calculation of sample size n*

$$n = \left[\frac{z(1-\beta) - z(\alpha)}{\mu_I - \mu_{II}}\right]^2 \sigma^2$$

Planning value for σ is used. Resultant value of n must be large.

(f) *Calculation of action limit A*

$$A = \mu_I + z(\alpha)\frac{s}{\sqrt{n}}$$

$\sigma\{\overline{X}\} = \sigma/\sqrt{n}$ estimated by $s\{\overline{X}\} = s/\sqrt{n}$.

(g) *Decision rule*

If $\overline{X} \geq A$, conclude H_0
If $\overline{X} < A$, conclude H_1

Conclude appropriate alternative based on observed $\overline{X}$.

11.8   TWO-SIDED TESTS

We now turn to tests on μ involving two-sided alternatives. Recall from (11.1c) that two-sided alternatives have the form:

$$H_0: \mu = \mu_I$$

$$H_1: \mu \neq \mu_I$$

Interest in this case centers on whether μ differs from μ_I in either direction.

◻ Examples

1. A company has required sales representatives to make forecasts each month of the orders they expect from their customers in the following month. The company is concerned whether or not the representatives' forecasts systematically overestimate or underestimate actual orders. Let μ denote the mean forecast error. The company then wishes to consider the alternatives $H_0: \mu = 0$ and $H_1: \mu \neq 0$. Here, $\mu_I = 0$.
2. A state demographer wishes to test whether or not the mean age of the nonresident population (μ) differs from the mean age of the resident population, which is 27.2 years according to census data. The nonresident population is to be studied by means of a sample survey, and the two alternatives will be $H_0: \mu = 27.2$ and $H_1: \mu \neq 27.2$. Here, $\mu_I = 27.2$. ◻

Evaluating Risks Inherent in Decision Rule

We shall use an illustration to explain how the risks inherent in a decision rule for two-sided alternatives are evaluated.

◻ Illustration

The research department of an art-publishing house is interested in the ability of "naive" subjects to judge the market value of art objects. In one test, subjects were asked to guess the market value of a thirteenth century Andalusian vase which (unknown to the subjects) recently sold for $550. It is desired to test whether or not the mean dollar guess, denoted by μ, differs from the actual market value of $550. If it were to differ, naive subjects would be biased assessors of the market value of the vase. Thus, the following alternatives are to be considered:

(11.19)
$$H_0: \mu = 550$$
$$H_1: \mu \neq 550$$

Here, $\mu_I = 550$. ◻

Decision Rule.   The choice between the alternatives in (11.19) will be based on the sample mean $\bar{X}$ of the dollar guesses for a random sample of naive subjects. Clearly, if $\bar{X}$ falls in an interval about $550, H_0 should be concluded. If, on the other hand, $\bar{X}$ lies some distance above or below $550, H_1 should be concluded. Based on this reasoning, the appropriate type of decision rule has the form:

$$\text{If } A_1 \leq \bar{X} \leq A_2, \text{ conclude } H_0 \ (\mu = 550)$$

$$\text{If } \bar{X} < A_1 \text{ or } \bar{X} > A_2, \text{ conclude } H_1 \ (\mu \neq 550)$$

Here, A_1 and A_2 are two action limits that straddle 550.

Suppose the action limits decided upon are $A_1 = 400$ and $A_2 = 700$. Thus, the decision rule employed is:

(11.20)
$$\text{If } 400 \leq \bar{X} \leq 700, \text{ conclude } H_0 \ (\mu = 550)$$
$$\text{If } \bar{X} < 400 \text{ or } \bar{X} > 700, \text{ conclude } H_1 \ (\mu \neq 550)$$

For the test, $n = 100$ subjects are to be chosen from a large population of naive subjects.

Power Curve. We evaluate the effectiveness of a decision rule for two-sided alternatives in the same way as for decision rules for one-sided alternatives.

Recall that the power curve is a graphic representation of $P(H_1; \mu)$ for various values of μ. Figure 11.8 illustrates the calculation of $P(H_1; \mu)$ for μ values of 350, 450, 550, 650, and 750 for our illustrative decision rule. We assume that the population standard deviation of subjects' guesses is $\sigma = 750$. Consequently, since $n = 100$, we have $\sigma\{\bar{X}\} = \sigma/\sqrt{n} = 750/\sqrt{100} = 75$.

To illustrate one of the calculations, consider the case $\mu = 450$. Since H_1 is concluded if $\bar{X} < 400$ or $\bar{X} > 700$, we must calculate the probabilities represented by the two shaded areas in Figure 11.8b. The z values corresponding to the two action limits are:

$$z = \frac{400 - 450}{75} = -.67 \qquad z = \frac{700 - 450}{75} = 3.33$$

Hence, the desired probability is:

$$P(H_1; 450) = P(Z < -.67) + P(Z > 3.33) = .2514 + .0004 = .2518$$

Note in the preceding calculation that we have assumed n is sufficiently large so that the sampling distribution of $\bar{X}$ is approximately normal.

The power curve for our decision rule is shown in Figure 11.9a for μ values between 250 and 850.

Error Curve. The error curve for our decision rule is plotted in Figure 11.9b. In the case of two-sided alternatives, a Type I error is made when H_1 is concluded at $\mu = \mu_I$ ($\mu_I = 550$ in our illustration), and a Type II error is made when H_0 is concluded at values of μ other than μ_I. For this reason, the error curve has a discontinuity at $\mu = 550$ in Figure 11.9b. The α risk at $\mu = 550$ is given by $P(H_1; 550)$ and was found in Figure 11.8 to be .0456. This α risk is shown in Figure 11.9b by a dot. The β risks are obtained by using relation (11.7). Thus, $P(H_0; 450) = 1 - P(H_1; 450) = 1 - .2518 = .7482$, which is the β risk when $\mu = 450$.

Effects of Action Limits and Sample Size on Risks. As with one-sided alternatives, the risks of incorrect decisions associated with decision rules for two-sided alternatives are affected by the sample size and action limits. Our earlier observations in (11.10) and (11.11) still apply:

FIGURE 11.8 *Computation of $P(H_1; \mu)$ for several values of μ*

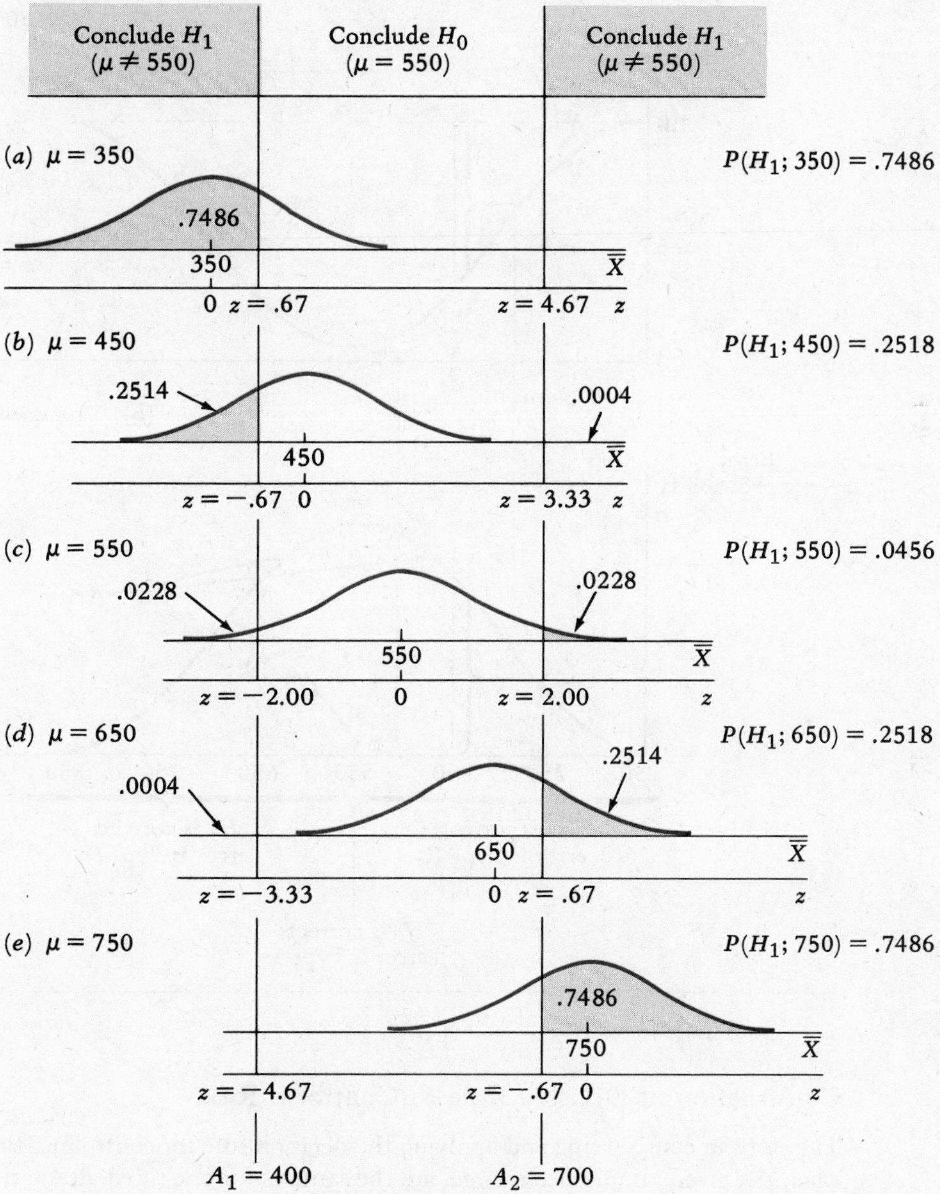

Conclude H_1 ($\mu \neq 550$)	Conclude H_0 ($\mu = 550$)	Conclude H_1 ($\mu \neq 550$)

(a) $\mu = 350$

.7486

350

0 $z = .67$ $z = 4.67$ z

$\overline{X}$

$P(H_1; 350) = .7486$

(b) $\mu = 450$

.2514

.0004

450

$z = -.67$ 0 $z = 3.33$ z

$\overline{X}$

$P(H_1; 450) = .2518$

(c) $\mu = 550$

.0228

.0228

550

$z = -2.00$ 0 $z = 2.00$ z

$\overline{X}$

$P(H_1; 550) = .0456$

(d) $\mu = 650$

.0004

.2514

650

$z = -3.33$ 0 $z = .67$ z

$\overline{X}$

$P(H_1; 650) = .2518$

(e) $\mu = 750$

.7486

750

$z = -4.67$ $z = -.67$ 0 z

$\overline{X}$

$P(H_1; 750) = .7486$

$A_1 = 400$ $A_2 = 700$

1. For a given random sample size, one type of error probability can be reduced only at the expense of increasing the other type.
2. A simultaneous reduction in both types of error probabilities in a given test situation can only be achieved by increasing the random sample size.

FIGURE 11.9 *Power curve and error curve for art example*

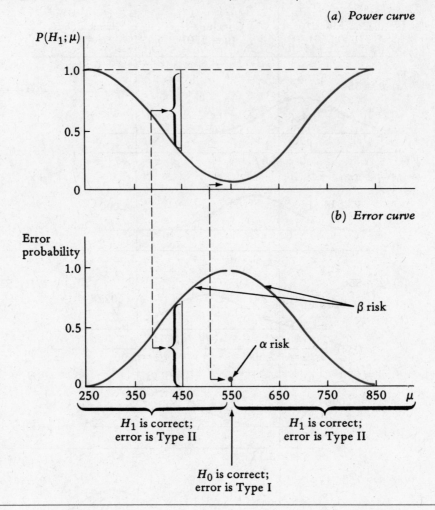

(a) Power curve

(b) Error curve

Construction of Decision Rule to Control α Risk

The steps in constructing and applying the decision rule for controlling the α risk, when the given sample size is large, are the same as for one-sided alternatives. We continue with the illustration dealing with the assessment of the market value of art objects by naive subjects. Recall that the alternatives are:

$$H_0: \mu = \mu_I = 550$$

$$H_1: \mu \neq \mu_I = 550$$

and that a sample of 100 subjects is to be used. It is desired to control the α risk at

FIGURE 11.10 *Construction of decision rule to control α risk—Large sample, two-sided test for μ*

(a) *Appropriate type of decision rule*

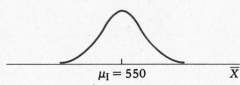

Conclude H_1	Conclude H_0	Conclude H_1
$(\mu \neq \mu_I = 550)$	$(\mu = \mu_I = 550)$	$(\mu \neq \mu_I = 550)$

$\overline{X}$ values differing substantially from μ_I imply H_1 is correct.

A_1 A_2 $\overline{X}$

(b) *Sampling distribution of $\overline{X}$ located at $\mu_I = 550$*

$\mu_I = 550$ $\overline{X}$

Distribution of $\overline{X}$ is centered at μ_I since α risk is controlled there.

(c) *Placement of action limits A_1 and A_2*

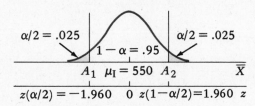

$\alpha/2 = .025$ $1 - \alpha = .95$ $\alpha/2 = .025$

A_1 $\mu_I = 550$ A_2 $\overline{X}$

$z(\alpha/2) = -1.960$ 0 $z(1-\alpha/2) = 1.960$ z

A_1 is placed $z(\alpha/2)$ deviations from μ_I; A_2 is placed $z(1 - \alpha/2)$ deviations from μ_I. Area in each tail equals $\alpha/2$.

(d) *Calculation of action limits A_1 and A_2*

$$A_1 = \mu_I + z(\alpha/2)s\{\overline{X}\}$$
$$= 550 - 1.960(78.7) = 396$$
$$A_2 = \mu_I + z(1 - \alpha/2)s\{\overline{X}\}$$
$$= 550 + 1.960(78.7) = 704$$

$\sigma\{\overline{X}\}$ estimated by $s\{\overline{X}\}$.

(e) *Decision rule*

If $396 = A_1 \leq \overline{X} \leq A_2 = 704$, conclude H_0

If $\overline{X} < A_1 = 396$ or $\overline{X} > A_2 = 704$, conclude H_1

Conclude appropriate alternative based on observed $\overline{X}$.

.05; that is, the probability of concluding H_1 (the subjects are biased assessors) when, in fact, H_0 is correct (the subjects are unbiased assessors) is to be controlled at .05.

Determination of Action Limits.   Figure 11.10a shows the type of decision rule appropriate here. Figure 11.10b shows the sampling distribution of $\overline{X}$ for the case $\mu = \mu_I = 550$, the case for which the α risk is specified. Note that the sampling distribution is approximately normal because the sample size is reasonably large here. The two action limits on either side of $\mu_I = 550$ are shown in Figure 11.10c. They are equidistant from 550, as is the customary practice for the two-sided case. Since A_1 and A_2 are set at equal distances from μ_I, the tail areas of the sampling distribution are both equal to $\alpha/2 = .025$ (and, of course, their sum is $\alpha = .05$).

Hence, the action limits are obtained as follows:

$$A_1 = \mu_I + z(\alpha/2)\sigma\{\bar{X}\} \qquad A_2 = \mu_I + z(1 - \alpha/2)\sigma\{\bar{X}\}$$
$$= 550 - 1.960\sigma\{\bar{X}\} \qquad\qquad = 550 + 1.960\sigma\{\bar{X}\}$$

where $z(1 - \alpha/2) = z(.975) = 1.960$ and $z(\alpha/2) = z(.025) = -1.960$.

Usually, we do not know $\sigma\{\bar{X}\}$ and replace it by $s\{\bar{X}\}$, the estimate based on the sample. For large sample sizes, this replacement has no material effect on the z multiple, as we have seen before. We now summarize the determination of the decision rule for two-sided alternatives:

(11.21) When the alternatives are:

$$H_0\colon \mu = \mu_I$$

$$H_1\colon \mu \neq \mu_I$$

and the sample size is reasonably large, the appropriate decision rule to control the α risk is:

$$\text{If } A_1 \leq \bar{X} \leq A_2, \text{ conclude } H_0$$

$$\text{If } \bar{X} < A_1 \text{ or } \bar{X} > A_2, \text{ conclude } H_1$$

where: $A_1 = \mu_I - zs\{\bar{X}\}$
$A_2 = \mu_I + zs\{\bar{X}\}$
$z = z(1 - \alpha/2)$
$s\{\bar{X}\}$ is given by (10.8)

☐   **Example**

In our art example, the alternatives are:

$$H_0\colon \mu = 550$$

$$H_1\colon \mu \neq 550$$

The α risk is to be controlled at .05; hence, $z(.975) = 1.960$. A sample of $n = 100$ subjects yielded the following results:

$$n = 100 \qquad \bar{X} = 843 \qquad s = 787$$

Hence, $s\{\bar{X}\} = s/\sqrt{n} = 787/\sqrt{100} = 78.7$ and the action limits are:

$$A_1 = 550 - 1.960(78.7) = 396 \qquad A_2 = 550 + 1.960(78.7) = 704$$

The decision rule therefore is:

$$\text{If } 396 \leq \bar{X} \leq 704, \text{ conclude } H_0 \ (\mu = 550)$$

$$\text{If } \bar{X} < 396 \text{ or } \bar{X} > 704, \text{ conclude } H_1 \ (\mu \neq 550)$$

Since $\bar{X} = 843$, the decision rule leads to conclusion H_1—that naive subjects are biased assessors of the value of the Andalusian vase.

Figures 11.10d and 11.10e show the calculation of the action limits and the final form of the decision rule, respectively.   ☐

Comment

The same conclusion could have been obtained from a 95 percent *two-sided* confidence interval, corresponding to the specified α risk of .05. We would have:

$$689 = 843 - 1.960(78.7) \leq \mu \leq 843 + 1.960(78.7) = 997$$

If μ_I falls within the interval, we conclude H_0; otherwise, we conclude H_1. Here, $\mu_I = 550$ does not fall within the confidence interval; hence, we conclude H_1, just as we did from the decision rule.

Control of Both Types of Risks through Determination of Sample Size

We continue with our art illustration to explain how both the α and β risks can be controlled when n is not predetermined. Suppose that in our art illustration no decision has yet been made on the number of subjects for the study. Figure 11.11 illustrates the necessary steps. The appropriate kind of decision rule is the same as before, as shown in Figure 11.11a. Figure 11.11b shows the sampling distribution of $\overline{X}$ when $\mu = \mu_I = 550$, where the α risk is to be controlled at .05. The control of this risk is the same as before.

Suppose it is also important to detect a bias of $200. In view of the symmetry of the problem, we can plan for either an upward or a downward bias. As the error curve in Figure 11.9b shows, we will receive the same protection against Type II errors on both sides of μ_I. Suppose we consider an upward bias of $200; hence, $\mu_{II} = 550 + 200 = 750$. The research department would like to control the β risk at .10 here. Figure 11.11c shows the sampling distribution of $\overline{X}$ for $\mu = \mu_{II} = 750$. The shaded area represents:

$$P(A_1 \leq \overline{X} \leq A_2; \mu_{II} = 750) = P(H_0; 750) = \beta$$

Observe that we use μ_{II} to denote that level of μ at which the Type II error is to be controlled. The shaded area in Figure 11.11c is .10, the specified level of β.

The two simultaneous specifications of the action limit A_2 based on the two risk specifications are shown in Figure 11.11d. In solving these two equations for the required sample size n, we obtain an expression that involves the unknown population standard deviation σ. We therefore require a planning value of σ in order to determine the needed sample size:

(11.22) For the two-sided test (11.21), the necessary sample size to control both the α and β risks, for a given planning value of σ, is:

$$n = \left[\frac{z(\beta) - z(1 - \alpha/2)}{\mu_{II} - \mu_I} \right]^2 \sigma^2$$

Once the sample size has been determined and the sample selected, the action limits A_1 and A_2 are obtained as for the case where n is given, using the sample standard deviation s to estimate σ.

FIGURE 11.11 *Construction of decision rule to control α and β risks—Large sample, two-sided test for μ*

(a) *Appropriate type of decision rule*

Conclude H_1 ($\mu \neq 550$)	Conclude H_0 ($\mu = 550$)	Conclude H_1 ($\mu \neq 550$)
A_1	A_2	$\overline{X}$

$\overline{X}$ values differing substantially from μ_I imply H_1 is correct.

(b) *Sampling distribution of $\overline{X}$ located at $\mu_I = 550$*

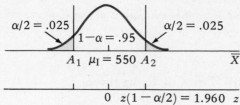

$\alpha/2 = .025$ $1 - \alpha = .95$ $\alpha/2 = .025$

$A_1 \quad \mu_I = 550 \quad A_2 \qquad \overline{X}$

$0 \quad z(1 - \alpha/2) = 1.960 \quad z$

Distribution of $\overline{X}$ is centered at μ_I since α risk is controlled there. Area in each tail equals $\alpha/2$.

(c) *Sampling distribution of $\overline{X}$ located at $\mu_{II} = 750$*

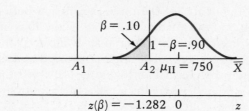

$\beta = .10$ $1 - \beta = .90$

$A_1 \qquad A_2 \; \mu_{II} = 750 \quad \overline{X}$

$z(\beta) = -1.282 \quad 0 \qquad z$

Distribution of $\overline{X}$ is centered at μ_{II} since β risk is controlled there.

(d) *Simultaneous specifications of action limit*

$$A_2 = \mu_I + z(1 - \alpha/2)\frac{\sigma}{\sqrt{n}}$$

$$A_2 = \mu_{II} + z(\beta)\frac{\sigma}{\sqrt{n}}$$

Simultaneous determination of action limit A_2 and sample size n, where $\sigma\{\overline{X}\} = \sigma/\sqrt{n}$.

(e) *Calculation of sample size n*

$$n = \left[\frac{z(\beta) - z(1 - \alpha/2)}{\mu_{II} - \mu_I}\right]^2 \sigma^2$$

$$= \left(\frac{-1.282 - 1.960}{750 - 550}\right)^2 (750)^2 = 148$$

Planning value for σ is used. Resultant value of n must be large.

(f) *Calculation of action limits A_1 and A_2*

$$A_1 = \mu_I - z\frac{s}{\sqrt{n}}$$

$$= 550 - 1.960\frac{710}{\sqrt{148}} = 436$$

$$A_2 = \mu_I + z\frac{s}{\sqrt{n}}$$

$$= 550 + 1.960\frac{710}{\sqrt{148}} = 664$$

$\sigma\{\overline{X}\} = \sigma/\sqrt{n}$ estimated by $s\{\overline{X}\} = s/\sqrt{n}$, and $z = z(1 - \alpha/2)$.

(g) *Decision rule*

If $436 = A_1 \leq \overline{X} \leq A_2 = 664$, conclude H_0

If $\overline{X} < A_1 = 436$ or $\overline{X} > A_2 = 664$, conclude H_1

Conclude appropriate alternative based on observed $\overline{X}$.

☐   Example

In our art illustration, the risk specifications are:

$$\mu_I = 550 \qquad \alpha = .05$$

$$\mu_{II} = 750 \qquad \beta = .10$$

Hence, $z(1 - \alpha/2) = z(.975) = 1.960$ and $z(\beta) = z(.10) = -1.282$. (Note that we are assuming a negligible area to the left of action limit A_1 for the sampling distribution centered at μ_{II}.) Suppose that a reasonable planning value for the population standard deviation is $\sigma = 750$. The sample size is then computed from (11.22) as follows:

$$n = \left(\frac{-1.282 - 1.960}{750 - 550}\right)^2 (750)^2 = 148$$

Suppose a random sample of 148 subjects was used and yielded the results:

$$n = 148 \qquad \bar{X} = 784 \qquad s = 710$$

Thus, $s\{\bar{X}\} = 710/\sqrt{148} = 58.4$ and the action limits are, by (11.21):

$$A_1 = 550 - 1.960(58.4) = 436 \qquad A_2 = 550 + 1.960(58.4) = 664$$

The decision rule then is:

If $436 \leq \bar{X} \leq 664$, conclude H_0 ($\mu = 550$)

If $\bar{X} < 436$ or $\bar{X} > 664$, conclude H_1 ($\mu \neq 550$)

Since $\bar{X} = 784$, the decision rule leads to conclusion H_1—that naive subjects are biased assessors of the value of the Andalusian vase.

Figures 11.11e, 11.11f, and 11.11g present the calculation of the sample size and action limits and the final form of the decision rule. ☐

Comment

The procedure that we have just described for controlling the probabilities of both types of errors is appropriate if the resultant sample size is reasonably large. Modifications in the procedure are needed if the sample size is small.

11.9   TESTS FOR μ—SMALL SAMPLE

The test procedures for μ discussed so far in this chapter have assumed that the sample size is reasonably large. This has permitted us to invoke central limit theorem extension (10.13), which states that the sampling distribution of $(\bar{X} - \mu)/s\{\bar{X}\}$ is then approximately normal.

When the sample size is small, approximate normality may no longer hold. If, however, the population is normal, we know from (10.15) that $(\bar{X} - \mu)/s\{\bar{X}\}$ is distributed as $t(n -- 1)$. Further, in view of the robustness of the t distribution, $(\bar{X} - \mu)/s\{\bar{X}\}$ is approximately distributed as $t(n - 1)$ as long as the population does not depart too markedly from a normal distribution. This means that an appropriate t value can be used in place of the corresponding z value in each type of decision rule that we have considered. Table 11.2 shows the form of the decision

TABLE 11.2 *Decision rules to control α risk for small-sample case, population normal or departure not marked*

Alternatives	Decision Rule
(a) H_0: $\mu \leq \mu_{\mathrm{I}}$ H_1: $\mu > \mu_{\mathrm{I}}$	If $\bar{X} \leq A$, conclude H_0 If $\bar{X} > A$, conclude H_1 where: $A = \mu_{\mathrm{I}} + t(1 - \alpha; n - 1)s\{\bar{X}\}$
(b) H_0: $\mu \geq \mu_{\mathrm{I}}$ H_1: $\mu < \mu_{\mathrm{I}}$	If $\bar{X} \geq A$, conclude H_0 If $\bar{X} < A$, conclude H_1 where: $A = \mu_{\mathrm{I}} + t(\alpha; n - 1)s\{\bar{X}\}$
(c) H_0: $\mu = \mu_{\mathrm{I}}$ H_1: $\mu \neq \mu_{\mathrm{I}}$	If $A_1 \leq \bar{X} \leq A_2$, conclude H_0 If $\bar{X} < A_1$ or $\bar{X} > A_2$, conclude H_1 where: $A_1 = \mu_{\mathrm{I}} - ts\{\bar{X}\}$ $A_2 = \mu_{\mathrm{I}} + ts\{\bar{X}\}$ $t = t(1 - \alpha/2; n - 1)$

rule for each of the three types of tests we have covered in this chapter. The rules are for the case where the sample size n is given and the α risk is to be controlled.

Example

In Chapter 10, we considered a quality control example in which a sample of five cans of standard-grade tomatoes in puree was taken at random from a production line immediately after filling, and the solid content of each can was weighed. Suppose it were desired to test whether the mean drained weight per can for the process is at the specified level $\mu = 22$ ounces. Past experience indicates that the distribution of the drained weights per can is approximately normal. The α risk for the test is to be controlled at .01.

The alternatives here are:

$$H_0: \mu = 22$$

$$H_1: \mu \neq 22$$

so $\mu_{\mathrm{I}} = 22$. The test procedure is that of Table 11.2c.

Recall that the sample results are:

$$n = 5 \qquad \bar{X} = 23.9 \qquad s = .822$$

Hence, $s\{\bar{X}\} = s/\sqrt{n} = .822/\sqrt{5} = .37$. We require, for $\alpha = .01$, $t(1 - \alpha/2; n - 1) = t(.995; 4) = 4.604$. Hence:

$$A_1 = 22 - 4.604(.37) = 20.3 \qquad A_2 = 22 + 4.604(.37) = 23.7$$

The decision rule then is:

$$\text{If } 20.3 \leq \bar{X} \leq 23.7, \text{ conclude } H_0 \ (\mu = 22)$$

$$\text{If } \bar{X} < 20.3 \text{ or } \bar{X} > 23.7, \text{ conclude } H_1 \ (\mu \neq 22)$$

Since $\bar{X} = 23.9$, conclusion H_1 should be accepted—that the process mean μ is not 22 ounces.

Comment

The 99 percent confidence interval for μ which was calculated for this same sample in the example of Section 10.5 is $22.2 \leq \mu \leq 25.6$. Note that $\mu_I = 22$ is not in the confidence interval, a result which is consistent with our test conclusion.

11.10   OPTIONAL TOPIC—STANDARDIZED TEST STATISTICS

The statistical tests which have been discussed in this chapter can be expressed in an alternative fashion by restating the test statistic $\bar{X}$ in standardized form. A number of statistical computer packages provide such standardized test statistics.

To illustrate the use of standardized test statistics, let us return to the machine component illustration, where the alternatives were:

$$H_0: \mu \leq \mu_I = 1{,}100$$

$$H_1: \mu > \mu_I = 1{,}100$$

and the statistical decision rule was of the form:

$$\text{If } \bar{X} \leq \mu_I + z(1 - \alpha)s\{\bar{X}\}, \text{ conclude } H_0$$

$$\text{If } \bar{X} > \mu_I + z(1 - \alpha)s\{\bar{X}\}, \text{ conclude } H_1$$

This decision rule can be restated by subtracting μ_I from each side and then dividing by $s\{\bar{X}\}$. Thus, we obtain:

$$\text{If } \frac{\bar{X} - \mu_I}{s\{\bar{X}\}} \leq z(1 - \alpha), \text{ conclude } H_0$$

$$\text{If } \frac{\bar{X} - \mu_I}{s\{\bar{X}\}} > z(1 - \alpha), \text{ conclude } H_1$$

The test statistic on the left is called a *standardized test statistic*. Note that the numerator is the deviation of $\bar{X}$ from μ_I, the level of μ where the α risk is controlled, and that this deviation is then expressed in units of the relevant standard deviation, here $s\{\bar{X}\}$. We shall denote this test statistic for large samples by z^*:

$$(11.23) \qquad z^* = \frac{\bar{X} - \mu_I}{s\{\bar{X}\}}$$

Figure 11.12 illustrates the equivalence of the test for the machine component example based on $\bar{X}$ and the test based on the standardized test statistic z^*.

When the sample size is small and the population is normal or does not depart too markedly from a normal distribution, we shall denote the test statistic by t^*:

$$(11.24) \qquad t^* = \frac{\bar{X} - \mu_I}{s\{\bar{X}\}}$$

The appropriate decision rules for use with standardized test statistics are shown in Table 11.3 for all cases considered in this chapter. The decision rules in

FIGURE 11.12 *Equivalence of decision rules based on $\bar{X}$ and z^*—Large sample, one-sided upper-tail test for μ.* The standardizing transformation changes values on the $\bar{X}$ scale to equivalent values on the z^* scale.

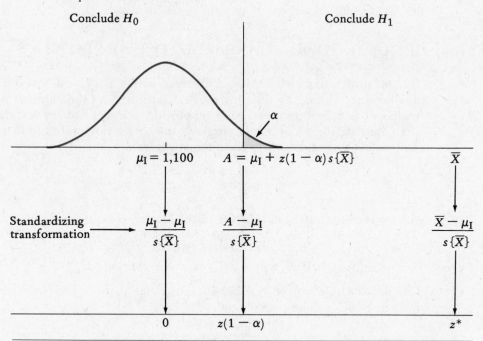

Table 11.3 are algebraic equivalents to the earlier decision rules based on $\bar{X}$, as illustrated in Figure 11.12 for the one-sided upper-tail test.

☐ Examples

1. For the machine component illustration, $\mu_I = 1{,}100$, $n = 36$, $\bar{X} = 1{,}121$, $s\{\bar{X}\} = 37.0$, and $\alpha = .01$. Hence, $z(1 - \alpha) = z(.99) = 2.326$ and:

$$z^* = \frac{1{,}121 - 1{,}100}{37.0} = .57$$

TABLE 11.3 *Decision rules for standardized test statistics*

Alternatives	Large Samples	Small Samples
(a) $H_0: \mu \leq \mu_I$ $H_1: \mu > \mu_I$	If $z^* \leq z(1 - \alpha)$, conclude H_0 Otherwise, conclude H_1	If $t^* \leq t(1 - \alpha; n - 1)$, conclude H_0 Otherwise, conclude H_1
(b) $H_0: \mu \geq \mu_I$ $H_1: \mu < \mu_I$	If $z^* \geq z(\alpha)$, conclude H_0 Otherwise, conclude H_1	If $t^* \geq t(\alpha; n - 1)$, conclude H_0 Otherwise, conclude H_1
(c) $H_0: \mu = \mu_I$ $H_1: \mu \neq \mu_I$	If $|z^*| \leq z(1 - \alpha/2)$, conclude H_0 Otherwise, conclude H_1	If $|t^*| \leq t(1 - \alpha/2; n - 1)$, conclude H_0 Otherwise, conclude H_1

NOTE: | | denotes the absolute value. Thus, if $z^* = -3.4$, $|z^*| = 3.4$.

The decision rule is:

$$\text{If } z^* \leq 2.326, \text{ conclude } H_0$$

$$\text{If } z^* > 2.326, \text{ conclude } H_1$$

Since $z^* = .57 \leq 2.326$, we conclude H_0, just as we did when we used the test statistic $\bar{X}$.

2. For the illustration dealing with the assessment of the market value of art objects by naive subjects, the alternatives were:

$$H_0: \mu = 550$$

$$H_1: \mu \neq 550$$

Here, $\mu_I = 550$ and the risk specification was $\alpha = .05$; also, we had obtained $\bar{X} = 843$ and $s\{\bar{X}\} = 78.7$. Hence, $z(1 - \alpha/2) = z(.975) = 1.960$ and the decision rule according to Table 11.3c is:

$$\text{If } |z^*| \leq 1.960, \text{ conclude } H_0$$

$$\text{If } |z^*| > 1.960, \text{ conclude } H_1$$

Since $z^* = (843 - 550)/78.7 = 3.72$, we have $|z^*| = |3.72| = 3.72 > 1.960$, and hence we conclude H_1. Again, this is the same conclusion that we obtained earlier based on the statistic $\bar{X}$.

Note that the decision rule statement:

$$\text{If } |z^*| \leq 1.960, \text{ conclude } H_0$$

is equivalent to:

$$\text{If } -1.960 \leq z^* \leq 1.960, \text{ conclude } H_0$$

Therefore, the test with the standardized statistic is a two-sided test, just as the one explained earlier based on the statistic $\bar{X}$.

3. In the quality control example involving the production process for canned tomatoes, the alternatives were:

$$H_0: \mu = 22$$

$$H_1: \mu \neq 22$$

Thus, $\mu_I = 22$. The α risk was to be controlled at .01 and $n = 5$, $\bar{X} = 23.9$, and $s\{\bar{X}\} = .37$. Hence, we require $t(1 - \alpha/2; n - 1) = t(.995; 4) = 4.604$. The standardized test statistic is $t^* = (23.9 - 22)/.37 = 5.14$ and the decision rule is:

$$\text{If } |t^*| \leq 4.604, \text{ conclude } H_0$$

$$\text{If } |t^*| > 4.604, \text{ conclude } H_1$$

Since $|t^*| = |5.14| = 5.14 > 4.604$, we conclude H_1. This, of course, is the same conclusion as the one based on the statistic $\bar{X}$. ☐

11.11 OPTIONAL TOPIC—*P*-VALUES FOR STATISTICAL TESTS

Statistical computer packages frequently provide information about the *P*-value of a statistical test in order that the user can draw the appropriate conclusion for any particular α risk level desired. For a test of a population mean, the P-value is defined as follows:

(11.25) The *P-value* of a statistical test for μ is the probability that, when $\mu = \mu_I$, the sample outcome could have been more extreme than the observed outcome.

Consider again the machine component illustration, where the alternatives were:

$$H_0: \mu \leq 1,100$$

$$H_1: \mu > 1,100$$

The sample mean was $\bar{X} = 1,121$ and the estimated standard deviation of $\bar{X}$ was $s\{\bar{X}\} = 37.0$. To find the P-value for this test, we consider the probability that $\bar{X}$ could have been greater than 1,121 when $\mu = \mu_I = 1,100$. Making use of the fact that the sampling distribution of $\bar{X}$ is approximately normal because of the large sample size, we find $P(\bar{X} > 1,121; \mu = 1,100)$ in the usual fashion. We obtain the z value, i.e., the standardized test statistic denoted by z^* in Section 11.10, as follows:

$$z^* = \frac{1,121 - 1,100}{37.0} = .57$$

Hence, $P(\bar{X} > 1,121; \mu = 1,100) = P(Z > z^* = .57) = .28$, so the P-value is .28.

When the P-value for the test statistic is sufficiently low, it indicates that such a sample outcome is most unlikely when $\mu = \mu_I$ and hence suggests that H_1 should be concluded. When the P-value is reasonably large, it indicates that the sample result is consistent with $\mu = \mu_I$ and hence suggests that H_0 should be concluded.

The determination of when the P-value is sufficiently small to suggest that H_1 holds is made by comparing the P-value with the specified α risk level:

(11.26) When the P-value is less than the specified α risk level, alternative H_1 is concluded, and when the P-value is greater than or equal to the specified α risk level, alternative H_0 is concluded.

☐ Examples

1. In the machine component illustration, the probability that the sample mean could have been greater than 1,121 when $\mu = \mu_I = 1,100$ is $P(Z > .57) = .28$, since the standardized test statistic is $z^* = .57$. For the specified α risk of .01, we find that the P-value of .28 is greater than .01; hence, we conclude H_0. This conclusion is identical to the one obtained with either the statistic $\bar{X}$ or the standardized test statistic z^*.

2. For the illustration about the assessment of the market value of art objects by naive subjects, the alternatives were:

$$H_0: \mu = 550$$

$$H_1: \mu \neq 550$$

and $\bar{X} = 843$, $s\{\bar{X}\} = 78.7$, and $\alpha = .05$. We now need to obtain the probability that the sample mean could have been more extreme than $\bar{X} = 843$ when $\mu = \mu_I = 550$. We found earlier that the standardized test statistic in this case is $z^* = 3.72$. Hence,

$P(\bar{X} > 843; \mu = 550) = P(Z > z^* = 3.72) = .0001$. In view of the fact that this is a two-sided, symmetric test, we need to consider extreme outcomes in both directions from $\mu = \mu_{\text{I}}$, so the P-value here is twice the one-sided value, or $2(.0001) = .0002$. For the specified α risk of .05, we thus find that the P-value of .0002 is smaller than .05; hence, we conclude H_1, as before. □

Comments

1. When the standardized test statistic t^* is employed for small samples, Table C–3 for the t distribution does not permit us to obtain exact P-values. We can only ascertain, for example, that the P-value is between .01 and .025, or between .05 and .10. However, many computer packages calculate the actual P-values in these cases and provide them for the user.
2. The P-value can be interpreted as the α risk level of the test when the action limit of the decision rule based on the sample mean is placed exactly at the value of the observed sample mean $\bar{X}$.
3. P-values are applicable for statistical tests other than for a population mean, as will be seen in later chapters.

PROBLEMS

11.1 For each of the following test situations: (1) specify the population mean; (2) state the alternatives H_0 and H_1; and (3) describe the Type I and Type II errors that are possible. Use the Text convention of labeling the alternative containing the equality sign as H_0.

    a. The mean donation per contributor to the Community Appeal was $11.83 prior to the initiation of a new public relations program aimed at making the community more aware of the social services sponsored by the Appeal. A random sample of donations received while the new public relations program is in effect is to be used to determine whether the mean contribution now is larger.

    b. The mean duration of marriages that ended in divorce or annulment during the past few years in a certain state was 8.1 years. A sociologist wishes to test whether new divorce legislation has changed the mean duration, based on a random sample of divorce records accumulated since the legislation was enacted.

11.2 For each of the following test situations: (1) specify the population mean; (2) state the alternatives H_0 and H_1; and (3) describe the Type I and Type II errors that are possible. Use the Text convention of labeling the alternative containing the equality sign as H_0.

    a. An engineer has developed a leaching process for reducing the amount of sulfur in coal. Sixty-four one-kilogram samples are selected randomly from coal treated by this process to test if the mean sulfur content of the treated coal is less than 10 grams per kilogram.

    b. Bilingual children achieve a mean score of 60 on a mathematics proficiency test when the subject is taught to them in their primary language. A random sample of 400 children are taught the subject in their second language. An equivalent proficiency test is then administered to see if the mean score differs from 60.

*11.3 A market study for a new industrial product indicates that the product should be launched by the firm if the mean number of units sold per customer in the first solicitation of the firm's customers is more than 2.0. Initial orders for the product are to be solicited from a random sample of the firm's large number of customers. It is anticipated that the standard deviation of the numbers of units purchased in initial orders will be $\sigma = 1.6$ units.

    a. Define the parameter μ and specify the alternatives H_0 and H_1 in this test situation.

    b. Suppose the following decision rule, based on a random sample of 100 customers, is adopted: If $\bar{X} \leq 2.30$, conclude H_0; if $\bar{X} > 2.30$, conclude H_1. Calculate the power of this decision rule for μ values of 2.00, 2.15, 2.30, 2.45, and 2.60. Sketch the power curve for this decision rule on a graph.

    c. Define Type I and Type II errors in this situation. Sketch the error curve for the decision rule in b on a separate graph. What is the maximum α risk associated with this decision rule? What is the maximum β risk?

    d. Management of the firm knows that if μ were 2.5 or more units and the product is not introduced, a competitor will successfully market a similar product to the firm's detriment. What is the probability with the decision rule in b that management would not introduce the product to the market if, in fact, $\mu = 2.5$ units per customer?

11.4 Refer to Problem 11.3. Answer all parts of this problem assuming that a random sample of 144 customers is to be selected and the following decision rule adopted: If $\bar{X} \leq 2.20$, conclude H_0; if $\bar{X} > 2.20$, conclude H_1.

11.5 Refer to Problem 11.3. Without making any calculations, show on a graph where the power curve which is obtained if the action limit of the decision rule is shifted from 2.30 to 2.10 would lie in relation to the original power curve. Make the same comparison for the error curves.

11.6 Refer to Problem 11.3. Sketch the power curve which is obtained if the sample size is decreased from 100 to 36; calculate the power for $\mu = 2.00$, 2.30, 2.50, and 2.70. Is the β risk when $\mu = 2.5$ higher with a sample size of 36 than with a sample size of 100? Explain.

*11.7 Refer to Problem 11.1a. The alternative conclusions are: H_0: $\mu \leq 11.83$, H_1: $\mu > 11.83$. A random sample consisting of 225 donations received while the new public relations program is in effect yields the following results: $\bar{X} = \$12.90$, $s = \$8.25$.

    a. Construct the decision rule which controls the α risk at .05 when $\mu_I = 11.83$. Which conclusion do you reach?

    b. For your decision rule in a, estimate the risk of concluding H_0 when $\mu = 13.00$.

11.8 It is known that the mean size of commercial loans made by a certain bank has been $60 thousand in the recent past. A new change in credit policy allows larger amounts to be borrowed under the same terms and the bank wants to test if, in fact, the mean loan size (μ) has increased as a result of the policy change. A random sample of $n = 144$ commercial loans made since the change in policy yields the results (in $000) $\bar{X} = 68.1$ and $s = 45.0$.

    a. Specify the alternatives H_0 and H_1 in this test situation.

    b. Construct the decision rule which controls the α risk at .01 when $\mu_I = 60$. Based on the sample results, has the mean loan size increased?

    c. If the α risk in b had been controlled at .05 rather than at .01, would your conclusion have been different? Explain.

*11.9 Refer to Problem 11.3. Management wishes to determine the required sample size for its customer survey to meet the following two specifications: (1) The α risk at $\mu_I = 2.0$ is to be .05. (2) The β risk at $\mu_{II} = 2.5$ is to be .01. What sample size is required for the survey if $\sigma = 1.6$ is a reasonable planning value?

11.10 A metal alloy manufacturer will produce a filament to be shipped in large batches to a customer. Specifications call for this filament to have a mean melting point (μ) of 300°C or less. On receiving each batch, the customer will select a random sample of filaments for a melting point test. The sample size must meet both the manufacturer's requirement that there is only a .01 chance of rejecting a batch if its mean melting point is $\mu_I = 300$°C and the customer's requirement that there is only a .01 chance of accepting a batch if its mean melting point is $\mu_{II} = 320$°C. Assuming the variability of filament melting points is $\sigma = 35$°C, what number of filaments from a batch must be tested to meet the requirements of both the manufacturer and customer?

11.11 Refer to Problem 11.10. After further negotiations, it was decided to use samples of 80 filaments.
  a. For one batch, the sample results were $\bar{X} = 311$°C and $s = 33$°C. What decision should be made, controlling the α risk at the specified level? State the alternatives, decision rule, and conclusion.
  b. Estimate the β risk at $\mu_{II} = 320$ for your decision rule in a.
  c. Assuming that filament melting points are approximately normally distributed, estimate the percent of filaments in the batch in a that have melting points of 300°C or less.

*11.12 Travelers returning from abroad are asked to declare the value of goods they are bringing back to their country. Customs authorities are interested in whether or not the mean reporting error (difference between declared value and actual value) is negative. Information will be obtained by an audit of the personal effects of a random sample of homecoming travelers. The authorities want to avoid concluding that travelers tend to understate the actual value of their goods unless the evidence confirms this fact quite clearly. Letting μ denote the mean reporting error, the alternatives of interest therefore are H_0: $\mu \geq 0$, H_1: $\mu < 0$.
  a. A study of the reporting errors for 100 randomly selected travelers produced the results $\bar{X} = -\$10.00$ and $s = \$15.00$. Conduct the appropriate test, controlling the α risk at .001 when $\mu_I = 0$. State the decision rule and conclusion.
  b. For your decision rule, estimate the probability of concluding H_0 when, in fact, $\mu = -\$5.00$.

11.13 Refer to Problem 11.12. Answer both parts assuming that the sample size is 225.

11.14 In a test of the alternatives H_0: $\mu \geq 100$, H_1: $\mu < 100$, a reasonable planning value of σ is 20.
  a. What sample size is needed so that the α risk is .001 at $\mu_I = 100$ and the β risk is .05 at $\mu_{II} = 90$?
  b. A random sample of the size determined in a has been selected and the sample results are $\bar{X} = 101$ and $s = 25$. Conduct the appropriate test, controlling the α risk at .001 when $\mu_I = 100$. State the decision rule and conclusion.
  c. Estimate the β risk at $\mu_{II} = 90$ for your decision rule in b. Why is it not .05 as planned?

*11.15 An auto parts company laboriously determined from a study of all sales invoices for the previous fiscal year that the mean gross profit per invoice was $16.50. For the current fiscal year, the company has decided to obtain this information by examining a random sample of sales invoices. The company wants to know if the current year mean gross profit per invoice (μ) has changed from that for last year.

a. Specify the alternatives H_0 and H_1 and describe the Type I and Type II errors that are possible in this test situation.

b. Suppose the following decision rule is adopted, based on a simple random sample of 196 invoices: If $15.00 \leq \bar{X} \leq 18.00$, conclude H_0; if $\bar{X} < 15.00$ or $\bar{X} > 18.00$, conclude H_1. From experience in previous years, it is assumed that the standard deviation of invoice gross profits in the current year is $\sigma = \$17.50$. Obtain the power and error probabilities for this decision rule for μ values of 13.50, 15.00, 16.50, 18.00, and 19.50.

c. Sketch the power and error curves for the decision rule in b, using separate graphs.

d. (1) What is the α risk associated with this decision rule? (2) What is the β risk if the mean gross profit has increased by $1.50? Has decreased by $3.00?

11.16 Refer to Problem 11.15. One company executive, after learning about the sampling plan, suggested that since $\bar{X}$ is a point estimate of μ, it should be concluded that the mean gross profit per invoice has changed from $16.50 if it is found that $\bar{X} \neq 16.50$. Discuss the reasonableness of the executive's suggestion.

11.17 A trade agreement governing the movement of agricultural products between countries in a common market stipulates that the mean weight of boxes of butter (μ) must be 25 kilograms. A large shipment of butter is to be tested to determine if it meets this requirement by selecting a random sample of 36 boxes and using the decision rule: If the sample mean weight per box lies within .05 kilogram of 25, consider the shipment to be acceptable; otherwise, consider the shipment to be unacceptable.

a. Specify the alternatives H_0 and H_1.

b. It is anticipated that the standard deviation of weights for boxes of butter is $\sigma = .12$ kilogram. Calculate the power of the decision rule at μ values of 24.90, 24.95, 25.00, 25.05, and 25.10.

c. Sketch the power curve for the decision rule based on the values calculated in b. Sketch the error curve on a separate graph.

d. What proportion of shipments containing boxes of butter which have a mean weight of exactly 25 kilograms will be judged unacceptable by the decision rule? What is the probability that a shipment with a mean weight per box which differs by .10 kilogram from 25 will be judged acceptable by the decision rule?

*11.18 Refer to Problem 11.15. A random sample of 225 invoices was actually selected and the gross profit computed for each. The sample results were $\bar{X} = \$17.14$ and $s = \$18.80$. Construct the decision rule appropriate for controlling the probability of concluding that the mean invoice gross profit has changed from $16.50 when, in fact, it has not, using $\alpha = .01$. State the decision rule and conclusion.

11.19 In the past, mean library usage per cardholder was 8.5 books during the summer. A random sample of 100 cardholders showed the following results for summer usage this year: $\bar{X} = 9.34$ books, $s = 3.31$ books. The library administration would like to know whether this year's mean usage (μ) has changed from that for past years. Conduct the appropriate test, controlling the α risk at .05. State the alternatives, decision rule, and conclusion.

11.20 Refer to Problem 11.19. Construct a two-sided 95 percent confidence interval for μ

based on the sample results. Explain how you can draw the same conclusion from this interval as from the test in Problem 11.19. What additional information does the interval estimate provide that the test does not?

*11.21   Refer to Problem 11.15. Suppose that the sample size has not yet been determined, and it is desired to control the α risk at .01 and the β risk at $\mu_{II} = \$18.00$ at .05.

   a.   How many invoices should be sampled, assuming that a reasonable planning value for the standard deviation of invoice gross profits in the current year is $\sigma = \$17.50$?

   b.   Suppose a random sample of the size determined in **a** has been selected and the results are $\bar{X} = \$16.93$ and $s = \$17.26$. Conduct the appropriate test, controlling the α risk at .01. State the decision rule and conclusion.

   c.   In constructing your decision rule in **b**, did you use $s = \$17.26$ or the planning value $\sigma = \$17.50$? Why?

11.22   Refer to Problem 11.17. Suppose that the sample size is still to be determined, and it is desired to control the α risk at .01 and the β risk at $\mu_{II} = 25.08$ kilograms at .005.

   a.   Assuming that $\sigma = .12$ kilogram is a reasonable planning value, what number of boxes should be sampled from a shipment of butter?

   b.   Suppose a random sample of the size determined in **a** is selected and the results are $\bar{X} = 25.053$ kilograms and $s = .10$ kilogram. Conduct the appropriate test, controlling the α risk at .01. State the decision rule and conclusion.

   c.   Estimate the β risk at $\mu_{II} = 25.08$ kilograms for your decision rule in **b**. Why is it not .005 as planned?

*11.23   In a study of risk-taking behavior, a researcher gave each subject a lottery ticket which offered the subject $100 with probability .1 and $0 with probability .9. By subsequent questioning, the researcher determined the minimum price at which the subject would sell the lottery ticket. Since the expected value of the lottery ticket is $10, the researcher wanted to test whether or not the mean minimum selling price is $10. Sixteen subjects participated in the study and the sample results were $\bar{X} = \$11.25$ and $s = \$1.52$. The researcher believes it is reasonable to assume that the minimum selling prices are approximately normally distributed. What conclusion should the researcher draw? Conduct the appropriate test, controlling the α risk at .01. State the alternatives, decision rule, and conclusion.

11.24   A random sample of 21 children from the same age group participated in a special language-skills program designed to build vocabulary. Measurements at the end of the program showed the children's vocabularies had a mean of $\bar{X} = 1,060.1$ words and a standard deviation of $s = 109.0$ words. It is known that the mean vocabulary of children in this age group ordinarily is 1,000 words. Does the program enhance the vocabulary of children in this age group? Conduct an appropriate test, controlling the α risk at $\mu_{I} = 1,000$ at .05. Assume that children's vocabularies are nearly normally distributed. State the alternatives, decision rule, and conclusion.

11.25   Refer to Problem 10.13. Test whether or not the mean blood-sugar level (μ) is above 110 milligrams. Control the α risk at $\mu_{I} = 110$ at .025. State the alternatives, decision rule, and conclusion.

*11.26   Refer to Problem 11.8.

   a.   Controlling the α risk at .01 when $\mu_{I} = 60$, give the decision rule for the test in terms of the standardized test statistic z^* defined in (11.23).

   b.   Calculate z^* for the given sample results. What conclusion should be drawn?

11.27 Refer to Problem 11.23.
    a. Give the decision rule for the test in terms of the standardized test statistic t^* defined in (11.24).
    b. Calculate t^* for the given sample results. What conclusion should be drawn?

11.28 If a sample yields the standardized test statistic $z^* = -2.260$ in a test of the alternatives H_0: $\mu = 0$, H_1: $\mu \neq 0$, what conclusion should be drawn when the α risk of the test is to be controlled at .01? At .05?

*11.29 Refer to Problem 11.8.
    a. Calculate the P-value for this test.
    b. In Problem 11.8 the appropriate conclusion for this test is H_1 when α is controlled at .05 and is H_0 when α is controlled at .01. Explain how these same conclusions can be drawn directly from knowledge of the P-value calculated in **a**.

11.30 Refer to Problem 11.19. Calculate the two-sided P-value for this test. Is the fact that H_1 was concluded in Problem 11.19 when the α risk is controlled at .05 consistent with this P-value? Comment.

11.31 Refer to Problem 11.24. Calculate the P-value for this test. What conclusion should be drawn if the α risk at $\mu_I = 1,000$ is to be controlled at .05? At .005?

EXERCISES

11.32 Briefly explain the term *level of significance*. In a statistical test with a predetermined sample size, does one achieve any control over the β risks by specifying the level of significance to be used? Discuss.

11.33 The Text illustrates how the two-sided alternatives H_0: $\mu = \mu_I$, H_1: $\mu \neq \mu_I$, may be tested at a specified α risk by noting whether or not μ_I lies between the confidence limits $\bar{X} \pm z(1 - \alpha/2)s\{\bar{X}\}$. Show how a lower confidence interval for μ may be used to test the one-sided alternatives (11.1a). Assume the sample size is large.

11.34 When testing the alternatives H_0: $\mu = \mu_I$, H_1: $\mu \neq \mu_I$ with the following pairs of action limits, what must be known about or given for the population and the sample size in order that the actual Type I error probability be exactly or nearly α?
(1) $\mu_I \pm z(1 - \alpha/2)s\{\bar{X}\}$, (2) $\mu_I \pm t(1 - \alpha/2; n - 1)s\{\bar{X}\}$,
(3) $\mu_I \pm z(1 - \alpha/2)\sigma\{\bar{X}\}$.

11.35 Demonstrate the equivalence of decision rules (11.12) and (11.26) for testing H_0: $\mu \leq \mu_I$, H_1: $\mu > \mu_I$ for a given α risk when the sample size is large.

STUDIES

11.36 A manufacturer has just marketed a new industrial appliance with a one-year warranty. The product development plan anticipates that the cost of meeting the warranty terms will be $50 per appliance, on the average, with a standard deviation of $40. As the warranty period expires for the first appliances sold, the manufacturer will note the actual costs of warranty servicing (X). Two testing approaches have been suggested for deciding whether or not average warranty costs are exceeding the $50 target: (1) Wait for warranty data on the first 100 appliances sold and conclude that the average cost will exceed the target if $\bar{X} > \$58$. (2) Wait for warranty data on the first 400 appliances sold and conclude that the average cost will exceed the target if $\bar{X} > \$54$. A timely conclusion about actual warranty costs is desirable because the manufacturer wishes to reduce the warranty terms if these are proving too generous to be profitable. On the other hand, an unnecessary reduction of warranty terms would adversely affect future sales of the industrial appliance. Assume that the warranty data on the initial sales will constitute random sample observations.

a. Define the parameter μ and the alternatives H_0 and H_1 which are of interest in this situation.

b. Sketch the error curves for the two test approaches on the same graph, assuming that the anticipated standard deviation of warranty costs is accurate. How do the risks of the two approaches compare if the mean warranty cost is on target? If it is $10 higher than the target?

c. The test approach with $n = 400$ was eventually chosen and the sample results were $\bar{X} = \$53.40$ and $s = \$38.40$. Using the specified decision rule for this approach, what conclusion should be drawn about average warranty costs? Estimate the α risk of your test when $\mu_1 = \$50$. Estimate the β risk of your test when the mean warranty cost is $10 higher than the target.

11.37 A market survey is being designed to assess consumer reactions to a new household product. The main factors to be taken into account in the design are: (1) the degree (X_1) to which a respondent prefers the new product to its principal competitor, and (2) the respondent's subjective assessment of the dollar value of the new product (X_2). The preference variable X_1 will be measured on an 11-point scale ranging from -5 (new product much worse than that of principal competitor) to $+5$ (new product much better than that of principal competitor). The test alternatives of interest in connection with the preference variable are H_0: $\mu_1 \leq 0$, H_1: $\mu_1 > 0$, where μ_1 denotes the mean of X_1. The mean subjective dollar value, denoted by μ_2, is to be estimated by a 99 percent confidence interval. In planning the sample size for the survey, it is desired for the test on the mean preference to control the α and β risks at .01 each when $\mu_1 = 0$ and $\mu_1 = +1$, respectively. It is also desired that the half-width of the confidence interval for μ_2 not exceed $5. Reasonable planning values for the standard deviations of X_1 and X_2 are, respectively, $\sigma_1 = 2.5$ and $\sigma_2 = \$20$.

a. What number of respondents is adequately large to satisfy both the testing and estimation requirements?

b. Consider the following two events: (1) The test on μ_1 leads to the incorrect conclusion. (2) The confidence interval for μ_2 is incorrect. In advance of conducting the survey, are these two events necessarily statistically independent? Explain why or why not.

c. Suppose the survey is conducted using the sample size determined in a and the following sample results are obtained:

Variable	Sample Mean	Sample Standard Deviation
Preference	1.32	2.38
Dollar value	$183.40	$18.10

Construct the desired interval estimate for μ_2 and conduct the test for μ_1, controlling the α risk at .01 when $\mu_1 = 0$. For the test, give your decision rule and conclusion. Interpret both your interval estimate and test conclusion.

In Chapters 10 and 11, we considered inferences about a population mean. In this chapter, we take up inferences about a population proportion.

12.1 POPULATION PROPORTION

In many applications of sampling, the characteristic of interest in the population elements is qualitative, with two possible outcomes.

☐ Examples

1. A government policy maker is concerned with the outlook by U.S. households about economic conditions. The relevant population consists of all U.S. households, and the characteristic of interest is whether or not the household expects an economic upturn within the next three months. The population parameter is taken to be the proportion p of U.S. households expecting an economic upturn within three months. (The parameter could equally well have been the proportion of households that do not expect an economic upturn within three months.)

2. In the processing of insurance applications, a characteristic of concern is whether or not the prepared application provides complete information. The population of interest in this case is an infinite one, because it pertains to the process of preparing insurance applications. The population proportion p of relevance is the proportion of all applications in the infinite population that are complete. (The proportion of applications that are incomplete could also have been considered to be the population proportion of interest.) Equivalently, the act of preparing an insurance application may be viewed as a Bernoulli random trial where the random variable X equals 1 when the application is complete and 0 otherwise, and the outcome 1 occurs with probability p. ☐

12.2 SAMPLE STATISTICS

When the characteristic of interest is qualitative with two possible outcomes, a sample statistic of interest is the *number of occurrences* among the n sample observations consisting of the particular outcome reflected in the population proportion. This number of occurrences is denoted by X.

☐   **Examples**

1. p is the population proportion of households that expect an economic upturn shortly; X then is the number of households in the sample that expect an economic upturn shortly.
2. p is the probability that an insurance application is complete; X then is the number of complete applications in the sample.   ☐

Another sample statistic of interest is the *sample proportion*, denoted by $\bar{p}$. It is defined as follows:

(12.1)
$$\bar{p} = \frac{X}{n}$$

Thus, $\bar{p}$ in the first example represents the proportion of households in the sample that expect an economic upturn, and in the second example it is the proportion of complete insurance applications in the sample.

12.3   SAMPLING DISTRIBUTIONS OF X AND $\bar{p}$

Binomial Probability Distribution as Sampling Distribution

When a simple random sample of size n is selected from an infinite population, we already know the exact sampling distribution of the number of occurrences X. Recall from Chapter 6 that when n Bernoulli trials are conducted, we have n Bernoulli random variables $X_1, X_2, \ldots, X_n$, each of which can take on the values 1 and 0. If the n trials are independent and $P(X_i = 1) = p$ for all trials, the sum $X = X_1 + X_2 + \cdots + X_n$ is a binomial random variable whose probability function is given by (6.3). The sum X is, of course, the number of occurrences in the sample that consist of the outcome denoted by 1.

Since the conditions of independence and constant probability p for each trial are precisely the conditions (8.11) for simple random sampling from an infinite population, it follows that the number of occurrences X in a simple random sample from an infinite population is a binomial random variable:

(12.2)   When a simple random sample of size n is selected from an infinite population (i.e., a Bernoulli process), the sum of the n Bernoulli variables, denoted by X:

$$X = X_1 + X_2 + \cdots + X_n$$

is a binomial random variable whose sampling distribution is given by the *binomial probability function*:

$$P(X) = \binom{n}{X} p^X (1 - p)^{n-X}$$

where: $\binom{n}{X} = \dfrac{n!}{X!(n-X)!}$

$X = 0, 1, \ldots, n$

Since the sample proportion $\bar{p}$, as defined in (12.1), is a constant multiple of the number of occurrences X, probabilities for $\bar{p}$ correspond to those for X. Thus, when $n = 3$, the probability that $\bar{p} = 2/3$ is the same as the probability that $X = 2$. Consequently, Table C–5 for the binomial probability distribution may be used to obtain the sampling distributions of both X and $\bar{p}$ for simple random sampling from an infinite population.

☐ Examples

1. Table 12.1 shows the sampling distributions of X and $\bar{p}$ when $n = 6$ and $p = .3$. Figure 12.1 shows these distributions graphically, the only difference between the sampling distributions of X and $\bar{p}$ being in the scaling on the horizontal axis. The probabilities were obtained from Table C–5.

2. For the previous example, we wish to find $P(\bar{p} \leq .5)$. Using Table 12.1, we find the desired probability to be:

$$P(\bar{p} \leq .5; n = 6, p = .3) = P(X \leq 3; n = 6, p = .3)$$
$$= .1176 + .3025 + .3241 + .1852 = .9294$$

Thus, the probability is .9294 that $\bar{p}$ is .5 or less.

3. For $n = 12$ and $p = .5$, we wish to find $P(X \geq 2)$. We obtain, using Table C–5:

$$P(X \geq 2; n = 12, p = .5) = 1 - P(X \leq 1; n = 12, p = .5)$$
$$= 1 - (.0002 + .0029) = .9969$$

4. For $n = 8$ and $p = .2$, we wish to find $P(\bar{p} \leq .25)$. We obtain:

$$P(\bar{p} \leq .25; n = 8, p = .2) = P(X \leq 2; n = 8, p = .2)$$
$$= .1678 + .3355 + .2936 = .7969 \qquad ☐$$

TABLE 12.1 *Sampling distributions of* X *and* $\bar{p}$ *for* n = 6 *and* p = .3. Since $\bar{p}$ is a constant multiple of X, the corresponding probabilities are the same.

Sampling Distribution of X		Sampling Distribution of $\bar{p}$	
X	P(X)	$\bar{p}$	$P(\bar{p})$
0	.1176	0	.1176
1	.3025	1/6	.3025
2	.3241	2/6	.3241
3	.1852	3/6	.1852
4	.0595	4/6	.0595
5	.0102	5/6	.0102
6	.0007	1	.0007
Total	1.00	Total	1.00

Comment

The probability function for $\bar{p}$ can be written explicitly by replacing X in (12.2) by $n\bar{p}$:

(12.3)
$$P(\bar{p}) = \binom{n}{n\bar{p}} p^{n\bar{p}} (1-p)^{n(1-\bar{p})}$$

where: $\bar{p} = 0, \dfrac{1}{n}, \dfrac{2}{n}, \ldots, 1$

Characteristics of Sampling Distributions of X and $\bar{p}$

Mean and Variance. The mean and variance of the sampling distribution of X have already been given in Chapter 6 for the binomial probability distribution. We now repeat these results for completeness and also give the mean and variance for the sampling distribution of $\bar{p}$:

(12.4) For simple random sampling from an infinite population (i.e., for a Bernoulli process):

Sampling Distribution of X	*Sampling Distribution of $\bar{p}$*
$E\{X\} = np$	$E\{\bar{p}\} = p$
$\sigma^2\{X\} = np(1-p)$	$\sigma^2\{\bar{p}\} = \dfrac{p(1-p)}{n}$
$\sigma\{X\} = \sqrt{np(1-p)}$	$\sigma\{\bar{p}\} = \sqrt{\dfrac{p(1-p)}{n}}$

FIGURE 12.1 *Sampling distributions of X and $\bar{p}$ for n = 6 and p = .3*

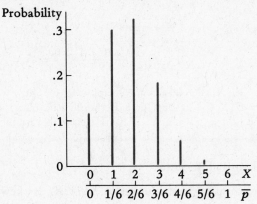

□   **Examples**

1. When $n = 6$ and $p = .3$, we wish to find the mean and variance of the sampling distribution of $\bar{p}$. We obtain:

$$E\{\bar{p}\} = .3 \qquad \sigma^2\{\bar{p}\} = \frac{.3(.7)}{6} = .035$$

The standard deviation of $\bar{p}$ therefore is $\sigma\{\bar{p}\} = \sqrt{.035} = .187$.

2. When $n = 100$ and $p = .8$, we wish to find the mean and variance of the sampling distribution of X. We obtain:

$$E\{X\} = 100(.8) = 80 \qquad \sigma^2\{X\} = 100(.8)(.2) = 16$$

The standard deviation of X therefore is $\sigma\{X\} = \sqrt{16} = 4$.   □

Shape of Distribution.   We know from Chapter 6 that the binomial probability distribution is skewed to the right if $p < .5$, skewed to the left if $p > .5$, and is symmetrical if $p = .5$. Hence, the sampling distribution of $\bar{p}$ has these same characteristics and is symmetrical only when $p = .5$.

Comments

1. To find the mean and variance of $\bar{p}$, we utilize the fact that $\bar{p} = X/n$ and the known results for the mean and variance of the random variable X:

$$E\{\bar{p}\} = E\left\{\frac{X}{n}\right\} = \frac{1}{n}E\{X\} = \frac{np}{n} = p \qquad \text{by (5.6b)}$$

$$\sigma^2\{\bar{p}\} = \sigma^2\left\{\frac{X}{n}\right\} = \frac{1}{n^2}\sigma^2\{X\} = \frac{np(1-p)}{n^2} = \frac{p(1-p)}{n} \qquad \text{by (5.11b)}$$

2. The sample proportion $\bar{p}$ is actually a special case of a sample mean $\bar{X}$. Recall that the binomial variable is $X = X_1 + X_2 + \cdots + X_n$, where $X_i = 0, 1$. Hence:

$$\bar{p} = \frac{X}{n} = \frac{X_1 + X_2 + \cdots + X_n}{n} = \bar{X}$$

For example, for $n = 4$ and $X_1 = 0$, $X_2 = 1$, $X_3 = 1$, and $X_4 = 0$, we have:

$$\bar{p} = \frac{0 + 1 + 1 + 0}{4} = .5 = \bar{X}$$

All of the earlier results for $\bar{X}$ apply to $\bar{p}$. Thus, we know from Chapter 9 that $E\{\bar{X}\} = E\{X_i\} = \mu$. But for a Bernoulli variable, $E\{X_i\} = 1(p) + 0(1 - p) = p$. Hence, $E\{\bar{X}\} = p$ when the X_i's are Bernoulli variables.

Central Limit Theorem

While the sampling distribution of $\bar{p}$ (or X) is skewed if $p \neq .5$, the skewness decreases as the sample size increases. Figure 12.2 illustrates this, showing the

FIGURE 12.2 *Sampling distributions of $\bar{p}$ for $n = 5$ and $n = 50$ when $p = .4$*

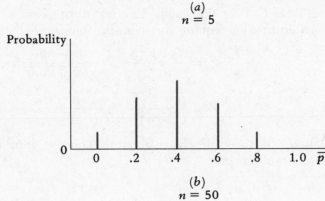

(a)
$n = 5$

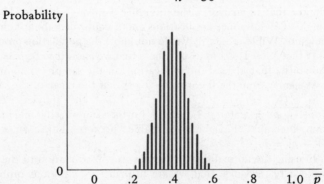

(b)
$n = 50$

sampling distribution of $\bar{p}$ when $p = .4$ for two sample sizes—$n = 5$ and $n = 50$. The sampling distribution of $\bar{p}$ for $n = 50$ in Figure 12.2b possesses very little skewness indeed. In fact, the outline of this sampling distribution appears to be approximately normal. This brings us to another encounter with the central limit theorem:

(12.5) The sampling distributions of X and $\bar{p}$ are approximately normal when the simple random sample size is sufficiently large.

Thus, the binomial probability distribution can be approximated by a normal probability distribution when the random sample size is reasonably large.

We shall use the *working rule: the normal approximation can be employed if both $np \geq 5$ and $n(1 - p) \geq 5$.* Suppose $n = 50$ and $p = .4$. Then $np = 50(.4) = 20$ and $n(1 - p) = 50(.6) = 30$. Since both quantities exceed 5, our working rule suggests that the normal approximation can be used here.

Probability Statements about Behavior of X and $\bar{p}$ Using Normal Approximation

Use of Table C-1 for the standard normal distribution for approximating the binomial distribution requires the following standardized variables:

(12.6)
$$Z = \frac{X - E\{X\}}{\sigma\{X\}} = \frac{X - np}{\sqrt{np(1 - p)}}$$

$$Z = \frac{\bar{p} - E\{\bar{p}\}}{\sigma\{\bar{p}\}} = \frac{\bar{p} - p}{\sqrt{p(1 - p)/n}}$$

We illustrate the theory developed about the sampling distributions of X and $\bar{p}$ by presenting two examples.

☐ Examples

1. Suppose that an analyst of the insurance company mentioned earlier selects a random sample of 50 insurance applications and is concerned about how many of these will be complete. While ordinarily we do not know the population proportion (i.e., the probability that $X_i = 1$), let us assume here that we know it to be $p = .7$. We wish to find the probability that at least 30 applications in the sample of 50 are complete.

   We know from the theorem in (12.4) that the mean of the sampling distribution of X is $E\{X\} = np = 50(.7) = 35$, and that the standard deviation is $\sigma\{X\} = \sqrt{np(1 - p)} = \sqrt{50(.7)(.3)} = 3.24$. Further, our working rule indicates that the central limit theorem (12.5) is applicable because both $np = 50(.7) = 35 \geq 5$ and $n(1 - p) = 50(.3) = 15 \geq 5$.

   In using the normal distribution as an approximation to the binomial distribution, we need to recognize that we are utilizing a continuous probability distribution to approximate a discrete one. Figure 12.3 illustrates the basic problem. Figure 12.3a shows the binomial probability that $X = 30$. Since the normal distribution is continuous and area under the normal curve corresponds to probability, we convert the discrete binomial probability to an area. We do this by treating $X = 30$ as extending from 29.5 to 30.5 and constructing a rectangle to represent the probability at $X = 30$, as shown in Figure 12.3b. Note that the area of the rectangle in Figure 12.3b is the same as the discrete probability in Figure 12.3a, since the height of the rectangle is the same and its width is 1.

   Thus, to find $P(X \geq 30)$, we will actually ascertain $P(X \geq 29.5)$ by the normal approximation since $X = 30$ is assumed to go down to 29.5 when discrete probability is converted to area. The desired probability is shown in Figure 12.3c.

   The z value for our example is, by (12.6):

   $$z = \frac{29.5 - 35}{3.24} = -1.70$$

   We thus find that $P(X \geq 29.5) = P(Z \geq -1.70) = .955$. Incidentally, the true probability according to the binomial distribution is .952, so the normal approximation here is very close.

   Using 29.5 rather than 30 in finding $P(X \geq 30)$ by means of the normal approximation involves a *correction for continuity*. The correction is so named because it arises when a continuous distribution is used to approximate a discrete one. The correction for continuity becomes trivial when the sample size is large, and we shall ordinarily omit it, as in the next example.

FIGURE 12.3 *Normal approximation to binomial distribution for n = 50, p = .7, using correction for continuity. X = 30 is treated as extending from 29.5 to 30.5.*

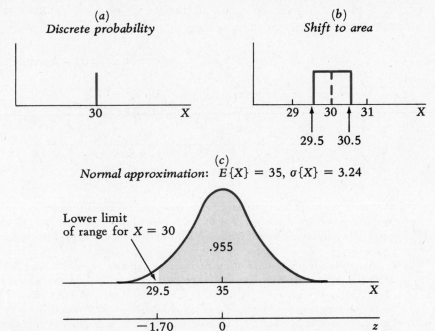

2. Consider again the case where the insurance analyst is selecting a random sample of applications. Suppose the sample size now is $n = 200$ and that, as before, $p = .7$. The analyst wishes to find the probability that $\bar{p}$ falls within 5 percent points (i.e., .05) of the population proportion p.

Figure 12.4 presents the normal approximation to the sampling distribution of $\bar{p}$. The mean and standard deviation of this distribution are, by (12.4):

$$E\{\bar{p}\} = .7 \qquad \sigma\{\bar{p}\} = \sqrt{\frac{.7(.3)}{200}} = .032$$

The desired probability is shaded in Figure 12.4. Because of the large sample size, we do not utilize the correction for continuity. The z values are:

$$z = \frac{.65 - .70}{.032} = -1.56 \qquad z = \frac{.75 - .70}{.032} = 1.56$$

The desired probability therefore is:

$$P(.65 \leq \bar{p} \leq .75) = P(-1.56 \leq Z \leq 1.56) = .881 \qquad \square$$

Comments

1. As we noted earlier, the sample proportion $\bar{p}$ is a special case of the sample mean $\bar{X}$. Hence, central limit theorem (12.5) for $\bar{p}$ is a special case of central limit theorem (9.5) for $\bar{X}$.

FIGURE 12.4 *Normal approximation to binomial distribution for n = 200, p = .7, omitting correction for continuity*

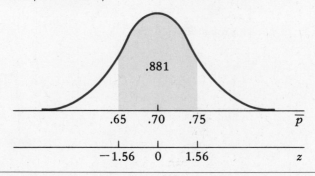

2. When p is small and n is large, the binomial probability distribution can also be approximated by the Poisson probability distribution (6.8) with $\lambda = np$. For many practical purposes, a useful approximation is obtained when n is reasonably large and $np < 5$.

Suppose the probability of a computer breakdown in a day is $p = .01$. We wish to find the probability that, in a random sample of $n = 20$ days, there are no days with a breakdown. We assume that the total number of days with breakdowns (X) is a binomial random variable. We have $\lambda = np = 20(.01) = .20$. From the Poisson Table C–6, we find for $\lambda = .20$ that $P(0) = .8187$. Hence, the probability is approximately .8187 that there will be no days with computer breakdowns in the 20 days. The exact binomial probability from Table C–5 is $P(0) = .8179$, so the Poisson approximation is quite close even though n is not especially large here.

Sampling Finite Populations

The earlier results about the sampling distributions of X and $\bar{p}$ for infinite populations are also applicable to finite populations whenever the finite population size N is large relative to the sample size n. Throughout this Text, we assume that the finite population size is large relative to the sample size unless otherwise noted. Thus, whenever the sample size is reasonably large (i.e., both $np \geq 5$ and $n(1 - p) \geq 5$), the normal approximation to the sampling distributions of X and $\bar{p}$ described earlier for infinite populations can be employed as well for large finite populations.

☐ Example

A tax analyst wishes to sample a population of $N = 500{,}000$ tax returns. Suppose that in this population the proportion of tax returns authorizing \$1 of support for political parties is $p = .45$. If the analyst randomly selects a sample of $n = 1{,}000$ returns, what is the probability that the sample proportion $\bar{p}$ falls within 3 percent points (i.e., .03) of the population proportion p?

Here, the population is large relative to the sample size. Further, $np = 450$ and

$n(1 - p) = 550$, both exceeding 5, so the normal distribution can be employed to approximate the sampling distribution. We know from (12.4) that:

$$E\{\bar{p}\} = .45 \qquad \sigma\{\bar{p}\} = \sqrt{\frac{.45(.55)}{1,000}} = .016$$

To find $P(.42 \leq \bar{p} \leq .48)$, we obtain the z values as usual:

$$z = \frac{.42 - .45}{.016} = -1.88 \qquad z = \frac{.48 - .45}{.016} = 1.88$$

Hence, $P(.42 \leq \bar{p} \leq .48) = P(-1.88 \leq Z \leq 1.88) = .94$. □

Comments

1. When a finite population is sampled, the sampling distributions of X and $\bar{p}$ are given exactly by the hypergeometric probability function (6.15). However, as long as the sample size is a small fraction of the population size, the binomial probability function provides a good approximation to the hypergeometric distribution.
2. In Chapter 14, we consider the case when the sample size is not small relative to the population size.

12.4 ESTIMATION OF POPULATION PROPORTION

It is frequently desired to estimate a population proportion by means of a confidence interval.

□ Illustration

An executive of a retail firm wished to estimate the proportion p of the firm's current credit card customers who make regular use of the firm's mail order services. A file check on a random sample of 1,240 credit card customers showed that $X = 372$ of these customers used the firm's mail order services regularly. Hence, the sample proportion is $\bar{p} = 372/1,240 = .30$. A confidence interval for the population proportion is now desired. □

Approximate Confidence Interval for p—Large Sample

When n is large, we know from central limit theorem (12.5) that the sampling distribution of $\bar{p}$ is approximately normal. We also know from (12.4) that the mean of the sampling distribution of $\bar{p}$ is $E\{\bar{p}\} = p$ and that the variance is $\sigma^2\{\bar{p}\} = p(1 - p)/n$.

To obtain approximate confidence limits for p when the sample size is large, we need to estimate $\sigma\{\bar{p}\}$. This can be done easily since, as we just noted, $\sigma^2\{\bar{p}\}$ depends only on p and n, and the sample proportion $\bar{p}$ is an unbiased point estimator of p.

Estimated Standard Deviation of $\bar{p}$. The estimated standard deviation of $\bar{p}$ is denoted by $s\{\bar{p}\}$ and is defined as follows for infinite or large finite populations:

(12.7)
$$s^2\{\bar{p}\} = \frac{\bar{p}(1 - \bar{p})}{n - 1} \qquad s\{\bar{p}\} = \sqrt{\frac{\bar{p}(1 - \bar{p})}{n - 1}}$$

It can be shown that $s^2\{\bar{p}\}$ is an unbiased estimator of $\sigma^2\{\bar{p}\}$, the variance of the sampling distribution of $\bar{p}$, for infinite populations.

Development of Confidence Interval. We utilize an extension of the central limit theorem for developing an approximate confidence interval for p when n is large:

(12.8) For infinite or large finite populations, $(\bar{p} - p)/s\{\bar{p}\}$ follows approximately a standard normal distribution when the random sample size is sufficiently large.

Using the same reasoning as in finding the confidence interval for μ, we obtain an analogous confidence interval for p:

(12.9) The confidence limits for the population proportion p with approximate confidence coefficient $1 - \alpha$, when the sample size is reasonably large, are:

$$\bar{p} \pm zs\{\bar{p}\}$$

where: $z = z(1 - \alpha/2)$
$\quad\quad s\{\bar{p}\}$ is given by (12.7)

☐ **Example**

In the mail order illustration described previously, the sample results are $n = 1,240$ and $\bar{p} = .30$, and a 99 percent confidence interval for the population proportion p of credit card customers who make regular use of the mail order services is desired. The population is large relative to the sample size, and we calculate by (12.7):

$$s^2\{\bar{p}\} = \frac{.30(1 - .30)}{1,240 - 1} = .0001695$$

so $s\{\bar{p}\} = \sqrt{.0001695} = .0130$. For 99 percent confidence limits, we require $z(1 - \alpha/2) = z(.995) = 2.576$. Thus, the limits are $.30 \pm 2.576(.0130)$ or $.30 \pm .03$, and the 99 percent confidence interval is:

$$.27 \leq p \leq .33$$

We can report, with 99 percent confidence, that between 27 and 33 percent of current credit card customers make regular use of the firm's mail order services. ☐

Comment

One-sided confidence intervals for a population proportion can be obtained by using the appropriate limit in (12.9), with $z = z(1 - \alpha)$. For example, suppose we wish to obtain a 90 percent lower confidence interval for p in the preceding mail order study. We have $\bar{p} = .30$ and $s\{\bar{p}\} = .0130$. We require $z(1 - \alpha) = z(.90) = 1.282$, so the lower confidence limit is $.30 - 1.282(.0130) = .283$ and the desired confidence interval is:

$$p \geq .283$$

Planning of Sample Size

Often, the sample size for estimating a population proportion p is not predetermined and can therefore be planned to provide a specified precision for a given confidence coefficient.

☐   Illustration

A commission is studying the spacial distribution of the labor force in a large metropolitan area. One parameter of interest is the proportion of workers (p) whose places of employment are within 15 miles of their residences. It is desired to select a random sample of workers of sufficient size to provide a 95 percent confidence interval for p with a half-width of $h = .02$.   ☐

The procedure to be described assumes that: (1) the population under study is large relative to the sample size, and (2) the sample size ultimately determined is reasonably large. As usual, the desired half-width of the confidence interval is denoted by h, and $z = z(1 - \alpha/2)$ is the standard normal value corresponding to the specified confidence coefficient. It is then seen from (12.9) that $h = z\sigma\{\bar{p}\}$. We use the notation for the true standard deviation $\sigma\{\bar{p}\}$ here because we are in the planning stage. We know from (12.4) that $\sigma^2\{\bar{p}\} = p(1 - p)/n$. Hence, it follows that:

$$h = z\sigma\{\bar{p}\} = z\sqrt{\frac{p(1 - p)}{n}}$$

In solving this equation for n, we obtain an expression involving the population proportion p, which is unknown. Hence, we require a planning value of p. Often, this planning value can be based on past experience with the same or a similar problem.

(12.10) The needed sample size, based on the desired half-width h of the confidence interval and the specified confidence coefficient $1 - \alpha$, for a given planning value of p, is:

$$n = \frac{z^2 p(1 - p)}{h^2}$$

where: $z = z(1 - \alpha/2)$

☐   Example

In our worker location study, we have $h = .02$ and $z(1 - \alpha/2) = z(.975) = 1.960$ because a 95 percent confidence interval is desired. Suppose it is expected that p is in the neighborhood of .9, so the planning value $p = .9$ will be used. Substituting into (12.10) yields:

$$n = \frac{(1.960)^2(.9)(.1)}{(.02)^2} = 864$$

A random sample of 864 workers from the labor force should provide an interval estimate of p with approximately the desired precision for the specified confidence coefficient if the planning value of p is reasonably accurate.   ☐

Comment

In (12.10), the term $p(1 - p)$ takes on its largest value when $p = .5$, and becomes smaller as p approaches 0 or 1. This implies that if no reliable planning value of p can be specified and a conservatively large sample size is desired, the planning value of p should be set equal to .5 in computing n from (12.10). This sample size may be quite conservative (i.e., larger than necessary), however, because the farther p lies from .5 (toward either 0 or 1), the smaller the sample size actually needed to provide a given degree of precision. In our illustration, $n = 864$ is adequate if $p = .9$, whereas $n = 2,401$ would be obtained if $p = .5$ were used as the planning value.

12.5 STATISTICAL TESTS FOR p

We now consider statistical tests about a population proportion p. These are conducted in the same manner as tests on the population mean μ.

Construction of Decision Rule to Control α Risk

We shall illustrate a one-sided lower-tail test *when the given sample size is large.*

☐ Example

A British food organization presently uses cod for "fish and chips," the fish traditionally used in making the dish. However, whiting is a cheaper and more plentiful fish. Management has decided to switch to whiting if at least 50 percent of consumers prefer whiting to cod. Let p denote the proportion who prefer whiting. Hence, the alternatives are:

$$H_0: p \geq p_I = .5$$

$$H_1: p < p_I = .5$$

Here, $p_I = .5$ is the level where the α risk is to be controlled. In other words, management is anxious to switch to whiting if $p = .5$ because of cost savings and better availability, and wishes to control the risk of making a Type I error at this level of p.

A random sample of $n = 265$ consumers are each subjected to a blind taste test of both types of fish prepared in the same fashion. Let $\bar{p}$ denote the sample proportion who prefer whiting. It is desired to control the α risk at .05 when $p = p_I = .5$. The test now proceeds as shown in Figure 12.5.

1. A small value of $\bar{p}$ is indicative of $p < .5$ (i.e., H_1), so the appropriate type of decision rule is the one shown in Figure 12.5a.
2. Because the sample size is large, the sampling distribution of $\bar{p}$ is approximately normal. The distribution of $\bar{p}$ is centered at $p_I = .5$ because the α risk is to be controlled there. See Figure 12.5b. When $p = p_I$, the sampling distribution has standard deviation $\sigma\{\bar{p}\} = \sqrt{p_I(1 - p_I)/n}$.
3. The action limit A must be positioned so the left-tail area is $\alpha = .05$. Hence, we require $z(\alpha) = z(.05) = -1.645$ and obtain:

$$A = p_I + z(\alpha) \sqrt{\frac{p_I(1 - p_I)}{n}} = .5 - 1.645 \sqrt{\frac{.5(.5)}{265}} = .449$$

Note that because the distribution is centered at $p_I = .5$, its standard deviation equals $\sigma\{\bar{p}\} = \sqrt{.5(1 - .5)/265}$. See Figures 12.5c and 12.5d.
4. The decision rule for the test is (see Figure 12.5e):

$$\text{If } \bar{p} \geq .449, \text{ conclude } H_0 \ (p \geq .5)$$

$$\text{If } \bar{p} < .449, \text{ conclude } H_1 \ (p < .5)$$

FIGURE 12.5 *Construction of decision rule to control α risk—Large sample, lower-tail test for p*

(a) *Appropriate type of decision rule*

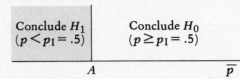

Small values of $\overline{p}$ imply H_1 is correct.

(b) *Sampling distribution of $\overline{p}$ located at $p_I = .5$*

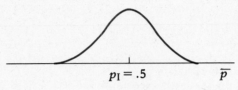

Distribution of $\overline{p}$ is centered at p_I since α risk is controlled there.

(c) *Placement of action limit A*

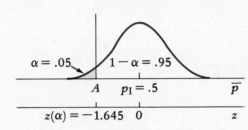

A is placed $z(\alpha)$ deviations from p_I so lower–tail area equals α.

(d) *Calculation of action limit A*

$$A = p_I + z(\alpha)\sqrt{\frac{p_I(1 - p_I)}{n}}$$

$\sigma\{\overline{p}\}$ is known for sampling distribution when $p = p_I$.

$$= .5 - 1.645\sqrt{\frac{.5(.5)}{265}} = .449$$

(e) *Decision rule*

If $\overline{p} \geq A = .449$, conclude H_0
If $\overline{p} < A = .449$, conclude H_1

Conclude appropriate alternative based on observed $\overline{p}$.

5. In the taste test, 144 of the 265 consumers preferred the whiting. Hence, $\overline{p} = 144/265 = .543$. Since $.543 \geq .449$, alternative H_0 is concluded—that at least half of the consumers prefer whiting. □

Decision Rule.   We summarize now the decision rule for the one-sided lower-tail test for p:

(12.11) When the alternatives are:

$$H_0: p \geq p_I$$
$$H_1: p < p_I$$

and the sample size is reasonably large, the appropriate decision rule to control the α risk at p_I is:

$$\text{If } \bar{p} \geq A, \text{ conclude } H_0$$

$$\text{If } \bar{p} < A, \text{ conclude } H_1$$

where: $A = p_I + z(\alpha) \sqrt{\dfrac{p_I(1 - p_I)}{n}}$

Comments

1. Note that the decision rule for a test concerning p can be completed in advance of the sample results, unlike tests for μ. The reason is that $\sigma\{\bar{p}\}$ is known when $p = p_I$, since it depends only on p_I and n.
2. The test for the fish and chips example can also be conducted by means of the standardized test statistic:

(12.12)
$$z^* = \frac{\bar{p} - p_I}{\sqrt{\dfrac{p_I(1 - p_I)}{n}}}$$

For test (12.11), small values of z^* lead to conclusion H_1. In our example, we have $\bar{p} = .543$, $p_I = .5$, and $\sqrt{p_I(1 - p_I)/n} = .031$; hence, $z^* = (.543 - .5)/.031 = 1.39$. For $\alpha = .05$, we require $z(.05) = -1.645$, and the decision rule is:

$$\text{If } z^* \geq -1.645, \text{ conclude } H_0$$

$$\text{If } z^* < -1.645, \text{ conclude } H_1$$

Here, $z^* = 1.39 \geq -1.645$; hence, we conclude H_0, just as we did with the decision rule based on $\bar{p}$.

Control of Both Types of Risks through Determination of Sample Size

To control both the α and β risks, the sample size and action limit must be jointly determined. Consider the fish preference test again, but this time assume that the sample size has not yet been determined. Suppose that the α risk is to be controlled as before ($\alpha = .05$ at $p_I = .5$). In addition, the β risk is to be controlled at .05 when $p_{II} = .4$. In other words, when only 40 percent of the consumers prefer whiting, the probability of concluding H_0: $p \geq .5$ is to be controlled at $\beta = .05$. Figure 12.6 shows the positioning of the two sampling distributions for controlling the α and β risks and the two simultaneous specifications of the action limit A. Note how $\sigma\{\bar{p}\}$ depends on the value of p in each case. Solving these two equations for the sample size n, we obtain the following:

(12.13) For the one-sided lower-tail test (12.11), the necessary sample size to control both the α and β risks is:

$$n = \left[\frac{z(1 - \beta) \sqrt{p_{II}(1 - p_{II})} - z(\alpha) \sqrt{p_I(1 - p_I)}}{p_I - p_{II}} \right]^2$$

FIGURE 12.6   *Simultaneous determination of action limit and sample size to control α and β risks—Large sample, lower-tail test for p*

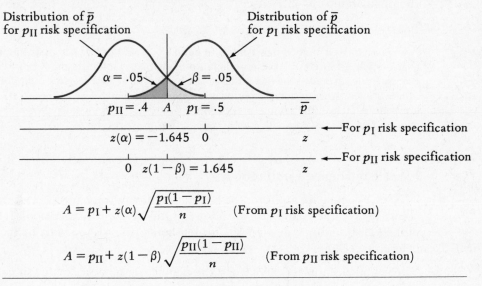

$$A = p_{\mathrm{I}} + z(\alpha)\sqrt{\frac{p_{\mathrm{I}}(1-p_{\mathrm{I}})}{n}} \qquad \text{(From } p_{\mathrm{I}} \text{ risk specification)}$$

$$A = p_{\mathrm{II}} + z(1-\beta)\sqrt{\frac{p_{\mathrm{II}}(1-p_{\mathrm{II}})}{n}} \qquad \text{(From } p_{\mathrm{II}} \text{ risk specification)}$$

Once the sample size has been determined, the decision rule is obtained from (12.11).

☐   **Example**

For our fish preference example, the specifications are:

$$p_{\mathrm{I}} = .5 \qquad \alpha = .05$$

$$p_{\mathrm{II}} = .4 \qquad \beta = .05$$

Hence, we require $z(\alpha) = z(.05) = -1.645$ and $z(1-\beta) = z(.95) = 1.645$. We obtain from (12.13):

$$n = \left[\frac{1.645\sqrt{.4(.6)} + 1.645\sqrt{.5(.5)}}{.5 - .4}\right]^2 = 265$$

This is, in fact, the sample size that was used in the study.   ☐

Comment

The determination of the required sample size to control both the α and β risks for one-sided upper-tail tests and for two-sided tests proceeds in corresponding fashion. The two simultaneous equations in A and n should be set up and solved for the required sample size.

12.6   OPTIONAL TOPIC—EXACT CONFIDENCE INTERVAL FOR p

When the sample size is small, so that the large-sample confidence interval (12.9) for the population proportion p is not applicable, an exact confidence interval for

p can be utilized. This interval is exact when the population is infinite, but can also be used for a finite population as long as the sample size is not large relative to the population size.

☐ **Illustration**

A financial analyst studied new corporate bond issues sold during the past five years. One characteristic of interest is the proportion p of the issues sold at a premium—i.e., where the issue price exceeded the face value of the bond. The analyst selected a random sample of $n = 15$ new issues during the period and determined that $X = 6$ of these were issued at a premium. Hence, the sample proportion is $\bar{p} = 6/15 = .40$. The analyst now wants to construct an interval estimate for the population proportion p. ☐

Exact Confidence Limits for p

The exact confidence limits for the population proportion p involve the number of occurrences X, the sample size n, and percentiles of the F distribution. (If a review of the F distribution is needed, the reader should turn to Appendix B, Section B.3 before proceeding.)

(12.14) The confidence limits L and U for the population proportion p with confidence coefficient $1 - \alpha$, when the population is infinite or large relative to n, are:

$$L = \frac{X}{X + (n - X + 1)F_1} \qquad U = \frac{(X + 1)F_2}{(X + 1)F_2 + (n - X)}$$

where: $F_1 = F[1 - \alpha/2; \, 2(n - X + 1), \, 2X]$

$F_2 = F[1 - \alpha/2; \, 2(X + 1), \, 2(n - X)]$

☐ **Example**

In our bond issue illustration, suppose that the financial analyst wishes to construct a 90 percent confidence interval for p. The sample results are $n = 15$ and $X = 6$, and $1 - \alpha = .90$. Consulting Table C-4 for the F distribution, we find:

$$F_1 = F[.95; \, 2(15 - 6 + 1), \, 2(6)] = F(.95; \, 20, \, 12) = 2.54$$

$$F_2 = F[.95; \, 2(6 + 1), \, 2(15 - 6)] = F(.95; \, 14, \, 18) = 2.29$$

where the latter value is obtained by interpolation. Substituting into (12.14) gives:

$$L = \frac{6}{6 + (15 - 6 + 1)(2.54)} = .19 \qquad U = \frac{(6 + 1)(2.29)}{(6 + 1)(2.29) + (15 - 6)} = .64$$

Therefore, a 90 percent confidence interval for the proportion of bond issues sold at a premium is:

$$.19 \le p \le .64$$

The analyst can conclude, with 90 percent confidence, that the population proportion of new issues selling at a premium is between 19 and 64 percent. These limits are quite wide, because the sample size is so small. Note that the point estimate $\bar{p} = .40$ gives no indication of this lack of precision. ☐

Comments

1. Unlike the large-sample confidence limits for p in (12.9), the exact confidence limits for p in (12.14) are not equally spaced about $\bar{p}$. In the example, the limits .19 and .64 are not equidistant from $\bar{p} = .40$.
2. The confidence limits (12.14) are based on two facts. First, the sampling distribution of $\bar{p}$ is given by the binomial probability function (12.2) when the population is infinite. Second, there exists a mathematical relation between the binomial distribution and the F distribution.

PROBLEMS

*12.1 The probability that an invoice is paid within 10 days is $p = .90$. Consider a random sample of $n = 9$ invoices.
   a. State briefly what X and $\bar{p}$ represent in this situation.
   b. What is the functional form of the sampling distribution of X? Of $\bar{p}$? How do the two distributions differ?
   c. Obtain the sampling distribution of $\bar{p}$ and calculate its mean and standard deviation. Confirm these results using (12.4).
   d. Plot the sampling distributions of X and $\bar{p}$ in one graph, as in Figure 12.1.
   e. (1) What is the probability that $X = 9$? (2) That $\bar{p} = 1$? (3) That $\bar{p} \geq 7/9$?

12.2 A customer entering a store makes a purchase with probability $p = .30$. A random sample of $n = 15$ customers is selected.
   a. State briefly what X and $\bar{p}$ represent in this situation.
   b. What is the functional form of the sampling distribution of X? Of $\bar{p}$? How do the two distributions differ?
   c. Obtain the sampling distributions of X and $\bar{p}$.
   d. Obtain the means and standard deviations of the sampling distributions of X and $\bar{p}$ using (12.4). How are these measures for X and $\bar{p}$ related?
   e. (1) What is the most probable value of X? Of $\bar{p}$? (2) What is the probability that $\bar{p} = .20$? (3) That $\bar{p} \leq .20$? (4) That $X \geq 14$?

12.3 The probability of a defective part is $p = .05$. Let X denote the number of defective parts in a random sample of $n = 5$ parts. Assume that the binomial distribution is applicable.
   a. Plot the sampling distributions of X and $\bar{p}$ in one graph, as in Figure 12.1.
   b. Obtain the means and standard deviations of the sampling distributions of X and $\bar{p}$ using (12.4). How are these measures for X and $\bar{p}$ related?
   c. (1) What is the probability that $X < 2$? (2) That $\bar{p} < 2/5$? (3) That $X = 5$?

12.4 For each of the following cases, indicate whether the working rule presented in the Text permits use of the normal approximation to the binomial distribution:
   a. $n = 100$, $p = .02$
   b. $n = 80$, $p = .15$
   c. $n = 40$, $p = .90$
   d. $n = 20$, $p = .60$

*12.5 A machine is in an unproductive state 20 percent of the time (i.e., $p = .20$). Let X denote the number of times that the machine is in an unproductive state when it is observed at $n = 50$ random times.

a. Verify that the values of n and p here satisfy the working rule for approximating the binomial distribution by a normal distribution.

b. Obtain the mean and standard deviation of the sampling distribution of X.

c. Using the normal approximation and the correction for continuity, obtain: (1) $P(X > 15)$, (2) $P(X \leq 4)$, (3) $P(X = 10)$.

d. Let $\bar{p}$ denote the proportion of times in the sample the machine is in an unproductive state. Use your results in c to obtain the following probabilities for the sample proportion: (1) $P(\bar{p} > .30)$, (2) $P(\bar{p} \leq .08)$, (3) $P(\bar{p} = .20)$.

12.6 The probability of a delayed shipment is $p = .10$. Let X denote the number of delayed shipments in a random sample of $n = 60$ shipments.

a. Verify that the values of n and p here satisfy the working rule for approximating the binomial distribution by a normal distribution.

b. Obtain the mean and standard deviation of the sampling distribution of X.

c. Using the normal approximation and the correction for continuity, obtain: (1) $P(X \leq 1)$, (2) $P(X > 6)$.

d. Use your results in c to obtain the following probabilities for the sample proportion: (1) $P(\bar{p} \leq 1/60)$, (2) $P(\bar{p} > .10)$.

*12.7 A large population of residential water meters is to be sampled to estimate the proportion p due for replacement within the next five years. Suppose $p = .40$, and that $n = 250$ meters are to be selected randomly.

a. Find the mean and standard deviation of the sampling distribution of $\bar{p}$.

b. Using a normal approximation without a correction for continuity, find the probability that: (1) $\bar{p}$ is less than .35, (2) $\bar{p}$ lies within $\pm.05$ of the population proportion p.

c. How much larger would be the probability in b(2) if the sample size were 1,000 instead of 250?

12.8 A nationwide survey is to be conducted to estimate the proportion p of food stores carrying fertile eggs. Suppose $p = .30$, and that $n = 225$ food stores are to be selected randomly.

a. Find the mean and standard deviation of the sampling distribution of $\bar{p}$.

b. Using a normal approximation without a correction for continuity, find: (1) the probability that $\bar{p}$ will exceed .35, (2) the interval about p within which $\bar{p}$ will fall 95 percent of the time (use symmetrical limits).

12.9 A public opinion poll of $n = 1,000$ persons is to be conducted to estimate the proportion p in the population who believe business conditions will worsen in the next 12 months. Assume that the binomial distribution is applicable and that it can be approximated by a normal distribution. Ignore the correction for continuity. What is the probability that the proportion of persons polled who believe business conditions will worsen is within $\pm.04$ of p if: (1) $p = .20$, (2) $p = .50$, (3) $p = .80$? What generalization is suggested by your results?

*12.10 A savings bank noted that 180 of the 225 new accounts opened last month involved a transfer of savings from an account at another bank. Assume that the data for the new accounts represent a random sample from a Bernoulli process.

a. Calculate $\bar{p}$, the sample proportion of new accounts at this savings bank which involve a transfer from another bank. Also calculate the estimated standard deviation $s\{\bar{p}\}$.

b. Obtain a 95 percent confidence interval for the population proportion p. Interpret your confidence interval.

c. An executive of this savings bank had claimed that three-quarters of all new accounts involve a transfer from another bank. Do your confidence limits in **b** suggest that the executive's claim is plausible? Comment.

12.11 A sample survey of $n = 400$ persons over 40 years of age revealed that 65 percent use some hair-conditioning product on a regular basis. The sample was a random one from a very large population of persons in this age category.

    a. Estimate the standard deviation of the sampling distribution of $\bar{p}$. Why is this standard deviation not known exactly and must be estimated?

    b. Construct a 99 percent confidence interval for the population proportion of persons in this age category who use a hair-conditioning product regularly. Interpret your interval estimate.

12.12 In a random sample of $n = 1,000$ workers from a certain occupation, it was found that 153 of the workers had some chronic illness. Construct a 90 percent confidence interval for the population proportion of workers with a chronic illness. Interpret your confidence interval.

12.13 (Based on Optional Topic Section 10.7.) Refer to Problem 12.12. Construct a 90 percent upper confidence interval for the population proportion of workers with a chronic illness. Can it be confidently concluded that this proportion does not exceed 18 percent? Discuss.

*12.14 A lathe operator is to be observed at random times to estimate p, the proportion of time he is in a productive state. It is desired to estimate this proportion within $\pm.05$, with a 95 percent confidence coefficient. It is expected that p is at least .8. What sample size will be large enough to assure that the estimate has the desired precision?

12.15 A sample survey of persons with arthritis is to be conducted to estimate the proportion who are taking a new antiarthritic drug. Preliminary sales for the new drug suggest that the proportion is in the neighborhood of .10. Using $p = .10$ as the planning value, determine the sample size which will give a confidence interval for the population proportion having a half-width of .02 at a confidence level of .90. Would the required sample size be substantially different if a planning value of $p = .12$ had been used?

12.16 As part of a study of postsecondary education, a random sample of women in the graduating classes of colleges and universities is to be selected to estimate the proportion who expect to enter the labor force upon graduation. It is desired to estimate this labor force participation rate within $\pm.03$, with a confidence coefficient of 99 percent. No reliable planning value for the participation rate is available. What sample size will be large enough to assure that the estimate has the desired precision?

12.17 Refer to Problem 12.16. It was finally decided to include 1,225 women in the study. Of these women, 950 stated that they would be entering the labor force upon graduation. Construct a 99 percent confidence interval for the population participation rate. Interpret the meaning of your interval estimate.

12.18 Refer to Problem 12.14. If the planning value for the proportion used in calculating the required sample size is incorrect, does this affect the validity of the confidence interval obtained from the sample results? The precision of the confidence interval? Comment.

*12.19 A large stock of items has been stored under unfavorable conditions. If fewer than 70 percent of the items are serviceable, the stock should be scrapped. A random sample of 100 items from the stock is to be selected and their condition ascertained. It is desired to test whether or not the stock should be scrapped.

a. Describe the parameter p and the alternatives H_0 and H_1 in this situation.

b. Calculate $\sigma\{\hat{p}\}$ when $p_I = .70$. Construct the appropriate decision rule so as to control the α risk at .01 when $p_I = .70$.

c. Suppose the sample contains 65 serviceable items. What conclusion should be drawn?

12.20 Refer to Problem 12.19. For your decision rule in **b**, what is the probability of concluding the stock should not be scrapped if, in fact, 60 percent of the items are serviceable? Why is this probability not exact? Is this probability a Type I or Type II error probability?

12.21 Nationally, the proportion of marriages ending in divorce or annulment which involve no children is .40. In one state, a random sample of 1,000 divorces and annulments showed that 437 involved no children. It is desired to test whether or not the state proportion is equal to .40.

a. Describe the parameter p and the alternatives H_0 and H_1 in this situation. Do your alternatives involve a one-sided or two-sided test?

b. Construct the appropriate decision rule so as to control the α risk at .05.

c. What conclusion should be drawn? Would your conclusion have been different if the α risk were controlled at .01?

12.22 Refer to Problem 12.12. A public health officer wishes to test whether or not the population proportion of workers in this occupation with a chronic illness is no greater than .11, the population proportion for workers in a different occupation. Conduct the test, controlling the α risk at .05 when $p_I = .11$. State the alternatives, decision rule, and conclusion.

*12.23 Refer to Problem 12.19. Suppose that the sample size has not yet been specified. In addition to requiring the α risk at $p_I = .70$ to be .01, it is also required that the β risk at $p_{II} = .50$ be .01.

a. What number of items in the stock must be sampled to give the desired control over the error probabilities?

b. Given the sample size determined in **a**, what is the appropriate decision rule for the test?

12.24 Refer to Problem 12.21. Suppose the sample size has not yet been specified. It is desired to control the α risk at $p_I = .40$ at level .05 and the β risk at $p_{II} = .45$ at level .01. What random sample size of divorces and annulments in the state is required to give the specified control over the error probabilities? What is the appropriate decision rule for the test?

12.25 A political analyst is planning a poll of eligible voters to ascertain their voting intentions for a forthcoming election between the incumbent party A and the opposition party B. Let p denote the population proportion of voters who would currently express an intention to vote for A. The analyst wants to test the alternatives $H_0: p \geq .50$, $H_1: p < .50$. She wishes to employ a sample size which will control the α risk at $p_I = .50$ at .02 and the β risk at $p_{II} = .45$ at .02.

a. What number of respondents should she select for the poll, and what decision rule should be employed?

b. For your decision rule, what is the probability of concluding H_0 when $p = .55$?

c. The analyst used the sample size determined in **a** for the poll and obtained $\hat{p} = .463$. What conclusion should be reached?

12.26 (Based on Optional Topic Section 11.10.) Refer to Problem 12.19.

  a. Calculate the standardized test statistic z^* corresponding to the sample result given in Problem 12.19c.

  b. Controlling the α risk at .01 when $p_I = .70$, give the decision rule for the test in terms of z^*. Is the conclusion reached with this decision rule the same as the one reached in Problem 12.19c? Must this always be the case? Comment.

12.27 (Based on Optional Topic Section 11.11.) Refer to Problem 12.19.

  a. Calculate the P-value for the sample result given in Problem 12.19c.

  b. What conclusion should be drawn if the α risk at $p_I = .70$ is to be controlled at .01? Must this be the same conclusion as the one reached in Problem 12.19c?

*12.28 An auditor examined a simple random sample of 15 accounts from a large population of accounts receivable. He discovered that 1 account in the sample was in error.

  a. Estimate the error rate in the population with an exact 98 percent confidence interval. Interpret your interval estimate.

  b. (Based on Optional Topic Section 10.7.) Estimate the error rate in the population with an exact upper 99 percent confidence interval.

12.29 A political scientist has conducted in-depth interviews with a random sample of 16 lawyers and has learned, among other things, that 11 favor the present system of judicial appointment. Construct an exact 90 percent confidence interval for the population proportion in favor of the present system.

EXERCISES

12.30 The probability that a garnet, when cut and polished, has a flaw which classes it as a "second" is .10. Assume that the production process is a Bernoulli process and consider a sample of $n = 3$ garnets. Let X denote the number of seconds in the sample.

  a. Obtain the sampling distribution of X by direct enumeration of all possible sample outcomes. Verify that it is the same as that given by (12.2).

  b. Calculate the mean and standard deviation of your sampling distribution in **a**. Verify that these results agree with those given by (12.4).

12.31 Explain why it is unnecessary to estimate $\sigma\{\bar{p}\}$ from the sample in constructing a decision rule to test a population proportion p, although it is necessary to estimate $\sigma\{\bar{X}\}$ from the sample in constructing a decision rule to test a population mean μ.

12.32 Refer to Problem 12.4. In which of the situations could a Poisson approximation to the binomial distribution be used?

12.33 An extensive record of airline flights shows that proportion $p = .008$ or .8 percent of them encounter mechanical problems on takeoff, landing, or in flight.

  a. What is the expected number of flights which will encounter mechanical problems in a random sample of $n = 500$ flights?

  b. Use the Poisson approximation to the binomial distribution to find the probability that 10 or more flights in a random sample of 500 flights will experience mechanical problems.

12.34 Show that $s^2\{\bar{p}\}$ as defined in (12.7) is an unbiased estimator of $\sigma^2\{\bar{p}\}$ as defined in (12.4). (Hint: $E\{\bar{p}^2\} = p^2 + \sigma^2\{\bar{p}\}$.)

12.35 (Based on Optional Topic Section 10.8.) A market researcher finds that $X = 3$

consumers from a random sample of $n = 15$ consumers have tried a new product recently introduced in food stores. Assume the binomial probability distribution applies.

a. From Table C–5, find the likelihoods of this sample outcome for $p = .10, .15, .20$, and $.25$.

b. Among the p values tabulated in Table C–5, for which is the likelihood of producing this sample outcome greatest?

c. The maximum likelihood estimator of p here can be shown to be $\bar{p} = X/n$. Is your finding in b consistent with this fact?

12.36 (Based on Optional Topic Section 10.8; calculus needed.) If X is a binomial random variable with parameters n and p, show that the sample proportion $\bar{p} = X/n$ is the maximum likelihood estimator of p.

STUDIES

12.37 a. An auditor will verify all accounts on a 100 percent basis if she finds one or more accounts in the sample are in error. Assuming that the binomial probability distribution is applicable and that she is utilizing a random sample of 120 accounts, what is the probability that the auditor will verify all accounts on a 100 percent basis if the population error rate is $p = .01$? If it is $p = .05$? Assuming that a 5 percent error rate is serious, is the auditor's sampling plan effective when $p = .05$?

b. Find the probability of no accounts in error in the sample if $p = .01$ by means of the Poisson approximation. How good is the approximation in this instance?

c. (Based on Optional Topic Sections 10.7 and 12.6.) Suppose that no accounts are in error in a random sample of 120 accounts. Construct an exact 99 percent upper confidence interval for p. Interpret your interval estimate. [Hint: $F(.99; 2, 240) = 4.69$.]

12.38 A trade association of small businesses plans to select a random sample of member firms to estimate, among other things: (1) the proportion of firms with no full-time employees, (2) the proportion of firms in which the owner works more than 60 hours per week, and (3) the proportion of firms which carry no liability insurance. The sample size should be adequate to provide 95 percent confidence intervals for each of these three population proportions with half-widths $h = .02$. Planning values for the three population proportions deemed appropriate by the research director of the trade association are .4, .7, and .2, respectively.

a. What sample size should the trade association employ so that the precision requirements for all three estimates can be met?

b. The trade association finally decided to sample only 1,000 firms to expedite the completion of the study. The final report will contain many other estimates of population proportions besides the three mentioned previously. To simplify the presentation of the precisions of these many estimates, prepare a graph which shows the half-width $zs\{\bar{p}\}$ of a 95 percent confidence interval for this survey as a function of the sample proportion $\bar{p}$, for $.01 \leq \bar{p} \leq .99$. Prepare a brief statement which explains how the graph is to be used.

13
Comparisons of Two
Populations and Other Inferences

Thus far, we have discussed statistical inferences (i.e., interval estimates and tests) about a population mean and a population proportion. In this chapter, we take up comparative studies where a comparison of two populations is of interest. Specifically, we describe the construction of interval estimates and tests for the difference between two population means and the difference between two population proportions. In addition, we consider as an optional topic inferences about the population variance.

13.1 COMPARATIVE STUDIES

Nature of Comparative Studies

The general purpose of comparative studies is to establish similarities or to detect and measure differences between populations. The populations involved in a comparative study may be of two basic types: (1) *existing* populations, and (2) *hypothetical* populations related to an experiment.

☐ Examples

1. An economist wishes to compare the proportions of families with incomes below the poverty level in two regions of the country, from sample data for each region. Specifically, he wishes to make inferences about the proportions p_1 and p_2 of the two populations from data based on the two samples. The populations here are existing populations.

2. A social psychologist wants to compare persons' attitudes toward a particular social behavior before and after they view an informational film about this behavior, based on a study of a sample of subjects. In this case, persons' attitudes "before" and "after" constitute two hypothetical populations related to an experiment in which the stimulus is the informational film. Using an established questionnaire to measure this attitude, the psychologist wishes to determine if the film causes a shift in the mean attitude level of subjects, where μ_1 and μ_2 denote the before and after stimulus means. That is, the psychologist wishes to ascertain whether or not μ_1 and μ_2 are equal and, if not, how great the difference is and in what direction. ☐

Design Considerations—Independent versus Matched Samples

In comparative sample studies, there is often the opportunity to choose between two alternative designs, called *independent samples* and *matched samples*. Clearly, one would like to choose that design which, for given cost, leads to smaller sampling errors. The following example highlights the differences between the two designs.

Suppose it is desired to obtain an estimate of the weight loss in a shipment of bananas during transit. Contrast the following two procedures:

1. *Independent samples* A random sample of banana bunches is selected from the lot before loading and weighed. After shipment, an independent random sample of bunches is selected and weighed during the unloading. The difference in the sample mean weight per bunch is used as the estimate of weight loss per bunch.
2. *Matched samples* Again, a random sample of bunches is selected and weighed before loading. After transit, the same bunches selected before loading are weighed again, and the difference in weight for each bunch is noted. The mean of these differences is used as an estimate of the weight loss per bunch.

Often, the matched samples method will lead to a smaller sampling error. To illustrate this, Table 13.1 contains hypothetical data on weights for a population of 10 banana bunches, giving the weights before and after shipment. In this idealized example, observe that each of the 10 bunches experiences the same weight loss of 2.0 kilograms. If independent samples of size 2, say, are drawn before and after transit, the uniform weight loss per bunch is hidden by the variation in the weights of bunches. The first sample might consist of bunches 1 and 8, with a mean weight of 28.5 kilograms, while the second sample might contain bunches 4 and 10, with a mean weight of 23.5 kilograms. The sample difference of 5.0 kilograms is far greater than the true difference of 2.0 kilograms. In contrast, the use of matched samples will always reveal exactly the 2.0 kilogram weight loss per bunch.

The example just presented is an extreme and simplified one, but it does illus-

TABLE 13.1 *Hypothetical weights for a population of banana bunches before and after shipment*

Bunch	Weight before Shipment (kilograms)	Weight after Shipment (kilograms)
1	34	32
2	29	27
3	31	29
4	28	26
5	23	21
6	32	30
7	27	25
8	23	21
9	28	26
10	23	21

trate the following general principle: if the observations before and after are positively related, the method involving matched observations will lead to smaller sampling errors than the use of independent samples. Note from Table 13.1 that the observations in our example are positively related. The heaviest bunches before shipment are still the heaviest bunches after shipment, and similarly for the lightest bunches. By weighing the same sample of banana bunches before and after shipment, variation in weights between banana bunches is eliminated as a source of sampling error and only the variation in the amounts of weight loss remains. On the other hand, the use of independent samples involves both of these sources of variation.

Matched samples need not always involve the same sample elements; instead, elements with similar characteristics may be paired in advance of the experiment. In a pharmacological research study involving the reaction of experimental animals to two drug dosages, two animals from the same litter and reared under identical conditions are randomly assigned the two dosages under study. By comparing the effects of the two drug dosages on the animals from the same litter, the effects of heredity, physiological condition, and environment are virtually eliminated from the sampling error.

In spite of the greater precision provided by matched samples when a positive relation exists between the paired observations, independent samples are used frequently. The cost or, perhaps, physical impossibility of obtaining matched observations often necessitates the use of independent samples. For instance, in the weight loss example, logistical difficulties may make it infeasible to tag banana bunches before shipment so that they can be identified for subsequent reweighing.

13.2 INFERENCES ABOUT DIFFERENCE BETWEEN TWO POPULATION MEANS—INDEPENDENT SAMPLES

As examples in the previous section have illustrated, we often want to estimate or test the difference between two population means. In this section, we discuss inferences of this nature for the case of two independent samples. First, we consider inferences when the populations are normal or approximately normal. Then we take up inferences for any populations when the sample sizes are large.

Confidence Interval for $\mu_2 - \mu_1$ —Normal Populations with Equal Variances

Assumptions

1. The two populations are normal or the departures are not too marked. The population means are denoted by μ_1 and μ_2.
2. The two populations have the same variance, denoted by σ^2.
3. Independent random samples of sizes n_1 and n_2 are drawn from the two populations, respectively.

☐   Illustration

An admissions officer of a small liberal arts college wishes to compare the mean admission test scores of applicants educated in rural high schools (population 1, X variable) and of applicants educated in urban high schools (population 2, Y variable). Independent random samples of sizes n_1 and n_2 from the two populations, respectively, will be used. It is reasonable to assume here that admission test scores in both populations are normally distributed with means μ_1 and μ_2, respectively, and common variance σ^2. Interest is in estimating the difference in mean test scores, $\mu_2 - \mu_1$, by a confidence interval. A point estimator of $\mu_2 - \mu_1$ is the difference between the two sample means:

(13.1)
$$\bar{Y} - \bar{X}$$

where $\bar{Y}$ is the mean of the Y observations from population 2 and $\bar{X}$ is the mean of the X observations from population 1. Let us now study the properties of the sampling distribution of $\bar{Y} - \bar{X}$.   ☐

Sampling Distribution of $\bar{Y} - \bar{X}$.   Since the populations sampled are assumed to be normal, we know from (9.7) that both $\bar{X}$ and $\bar{Y}$ have normal sampling distributions. The means of these sampling distributions are μ_1 and μ_2, from (9.3), and the variances are σ^2/n_1 and σ^2/n_2, from (9.4).

Turning now to the sampling distribution of $\bar{Y} - \bar{X}$, the following theorem tells us that this difference is also normal:

(13.2) Any linear combination of independent normal random variables is also normally distributed.

$\bar{Y} - \bar{X}$ is such a linear combination of independent normal random variables here. Remember that $\bar{X}$ and $\bar{Y}$ are independent because the two samples are independent.

Theorem (5.14a) tells us that the expected value of $\bar{Y} - \bar{X}$ is:

$$E\{\bar{Y} - \bar{X}\} = E\{\bar{Y}\} - E\{\bar{X}\} = \mu_2 - \mu_1$$

and theorem (5.14b) enables us to ascertain the variance:

$$\sigma^2\{\bar{Y} - \bar{X}\} = \sigma^2\{\bar{Y}\} + \sigma^2\{\bar{X}\} = \frac{\sigma^2}{n_2} + \frac{\sigma^2}{n_1} = \sigma^2\left(\frac{1}{n_2} + \frac{1}{n_1}\right)$$

We now summarize these results:

(13.3) When two independent random samples are selected from normal populations with common variance σ^2, the sampling distribution of $\bar{Y} - \bar{X}$ has the following properties:

$$\text{Mean: } E\{\bar{Y} - \bar{X}\} = \mu_2 - \mu_1$$

$$\text{Variance: } \sigma^2\{\bar{Y} - \bar{X}\} = \sigma^2\left(\frac{1}{n_2} + \frac{1}{n_1}\right)$$

Functional form: Normal distribution

Development of Confidence Interval. To construct a confidence interval for $\mu_2 - \mu_1$, we require an estimate of the common population variance σ^2. If s_1^2 and s_2^2 are the variances of the two samples, it can be shown that the following weighted average of s_1^2 and s_2^2 is the best unbiased estimator of σ^2:

$$(13.4) \qquad s_c^2 = \frac{(n_1 - 1)s_1^2 + (n_2 - 1)s_2^2}{(n_1 - 1) + (n_2 - 1)}$$

The estimator s_c^2 is called a *pooled* or *combined estimator* of σ^2.

With s_c^2 as an unbiased estimator of σ^2, it can be seen from (13.3) that the following is an unbiased estimator of $\sigma^2\{\bar{Y} - \bar{X}\}$:

$$(13.5) \qquad s^2\{\bar{Y} - \bar{X}\} = s_c^2\left(\frac{1}{n_2} + \frac{1}{n_1}\right)$$

A statistical theorem tells us:

(13.6) For two independent random samples from normal populations with common variance:

$$\frac{(\bar{Y} - \bar{X}) - (\mu_2 - \mu_1)}{s\{\bar{Y} - \bar{X}\}} = t(n_1 + n_2 - 2)$$

where: $s\{\bar{Y} - \bar{X}\}$ is given by (13.5)

Hence, the situation is parallel to that for interval estimation of μ for a normal population:

(13.7) The confidence limits for $\mu_2 - \mu_1$ with confidence coefficient $1 - \alpha$, when the populations are normal (or do not depart too markedly from normality) with common variance σ^2 and the samples are independent, are:

$$(\bar{Y} - \bar{X}) \pm ts\{\bar{Y} - \bar{X}\}$$

where: $t = t(1 - \alpha/2; n_1 + n_2 - 2)$
$s\{\bar{Y} - \bar{X}\}$ is given by (13.5)

The t value in this case is based on $n_1 + n_2 - 2$ degrees of freedom because the common variance σ^2 is estimated from two sample variances having a total number of degrees of freedom of $(n_1 - 1) + (n_2 - 1) = n_1 + n_2 - 2$.

☐ Example

Independent random samples of $n_1 = 15$ rural applicants and $n_2 = 17$ urban applicants were selected in the admission test scores case. The results were:

$$n_1 = 15 \qquad \bar{X} = 495 \qquad s_1 = 55$$
$$n_2 = 17 \qquad \bar{Y} = 545 \qquad s_2 = 50$$

A 95 percent confidence interval for $\mu_2 - \mu_1$ is desired.

Based on these sample results, we have:

$$\bar{Y} - \bar{X} = 545 - 495 = 50$$

$$s_c^2 = \frac{(15 - 1)(55)^2 + (17 - 1)(50)^2}{(15 - 1) + (17 - 1)} = 2{,}745$$

$$s^2\{\bar{Y} - \bar{X}\} = 2{,}745\left(\frac{1}{17} + \frac{1}{15}\right) = 344.47$$

$$s\{\bar{Y} - \bar{X}\} = \sqrt{344.47} = 18.56$$

For a 95 percent confidence interval, we require $t(1 - \alpha/2; \; n_1 + n_2 - 2) = t(.975;$ $30) = 2.042$. Hence, the desired confidence limits are $50 \pm 2.042(18.56)$ and the resulting confidence interval is:

$$12.1 \leq \mu_2 - \mu_1 \leq 87.9$$

Therefore, with 95 percent confidence, it can be stated that the mean admission test score for urban applicants for this college exceeds the mean score for rural applicants by between 12.1 and 87.9 points. Clearly, larger samples would be required to measure this difference more precisely. □

Confidence Interval for $\mu_2 - \mu_1$ —Large Samples from Arbitrary Populations

Assumptions

1. There are no restrictions on either population other than that the sample mean be approximately normally distributed for large samples.
2. Independent random samples of sizes n_1 and n_2 are selected, both of which are reasonably large.

□ Illustration

An economist wishes to estimate the difference in mean family income for households in two socioeconomic groups, neither of which contains households with extremely high incomes. Large random samples of n_1 and n_2 households are drawn from the respective groups and their incomes ascertained. The difference in the sample means, $\bar{Y} - \bar{X}$, will be used to construct an interval estimate for $\mu_2 - \mu_1$ (the difference in mean family incomes for the two groups). □

Sampling Distribution of $\bar{Y} - \bar{X}$. From the assumptions made in this case, we can determine the properties of the sampling distribution of $\bar{Y} - \bar{X}$:

(13.8) When two large independent random samples are selected from almost any populations, the sampling distribution of $\bar{Y} - \bar{X}$ has the following properties:

$$\text{Mean: } E\{\bar{Y} - \bar{X}\} = \mu_2 - \mu_1$$

$$\text{Variance: } \sigma^2\{\bar{Y} - \bar{X}\} = \sigma^2\{\bar{Y}\} + \sigma^2\{\bar{X}\}$$

Functional form: Approximately normal distribution

Because the samples are drawn independently, $\bar{Y} - \bar{X}$ has a variance by (5.14b) equal to the sum of the variances of $\bar{Y}$ and $\bar{X}$. Also, since both sample sizes are large, central limit theorem (9.5) guarantees that $\bar{Y}$ and $\bar{X}$ are each approximately normal and, consequently, by (13.2) their difference $\bar{Y} - \bar{X}$ is also approximately normal.

Development of Confidence Interval. To construct a confidence interval for $\mu_2 - \mu_1$, we estimate $\sigma^2\{\bar{Y} - \bar{X}\}$ with the following unbiased estimator:

(13.9)
$$s^2\{\bar{Y} - \bar{X}\} = s^2\{\bar{Y}\} + s^2\{\bar{X}\}$$

where: $s^2\{\bar{Y}\}$ and $s^2\{\bar{X}\}$ are given by (10.8)

Because we are dealing with large random samples, we can draw a parallel between this case and the case of interval estimation for μ based on a large random sample:

(13.10) The confidence limits for $\mu_2 - \mu_1$ with approximate confidence coefficient $1 - \alpha$, when the random samples are independent and reasonably large, are:

$$(\bar{Y} - \bar{X}) \pm zs\{\bar{Y} - \bar{X}\}$$

where: $z = z(1 - \alpha/2)$
$s\{\bar{Y} - \bar{X}\}$ is given by (13.9)

☐ **Example**

In our income study, samples of $n_1 = 200$ and $n_2 = 250$ families were selected from the respective populations. The results were:

$$n_1 = 200 \qquad \bar{X} = \$15,530 \qquad s_1 = \$5,160$$
$$n_2 = 250 \qquad \bar{Y} = \$16,910 \qquad s_2 = \$5,840$$

A 95 percent confidence interval for $\mu_2 - \mu_1$ is desired.
We have:

$$\bar{Y} - \bar{X} = 16,910 - 15,530 = 1,380$$

$$s^2\{\bar{Y} - \bar{X}\} = \frac{(5,160)^2}{200} + \frac{(5,840)^2}{250} = 269,550$$

$$s\{\bar{Y} - \bar{X}\} = \sqrt{269,550} = 519$$

We require $z(1 - \alpha/2) = z(.975) = 1.960$. Hence, the desired confidence limits are $1,380 \pm 1.960(519)$ and the resulting confidence interval is:

$$363 \le \mu_2 - \mu_1 \le 2,397$$

Therefore, it can be concluded, with 95 percent confidence, that mean family income in the second group exceeds that in the first group by between \$363 and \$2,397. ☐

Statistical Tests for $\mu_2 - \mu_1$

For both the case of normal populations and that of arbitrary populations with large sample sizes, decision rules for making statistical tests concerning $\mu_2 - \mu_1$ are constructed in the same manner as for tests concerning a single population mean μ. We present one example to show the completely parallel structure.

☐ **Example**

In our income study, it is desired to test if both socioeconomic groups have the same mean family income, i.e., to test:

$$H_0: \mu_2 - \mu_1 = 0$$

$$H_1: \mu_2 - \mu_1 \neq 0$$

The α risk is to be controlled at .05 when $\mu_2 - \mu_1 = 0$, i.e., when $\mu_2 = \mu_1$.

FIGURE 13.1 *Construction of decision rule to control α risk—Two-sided test for* $\mu_2 - \mu_1$, *large independent samples*

(a) *Appropriate type of decision rule*

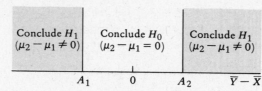

Large negative or positive values of $\overline{Y} - \overline{X}$ imply H_1 is correct.

(b) *Sampling distribution of $\overline{Y} - \overline{X}$ located at $\mu_2 - \mu_1 = 0$*

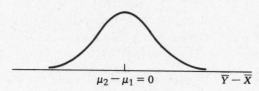

Distribution of $\overline{Y} - \overline{X}$ is centered at $\mu_2 - \mu_1 = 0$ because α risk is controlled there.

(c) *Placement of action limits*

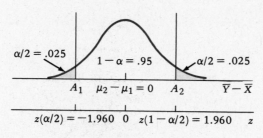

A_1 and A_2 are placed so area in each tail equals $\alpha/2$.

(d) *Calculation of action limits*

$$A_1 = 0 + z(\alpha/2)s\{\overline{Y} - \overline{X}\}$$
$$= -1.960(519) = -1,017$$

$$A_2 = 0 + z(1 - \alpha/2)s\{\overline{Y} - \overline{X}\}$$
$$= 1.960(519) = 1,017$$

$\sigma\{\overline{Y} - \overline{X}\}$ estimated by $s\{\overline{Y} - \overline{X}\}$.

(e) *Decision rule*

If $-1,017 \leq \overline{Y} - \overline{X} \leq 1,017$, conclude H_0

If $\overline{Y} - \overline{X} < -1,017$ or $\overline{Y} - \overline{X} > 1,017$, conclude H_1

Conclude appropriate alternative based on observed $\overline{Y} - \overline{X}$.

We proceed with the test following the model in Figure 13.1. The test is two-sided, as shown in Figure 13.1a. The sampling distribution centered around $\mu_2 - \mu_1 = 0$ is shown in Figure 13.1b. The two action limits are centered around $\mu_2 - \mu_1 = 0$, as shown in Figure 13.1c. Since each shaded tail area must be $\alpha/2 = .025$, the two action limits are:

$$A_1 = 0 + z(\alpha/2)s\{\bar{Y} - \bar{X}\} = -1.960(519) = -1{,}017$$

$$A_2 = 0 + z(1 - \alpha/2)s\{\bar{Y} - \bar{X}\} = 1.960(519) = 1{,}017$$

The decision rule is:

If $-1{,}017 \leq \bar{Y} - \bar{X} \leq 1{,}017$, conclude H_0 $(\mu_2 - \mu_1 = 0)$

If $\bar{Y} - \bar{X} < -1{,}017$ or $\bar{Y} - \bar{X} > 1{,}017$, conclude H_1 $(\mu_2 - \mu_1 \neq 0)$

Since $\bar{Y} - \bar{X} = 1{,}380$, it is concluded that the two groups do not have equal mean family income. □

Decision Rule. We summarize the decision rule for a two-sided test concerning $\mu_2 - \mu_1$ when the two sample sizes are large:

(13.11) When the alternatives are:

$$H_0: \mu_2 - \mu_1 = 0$$

$$H_1: \mu_2 - \mu_1 \neq 0$$

and the samples are independent and reasonably large, the appropriate decision rule to control the α risk is:

If $A_1 \leq \bar{Y} - \bar{X} \leq A_2$, conclude H_0

If $\bar{Y} - \bar{X} < A_1$ or $\bar{Y} - \bar{X} > A_2$, conclude H_1

where: $A_1 = 0 - zs\{\bar{Y} - \bar{X}\}$
$A_2 = 0 + zs\{\bar{Y} - \bar{X}\}$
$z = z(1 - \alpha/2)$
$s\{\bar{Y} - \bar{X}\}$ is given by (13.9)

Comment

The test of whether both socioeconomic groups have the same mean family income, i.e., $\mu_2 - \mu_1 = 0$, can also be conducted with the standardized test statistic:

(13.12)
$$z^* = \frac{(\bar{Y} - \bar{X}) - 0}{s\{\bar{Y} - \bar{X}\}}$$

The decision rule, for $\alpha = .05$, with this approach is:

If $|z^*| \leq 1.960$, conclude H_0

If $|z^*| > 1.960$, conclude H_1

Here, $z^* = 1{,}380/519 = 2.66$. Since $|z^*| = |2.66| = 2.66 > 1.960$, we reach conclusion H_1, as before.

13.3 INFERENCES ABOUT DIFFERENCE BETWEEN TWO POPULATION MEANS—MATCHED SAMPLES

When matched samples are employed for making inferences about $\mu_2 - \mu_1$, it turns out that the procedures simplify to those for a single population mean μ. The following example illustrates this.

☐ Illustration

We described earlier an example in which a social psychologist wishes to compare persons' attitudes toward a particular social behavior before and after they view a film about the behavior. For each subject, an attitude measurement is made before the film is viewed (X) and a second measurement is made after the film is viewed (Y). Letting μ_1 and μ_2 denote the population mean values of X and Y, respectively, the psychologist wishes to estimate $\mu_2 - \mu_1$.

We shall let X_i and Y_i denote the "before" and "after" measurements, respectively, for the ith subject, so (X_i, Y_i) is a "matched pair" of observations. To measure the change in attitude of the ith subject, the psychologist will use the difference:

$$(13.13) \qquad D_i = Y_i - X_i$$

If n subjects are studied, he will have n differences $D_1, D_2, \ldots, D_n$.

Now, these differences $D_1, D_2, \ldots, D_n$ may be thought of as a random sample of observations from a population of differences. We denote the mean of this population of differences by μ_D:

$$(13.14) \qquad E\{D_i\} = \mu_D$$

and the variance by σ_D^2. Using definition (13.13), we can express μ_D in terms of μ_1 and μ_2:

$$E\{D_i\} = E\{Y_i - X_i\} = E\{Y_i\} - E\{X_i\} = \mu_2 - \mu_1$$

Hence:

$$(13.14a) \qquad \mu_D = \mu_2 - \mu_1$$

Thus, we shall estimate μ_D by the earlier procedures for a single population mean, but we are really estimating $\mu_D = \mu_2 - \mu_1$, the difference between the means of the two populations. ☐

The procedure therefore is to first calculate the sample mean of the observed differences $D_1, D_2, \ldots, D_n$, denoted by $\bar{D}$. Next, we calculate the sample variance of the D_i's, denoted by s_D^2 as a reminder that it is a variance of differences. Finally, we calculate $s^2\{\bar{D}\}$, the estimated variance of $\bar{D}$. Thus, we have:

$$(13.15) \qquad \bar{D} = \frac{\sum_{i=1}^{n} D_i}{n}$$

$$(13.16) \qquad s_D^2 = \frac{\sum_{i=1}^{n} (D_i - \bar{D})^2}{n - 1}$$

$$(13.17) \qquad s^2\{\bar{D}\} = \frac{s_D^2}{n}$$

Then we proceed as earlier, depending on whether the population of differences is approximately normal or the population of differences is arbitrary and the number of differences (n) is large.

Confidence Interval for $\mu_2 - \mu_1$ —Normal Population of Differences

When the differences D_i are normally distributed, or when their distribution does not depart from normality too markedly, a confidence interval for $\mu_2 - \mu_1$ is equivalent to that in (10.16):

(13.18)   The confidence limits for $\mu_2 - \mu_1$ with confidence coefficient $1 - \alpha$, when the population of differences is normal or does not depart too markedly from normality, are:

$$\bar{D} \pm ts\{\bar{D}\}$$

where: $t = t(1 - \alpha/2; n - 1)$
$s\{\bar{D}\}$ is given by (13.17)

☐   Example

In the social behavior study, $n = 10$ subjects were studied and the attitude measures shown in Table 13.2 were obtained. It is desired to estimate $\mu_2 - \mu_1$ with a 90 percent confidence interval. It is reasonable here to assume that the population of differences is approximately normally distributed.

For the sample data, we have $\bar{D} = 7.64$ and $s_D = 12.57$. Therefore, $s^2\{\bar{D}\} = s_D^2/n = (12.57)^2/10 = 15.80$ and $s\{\bar{D}\} = \sqrt{15.80} = 3.97$. We require $t(.95; 9) = 1.833$. The confidence limits equal $7.64 \pm 1.833(3.97)$ and the desired interval is:

$$.36 \leq \mu_2 - \mu_1 \leq 14.92$$

The psychologist can assert, with 90 percent confidence, that the mean attitude measurement after viewing the film exceeds the mean attitude measurement before viewing by between .36 and 14.92 units on the measurement scale. Thus, a change in mean attitude occurred after viewing the film, but the small-scale study does not make clear whether the change is unimportant or large.   ☐

TABLE 13.2   *Sample data on attitudes before and after viewing an informational film*

Subject i	Before X_i	After Y_i	$D_i = Y_i - X_i$	Subject i	Before X_i	After Y_i	$D_i = Y_i - X_i$
1	41.0	46.9	+5.9	6	22.5	56.8	+34.3
2	60.3	64.5	+4.2	7	67.5	60.7	− 6.8
3	23.9	33.3	+9.4	8	50.3	57.3	+ 7.0
4	36.2	36.0	− .2	9	50.9	65.4	+14.5
5	52.7	43.5	−9.2	10	24.6	41.9	+17.3

$\bar{D} = 7.64$ $s_D = 12.57$

Comment

Matched sampling was highly effective here. To see why, we shall obtain $\sigma^2\{D_i\}$. Since $D_i = Y_i - X_i$, we can use (5.30) to find:

$$\sigma^2\{D_i\} = \sigma^2\{Y_i - X_i\} = \sigma^2\{Y_i\} + \sigma^2\{X_i\} - 2\sigma\{X_i, Y_i\}$$

This variance has been denoted by σ_D^2 earlier—i.e., $\sigma^2\{D_i\} = \sigma_D^2$. Thus, when $\sigma\{X_i, Y_i\}$, the covariance between X_i and Y_i, is positive, σ_D^2 will be smaller than when X_i and Y_i are independent. Since $\sigma^2\{\bar{D}\} = \sigma_D^2/n$, it follows that $\sigma^2\{\bar{D}\}$ is smaller when the X_i and Y_i observations are positively related than when they are independent. Here, the "before" and "after" attitude measures are strongly positively related, leading to a more precise estimate than if independent samples had been employed. Matched sample designs seek to assure a high positive relation between X and Y to attain this improvement in precision.

Confidence Interval for $\mu_2 - \mu_1$
—Large Sample from Arbitrary Population of Differences

When the sample of differences is reasonably large, the nature of the population of differences does not matter and the confidence interval for $\mu_2 - \mu_1$ is equivalent to that in (10.11):

(13.19)   The confidence limits for $\mu_2 - \mu_1$ with approximate confidence coefficient $1 - \alpha$, when the sample of differences is reasonably large, are:

$$\bar{D} \pm zs\{\bar{D}\}$$

where: $z = z(1 - \alpha/2)$
$s\{\bar{D}\}$ is given by (13.17)

Statistical Tests for $\mu_2 - \mu_1$

Like the construction of confidence intervals for $\mu_2 - \mu_1$, tests on $\mu_2 - \mu_1$ for the case of matched samples parallel those for a single population mean. $\bar{D}$ corresponds to $\bar{X}$, $\mu_D = \mu_2 - \mu_1$ corresponds to μ, and σ_D corresponds to σ.

☐   Example

In the social behavior study, suppose we wish to test whether or not viewing the film leads to an upward shift in the mean attitude level. Thus, the alternatives are:

$$H_0\!: \mu_2 - \mu_1 \leq 0$$
$$H_1\!: \mu_2 - \mu_1 > 0$$

Suppose the α risk is to be controlled at .05 when $\mu_2 - \mu_1 = 0$. We found earlier that $\bar{D} = 7.64$ and $s\{\bar{D}\} = 3.97$. The appropriate form of the test is given in Table 11.2a, since the population of differences is approximately normal and a one-sided upper-tail test is desired.

We require $t(.95; 9) = 1.833$. The action limit is (note the change from "X" to "D" notation and the substitution of $\mu_2 - \mu_1 = 0$ for μ_I):

$$A = 0 + t(1 - \alpha; n - 1)s\{\bar{D}\} = 1.833(3.97) = 7.28$$

The decision rule is:

$$\text{If } \bar{D} \leq 7.28, \text{ conclude } H_0 \ (\mu_2 - \mu_1 \leq 0)$$

$$\text{If } \bar{D} > 7.28, \text{ conclude } H_1 \ (\mu_2 - \mu_1 > 0)$$

Since $\bar{D} = 7.64$, H_1 is concluded—that there has been an upward shift in the mean attitude after viewing of the film.   □

Comment

The test statistic for the preceding example using the standardized test statistic approach is:

(13.20)
$$t^* = \frac{\bar{D} - 0}{s\{\bar{D}\}}$$

and the decision rule for $\alpha = .05$ with this approach is:

$$\text{If } t^* \leq 1.833, \text{ conclude } H_0$$

$$\text{If } t^* > 1.833, \text{ conclude } H_1$$

Since $t^* = 7.64/3.97 = 1.92 > 1.833$, we again conclude H_1.

13.4   INFERENCES ABOUT DIFFERENCE BETWEEN TWO POPULATION PROPORTIONS

□   **Illustration**

A market survey organization carried out a product taste study with consumers in two regions. In one region, a random sample of $n_1 = 400$ consumers was selected while in the other region an independent random sample of $n_2 = 300$ consumers was selected. Each person was asked to indicate which of two servings of product had a better taste. Unknown to the subject, one serving was a new high-protein breakfast cereal and the other was an existing cereal. In the first region, proportion $\bar{p}_1 = .55$ of the sample persons preferred the new cereal. In the second region, the proportion was $\bar{p}_2 = .65$. It is now desired to construct an interval estimate of $p_2 - p_1$, the difference in the population proportions of consumers in the two regions who prefer the new cereal.   □

Sampling Distribution of $\bar{p}_2 - \bar{p}_1$

As may be clear already, this type of estimation problem is very similar to the problem of estimating the difference between two population means. A point estimator of $p_2 - p_1$ is the difference in sample proportions:

(13.21)
$$\bar{p}_2 - \bar{p}_1$$

We now summarize the properties of the sampling distribution of $\bar{p}_2 - \bar{p}_1$ when *large and statistically independent random samples* are drawn from the two populations:

(13.22) When large independent random samples are selected from two populations, the sampling distribution of $\bar{p}_2 - \bar{p}_1$ has the following properties:

$$\text{Mean: } E\{\bar{p}_2 - \bar{p}_1\} = p_2 - p_1$$

$$\text{Variance: } \sigma^2\{\bar{p}_2 - \bar{p}_1\} = \sigma^2\{\bar{p}_2\} + \sigma^2\{\bar{p}_1\}$$

Functional form: Approximately normal distribution

Observe that the variance of $\bar{p}_2 - \bar{p}_1$ is the sum of the variances of $\bar{p}_1$ and $\bar{p}_2$ by (5.14b) because the samples are independent. In addition, since the individual sample sizes are large, $\bar{p}_1$ and $\bar{p}_2$ are each approximately normal by (12.5) and, by (13.2), their difference $\bar{p}_2 - \bar{p}_1$ is therefore also approximately normal.

Confidence Interval for $p_2 - p_1$

To construct an interval estimate for $p_2 - p_1$, we need an unbiased estimator of $\sigma^2\{\bar{p}_2 - \bar{p}_1\}$. This is provided by using $s^2\{\bar{p}_1\}$ and $s^2\{\bar{p}_2\}$ as given in (12.7) as estimators of the true variances $\sigma^2\{\bar{p}_1\}$ and $\sigma^2\{\bar{p}_2\}$. The resulting estimator is denoted by $s^2\{\bar{p}_2 - \bar{p}_1\}$. Thus:

(13.23)
$$s^2\{\bar{p}_2 - \bar{p}_1\} = s^2\{\bar{p}_2\} + s^2\{\bar{p}_1\}$$

where: $s^2\{\bar{p}_2\}$ and $s^2\{\bar{p}_1\}$ are given by (12.7)

The confidence interval for $p_2 - p_1$ takes the usual form for the large-sample case:

(13.24) The confidence limits for $p_2 - p_1$ with approximate confidence coefficient $1 - \alpha$, when the two samples are independent and reasonably large, are:

$$(\bar{p}_2 - \bar{p}_1) \pm zs\{\bar{p}_2 - \bar{p}_1\}$$

where: $z = z(1 - \alpha/2)$
$s\{\bar{p}_2 - \bar{p}_1\}$ is given by (13.23)

☐ Example

For our market survey case, we have:

$$n_1 = 400 \qquad \bar{p}_1 = .55$$

$$n_2 = 300 \qquad \bar{p}_2 = .65$$

A 90 percent confidence interval for $p_2 - p_1$ is desired.
We have $\bar{p}_2 - \bar{p}_1 = .65 - .55 = .10$ and:

$$s^2\{\bar{p}_1\} = \frac{\bar{p}_1(1 - \bar{p}_1)}{n_1 - 1} = \frac{.55(.45)}{400 - 1} = .0006203$$

$$s^2\{\bar{p}_2\} = \frac{\bar{p}_2(1 - \bar{p}_2)}{n_2 - 1} = \frac{.65(.35)}{300 - 1} = .0007609$$

$$s^2\{\bar{p}_2 - \bar{p}_1\} = .0007609 + .0006203 = .0013812$$

$$s\{\bar{p}_2 - \bar{p}_1\} = \sqrt{.0013812} = .0372$$

We require $z(.95) = 1.645$, and the confidence limits are $.10 \pm 1.645(.0372)$. The resulting confidence interval is:

$$.039 \leq p_2 - p_1 \leq .161$$

We conclude, with 90 percent confidence, that the proportion of consumers in the second region who prefer the new cereal exceeds the proportion for the first region by between 3.9 and 16.1 percent points. □

Statistical Tests for $p_2 - p_1$

Statistical tests regarding the difference between two population proportions parallel those for the difference between two population means. We illustrate the procedure for a two-sided test.

□ Example

In the market survey study, it is desired to test whether or not the proportions of consumers preferring the new high-protein cereal for the two regions are equal. Thus, the alternatives are:

$$H_0: p_2 - p_1 = 0$$

$$H_1: p_2 - p_1 \neq 0$$

The α risk at $p_2 - p_1 = 0$ is to be controlled at .10. Figure 13.2 displays the steps to be followed in the test.

1. Since large negative or positive values of $\bar{p}_2 - \bar{p}_1$ are indicative of H_1 being true, the appropriate decision rule is that shown in Figure 13.2a.
2. The α risk is controlled at $p_2 - p_1 = 0$; hence, the sampling distribution of $\bar{p}_2 - \bar{p}_1$ is centered on 0 in Figure 13.2b.
3. Figure 13.2c shows the positioning of the action limits so that the tail areas are each $\alpha/2$.
4. To calculate the action limits, we require an estimate of $\sigma\{\bar{p}_2 - \bar{p}_1\}$ when $p_2 - p_1 = 0$. Since $p_2 = p_1$ in this case, we shall let p denote their common value. To estimate p from the two samples, we employ the following *pooled* estimator of p:

(13.25)
$$\bar{p}' = \frac{n_1\bar{p}_1 + n_2\bar{p}_2}{n_1 + n_2}$$

Note that $\bar{p}'$ is simply a weighted average of $\bar{p}_1$ and $\bar{p}_2$.

When $p_1 = p_2 = p$, we know that:

$$\sigma^2\{\bar{p}_2 - \bar{p}_1\} = \frac{p_2(1 - p_2)}{n_2} + \frac{p_1(1 - p_1)}{n_1} = p(1 - p)\left(\frac{1}{n_2} + \frac{1}{n_1}\right)$$

Hence, a sample estimator of $\sigma^2\{\bar{p}_2 - \bar{p}_1\}$ based on the pooled estimator $\bar{p}'$ is:

(13.26)
$$s^2\{\bar{p}_2 - \bar{p}_1\} = \bar{p}'(1 - \bar{p}')\left(\frac{1}{n_2} + \frac{1}{n_1}\right)$$

where: $\bar{p}'$ is given by (13.25).

FIGURE 13.2 *Construction of decision rule to control α risk—Two-sided test for* $p_2 - p_1$, *large independent samples*

(a) *Appropriate type of decision rule*

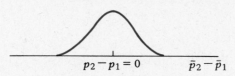

Conclude H_1 ($p_2 - p_1 \neq 0$)	Conclude H_0 ($p_2 - p_1 = 0$)	Conclude H_1 ($p_2 - p_1 \neq 0$)

$A_1 \qquad 0 \qquad A_2 \qquad \bar{p}_2 - \bar{p}_1$

Large negative or positive values of $\bar{p}_2 - \bar{p}_1$ imply H_1 is correct.

(b) *Sampling distribution of $\bar{p}_2 - \bar{p}_1$ located at $p_2 - p_1 = 0$*

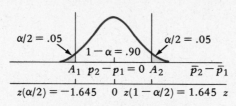

$p_2 - p_1 = 0 \qquad \bar{p}_2 - \bar{p}_1$

Distribution of $\bar{p}_2 - \bar{p}_1$ centered at $p_2 - p_1 = 0$ because α risk is controlled there.

(c) *Placement of action limits*

$\alpha/2 = .05 \qquad\qquad \alpha/2 = .05$

$1 - \alpha = .90$

$A_1 \quad p_2 - p_1 = 0 \quad A_2 \qquad \bar{p}_2 - \bar{p}_1$

$z(\alpha/2) = -1.645 \quad 0 \quad z(1 - \alpha/2) = 1.645 \quad z$

A_1 and A_2 are placed so area in each tail equals $\alpha/2$.

(d) *Calculation of action limits*

$$\bar{p}' = \frac{n_1\bar{p}_1 + n_2\bar{p}_2}{n_1 + n_2} = \frac{400(.55) + 300(.65)}{400 + 300} = .593$$

$\bar{p}'$ used to estimate $p_1 = p_2 = p$.

$$s^2\{\bar{p}_2 - \bar{p}_1\} = \bar{p}'(1 - \bar{p}')\left(\frac{1}{n_2} + \frac{1}{n_1}\right)$$

$$= .593(.407)\left(\frac{1}{300} + \frac{1}{400}\right)$$

$$= .001408$$

$\bar{p}'$ used in calculation of $s\{\bar{p}_2 - \bar{p}_1\}$.

$$s\{\bar{p}_2 - \bar{p}_1\} = \sqrt{.001408} = .0375$$

$$A_1 = 0 + z(\alpha/2)s\{\bar{p}_2 - \bar{p}_1\} = -1.645(.0375) = -.062$$

$$A_2 = 0 + z(1 - \alpha/2)s\{\bar{p}_2 - \bar{p}_1\} = 1.645(.0375) = .062$$

$s\{\bar{p}_2 - \bar{p}_1\}$ used as estimate of $\sigma\{\bar{p}_2 - \bar{p}_1\}$.

(e) *Decision rule*

If $-.062 = A_1 \leq \bar{p}_2 - \bar{p}_1 \leq A_2 = .062$, conclude H_0

If $\bar{p}_2 - \bar{p}_1 < A_1 = -.062$ or $\bar{p}_2 - \bar{p}_1 > A_2 = .062$, conclude H_1

Conclude appropriate alternative based on observed $\bar{p}_2 - \bar{p}_1$.

Figure 13.2d shows the calculation of the action limits using $s\{\bar{p}_2 - \bar{p}_1\}$ to estimate $\sigma\{\bar{p}_2 - \bar{p}_1\}$.

5. Applying the decision rule in Figure 13.2e, since $\bar{p}_2 - \bar{p}_1 = .10$, we conclude H_1—that the proportions preferring the new cereal in the two regions differ. ☐

Decision Rule. We summarize the decision rule for a two-sided test concerning $p_2 - p_1$:

(13.27) When the alternatives are:

$$H_0: p_2 - p_1 = 0$$

$$H_1: p_2 - p_1 \neq 0$$

and the samples are independent and reasonably large, the appropriate decision rule to control the α risk is:

If $A_1 \leq \bar{p}_2 - \bar{p}_1 \leq A_2$, conclude H_0

If $\bar{p}_2 - \bar{p}_1 < A_1$ or $\bar{p}_2 - \bar{p}_1 > A_2$, conclude H_1

where: $A_1 = 0 - zs\{\bar{p}_2 - \bar{p}_1\}$
$A_2 = 0 + zs\{\bar{p}_2 - \bar{p}_1\}$
$z = z(1 - \alpha/2)$
$s\{\bar{p}_2 - \bar{p}_1\}$ is given by (13.26)

Comment

The standardized test statistic for the preceding two-sided test is:

(13.28)

$$z^* = \frac{(\bar{p}_2 - \bar{p}_1) - 0}{s\{\bar{p}_2 - \bar{p}_1\}}$$

13.5 OPTIONAL TOPIC—INFERENCES ABOUT POPULATION VARIANCE AND RATIO OF TWO POPULATION VARIANCES

Interest in Population Variance

On various previous occasions, we have used the sample variance s^2 as an unbiased point estimator of the population variance σ^2. In some applications, we are interested in constructing an interval estimate for σ^2 or in testing the value of σ^2. We discuss these types of inferences now for the special case where *the population is normal or approximately normal*. In Chapter 15, we consider inferences about the population variance when the population is not normal or approximately normal.

☐ Illustration

The population variance is often of interest in studies where the extent of dispersion in the population is of concern. Consider the following quarterly rates of return (dividends plus price change as a percent of opening price) on a common stock investment over a two-year period:

Quarter:	1	2	3	4	5	6	7	8
Rate of return:	4.8	2.8	9.9	7.6	9.5	6.0	8.4	−5.0

Suppose that these eight observations can be considered as a random sample from the infinite population of all quarterly rates of return for the stock, and that this population is approximately normal. The population variance σ^2 may be viewed as a measure of uncertainty for the common stock investment, and we may therefore be interested in constructing an interval estimate for σ^2. ☐

Confidence Interval for σ^2

Sampling Distribution of s^2. Statistical theory provides us with an important theorem:

(13.29) If a random sample of size n is selected from a normal population with variance σ^2, then:

$$\frac{(n-1)s^2}{\sigma^2} = \chi^2(n-1)$$

where: s^2 is the sample variance defined in (8.12b)

(If a review of the χ^2 distribution is needed, the reader should turn to Appendix B, Section B.1 before proceeding.)

Figure 13.3a contains a representative χ^2 distribution which illustrates theorem (13.29). A χ^2 distribution has its mean at v, its degrees of freedom. Here, $v = n - 1$, so the distribution of $\chi^2(n-1)$ is located around $n - 1$. Since a χ^2 distribution is always right-skewed, it is possible for s^2 to be substantially larger than σ^2 on some occasions.

FIGURE 13.3 *Sampling distribution of $(n-1)s^2/\sigma^2$ for a normal population*

(a)

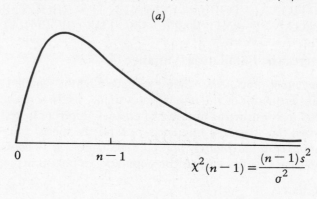

(b)

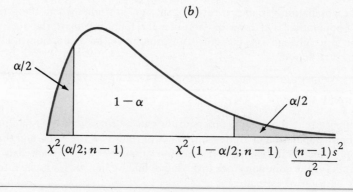

Confidence Limits. The confidence limits for σ^2 are based on theorem (13.29) and make use of the percentiles of the χ^2 distribution shown in Figure 13.3b:

(13.30) The confidence limits L and U for the population variance σ^2 with confidence coefficient $1 - \alpha$, when the population is normal or approximately so, are:

$$L = \frac{(n - 1)s^2}{\chi^2(1 - \alpha/2; n - 1)} \qquad U = \frac{(n - 1)s^2}{\chi^2(\alpha/2; n - 1)}$$

☐   Example

For our investment illustration, $n = 8$ and $s^2 = 23.78$ (calculation not shown). A 95 percent confidence interval is desired; hence:

$$\chi^2(1 - \alpha/2; n - 1) = \chi^2(.975; 7) = 16.01$$

$$\chi^2(\alpha/2; n - 1) = \chi^2(.025; 7) = 1.69$$

Substituting these values into (13.30) yields:

$$L = \frac{(8 - 1)23.78}{16.01} = 10.4 \qquad U = \frac{(8 - 1)23.78}{1.69} = 98.5$$

Hence, the 95 percent confidence interval for σ^2 is:

$$10.4 \leq \sigma^2 \leq 98.5$$

Note that the magnitude of σ^2 cannot be ascertained precisely here with such a small sample size.   ☐

Development of Confidence Interval. Since theorem (13.29) tells us that $(n - 1)s^2/\sigma^2$ follows a $\chi^2(n - 1)$ distribution, it is a consequence of the definition of percentiles for a χ^2 distribution that:

$$P\left[\chi^2(\alpha/2; n - 1) \leq \frac{(n - 1)s^2}{\sigma^2} \leq \chi^2(1 - \alpha/2; n - 1)\right] = 1 - \alpha$$

This probability statement is illustrated in Figure 13.3b. Observe that α is divided equally between the two tails, as has been our practice in constructing two-sided confidence intervals. Rearranging the inequalities, we obtain:

$$P\left[\frac{(n - 1)s^2}{\chi^2(1 - \alpha/2; n - 1)} \leq \sigma^2 \leq \frac{(n - 1)s^2}{\chi^2(\alpha/2; n - 1)}\right] = 1 - \alpha$$

Thus, we have $1 - \alpha$ confidence limits for σ^2.

Comments

1. Observe that the confidence limits for σ^2 are not equally spaced about s^2. This reflects the right skewness of the χ^2 sampling distribution.
2. A confidence interval for σ is obtained by taking the square roots of the confidence limits for σ^2 in (13.30). For our example, the 95 percent confidence interval for σ is:

$$3.2 = \sqrt{10.4} \leq \sigma \leq \sqrt{98.5} = 9.9$$

Hence, the standard deviation of the common stock's quarterly rate of return is between 3.2 and 9.9 percent, with 95 percent confidence.

3. For large n, the χ^2 percentiles in (13.30) may be approximated by standard normal percentiles using the approximation given in (B.4) in Appendix B.

4. One-sided confidence limits for σ^2, with confidence coefficient $1 - \alpha$, can be obtained from (13.30) by replacing $1 - \alpha/2$ by $1 - \alpha$ in the percentiles.

Statistical Tests for σ^2

Statistical tests for σ^2 may be constructed using theorem (13.29). We illustrate the construction of a two-sided test when the sample size is given and the α risk is controlled.

☐ Example

Consider the investment illustration again. An investment analyst has developed a model according to which σ^2 for the common stock, given the nature of the business, the risks involved, and other related factors, should be equal to 5. We wish to test whether or not the model is correct, i.e., whether or not $\sigma^2 = 5$.

$$H_0: \sigma^2 = \sigma_I^2 = 5$$

$$H_1: \sigma^2 \neq \sigma_I^2 = 5$$

Here, σ_I^2 is the value of σ^2 at which the α risk will be controlled. In this case, $\sigma_I^2 = 5$, and we wish to control α at .05. Figure 13.4 displays the steps in the test procedure graphically.

1. It is convenient to work with the test statistic $(n - 1)s^2/\sigma_I^2$ rather than with s^2 itself. Since exceptionally large or small values of s^2 and, hence, of $(n - 1)s^2/\sigma_I^2$, suggest that H_1 ($\sigma^2 \neq \sigma_I^2$) is correct, the appropriate type of decision rule is two-sided as shown in Figure 13.4a.

2. When $\sigma^2 = \sigma_I^2$ (the level where the α risk is controlled), theorem (13.29) states that the sampling distribution of $(n - 1)s^2/\sigma_I^2$ is $\chi^2(n - 1)$. This sampling distribution is shown in Figure 13.4b. In this case, $\chi^2(n - 1) = \chi^2(7)$ because $n = 8$.

3. The action limits are positioned on the sampling distribution in Figure 13.4c. In order to give equal tail areas of $\alpha/2$, the action limits must correspond to the following χ^2 percentiles (recall that $\alpha = .05$ in this example):

$$A_1 = \chi^2(\alpha/2; n - 1) = \chi^2(.025; 7) = 1.69$$

$$A_2 = \chi^2(1 - \alpha/2; n - 1) = \chi^2(.975; 7) = 16.01$$

4. The decision rule for the test therefore is (see Figure 13.4d):

$$\text{If } 1.69 \leq \frac{(n - 1)s^2}{\sigma_I^2} \leq 16.01, \text{ conclude } H_0 \ (\sigma^2 = 5)$$

$$\text{If } \frac{(n - 1)s^2}{\sigma_I^2} < 1.69 \text{ or } \frac{(n - 1)s^2}{\sigma_I^2} > 16.01, \text{ conclude } H_1 \ (\sigma^2 \neq 5)$$

Since $s^2 = 23.78$ in our sample, $(n - 1)s^2/\sigma_I^2 = (8 - 1)23.78/5 = 33.3$ and hence H_1 should be concluded. Thus, the variance of quarterly returns for the stock differs from the magnitude projected by the model. If the model were to perform as badly for other stocks, the analyst should abandon the model. ☐

FIGURE 13.4   *Construction of decision rule to control α risk—Two-sided test for σ^2*

(a) *Appropriate type of decision rule*

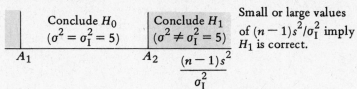

Small or large values of $(n-1)s^2/\sigma_I^2$ imply H_1 is correct.

(b) *Sampling distribution of $(n-1)s^2/\sigma_I^2$ when $\sigma^2 = \sigma_I^2$*

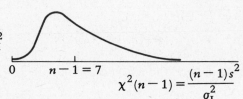

Distribution of $(n-1)s^2/\sigma_I^2$ is $\chi^2(n-1)$ when $\sigma^2 = \sigma_I^2$.

(c) *Placement of action limits*

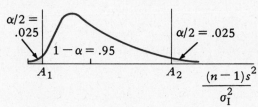

A_1 and A_2 are placed so area in each tail is $\alpha/2$.

where: $A_1 = \chi^2(\alpha/2; n-1) = 1.69$
$A_2 = \chi^2(1 - \alpha/2; n-1) = 16.01$

(d) *Decision rule*

If $1.69 = A_1 \leq \dfrac{(n-1)s^2}{\sigma_I^2} \leq A_2 = 16.01$,

conclude H_0

If $\dfrac{(n-1)s^2}{\sigma_I^2} < A_1 = 1.69$

or $\dfrac{(n-1)s^2}{\sigma_I^2} > A_2 = 16.01$,

conclude H_1

Conclude appropriate alternative based on observed s^2.

Decision Rule.   We present the two-sided test for σ^2 in general:

(13.31) When the alternatives are:

$$H_0: \sigma^2 = \sigma_I^2$$
$$H_1: \sigma^2 \neq \sigma_I^2$$

and the population is normal or approximately so, the appropriate decision rule to control the α risk at σ_I^2 is:

$$\text{If } A_1 \leq \frac{(n-1)s^2}{\sigma_I^2} \leq A_2, \text{ conclude } H_0$$

$$\text{If } \frac{(n-1)s^2}{\sigma_I^2} < A_1 \text{ or } \frac{(n-1)s^2}{\sigma_I^2} > A_2, \text{ conclude } H_1$$

where: $A_1 = \chi^2(\alpha/2; n-1)$
   $A_2 = \chi^2(1 - \alpha/2; n-1)$

Comparison of Two Population Variances

In some comparative studies, interest lies in the dispersions of two populations. For instance, a manufacturer may wish to know whether two versions of a product have the same variability in shelf life. We now discuss interval estimates and tests for the ratio of two population variances. Again, we assume that *the two populations are normal or approximately so* and that *the two random samples are independent.*

☐   Illustration

A behavioral scientist is interested in determining whether group discussion tends to affect the closeness of judgments by property appraisers. She obtained the cooperation of 16 real estate appraisers. Randomly, she selected 6 and asked them to make individual appraisals of a piece of commercial property after all 6 had participated in a group discussion of relevant appraisal factors. The other 10 appraisers were also asked to make individual appraisals of the same property, but without any prior group discussion.

   As a measure of the closeness of agreement among appraisers, the scientist is using the variance. We shall let σ_1^2 denote the variance of appraisals made with group discussion and σ_2^2 the variance of appraisals made without group discussion. The scientist wants to compare the magnitudes of σ_1^2 and σ_2^2. Statistical theory provides a ready way for her to do this by computing an interval estimate of the ratio σ_2^2/σ_1^2.   ☐

Confidence Interval for σ_2^2/σ_1^2

To construct a confidence interval for σ_2^2/σ_1^2, we employ an important statistical theorem:

(13.32) If independent random samples of sizes n_1 and n_2 are selected from normal populations with variances σ_1^2 and σ_2^2, respectively, then:

$$\frac{s_1^2/\sigma_1^2}{s_2^2/\sigma_2^2} = F(n_1 - 1, n_2 - 1)$$

where: s_1^2 and s_2^2 are the two sample variances, respectively

(If a review of the F distribution is needed, the reader should turn to Appendix B, Section B.3 before proceeding.)

Confidence Limits. The confidence limits for σ_2^2/σ_1^2 are based on theorem (13.32) and make use of percentiles of the F distribution:

(13.33) The confidence limits L and U for σ_2^2/σ_1^2 with confidence coefficient $1 - \alpha$, when the populations are normal or approximately so and the samples are independent, are:

$$L = F(\alpha/2; n_1 - 1, n_2 - 1)\frac{s_2^2}{s_1^2} \qquad U = F(1 - \alpha/2; n_1 - 1, n_2 - 1)\frac{s_2^2}{s_1^2}$$

Note that the numerator degrees of freedom for the F distribution are those of s_1^2, and the denominator degrees of freedom are those of s_2^2.

☐ Example

The data for our appraisal study are presented in Table 13.3. A 90 percent confidence interval for σ_2^2/σ_1^2 is desired. The scientist states that it is reasonable to assume that the respective populations of appraisals are approximately normal.

We note first that $s_2^2/s_1^2 = 194.2/77.9 = 2.49$. For a 90 percent confidence interval, we require:

$$F(1 - \alpha/2; n_1 - 1, n_2 - 1) = F(.95; 5, 9) = 3.48$$

$$F(\alpha/2; n_1 - 1, n_2 - 1) = F(.05; 5, 9) = 1/4.77 = .210$$

Therefore, the confidence limits are $.210(2.49) = .523$ and $3.48(2.49) = 8.67$, and the desired confidence interval is:

$$.523 \leq \frac{\sigma_2^2}{\sigma_1^2} \leq 8.67$$

Since the confidence interval is so wide and straddles 1, no clear conclusion about the effect of group discussions on the variability of appraisals is possible. The scientist will have to enlarge her study to draw any firmer conclusions. ☐

TABLE 13.3 *Experimental data on assessed values for a commercial property*

Appraisals with Group Discussion ($000)	Appraisals without Group Discussion ($000)
97	118
111	109
102	84
99	85
88	100
111	121
	115
	93
	91
	112
$n_1 = 6$	$n_2 = 10$
$s_1^2 = 77.9$	$s_2^2 = 194.2$

Development of Confidence Interval. From theorem (13.32) and the definition of percentiles of the F distribution, we can make the following probability statement:

$$P\left[F(\alpha/2; n_1 - 1, n_2 - 1) \leq \frac{s_1^2/\sigma_1^2}{s_2^2/\sigma_2^2} \leq F(1 - \alpha/2; n_1 - 1, n_2 - 1)\right] = 1 - \alpha$$

Observe that α is equally divided between the two tails, as has been our practice in constructing two-sided confidence intervals. Rearranging the inequalities, we obtain:

$$P\left[F(\alpha/2; n_1 - 1, n_2 - 1)\frac{s_2^2}{s_1^2} \leq \frac{\sigma_2^2}{\sigma_1^2} \leq F(1 - \alpha/2; n_1 - 1, n_2 - 1)\frac{s_2^2}{s_1^2}\right] = 1 - \alpha$$

Hence, we have $1 - \alpha$ confidence limits for σ_2^2/σ_1^2.

Comment

A confidence interval for the ratio of two standard deviations, σ_2/σ_1, is obtained by taking the square roots of the confidence limits for the variance ratio in (13.33). For our example, the 90 percent confidence interval is:

$$.72 = \sqrt{.523} \leq \frac{\sigma_2}{\sigma_1} \leq \sqrt{8.67} = 2.94$$

Statistical Tests for σ_2^2/σ_1^2

A variety of statistical tests concerning σ_2^2/σ_1^2 may be constructed. We illustrate a one-sided upper-tail test.

Example

Suppose the behavioral scientist had hypothesized that group discussion would reduce the variability of appraisals. The alternatives then would be:

$$H_0: \frac{\sigma_2^2}{\sigma_1^2} \leq 1$$

$$H_1: \frac{\sigma_2^2}{\sigma_1^2} > 1$$

The scientist wishes to control the α risk at .05 when $\sigma_2^2/\sigma_1^2 = 1$. Figure 13.5 displays the steps in the test procedure.

1. The statistic of interest is the ratio of the sample variances s_2^2/s_1^2. Values of this statistic substantially larger than 1 suggest that σ_1^2 is smaller than σ_2^2 and hence that H_1 is true. Thus, the appropriate type of decision rule is the one shown in Figure 13.5a.
2. When $\sigma_2^2 = \sigma_1^2$, the case for which the α risk is to be controlled, theorem (13.32) states that the sampling distribution of s_2^2/s_1^2 is $F(n_2 - 1, n_1 - 1)$. This sampling distribution is shown in Figure 13.5b.

FIGURE 13.5 *Construction of decision rule to control α risk—Upper-tail test for* σ_2^2/σ_1^2, *independent samples*

(a) *Appropriate type of decision rule*

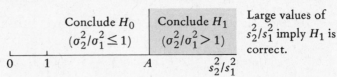

Large values of s_2^2/s_1^2 imply H_1 is correct.

(b) *Sampling distribution of s_2^2/s_1^2 when $\sigma_2^2 = \sigma_1^2$*

Distribution of s_2^2/s_1^2 when $\sigma_2^2 = \sigma_1^2$ is $F(n_2 - 1, n_1 - 1)$.

(c) *Placement of action limit*

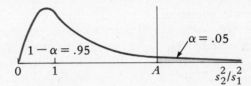

A is placed so upper–tail area is α.

where: $A = F(1 - \alpha; n_2 - 1, n_1 - 1) = 4.77$

(d) *Decision rule*

If $s_2^2/s_1^2 \leq A = 4.77$, conclude H_0

If $s_2^2/s_1^2 > A = 4.77$, conclude H_1

Conclude appropriate alternative based on observed ratio s_2^2/s_1^2.

3. The action limit is positioned at $A = F(.95; 9, 5) = 4.77$ in Figure 13.5c in order that the upper-tail area is $\alpha = .05$. (Recall that $n_2 - 1 = 10 - 1 = 9$ and $n_1 - 1 = 6 - 1 = 5$.)
4. The decision rule for the test is shown in Figure 13.5d. Since $s_1^2 = 77.9$ and $s_2^2 = 194.2$ in the scientist's study (see Table 13.3), it follows that $s_2^2/s_1^2 = 194.2/77.9 = 2.49$. Since $2.49 \leq 4.77$, we conclude H_0—that the variability of appraisals with group discussion is not smaller than that without it. □

Decision Rule. We now summarize the decision rule for the one-sided upper-tail test:

(13.34) When the alternatives are:

$$H_0: \frac{\sigma_2^2}{\sigma_1^2} \leq 1$$

$$H_1: \frac{\sigma_2^2}{\sigma_1^2} > 1$$

and the populations are normal or approximately so and the samples are independent, the appropriate decision rule to control the α risk is:

$$\text{If } \frac{s_2^2}{s_1^2} \leq F(1 - \alpha; n_2 - 1, n_1 - 1), \text{ conclude } H_0$$

$$\text{If } \frac{s_2^2}{s_1^2} > F(1 - \alpha; n_2 - 1, n_1 - 1), \text{ conclude } H_1$$

Need for Normality

The inferential procedures for a population variance and the comparison of two population variances described in this section are exact when the underlying population is normal. These procedures yield approximately correct results when departures from normality are small, but under moderate and large departures the results can be quite unreliable.

Fortunately, when nonnormal populations are encountered, several alternative techniques are available. One of these is to use mathematical transformations on the sample data. For example, a logarithmic transformation is frequently employed on income and other right-skewed data because such data in the form of $\log X_i$ generally are more normal than the untransformed data X_i. Figure 13.6 illustrates this point. In Figure 13.6a is a histogram of the income distribution for persons under 35 years old in a community. Figure 13.6b presents the same data on a log X scale. Clearly, use of the logarithmic transformation of income has resulted here in a much more symmetric distribution.

A second alternative is to employ an inferential procedure which does not require normality of the population. One such procedure is the jackknife procedure, which is discussed in Chapter 15.

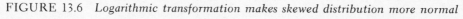

FIGURE 13.6 *Logarithmic transformation makes skewed distribution more normal*

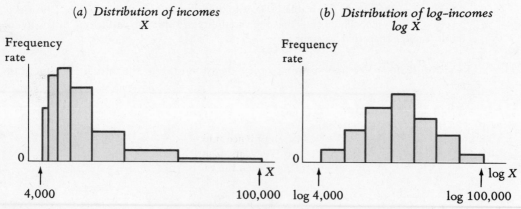

(a) *Distribution of incomes*
X

(b) *Distribution of log-incomes*
log X

PROBLEMS

*13.1 Annual snowfall in a northern city averaged 284 centimeters in the last 21 years (period 2), with a standard deviation of 66 centimeters. In the 21 years before that (period 1), the average annual snowfall was 264 centimeters, with a standard deviation of 65 centimeters. Assume that the two sets of observations constitute independent random samples from the populations defined by the weather environments prevailing in the respective 21-year periods. Also assume that each of these populations is normal, with the same variance. Estimate the difference in mean annual snowfall for the two weather environments, using a 90 percent confidence interval. Interpret your interval estimate.

13.2 An operations analyst wished to estimate the difference in mean tread life for a certain make of automobile tire when it is inflated to higher pressure (to improve gas mileage) instead of to the standard pressure. She selected two independent random samples of 15 tires each from the production process, inflated the tires in sample 1 to the standard pressure and the tires in sample 2 to the higher pressure, and conducted tread-life tests for all tires. The results (in thousand miles of tread life) follow (n_1 is 14 because one of the tires was defective and the testing could not be completed).

Standard Pressure	Higher Pressure
$n_1 = 14$	$n_2 = 15$
$\bar{X} = 43.0$	$\bar{Y} = 40.7$
$s_1 = 1.1$	$s_2 = 1.3$

Assume that each of the underlying populations is normal, with the same variance.
a. Identify the populations here, taking into account the exclusion of the defective tire from the study.
b. Estimate the difference in mean tread life for the two tire pressures, using a 95 percent confidence interval. Interpret your interval estimate.

*13.3 A simple random sample of 400 households from a large community was selected to estimate the mean residential electricity usage per household during February of last year. Another simple random sample of 450 households was selected, independently of the first, to estimate mean residential electricity usage during February of this year. The sample results (expressed in kilowatt hours) were:

Last Year	This Year
$n_1 = 400$	$n_2 = 450$
$\bar{X} = 1,252$	$\bar{Y} = 1,330$
$s_1 = 257$	$s_2 = 251$

a. Construct a 99 percent confidence interval for the change in mean usage per household between the two years. Can it be concluded that mean usage per household has changed between the two years? If so, was the change small or large? Explain.
b. If 300 of the households selected last year were also included in the sample for this year, why would the estimation procedure used in **a** not be appropriate?

13.4 In a study of company salaries, random samples of 150 employees were selected independently from each of two large departments. The following salary statistics were obtained:

Accounting Department	Engineering Department
$n_1 = 150$	$n_2 = 150$
$\bar{X} = \$27,250$	$\bar{Y} = \$29,212$
$s_1 = \$5,541$	$s_2 = \$5,356$

a. Construct a 90 percent confidence interval for the difference in mean salaries for the two departments. Interpret your confidence interval.

b. What information is provided by this confidence interval which would not be provided by a test of whether or not mean salaries in the two departments are the same?

*13.5 Refer to Problem 13.1.

a. Test whether or not the mean annual snowfall is the same for the two weather environments, controlling the α risk at .10. State the alternatives, decision rule, and conclusion.

b. Is your conclusion in **a** consistent with the confidence interval obtained in Problem 13.1? Explain.

13.6 Refer to Problem 13.2. Test whether or not the mean tread life for tires inflated to the higher pressure is less than that for tires inflated to the standard pressure. Control the α risk at .025 when $\mu_2 - \mu_1 = 0$. State the alternatives, decision rule, and conclusion.

*13.7 Refer to Problem 13.4. Test whether or not the mean salaries in the two departments are equal, controlling the α risk at .10. State the alternatives, decision rule, and conclusion.

13.8 Refer to Problem 13.3. Test whether or not mean residential usage of electricity increased from February of last year to February of this year. Control the α risk at .005 when $\mu_2 - \mu_1 = 0$. State the alternatives, decision rule, and conclusion.

*13.9 In a test involving highway cruising at 55 mph, eight new cars of a certain make were each driven 1,000 miles without an oil additive and 1,000 miles with an oil additive. The order of the two runs was randomized. Let X_i and Y_i denote, respectively, the miles per gallon for the ith car without and with the oil additive. Data for the differences $D_i = Y_i - X_i$ follow.

i:	1	2	3	4	5	6	7	8
D_i:	1.87	1.71	2.38	2.19	1.89	1.96	1.76	2.00

Assume that the population of differences is normal. Construct a 90 percent confidence interval for the mean difference per car in miles per gallon with and without the additive. Interpret your interval estimate. Does the additive appear to improve gas economy?

13.10 An educational psychologist is studying differences in mathematics achievment scores produced by two tests (A and B). Eighteen subjects randomly selected were each given both tests, half taking test A first and the remainder taking test B first. The results were as follows, where X and Y denote the scores for tests A and B, respectively:

Subject i	Y_i	X_i	Subject i	Y_i	X_i	Subject i	Y_i	X_i
1	81	93	7	103	94	13	123	97
2	103	70	8	102	87	14	98	73
3	88	79	9	111	119	15	124	99
4	85	91	10	114	111	16	113	108
5	84	83	11	109	95	17	80	77
6	100	106	12	91	93	18	107	98

Assume that the population of test score differences is approximately normal.

a. Construct a 95 percent confidence interval for the difference in mean scores for the two test procedures. Interpret your confidence interval.

b. Can it be concluded from your interval estimate that the mean scores for the two test procedures differ? If so, is the difference large? Explain.

*13.11 Refer to Problem 13.4. In further analysis of the data, the 150 employees of each department were ranked by age, then the salaries of employees with the same rank were matched, and the salary difference was calculated for each pair. The average salary difference for the 150 pairs was found to be $1,962 in favor of the engineering department, with the standard deviation of the differences being $4,383.

a. Based on these data on salary differences, construct a 90 percent confidence interval for the difference in mean salaries for the two departments.

b. What accounts for the confidence interval in a being tighter than the one in Problem 13.4a? Explain.

13.12 A financial analyst wished to investigate whether bond yields are affected by the presence of a certain sinking fund provision in the bond indenture. He gathered data on 160 pairs of bond issues selected randomly from bond issues in the last 10 years. Each pair is identical with respect to coupon rate, year of maturity, and investment quality rating, but only one bond issue in each pair has the sinking fund provision. Taking the difference $D_i = Y_i - X_i$ for each pair between the yield of the bond issue with the provision (Y) and the one without it (X), the analyst obtained a mean difference in yields of $-.18$ percent point and a standard deviation of differences of .46 percent point.

a. Estimate by means of a 98 percent confidence interval the mean difference in yields with and without the provision. Interpret your interval estimate.

b. The analyst would obtain the same mean yield difference if he simply averaged the yields of the 160 issues with the provision, then averaged the yields of the 160 issues without the provision, and finally took the difference between the two averages. What would the analyst be likely to gain by matching pairs of bond issues?

c. (Based on Optional Topic Section 10.7.) Construct an upper 99 percent confidence interval for the mean difference in yields. Interpret your interval estimate.

*13.13 Refer to Problem 13.9. Test whether or not the oil additive increases the mean gas mileage, controlling the α risk at .05 when $\mu_2 - \mu_1 = 0$. State the alternatives, decision rule, and conclusion.

13.14 Refer to Problem 13.10.

a. Test whether or not the mean achievement score for test B is higher than that for test A, controlling the α risk at .025 when $\mu_2 - \mu_1 = 0$. State the alternatives, decision rule, and conclusion.

  b. If all the subjects had taken the two tests in the same sequence, could one conclude that a difference in mean test scores is necessarily due to differences in the two tests? Explain.

*13.15 Refer to Problem 13.12. The analyst believes the sinking fund provision should reduce bond yields, on the average. Test this hypothesis, controlling the α risk at .01 when $\mu_2 - \mu_1 = 0$. State the alternatives, decision rule, and conclusion.

*13.16 A random sample of $n_1 = 1,000$ recent divorces and annulments in a state contained 437 which involved no children. In a neighboring state, an independent random sample of $n_2 = 1,200$ divorces and annulments contained 538 which did not involve any children.

  a. Construct a 90 percent confidence interval for $p_2 - p_1$, where p_1 and p_2 denote, respectively, the proportions of divorces and annulments involving no children in the first and second states.

  b. Does it appear from your interval in **a** that the two state proportions are close to each other? Explain.

13.17 The following data show the number of business school graduates hired by a firm three years ago who are still with the firm, classified by degree:

Degree	Number Hired	Number Remaining
1. Bachelor's	210	141
2. Master's	150	48

  Assume that the numbers remaining, 141 and 48, are outcomes of independent binomial random variables, i.e., that these are independent outcomes from two Bernoulli processes. Construct a 95 percent confidence interval for the difference in the probabilities of bachelor's and master's degree graduates remaining with the firm for three years. Interpret your confidence interval.

*13.18 Refer to Problem 13.16. Test whether or not the two state proportions are equal, controlling the α risk at .10. State the alternatives, decision rule, and conclusion.

13.19 Refer to Problem 13.17. Test whether or not the probability of a bachelor's degree graduate remaining with the firm for three years exceeds that for a master's degree graduate, controlling the α risk at .025 when $p_2 - p_1 = 0$. State the alternatives, decision rule, and conclusion.

*13.20 A pharmaceutical firm produces tablets that are supposed to contain a consistent amount of active ingredient. A random sample of 41 tablets taken from the process has a standard deviation $s = 1.09$ milligrams of active ingredient per tablet ($s^2 = 1.1881$). Assume that the amounts of active ingredient per tablet are normally distributed for the process.

  a. Construct a 95 percent confidence interval for σ^2, the process variance of amounts of active ingredient in tablets.

  b. Convert your interval estimate in **a** into a 95 percent confidence interval for σ. Interpret the meaning of this interval estimate.

13.21 The standard deviation of the lengths of 25 mature trout raised experimentally on a standard diet in a hatchery was $s = 4.35$ centimeters ($s^2 = 18.9225$). Assume that the population distribution is normal.

  a. Construct a 99 percent confidence interval for σ^2.

  b. Convert your interval estimate in **a** into a 99 percent confidence interval for σ. Interpret your interval estimate.

  c. Why are your confidence limits in **a** not equally spaced around s^2?

13.22 (Based on Optional Topic Section 10.7.) Refer to Problem 13.20. Obtain an upper 97.5 percent confidence interval for σ and interpret your interval estimate.

*13.23 Refer to Problem 13.21. Test whether or not the population variance of the lengths of trout equals 16.32, controlling the α risk at .01. State the alternatives, decision rule, and conclusion.

13.24 Refer to Problem 13.20. A process engineer had specified that the process variance for the amount of active ingredient should not exceed 1.10. Test whether or not the process variance meets this requirement, controlling the α risk at .025 when $\sigma^2 = 1.10$. State the alternatives, decision rule, and conclusion.

13.25 A food-distributing company has specified that the standard deviation in weights of broiling chickens received from vendors be $\sigma = 71$ grams. Management wishes to identify for special handling those shipments where the variability in weights departs materially from 71 grams in either direction. In a random sample of 81 broiling chickens taken from a large shipment, $s = 76.7$ grams. Test whether or not the specification is being met in this shipment, controlling the α risk at .02. Assume the weights in the shipment are approximately normally distributed. State the alternatives, decision rule, and conclusion.

*13.26 In a random sample of 25 candles of brand 1, the standard deviation of burning times was found to be $s_1 = 1.60$ minutes ($s_1^2 = 2.56$), while in an independent random sample of 21 similar candles of brand 2, the standard deviation of burning times was found to be $s_2 = 2.40$ minutes ($s_2^2 = 5.76$). Assume that the two populations of burning times are each normal.

    a. Construct a 90 percent confidence interval for the ratio of the two population variances.

    b. Convert your confidence interval in **a** into a 90 percent confidence interval for the ratio of the two population standard deviations. Interpret your interval estimate.

13.27 Refer to Problem 13.20. A random sample of $n_2 = 31$ tablets selected independently a week later has a standard deviation $s_2 = 1.21$ milligrams of active ingredient per tablet ($s_2^2 = 1.4641$). Assume that the distribution of amounts of active ingredient per tablet is still normal.

    a. Construct a 98 percent confidence interval for the ratio of the process variances for the two weeks.

    b. Convert your interval estimate in **a** into a 98 percent confidence interval for the ratio of the process standard deviations. Interpret your interval estimate.

*13.28 Refer to Problems 13.20 and 13.27. Test whether the process variance has increased between the two weeks, controlling the α risk at .01 when $\sigma_1^2 = \sigma_2^2$. State the alternatives, decision rule, and conclusion.

13.29 Refer to Problem 13.26. Test whether or not the two population variances are equal, controlling the α risk at .10. State the alternatives, decision rule, and conclusion.

13.30 Refer to Problem 13.21. It was hypothesized that the variability in lengths of mature trout could be reduced by increasing the protein in the diet. The standard deviation of the lengths of $n_2 = 25$ mature trout raised on a higher-protein diet was found to be $s_2 = 2.76$ centimeters ($s_2^2 = 7.6176$). Assume that the underlying population of lengths here also is normally distributed. Test whether the higher-protein diet leads to a smaller variance, controlling the α risk at .05 when $\sigma_1^2 = \sigma_2^2$. State the alternatives, decision rule, and conclusion.

EXERCISES

13.31 Demonstrate that: (1) s_c^2 in (13.4) is an unbiased estimator of σ^2, (2) $(n_1 + n_2 - 2)s_c^2/\sigma^2$ has a $\chi^2(n_1 + n_2 - 2)$ distribution.

13.32 (Based on Optional Topic Section 5.9.) Let T_1 and T_2 be unbiased estimators of total health expenditures by U.S. families in two successive years, which we denote by τ_1 and τ_2, respectively.

   a. Show that $T_2 - T_1$ is an unbiased estimator of the difference $\tau_2 - \tau_1$.

   b. Show that the more positive is the covariation between T_1 and T_2 (other factors being unchanged), the more efficient is $T_2 - T_1$ as an estimator of $\tau_2 - \tau_1$.

   c. Explain the relevance of the results in **a** and **b** to the use of matched samples to estimate the change in total health expenditures between the two years.

13.33 A student asks why a pooled estimator is employed in testing whether or not $p_2 - p_1 = 0$, using (13.27), whereas it is not employed in constructing a confidence interval for $p_2 - p_1$, using (13.24). Respond.

13.34 Examine whether the confidence limits (13.30) for the population variance σ^2 become more symmetrical about the point estimator s^2 as the sample size increases by considering the case $s^2 = 100$, $1 - \alpha = .95$, for $n = 5, 11, 21, 51$, and 101. What do you conclude?

STUDIES

13.35 A labor economist is studying the durations of the most recent strikes in the vehicles and construction industries to see if strikes in the two industries are equally difficult to settle. To achieve approximate normality and equal variances, the economist decided to work with the logarithms (to base 10) of the duration data (expressed in days), and obtained the following sample statistics:

Industry	Number of Strikes	Mean Log-Duration	Standard Deviation of Log-Duration
1. Vehicles	13	.593	.294
2. Construction	15	.973	.349

The economist believes it is reasonable to view the data as independent random samples.

   a. Construct a 90 percent confidence interval for the difference in the mean log-durations of strikes in the two industries. Interpret your interval estimate.

   b. (Based on Optional Topic Section 3.4.) What do the antilogarithms of the two confidence limits in **a** represent? [*Hint:* Recall formula (3.22) and the properties of logarithms.]

   c. Test whether strikes in the two industries have the same mean log-durations, controlling the α risk at .10. State the alternatives, decision rule, and conclusion.

   d. Test the economist's assumption that the log-durations of strikes in the two industries have equal variances, controlling the α risk at .10. What do you conclude?

13.36 Refer to Problem 13.20. Specifications require that at least 95 percent of tablets have an amount of active ingredient within ± 2.0 milligrams of the mean amount. Test whether or not the process meets this specification, controlling the α risk at .01. State the alternatives, decision rule, and conclusion. [*Hint:* What value of σ satisfies $2/\sigma = z(.975)$?]

13.37 Sixty-two persons who were seriously overweight were randomly assigned to one of two weight reduction regimens. During the study period, one person in regimen 2 moved out of town. All other persons remained in the study. At the end of the study period, the weight losses were ascertained. The data on weight losses (in kilograms) follow.

		Regimen 1						Regimen 2			
15.2	14.4	16.3	13.6	11.7	10.2	16.2	18.6	12.6	17.9	18.2	20.0
12.6	12.6	14.3	16.7	13.0	17.1	17.3	18.3	19.7	15.1	16.8	16.8
11.7	12.9	14.4	14.2	14.5	14.8	15.5	20.5	16.7	16.4	14.8	18.3
12.7	15.1	11.1	10.0	12.1	14.7	16.6	17.9	18.8	16.9	18.2	17.3
11.9	17.2	11.3	13.3	14.6	12.9	16.2	18.2	18.1	15.2	16.5	19.2
13.7											

a. Construct a frequency distribution for each sample and plot the polygons on the same graph. Do the two sets of data appear to have about the same amount of variability? Do the distributions suggest that the populations are normal? Discuss.

b. Assume that the two populations are approximately normal. Test whether or not the two population variances are equal, controlling the α risk at .02. State the alternatives, decision rule, and conclusion.

c. Assume that the two populations are normal, with equal variances. Test whether or not the mean weight loss for regimen 2 exceeds that for regimen 1, as had been expected. Control the α risk at .05 when $\mu_2 - \mu_1 = 0$. State the alternatives, decision rule, and conclusion. [*Hint:* $t(.95; 59) = 1.671$.]

d. Estimate the difference in the mean weight losses, using a 90 percent confidence interval. Interpret your interval estimate.

e. Obtain a point estimate (which relies on the normality of the two populations) of the probability of a weight loss of at least 16.0 kilograms for each of the two regimens. Does there appear to be a substantial difference between the two probabilities? Comment.

14
Sampling Procedures

In previous chapters, we have restricted our discussion to simple random sampling from infinite or large finite populations. In this chapter, we consider simple random sampling situations where the sample size is not small relative to the population size. We also describe several other probability sampling procedures which are of practical importance. Finally, we consider two applications of sampling methodology—namely, acceptance sampling and control charts.

14.1 SIMPLE RANDOM SAMPLING FROM FINITE POPULATIONS

Throughout the Text up to this point, we have assumed the sampling fraction n/N is small whenever the population sampled is finite. There are instances, however, when this condition does not hold, e.g., when a sample of 5 generators in a population of 15 is sampled to study the working condition of an important safety feature. In cases where the sampling fraction n/N is not small, recognition of the finite nature of the population is essential.

Estimation of Population Mean

The formula for the estimated standard deviation of $\bar{X}$ given by (10.8), i.e., $s\{\bar{X}\} = s/\sqrt{n}$, is exact only for the case of an infinite population. When the population is finite and the sampling fraction n/N is small, formula (10.8) still applies approximately. However, when the sampling fraction n/N is not small, the estimated standard deviation $s\{\bar{X}\}$ must be modified. For finite populations, the estimated standard deviation $s\{\bar{X}\}$ and the variance $s^2\{\bar{X}\}$ are as follows:

(14.1)
$$s^2\{\bar{X}\} = \left(1 - \frac{n}{N}\right)\frac{s^2}{n} \qquad s\{\bar{X}\} = \sqrt{1 - \frac{n}{N}}\,\frac{s}{\sqrt{n}}$$

The factor $\sqrt{1 - (n/N)}$ is called the *finite population correction* because the standard deviation $s/\sqrt{n}$ for an infinite population is "corrected" by this factor to yield the proper estimate when the population is finite.

Several characteristics of the finite population correction are noteworthy:

1. For a given sample size n, the finite population correction $\sqrt{1 - (n/N)}$ is nearly equal to 1 when N is large relative to n. It is for this reason that a finite population can be treated as an infinite one when the sampling fraction n/N is small, since $s\{\overline{X}\}$ then equals $s/\sqrt{n}$ for all practical purposes. A rule of thumb often employed is: *A finite population can be treated as infinite when $n/N \leq .05$, i.e., when no more than 5 percent of the population is sampled.* In all applications considered thus far in the Text, the sampling fraction n/N has been small enough for the correction to be ignored.

2. For a given finite population, the finite population correction approaches 0 as n approaches N. Hence, $s\{\overline{X}\} = 0$ when $n = N$. This is intuitively reasonable since a census of the population is taken when $n = N$ and hence there is no error due to sampling present in $\overline{X}$, i.e., $\overline{X} = \mu$ then.

Confidence Interval for μ. Whenever the sample size is not an exceptionally large portion of the finite population size, the estimation and testing procedures for a population mean μ recognizing the finite nature of the population are the same as when the population can be treated as infinite, except that (14.1) is used to obtain $s\{\overline{X}\}$. We now illustrate the construction of a confidence interval for μ when the sampling fraction n/N is large.

Example

A chemical company serves $N = 810$ dry cleaning establishments. It selected a simple random sample of $n = 140$ of these establishments to estimate the mean annual usage per establishment (μ) that would be made of a new variation of one of its current solvent products. The survey results gave $\overline{X} = 410$ gallons and $s = 215$ gallons. A 95 percent confidence interval for μ is desired.

The sample fraction n/N in this case is $140/810 = .1728$, which is large enough to make the finite population correction appropriate, but not so large as to affect the estimation procedure. Hence, the confidence interval for μ is of the same form as (10.11), except that $s\{\overline{X}\}$ is computed using (14.1).

We have from (14.1):

$$s\{\overline{X}\} = \sqrt{1 - \frac{140}{810}}\,\frac{215}{\sqrt{140}} = 16.526$$

For 95 percent confidence, $z(.975) = 1.960$ is required. Hence, the confidence limits are $410 \pm 1.960(16.526)$ and the desired confidence interval is $377.6 \leq \mu \leq 442.4$. □

Planning of Sample Size. When planning a sample from a finite population, the finite nature of the population must be taken into account when the resulting sample size is not a small fraction of the population size. The appropriate planning formula for the case when it is desired to estimate μ with a $1 - \alpha$ confidence interval with a half-width h is a modification of (10.14):

(14.2)
$$n = \frac{z^2\sigma^2}{h^2 + \dfrac{z^2\sigma^2}{N}}$$

where: $z = z(1 - \alpha/2)$

Here, σ, as in (10.14), denotes the planning value for the population standard deviation.

☐ Example

A random sample from a population of $N = 900$ Canadian general hospitals is to be taken to obtain a confidence interval for the mean number of pediatric beds per hospital, with a half-width of $h = 5$, and a 95 percent confidence coefficient. Results of a previous study suggest that an appropriate planning value of the population standard deviation is $\sigma = 40$ beds. Since $z(.975) = 1.960$, the needed sample size is:

$$n = \frac{(1.960)^2(40)^2}{(5)^2 + \dfrac{(1.960)^2(40)^2}{900}} = 193$$

Note that $n/N = 193/900 = .214$. Thus, over 21 percent of all hospitals must be sampled to obtain an estimate with the desired precision. ☐

Comment

As the population size N gets large, the second term in the denominator of (14.2) approaches zero, and (14.2) simplifies to (10.14).

Estimation of Population Proportion

Confidence Interval for p. When estimating a population proportion p where the finite nature of the population must be recognized, the estimated standard deviation $s\{\bar{p}\}$ needs to include the finite population correction, as follows:

(14.3)
$$s^2\{\bar{p}\} = \left(1 - \frac{n}{N}\right)\frac{\bar{p}(1 - \bar{p})}{n - 1} \qquad s\{\bar{p}\} = \sqrt{1 - \frac{n}{N}}\sqrt{\frac{\bar{p}(1 - \bar{p})}{n - 1}}$$

The construction of the confidence interval for p proceeds as for the case where the population is infinite, using (12.9), except that $s\{\bar{p}\}$ is obtained from (14.3).

☐ Example

For a population of $N = 1,200$ railroad tank cars, a simple random sample of $n = 200$ cars revealed that 33 had defective discharge valves. A 90 percent confidence interval for the population proportion of tank cars with defective valves is desired. We require $z(.95) = 1.645$, $\bar{p} = 33/200 = .165$, and:

$$s\{\bar{p}\} = \sqrt{1 - \frac{200}{1,200}}\sqrt{\frac{.165(1 - .165)}{200 - 1}} = .0240$$

The confidence limits therefore are $.165 \pm 1.645(.0240)$ and the desired confidence interval is $.13 \leq p \leq .20$. ☐

Planning of Sample Size. When planning a sample that will represent a fraction of the population which is not small, the planning formula in (12.10) must be modified as follows:

(14.4)
$$n = \frac{z^2 p(1 - p)}{h^2 + \dfrac{z^2 p(1 - p)}{N}}$$

where: $z = z(1 - \alpha/2)$

Here, p, as in (12.10), denotes the planning value for the population proportion.

☐ Example

Union leaders of a local with $N = 850$ members want to survey members' opinions on a proposed change in their labor contract. It is desired to estimate the proportion p supporting the change by a confidence interval with half-width $h = .05$ and confidence coefficient .99. A planning value of $p = .5$ is proposed. Since $z(.995) = 2.576$, the number of members needed to be included in the sample is:

$$n = \frac{(2.576)^2(.5)(1 - .5)}{(.05)^2 + \dfrac{(2.576)^2(.5)(1 - .5)}{850}} = 373$$

☐

Efficiency of a Sampling Procedure

Simple random sampling is not always the best sampling procedure to employ. Other procedures often are more efficient. The concept of efficiency of a sampling procedure is usually defined in terms of the precision of the estimator, as measured by the width of the confidence interval for a specified confidence coefficient. Every probability sampling procedure can be evaluated in terms of the cost of attaining a given statistical precision, or alternatively in terms of the precision furnished for a given cost:

(14.5) A probability sampling procedure is said to be more *efficient* than simple random sampling if it offers, at a given level of confidence, the same precision at less cost or greater precision at the same cost.

We now consider a number of probability sampling procedures which, under certain conditions, are more efficient than simple random sampling.

14.2   STRATIFIED SAMPLING

Stratified sampling is a sampling procedure that is widely used.

☐ Illustration

A state teachers' association is studying the educational qualifications of its membership, which consists of 50,000 elementary school, high school, and college teachers. A major characteristic of interest is the mean number of years of education of its members—i.e., the average number of years of formal education (including teacher training) of the association's teachers. The analyst conducting the study knew that the number of years of education tends to vary for teachers by level of school, being greater for teachers in higher-level educational institutions. Hence, the analyst used a sample design in which independent simple random samples of teachers are selected from among the member teachers in each of the three school levels. Column 1 of Table 14.1 shows the total number of member teachers in each school level, and column 2 shows the sample size for each group. □

Definition

The sampling procedure used here is called stratified random sampling:

(14.6) With *stratified random sampling*, the population is divided into a number of mutually exclusive subpopulations or *strata*, and independent simple random samples are selected from the strata.

In our illustration, the 25,000 teachers in elementary schools represent a stratum of the population of 50,000 teachers.

Advantages of Stratified Sampling

There are three major advantages of stratified sampling: (1) efficiency, (2) information about subpopulations, and (3) feasibility.

Efficiency.   Frequently, a stratified sample will be much more efficient than a simple random sample. To illustrate this, consider Table 14.2a, which contains a hypothetical population of nine teachers. The number of years of education is given for each. The population variance is 2.61. This is the variability affecting the precision of the sample mean $\bar{X}$ with simple random sampling.

Suppose we stratify the teachers by level of school. This is done in Table 14.2b, which also shows the strata variances. Note that the variability of years of education is much smaller within each stratum than in the entire population. Consequently, a stratified random sample of given size will yield a more precise estimate than a simple random sample of the same size, since much smaller variability is encountered within each stratum.

Thus, for stratified sampling to be efficient, it is necessary that the strata be designed to contain relatively homogeneous elements. This is accomplished when

TABLE 14.1 *An illustration of stratified sampling*

Stratum	Level of School	(1) Number of Members In Population	(2) In Sample
1	Elementary	25,000	150
2	High school	20,000	120
3	College	5,000	30
	Total	50,000	300

TABLE 14.2 *An illustration of the effects of stratification*

(a)
Population

Teacher	Level of School	Years of Education
1	Elementary	15.5
2	College	19.0
3	High school	16.5
4	Elementary	16.0
5	High school	18.0
6	College	20.5
7	High school	17.5
8	College	19.5
9	Elementary	16.5

Variance = 2.61

(b)
Strata

Stratum 1 Elementary		Stratum 2 High School		Stratum 3 College	
Teacher	Years of Education	Teacher	Years of Education	Teacher	Years of Education
1	15.5	3	16.5	2	19.0
4	16.0	5	18.0	6	20.5
9	16.5	7	17.5	8	19.5
Variance = .17		Variance = .39		Variance = .39	

the basis of stratification is related to the characteristic under study. In our illustration, level of school in which members teach is related to number of years of education. Similarly, in a survey of firms to estimate mean charitable contributions per firm, a stratification by size of firm is likely to be highly effective in view of the positive relation between size of firm and amount of contributions.

While the selection of a stratified random sample may cost somewhat more than the selection of a simple random sample of equal size (e.g., separate frames for each stratum are required), this is usually more than balanced by the substantially improved precision. Thus, a stratified sample to yield a specified precision can usually be significantly smaller than the corresponding simple random sample and consequently cost less.

Information about Subpopulations.   Stratified sampling can provide estimates of strata characteristics in addition to estimates of overall population characteristics. For example, a study by a university of the effects of tuition increases is to provide separate information for undergraduate and graduate students as well as information for all students. Stratification by student class status will permit precise estimates for undergraduate and graduate students separately, as well as for all students.

Feasibility. At times, stratified sampling is the most feasible type of sampling. Consider the sampling of welfare claims in a state. In the one large city of the state, the claims records are in computerized files, whereas elsewhere in the state, the records are kept locally in file cabinets. Administrative considerations here require separate sampling of claims in the city and elsewhere in the state.

Estimation of Population Mean

Notation. The total number of strata into which the population is divided is denoted by k. The size of the jth stratum (i.e., the number of population elements in it) is denoted by N_j. The total population size is denoted as usual by N, where:

$$(14.7) \qquad N = \sum_{j=1}^{k} N_j$$

Analogously, n_j denotes the size of the sample selected from the jth stratum, and n denotes the total sample size. Thus:

$$(14.8) \qquad n = \sum_{j=1}^{k} n_j$$

Finally, $\bar{X}_j$ and s_j^2 denote the mean and variance, respectively, of the sample drawn from the jth stratum.

In our teachers' association example (see Table 14.1), $k = 3$ strata are employed. The stratum size and sample size for the first stratum (elementary schools) are $N_1 = 25,000$ and $n_1 = 150$, respectively. The total population and sample sizes are $N = 50,000$ and $n = 300$.

Development of Confidence Interval. An unbiased point estimator of the population mean μ is as follows, where $\bar{X}_{st}$ denotes the stratified estimator:

$$(14.9) \qquad \bar{X}_{st} = \left(\frac{N_1}{N}\right)\bar{X}_1 + \left(\frac{N_2}{N}\right)\bar{X}_2 + \cdots + \left(\frac{N_k}{N}\right)\bar{X}_k = \sum_{j=1}^{k} \left(\frac{N_j}{N}\right)\bar{X}_j$$

Note that $\bar{X}_{st}$ is simply a weighted average of the strata sample means, with the weights being the proportions of the total population in each stratum.

We can use the theory of random variables developed in Chapter 5 to obtain $\sigma^2\{\bar{X}_{st}\}$, the variance of the sampling distribution of $\bar{X}_{st}$. Since the $\bar{X}_j$'s are based on independent random samples, we know from (5.15b) that the variances of the terms in $\bar{X}_{st}$ are additive. Further, by (5.11b), the variance of $(N_j/N)\bar{X}_j$ is $(N_j/N)^2\sigma^2\{\bar{X}_j\}$. Therefore:

$$(14.10) \qquad \sigma^2\{\bar{X}_{st}\} = \left(\frac{N_1}{N}\right)^2 \sigma^2\{\bar{X}_1\} + \left(\frac{N_2}{N}\right)^2 \sigma^2\{\bar{X}_2\} + \cdots + \left(\frac{N_k}{N}\right)^2 \sigma^2\{\bar{X}_k\}$$

We must still estimate. the variances $\sigma^2\{\overline{X}_j\}$. For this purpose, we employ $s^2\{\overline{X}_j\}$ given by (14.1). Hence, the following is a point estimator of $\sigma^2\{\overline{X}_{st}\}$:

(14.11)
$$s^2\{\overline{X}_{st}\} = \left(\frac{N_1}{N}\right)^2 s^2\{\overline{X}_1\} + \left(\frac{N_2}{N}\right)^2 s^2\{\overline{X}_2\} + \cdots + \left(\frac{N_k}{N}\right)^2 s^2\{\overline{X}_k\}$$

$$= \sum_{j=1}^{k} \left(\frac{N_j}{N}\right)^2 s^2\{\overline{X}_j\}$$

As usual, the finite population correction is omitted for any stratum for which the sample fraction n_j/N_j does not exceed 5 percent.

To complete the construction of the interval estimate of μ, we note that the sampling distribution of $(\overline{X}_{st} - \mu)/s\{\overline{X}_{st}\}$ is approximately normal when the sample size n is reasonably large. Thus, the confidence limits for μ have the usual large-sample form:

(14.12) The confidence limits for μ with approximate confidence coefficient $1 - \alpha$, when stratified random sampling is employed and the sample size is reasonably large, are:

$$\overline{X}_{st} \pm zs\{\overline{X}_{st}\}$$

where: $z = z(1 - \alpha/2)$
   $\overline{X}_{st}$ is given by (14.9)
   $s\{\overline{X}_{st}\}$ is given by (14.11)

☐   Example

For our teachers' association example, a 95 percent confidence interval for the mean number of years of education is desired. Table 14.3 contains the sample results and shows the calculation of $\overline{X}_{st}$. We find that $\overline{X}_{st} = 16.25$ years. This table also shows the calculation of $s^2\{\overline{X}_{st}\}$. Since the strata sizes N_j are large relative to the sample sizes n_j, we have ignored the finite population corrections and used s_j^2/n_j for calculating $s^2\{\overline{X}_j\}$. Since $s^2\{\overline{X}_{st}\} = .01545$, it follows that $s\{\overline{X}_{st}\} = \sqrt{.01545} = .1243$ year.

For a 95 percent confidence coefficient, we require $z(.975) = 1.960$. Hence, the 95 percent confidence limits for μ are $16.25 \pm 1.960(.1243)$ and the confidence interval is:

$$16.01 \leq \mu \leq 16.49$$

It can be asserted, with 95 percent confidence, that the mean number of years of education of the member teachers is between 16.0 and 16.5.   ☐

TABLE 14.3   *Stratified sample results*

j	Level of School	N_j	n_j	$\overline{X}_j$	s_j^2	$\left(\frac{N_j}{N}\right)\overline{X}_j$	$\left(\frac{N_j}{N}\right)^2\left(\frac{s_j^2}{n_j}\right)$
1	Elementary	25,000	150	14.8	6.40	7.40	.01067
2	High school	20,000	120	17.3	2.69	6.92	.00359
3	College	5,000	30	19.3	3.57	1.93	.00119
	Total	50,000	300			16.25	.01545
		↑ N	↑ n			↑ $\overline{X}_{st}$	↑ $s^2\{\overline{X}_{st}\}$

Comment

Stratified sampling can also be used to estimate a population proportion p. For example, an economically depressed district of a city might be stratified into groups of contiguous city blocks and a sample of households drawn from each stratum to estimate the proportion of households in the district with inadequate living space. The procedures for estimating the population proportion p are analogous to those for estimating the population mean μ.

Planning Strata Sample Sizes

An important design consideration in stratified sampling is the allocation of the total sample size to the individual strata.

Proportional Allocation.   Note from Table 14.3 that the sample size is the same proportion (.6 percent) of the stratum size for all strata. This method of allocation is called proportional allocation:

(14.13) With *proportional allocation*, the sample size for the jth stratum is:

$$n_j = \left(\frac{N_j}{N}\right)n$$

Proportional allocation is frequently used because it is simple and often quite effective.

Optimal Allocation.   This approach allocates the total sample size to the individual strata so as to minimize $\sigma^2\{\bar{X}_{st}\}$—i.e., so the estimate of μ is as precise as possible for the given total sample size. A statistical theorem provides the strata sample sizes for optimal allocation:

(14.14) With *optimal allocation*, the sample size for the jth stratum is:

$$n_j = \left(\frac{N_j\sigma_j}{\displaystyle\sum_{j=1}^{k} N_j\sigma_j}\right)n$$

where: σ_j is the standard deviation for the jth stratum

Note that (14.14) involves the strata standard deviations σ_j. Since these are generally unknown, planning values for the σ_j's are required. Approximate information about the order of magnitude of the σ_j's is usually adequate for planning purposes.

☐   Example

The analyst for the teachers' association considered optimal allocation for the stratified sample. Table 14.4 shows the planning values for the σ_j's, based on earlier surveys, together with the calculations for optimally allocating the total sample size $n = 300$ to the three strata. We see that the strata sample sizes for optimal allocation differ somewhat from those for proportional allocation in Table 14.3.

In the end, the analyst did not utilize the optimal allocation sample sizes. The reason

TABLE 14.4   *Optimal allocation of total sample size in stratified sampling*

j	Level of School	N_j	Planning Value of σ_j	$N_j\sigma_j$	$\dfrac{N_j\sigma_j}{\Sigma N_j\sigma_j}$	n_j
1	Elementary	25,000	2.5	62,500	.6098	183
2	High school	20,000	1.5	30,000	.2927	88
3	College	5,000	2.0	10,000	.0975	29
	Total	50,000		$\Sigma N_j\sigma_j = 102,500$	1.0000	300

was that the sample was intended to provide information about various other characteristics besides the mean number of years of education. It turned out that the optimal allocations for estimating other major characteristics differed markedly from the one here, because the relative magnitudes of the σ_j's for the other characteristics were not the same as for years of education. The analyst then decided that proportional allocation represented a reasonable compromise. ☐

14.3   CLUSTER SAMPLING

Single-Stage Cluster Sampling

Sometimes, it is efficient to sample population elements in groups rather than individually.

☐   Illustration

An inspector needs to examine a shipment to determine the extent of deterioration, if any, of the parts in the shipment. The shipment consists of 2,000 sealed cartons, each containing 5 parts. If a simple random sample of 100 parts is to be examined, it is conceivable that as many as 100 cartons may have to be unsealed. Since unsealed cartons are difficult to store and handle, and items in unsealed cartons may be stolen, the inspector wishes to open fewer cartons. His sampling procedure calls for the selection of 20 of the 2,000 cartons at random and inspection of all the parts in each selected carton. ☐

Definition.   The procedure just described illustrates single-stage cluster sampling. Each carton represents a group or cluster of elements, and a probability sample of these clusters is selected. All elements in the selected clusters are then included in the sample.

(14.15)   With *single-stage cluster sampling*, the population is divided into *clusters* of elements, a probability sample of clusters is selected, and all elements in the selected clusters are included in the sample.

Efficiency.   In general, cluster sampling requires a larger number of elements to yield a specified precision than simple random sampling. The reason is that the elements within a cluster (persons in a family, parts in a carton) tend to be more homogeneous than is the case in the population at large. Nevertheless, under some circumstances cost considerations make cluster sampling more efficient than simple random sampling:

1. When the cost of constructing a frame of elements is high, cluster sampling may be more efficient. For example, in a survey of eligible voters in a city, a current listing of city blocks is available but a current listing of eligible voters is not. Here, cluster sampling of city blocks has the advantage of requiring the preparation of voters' lists only for city blocks actually sampled. On the other hand, simple random sampling of eligible voters would require a voters' list for the entire city, which is an expensive undertaking.

2. Cluster sampling may also be efficient when it is more expensive to sample items scattered throughout the population than items that are "close" to one another. In our inspection example, it is much more costly to open 100 cartons to inspect 100 parts than to inspect the same number of parts in 20 cartons. As another example, a block sample of households in a large city requires substantially less travel time by interviewers than a random sample of households scattered throughout the city.

Multistage Cluster Sampling

In some applications of cluster sampling, it is desirable to sample in two or more stages.

☐   Illustration

In a statewide survey of urban households to obtain data on food expenditures, a probability sample of cities is first selected. Within the chosen cities, probability samples of city blocks are selected, and within the selected blocks, random samples of households are chosen.   ☐

Definition. The procedure just described illustrates a three-stage cluster sample. Cities are sampled in the first stage and are called primary sampling units. Blocks are sampled in the second stage and are called secondary sampling units. Households within blocks are sampled in a third stage and are called tertiary sampling units.

(14.16) With *multistage cluster sampling*, a probability sample of *primary sampling units* is selected. From each chosen primary sampling unit, a probability sample of *secondary sampling units* is selected, and so on, until the final stage of sampling.

Efficiency. Cluster sampling in more than one stage is efficient when the primary sampling units are large and heterogeneous, such as cities or counties. It is also efficient when the cost of frame construction is low for the primary sampling units and much higher for secondary and other subsequent sampling units. For instance, a frame of cities in a state is readily available, as are frames for blocks in most cities, but the cost of constructing a current frame of households in a state is very large. Also, when clusters at any stage of sampling are homogeneous, precision will be enhanced if the budget is devoted to the selection of more clusters rather than to the selection of more elements within a cluster. The reason is that selecting more elements in a cluster will not provide much more information than is provided by a few elements, in view of the homogeneity of the clusters. As an extreme example, if all persons in a household have the same preference among presidential candidates, a sample of one person from a household provides full information about that cluster.

Comments

1. In both single- and multistage cluster sampling, probability sampling procedures other than simple random sampling may be advantageous for selecting the sampling units at the different stages. In our food expenditures survey illustration, cities might be selected using probabilities proportional to their population sizes, and blocks within a city might be selected similarly.
2. When the clusters refer to geographical areas such as counties or city blocks, cluster sampling is often referred to as *area sampling*.
3. Cluster sampling can be used in conjunction with stratified sampling. For example, a statewide sample of urban households might involve in the first stage of sampling a stratified sample of cities. In the second stage, city blocks in a selected city might be sampled using stratification based on city districts.

14.4   SYSTEMATIC SAMPLING

Systematic selection of sample elements is widely employed because of its ease and convenience.

☐   Illustration

An auditor wishes to sample 200 sales receipts from a population of 10,000 receipts issued during the past quarter. The receipts are serialized and stacked in order of the serial numbers. To select the sample quickly, the auditor decides to select every $10,000/200 = 50$th receipt. He chooses one of the first 50 receipts at random (he happens to choose random number 39) and then selects every 50th receipt thereafter. Thus, the sample consists of the 39th, 89th, 139th, . . . , 9,989th receipts. The actual selection is extremely easy because of the serialized ordering of the population.   ☐

Definition

The sampling procedure described here is called systematic random sampling:

(14.17) With *systematic random sampling*, every kth element in the frame is selected for the sample, with the starting point among the first k elements determined at random.

If every kth element is selected, then $100/k$ percent of the population is sampled. The resulting sample is called a *$100/k$ percent systematic sample*. The auditor, for instance, selected a $100/50 = 2$ percent systematic sample.

Efficiency

Systematic sampling has the advantage of being simple to execute and hence is economical and convenient to use. A systematic random sample frequently will also provide more precise estimates than a simple random sample of the same size. This occurs when the frame listing in effect stratifies the population elements. For example, the sales receipts in our audit example are issued consecutively during the quarter. Thus, a systematic sample will automatically contain receipts issued at different times during the quarter. In essence, the auditor's systematic sample is stratified by time.

Comments

1. In using systematic sampling, one must avoid a sampling interval that corresponds to any natural period in the frame. Thus, in sampling average monthly stock prices for a 30-year period, one should not sample every 12th month, or some multiple thereof. The resulting price data would always correspond to the same month of the year and might be systematically different from the data for the other months.
2. Systematic sampling can be used in combination with other sampling procedures. For example, a stratified sample may involve systematic selection within each stratum.

14.5 OPTIONAL TOPIC—ACCEPTANCE SAMPLING

Acceptance sampling is an application of statistical testing to quality control. The choice is between accepting or rejecting a lot of items, such as an incoming shipment of material, an outgoing shipment of finished product, or a batch of clerical work. The lots constitute finite populations; hence, they could be given 100 percent inspection unless the testing is destructive. However, 100 percent inspection can be prohibitively expensive and usually would not detect all defects and errors anyway in view of inspection fatigue and other sources of error. Hence, sampling is frequently used to decide whether the lot should be accepted (H_0) or rejected (H_1). The latter decision may lead to 100 percent inspection of the items, a return of the lot to the vendor, complete verification of the batch of clerical work, or some other action of this nature.

Single Sampling Plans

In acceptance sampling, the quality of an item often is expressed as satisfactory or defective. A single sampling plan for such a quality characteristic specifies the random sample size and the action limit for the number of defective items in the sample.

☐ Example

A firm receives bushings machined to fine tolerances in lots of 10,000 from a supplier. The alternative conclusions for any particular lot are:

$$H_0: p \leq p_I = .04 \qquad \text{(Accept lot)}$$

$$H_1: p > p_I = .04 \qquad \text{(Reject lot)}$$

where p denotes the proportion of defective bushings in the lot and $p_I = .04$ denotes the maximum acceptable proportion of defectives. In acceptance sampling terminology, p_I is referred to as the *acceptable quality level,* or AQL.

The firm and the supplier have reached an agreement that the following single sampling plan, stated in condensed form in Table 14.5a, be employed:

Select a random sample of 125 bushings from the lot. If 10 or fewer bushings in the sample are defective, accept the lot; if 11 or more are defective, reject the lot.

The maximum number of defectives that leads to acceptance of the lot is 10 here. This value is called the *acceptance number* of the sampling plan. The minimum number of defectives leading to rejection, here 11, is called the *rejection number.* □

Theory. We can see immediately that this sampling plan is the equivalent of a decision rule for a sample proportion with an associated sample size of $n = 125$. Since $10/125 = .08$, the rule can be written:

$$\text{If } \bar{p} \leq .08, \text{ conclude } H_0 \qquad \text{(Accept lot)}$$

$$\text{If } \bar{p} > .08, \text{ conclude } H_1 \qquad \text{(Reject lot)}$$

Using the principles set out in Chapter 12, we can compute the probability of accepting the lot for different values of p. These probabilities, which we denote by $P(H_0; p)$, are plotted in Figure 14.1. The plot of $P(H_0; p)$ in Figure 14.1 is called the *operating characteristic curve* of the sampling plan. The operating characteris-

TABLE 14.5 *Matched single and multiple sampling plans*

(a)
Single sampling plan

Sample Size	Acceptance Number	Rejection Number
125	10	11

(b)
Multiple sampling plan

Sample	Sample Size	Cumulative Sample Size	Acceptance Number	Rejection Number
First	32	32	0	5
Second	32	64	3	8
Third	32	96	6	10
Fourth	32	128	8	13
Fifth	32	160	11	15
Sixth	32	192	14	17
Seventh	32	224	18	19

SOURCE: Military Standard 105D Tables (Ref. 14.1).

FIGURE 14.1 *Operating characteristic curve of sampling plan*

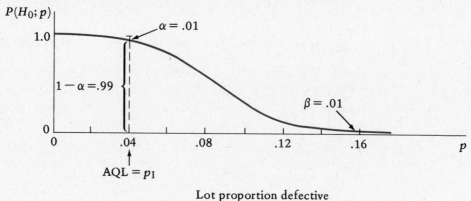

Lot proportion defective

tic curve is simply the complement of the power curve, as can be seen from (11.7). Since the α risk is the probability of rejecting an acceptable shipment, this risk is often called the *producer's* or *supplier's risk*. Figure 14.1 shows that this risk is approximately .01 when $p = p_\mathrm{I} = .04$.

On the other hand, the β risk is the probability of accepting an unacceptable shipment, and hence is often called the *buyer's* or *purchaser's risk*. Figure 14.1 shows, for example, that the β risk is approximately .01 when $p = .16$.

Multiple Sampling Plans

Multiple sampling plans involve sampling in two or more stages. At each stage a decision is made whether to accept or reject the lot or to continue sampling. Multiple sampling plans usually result in smaller total sample sizes than corresponding single sampling plans with the same α and β risks.

Example

Table 14.5b contains a multiple sampling plan for our bushings illustration which has approximately the same operating characteristic curve as the single sampling plan in Table 14.5a. With this sampling plan, an initial sample of 32 bushings is selected. If 0 bushings are defective (the acceptance number), the lot is accepted. If 5 (the rejection number) or more are defective, the lot is rejected. If the number of defectives is 1, 2, 3, or 4, sampling continues and a second sample of 32 bushings is selected. We now consider the *combined* sample of 64 items. If 3 or fewer bushings in the combined sample are defective, we accept the lot. If 8 or more are defective, we reject it. If the number of defectives is 4, 5, 6, or 7, a third sample of 32 is selected. This continues, if necessary, until the seventh sample of 32 is selected. At this stage, the lot is either accepted (18 or fewer defectives) or rejected (19 or more defectives).

It is intuitively clear that the total sample size depends on the lot proportion defective p. The multiple sampling plan will detect exceptionally good or bad lots at the first stage, with a resulting sample size of 32 compared with the sample size of 125 used in the single sampling plan. Lots of intermediate quality may require two or more samples before a final decision is reached. Only infrequently, however, will the combined sample size exceed 128—i.e., go beyond the fourth sample. □

Comments

1. When multiple sampling plans provide for sampling one item at each stage, so that an accept, reject, or continue decision is made with each sample observation, they are called *sequential sampling plans*.
2. Extensive tables of sampling plans have been published. One set is the *Military Standard 105D Tables* (Ref. 14.1) from which the sampling plans in Table 14.5 have been extracted. These tables enable the user to select readily a plan to fit the circumstances.

14.6   OPTIONAL TOPIC—CONTROL CHARTS

Control charts are a widely used application of statistical tests to the quality control of an ongoing process. Consider the production of a certain tablet by a pharmaceutical firm. The weight of the tablet is an important quality characteristic to be controlled. Management wishes to know when a change in the process mean weight takes place so that appropriate action can be taken. The use of control charts can be very helpful here, as we shall now see.

Constructing a Control Chart

To set up a control chart for tablet weight, information must be obtained about the process mean μ and standard deviation σ of the weight of tablets. Extensive past experience indicates that the distribution of tablet weights is approximately normal, with $\mu = .441$ gram and $\sigma = .008$ gram. These levels are considered satisfactory by management.

Random samples of $n = 5$ tablets will be taken from the process periodically and decisions made, based on the sample mean $\bar{X}$, whether or not the process mean has remained stable. The sampling distribution of $\bar{X}$ will be approximately normal here, even though the sample size is small, because of the approximate normality of the population. Furthermore, we know from (9.3) and (9.4) that the mean and standard deviation of $\bar{X}$ are $E\{\bar{X}\} = \mu = .441$ gram and $\sigma\{\bar{X}\} = \sigma/\sqrt{n} = .008/\sqrt{5} = .00358$ gram, respectively.

The test to be performed for each sample of $n = 5$ tablets from the process involves the following two alternatives:

(14.18)
$$H_0: \mu = \mu_I \quad \text{(Let process alone)}$$
$$H_1: \mu \neq \mu_I \quad \text{(Look for cause of change in } \mu)$$

Here, μ_I denotes the process mean during the past. We have $\mu_I = .441$ gram, as noted earlier.

It is customary in the United States and Canada to construct a decision rule using action limits set at three standard deviations above and below the process mean μ_I. Thus, the decision rule has the form:

(14.19)   If $\mu_I - 3\sigma\{\bar{X}\} \leq \bar{X} \leq \mu_I + 3\sigma\{\bar{X}\}$, conclude H_0 (Let process alone)

If $\bar{X} < \mu_I - 3\sigma\{\bar{X}\}$ or $\bar{X} > \mu_I + 3\sigma\{\bar{X}\}$,

conclude H_1 (Look for cause of change in μ)

For our example, the action limits are $.441 \pm 3(.00358)$, or .430 and .452 gram, respectively.

The test procedure just described is analogous to that summarized in Figure 11.10. Because the sampling distribution of $\bar{X}$ is approximately normal, we know that the probability of $\bar{X}$ lying outside the action limits $\mu_I \pm 3\sigma\{\bar{X}\}$ is negligibly small when $\mu = \mu_I$. Thus, the decision rule has a small α risk—i.e., it is unlikely to lead one to look for a cause of change in the process mean when, in fact, no change has occurred.

To facilitate the use of rule (14.19) by plant personnel, as well as for other reasons discussed shortly, the rule is usually presented graphically as a *control chart*. Figure 14.2 presents the control chart for our example. Note that the process mean μ_I and upper and lower action limits are plotted as horizontal lines. The vertical axis denotes sample mean weight of tablets; the horizontal axis represents time over which the periodic samples are taken. The upper and lower action limits are called the *upper control limit* (UCL) and *lower control limit* (LCL), respectively.

Using a Control Chart

Plotting Procedure.   Periodic samples—in this instance, of five tablets—are taken from the process, and a point representing $\bar{X}$ for each sample is plotted on the control chart. As long as the points stay within the control limits, the process is to be let alone. If a point falls outside the control limits, however, it is concluded that the process mean has changed as the result of an assignable cause that was not present previously. A search for the cause is then undertaken, on the basis of which appropriate remedial action is initiated. Points inside and outside the control limits indicate that the process is *in control* and *out of control*, respectively. For instance, in Figure 14.2 the process was out of control on Monday at 4 P.M.

Information Provided by Control Chart.   One important type of information furnished by the control chart is the approximate time when a process goes out of control. This often provides a valuable clue to the assignable cause. For instance, a process may have gone out of control at about the same time that raw material from a new supplier was put into the process. In another instance, a point may have fallen outside the control limits shortly after an operator adjusted his or her machine.

The periodic sampling of the process assures management that the process will not operate too long after it has gone out of control without this condition being detected. If inspection were undertaken only after the production of large batches

FIGURE 14.2 *Control chart for process mean weight of drug tablets*

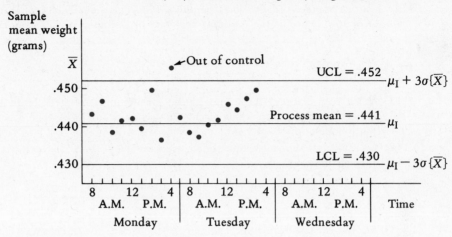

of items, the process might turn out many defective items before the condition is discovered.

The control chart furnishes other valuable clues in addition to indicating when a process is out of control. Through the plotting of sample results over a period of time, one can study trends or other changes in the quality of performance. For instance, Figure 14.2 indicates that there has been an upward trend in the mean weights of the Tuesday samples. This raises the possibility that the process soon might go out of control. Remedial action might therefore be taken even before the process actually goes out of control. Statistical tests should be used, of course, to decide whether the trend in sample means is a real one or simply a chance phenomenon. In the next chapter, we present a test that is useful for determining if runs of points upward or downward on a control chart are due to chance or not.

When an assignable cause of a process change has been discovered and appropriate action taken, changes in the process mean μ, process standard deviation σ, or both, frequently result. The control chart must then be revised to reflect the new conditions.

Comments

1. The quality of a product depends upon the process variability as well as the process mean. Should σ increase, for instance, the product will be more variable and a larger proportion of items will not meet specifications. Hence, it is important to control the process variability as well as the process mean. This is done by another control chart similar to the one described for the process mean. Control charts also can be constructed for the process proportion in cases where quality is measured in terms of the proportion of defectives or errors.
2. Our discussion of control charts has been in terms of industrial production processes. However, clerical work, accident experience, sales performance, and other such processes may be controlled successfully by this statistical approach.

Cited Reference

14.1 *Military Standard 105D Tables: Sampling Procedures and Tables for Inspection by Attributes.* Office of the Assistant Secretary of Defense, Washington, D.C., 1963.

PROBLEMS

*14.1 What is the ratio of the value of $s\{\bar{X}\}$ calculated without the finite population correction to the exact value calculated according to (14.1) when 3 percent of a finite population is sampled, i.e., when $n/N = .03$?

14.2 a. Calculate the finite population correction $\sqrt{1 - (n/N)}$ for each of the following three cases:

N:	1,000	10,000	100,000
n:	500	500	500

   b. Use the results in **a** to calculate $s\{\bar{X}\}$ for each of the three cases when $s^2 = 16$. Is there much of a change in $s\{\bar{X}\}$ when N goes from 1,000 to 10,000? From 10,000 to 100,000?

*14.3 A simple random sample of 50 bank branches was selected from a population of 400 branches to estimate the mean number of savings certificates issued per branch during the past month. The sample results were $\bar{X} = 200$ and $s = 15$. Construct a 95 percent confidence interval for the population mean.

14.4 An auditor selected a random sample of 100 expense claims from among the 500 filed by company executives during the first quarter of the year to ascertain the expense amount supported by acceptable receipts for each claim. The sample results were $\bar{X} = \$418$ and $s = \$180$. Construct a 90 percent confidence interval for the mean supported expenses per claim for all claims filed during the quarter.

14.5 Refer to Problem 14.3. Estimate the total number of savings certificates issued by all 400 branches during the past month, using a 95 percent confidence interval.

*14.6 What random sample size is needed to obtain a 95 percent confidence interval for the mean of a population containing 1,200 items if the interval is to have a half-width of 2.5? Assume $\sigma = 30$ is a suitable planning value for the population standard deviation.

14.7 Refer to Problem 14.4. The auditor is now planning to review the 650 expense claims filed by company executives in the second quarter of the year. What random sample size is required if the mean supported expenses per claim is to be estimated with a 90 percent confidence interval having a half-width of $25? Assume the sample standard deviation from the first quarter's examination, namely $180, is a suitable planning value for σ.

14.8 Refer to Problem 14.6. Recompute the required sample size assuming that the population is infinite. Does the assumption of an infinite population increase the required sample size substantially here? Comment.

*14.9 In a random sample of 60 garment manufacturers selected from the 300 garment manufacturers operating in a state, 33 were found to pay full-time employees more than the state's minimum wage rate. Construct a 90 percent confidence interval for the proportion of garment manufacturers in the state who pay employees more than the minimum wage rate.

14.10 A pharmaceutical company reports that in a recent survey of 250 physicians from among the 2,000 physicians practicing in one of the company's sales districts, 60 percent wrote more than 20 prescriptions per working day. Assuming that the surveyed physicians constitute a simple random sample, estimate by means of a 95 percent confidence interval the proportion of physicians in the district who write more than 20 prescriptions per working day.

*14.11 What random sample size is needed to estimate a population proportion p by means of a 99 percent confidence interval with a half-width of .05 when $N = 3,000$ and $p = .3$ is a suitable planning value?

14.12 A population of 900 firms is to be surveyed to estimate the proportion of firms that have major union contracts which come up for renewal in the next twelve months. What random sample size is needed to estimate the proportion within ±.05, with 95 percent confidence? Assume $p = .5$ is a suitable planning value.

*14.13 A stratified random sample of 700 households in a community was selected to estimate mean home improvement expenditures last year. The strata employed were geographic areas. The strata sizes and sample results were:

Stratum	N_j	n_j	$\overline{X}_j$	s_j
1	12,000	300	480	200
2	6,000	150	380	150
3	10,000	250	510	300
Total	28,000	700		

Obtain a 95 percent confidence interval for the mean home improvement expenditures per household in the community.

14.14 A dairy association with 9,000 member producers selected a stratified random sample of 400 members to estimate the mean herd size per producer for the entire association. The strata were set up according to the amount of milk shipped in the last calendar year, which was known for each member. The strata sizes and sample results on herd sizes follow.

Stratum	N_j	n_j	$\overline{X}_j$	s_j
1	4,500	170	15.7	8.3
2	2,700	90	25.1	7.7
3	1,800	140	58.2	18.0
Total	9,000	400		

Obtain a 99 percent confidence interval for the population mean herd size.

14.15 Refer to Problem 14.14. Suppose that the population standard deviation is $\sigma = 19.0$. How would the precision for a simple random sample of $n = 400$ members tend to compare with that obtained for the stratified random sample? What does this imply about the relative advantage of a stratified random sample over a simple random sample of the same size in this case?

14.16 Stratified random sampling is to be employed in each of the following cases. Describe a basis of stratification likely to be useful for each.
   a. A sample of personal checking accounts of a bank, to estimate the mean number of transactions per account for May.
   b. A sample of logs in a shipment, to estimate the mean dollar yield of lumber per log.

c. A sample of school bus drivers from all those licensed in Iowa, to estimate the mean number of years of driving experience per driver.

*14.17 A company wishes to estimate the mean number of days elapsed between the date an invoice is sent to a customer and the date payment is received. The study is to be confined to the 18,000 invoices issued in the last fiscal quarter. Three strata, according to the amount of the invoice, are to be employed. Strata sizes and the anticipated strata standard deviations are:

Stratum	N_j	Anticipated σ_j (days)
1	9,000	8
2	6,000	5
3	3,000	3
Total	18,000	

The budget permits a sample of 600 invoices to be examined.
a. What are the strata sample sizes with proportional allocation?
b. What are the strata sample sizes with optimal allocation?
c. Do the two methods of allocation lead to substantially different strata sample sizes here? Discuss.

14.18 Refer to Problem 14.14. If proportional allocation had been used in assigning the sample of 400 members to the strata, what would have been the strata sample sizes?

14.19 a. What information about the population strata is required for optimal allocation but not for proportional allocation?
b. Under what special circumstances would optimal allocation and proportional allocation lead to the same allocation of strata sample sizes, for a given total sample size?

14.20 Refer to Problem 14.17.
a. What is the anticipated standard deviation of $\bar{X}_{st}$ with proportional allocation? With optimal allocation?
b. Is optimal allocation expected to lead to a substantial gain in precision over proportional allocation here? Comment.
c. If the sample survey were also to estimate the mean delivery charge per invoice, would the strata sample sizes obtained in Problem 14.17b also be optimal for this purpose? Discuss.

14.21 Explain how cluster sampling might be employed in each of the following situations:
a. Sampling of peaches in a fruit orchard to estimate the percent of peaches that suffered frost damage.
b. Sampling of fifth-grade students throughout a state to estimate the mean reading achievement.
c. Sampling of prisoners in federal penal institutions to estimate the mean number of hours spent last month in formal education and training programs.

14.22 An auditor must check a company's inventory of raw materials and finished goods located in a large number of warehouses throughout the country. The auditor will perform the check by selecting a sample of inventory items, counting the actual inventory on hand for these items, and comparing these counts with company records. The total error in the book inventory is to be estimated.

      a. Explain how the auditor might employ cluster sampling in this investigation. Does your proposal involve single-stage or multistage cluster sampling?

      b. What advantages does cluster sampling offer here relative to simple random sampling? What disadvantages?

*14.23 A 4 percent systematic random sample is selected from a computer listing of 6,000 employee payroll records. What is the value of k here? How many records are selected for the sample in this case?

14.24 A population of 5,000 physicians specializing in plastic surgery is listed in order of age. A systematic random sample of every 50th physician was selected from this frame, to estimate the mean number of years of practice in this specialty. The analyst has evaluated the sample results as if the sample were a simple random one. Will the calculated precision so obtained likely overstate, understate, or be the same as the actual precision of the systematic random sample estimate? Explain.

*14.25 Consider the following single sampling plan for examining incoming lots:

Sample Size	Acceptance Number	Rejection Number
20	1	2

      a. What action should be taken if the sample contains 3 defectives?

      b. What action should be taken if the sample contains 1 defective?

14.26 Consider the following multiple sampling plan with two stages for examining incoming lots:

Sample	Sample Size	Cumulative Sample Size	Acceptance Number	Rejection Number
First	15	15	1	3
Second	15	30	3	4

      a. What action should be taken if the first sample of 15 items from a lot contains: (1) 0 defectives, (2) 3 defectives, (3) 2 defectives?

      b. What action should be taken if the first sample contains 2 defectives and the second contains 2 defectives?

*14.27 Samples of nine alloy filaments are taken daily from a production process and the melting point of each is determined. Extensive past experience has shown that the melting points of filaments produced by this process are normally distributed with $\mu = 335°C$ and $\sigma = 21°C$. The sample means for the 28 days of February are:

Day	$\bar{X}$	Day	$\bar{X}$	Day	$\bar{X}$	Day	$\bar{X}$	Day	$\bar{X}$	Day	$\bar{X}$	Day	$\bar{X}$
1	340	5	351	9	336	13	339	17	325	21	340	25	351
2	328	6	340	10	317	14	355	18	338	22	345	26	344
3	342	7	323	11	337	15	334	19	326	23	330	27	328
4	335	8	339	12	350	16	310	20	328	24	332	28	333

      a. Construct a control chart for the process mean melting point. What are the upper and lower control limits?

      b. Plot the data for February on your chart, identifying any points where the process was out of control with respect to the process mean melting point.

14.28 Refer to Problem 14.27.

    a. Under what conditions must the upper and lower control limits constructed in Problem 14.27a be changed?

    b. What process characteristic other than the mean would you also be interested in controlling with respect to filament melting points? Explain.

14.29 A credit card company sends bills to its credit card holders monthly. Extensive past experience has shown that 4 percent of the bills contain a billing error. Random samples of 300 billings are selected monthly and checked for billing errors. The numbers of billing errors found in the samples for the past 12 months are, chronologically: 10, 15, 10, 13, 13, 10, 11, 13, 14, 8, 9, 3.

    a. Construct a control chart for the process proportion of incorrect billings and plot the data for the past 12 months on your chart. What are the upper and lower control limits?

    b. Was billing quality in control during the last 12 months? Has experience during this period exhibited any noteworthy patterns? Discuss.

EXERCISES

14.30 In estimating a population mean from a stratified random sample, why is the estimated variance (14.11) not applicable when the strata samples are not drawn independently? Explain.

14.31 An observer has stated that if the strata standard deviations σ_j are the same for all strata, stratified random sampling offers no greater precision than simple random sampling for the same sample size. Do you agree? Discuss. (*Hint:* If the strata standard deviations are all equal, must the population standard deviation also be the same?)

STUDIES

14.32 Refer to Appendix D.

    a. Select a stratified random sample of 64 firms from the population of all firms. Use industry groupings as strata and employ proportional allocation.

    b. Estimate the mean 1981 net income per firm for the population using a 95 percent confidence interval. Use the finite population correction in your calculations. Interpret your interval estimate.

    c. How would a 16.67 percent systematic random sample here be selected? Why might this systematic sample be considered, for purposes of analysis, as approximately equivalent to the stratified sample selected in **a**?

14.33 Refer to Problem 14.32. Do this problem using the data for 1980 net income.

14.34 Refer to Problem 14.25.

    a. Suppose the acceptable quality level (AQL) for a lot is $p_{\mathrm{I}} = .05$. What is the probability that the plan will accept a very large lot which contains 5 percent defectives? (*Hint:* Use Table C-5.) What is the supplier's risk in this case?

    b. What is the buyer's risk (i.e., the β risk) in using this plan when a very large lot contains 10 percent defectives?

14.35 Refer to Problem 14.26.

    a. If a large incoming lot contains 10 percent defectives, what is the probability that it will be accepted by the sampling plan in the first stage? In either stage? (*Hint:* Use Table C–5.)

    b. If a large incoming lot contains 10 percent defectives, what is the expected total sample size needed to reach an accept-reject decision? (*Hint:* The total sample size will be either 15 or 30.)

14.36 A company receives raw material in lots of 8,000 units. A random sample of 50 units is selected from each lot, on the basis of which a decision is made to accept (H_0) or reject (H_1) the lot. A rejected lot is inspected 100 percent, all defective units are replaced with acceptable ones, and the supplier is charged for the work and defective materials. Suppose the decision rule used has the following properties:

Lot Proportion Defective p	$P(H_0; p)$	$P(H_1; p)$
.01	.91	.09
.03	.56	.44
.10	.03	.97

    a. What is the average quality of lots after inspection if the incoming lot proportion defective is $p = .01$? .03? .10? Ignore defective units found in samples from accepted lots.

    b. Explain why the average quality of lots after inspection is better if the lot proportion defective is .10 than if it is .03. Is it therefore desirable for the supplier to submit poor-quality lots? Explain.

    c. What is the average number of units inspected per lot if the incoming lot proportion defective is $p = .01$? .03? .10?

UNIT FOUR

———◆———

Estimation
and
Testing—II

15

Nonparametric Procedures

In the estimation and testing procedures considered thus far, heavy reliance has been placed either on the assumption that the underlying population is of a particular form (e.g., normal) or, when little could be assumed about the nature of the population, on large sample sizes and the central limit theorem. In this chapter, we discuss some procedures that require only minimal assumptions about the population and can be used for small as well as large sample sizes. These procedures are called *nonparametric* or *distribution-free* procedures, and they are typically very simple to implement.

15.1 SINGLE POPULATION STUDIES

We first discuss nonparametric procedures that are concerned with the median of a single population. The estimation of the population median is important in many applications, for two major reasons. First, when the population is highly skewed (e.g., in studies of family incomes, store sales, and manufacturers' inventories), the population median is located more in the center of the distribution than the population mean and thus may be a more meaningful measure of location. Second, the population median is an equally meaningful measure as the mean for symmetrical populations because the two parameters coincide in this case.

Assumptions

The inference procedures for the population median which we now consider require only the following two assumptions:

1. The population is continuous.
2. The sample is a random one.

Recall that a random variable for a continuous population can take on any value in an interval. Figure 5.4 (p. 129) illustrates two continuous populations.

Point Estimation of Population Median

We denote the population median by η (Greek eta). We shall use the sample median Md as defined in (3.8) as the point estimator of η. Recall that the sample median Md is the middle value when all items in the sample are arrayed in ascending or descending order. When the sample size is even, Md is taken to be the mean of the middle two values.

The choice of the sample median Md as a point estimator of η is appropriate for several reasons:

1. For a symmetrical population, it can be shown that Md is an unbiased estimator of η.
2. For any continuous population, whether symmetrical or not, the sample median, when n is odd, has an equal chance of lying above η or below it. In other words, the median of the sampling distribution of Md equals η. The same result holds approximately when n is even. The reason why the result is not exact when n is even is our convention of taking Md as the mean of the middle two values in this case.
3. The sample median is a consistent estimator of η when the population is continuous, but it is not always the most efficient.

Example

An administrator is studying how long it takes to settle claims made under a workers' compensation insurance plan for injuries received from on-the-job accidents. For one category of claims, case durations for a random sample of nine settled cases are as follows:

Case:	1	2	3	4	5	6	7	8	9
Case duration (*months*):	4.5	5.8	2.6	4.3	2.7	10.5	5.6	1.9	4.1

Case duration refers to the length of time from the occurrence of the accident to the closing date for the case. We are to estimate the median duration η for claims in this category.

Arraying the nine case durations in ascending order, it is seen that the sample median is $Md = 4.3$ months:

$$1.9 \quad 2.6 \quad 2.7 \quad 4.1 \quad \boxed{4.3} \quad 4.5 \quad 5.6 \quad 5.8 \quad 10.5$$

Therefore, $Md = 4.3$ months is a point estimate of η for the population of case durations under study here.

Interval Estimation of Population Median

A confidence interval for the population median η is very easy to construct. Consider a random sample of size $n = 2$, and denote the two sample observations by X_1 and X_2 as usual. Suppose that X_1 is the smaller of the two observations. A confidence interval for η, with a 50 percent confidence coefficient, is then simply:

$$X_1 \leq \eta \leq X_2$$

The reason why the confidence coefficient is 50 percent here is as follows: For any continuous population, the probability that an observation X_i falls below the median η is $P(X_i < \eta) = .5$, and the probability that it falls above it is $P(X_i > \eta) = .5$. Figure 15.1 illustrates this. Remember that for a continuous population, we have $P(X_i = \eta) = 0$.

Now our confidence interval $X_1 \leq \eta \leq X_2$ will be correct if $X_1 \leq \eta$ and if $X_2 \geq \eta$. Alternatively, it will be incorrect if both X_1 and X_2 fall below η or if both fall above η. Because of the independence of the observations in a random sample, these latter two probabilities are:

$$P[(X_1 < \eta) \cap (X_2 < \eta)] = .5(.5) = .25$$

$$P[(X_1 > \eta) \cap (X_2 > \eta)] = .5(.5) = .25$$

But these are mutually exclusive occurrences. Hence, by the complementation theorem (4.19), we obtain:

$$P(X_1 \leq \eta \leq X_2) = 1 - (.25 + .25) = .50$$

Clearly, if the sample size is larger, the use of the smallest and largest sample observations as confidence limits for η will involve a higher confidence coefficient. Indeed, we do not need to use the two extreme sample outcomes. Let us denote the rth smallest sample outcome by L_r and the rth largest by U_r. Then we can use L_r and U_r as confidence limits for the population median η:

(15.1) A confidence interval for the population median η, when the population is continuous, is:

$$L_r \leq \eta \leq U_r$$

where: confidence coefficient $1 - \alpha$ is given in Table C-10

Small Sample. Table C-10 provides the confidence coefficient $1 - \alpha$ for different L_r and U_r, for sample sizes up to 15. We now illustrate the use of this table:

1. If $n = 10$, use of the confidence limits L_1 and U_1 (i.e., the two extreme sample observations) involves a confidence coefficient of .998.
2. If $n = 11$, use of the confidence limits L_2 and U_2 (i.e., the second smallest and second largest sample observations) involves a confidence coefficient of .988.

FIGURE 15.1 *Position of population median in a continuous population*

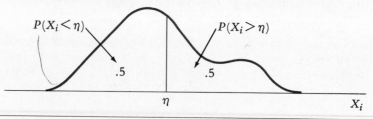

3. If $n = 15$, use of the confidence limits L_4 and U_4 (i.e., the fourth smallest and fourth largest sample observations) involves a confidence coefficient of .965.

☐ Examples

1. In our case duration study, a confidence interval for η is desired with a 95 percent confidence coefficient. Table C–10 shows that for $n = 9$, the closest we can come to 95 percent is by use of $r = 2$, which involves a 96.1 percent confidence coefficient. Hence, we require L_2 and U_2. We see from the sample results presented earlier that $L_2 = 2.6$ and $U_2 = 5.8$. Hence, our confidence interval is:

$$2.6 \leq \eta \leq 5.8$$

Thus, we can conclude, with 96.1 percent confidence, that the median case duration is between 2.6 and 5.8 months.

2. A random sample of $n = 8$ households showed the following family incomes (in $ thousands) in ascending order:

$$5.9 \quad 6.4 \quad 8.3 \quad 11.2 \quad 14.9 \quad 16.3 \quad 17.6 \quad 31.7$$

A 90 percent confidence interval is desired for the population median family income. We see from Table C–10 that for $n = 8$, $r = 2$ involves a 93.0 percent confidence coefficient, which exceeds the desired 90 percent level just a little. Hence, we use $L_2 = 6.4$ and $U_2 = 17.6$ as confidence limits, and the 93 percent confidence interval for the population median family income is:

$$6.4 \leq \eta \leq 17.6$$

The limits are wide here because of the very small sample size and the large variability in the observations. ☐

Sample Size Not Small.   For sample sizes larger than 15, the value of r required for the confidence interval $L_r \leq \eta \leq U_r$ can be obtained by an approximation formula:

(15.2) When the sample size n exceeds 15, for a desired $1 - \alpha$ confidence coefficient for the confidence interval $L_r \leq \eta \leq U_r$, select the largest integer r that does not exceed:

$$.5[n + 1 - z(1 - \alpha/2) \sqrt{n}]$$

☐ Examples

1. A random sample of 43 tax returns has been selected. It is desired to construct a 95 percent confidence interval for the population median amount of interest income per return. Since $1 - \alpha = .95$, we require $z(1 - \alpha/2) = z(.975) = 1.960$ and we obtain:

$$.5(43 + 1 - 1.960 \sqrt{43}) = 15.57$$

The largest integer not exceeding 15.57 is 15. Hence, $r = 15$, and the required confidence limits for η are the 15th smallest (L_{15}) and the 15th largest (U_{15}) interest incomes in the sample of 43 returns.

2. A random sample of 75 vouchers has been selected. It is desired to construct a 90

percent confidence interval for the population median amount expended per voucher. Since $1 - \alpha = .90$, we require $z(.95) = 1.645$ and we obtain:

$$.5(75 + 1 - 1.645\sqrt{75}) = 30.88$$

The largest integer not exceeding 30.88 is 30. Hence, $r = 30$, and we use L_{30} and U_{30} (the 30th smallest and 30th largest voucher amounts) for the confidence limits for η.

□

Evaluation of Confidence Coefficient. We explain now how the confidence coefficients in Table C–10 are obtained. We return to our case duration study where a random sample of $n = 9$ cases was studied. Suppose we wish to use L_2 and U_2 as the confidence limits for η. In other words, we plan to use the second smallest and second largest sample durations for the confidence limits.

Figure 15.2 shows schematically seven different possible locations for the nine sample observations. Note that the confidence interval $L_2 \leq \eta \leq U_2$ is correct whenever at least two observations fall below η and at least two observations fall above it (cases 1, 2, and 3). The confidence interval is incorrect when none or only one observation is above η (cases 4 and 5) and when eight or nine observations are above η (cases 6 and 7). (Remember that for a continuous population, the probability is zero that an observation equals the population median.)

Thus, the confidence interval $L_2 \leq \eta \leq U_2$ is correct here whenever at least two but not more than seven observations fall above η. Let B denote the number of sample observations above the median. We can then express the probability that the confidence interval will be correct as follows:

$$P(2 \leq B \leq 7)$$

FIGURE 15.2 *Several confidence intervals of the form* $L_2 \leq \eta \leq U_2$ *for a sample of size* $n = 9$. The confidence interval is correct whenever at least two but no more than seven observations fall above η.

Case	Population median η		Interval spans η	Number of observations above η
1	$\_L_2$	$_____U_2\_\_$	Yes	7
2	$\_L_2_____$	$U_2\_$	Yes	2
3	$\_\_L_2\_\_\_\_$	$\_\_U_2\_$	Yes	4
4	$L_2_____U_2\_$		No	0
5	$\_L_2_____U_2\_$		No	1
6	$\_$	$L_2_____U_2\_$	No	8
7		$\_L_2_____U_2\_$	No	9

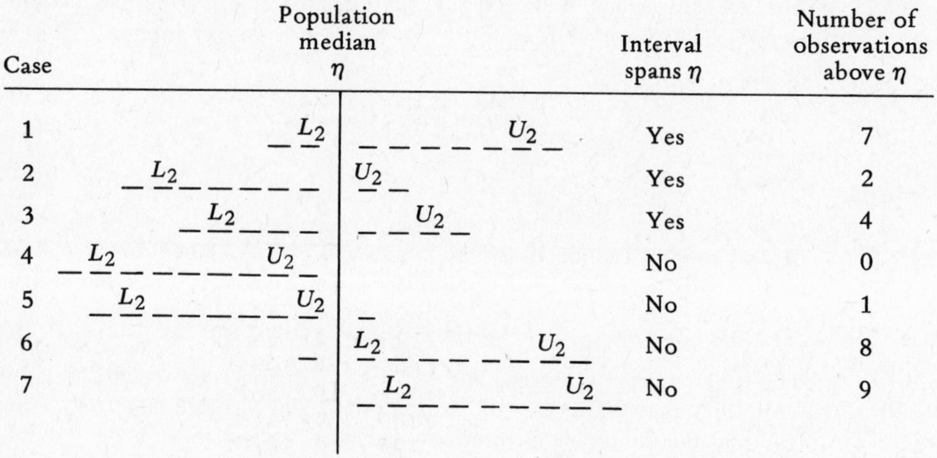

But B is a binomial random variable with $p = .5$ because: (1) for each observation X_i, $P(X_i > \eta) = .5$ since the population is continuous, and (2) the sample observations are independent because a random sample is assumed. Hence, we have by (6.3) for our example where $n = 9$:

$$P(2 \leq B \leq 7) = \sum_{i=2}^{7} \binom{9}{i}(.5)^i(.5)^{9-i}$$

Consulting the binomial tables (Table C-5) for $n = 9$ and $p = .50$, reproduced in Figure 15.3, we add the probability values associated with outcomes $2, 3, \ldots, 7$ and obtain $.0703 + .1641 + \cdots + .0703 = .961$. This is the confidence coefficient shown in Table C-10 for $n = 9$ and $r = 2$.

In general, the confidence interval $L_r \leq \eta \leq U_r$ will be correct if at least r and not more than $n - r$ out of the n sample observations fall above the median. Hence, the confidence coefficient is equivalent to the probability $P(r \leq B \leq n - r)$:

(15.3) The confidence interval $L_r \leq \eta \leq U_r$ for the population median η has a confidence coefficient given by:

$$P(r \leq B \leq n - r) = \sum_{i=r}^{n-r} \binom{n}{i}(.5)^i(.5)^{n-i}$$

where: B is the number of sample observations above η

Comments

1. In general, the confidence limits L_r and U_r will not be equidistant from the sample median Md. Recall that in our case duration example, we had $Md = 4.3$, $L_2 = 2.6$, and $U_2 = 5.8$.

FIGURE 15.3 *Fragment of binomial table showing probability that L_2 and U_2 will span η when $n = 9$*

$$p = .50$$

		0164	0.0312
		0017	0.0039
n	x		
9	0	C46	0.0020
	1	139	0.0176
	2	10	0.0703
	3	19	0.1641
	4	500	0.2461
	5	128	0.2461
	6	160	0.1641
	7	0407	0.0703
	8	0083	0.0176
	9	0008	0.0020

.961 ← (arrow pointing to boxed values 0.0703 through 0.0703)

2. The procedure just described does not usually enable us to achieve exactly a desired confidence coefficient such as 90 or 95 percent. The reason is that the binomial probability distribution, which determines the confidence coefficient, is discrete. However, we can usually get quite close to the desired confidence coefficient when the sample size is not exceedingly small.

3. Formula (15.2) for r when n is not small is obtained from the normal approximation to the binomial probability in (15.3), using the correction for continuity discussed in Chapter 12.

4. The rth smallest sample value, L_r, is often referred to as the *rth order statistic* of the sample.

5. Confidence intervals for other population percentiles, such as the 25th or 75th percentiles, can be obtained by a procedure similar to the one discussed here for the median.

Test for Population Median via Confidence Interval

Confidence interval (15.1) for η can be used in the usual way to test the following alternatives:

$$(15.4) \qquad H_0: \eta = \eta_I$$

$$H_1: \eta \neq \eta_I$$

Here, η_I denotes the value of the population median at which the α risk is to be controlled. To control this risk at α, the confidence interval must have a confidence coefficient of $1 - \alpha$. If η_I falls within the confidence interval, we conclude H_0; otherwise, we conclude H_1.

☐ Example

In our case duration study, the administrator wishes to test whether or not the median case duration for the current process is $\eta = \eta_I = 6$ months (6 months being the historical median). Our 96.1 percent confidence interval for η is $2.6 \leq \eta \leq 5.8$ months. Since $\eta_I = 6$ months lies outside this interval, H_1 must be concluded—that the median case duration has changed from its historical level. The α risk implicit in the administrator's test is $\alpha = 1 - .961 = .039$. ☐

Sign Test for Population Median

The alternatives in (15.4) can also be tested by a statistical test called the *sign test*. This test is equivalent to using a confidence interval for η for testing these alternatives.

The sign test is based on the number of sample observations that fall above η_I. Let p denote the probability that an observation falls above η_I, i.e., $p = P(X_i > \eta_I)$. Then an equivalent way of expressing the alternatives in (15.4) is:

$$(15.5) \qquad H_0: p = P(X_i > \eta_I) = .5$$

$$H_1: p = P(X_i > \eta_I) \neq .5$$

Observe from Figure 15.1 that if η_I is below η, then $p = P(X_i > \eta_I) > .5$, and if η_I is above η, then $p = P(X_i > \eta_I) < .5$. Only when η_I coincides with η is $p = .5$.

We shall let B denote the number of sample observations larger than η_I and

$\bar{p} = B/n$ the proportion of sample observations larger than η_I. If H_0 holds so $\eta = \eta_I$, then B and $\bar{p}$ follow the binomial probability distribution with $p = .5$.

Large Sample. For large n, the sign test is nothing more than a test for a population proportion as described in Chapter 12, where $p = .5$ is specified in H_0.

☐ Example

A random sample of $n = 250$ travelers checks cashed in the past year was selected and the length of time outstanding (time from issue to redemption) was obtained for each check. In the past, the median time outstanding has been $\eta_I = 2.1$ months. Management would like to test:

$$H_0: \eta = \eta_I = 2.1$$

$$H_1: \eta \neq \eta_I = 2.1$$

The α risk is to be controlled at .05. The sample data, in part, are:

Check i	Months Outstanding X_i	$X_i - \eta_I$	Sign
1	1.6	$-\ .5$	$-$
2	4.2	$+\ 2.1$	$+$
$\vdots$	$\vdots$	$\vdots$	$\vdots$
250	12.5	$+10.4$	$+$

We can obtain B, the number of observations larger than η_I, by taking the differences $X_i - \eta_I$ and counting the number of plus signs (hence, the name sign test). In this sample, it was found that $B = 89$ and hence $\bar{p} = B/n = 89/250 = .356$.

We now follow the procedure in Figure 15.4. The appropriate decision rule is two-sided, as shown in Figure 15.4a. The sampling distribution of $\bar{p}$ when $p = p_I = .5$ is shown in Figure 15.4b. It is approximately normal because of the large sample size. The standard deviation of $\bar{p}$ when $p = p_I = .5$ is:

$$\sigma\{\bar{p}\} = \sqrt{\frac{.5(.5)}{250}} = .032$$

For $\alpha = .05$, we require $z(.975) = 1.960$, so the action limits, as illustrated in Figures 15.4c and 15.4d, are $.5 \pm 1.960(.032)$. The decision rule then is:

If $.437 \leq \bar{p} \leq .563$, conclude H_0

If $\bar{p} < .437$ or $\bar{p} > .563$, conclude H_1

For our sample, $\bar{p} = .356$. Our decision rule therefore leads to conclusion H_1—that the population median duration outstanding for travelers checks is no longer 2.1 months.

Equivalently, the standardized test statistic $z^* = (.356 - .5)/.032 = -4.50$ could have been employed. Since $|z^*| = 4.50 > 1.960$, conclusion H_1 is indicated. ☐

Small Sample. When n is small, the decision rule for the sign test is set up by using the exact sampling distribution (i.e., the binomial distribution) and the statistic B.

FIGURE 15.4 *Construction of decision rule for sign test—Large sample, two-sided test for η*

(a) *Appropriate type of decision rule*

Conclude H_1 $(\eta \neq \eta_{\mathrm{I}})$	Conclude H_0 $(\eta = \eta_{\mathrm{I}})$	Conclude H_1 $(\eta \neq \eta_{\mathrm{I}})$

$A_1 \quad p_{\mathrm{I}} = .5 \quad A_2 \qquad \bar{p}$

$\bar{p}$ values differing substantially from .5 imply H_1 is correct.

(b) *Sampling distribution of $\bar{p}$ located at $p_{\mathrm{I}} = .5$*

$p_{\mathrm{I}} = .5 \qquad \bar{p}$

Distribution of $\bar{p}$ centered at $p_{\mathrm{I}} = .5$ since α risk is controlled there.

(c) *Placement of action limits*

$\alpha/2 \qquad 1 - \alpha \qquad \alpha/2$

$A_1 \quad p_{\mathrm{I}} = .5 \quad A_2 \qquad \bar{p}$

$z(\alpha/2) \quad 0 \quad z(1 - \alpha/2) \qquad z$

A_1 and A_2 are placed at equal distance from $p_{\mathrm{I}} = .5$ to give tail areas of $\alpha/2$ each.

(d) *Calculation of action limits*

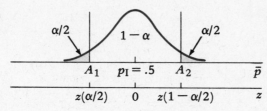

$$A_1 = .5 + z(\alpha/2) \sqrt{\frac{.5(.5)}{n}}$$

$$A_2 = .5 + z(1 - \alpha/2) \sqrt{\frac{.5(.5)}{n}}$$

$\sigma\{\bar{p}\}$ is known for sampling distribution when $p = .5$.

(e) *Decision rule*

If $A_1 \leq \bar{p} \leq A_2$, conclude H_0

If $\bar{p} < A_1$ or $\bar{p} > A_2$, conclude H_1

Conclude appropriate alternative based on $\bar{p}$.

☐ Example

In our case duration study, we wish to test whether or not $\eta = \eta_{\mathrm{I}} = 6$ months. The α risk is to be as close as possible to .05 without exceeding it. Figure 15.5a shows the appropriate type of decision rule. The sampling distribution of B and the placement of the action limits are illustrated in Figure 15.5b. Since $\alpha = .05$, the probability to the left of action limit A_1 in Figure 15.5b must not exceed $\alpha/2 = .025$. Similarly, the probability to the right of A_2 must not exceed .025. Referring to the binomial distribution when $p = p_{\mathrm{I}} = .5$ and $n = 9$, reproduced in Figure 15.3, we see that the lower tail must consist of outcomes 0 and 1 so that the probability will not exceed .025. By symmetry of the binomial distribution when $p = .5$, we know that the upper tail must then consist of outcomes 8 and 9. Hence, the action limits A_1 and A_2 are located as shown in Figure 15.5b and the decision rule is:

$$\text{If } 2 \leq B \leq 7, \text{ conclude } H_0$$

$$\text{If } B < 2 \text{ or } B > 7, \text{ conclude } H_1$$

FIGURE 15.5   *Construction of decision rule for sign test—Small sample, two-sided test for η*

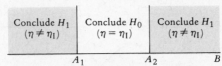

(a) *Appropriate type of decision rule*

Conclude H_1 ($\eta \neq \eta_I$)	Conclude H_0 ($\eta = \eta_I$)	Conclude H_1 ($\eta \neq \eta_I$)

A_1 A_2 B

B values close to 0 or n imply H_1 is correct.

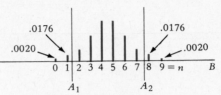

(b) *Sampling distribution of B when H_0 holds, and placement of action limits*

.0176   .0176

.0020   .0020

0   1 | 2   3   4   5   6   7 | 8   9 = n B

A_1 A_2

Distribution of B is binomial with $p = p_1 = .5$ since α risk is controlled there.

A_1 and A_2 are symmetrically placed so probability in each tail does not exceed $\alpha/2$.

In our example, $B = 1$ because one case duration exceeded $\eta_I = 6$ (see p. 373). Hence, we conclude H_1, just as we did in the test via the confidence interval.

Note that the exact α risk here is $2(.0020 + .0176) = .039$. Also note that $A_1 = 2 = r$ and $A_2 = 7 = n - r$. This shows the equivalence of the sign test and the test via the confidence interval. Recall that the confidence interval had confidence coefficient $1 - \alpha = 1 - .039 = .961$ and that a correct interval is produced when B is between 2 and 7 inclusive.

Comments

1. The statistic used in the sign test can be the number of observations *below* η_I rather than the number *above*. Either statistic gives the same result.
2. In theory, sample values should not exactly equal η_I because the population is assumed to be continuous. In practice, zero differences $(X_i - \eta_I)$ may occur occasionally because of rounding. It is then conventional to disregard the zero differences for the sign test and reduce the sample size accordingly.

15.2   COMPARATIVE STUDIES OF TWO POPULATIONS—INDEPENDENT SAMPLES

In this section, we describe a nonparametric procedure for comparing two populations on the basis of independent random samples. This procedure involves a test of whether two population distributions have different locations, as illustrated in Figure 15.6.

Illustration

An instructor, wishing to compare the effectiveness of two different pedagogical approaches to the introductory marketing course, randomly split the total enrollment of 24 students into two equal course sections. One section was given live lectures by the instructor, while the other viewed video-taped lectures prepared by the same instructor. In other respects, the two course sections were handled in identical fashion. The instructor now

FIGURE 15.6 *Two population distributions of identical shape differing in location by amount δ*

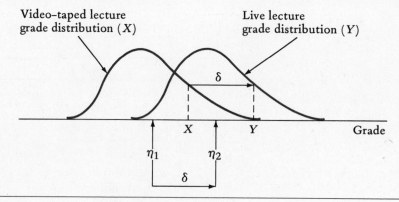

wants to test whether or not the final grade distributions for students exposed to video-taped lectures (X population) and to live lectures (Y population) are located at the same position. He believes that the type of instruction should not affect the shape of the grade distribution but could affect its location. Figure 15.6 illustrates the situation for the case when live lectures lead to higher grades. □

Shift Parameter

The two distributions for X and Y in Figure 15.6 are identical in shape but differ in location. We denote the difference in location by δ (Greek delta) and call it the *shift parameter*. If $\delta > 0$, as in Figure 15.6, the distribution for Y lies a distance δ to the right of the distribution for X. If $\delta < 0$, the distribution for Y lies to the left of the distribution for X. Finally, if $\delta = 0$, the two distributions are at the same location and hence are identical.

Mann-Whitney-Wilcoxon (M-W-W) Test

The Mann-Whitney-Wilcoxon test (the M-W-W test, for short) is a widely used nonparametric procedure for determining whether or not two population distributions of the same shape differ in location. In our subsequent discussion, populations 1 and 2 correspond to the X and Y populations, respectively.

Assumptions

1. The two population distributions are continuous and have the same shape, but may differ by the distance δ in location.
2. Independent random samples of sizes n_1 and n_2 are drawn from the two populations.

Note that assumption 1 does *not* specify the shape of the distributions.

Alternatives. The M-W-W test can be used for testing two-sided or one-sided alternatives concerning the shift parameter δ:

(15.6)
$$H_0: \delta = 0 \qquad H_0: \delta \leq 0 \qquad H_0: \delta \geq 0$$
$$H_1: \delta \neq 0 \qquad H_1: \delta > 0 \qquad H_1: \delta < 0$$

Since the two population distributions are assumed to be identical in shape, the shift parameter δ measures the distance between the two population means, $\mu_2 - \mu_1$, or the distance between the two population medians, $\eta_2 - \eta_1$. Figure 15.6 illustrates this for the two population medians. Hence, the tests on δ may also be viewed as tests on $\mu_2 - \mu_1$ or on $\eta_2 - \eta_1$.

Test Statistic. The test statistic for the M-W-W test is simple. We combine the n_1 sample observations from the X population and the n_2 sample observations from the Y population, and array them in ascending order. Then we assign ranks to the observations (starting with 1 for the smallest observation) and sum the ranks for the n_2 sample observations from the Y population. This sum is denoted by S_2.

☐ **Example**

The following are the sample results when $n_1 = 2$ and $n_2 = 3$ sample observations were selected from two populations, respectively.

Sample from X Population		Sample from Y Population		
$n_1 = 2$		$n_2 = 3$		
$X_1 = 4$ $\quad$ $X_2 = 12$		$Y_1 = 14$	$Y_2 = 10$	$Y_3 = 17$

The array and ranking of the combined sample are:

Sample value:	4	10	12	14	17
Rank:	1	2	3	4	5
Population:	X	Y	X	Y	Y

The ranks shown in boxes are for the three sample observations from the Y population. We see that the test statistic is:

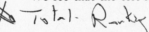 Total Ranks

$$S_2 = 2 + 4 + 5 = 11 \qquad\qquad ☐$$

Since there were $n_2 = 3$ sample observations from the Y population in our example, the smallest value that S_2 could have assumed is the sum of the first three ranks, namely, $1 + 2 + 3 = 6$. The largest possible value of S_2 is the sum of the last three ranks, namely, $3 + 4 + 5 = 12$.

In general, the smallest possible value of S_2 is the sum of the first n_2 ranks, i.e., $1 + 2 + \cdots + n_2$, which can be shown to equal $n_2(n_2 + 1)/2$. The largest possible value of S_2 is the sum of the last n_2 ranks, i.e., $(n_1 + 1) + (n_1 + 2) + \cdots + (n_1 + n_2)$. This sum equals $n_1 n_2 + [n_2(n_2 + 1)/2]$. For our example, $n_1 = 2$ and $n_2 = 3$, so:

$$\frac{n_2(n_2 + 1)}{2} = \frac{3(3 + 1)}{2} \qquad\qquad n_1 n_2 + \frac{n_2(n_2 + 1)}{2} = 2(3) + \frac{3(3 + 1)}{2}$$
$$= 6 \qquad\qquad\qquad\qquad\qquad\qquad = 12$$

If the Y distribution lies to the right of the X distribution (so $\delta > 0$ as in Figure 15.6), the Y's will tend to be larger than the X's and, consequently, S_2 will be close to its largest value. Conversely, if the Y distribution is to the left of the X distribution (so $\delta < 0$), the Y's will tend to be smaller than the X's and S_2 will tend to be close to its smallest possible value. Finally, if the X and Y distributions are identical (so $\delta = 0$), all possible orderings of the combined samples will be equally likely and S_2 will tend to have a value near midway between the two extremes.

Thus, the appropriate type of decision rule for each set of alternatives in (15.6) is intuitively clear. To test, for example:

$$H_0: \delta = 0$$

$$H_1: \delta \neq 0$$

very small and very large values of S_2 will lead to H_1, and intermediate values will lead to H_0.

Sampling Distribution of S_2 When $\delta = 0$. To control the α risk when $\delta = 0$, we need to know the sampling distribution of S_2 when $\delta = 0$. As we mentioned earlier, all possible orderings of the combined $n_1 + n_2$ observations are equally likely when $\delta = 0$, and the exact sampling distribution of S_2 can be worked out. It depends only on the sample sizes n_1 and n_2. Tables of the exact sampling distribution of S_2 are available but are not needed unless n_1 and n_2 are very small.

(15.7) When independent random samples of n_1 and n_2 are selected from two identical populations (i.e., identical shape and $\delta = 0$), the sampling distribution of S_2 has mean and variance:

$$\mathrm{E}\{S_2\} = \frac{n_2(n_1 + n_2 + 1)}{2} \qquad \sigma^2\{S_2\} = \frac{n_1 n_2(n_1 + n_2 + 1)}{12}$$

Further, when n_1 and n_2 are sufficiently large, the sampling distribution of S_2 is approximately normal.

The normal approximation is accurate enough when n_1 and n_2 are both 10 or more. Thus, the sample sizes need not be particularly large for the normal approximation to be applicable.

Decision Rule for Two-Sided Test. Given the approximate normality of the sampling distribution of S_2, the construction of the decision rule parallels earlier large-sample tests. Figure 15.7 illustrates the procedure for a two-sided test.

(15.8) When the alternatives are:

$$H_0: \delta = 0$$

$$H_1: \delta \neq 0$$

and the samples are independent and reasonably large from populations with

FIGURE 15.7   *Construction of decision rule for the M-W-W test—Large sample, two-sided test for δ*

(a) *Appropriate type of decision rule*

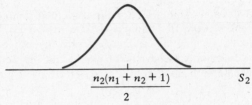

Conclude H_1 ($\delta \neq 0$)	Conclude H_0 ($\delta = 0$)	Conclude H_1 ($\delta \neq 0$)

$A_1 \quad \dfrac{n_2(n_1 + n_2 + 1)}{2} \quad A_2 \qquad S_2$

Extreme values of S_2 imply H_1 is correct.

(b) *Sampling distribution of S_2 when populations identical (i.e., δ = 0)*

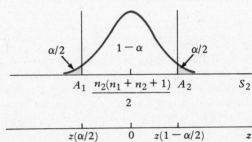

$\dfrac{n_2(n_1 + n_2 + 1)}{2} \qquad S_2$

Distribution of S_2 is approximately normal when n_1 and n_2 are large and $\delta = 0$.

(c) *Placement of action limits*

$\alpha/2 \qquad 1 - \alpha \qquad \alpha/2$

$A_1 \quad \dfrac{n_2(n_1 + n_2 + 1)}{2} \quad A_2 \qquad S_2$

$z(\alpha/2) \qquad 0 \qquad z(1 - \alpha/2) \qquad z$

A_1 and A_2 are placed at equal distances from $E\{S_2\}$ to give tail areas of $\alpha/2$ each.

where:

$$A_1 = \frac{n_2(n_1 + n_2 + 1)}{2} + z(\alpha/2) \sqrt{\frac{n_1 n_2(n_1 + n_2 + 1)}{12}}$$

$$A_2 = \frac{n_2(n_1 + n_2 + 1)}{2} + z(1 - \alpha/2) \sqrt{\frac{n_1 n_2(n_1 + n_2 + 1)}{12}}$$

(d) *Decision rule*

If $A_1 \leq S_2 \leq A_2$, conclude H_0
If $S_2 < A_1$ or $S_2 > A_2$, conclude H_1

Conclude appropriate alternative based on observed S_2.

identically shaped distributions, the appropriate decision rule to control the α risk is:

$$\text{If } A_1 \leq S_2 \leq A_2, \text{ conclude } H_0$$

$$\text{If } S_2 < A_1 \text{ or } S_2 > A_2, \text{ conclude } H_1$$

where: $A_1 = E\{S_2\} - z\sigma\{S_2\}$
$A_2 = E\{S_2\} + z\sigma\{S_2\}$
$z = z(1 - \alpha/2)$
$E\{S_2\}$ and $\sigma\{S_2\}$ are given by (15.7)

☐   **Examples**

1. We return to the marketing instruction case. Table 15.1a contains the sample results. Note that $n_1 = 11$ because one student dropped out of the experimental video-lectures class. The instructor wishes to test:

$$H_0: \delta = 0$$

$$H_1: \delta \neq 0$$

and wants to control the α risk at .05.

Table 15.1b contains the array for the combined samples and shows the rank for each sample observation (the Y ranks are shown in boxes). Note that the one case of ties is handled by assigning the mean rank to each tied observation. Thus, since the two grades of 78 would normally have had ranks 14 and 15, each has been given rank 14.5. Table 15.1c assembles the ranks for the sample from the Y population. The test statistic is $S_2 = 149$.

The minimum and maximum values of S_2 here are:

$$\frac{12(13)}{2} = 78 \qquad 11(12) + \frac{12(13)}{2} = 210$$

so we see that $S_2 = 149$ is not an extreme value.

Figure 15.7 illustrates the construction of the decision rule. The appropriate decision rule is shown in Figure 15.7a. The sampling distribution of S_2 when H_0 holds is presented in Figure 15.7b, and Figure 15.7c shows the calculation of the action limits. When H_0 holds, we have by (15.7):

$$E\{S_2\} = \frac{12(11 + 12 + 1)}{2} = 144 \qquad \sigma^2\{S_2\} = \frac{11(12)(11 + 12 + 1)}{12} = 264$$

Thus, $\sigma\{S_2\} = \sqrt{264} = 16.25$. To control the α risk at .05, we require $z(.975) = 1.960$.

TABLE 15.1   *Marketing instruction experiment results*

(a)
Final grades

Video X_i: ($n_1 = 11$)	87	72	94	49	56	88	74	61	80	52	75	
Live Y_i: ($n_2 = 12$)	71	42	69	97	78	84	57	64	78	73	85	91

(b)
Combined array (Y ranks boxed)

Value:	42	49	52	56	57	61	64	69	71	72	73	74
Rank:	☐1	2	3	4	☐5	6	☐7	☐8	☐9	10	☐11	12

Value:	75	78	78	80	84	85	87	88	91	94	97
Rank:	13	☐14.5	☐14.5	16	☐17	☐18	19	20	☐21	22	☐23

(c)
Y ranks

1	5	7	8	9	11	14.5	14.5	17	18	21	23	$S_2 = 149$

Hence, the action limits are $144 \pm 1.960(16.25)$ and the appropriate decision rule is, as shown in Figure 15.7d:

$$\text{If } 112.2 \leq S_2 \leq 175.9, \text{ conclude } H_0$$

$$\text{If } S_2 < 112.2 \text{ or } S_2 > 175.9, \text{ conclude } H_1$$

Since $S_2 = 149$, we conclude H_0—that $\delta = 0$. Thus, this small-scale experiment does not indicate any difference in students' grades for live and video-taped lectures.

This test could also have been conducted by means of the standardized test statistic $z^* = (149 - 144)/16.25 = .31$. Since $|z^*| = .31 \leq 1.960$, conclusion H_0 is indicated.

2. Suppose the instructor had hypothesized that live lectures should lead to better performance. The alternatives then would be:

$$H_0: \delta \leq 0$$

$$H_1: \delta > 0$$

and an upper-tail test is required. (Remember that S_2 will tend to be large if $\delta > 0$, i.e., if the Y distribution lies to the right of the X distribution.) Assuming that the α risk is to be .05, we require $z(1 - \alpha) = z(.95) = 1.645$. As before:

$$E\{S_2\} = 144 \qquad \sigma\{S_2\} = 16.25$$

Hence, the upper-tail action limit is $144 + 1.645(16.25) = 170.7$ and the decision rule is:

$$\text{If } S_2 \leq 170.7, \text{ conclude } H_0$$

$$\text{If } S_2 > 170.7, \text{ conclude } H_1$$

Since $S_2 = 149$, we conclude H_0. □

Comments

1. When the sample sizes n_1 and n_2 are small, the normal approximation for the sampling distribution of S_2 is not appropriate for the M-W-W test, and tables of the exact sampling distribution should be used. These are available in specialized texts on nonparametric procedures.
2. The M-W-W test procedure could equally well have been based on the sum of the ranks of the sample from the X population, which we might denote by S_1. However, the expected value of S_1 when $\delta = 0$ is not the same as $E\{S_2\}$ unless $n_1 = n_2$. It is:

(15.9)
$$E\{S_1\} = \frac{n_1(n_1 + n_2 + 1)}{2}$$

The variance of S_1 when $\delta = 0$ is the same as that of S_2.
3. The procedure we have described for handling ties is suitable provided they are not numerous. The decision rule for the M-W-W test must be modified if ties are numerous.
4. Although the M-W-W test uses only the ranks rather than the values of the observations, it can be shown that for reasonably large samples, the M-W-W test is at times almost as powerful as tests that depend on specific knowledge about the form of the underlying populations. For instance, it compares favorably with the t test described in Section 13.2 for comparing the locations of two normal distributions. That is, when

both tests have the same α risk, the probability of correctly concluding H_1 when $\delta \neq 0$ is almost as large for the M-W-W test as for the t test.

5. The statistic S_2 is also called the *Wilcoxon rank sum statistic* because it was originally developed by Wilcoxon. Mann and Whitney independently formulated an alternative test statistic which is a simple function of S_2 and leads to identical tests.

15.3 COMPARATIVE STUDIES OF TWO POPULATIONS—MATCHED SAMPLES

In this section, we describe several nonparametric procedures for comparing two populations based on matched samples. Recall from Chapter 13 that matched samples involve the pairing of sample observations X_i and Y_i from two populations as, for example, when male and female employees are paired on the basis of their job qualifications in order to study possible salary differentials. Like the procedures discussed in Chapter 13, the nonparametric procedures are based on the differences $D_i = Y_i - X_i$ for matched pairs.

We shall focus in this section on the median of the population of differences, denoted by η_D. When the Y_i's tend to be larger than the X_i's, the D_i's tend to be positive and η_D is positive. Similarly, when the Y_i's tend to be smaller than the X_i's, the D_i's tend to be negative and η_D is negative. Finally, if the X and Y distributions do not differ in location, the D_i's typically tend to be positive and negative with approximately equal frequency and η_D then is near zero. Thus, η_D may be thought of as a parameter that measures how far apart the X and Y distributions are. This is particularly evident in the case of a symmetrical distribution of differences because then η_D equals μ_D, the mean of the population of differences, and hence $\eta_D = \mu_D = \mu_2 - \mu_1$.

Confidence Interval for η_D

As in Chapter 13, nonparametric inferences for matched samples are based on inference procedures for single populations, the population being that of the differences D_i. Thus, the setting up of a confidence interval for the median difference η_D utilizes the procedures of Section 15.1.

Assumptions

1. The population of differences D_i is continuous.
2. The n differences are a random sample from the population of differences.

☐ Example

A business researcher is studying the impact of information on credit decisions. Forty volunteer credit managers are grouped into $n = 20$ pairs, each pair having similar backgrounds with respect to business experience, education, etc. Two case descriptions are prepared of a hypothetical business that is requesting trade credit. The first one provides limited information about the business (e.g., product line, years in business, total assets). The second case description provides detailed information, including financial and operating data for the past several years. In each pair, a random number is used to decide which manager receives description 1 and which one receives description 2. Each manager then studies the description and determines the maximum amount of credit he or she would extend to the business.

The sample results are presented in Table 15.2. The credit amounts with limited information are denoted by X and those with detailed information are denoted by Y. The researcher would like to estimate η_D of the population of differences $D_i = Y_i - X_i$, with a 95 percent confidence interval.

The n sample differences are shown in Table 15.2. Since $n = 20$ is a reasonably large sample, we use formula (15.2) for finding r. For $1 - \alpha = .95$, we require $z(1 - \alpha/2) = z(.975) = 1.960$. Hence:

$$.5[n + 1 - z(1 - \alpha/2)\sqrt{n}] = .5(20 + 1 - 1.960\sqrt{20}) = 6.12$$

so $r = 6$. We therefore choose L_6 and U_6, the 6th smallest and 6th largest sample differences, for our confidence limits. A study of Table 15.2 shows $L_6 = 10$ and $U_6 = 28$. Thus, the desired confidence interval is:

$$10 \leq \eta_D \leq 28$$

The researcher can state, with 95 percent confidence, that detailed information in the case description has resulted in a median increase of between \$10 and \$28 thousand in trade credit extended. These limits clearly indicate that the maximum credit tends to be larger here with detailed than with limited information.   □

TABLE 15.2   *Amounts of credit extended with limited and detailed information*

Pair of Managers i	Credit Amount ($000)		Difference D_i
	Limited Information X_i	Detailed Information Y_i	
1	85	90	+ 5
2	50	60	+10
3	85	113	+28
4	70	100	+30
5	30	45	+15
6	75	99	+24
7	67	86	+19
8	60	85	+25
9	60	85	+25
10	85	87	+ 2
11	70	50	−20
12	85	90	+ 5
13	70	93	+23
14	77	115	+38
15	70	62	− 8
16	50	85	+35
17	55	72	+17
18	75	125	+50
19	81	103	+22
20	20	60	+40

Sign Test for η_D

As we noted in Section 15.1, we can conduct a sign test for a population median η directly or via a confidence interval for η. Both of these approaches can be used for tests about the median difference η_D with matched samples.

☐ Example

In our credit study, suppose the researcher wishes to test whether or not $\eta_D = 0$. The alternatives therefore are:

$$H_0\colon \eta_D = 0$$

$$H_1\colon \eta_D \neq 0$$

The α risk is to be controlled at .05. Using the 95 percent confidence interval obtained before:

$$10 \leq \eta_D \leq 28$$

we see that it does not include 0, hence we conclude H_1.   ☐

Wilcoxon Signed Rank Test for η_D

When it can be assumed that the population of differences is symmetrical, the Wilcoxon signed rank test is generally more powerful than the sign test for making inferences about the population median difference η_D.

In practice, and especially in experimental settings, the population of differences between matched pairs frequently will be symmetrical, or approximately so. In the case of experiments where matched subjects are each assigned randomly to two different treatments, a symmetrical distribution is to be expected when the two treatments have no differential effects because each subject has an equal chance of being assigned to either treatment.

Assumptions

1. The population of differences D_i is continuous and symmetrical.
2. The n differences are a random sample from the population of differences.

Alternatives. The Wilcoxon signed rank test can be employed for both two-sided and one-sided alternatives concerning η_D.

Test Statistic. The test statistic for the Wilcoxon test is simple. We consider the absolute differences $|D_i|$. We rank these. To each rank, we attach a plus sign or a minus sign according to whether D_i is positive or negative. We sum the signed ranks and denote the sum by T. If a difference D_i should happen to equal zero, we discard it and reduce the sample size accordingly. When absolute differences are tied, they are assigned the average value of the corresponding ranks, as in the M-W-W test.

☐ Example

Table 15.3 shows the calculation of the test statistic T for our credit study. We repeat in column 1 the differences D_i from Table 15.2. The absolute values of the differences are

TABLE 15.3 *Wilcoxon signed rank test for credit example*

Pair of Managers i	(1) Difference D_i	(2) Absolute Difference $\|D_i\|$	(3) Rank	(4) Signed Rank
1	+ 5	5	2.5	+ 2.5
2	+10	10	5	+ 5
3	+28	28	15	+15
4	+30	30	16	+16
5	+15	15	6	+ 6
6	+24	24	12	+12
7	+19	19	8	+ 8
8	+25	25	13.5	+13.5
9	+25	25	13.5	+13.5
10	+ 2	2	1	+ 1
11	−20	20	9	− 9
12	+ 5	5	2.5	+ 2.5
13	+23	23	11	+11
14	+38	38	18	+18
15	− 8	8	4	− 4
16	+35	35	17	+17
17	+17	17	7	+ 7
18	+50	50	20	+20
19	+22	22	10	+10
20	+40	40	19	+19
				$T = 184$

shown in column 2. The ranks from smallest absolute difference (rank 1) to largest (rank n) are shown in column 3. The signed ranks are shown in column 4. It can be seen from Table 15.3 that $T = 184$. □

If all the differences are positive, T is the sum of the ranks $1, 2, \ldots, n$, which is $n(n + 1)/2$. For $n = 20$, $n(n + 1)/2 = 210$. If all the differences are negative, T equals $-n(n + 1)/2$. For $n = 20$, this is -210. Finally, if the distribution of differences is symmetrical about 0 (i.e., if $\eta_D = 0$), the rank associated with any absolute difference has an equal probability of being positive or negative. In this case, T will tend to have a value midway between the two extremes, i.e., near 0. Thus, the appropriate decision rules are clear. For a two-sided test, for example, values of T close to $-n(n + 1)/2$ or close to $n(n + 1)/2$ lead us to conclude H_1 ($\eta_D \neq 0$); otherwise, we conclude H_0 ($\eta_D = 0$). Figure 15.8a shows this decision rule.

Sampling Distribution of T When $\eta_D = 0$. To control the α risk at a specified level when $\eta_D = 0$, we need to know the sampling distribution of T when $\eta_D = 0$. As we mentioned earlier, when $\eta_D = 0$, the rank associated with any absolute difference has an equal probability of being positive or negative. Hence, the sampling distribution of T can be determined exactly for all sample sizes. It depends only on the number of differences n. Tables of the exact sampling distribution of T for small sample sizes are available but are not needed unless n is very small.

FIGURE 15.8 *Construction of decision rule for Wilcoxon signed rank test—Large sample, two-sided test for η_D*

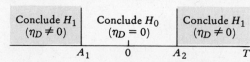

(a) *Appropriate type of decision rule*

Conclude H_1 ($\eta_D \neq 0$)	Conclude H_0 ($\eta_D = 0$)	Conclude H_1 ($\eta_D \neq 0$)
A_1	0	A_2 $\quad T$

Extreme values of T imply H_1 is correct.

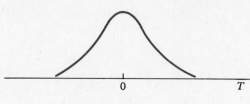

(b) *Sampling distribution of T when $\eta_D = 0$*

Distribution of T is approximately normal when n is large and $\eta_D = 0$

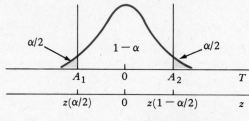

(c) *Placement of action limits*

$\alpha/2$ $1-\alpha$ $\alpha/2$

$A_1 \quad 0 \quad A_2 \quad\quad T$

$z(\alpha/2) \quad 0 \quad z(1-\alpha/2) \quad\quad z$

A_1 and A_2 are placed at equal distances from 0 to give tail areas of $\alpha/2$ each.

$$\text{where: } A_1 = 0 + z(\alpha/2)\sqrt{\frac{n(n+1)(2n+1)}{6}}$$

$$A_2 = 0 + z(1-\alpha/2)\sqrt{\frac{n(n+1)(2n+1)}{6}}$$

(d) *Decision rule*

If $A_1 \leq T \leq A_2$, conclude H_0
If $T < A_1$ or $T > A_2$, conclude H_1

Conclude appropriate alternative based on observed T.

(15.10) When a random sample of n differences is selected from a symmetrical population of differences with $\eta_D = 0$, the sampling distribution of T has mean and variance:

$$E\{T\} = 0 \qquad \sigma^2\{T\} = \frac{n(n+1)(2n+1)}{6}$$

Further, when n is sufficiently large, the sampling distribution of T is approximately normal.

The normal approximation is applicable when n, the number of differences, is 10 or more. Figure 15.8b shows the sampling distribution of T when $\eta_D = 0$.

Decision Rule for Two-sided Test.   Given the approximate normality of T, the construction of the decision rule proceeds as usual for large-sample tests. Figure 15.8 illustrates the procedure for a two-sided test.

(15.11) When the alternatives are:

$$H_0: \eta_D = 0$$

$$H_1: \eta_D \neq 0$$

and the sample of differences is reasonably large from a symmetrical population of differences, the appropriate decision rule to control the α risk is:

$$\text{If } A_1 \leq T \leq A_2, \text{ conclude } H_0$$

$$\text{If } T < A_1 \text{ or } T > A_2, \text{ conclude } H_1$$

where: $A_1 = 0 - z\sigma\{T\}$
   $A_2 = 0 + z\sigma\{T\}$
   $z = z(1 - \alpha/2)$
   $\sigma\{T\}$ is given by (15.10)

Examples

1. The sample results for the credit study are presented in Table 15.3. The test statistic is $T = 184$, which is close to the maximum value of 210. The researcher wishes to test whether or not $\eta_D = 0$:

$$H_0: \eta_D = 0$$

$$H_1: \eta_D \neq 0$$

When $\eta_D = 0$, we know that the mean and variance of T are:

$$E\{T\} = 0 \qquad \sigma^2\{T\} = \frac{n(n + 1)(2n + 1)}{6} = \frac{20(21)(41)}{6} = 2{,}870$$

and hence $\sigma\{T\} = \sqrt{2{,}870} = 53.57$. The researcher would like to control the α risk at .05. Hence, we require $z(.975) = 1.960$, and the action limits are $0 \pm 1.960(53.57)$. The decision rule therefore is:

$$\text{If } -105.00 \leq T \leq 105.00, \text{ conclude } H_0$$

$$\text{If } T < -105.00 \text{ or } T > 105.00, \text{ conclude } H_1$$

Since $T = 184$, we conclude H_1—that the median difference in the credit amounts is not zero.

This test could also be carried out by the standardized test statistic $z^* = (184 - 0)/53.57 = 3.43$. Since $|z^*| = 3.43 > 1.960$, conclusion H_1 is indicated.

2. Suppose the following one-sided test had been desired:

$$H_0: \eta_D \leq 0$$

$$H_1: \eta_D > 0$$

and the α risk were to be held at .05. We require $z(.95) = 1.645$, and the action limit is $0 + 1.645(53.57) = 88.12$. The decision rule then is:

$$\text{If } T \leq 88.12, \text{ conclude } H_0$$

$$\text{If } T > 88.12, \text{ conclude } H_1$$

Since $T = 184$, we would conclude H_1—that $\eta_D > 0$.

Comments

1. The Wilcoxon signed rank test accommodates tied ranks if they are not too numerous. If they are numerous, modifications must be made in the decision rule.
2. The Wilcoxon signed rank test may also be used to test if a single population (which is symmetrical and continuous) has a median equal to some specified value, say, $\eta = \eta_I$. If $X_1, X_2, \ldots, X_n$ are the random sample observations from the population, then the differences $D_i = X_i - \eta_I$ are computed and the test whether or not $\eta_D = 0$ is applied to these differences in the standard fashion. Clearly, testing if the differences D_i have a zero median is equivalent to testing if the X_i's have a median equal to η_I.
3. Like the M-W-W test, the Wilcoxon signed rank test often compares favorably with parametric tests based on stronger assumptions about the underlying population (such as the t test for a normal population of differences discussed in Section 13.3). For this reason, the test is widely used in practice.

15.4 OPTIONAL TOPIC—TEST FOR RANDOMNESS

In this section, we take up a nonparametric procedure for examining whether or not a set of observations constitutes a random sample from an infinite population.

Nature of Randomness

We saw in Chapter 8 from definition (8.11) that a random sample from an infinite population is one in which the observations are statistically independent and have the same probability distribution. This means that the process generating the sample data is independent and stationary as defined in (6.1). Figure 15.9a illustrates a sequence of observations generated by an independent and stationary process from the standard normal distribution $N(0, 1)$.

Departures from randomness can occur in many ways, some of which are illustrated in Figure 15.9. The sequence in Figure 15.9b comes from a process with an upward trend, which is inconsistent with the assumption that all outcomes of a random sequence are from the same distribution. In Figure 15.9c the sequence of observations exhibits increasing variability, which again illustrates that the observations are not from the same distribution. Figures 15.9d and 15.9e contain sequences in which consecutive observations are dependent. In Figure 15.9d the process involves a cyclical pattern, while in Figure 15.9e the process alternates regularly.

Tests for randomness are of major importance because the assumption of randomness underlies statistical inference. In addition, tests for randomness are important for time series analysis, discussed in Chapters 24 and 25.

Test of Runs Up and Down

Departures from randomness can take so many forms that no single test for randomness is best for all situations. One of the most common departures from randomness is the tendency of a sequence to persist in its direction of movement. This is the case in Figure 15.9b, where the sequence contains a trend, and in Figure

FIGURE 15.9   *Sequences from random and nonrandom processes*

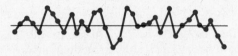

(a) *Random normal process*

(b) *Process involving an upward trend*

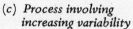

(c) *Process involving increasing variability*

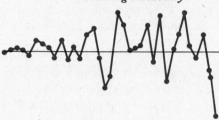

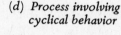

(d) *Process involving cyclical behavior*

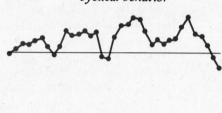

(e) *Process involving regular and frequent changes in direction*

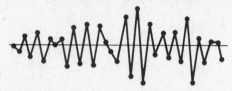

15.9d, where the persistence of movement produces a pronounced cyclical pattern. We now present a nonparametric test that is particularly effective when a process contains this persistence, as well as when a process contains excessively frequent changes in direction, as in Figure 15.9e.

Assumption.   The test assumes only that all sample observations come from continuous populations.

Alternatives.   The runs-up-and-down test, as already noted, can be used for detecting trend and other forms of persistence in a sequence of sample observations. Here, the alternatives are:

(15.12) H_0: Sequence generated by a random process

H_1: Sequence generated by a process containing persistence

When the test is used to detect departures from randomness consisting of either frequent or infrequent changes in direction, the alternatives are:

(15.13)

H_0: Sequence generated by a random process

H_1: Sequence generated by a nonrandom process

☐ Illustration

Table 15.4 shows the maximum level of a lake each year for a 30-year period. It is desired to test whether the 30 observations have been generated by a random process or whether the process contains a persistent trend. The presence of a trend would have significant environmental policy implications.　☐

Test Statistic.　The test statistic is a simple one, namely the total number of runs up and down. In Table 15.4, the direction of change in the sequence of maximum lake levels is shown by plus and minus signs. Thus, since $X_2 = 6.59$ is less than $X_1 = 6.63$, the change is $-$. A *run* is a succession of plus or minus signs, surrounded by the opposite sign. Thus, we see in Table 15.4 that the first run consists of two minus signs and the next run consists of one plus sign. Let R denote the total number of runs in the sequence. We have $R = 12$ for the data in Table 15.4.

When a trend or other such form of persistence is present, R will clearly be small. The smallest possible value of R is 1 (a single run). On the other hand, if the process involves frequent changes in direction, R will be large. The largest possible value of R is $n - 1$, when the signs alternate with each observation. Thus, when testing the alternatives in (15.12), small values of R will lead to H_1, and when testing the alternatives in (15.13), both small and large values of R will lead to H_1.

TABLE 15.4　*Maximum level of lake in each of 30 consecutive years*

Year	Level (meters above 190)	Change	Year	Level (meters above 190)	Change
1	6.63		16	5.91	−
2	6.59	− } Run 1	17	5.81	−
3	6.46	−	18	5.64	− } Run 7
4	6.49	+ } Run 2	19	5.51	−
5	6.45	−	20	5.31	−
6	6.41	−	21	5.36	+ } Run 8
7	6.38	−	22	5.17	−
8	6.26	− } Run 3	23	5.07	− } Run 9
9	6.09	−	24	4.97	−
10	5.99	−	25	5.00	+ } Run 10
11	5.92	−	26	5.01	+
12	5.93	+ } Run 4	27	4.85	−
13	5.83	− } Run 5	28	4.79	− } Run 11
14	5.82	−	29	4.73	−
15	5.95	+ } Run 6	30	4.76	+ } Run 12

Sampling Distribution of R When H_0 Holds. To control the α risk at a specified level, the sampling distribution of R must be known when H_0 is true, i.e., when the sequence is generated by a random process. This sampling distribution can be determined exactly for all sample sizes, since under H_0 all possible permutations of the observed values are equally likely. Tables of this sampling distribution for small sample sizes are available, but for most practical applications they are not needed since a normal approximation is applicable for all but very small sample sizes.

(15.14) When a sequence of n observations is generated by a random process (i.e., stationary and independent) and the distribution is continuous, the sampling distribution of R has mean and variance:

$$E\{R\} = \frac{2n-1}{3} \qquad \sigma^2\{R\} = \frac{16n-29}{90}$$

Further, when n is sufficiently large, the sampling distribution of R is approximately normal.

The normal approximation is applicable when the sample size n is 20 or more.

Decision Rule for One-sided Test. Figure 15.10 illustrates the construction of the decision rule for testing the alternatives in (15.12). The procedure follows that of previous large-sample tests:

(15.15) When the alternatives are:

H_0: Sequence generated by a random process

H_1: Sequence generated by a process containing persistence

and the number of observations is reasonably large, the appropriate decision rule to control the α risk is:

If $R \geq A$, conclude H_0

If $R < A$, conclude H_1

where: $A = E\{R\} + z(\alpha)\sigma\{R\}$
$E\{R\}$ and $\sigma\{R\}$ are given by (15.14)

☐   Examples

1. In our lake study, the alternatives to be tested are those in (15.12). The α risk is to be controlled at .01.

   First, we obtain the mean and variance of R for $n = 30$ when H_0 holds:

$$E\{R\} = \frac{2(30)-1}{3} = 19.67 \qquad \sigma^2\{R\} = \frac{16(30)-29}{90} = 5.01$$

so $\sigma\{R\} = \sqrt{5.01} = 2.24$. The test is a lower-tail test as illustrated in Figure 15.10. For

FIGURE 15.10 *Construction of decision rule for test of runs up and down—Large sample, lower-tail test for randomness*

(a) *Appropriate type of decision rule*

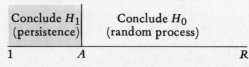

Small values of R imply H_1 is correct.

(b) *Sampling distribution of R when process random*

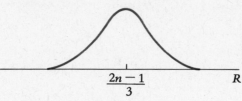

Distribution of R is approximately normal when n is large and process is random.

(c) *Placement of action limit*

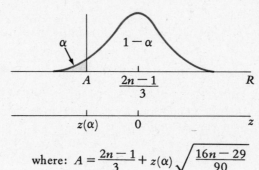

A is placed $z(\alpha)\sigma\{R\}$ below $E\{R\}$ to give lower–tail area of α.

where: $A = \dfrac{2n-1}{3} + z(\alpha)\sqrt{\dfrac{16n-29}{90}}$

(d) *Decision rule*

If $R \geq A$, conclude H_0

If $R < A$, conclude H_1

Conclude appropriate alternative based on observed R.

$\alpha = .01$, we require $z(.01) = -2.326$. Hence, the action limit is $19.67 - 2.326(2.24)$ and the decision rule is:

$$\text{If } R \geq 14.46, \text{ conclude } H_0$$

$$\text{If } R < 14.46, \text{ conclude } H_1$$

Since $R = 12 < 14.46$, we conclude H_1—that the process has not been a random one over the 30-year period and has contained persistence. Thus, the downward tendency in the data in Table 15.4 is indicative of a real trend.

2. If the two-sided alternatives in (15.13) had been of interest, with the α risk controlled at .01, we would use $z(.995) = 2.576$ for setting up the action limits $19.67 \pm 2.576(2.24)$, or 13.90 and 25.44. Since $R = 12$ is less than the lower action limit $A_1 = 13.90$, we would conclude H_1—that the process is not random. □

Comment

If consecutive observations are tied, we suggest that the tied observation be treated as a continuation of the existing run. Thus, tied observations following a minus sign should be treated as a minus, and similarly if the preceding run consists of plus signs. If ties are too numerous, the decision rule must be modified.

15.5   OPTIONAL TOPIC—JACKKNIFE ESTIMATION OF POPULATION STANDARD DEVIATION

In Section 13.5, we considered confidence interval estimation of the population variance and standard deviation when the population is normal. We cautioned there that the methodology does not yield reliable results when the population distribution is not close to normal. We now consider a robust procedure for estimating the population standard deviation, i.e., a procedure applicable under a wide variety of conditions. This procedure is called the *jackknife procedure*. It is so named because the procedure is a multipurpose estimation tool which is reliable under quite general conditions.

The jackknife procedure for estimating the population standard deviation σ (or some other parameter) is conceptually quite simple, involving the following three steps for estimating σ:

1. Based on the n sample observations, n standard deviations are calculated, each time omitting one sample observation. We shall let s_{-i} denote the sample standard deviation obtained when X_i is omitted. Clearly, each s_{-i} is an estimator of the population standard deviation σ.
2. Next, n *pseudo-values* J_i are calculated, utilizing the standard deviation s for the full sample and the standard deviations s_{-i} obtained by omitting one X_i observation at a time:

$$(15.16) \qquad J_i = ns - (n-1)s_{-i} \qquad i = 1, 2, \dots, n$$

Since both s and s_{-i} are estimators of the population standard deviation σ, each J_i is also an estimator of σ.

3. Now we treat the n pseudo-values $J_1, J_2, \dots, J_n$ as if they were a random sample from a normal population. We compute the mean and standard deviation of the pseudo-values in the usual way—here, we denote them by $\overline{J}$ and s_J, respectively:

$$(15.17) \qquad \overline{J} = \frac{\sum\limits_{i=1}^{n} J_i}{n} \qquad s_J = \sqrt{\frac{\sum\limits_{i=1}^{n} (J_i - \overline{J})^2}{n-1}}$$

Next, we compute the estimated standard deviation of $\overline{J}$, denoted here by $s\{\overline{J}\}$, just as is done for the sample mean $\overline{X}$:

$$(15.18) \qquad s\{\overline{J}\} = \frac{s_J}{\sqrt{n}}$$

$\overline{J}$ is an estimator of the population standard deviation σ because each J_i is an estimator of σ. To estimate the population standard deviation σ by means of an interval estimate, we employ confidence interval (10.16) based on a sample mean. This interval, which utilizes the t distribution, is appropriate here because the pseudo-values J_i are treated as if they were a random sample from a normal population.

(15.19) The confidence limits for σ, with approximate confidence coefficient $1 - \alpha$, are:

$$\bar{J} \pm ts\{\bar{J}\}$$

where: $t = t(1 - \alpha/2; n - 1)$
$\bar{J}$ is given by (15.17)
$s\{\bar{J}\}$ is given by (15.18)

This procedure turns out to be reliable under many circumstances, thus making the jackknife estimator a robust estimator.

☐   **Example**

A random sample of seven fire insurance claims under a particular type of residential policy consists of the following claims (in thousands of dollars):

$$6.2 \quad 3.1 \quad 1.0 \quad 9.3 \quad 4.2 \quad .7 \quad 11.7$$

It is desired to obtain a 95 percent confidence interval for σ, the population standard deviation for claims. Since the population of claims is quite skewed, confidence interval (13.30) for a normal population is not appropriate, and the jackknife procedure is to be employed. We shall follow the three-step procedure:

1. The sample mean for the $n = 7$ claims is $\bar{X} = 5.17$. Hence, the sample variance based on the full sample is:

$$s^2 = \frac{1}{7 - 1}[(6.2 - 5.17)^2 + (3.1 - 5.17)^2 + \cdots + (11.7 - 5.17)^2] = 17.226$$

so $s = \sqrt{17.226} = 4.15$. This value is shown in the last line of Table 15.5.

Next, we calculate the standard deviations s_{-i} by omitting one of the sample observations at a time. When X_1 is omitted, for instance, the mean of the remaining six observations is 5.00 and we obtain:

$$s_{-1}^2 = \frac{1}{6 - 1}[(3.1 - 5.00)^2 + (1.0 - 5.00)^2 + \cdots + (11.7 - 5.00)^2] = 20.424$$

TABLE 15.5   *Calculations for a jackknife estimate of σ*

i	(1) Omitted Observation X_i	(2) Sample Standard Deviation s_{-i}	(3) Pseudo-Value J_i
1	6.2	4.52	1.93
2	3.1	4.44	2.41
3	1.0	4.08	4.57
4	9.3	4.09	4.51
5	4.2	4.52	1.93
6	.7	4.00	5.05
7	11.7	3.28	9.37
	None	$s = 4.15$	Total 29.77

so $s_{-1} = \sqrt{20.424} = 4.52$. This standard deviation is shown in Table 15.5, column 2, first line. The other standard deviations s_{-i} are also shown in this table.

2. Next, we calculate the pseudo-values J_i using (15.16). For $i = 1$, we obtain:

$$J_1 = ns - (n - 1)s_{-1} = 7(4.15) - 6(4.52) = 1.93$$

This and the other pseudo-values are shown in column 3 of Table 15.5.

3. Now we treat the n pseudo-values J_i as if they were a random sample from a normal population. Using (15.17), we obtain:

$$\bar{J} = \frac{1}{7}(1.93 + 2.41 + \cdots + 9.37) = \frac{29.77}{7} = 4.25$$

and:

$$s_J^2 = \frac{1}{7 - 1}[(1.93 - 4.25)^2 + (2.41 - 4.25)^2 + \cdots + (9.37 - 4.25)^2] = 6.862$$

so $s_J = \sqrt{6.862} = 2.62$. Finally, the estimated standard deviation of $\bar{J}$, using (15.18), is:

$$s\{\bar{J}\} = \frac{s_J}{\sqrt{n}} = \frac{2.62}{\sqrt{7}} = .99$$

For $1 - \alpha = .95$ and $n - 1 = 6$, we require $t(.975; 6) = 2.447$. Hence, the confidence limits using (15.19) are $4.25 \pm 2.447(.99)$ and the confidence interval is:

$$1.83 \leq \sigma \leq 6.67$$

We conclude, with 95 percent confidence, that the standard deviation of the population of claims is between \$1.8 thousand and \$6.7 thousand. ☐

Comments

1. In addition to being robust, jackknife estimators have the desirable property that they avoid some types of bias which may be present in the original estimator. For example, s is a biased estimator of σ but $\bar{J}$ has almost no bias. Jackknife estimators are also very useful for obtaining confidence intervals for parameters that must be estimated by very complicated statistics. Since this situation is common in large-scale surveys, jackknife estimation is widely used in analyzing survey data.

2. The appropriateness of confidence intervals based on the jackknife method can sometimes be improved by employing a mathematical transformation to reduce possible skewness in the distribution of the pseudo-values. In our example, for instance, we could have used log s_{-i} and log s instead of s_{-i} and s for obtaining the pseudo-values. If the population of claims were an extremely skewed distribution, this transformation would help make the pseudo-values more nearly normally distributed.

3. When n is large, computing n pseudo-values by omitting one observation at a time involves much computation. To remedy this situation, the sample can be divided randomly into groups. For example, a sample of size 300 might be divided into 10 groups of size 30 each. To estimate the population standard deviation, 10 sample standard deviations s_{-i} would be computed, dropping out one group of observations at a time. Ten pseudo-values would then be calculated according to (15.16) and the confidence interval for σ would be obtained using (15.19), but now with $n = 10$.

PROBLEMS

*15.1 Consider a random sample of size $n = 8$ from a continuous population, and use Table C–10 to answer the following questions:

    a. What is the value of r for which confidence interval (15.1) has a confidence coefficient closest to .99?

    b. What is the confidence coefficient of the interval estimate $L_4 \leq \eta \leq U_4$?

15.2 Refer to Problem 15.1. Answer both parts for a random sample of size $n = 15$.

*15.3 In 12 recent automobile damage suits where the accidents were similar, the amounts awarded to plaintiffs (in thousands of dollars) were: 5.2, 5.5, 3.8, 12.5, 8.3, 2.1, 1.7, 20.0, 4.8, 6.9, 7.5, 10.6. Assume that these data represent a random sample from a continuous population.

    a. Array the observations and obtain a point estimate of η, the population median amount awarded in this type of suit.

    b. Construct a confidence interval for η with a confidence coefficient near .99.

15.4 Refer to Problem 15.3. Answer both parts assuming that the sample consists of the following amounts awarded in 15 suits: 6.0, 23.7, 8.6, 5.1, 2.4, 11.7, 33.6, 10.9, 5.6, 8.8, 4.2, 1.9, 10.1, 15.4, 19.1.

*15.5 Consider a random sample of size $n = 36$ from a continuous population. Using approximation (15.2), give the value of r for which the interval estimate (15.1) has a confidence coefficient of about .95.

15.6 Refer to Problem 15.5. Do the problem for sample size $n = 25$.

*15.7 Refer to Appendix D. Assume that the 1981 net assets data for paper companies represent a random sample from a continuous population. Construct a 95 percent confidence interval for the population median net assets of paper companies in 1981. Interpret your confidence interval.

15.8 Refer to Problem 15.7. Do the problem using 1981 net assets data for electronics companies.

15.9 An aluminum company experienced the following failure ages (in days) with a group of 21 electrolytic cells of design A that were installed in one of its smelter potrooms (in order of failure):

518	903	1,192	1,477	1,814	2,060	2,421
775	1,015	1,354	1,604	1,826	2,274	2,591
888	1,189	1,367	1,708	2,040	2,330	2,716

The 21 observations may be considered a random sample from a continuous population of failure ages. Construct a 90 percent confidence interval for the population median failure age of cells of design A. Interpret your interval estimate.

*15.10 Refer to Problem 15.3. Use your confidence interval to test whether or not $\eta = \$13$ thousand. State your conclusion. What is the α risk inherent in the test?

15.11 Refer to Problem 15.7. Use your confidence interval to test whether or not $\eta = \$300$ million. State your conclusion. What is the α risk inherent in the test?

15.12 Refer to Problem 15.9. Cells of a standard design are known to have a median failure age of 1,300 days. Use your confidence interval to test whether or not the population median failure age of cells of design A equals 1,300 days. State your conclusion. What is the α risk inherent in the test?

*15.13 In mail fund raising by a charitable foundation in the recent past, the median contribution from regular contributors has been $62.50. As an experiment, the foundation prepared a glossy brochure describing the foundation's work and distributed it to a random sample of 250 regular contributors. Subsequent analysis indicated that 140 of the 250 contributors gave in excess of $62.50 and 110 gave less than $62.50.

   a. Use the sign test to determine whether or not the median contribution with inclusion of the brochure has changed, controlling the α risk at .10. State the alternatives, decision rule, and conclusion.

   b. Redo your test in **a** on the assumption that 140 contributions were in excess of $62.50, 108 were less than $62.50, and 2 were exactly $62.50.

15.14 The median score of students enrolled in a community school system on a chemistry aptitude test in the past few years was 512. Of 800 students who took the aptitude test this year, 493 had scores above 512 and 307 had scores below 512. Use the sign test to test whether the population median score now is above 512, controlling the α risk at .025. State the alternatives, decision rule, and conclusion.

*15.15 Refer to Problem 15.3. Use the sign test to test whether or not the population median amount awarded equals $13 thousand, controlling the α risk near .01. State the alternatives, decision rule, and conclusion. Is your sign test equivalent to the test in Problem 15.10 based on the confidence interval for the population median? Explain.

15.16 A gas industry report states that half of residential customers pay their gas bills within 14 days after billing. In a recent random sample of 10 residential customers of one gas company, the bills were paid within the following numbers of days after billing (arrayed by magnitude): 2, 4, 8, 13, 15, 28, 29, 35, 36, 46. Use the sign test to determine whether or not the median payment period for this company is longer than that for the industry; control the α risk near .02. State the alternatives, decision rule, and conclusion.

15.17 The median score of the general population on a standard psychological test designed to measure extrinsic-reward motivation is 115. A random sample of eight successful sales managers who took the test obtained the following scores (arrayed by magnitude): 116, 120, 121, 128, 131, 138, 139, 145. Use the sign test to determine if the median test score of successful sales managers exceeds that of the general population, controlling the α risk near .05. State the alternatives, decision rule, and conclusion. What is the exact α risk of your test?

*15.18 Participants in a management training seminar play a computerized business game intended to develop their decision-making skills in a simulated business environment. The game is played independently by each participant, and success is measured by total profits earned. In one seminar having 15 participants, the game was played very early in the seminar period. In a second seminar having 13 participants, the game was played toward the end of the seminar period. In all other respects, the two seminars were identical. The following profits (in thousands of dollars) were earned by the participants (ordered by magnitude):

Game Scheduled Early in Seminar X_i			Game Scheduled Late in Seminar Y_i		
−4.4	2.2	6.8	.1	6.1	9.6
−2.1	4.4	8.2	.5	6.2	11.2
−1.2	4.7	8.3	1.0	7.6	15.7
0.0	5.5	8.5	4.3	8.6	
.7	5.9	9.1	4.5	8.7	

Use the M-W-W test to determine whether or not the location of the perform-ance distribution is affected by when the game is played in the seminar period; control the α risk at .10. State the alternatives, decision rule, and conclusion.

15.19 The numbers of units produced by 25 employees randomly selected from shift 1 and by 25 employees randomly and independently selected from shift 2 follow (in order of magnitude).

Shift 1 X_i						Shift 2 Y_i				
169	185	198	202	213	172	186	203	212	231
176	188	198	204	215	175	190	206	218	235
177	189	199	205	220	178	191	207	221	237
182	192	200	208	225	181	196	209	227	246
183	195	201	211	230	184	197	210	229	253

a. Use the M-W-W test to determine whether or not employee productivity for shift 2 is higher than that for shift 1. Control the α risk at .05. State the alternatives, decision rule, and conclusion.

b. What assumptions did you make in **a** in employing the M-W-W test proce-dure?

c. What additional assumptions would be required to use the t test procedure based on (13.6) to test whether the population mean for shift 2 is higher than that for shift 1?

15.20 Refer to Problem 15.19.

a. Construct a frequency distribution for each of the two sample data sets and plot the frequency polygons of the distributions on the same graph. Do the frequency polygons support the applicability of the M-W-W test here? Explain.

b. Interpret the shift parameter δ here.

15.21 A new method of swathing grain is under experimentation in a dry northern re-gion. The method leaves strips of standing grain to increase snow retention on fields during winter and thereby to increase soil moisture and crop yields in the following growing season. In one experiment, 34 similar fields were divided ran-domly into two equal groups. Fields in one group were swathed conventionally and fields in the other group were swathed using the snow-strip technique. Grain yields (in bushels per acre) for the fields in the following year were (arrayed by magni-tude):

Conventional Swathing X_i				Snow-Strip Swathing Y_i			
15.6	19.7	23.5	27.1	21.4	26.3	30.2	33.2
16.1	20.8	24.4	28.3	23.3	27.5	30.3	36.3
17.7	21.0	25.7		25.7	28.0	30.7	
18.6	22.5	26.5		25.9	28.4	31.4	
19.7	22.8	26.6		26.2	29.4	32.9	

a. Use the M-W-W test to determine whether or not snow-strip swathing leads to higher crop yields in the following year. Control the α risk at .01. State the alternatives, decision rule, and conclusion.

b. Under the assumptions of the M-W-W test, $\delta = \mu_2 - \mu_1$. Estimate δ by calcu-lating $\bar{Y} - \bar{X}$. Why can δ also be estimated by the difference of the two sample medians if the M-W-W test assumptions hold? Obtain this estimate. How does it compare with the estimate based on $\bar{Y} - \bar{X}$?

*15.22 A company's personnel policy provides a transferred executive with a guaranteed purchase price for his or her home if it cannot be sold for a larger amount within a fixed period from the transfer notification date. The guaranteed price is the average of two appraisals made independently by company-appointed real estate appraisers. The following are appraisal data (in thousands of dollars) for a random sample of 15 houses appraised by the same pair of appraisers:

House	Appraiser No. 1 X_i	Appraiser No. 2 Y_i	House	Appraiser No. 1 X_i	Appraiser No. 2 Y_i	House	Appraiser No. 1 X_i	Appraiser No. 2 Y_i
1	40.3	40.6	6	59.7	60.7	11	74.4	75.5
2	54.6	55.1	7	61.5	61.2	12	64.4	67.0
3	53.1	53.5	8	50.1	50.8	13	45.2	46.9
4	65.1	67.8	9	63.7	63.3	14	80.1	82.3
5	54.8	55.2	10	58.6	60.4	15	78.5	77.7

Construct a confidence interval for the population median difference in appraisals for these two appraisers, with a confidence coefficient near 95 percent. Interpret your interval estimate. Does it indicate that the appraisals by the two appraisers tend to differ systematically?

*15.23 Refer to Problem 15.22. Use the sign test to test whether or not the population median difference in appraisals is zero, controlling the α risk near .05. State the alternatives, decision rule, and conclusion.

15.24 Refer to Table 15.2, which pertains to the credit study example.
   a. Use the sign test to determine whether or not the population median difference in the amount of credit extended with limited and detailed information is zero. Control the α risk near .005. State the alternatives, decision rule, and conclusion.
   b. The Wilcoxon signed rank procedure was used in the Text to perform the test in a. What assumption does it require that the sign test does not?

*15.25 Refer to Problem 15.22. Use the Wilcoxon signed rank test to determine whether or not the population median difference in appraisals for the two appraisers is zero. Control the α risk at .05. State the alternatives, decision rule, and conclusion.

15.26 Refer to Problems 15.22 and 15.25. Make a frequency distribution of the differences in appraisals for the 15 houses and plot it in the form of a histogram. Does the histogram support the appropriateness of the Wilcoxon signed rank test here? Explain.

15.27 Refer to Problem 13.10.
   a. Construct a confidence interval for the population median test score difference, with a confidence coefficient of 99 percent. Interpret your interval estimate.
   b. Use the Wilcoxon signed rank test to determine whether or not the population median test score difference is zero. Control the α risk at .01. State the alternatives, decision rule, and conclusion.
   c. What assumptions did you make in employing the Wilcoxon signed rank test in b? What additional assumptions would be required to use the t test procedure of Section 13.3 for testing whether or not the two population means are equal?

*15.28 The numbers of marriages (in thousands) in the United States each month during the period July 1976–June 1980 follow.

1976		1977			1978			1979			1980	
J	219	J	121	J	228	J	114	J	231	J	131	J	227	J	140
A	226	F	131	A	226	F	128	A	240	F	135	A	254	F	155
S	185	M	140	S	195	M	151	S	223	M	150	S	204	M	153
O	186	A	176	O	183	A	178	O	185	A	189	O	199	A	190
N	162	M	187	N	165	M	202	N	174	M	221	N	187	M	224
D	180	J	245	D	182	J	262	D	194	J	277	D	181	J	278

SOURCE: U.S. Department of Health and Human Services.

  a. Do you observe any departure from randomness in this monthly time series?
  b. Perform the runs-up-and-down test for persistence, controlling the α risk at .01. State the alternatives, decision rule, and conclusion.

15.29 Refer to Problem 14.27.
  a. Use the runs-up-and-down test to determine whether or not the sequence of sample means for the 28 days of February was generated randomly, controlling the α risk at .05. State the alternatives, decision rule, and conclusion.
  b. Did you use a lower-tail or a two-sided test in **a**? Why?

15.30 Refer to Problem 16.9. We wish to test whether the computer program written by the graduate student generates numbers in a random sequence.
  a. Use the runs-up-and-down test to determine whether or not the computer program generates numbers in a random sequence. Control the α risk at .10. State the alternatives, decision rule, and conclusion.
  b. Did you use a one-sided or two-sided test in **a**? Why?
  c. If the runs test in **a** were performed on 100 different sequences of 50 numbers, each generated by the computer program and each test controlling α at .10, how many times would the conclusion of nonrandomness be expected if, indeed, the computer program generates numbers in a random sequence?

*15.31 The durations (in days) of the six most recent transit strikes in a city are: 5, 2, 7, 6, 4, 10. As a measure of the variability of durations of transit strikes in this city, a 95 percent confidence interval for the standard deviation of the probability distribution of strike durations is to be constructed. Assume the six observations constitute a random sample from this distribution.
  a. Verify that $s = 2.733$.
  b. Obtain the desired confidence interval using the jackknife procedure. Some of the required calculational results follow.

i:	1	2	3	4	5	6
X_i:	5	2	7	6	4	10
s_{-i}:	3.033	2.302	2.966	—	—	1.924
J_i:	1.233	4.888	1.568	—	—	6.778

15.32 Rates of return (in percent) earned on the common stock of Polyphase, Inc., during each of the past five quarters are: -5, 10, 2, 19, 15. Construct a 90 percent confidence interval for the population standard deviation of quarterly rates of return on this stock using the jackknife procedure. Assume the five observations constitute a random sample from the population of all quarterly rates of return. Some of the required calculational results follow.

i:	1	2	3	4	5	
X_i:	−5	10	2	19	15	
s_{-i}:	7.326	11.177	10.500	—	—	$s = 9.731$
J_i:	19.351	3.947	6.655	—	—	

15.33 (Based on Optional Topic Section 13.5.) Refer to Problem 15.32. Use (13.30) to construct a 90 percent confidence interval for the population standard deviation. What assumption is required for this confidence interval that is not required by the jackknife procedure?

EXERCISES

15.34 For a random sample from a symmetrical population, explain why $E\{Md\} = \mu = \eta$.

15.35 For an odd-sized random sample from a continuous population, show that η is the median of the sampling distribution of Md. (*Hint:* Does Md have an equal probability of falling on either side of η?)

15.36 Use Table C–5 to evaluate the confidence level of the interval $L_8 \leq \eta \leq U_8$ when $n = 20$.

15.37 Verify (15.2). [*Hint:* Use the normal approximation to the binomial probability in (15.3), together with the correction for continuity discussed in Section 12.3.]

15.38 Derive (15.9) from (15.7). [*Hint:* For the M-W-W test statistics: $S_1 + S_2 = (n_1 + n_2)(n_1 + n_2 + 1)/2$.]

15.39 Derive $E\{T\}$ and $\sigma^2\{T\}$ in (15.10) when $\eta_D = 0$.

$$[Hint: \sum_{i=1}^{n} i^2 = n(n + 1)(2n + 1)/6.]$$

STUDIES

15.40 Refer to Problem 15.9.
  a. Suppose the population of cell failure ages is approximately normally distributed. Does your confidence interval in Problem 15.9 then serve as a confidence interval for the mean failure age of cells of design A? Explain.
  b. Construct a "better" 90 percent confidence interval for the median failure age, knowing that the population is normal. In what sense is this confidence interval "better"?

15.41 Refer to Problem 15.9. Cells of a standard design are known to have a median failure age of 1,300 days. Assuming that the population is symmetrical, use the Wilcoxon signed rank test to determine whether or not the population median failure age of cells of design A exceeds 1,300 days. Control the α risk at .05. State the alternatives, decision rule, and conclusion.

15.42 Refer to Problem 15.9. The company also has installed 13 cells of a new design B. The distribution of cell failure ages for this design is known to be approximately normal. All 13 cells were installed on the same day. The first 10 failures among the

13 cells in this sample have occurred, and their failure ages (in days) are (in order of failure): 808, 980, 1,351, 1,543, 1,627, 1,666, 1,726, 1,993, 2,168, 2,271. The company wishes to estimate the mean failure age of cells of this design based on these incomplete sample data.

a. Can any of the estimation procedures for the mean of a normal population discussed earlier in the Text be employed in this case? Explain.

b. Since the mean and median of a normal population coincide, it is decided to construct a confidence interval for the population median failure age, with a confidence coefficient near 90 percent. Obtain this interval estimate and interpret it.

c. Use the Wilcoxon signed rank test to determine whether or not the median failure age of cells of design B exceeds 1,300 days, controlling the α risk at .05. Explain why the test procedure can be applied even though the failure ages of the last three cells are not known yet.

15.43 Refer to Problem 10.14. It is desired to estimate the population standard deviation of operating costs per mile for motor vehicles in the fleet. Use the jackknife procedure for obtaining a 99 percent confidence interval. Interpret your interval estimate.

15.44 A random sample of 10 sales executives in smaller firms showed the following annual incomes (in thousands of dollars): 37.0, 45.7, 43.1, 56.8, 94.1, 36.9, 51.4, 38.3, 64.1, 142.9. The analyst suspected that the income distribution for sales executives in smaller firms is quite skewed, so inference procedures that do not depend on approximate normality should be used.

a. Estimate the population median income by means of an interval estimate, with a confidence coefficient near 95 percent.

b. The analyst wishes to test whether or not the population median income of sales executives in smaller firms exceeds $60 thousand. Conduct such a test, controlling the α risk near .025. State the alternatives, decision rule, and conclusion.

c. The analyst also wishes to estimate the population standard deviation σ. To improve the appropriateness of the jackknife procedure, the analyst believes it would be desirable to use the logarithms of the standard deviations s and s_{-i}, as described in Comment 2, p. 401. Construct a 95 percent confidence interval for σ by first obtaining a confidence interval for log σ, using the pseudo-values $J_i = n \log(s) - (n - 1)\log(s_{-i})$, and then converting this interval into one for σ.

16
Goodness of Fit

16.1   GOODNESS OF FIT TESTS

Frequently in investigations, it is important to obtain information from a sample about the functional form of the population of interest. This information may be needed for: (1) confirming theories or validating models, or for (2) determining whether the population assumptions of a statistical inference procedure are appropriate for the given circumstances. We illustrate these types of situations with three examples.

☐   Examples

1. An operations analyst is developing a model of replacement demand for a spare part. She knows that for many spare parts, replacement demand follows a Poisson distribution. The analyst wishes to decide whether the weekly demand for this particular part follows a Poisson distribution. If so, the analyst will utilize Poisson probabilities in the model to ascertain the optimal inventory level. She has collected data on the number of replacement units of the spare part demanded per week for a sample of 52 weeks. She now wants to use these data to test which of the following alternatives should be accepted:

(16.1) $$H_0: \text{The probability distribution is Poisson}$$
$$H_1: \text{The probability distribution is not Poisson}$$

2. A researcher for a milk producers' association has production data for a sample of 40 milk cows of the same age and breed. He wishes to estimate the variability in output among cows in the relevant population and would like to use confidence interval (13.30) which requires the population to be approximately normal. A test of the assumption of normality involves the following alternatives:

(16.2) $$H_0: \text{The probability distribution is normal}$$
$$H_1: \text{The probability distribution is not normal}$$

3. Extensive experience in another country has shown that the length of hospital stay (in days) for a certain psychiatric condition has an exponential distribution with parameter value $\lambda = .036$. A medical researcher has collected comparable data for 10 cases in this

country and wishes to test if his data also came from an exponential distribution with $\lambda = .036$; i.e., he wishes to test the alternatives:

(16.3) H_0: The probability distribution is exponential with $\lambda = .036$

H_1: The probability distribution is not exponential with $\lambda = .036$ ☐

Alternatives in Goodness of Fit Tests

Tests for deciding between the alternatives in our three examples are called *goodness of fit tests*. The name arises because the alternatives involve the question whether or not a particular probability distribution is a good model for the population sampled, and this will be judged by whether or not the probability distribution specified in H_0 is a good fit for the sample data.

The alternatives in our three examples differ in some important ways. In (16.1), the alternatives relate to a discrete probability distribution, whereas in (16.2) and (16.3), they relate to continuous probability distributions. Another difference is whether or not the parameter values are specified. The alternatives in (16.1) and (16.2) pertain to a family of probability distributions (Poisson, normal) without specifying the parameter value(s). On the other hand, H_0 in (16.3) specifies not only the family (exponential) but also the parameter value ($\lambda = .036$). Thus, conclusion H_1 in (16.1) and (16.2) implies that the probability distribution is not that specified in H_0 (Poisson, normal). On the other hand, conclusion H_1 in (16.3) implies either that the probability distribution is not exponential or that it is exponential but λ is not equal to .036.

16.2   CHI-SQUARE TEST

One of the most widely used goodness of fit tests is the *chi-square test*.

Assumptions

There are only two assumptions for the chi-square test:

1. The sample is a simple random one from the population.
2. The sample size is reasonably large.

Nature of Test

The basic structure of the chi-square test is shown in Table 16.1. We first classify the sample data into k classes. For the ith class, we denote the observed number or frequency of sample values falling into that class by f_i. Of course, the f_i's must sum to the total sample size n:

(16.4) $$\sum_{i=1}^{k} f_i = n$$

TABLE 16.1   *Basic structure of chi-square test for goodness of fit*

Class	Observed Frequency	Expected Frequency under H_0	Relative Squared Residual
1	f_1	F_1	$(f_1 - F_1)^2/F_1$
2	f_2	F_2	$(f_2 - F_2)^2/F_2$
$\vdots$	$\vdots$	$\vdots$	$\vdots$
k	f_k	F_k	$(f_k - F_k)^2/F_k$
Total	n	n	$X^2 = \Sigma\ (f_i - F_i)^2/F_i$

Now, given the form of the distribution specified in H_0, we can determine the expected frequency for each class when H_0 is true. The expected frequency for the ith class when H_0 holds is denoted by F_i. Naturally:

$$(16.5) \qquad \sum_{i=1}^{k} F_i = n$$

The test is based on a comparison of the observed frequencies f_i with the corresponding expected frequencies F_i under H_0. The difference between the observed frequency and the expected frequency, $f_i - F_i$, is a *residual*, as explained in Chapter 2 (Section 2.7). The closer the observed frequencies are to the expected frequencies, the greater the weight of evidence in favor of H_0. The farther apart the observed and expected frequencies, the greater the weight of evidence in favor of H_1. The specific measure of closeness used for each class is the squared residual expressed relative to the expected frequency, i.e, $(f_i - F_i)^2/F_i$, as shown in Table 16.1. We shall call this ratio the *relative squared residual*. The test statistic is the sum of the relative squared residuals for all classes.

Test Involving Discrete Probability Distribution

☐   Illustration

In our Example 1 dealing with the demand for a spare part, the alternatives are:

> H_0: The probability distribution is Poisson

> H_1: The probability distribution is not Poisson

The weekly demand data for a random sample of 52 weeks are presented in Table 16.2a. In Table 16.2b, the 52 sample observations are classified into $k = 8$ classes and the observed frequency for each class (f_i) is shown. The reason for using these eight classes will be explained shortly.

The observed frequencies under H_0 are hypothesized to have arisen from a Poisson distribution. We must now obtain the expected frequencies if the population is a Poisson distribution. We know from (6.8) that the Poisson distribution has one parameter—its mean λ. Since H_0 does not specify the value of this parameter, we estimate it by the sample mean $\bar{X}$. For the 52 sample values in Table 16.2a, $\bar{X} = 4.0$ units (calculation not shown). Now we use the Poisson distribution with $\lambda = 4.0$ in Table C-6 to obtain the probabilities that a sample observation will fall into each of the k classes in Table 16.2b.

TABLE 16.2 *Chi-square test for goodness of fit of Poisson distribution—Demand for spare part example*

(a)
Weekly replacement demands

Weeks	Demand											
1–13	5	5	2	1	3	3	3	3	8	6	5	3	6
14–26	4	6	5	3	2	6	2	5	3	2	8	2	2
27–39	6	3	6	2	5	4	6	4	5	3	4	2	2
40–52	6	6	3	7	5	0	4	4	10	1	3	4	0

(b)
Frequency distribution and calculations for chi-square test

Class i	Number Demanded X	Observed Frequency f_i	Probability under H_0	Expected Frequency under H_0 F_i	Residual $f_i - F_i$	Relative Squared Residual $(f_i - F_i)^2/F_i$
1	0–1	4	.0916	4.763	−.763	.122
2	2	9	.1465	7.618	1.382	.251
3	3	11	.1954	10.161	.839	.069
4	4	7	.1954	10.161	−3.161	.983
5	5	8	.1563	8.128	−.128	.002
6	6	9	.1042	5.418	3.582	2.368
7	7	1	.0595	3.094	−2.094	1.417
8	8 or more	3	.0511	2.657	.343	.044
	Total	52	1.0000	52.000	0.000	$X^2 = 5.256$

For instance, the probability that a week's demand X will be 0 or 1 and hence fall in the first class is $P(0) + P(1) = .0183 + .0733 = .0916$, and that X will be 2 and fall in the second class is $P(2) = .1465$. These probabilities are shown in Table 16.2b.

Next, we obtain the expected frequencies if H_0 holds (F_i) by multiplying each probability by the sample size n ($n = 52$ in our illustration). For the first class we obtain $F_1 = 52(.0916) = 4.763$, and for the second class $F_2 = 52(.1465) = 7.618$. These expected frequencies are also shown in Table 16.2b. We now have the observed frequencies f_i and the expected frequencies F_i based on the assumption that weekly demand is Poisson-distributed. At this point, we are ready to calculate the test statistic. ☐

Test Statistic and Decision Rule. Our test statistic is the sum of the relative squared residuals, which we shall denote by X^2, i.e.:

$$(16.6) \qquad X^2 = \sum_{i=1}^{k} \frac{(f_i - F_i)^2}{F_i}$$

Note that the farther the observed frequency f_i departs in either direction from the expected frequency F_i, the larger is $(f_i - F_i)^2$ and hence the larger is X^2. On the other hand, if f_i and F_i are identical for all classes, $X^2 = 0$ because each $(f_i - F_i)^2 = 0$. It follows, therefore, that large values of X^2 are indicative of H_1

being true while small values of X^2 are indicative of H_0 being true. Thus, an upper-tail test as shown in Figure 16.1a is appropriate.

To control the α risk, we require the sampling distribution of X^2 when H_0 is true. The following theorem applies:

(16.7) When the population sampled has the probability distribution specified in H_0 and the sample size is reasonably large:

$$X^2 \simeq \chi^2(k - m - 1)$$

where: k is the number of classes

m is the number of parameters estimated from the sample data

Thus, when n is reasonably large, X^2 is distributed approximately as a χ^2 random variable with $k - m - 1$ degrees of freedom if H_0 holds. (If a review of the χ^2 distribution is needed, the reader should turn to Appendix B, Section B.1 before proceeding.)

Statistical research has suggested that the sample size need not be very large before the χ^2 approximation is appropriate. We shall use the rule: *n is large*

FIGURE 16.1 *Construction of decision rule for chi-square goodness of fit test*

(*a*) *Appropriate type of decision rule*

Conclude H_0 (population has specified form) | Conclude H_1

$$\begin{array}{ccc} 0 & & A \quad X^2 \end{array}$$

Large values of X^2 imply H_1 is correct.

(*b*) *Sampling distribution of X^2 when population has form specified in H_0*

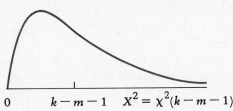

$$\begin{array}{cc} 0 & k - m - 1 \quad X^2 = \chi^2(k - m - 1) \end{array}$$

Distribution of X^2 is approximately $\chi^2(k - m - 1)$ when H_0 is true.

(*c*) *Placement of action limit*

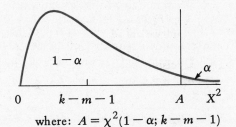

$1 - \alpha$

α

$$\begin{array}{cccc} 0 & k - m - 1 & A & X^2 \end{array}$$

where: $A = \chi^2(1 - \alpha; k - m - 1)$

A is placed so right–tail area equals α.

(*d*) *Decision rule*

If $X^2 \leq \chi^2(1 - \alpha; k - m - 1)$, conclude H_0
If $X^2 > \chi^2(1 - \alpha; k - m - 1)$, conclude H_1

Conclude appropriate alternative based on observed X^2.

enough for the χ^2 approximation in (16.7) *if the expected frequency F_i for each class is at least 2 and if most F_i's are at least 5.*

Figure 16.1b illustrates the sampling distribution of X^2 when H_0 is true. Figure 16.1c shows the position of the action limit when the risk of a Type I error is controlled at α. The area in the right tail must be α, and hence the action limit must be the $100(1 - \alpha)$ percentile of the χ^2 distribution with $k - m - 1$ degrees of freedom; i.e., $A = \chi^2(1 - \alpha; k - m - 1)$. Hence, the appropriate decision rule takes the form in Figure 16.1d.

(16.8) When the alternatives are:

$$H_0: \text{Population has specified form}$$

$$H_1: \text{Population does not have specified form}$$

and the sample is reasonably large, the appropriate decision rule to control the α risk is:

$$\text{If } X^2 \leq \chi^2(1 - \alpha; k - m - 1), \text{ conclude } H_0$$

$$\text{If } X^2 > \chi^2(1 - \alpha; k - m - 1), \text{ conclude } H_1$$

☐   **Example**

In our replacement part case, the α risk is to be controlled at .05. Since there are $k = 8$ classes and $m = 1$ parameter was estimated (namely, λ), the required χ^2 distribution has $k - m - 1 = 8 - 1 - 1 = 6$ degrees of freedom. We therefore use Table C-2 to find $\chi^2(.95; 6) = 12.59$. Note that the χ^2 approximation is appropriate because all F_i values exceed 2 and most exceed 5.

The decision rule therefore is:

$$\text{If } X^2 \leq 12.59, \text{ conclude } H_0$$

$$\text{If } X^2 > 12.59, \text{ conclude } H_1$$

From Table 16.2b, we see that $X^2 = 5.256 \leq 12.59$. Hence, we conclude H_0—that a Poisson distribution adequately fits the observed frequencies for weekly demand.   ☐

Comment

When the expected frequency of a class is less than 2, it is conventional to pool or combine adjacent classes to bring the combined expected frequency to 2 or more to satisfy the χ^2 approximation requirements. Thus, the first class in Table 16.2b was formed as follows:

	Original Classes				Pooled Class		
X	f_i	Probability under H_0	F_i		X	f_i	Probability under H_0	F_i
0	2	.0183	.952		0-1	4	.0916	4.763
1	2	.0733	3.812					

Note that if $X = 0$ and $X = 1$ had not been pooled, the expected frequency in the first class would have been less than 2. Pooling was also employed to form the class $X = 8$ or more in Table 16.2b.

Test Involving Continuous Probability Distribution

When the probability distribution specified in H_0 is continuous (e.g., normal or exponential), the class intervals used to classify the sample data can be chosen in many ways. To strengthen the discriminating capability of the test, we employ the following procedure: *The classes of the continuous probability distribution speci-fied in H_0 shall have equal probabilities; i.e., the expected frequencies for all classes shall be equal.*

Figure 16.2 presents the procedure graphically. If, say, $k = 10$ classes are to be used, each class has probability $1/k = .10$. The percentile separating classes 1 and 2 then is the $100(1/k) = 100(1/10) = 10$th percentile. The percentile separating classes 2 and 3 is the 20th percentile, and so on.

☐   Example

In our Example 2, the researcher wishes to determine whether or not the milk production of cows of a certain breed and age is normally distributed. His production data for the sample of 40 cows are presented in Table 16.3a. The alternatives are:

$$H_0: \text{The probability distribution is normal}$$

$$H_1: \text{The probability distribution is not normal}$$

The researcher wishes to control the α risk at .01.

To apply the chi-square test, we first must estimate the unknown parameters μ and σ of the normal distribution since these are not specified in H_0. As usual, we employ the sample estimators $\bar{X}$ and s, respectively. For the sample data in Table 16.3a, $\bar{X} = 15.96$ and $s = 2.144$ (calculations not shown).

We shall utilize eight class intervals. Since $n = 40$, each class will then have an expected frequency of 5. If we were to use more than eight classes, the expected frequencies would be less than 5 (remember that they are to be equal for all classes) and this would violate our requirement that most of the expected frequencies be at least 5. If we were to use fewer than eight classes, the test would have less discriminating capability for detecting nonnormality (i.e., the β risks would be greater).

FIGURE 16.2   *Partitioning a continuous probability distribution into classes of equal probability*

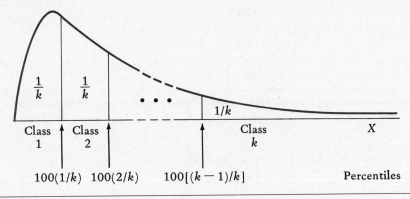

TABLE 16.3 *Chi-square test for goodness of fit of normal distribution—Milk production example*

(a)
Milk production per cow (thousands of pounds)

16.93	18.79	14.62	13.98	15.79	12.39	13.20	16.08	13.97	16.16
16.12	17.81	18.74	15.99	13.32	13.63	16.40	13.76	16.58	15.25
18.97	18.36	15.04	18.79	18.08	17.32	16.32	17.54	18.05	14.20
18.04	13.00	13.25	12.43	16.56	14.12	20.55	16.75	13.29	18.23

(b)
Frequency distribution and calculations for chi-square test

Milk Production X	Observed Frequency f_i	Probability under H_0	Expected Frequency under H_0 F_i	Residual $f_i - F_i$	Relative Squared Residual $(f_i - F_i)^2/F_i$
Under 13.49	7	.125	5	2	.80
13.49–14.51	6	.125	5	1	.20
14.52–15.26	3	.125	5	−2	.80
15.27–15.95	1	.125	5	−4	3.20
15.96–16.64	8	.125	5	3	1.80
16.65–17.39	3	.125	5	−2	.80
17.40–18.42	7	.125	5	2	.80
18.43 and over	5	.125	5	0	0.00
Total	40	1.000	40	0	$X^2 = 8.40$

Figure 16.3 illustrates the construction of the classes. The probability of each class is to be the same; hence, it must be $1/8 = .125$. The first class limit corresponds therefore to the 12.5th percentile, the second to the 25th percentile, and so on. These percentiles are estimated, using $\bar{X} = 15.96$ and $s = 2.144$, as follows:

$$\bar{X} + z(.125)s = 15.96 - 1.15(2.144) = 13.49$$

$$\bar{X} + z(.250)s = 15.96 - 0.67(2.144) = 14.52$$

$$\vdots \qquad\qquad \vdots \qquad\qquad \vdots$$

$$\bar{X} + z(.875)s = 15.96 + 1.15(2.144) = 18.43$$

We see, for example, that the 12.5th percentile of the standard normal distribution is $z(.125) = -1.15$, and hence the corresponding estimated 12.5th percentile of the distribution of milk production must be 1.15 standard deviations ($s = 2.144$) below the mean ($\bar{X} = 15.96$). With this method of constructing the class limits, five sample observations are expected to lie in each class—five observations in the class under 13.49, five observations in the class 13.49–14.51, etc.—as shown in Table 16.3b.

The 40 sample observations are now classified into these eight classes, as shown in Table 16.3b. The testing then proceeds as in the previous example. If H_0 holds (i.e., milk production is normally distributed), X^2 follows an approximate χ^2 distribution with $k - m - 1 = 8 - 2 - 1 = 5$ degrees of freedom. Remember that $m = 2$ parameters (namely, μ and σ) had to be estimated here. For $\alpha = .01$, we require $\chi^2(.99; 5) = 15.09$. Hence, the decision rule is:

$$\text{If } X^2 \leq 15.09, \text{ conclude } H_0$$

$$\text{If } X^2 > 15.09, \text{ conclude } H_1$$

FIGURE 16.3   *Partitioning a normal distribution into eight intervals of equal probability for milk production example*

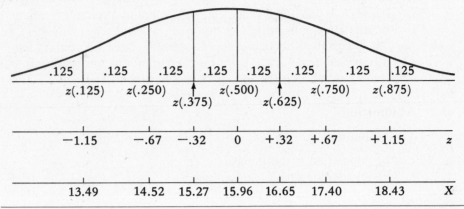

From Table 16.3b, we see that $X^2 = 8.40 \leq 15.09$. Hence, we conclude H_0—that the fit of a normal distribution to the sample data is adequate.   □

Comments

1. As we have seen in earlier chapters, many statistical tests that assume a normal population are quite robust and hence are satisfactory even in the absence of exact normality. Similarly in the development of models, exact normality is often not essential. In these cases, use of a low α risk with the chi-square test may be reasonable. The risk of a Type II error (concluding population is normal when it is not) will then be high if the departure from normality is small, but the normal model is still a reasonable one. However, the risk of a Type II error will be low when the departure from normality is great and the normal model is not reasonable.

2. When H_0 is rejected because of sharp departures between observed and expected frequencies, an examination of the residuals $f_i - F_i$ often yields clues to the true form of the probability distribution.

3. When sample data from a continuous probability distribution are already classified in a frequency distribution (e.g., in published data), one may not be able to construct equal-probability classes for the goodness of fit test. The chi-square procedure can still be used with the frequency classes already provided, pooling them where needed to achieve F_i's of adequate size. The resulting loss in the discriminating ability of the test will, in most cases, be small.

4. The parameter estimates on which theorem (16.7) is based employ the observed frequencies $f_1, f_2, \ldots, f_k$. As a practical matter, however, it is usually more convenient to estimate the parameters from the original data, as we did when we obtained $\bar{X}$ and s in our milk production example from the original data in Table 16.3a. It can be shown that use of the original data to estimate the parameters leads to an actual α risk level that may be somewhat larger than the specified one. However, when the number of classes utilized in the test is reasonably large and the expected frequencies F_i satisfy the conditions stated earlier, the actual α risk level will be close enough to the specified one to be satisfactory.

16.3 OPTIONAL TOPIC—CONFIDENCE BAND AND TEST FOR CUMULATIVE PROBABILITY FUNCTION

In Chapter 5, it was noted that a continuous random variable X can be described by its cumulative probability function $F(x)$, where $F(x) = P(X \leq x)$. In this section, we discuss the Kolmogorov-Smirnov (abbreviated K-S) estimation procedure, which provides a confidence band for the cumulative probability function $F(x)$. This procedure also can be used as a goodness of fit test.

Assumptions

The assumptions for the K-S estimation procedure are:

1. The population is continuous.
2. The sample is a simple random one.

K-S Statistic

Cumulative Sample Function. The basis of the K-S estimation procedure is the cumulative sample function, which we denote by $S(x)$:

(16.9) The *cumulative sample function*, denoted by $S(x)$, specifies for each value of x the proportion of sample values less than or equal to x.

From this definition, it is seen that $S(x)$ is simply a step-function ogive, such as the one illustrated in Figure 3.3b.

☐ Example

Figure 16.4a contains the sample results in array form for our Example 3 in which a researcher is studying length of hospital stay for certain psychiatric patients. The step-function ogive $S(x)$ for these data is plotted in Figure 16.4b, based on the cumulative data:

x:	16	17	20	27	31	34	37	55
$S(x)$:	.3	.4	.5	.6	.7	.8	.9	1.0

The step function $S(x)$ has steps of $1/n = .1$ except at $x = 16$, where the step is .3 because there are three observations there. ☐

$D(n)$ Statistic. The K-S estimation procedure utilizes a statistic, denoted by $D(n)$, which is based on the differences between the cumulative sample function $S(x)$ and the cumulative probability function $F(x)$.

(16.10)
$$D(n) = \max_x |S(x) - F(x)|$$

In other words, $D(n)$ equals the largest absolute deviation of $S(x)$ from $F(x)$ when all values of x are considered.

$D(n)$ is shown as a function of n because it depends on the sample size. Surprisingly, however, it does not depend on the specific form of $F(x)$. Thus, the sampling distribution of $D(n)$ can be tabulated for different values of n, irrespective of $F(x)$.

FIGURE 16.4   *Hospital stay data and cumulative sample function for 10 psychiatric cases*

(a)
Array of hospital stays

Case	Stay (days)	Case	Stay (days)
TAE	16	RJV	27
MSC	16	AAM	31
AGB	16	JJT	34
KFG	17	WKR	37
JCM	20	RAM	55

(b)
Cumulative sample function

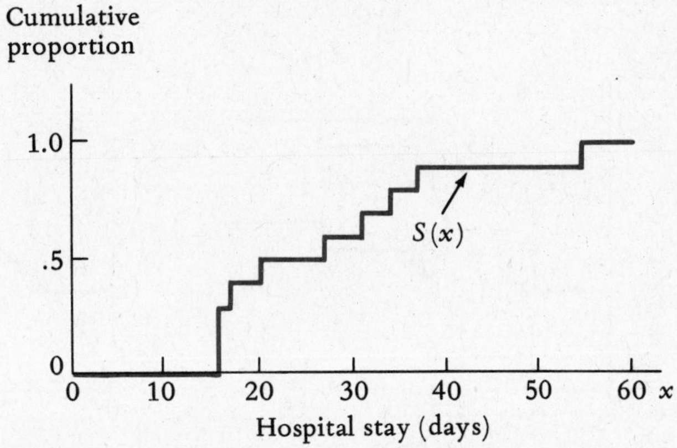

In Table C-11 are given percentiles of the $D(n)$ distribution for selected percentages and values of n, up to $n = 50$. For $n > 50$, the table shows formulas for computing the percentiles. Letting $D(a; n)$ denote the $100a$ percentile of the $D(n)$ distribution, we note, for instance, that $D(.90; 10) = .37$. This means that the probability is .90 that the largest absolute deviation of $S(x)$ from $F(x)$ is less than or equal to .37 when $n = 10$, no matter what is the cumulative probability function $F(x)$.

Confidence Band for $F(x)$

Development of Confidence Band.   The statistic $D(n)$ provides a simple way of constructing a confidence band from $S(x)$ for the cumulative probability function $F(x)$. Since $D(n)$ is the maximum *absolute* deviation between $S(x)$ and $F(x)$ and since $D(n)$ is less than $D(1 - \alpha; n)$ with probability $1 - \alpha$, it follows that $S(x)$ has a probability of $1 - \alpha$ of lying within distance $D(1 - \alpha; n)$ of $F(x)$, i.e.:

$$P[S(x) - D(1 - \alpha; n) \leq F(x) \leq S(x) + D(1 - \alpha; n)] = 1 - \alpha$$

This probability statement provides us with a confidence band for $F(x)$:

(16.11)   The confidence band for the cumulative probability function $F(x)$ with confidence coefficient $1 - \alpha$ is of the form $L(x) \leq F(x) \leq U(x)$, where:

$$L(x) = S(x) - D(1 - \alpha; n)   \qquad U(x) = S(x) + D(1 - \alpha; n)$$

☐   **Example**

For the psychiatric study, a 90 percent confidence band for $F(x)$ is desired. Figure 16.5a shows the construction of the confidence band. We require from Table C-11, for $n = 10$

FIGURE 16.5   *K-S procedures for psychiatric hospital stays example*

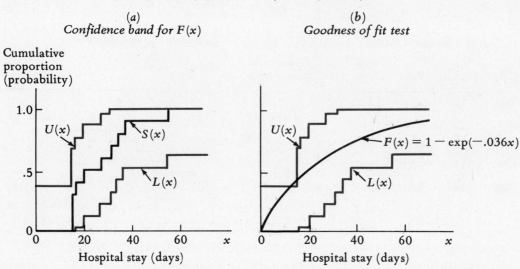

(a)
Confidence band for F(x)

(b)
Goodness of fit test

and $1 - \alpha = .90$, $D(.90; 10) = .37$. Thus, $L(x)$ and $U(x)$ are obtained by constructing a pair of step functions that are identical to $S(x)$ but at vertical distances of .37 below and above it, respectively. We obtain:

x:	16	17	20	27	31	34	37	55
$S(x)$:	.30	.40	.50	.60	.70	.80	.90	1.0
$L(x)$:	0	.03	.13	.23	.33	.43	.53	.63
$U(x)$:	.67	.77	.87	.97	1.0	1.0	1.0	1.0

Note that there is no need to extend the confidence band above 1 or below 0 since $F(x)$ cannot be outside these limits by definition. In conclusion, we can state with 90 percent confidence that $F(x)$ lies *entirely* within the $L(x)$ and $U(x)$ band in Figure 16.5a.   □

Comment

When the underlying population is discrete, the confidence band in (16.11) can be applied to the cumulative probability function but the actual confidence coefficient will then be greater than $1 - \alpha$.

Goodness of Fit Test for $F(x)$

The confidence band for $F(x)$ constructed in the previous section affords a convenient goodness of fit test for $F(x)$. We simply plot the cumulative probability function $F(x)$ specified in H_0 together with the confidence band. If the specified $F(x)$ falls entirely within the confidence band, we conclude H_0; otherwise, we conclude H_1.

◻ Example

In the psychiatric study, the researcher wishes to test whether or not the probability distribution is exponential with $\lambda = .036$:

H_0: The probability distribution is exponential with $\lambda = .036$

H_1: The probability distribution is not exponential with $\lambda = .036$

The α risk is to be controlled at .10.

Figure 16.5b repeats the 90 percent confidence band for $F(x)$ from Figure 16.5a and also shows the exponential cumulative probability function under H_0—see (7.16):

$$F(x) = 1 - \exp(-\lambda x) = 1 - \exp(-.036x)$$

We can use Table C-7 for plotting the specified $F(x)$. For example, when $x = 20$, $\lambda x = .036(20) = .72$ and $F(x) = .5132$.

We note from Figure 16.5b that the specified $F(x)$ does not fall entirely within the confidence band. Hence, we conclude H_1—that the probability distribution is not exponential with $\lambda = .036$. Thus, either the distribution is not exponential or it is exponential but $\lambda \neq .036$.

The researcher noted that the sample mean is $\bar{X} = 26.9$, so $1/26.9 = .037$. Since this is so close to the specified $\lambda = .036$, he concluded that the specified λ value must be applicable here and therefore the probability distribution is likely not exponential. The major lack of fit is in the neighborhood of $X = 16$, where three sample observations occur. Upon further analysis, the researcher found that two of these were actually somewhat shorter hospital stays but were incorrectly recorded at the usual minimum stay of 16 days. With the corrected data, the researcher found that an exponential distribution with $\lambda = .036$ is an adequate model. ◻

Comments

1. When H_0 does not specify the parameter value(s), the sample data may be used for estimating the parameter(s) and then obtaining the specified $F(x)$. However, the test with the K-S procedure now will involve an α risk less than the specified one and the power of the test (ability to detect if distribution is different from the one specified in H_0) will be smaller.
2. In situations where both the chi-square test and the K-S test might be applied, it has been found that the latter is often more powerful when the true probability distribution is not the one specified in H_0, where both tests control the α risk at the same level.
3. The K-S goodness of fit test may be employed when the probability distribution specified in H_0 is discrete, but the actual α risk level for the test will then be less than the one specified. The test is performed exactly as in the continuous case with the hypothesized discrete cumulative probability function plotted in lieu of a continuous one.

PROBLEMS

*16.1 For each of the following applications of the chi-square goodness of fit test, state: (1) H_1, (2) the number of degrees of freedom associated with X^2 under H_0, (3) the value of the action limit in the decision rule.

    a. H_0 specifies that the distribution is Poisson. There are 9 classes. The α risk is to be controlled at .01.

b. H_0 specifies that the distribution is uniform with $a = 2$ and $b = 10$. There are 6 classes. The α risk is to be controlled at .10.

c. H_0 specifies that the distribution is normal. There are 12 classes. The α risk is to be controlled at .05.

16.2 For each of the following applications of the chi-square goodness of fit test, state: (1) H_1, (2) the number of degrees of freedom associated with X^2 under H_0, (3) the value of the action limit in the decision rule.

a. H_0 specifies that the distribution is Poisson with $\lambda = 3$. There are 8 classes. The α risk is to be controlled at .10.

b. H_0 specifies that the distribution is binomial with $n = 4$. There are 5 classes. The α risk is to be controlled at .01.

c. H_0 specifies that the distribution is $N(0, 1)$. There are 6 classes. The α risk is to be controlled at .05.

*16.3 A city report described the results of an intensive promotional campaign concerned with public fire safety. Among other facts, the report provided the following frequency distribution of the number of fire calls answered by one fire station between 12 midnight and 8 A.M. each day during a 60-day period following the campaign:

Number of calls:	0	1	2	3 or more
Number of days:	28	20	7	5

Extensive data accumulated for this station showed that, prior to the promotional campaign, the mean number of calls during this eight-hour period each day had been 1.5. Treating the 60 days of data as a random sample, test if the number of daily calls during this period is Poisson-distributed with $\lambda = 1.5$. Control the α risk at .05. State the alternatives, decision rule, and conclusion.

16.4 Refer to Problem 16.3. The sample mean number of calls for the 60 days of data is $\bar{X} = .9$.

a. Test whether the number of calls follows a Poisson distribution (i.e., with no specification of the value of λ). Use $\alpha = .05$. State the alternatives, decision rule, and conclusion.

b. The test in Problem 16.3 leads to the conclusion that the population is not Poisson with $\lambda = 1.5$. Is the conclusion in **a** inconsistent with the conclusion of the test in Problem 16.3? Explain.

16.5 A researcher observed the final digit in the daily volume of stock transactions during the past 50 days. The frequency distribution was as follows:

Final digit:	0	1	2	3	4	5	6	7	8	9
Frequency:	5	2	4	4	7	5	6	5	7	5

Test if the final digits were generated from a discrete uniform probability distribution for integer outcomes $0, 1, \ldots, 9$, controlling the α risk at .05. State the alternatives, decision rule, and conclusion.

*16.6 Refer to Problem 10.7. The airline researcher conjectures that the number of no-shows per flight can be treated as Poisson-distributed. She has calculated the sample mean for the 100 observations and found it to be $\bar{X} = 1.500$.

a. Test the conjecture, controlling the α risk at .10. State the alternatives, decision rule, and conclusion. Pool the classes 4, 5, and 6 in performing your test.

b. What is the reason for pooling the classes in **a**?

16.7 In a product-testing experiment, each of 150 subjects independently examined three patches of fabric. For each patch, the subject was asked whether the fiber is synthetic or natural. The numbers of correct identifications by subjects were as follows:

Number of correct identifications by subject:	0	1	2	3
Number of subjects:	9	40	36	65

If the subjects simply make a separate guess for each patch, the number of correct identifications by a subject will follow a binomial probability distribution with $n = 3$ and $p = .5$.

a. Verify that if all subjects guess, the expected numbers of subjects giving 0, 1, 2, and 3 correct identifications are 18.75, 56.25, 56.25, and 18.75, respectively.

b. Test whether or not the subjects in this experiment were guessing, controlling the α risk at .01. State the alternatives, decision rule, and conclusion.

*16.8 Refer to Problem 10.32. Test whether or not service lives for electrolytic cells of this design are normally distributed. Use a chi-square test with five equal-probability classes, and control the α risk at .10. State the alternatives, decision rule, and conclusion. (Hint: $\bar{X} = 1,379.13$, $s = 399.01$.)

16.9 The following 50 numbers were generated by a computer program written by a graduate student to generate standard normal random numbers:

−0.957	−0.238	−1.029	0.551	−0.298
0.525	−0.869	0.479	0.418	1.064
−1.865	−1.016	2.709	0.074	0.162
−0.273	0.417	−0.057	0.524	−0.129
−0.035	0.056	−0.300	0.479	−1.204
0.371	0.561	−0.594	0.326	1.097
−0.702	−2.357	−1.047	1.114	−0.916
−0.432	1.956	−1.347	1.068	1.222
−0.465	−0.281	0.996	0.772	−1.153
0.120	0.932	−1.023	−0.226	1.298

Use a chi-square procedure to test if the random-number generator is generating numbers from a standard normal distribution. Control the α risk at .01, and use 10 equal-probability classes. State the alternatives, decision rule, and conclusion.

16.10 An industrial machine has a 1.5-meter hydraulic hose which ruptures occasionally. The manufacturer has recorded the locations of these ruptures for a random sample of 25 ruptured hoses. These locations (measured in meters from the pump end of the hose) are:

1.32	1.19	1.21	1.36	.64	1.46	.80	1.37	.38
1.07	.13	1.16	.33	1.42	1.27	.08	.75	
1.22	1.37	1.14	1.43	.97	1.12	.27	1.46	

a. Use a chi-square procedure with five equal-probability classes to test if rupture location is uniformly distributed along the length of the hose. Control the α risk at .10. State the alternatives, decision rule, and conclusion.

b. Examine the residuals $f_i - F_i$ of your test in a. In which section of the hose is the rupture frequency most out of line with the expectations under H_0? What is the implication of your finding for the engineering design of this hose?

*16.11 Rates of return (in percent) realized on a sample of 10 oil stocks last year were: 3, 18, 21, 6, 25, 12, −1, 35, 9, 15.

    a. Construct the cumulative sample function and plot it on a graph.
    b. Treat the 10 rates as a random sample from a continuous probability distribution and obtain the 90 percent confidence band for the cumulative probability function. Plot the confidence band on your graph in **a**.

16.12 Fifteen consumers were selected randomly to test a new home garden tool and then each was asked what would be a reasonable price for this tool. The answers (in dollars) were: 18, 23, 36, 33, 21, 26, 26, 16, 31, 30, 41, 27, 29, 40, 38.
    a. Construct a graph showing the 95 percent confidence band for the cumulative probability function of the price considered reasonable by consumers.
    b. From your confidence band, what is the maximum proportion of consumers who consider a price of $20 or less reasonable?

*16.13 Refer to Problem 16.9. Consider the 10 values in the first two rows. Use the K-S procedure to test if these were generated from a standard normal distribution. Control the α risk at .10. State the alternatives, decision rule, and conclusion.

16.14 Refer to Problem 16.12. Before collecting the data, the market researcher in charge of the study hypothesized that the prices would follow a uniform distribution with $a = \$15$ and $b = \$45$. Use the K-S procedure to test the researcher's hypothesis, controlling the α risk at .05. State the alternatives, decision rule, and conclusion.

EXERCISES

16.15 Show that test statistic X^2 in (16.6) can be written as the following weighted sum of the residuals $f_i - F_i$:

$$X^2 = \sum_{i=1}^{k} \frac{f_i}{F_i}(f_i - F_i)$$

16.16 Consider the test statistic X^2 in (16.6). Assume that H_0 fully specifies the population distribution and that, consequently, no parameters need to be estimated from the sample data.
    a. Show that for any class i, $E\{(f_i - F_i)^2/F_i\} = 1 - (F_i/n)$ when H_0 holds. [*Hint:* f_i is a binomial random variable with mean F_i and variance $F_i(n - F_i)/n$ in this case.]
    b. Use the result in **a** to show that $E\{X^2\} = k - 1$ when H_0 holds.

16.17 $D(1 - \alpha; n)$ approaches 0 as n increases for any α. What does this imply about the probability of a Type II error with the K-S goodness of fit procedure when n is large?

16.18 Show that for a continuous random variable, $S(x)$ is a consistent and unbiased estimator of $F(x)$ at any specified value of x. [*Hint:* Let $F(x) = p$ and $S(x) = \bar{p}$ and employ the Chebyshev inequality.]

STUDIES

16.19 Refer to the after-pollution-controls data set in Table 3.1a (p. 46). Consider these data to constitute a random sample and ignore the rounded nature of the data.
    a. Use the chi-square procedure to test whether the data were generated from an exponential probability distribution with mean 28. Use eight equal-probability

classes and $\alpha = .05$. State the alternatives, decision rule, and conclusion. (*Hint:* The first two classes are: 0–under 3.64, 3.64–under 8.12.)

b. The mean of the data set is $\bar{X} = 27.6$. Would the goodness of fit be much improved if you had tested only whether or not the probability distribution is exponential? Explain.

c. Plot the residuals $f_i - F_i$ obtained in your test in **a**. What does the plot show about the lack of fit of an exponential distribution?

16.20 Refer to Problem 16.7. A statistician has suggested an alternative model—namely, that a proportion P of the subjects know the identification for certain and the remainder guess.

a. Test the aptness of this model when $P = .3$, controlling the α risk at .01. State the alternatives, decision rule, and conclusion.

b. How might one determine an estimate of P here? (*Hint:* Utilize the X^2 statistic.)

16.21 A newspaper's sports page contained the following frequency distribution of times between consecutive goals for the city's soccer team in a recent season:

Time between Goals (*minutes*)	Number of Cases	Time between Goals (*minutes*)	Number of Cases
0–under 10	105	30–under 40	21
10–under 20	72	40–under 50	14
20–under 30	28	50 and over	20

For the 260 cases, $\bar{X} = 20.0$.

a. Test if the times between goals are exponentially distributed, controlling the α risk at .05. Since the original data are not available for setting up equal-probability classes, the classes given in the newspaper article must be used. State the alternatives, decision rule, and conclusion.

b. A sportswriter, commenting on the data, said they showed that goal scoring has a degree of contagion—once the team scores a goal, there is a high probability that another goal will follow shortly. Does the residual $f_i - F_i$ for the class 0–under 10 support this? Discuss.

16.22 A financial analyst has undertaken a computer simulation study of a complex and risky capital investment proposal. As part of the study, he has obtained in 20 simulation runs the following rates of return (in percent), which may be considered as a simple random sample from the true rate-of-return distribution for this investment proposal:

−4.7	22.7	12.3	5.1	20.4	16.7	11.1
−1.5	4.3	2.6	−11.3	1.0	13.4	23.0
12.1	10.3	−2.1	−3.4	1.1	10.1	

a. Construct a graph showing the 90 percent confidence band for the cumulative probability function $F(x)$. Interpret this band.

b. From your confidence band, obtain an interval estimate for $F(0)$, the probability that the investment proposal will produce a negative or zero rate of return.

c. How many rates of return would have to be simulated in order to have probability .99 that $S(x)$ will lie within a vertical distance of $\pm.04$ of $F(x)$ for all values x?

d. (Based on Optional Topic Section 12.6.) Letting $p = F(0)$, obtain an exact 90 percent confidence interval for p using (12.14). Why is this interval tighter than the one obtained in **b**?

17
Multinomial Populations

17.1 NATURE OF MULTINOMIAL POPULATIONS

In many problems, one is interested in populations containing elements classified into several categories or classes. Such populations are called multinomial populations:

(17.1) When each element of a population is assigned to one (and only one) of two or more attribute classes or categories, the population is called a *multinomial population.*

We encountered multinomial populations in Chapter 4. Table 4.1a provides an example. We now consider some other examples.

☐ Examples
1. A large lot of used machine components is the population of interest. Each machine component is classified according to its condition, the attribute categories being excellent, good, poor, and scrap.
2. All of the households in a city constitute the population of interest. Each household is classified as owning no cars, one car, or two or more cars.
3. The process of producing memory chips for computers is the population of interest. A chip is classified as acceptable, having minor defects, or having major defects. Observe that the population here is infinite.

☐

Parameters of Interest

We denote the number of attribute classes into which a population element can be classified by k. Thus, in Example 1 we have $k = 4$ and in Examples 2 and 3 we have $k = 3$. The probability that a population element selected at random comes from the ith attribute class is denoted by p_i $(i = 1, 2, \ldots, k)$. Of course, these probabilities must sum to 1.

For example, the probabilities for the lot of machine components might be:

i	Condition of Component	p_i
1	Excellent	$p_1 = .4$
2	Good	$p_2 = .3$
3	Poor	$p_3 = .2$
4	Scrap	$p_4 = .1$
	Total	1.0

The probabilities p_i are the parameters of interest in multinomial populations but are usually unknown. Sampling is then employed to make inferences about these parameters.

Comment

When only $k = 2$ categories are used for classifying a population element, a random outcome from the multinomial population is equivalent to a Bernoulli trial. For example:

i	Sex of Employee	p_i
1	Male	$p_1 = p$
2	Female	$p_2 = 1 - p$
	Total	1.0

17.2 MULTINOMIAL PROBABILITY DISTRIBUTIONS

When a sample of n elements is selected from a multinomial population, we denote the number of sample observations from the ith attribute category by f_i. Of course, the sum of the f_i's is equal to the sample size n.

For example, a sample of 10 machine components might yield the following results:

i	Condition of Component	f_i
1	Excellent	$f_1 = 3$
2	Good	$f_2 = 3$
3	Poor	$f_3 = 2$
4	Scrap	$f_4 = 2$
	Total	$n = 10$

When a simple random sample is selected from an infinite multinomial population, the probability of obtaining the sample frequencies $f_1, f_2, \ldots, f_k$, denoted by $P(f_1, f_2, \ldots, f_k)$, is given by the multinomial probability function:

(17.2) The *multinomial probability function* is:

$$P(f_1, f_2, \ldots, f_k) = \frac{n!}{f_1! \, f_2! \, \cdots \, f_k!} p_1^{f_1} \, p_2^{f_2} \cdots p_k^{f_k}$$

where: $\displaystyle\sum_{i=1}^{k} f_i = n$

$\displaystyle\sum_{i=1}^{k} p_i = 1$

Observe that the multinomial probability function is discrete, with parameters n and $p_1, p_2, \ldots, p_k$. Each different set of parameter values defines a different probability function in the multinomial family.

When the multinomial population is finite, the multinomial probability function (17.2) provides approximate probabilities of the sample outcomes as long as the sampling fraction n/N is not large. Throughout this chapter, we shall assume that finite populations can be treated as infinite ones.

☐ Examples

1. We wish to find the probability of the outcomes of the machine component sample when the probabilities p_i are those cited earlier. Thus, we have:

$$p_1 = .4 \qquad p_2 = .3 \qquad p_3 = .2 \qquad p_4 = .1$$
$$n = 10 \qquad f_1 = 3 \qquad f_2 = 3 \qquad f_3 = 2 \qquad f_4 = 2$$

We then obtain:

$$P(3, 3, 2, 2) = \frac{10!}{3!\ 3!\ 2!\ 2!}(.4)^3(.3)^3(.2)^2(.1)^2 = .01742$$

2. For the same parameter values, we wish to find the probability of the sample outcomes $f_1 = 8$, $f_2 = 2$, $f_3 = 0$, $f_4 = 0$:

$$P(8, 2, 0, 0) = \frac{10!}{8!\ 2!\ 0!\ 0!}(.4)^8(.3)^2(.2)^0(.1)^0 = .002654$$

☐

Comment

The binomial probability function (6.3) is a special case of the multinomial probability function (17.2). This is to be expected since we saw that a Bernoulli trial is a special case of a random outcome from a multinomial population. Specifically, when $k = 2$, (17.2) reduces to (6.3) with:

$$f_1 = x \qquad f_2 = n - x$$
$$p_1 = p \qquad p_2 = 1 - p$$

17.3   INFERENCES CONCERNING THE PARAMETERS p_i

Various types of inferences can be made concerning the parameters p_i of a multinomial population. We discuss here two basic types—inferences on a given parameter p_i and inferences on all the p_i.

Assumptions

1. A simple random sample of size n has been selected from an infinite or large multinomial population.
2. The sample size n is reasonably large.

Inferences for One p_i

The estimation or testing of any one of the multinomial parameters p_i is carried out by the procedures of Chapter 12 for a population proportion. To see why this is the case, consider our machine component example again. Suppose we wish to estimate p_1, the probability that a used component is in excellent condition. To estimate p_1, we do not care what the other categories are and thus can combine them into one class consisting of all components that are not in excellent condition. But then we have the binomial situation where there are two categories (excellent condition, not excellent condition), with probabilities p_1 and $1 - p_1$, respectively, and all of our earlier inference procedures for a population proportion are applicable.

☐   Examples

1. A railroad engineer is investigating maintenance problems with rolling stock. He is currently studying a certain type of bearing failure to ascertain whether occurrences of these failures are associated with operational levels only or whether other factors are also involved. Table 17.1 contains a record of 1,044 bearing failures on rolling stock over the past several years, classified by quarter of the year of occurrence.

   The engineer believes it is reasonable to treat these data as a random sample from an infinite multinomial population of bearing failures. Here, p_1 is the probability that a failure occurs in the first quarter, p_2 is the probability that it occurs during the second quarter, and p_3 and p_4 are similarly defined.

   To estimate the probability of a failure occurring in the first quarter, we use the sample proportion:

   $$\hat{p}_1 = \frac{f_1}{n} = \frac{249}{1,044} = .2385$$

   A 90 percent confidence interval for p_1 is desired. Since $n = 1,044$, we employ the large-sample confidence interval (12.9) for a population proportion. We require:

   $$s\{\hat{p}_1\} = \sqrt{\frac{\hat{p}_1(1 - \hat{p}_1)}{n - 1}} = \sqrt{\frac{.2385(1 - .2385)}{1,044 - 1}} = .0132$$

   $$z(1 - \alpha/2) = z(.95) = 1.645$$

TABLE 17.1   *Record of bearing failures in rolling stock, classified by quarter of occurrence*

Quarter of Year i	Number of Failures f_i
1	249
2	256
3	297
4	242
Total $n = 1,044$	

Hence, the 90 percent confidence limits for p_1 are $.2385 \pm 1.645(.0132)$ and the confidence interval is:

$$.217 \leq p_1 \leq .260$$

Thus, the engineer can conclude with 90 percent confidence that the probability of a bearing failure occurring during the first quarter is between .217 and .260.

2. The proportion of annual traffic carried in the first quarter is .210. The engineer would like to test whether the relative frequency of bearing failure occurrences in the first quarter is the same as the proportion of traffic. That is, he would like to test:

$$H_0: p_1 = .210$$

$$H_1: p_1 \neq .210$$

The α risk is to be controlled at .10. Since the 90 percent confidence interval obtained in Example 1 does not include .210, H_1 should be concluded. □

Inferences for All p_i

At times, it is desired to make inferences about all of the parameters p_i of a multinomial population.

□ Illustration

In our railroad maintenance example, the engineer would like to test whether the occurrences of bearing failures by quarter vary in direct proportion to the volume of railroad traffic. The quarterly proportions of annual traffic are as follows:

Quarter:	1	2	3	4
Proportion of annual traffic:	.21	.23	.30	.26

For example, about 21 percent of annual traffic is handled in the first quarter of each year. □

Alternatives. In this type of situation, the alternatives take the following form:

(17.3)

$$H_0: p_i = p_{iI} \quad \text{for all } i$$

$$H_1: p_i \neq p_{iI} \quad \text{for some } i$$

Here, p_{iI} denotes the value specified for the parameter p_i $(i = 1, 2, \ldots, k)$ in H_0. In the engineer's test, H_0 would be:

$$H_0: p_1 = p_{1I} = .21 \qquad p_3 = p_{3I} = .30$$

$$p_2 = p_{2I} = .23 \qquad p_4 = p_{4I} = .26$$

Note that H_1 states in formal terms that H_0 does not hold. This does not require that *all* $p_i \neq p_{iI}$. For instance, alternative H_0 in our example would be incorrect if $p_1 = .21$ and $p_2 = .23$ but $p_3 = .40$ and $p_4 = .16$, even though p_1 and p_2 do agree with the specified values.

Test Procedure. The test for the multinomial alternatives (17.3) is precisely of the same form as the goodness of fit test (16.8), and Figure 16.1 describes the test procedure. The correspondence to the earlier test can be seen as follows:

1. In each case, the sample outcomes are classified into one of k classes.
2. Alternative H_0 provides the expected frequencies F_i when H_0 holds.
3. The test then compares the observed frequencies f_i with the frequencies F_i expected when H_0 holds.

Since the multinomial alternative H_0 in (17.3) specifies all of the probabilities p_i so that no parameters must be estimated, the degrees of freedom for the χ^2 distribution here are $k - m - 1 = k - 0 - 1 = k - 1$.

(17.4) When the alternatives are:

$$H_0: p_i = p_{i\text{I}} \qquad \text{for all } i$$

$$H_1: p_i \neq p_{i\text{I}} \qquad \text{for some } i$$

and the sample is reasonably large, the appropriate decision rule to control the α risk is:

$$\text{If } X^2 \leq \chi^2(1 - \alpha; k - 1), \text{ conclude } H_0$$

$$\text{If } X^2 > \chi^2(1 - \alpha; k - 1), \text{ conclude } H_1$$

where: X^2 is given by (16.6)

Again, we shall use the rule that the sample size is large enough for the χ^2 approximation when no F_i is less than 2 and most F_i's are at least 5.

☐   **Example**

The railroad engineer wishes to test:

$$H_0: p_1 = p_{1\text{I}} = .21 \qquad p_3 = p_{3\text{I}} = .30$$

$$p_2 = p_{2\text{I}} = .23 \qquad p_4 = p_{4\text{I}} = .26$$

$$H_1: p_i \neq p_{i\text{I}} \qquad \text{for some } i$$

The α risk is to be controlled at .05.

Table 17.2 repeats the observed frequencies f_i from Table 17.1 and also shows the probabilities $p_{i\text{I}}$ under H_0. The expected frequencies F_i when H_0 holds are then obtained as usual by multiplying the probabilities $p_{i\text{I}}$ by n; i.e., $F_i = np_{i\text{I}}$. The test statistic X^2 is then calculated in the usual way; we see from Table 17.2 that $X^2 = 9.1208$. For $\alpha = .05$ and $k - 1 = 4 - 1 = 3$, we require $\chi^2(.95; 3) = 7.81$. Hence, the decision rule is:

$$\text{If } X^2 \leq 7.81, \text{ conclude } H_0$$

$$\text{If } X^2 > 7.81, \text{ conclude } H_1$$

Since $X^2 = 9.12 > 7.81$, we conclude H_1—that the occurrence of bearing failures by quarter is not entirely consistent with volume of traffic.

The engineer noted in Table 17.2 the particularly large residuals (in absolute terms) for the first and fourth quarters, and that the residuals were positive for the first two quarters

TABLE 17.2 *Chi-square test for multinomial population—Bearing failures example*

Quarter i	Observed Frequency f_i	Probability under H_0 p_{iI}	Expected Frequency under H_0 $F_i = np_{iI}$	Residual $f_i - F_i$	Relative Squared Residual $(f_i - F_i)^2/F_i$
1	249	.21	219.24	29.76	4.0397
2	256	.23	240.12	15.88	1.0502
3	297	.30	313.20	−16.20	.8379
4	242	.26	271.44	−29.44	3.1930
Total	$n = 1{,}044$	1.00	1,044.00	0.0	$X^2 = 9.1208$

and negative for the last two. These proved to be valuable clues to him in his search for factors in addition to operating volume which affect the occurrence of bearing failures by quarter, such as temperature and company maintenance policy. ☐

Comment

When there are $k = 2$ categories only, test (17.4) reduces to a two-sided test for a population proportion.

17.4 BIVARIATE MULTINOMIAL POPULATIONS

Often when considering multinomial populations, we are interested in two or more characteristics of the population elements and wish to study the relationships between the variables. In this section, we are concerned with the analysis of bivariate multinomial populations.

(17.5) A multinomial population having attribute categories arranged in a bivariate classification system is called a *bivariate multinomial population*.

We encountered bivariate multinomial populations in Chapter 4. Table 4.1b provides an example. We now consider several other examples.

☐ Examples

1. Persons in the labor force may be categorized according to the following bivariate classification scheme in a study of the relationship between age and employment status:

		Age Class			
		25 and Under	26–45	46–65	Over 65
Employment Status	Unemployed Employed part-time Employed full-time				

2. Electronic instrument failures may be categorized by the type of failure that is observed (T_1 or T_2) and the location of the failure in the instrument (L_1, L_2, or L_3) as follows:

	Location of Failure		
	L_1	L_2	L_3
Type of T_1			
Failure T_2			

This bivariate classification system will enable the analyst to study whether or not type and location of failure are related variables. If so, the nature of the relationship is important for maintenance trouble-shooting and for some aspects of instrument design. ☐

Comment

Note that each of the bivariate multinomial populations in the two examples can be viewed as an ordinary multinomial population with every cell in the bivariate classification constituting a class. Thus, the classification in Example 1 involves $k = 3(4) = 12$ classes, and in Example 2 there are $k = 2(3) = 6$ classes. The bivariate arrangement facilitates, of course, the analysis of the relationship between the two variables.

Parameters of Interest

The probability parameters of a bivariate multinomial population are displayed in a convenient form in Table 17.3a. The table shows a bivariate classification system with the row variable having r classes and the column variable having c classes so, in all, the multinomial population has $k = rc$ classes or cells. The symbol p_{ij} de-

TABLE 17.3 *Bivariate multinomial probability distribution*

(a)
Bivariate multinomial probability parameters

		Class for Column Variable				
		$j = 1$	$j = 2$	$\cdots$	$j = c$	Total
Class for Row Variable	$i = 1$	p_{11}	p_{12}	$\cdots$	p_{1c}	$p_{1.}$
	$i = 2$	p_{21}	p_{22}	$\cdots$	p_{2c}	$p_{2.}$
	$\vdots$	$\vdots$	$\vdots$	$\vdots$	$\vdots$	$\vdots$
	$i = r$	p_{r1}	p_{r2}	$\cdots$	p_{rc}	$p_{r.}$
	Total	$p_{.1}$	$p_{.2}$	$\cdots$	$p_{.c}$	1.0

(b)
Bivariate multinomial sample frequencies

		Class for Column Variable				
		$j = 1$	$j = 2$	$\cdots$	$j = c$	Total
Class for Row Variable	$i = 1$	f_{11}	f_{12}	$\cdots$	f_{1c}	$f_{1.}$
	$i = 2$	f_{21}	f_{22}	$\cdots$	f_{2c}	$f_{2.}$
	$\vdots$	$\vdots$	$\vdots$	$\vdots$	$\vdots$	$\vdots$
	$i = r$	f_{r1}	f_{r2}	$\cdots$	f_{rc}	$f_{r.}$
	Total	$f_{.1}$	$f_{.2}$	$\cdots$	$f_{.c}$	n

notes the joint probability that a population element selected at random falls in the ith class of the row variable and the jth class of the column variable. The symbols $p_{i.}$ and $p_{.j}$ denote the marginal probabilities for the ith row and jth column, respectively. An example of the display in Table 17.3a with probability values is found in Table 4.1b.

Sampling a Bivariate Multinomial Population

When a sample of size n is selected from a bivariate multinomial population, the sample frequencies will be denoted as shown in Table 17.3b. In the table, f_{ij} denotes the number of sample observations in the ith class of the row variable and the jth class of the column variable. We shall let $f_{i.}$ and $f_{.j}$ denote the totals of the sample frequencies for the ith row and jth column, respectively. An example of the display in Table 17.3b with sample values is found in Table 17.4, which we will discuss shortly.

Test for Statistical Independence (Contingency Table Test)

In analyzing bivariate multinomial populations, the first question of interest usually is whether or not the two variables are statistically independent. If they are independent, we know from Chapter 4 that there is no relationship between them. If it turns out that they are not independent and a relationship does exist between the two variables, the next step in the analysis then is to study the nature of the relationship. We begin with the first step of the analysis, testing whether or not the two variables are independent.

The test for statistical independence has a structure with which we are now well familiar: We compare the observed frequencies f_{ij} with the frequencies F_{ij} that are expected if the variables are independent, and calculate the statistic X^2. If X^2 is small, it indicates that the variables are statistically independent; if X^2 is large, the data support the conclusion that the variables are not independent.

Assumptions

1. A simple random sample of size n has been selected from an infinite or large bivariate multinomial population.
2. The sample size n is reasonably large.

☐ Illustration

For the study of electronic instrument failures classified by type and location of failure, historical data on $n = 200$ failures are available. These are displayed in Table 17.4. The analyst desires to test whether or not type and location of failure are statistically independent variables. She assumes that the 200 observations constitute a random sample from an infinite bivariate multinomial population of failures for the electronic instrument under study. ☐

Alternatives. For two variables to be statistically independent, we know from (4.24) that each joint probability in the bivariate distribution must equal the

TABLE 17.4 *Testing statistical independence of type and location of failure.* The numbers in parentheses are the estimated expected frequencies under H_0.

Type of Failure		Location of Failure			
		$j = 1$ L_1	$j = 2$ L_2	$j = 3$ L_3	Total
$i = 1$	T_1	50 (53.84)	16 (20.37)	31 (22.80)	97
$i = 2$	T_2	61 (57.17)	26 (21.63)	16 (24.21)	103
	Total	111	42	47	$n = 200$

product of the corresponding row and column marginal probabilities; i.e., in the notation of Table 17.3a:

$$p_{ij} = p_{i.}p_{.j} \qquad \text{for all } (i, j)$$

Thus, the appropriate alternatives for a test of statistical independence are:

(17.6)
$$H_0: p_{ij} = p_{i.}p_{.j} \qquad \text{for all } (i, j)$$
$$H_1: p_{ij} \neq p_{i.}p_{.j} \qquad \text{for some } (i, j)$$

Here, H_0 represents statistical independence and H_1 statistical dependence. Note in H_1 that statistical dependence exists even if only some (not all) $p_{ij} \neq p_{i.}p_{.j}$.

Computing Expected Frequencies. Under H_0 of (17.6), $p_{ij} = p_{i.}p_{.j}$ for each cell of the bivariate multinomial population. Hence, an estimate of p_{ij} under the assumption of statistical independence can be obtained by first estimating the marginal probabilities $p_{i.}$ and $p_{.j}$, and then forming their product. Estimates of $p_{i.}$ and $p_{.j}$, which we denote by $\bar{p}_{i.}$ and $\bar{p}_{.j}$, may be computed from the marginal sample frequencies as follows:

(17.7)
$$\bar{p}_{i.} = \frac{f_{i.}}{n} \qquad \bar{p}_{.j} = \frac{f_{.j}}{n}$$

Referring to Table 17.4, we find, for example, that $\bar{p}_{1.} = 97/200 = .4850$ and $\bar{p}_{.1} = 111/200 = .5550$. Hence, a point estimate of the joint probability p_{11} under the assumption of statistical independence is $\bar{p}_{11} = \bar{p}_{1.}\bar{p}_{.1} = .4850(.5550) = .2692$. Thus, point estimates of the joint probabilities if H_0 holds are obtained as follows:

(17.8)
$$\bar{p}_{ij} = \bar{p}_{i.}\bar{p}_{.j}$$

As another example, $\bar{p}_{23} = \bar{p}_{2.}\bar{p}_{.3} = (103/200)(47/200) = .1210$.

Finally, to obtain the expected frequencies F_{ij} if H_0 holds, we multiply the estimated joint probabilities by the sample size n; i.e.:

(17.9)
$$F_{ij} = n\bar{p}_{ij} = n\bar{p}_{i.}\bar{p}_{.j}$$

For example, the estimated expected frequency if H_0 holds for failure type T_1 in location L_1 is $F_{11} = n\bar{p}_{11} = 200(.2692) = 53.84$. The expected frequencies F_{ij} for our electronic instrument example are shown in parentheses in Table 17.4 alongside the corresponding observed frequencies f_{ij}.

Test Statistic and Decision Rule. As usual, we employ the X^2 test statistic:

$$(17.10) \qquad X^2 = \sum_{i=1}^{r} \sum_{j=1}^{c} \frac{(f_{ij} - F_{ij})^2}{F_{ij}}$$

The X^2 statistic is obtained as always by summing over all classes, but a double summation is required here since each class (i, j) is identified in terms of the categories of both variables. (For a review of the double summation notation, refer to Appendix A, Section A.1.)

To control the α risk, we need to know the sampling distribution of X^2 when H_0 holds:

(17.11) When the variables of a bivariate multinomial population are statistically independent and the sample size is reasonably large:

$$X^2 \simeq \chi^2[(r-1)(c-1)]$$

As usual, the sample size is considered reasonably large for the χ^2 approximation to hold if no F_{ij} is less than 2 and most are at least 5.

The decision rule then follows the standard form:

(17.12) When the alternatives are:

$$H_0: p_{ij} = p_{i.}p_{.j} \qquad \text{for all } (i, j)$$

$$H_1: p_{ij} \neq p_{i.}p_{.j} \qquad \text{for some } (i, j)$$

and the sample is reasonably large, the appropriate decision rule to control the α risk is:

$$\text{If } X^2 \leq \chi^2[1 - \alpha; (r-1)(c-1)], \text{ conclude } H_0$$

$$\text{If } X^2 > \chi^2[1 - \alpha; (r-1)(c-1)], \text{ conclude } H_1$$

where: X^2 is given by (17.10)

☐ **Example**

The analyst wishes to test whether or not type of failure and location of failure are statistically independent and specifies that the α risk is to be controlled at .05. Since $r = 2$ and $c = 3$, there are $(r-1)(c-1) = (2-1)(3-1) = 2$ degrees of freedom for the χ^2 distribution when H_0 holds. Hence, for $\alpha = .05$, we require $\chi^2(.95; 2) = 5.99$.

To complete the test, we compute test statistic (17.10) from the data in Table 17.4 as follows:

$$X^2 = \frac{(50 - 53.84)^2}{53.84} + \frac{(16 - 20.37)^2}{20.37} + \cdots + \frac{(16 - 24.21)^2}{24.21} = 8.08$$

Because $X^2 = 8.08 > 5.99$, we conclude H_1—that type and location of failure are statistically dependent. The next step in the investigation therefore will be to study the nature of the relationship. ☐

Comments

1. A shortcut formula for calculating F_{ij} can be obtained by combining (17.7) and (17.9):

(17.13)
$$F_{ij} = \frac{f_{i.}f_{.j}}{n}$$

2. A bivariate sample distribution (as shown in Table 17.3b) is often called a *contingency table*. For this reason, the test procedure described here is sometimes called a *contingency table test*.
3. The degrees of freedom for the χ^2 distribution for the statistical independence test are obtained as follows. Since there are rc multinomial classes, we have $k = rc$. The number of parameters estimated from the sample is $(r - 1) + (c - 1)$. The reason is that both sets of marginal probabilities must add to 1, hence only $r - 1$ of the row marginal probabilities and $c - 1$ of the column marginal probabilities are independently estimated from the sample data. Thus, the χ^2 distribution for the test has $k - m - 1 = rc - [(r - 1) + (c - 1)] - 1 = (r - 1)(c - 1)$ degrees of freedom.

Examination of Nature of Statistical Dependence

When the test for statistical independence leads to the conclusion of dependence, a simple way to examine the nature of the statistical dependence is to consider the conditional probability distributions, as discussed in Chapter 4. Of course, since only sample data are available here, we must work with estimated conditional probabilities.

Example

The analyst for the electronic failures study was particularly interested in the effect of location of failure on type of failure. Table 17.5 shows the estimated conditional probability distributions for type of failure for each of the three locations. The estimated probabilities are computed from the sample data in Table 17.4. Note how the conditional probability distribution for location L_3 differs from the distributions for locations L_1 and L_2. Indeed, the more likely failure type has reversed. Design engineers can use such insights to identify potential causes of failure and to recommend design modifications aimed at improved instrument reliability.

TABLE 17.5 *Estimated conditional probability distributions for type of instrument failure*

Type of Failure		Conditional upon Location of Failure		
		L_1	L_2	L_3
T_1		.45	.38	.66
T_2		.55	.62	.34
	Total	1.00	1.00	1.00

17.5 COMPARISONS OF SEVERAL MULTINOMIAL POPULATIONS

Frequently, it is of interest to compare several multinomial populations to see if they are identical.

☐ Examples

1. An educator wishes to compare the performance of three sections of the same course, each taught by a different instructional method, to see if the methods produce the same or different letter grade distributions.
2. A market researcher wishes to compare the preferences among a sample of men with those among a sample of women to determine whether the proportions of persons preferring each of four different syrup flavors are the same for men and women. ☐

Parameters of Interest

Table 17.6 illustrates the parameters of interest. There are c multinomial populations (e.g., 3 instructional methods, 2 sexes), each with the same r attribute classes (e.g., 5 letter grades, 4 syrup flavors). The probability that an element selected at random from population j falls into class i is denoted by p_{ij}. Of course, the p_{ij}'s sum to 1 for each population j.

Assumptions

1. Independent random samples of sizes $n_1, n_2, \ldots, n_c$ are selected from c infinite or large multinomial populations.
2. Each of the c sample sizes $n_1, n_2, \ldots, n_c$ is reasonably large.

Alternatives

If all of the multinomial populations are identical, $p_{11} = p_{12} = \cdots = p_{1c}$. Similarly, p_{2j} will be the same value for each population j; and so on. Thus, a test for the identity of the c populations involves the following alternatives:

(17.14)
$$H_0: p_{i1} = p_{i2} = \cdots = p_{ic} \quad \text{for all } i$$
$$H_1: \text{Not all equalities in } H_0 \text{ hold}$$

TABLE 17.6 *c multinomial populations with the same r classes*

		Multinomial Population		
Class	$j = 1$	$j = 2$	$\cdots$	$j = c$
$i = 1$	p_{11}	p_{12}	$\cdots$	p_{1c}
$i = 2$	p_{21}	p_{22}	$\cdots$	p_{2c}
$\vdots$	$\vdots$	$\vdots$	$\vdots$	$\vdots$
$i = r$	p_{r1}	p_{r2}	$\cdots$	p_{rc}
Total	1.0	1.0	$\cdots$	1.0

Test Procedure

Again, we shall employ the test statistic X^2, which compares the observed frequencies with the frequencies expected if H_0 holds. It turns out that the test statistic X^2 in (17.10) for testing statistical independence in a bivariate multinomial population is also the test statistic for comparing the identity of c multinomial populations, and that the decision rule follows the standard form:

(17.15) When the alternatives are:

$$H_0: p_{i1} = p_{i2} = \cdots = p_{ic} \qquad \text{for all } i$$

$$H_1: \text{Not all equalities in } H_0 \text{ hold}$$

and each of the c samples is reasonably large, the appropriate decision rule to control the α risk is:

$$\text{If } X^2 \leq \chi^2[1 - \alpha; (r - 1)(c - 1)], \text{ conclude } H_0$$

$$\text{If } X^2 > \chi^2[1 - \alpha; (r - 1)(c - 1)], \text{ conclude } H_1$$

where: X^2 is given by (17.10)

The sample sizes may be considered reasonably large when no expected frequency F_{ij} is less than 2 and most are at least 5.

The logic of deriving the expected frequencies for testing the identity of c multinomial populations is not the same as for testing statistical independence in a bivariate multinomial population, even though the algebraic results are identical. We shall explain the logic by means of the syrup flavor preference study mentioned earlier.

☐ Illustration

In the syrup flavor preference study, the market researcher selected a random sample of $n_1 = 250$ men and independently a random sample of $n_2 = 400$ women. Each person was given four syrup flavors to taste and then was asked for the preferred flavor. The sample results are presented in Table 17.7. We let f_{ij} denote the observed frequency in the ith class for the jth population. For our example, $i = 1, 2, 3, 4$ and $j = 1, 2$. Thus, we see from Table 17.7 that $f_{11} = 23$, $f_{12} = 174$, and so on.

TABLE 17.7 *Testing identity of syrup flavor preferences among men and women.* The numbers in parentheses are the estimated expected frequencies under H_0.

Multinomial Class	Preferred Syrup	Sex $j = 1$ Men	$j = 2$ Women	Total
$i = 1$	Flavor 1	23 (75.8)	174 (121.2)	197
$i = 2$	Flavor 2	169 (78.1)	34 (124.9)	203
$i = 3$	Flavor 3	48 (76.2)	150 (121.8)	198
$i = 4$	Flavor 4	10 (20.0)	42 (32.0)	52
	Total	$n_1 = 250$	$n_2 = 400$	$n_T = 650$

We denote the total number of observations in the study by n_T:

$$(17.16) \qquad n_T = \sum_{j=1}^{c} n_j$$

In our example, $n_T = n_1 + n_2 = 250 + 400 = 650$.

The researcher would like to test whether the preference distributions for men and women are the same:

$$H_0: p_{i1} = p_{i2} \qquad \text{for } i = 1, 2, 3, 4$$

$$H_1: \text{Not all equalities in } H_0 \text{ hold}$$

The α risk should be controlled at .01.

If H_0 holds, all multinomial populations are the same. Let $p_1, p_2, \ldots, p_r$ denote the parameters for the common multinomial distribution if H_0 holds. To estimate the frequencies that one expects when H_0 is true, we must first estimate these common parameters. An estimator for p_i is the proportion of observations falling in the ith class among all c samples combined. Denoting this estimator by $\bar{p}_i$, we have:

$$(17.17) \qquad \bar{p}_i = \frac{\sum_{j=1}^{c} f_{ij}}{n_T} = \frac{f_{i.}}{n_T}$$

From Table 17.7, we see that $f_{1.} = 197$ persons of the total of $n_T = 650$ in the study preferred syrup flavor 1 (i.e., fell in class $i = 1$). Thus, $\bar{p}_1 = 197/650 = .3031$. The parameter estimates for the other classes are obtained similarly.

We shall let F_{ij} denote the frequency expected in class i for the sample from the jth population when the multinomial distributions are, in fact, identical—i.e., when H_0 holds. It follows from our foregoing argument that:

$$(17.18) \qquad F_{ij} = n_j \bar{p}_i$$

For instance, in the sample of $n_1 = 250$ men, we would expect $F_{11} = n_1 \bar{p}_1 = 250(.3031) = 75.8$ men to fall in the first class if H_0 is true. Similarly, $F_{21} = n_1 \bar{p}_2 = 250(203/650) = 250(.3123) = 78.1$ men would be expected to fall in the second class. Table 17.7 shows the values of F_{ij} in parentheses immediately beside the corresponding observed frequencies f_{ij}.

The required action limit for the decision rule is $\chi^2[1 - \alpha; (r - 1)(c - 1)] = \chi^2[.99; (4 - 1)(2 - 1)] = \chi^2(.99; 3) = 11.34$. The test statistic (17.10) is calculated from the data in Table 17.7 as follows:

$$X^2 = \frac{(23 - 75.8)^2}{75.8} + \frac{(169 - 78.1)^2}{78.1} + \cdots + \frac{(42 - 32.0)^2}{32.0} = 256.8$$

Because $X^2 = 256.8 > 11.34$, we conclude H_1—that the preference patterns for men and women are different.

The market researcher thereupon investigated the nature of the differences. The estimated probabilities of preference for men and women, based on the data in Table 17.7, are as follows:

Preferred Flavor		Men	Women
1		.092	.435
2		.676	.085
3		.192	.375
4		.040	.105
	Total	1.000	1.000

Clearly, men show a strong preference for flavor 2 and women have strong preferences divided between flavors 1 and 3. We could use the procedures for comparing two population proportions when the samples are independent in Section 13.4 for analyzing further the differences in the probabilities between men and women for any of the flavors.   □

Comments

1. The degrees of freedom $(r - 1)(c - 1)$ for the χ^2 distribution when testing the identity of multinomial populations is arrived at as follows. For any one population sampled, there would be $r - 1$ degrees of freedom if H_0 specified the common probabilities p_1, $p_2, \ldots, p_r$ completely. It can be shown that independent χ^2 variables are additive, as are their degrees of freedom. Hence, there would be $c(r - 1)$ degrees of freedom for all c populations if the common probabilities were specified by H_0. But the common probabilities are not specified by H_0 and have to be estimated from the sample. Hence, $m = r - 1$ degrees of freedom are lost, and the number of degrees of freedom is $c(r - 1) - (r - 1) = (r - 1)(c - 1)$.

2. Formula (17.18) for F_{ij} is identical to formula (17.13) for the expected frequency in the statistical independence test, with $n_j = f_j$ and $n_T = n$, even though the logic in arriving at F_{ij} differs for the two cases. Also, the alternatives tested differ for the two cases.

3. The test for the identity of c multinomial populations is sometimes called a *test of homogeneity*.

4. When there are only $c = 2$ populations and each has $r = 2$ categories, test (17.15) reduces to the two-sided test for two population proportions in (13.27).

17.6   OPTIONAL TOPIC—INFERENCES FOR A RATIO OF MULTINOMIAL PROBABILITIES

Often, we are interested in comparing two probabilities p_i and p_j of a multinomial population in terms of their ratio p_i/p_j.

□   Examples

1. In an election poll, the following results were obtained:

i	Candidate Preference		p_i	f_i
1	A		p_1	$f_1 = 460$
2	B		p_2	$f_2 = 350$
3	Undecided		p_3	$f_3 = 190$
		Total	1.0	$n = 1,000$

It is desired to estimate the margin of preference for candidate A among the "decided" voters. That is , p_1/p_2 is to be estimated. A point estimate is $460/350 = 1.31$. This states that among the "decided" voters, we estimate that 131 favor candidate A for every 100 favoring candidate B. The ratio 1.31, when expressed in the form 1.31 to 1, is called the *odds*, as discussed in Section 4.2.

2. In our railroad maintenance example, the engineer knows that during the third quarter, 143 percent as much volume of traffic is carried as during the first quarter. He would like to estimate p_3/p_1 for the bearing failures to see if this ratio is near 1.43. □

Confidence Interval

Confidence limits for p_i/p_j can be constructed quite easily when both f_i and f_j are large:

(17.19) The confidence limits L and U for p_i/p_j with approximate confidence coefficient $1 - \alpha$, when f_i and f_j are reasonably large, are:

$$L = \frac{f_i}{f_j}\exp\left(-z\sqrt{\frac{1}{f_i} + \frac{1}{f_j}}\right) \qquad U = \frac{f_i}{f_j}\exp\left(z\sqrt{\frac{1}{f_i} + \frac{1}{f_j}}\right)$$

where: $z = z(1 - \alpha/2)$
$\exp(x)$ denotes e^x

(A review of exponentiation and the base e of natural logarithms can be found in Appendix A, Section A.2.)

□   Example

In our railroad maintenance example (data in Table 17.1), the engineer would like to estimate the ratio p_3/p_1 with a 95 percent confidence interval. Since $f_3 = 297$ and $f_1 = 249$, the large-sample confidence limits in (17.19) can be employed. For $1 - \alpha = .95$, $z(.975) = 1.960$. Hence, the limits are:

$$L = \frac{297}{249}\exp\left(-1.960\sqrt{\frac{1}{297} + \frac{1}{249}}\right) \qquad U = \frac{297}{249}\exp\left(1.960\sqrt{\frac{1}{297} + \frac{1}{249}}\right)$$

$$= 1.1928 \exp(-.1684) = 1.008 \qquad\qquad = 1.1928 \exp(.1684) = 1.412$$

and the 95 percent confidence interval for p_3/p_1 is:

$$1.008 \leq \frac{p_3}{p_1} \leq 1.412$$

Thus, the engineer can conclude with 95 percent confidence that the probability a bearing failure occurs in the third quarter is at least as great as the probability the failure occurs in the first quarter, and may be up to 1.41 times as great.

The engineer was interested in this result because the confidence interval does not include 1.43, the ratio of the volumes of traffic in the two quarters. Thus, the occurrence of bearing failures by quarter does not appear to be entirely consistent with traffic volume. □

Comments

1. The confidence limits in (17.19) can also be expressed in logarithms. Taking logarithms to base 10, we obtain for $\log L$ and $\log U$:

(17.19a)
$$\log\left(\frac{f_i}{f_j}\right) \pm z\left(.434294\sqrt{\frac{1}{f_i} + \frac{1}{f_j}}\right)$$

where $z = z(1 - \alpha/2)$ and .434294 (i.e., $\log_{10}e$) is the factor for converting from natural logarithms to common logarithms. Upon taking the antilogarithms, one obtains the limits L and U.

For the preceding example, the logarithms of the lower and upper confidence limits, L and U, are by (17.19a):

$$\log\left(\frac{297}{249}\right) \pm 1.960\left(.434294\sqrt{\frac{1}{297} + \frac{1}{249}}\right) = .07656 \pm .07314$$

so $\log L = .07656 - .07314 = .00342$ and $\log U = .07656 + .07314 = .14970$. Taking antilogarithms gives $L = $ antilog $.00342 = 1.008$ and $U = $ antilog $.14970 = 1.412$.

2. The confidence limits (17.19a) are in the usual form: point estimate $\pm z$ standard deviations. The reason is that $\log(f_i/f_j)$ is approximately normally distributed when f_i and f_j are reasonably large.

PROBLEMS

*17.1 A used machine component selected randomly from a very large lot has probabilities of .5, .3, .1, and .1 of being in excellent, good, poor, or scrap condition, respectively.
   a. What is the probability that a random sample of five components from the lot will contain: (1) two in excellent condition and one in each of the other conditions, (2) three in good condition and two in poor condition?
   b. Construct the probability distribution of all possible outcomes in a random sample of two components from the lot. Which outcome is most probable?

17.2 Refer to Problem 17.1. Answer both parts using respective probabilities .6, .2, .1, and .1.

17.3 The probability that an account will be paid within 30 days is .7, between 31 and 60 days is .2, and after 60 days is .1. The timings of payment for different accounts are statistically independent.
   a. What is the probability that in a random sample of three accounts, two will be paid within 30 days and one will be paid after 60 days? All three will be paid after 60 days?
   b. Construct the probability distribution of all possible payment timing outcomes in a random sample of three accounts. Which outcome is most probable?

17.4 Refer to Problem 17.3. What is the probability that in a random sample of four accounts, one will be paid within 30 days and three will not be paid within 30 days? (*Hint:* Redefine the attribute classes and obtain the appropriate probabilities.)

*17.5 In a random sample of 180 women who use a certain type of skin cream regularly, each was asked to state which of the three brands on the market (A, B, C) she preferred. The preference responses were:

Brand most preferred:	A	B	C
Number of women:	75	54	51

    a. Describe the parameters of the multinomial population being sampled.

    b. Construct a 99 percent confidence interval for the probability a regular skin cream user in the market studied prefers brand A. Interpret your interval estimate.

17.6 The manager of the customer services department of a large chain discount store compiled the following tabulation of the 250 most recent complaints:

Nature of Complaint	Number of Complaints
Quality of merchandise	69
Price	48
Service	122
Other	11
Total	250

    a. Describe the parameters of the relevant multinomial population.

    b. Construct a 95 percent confidence interval for the probability that a complaint deals with service. Interpret your confidence interval.

*17.7 Refer to Problem 17.5. Current market shares for the three brands of skin cream, reflecting purchases by all users (regular users as well as nonregular users), are:

Brand:	A	B	C
Market share (*percent*):	35	45	20

The study director wants to determine whether or not the probabilities of regular users preferring brands A, B, and C are the same as the respective current market shares based on purchases by all users.

    a. Perform the appropriate test, controlling the α risk at .01. State the alternatives, decision rule, and conclusion.

    b. Examine the residuals $f_i - F_i$ for the test and analyze the major differences between the preference distributions for regular users and all users.

17.8 Refer to Problem 17.6. In the entire chain, the distribution of complaints is known from extensive experience to be as follows:

Nature of Complaint	Proportion of Complaints
Quality of merchandise	.25
Price	.19
Service	.24
Other	.32
Total	1.00

    a. Test whether or not the store's complaint pattern is the same as that for the entire chain, controlling the α risk at .05. State the alternatives, decision rule, and conclusion.

    b. Examine the residuals $f_i - F_i$ for the test and analyze how the complaint pattern for the store differs from that for the chain.

17.9 Refer to Problem 17.5.

    a. Test whether or not the preference probabilities for regular users are the same for all three brands. Control the α risk at .10. State the alternatives, decision rule, and conclusion.

    b. Using a confidence interval, test whether or not the probability that a regular user prefers brand B is equal to 1/3. Control the α risk at .10. Is your conclusion inconsistent with the conclusion obtained in **a**? Explain.

*17.10 An investment analyst is studying the relation between stock price movements in two consecutive weeks in March. A random sample of 100 stocks was selected and the price movements of each stock during the two weeks were cross-classified, as follows:

		Movement in Second Week		
		Increase	No Change	Decrease
Movement	Increase	28	6	1
in First	No change	6	32	4
Week	Decrease	2	6	15

a. Test whether or not price movements in the two weeks are statistically independent, controlling the α risk at .10. State the alternatives, decision rule, and conclusion.

b. Examine the residuals $f_i - F_i$ for the test and analyze how the price movements depart from independence.

17.11 A pharmaceutical firm has surveyed a random sample of 120 persons suffering from Parkinson's disease. Among the facts obtained was the following bivariate frequency distribution of disease duration and degree of self-reliance:

		Disease Duration (years)			
		Under 5	5–9	10–14	15 or more
Self-	Considerable	36	25	19	10
Reliance	Little	4	7	9	10

a. Test whether or not disease duration and degree of self-reliance are statistically independent, controlling the α risk at .01. State the alternatives, decision rule, and conclusion.

b. For each disease duration category, obtain the estimated conditional probability distribution of the degree of self-reliance. What do these distributions show about the nature of the relationship between disease duration and self-reliance?

17.12 A savings bank surveyed a random sample of 400 heads of households to determine how many knew the current interest rate paid on six-month savings certificates. The following results were obtained when the survey data were cross-classified by occupational category and response:

Occupation	Did Know	Did Not Know
Wage earner, clerical worker	110	100
Manager, professional	60	20
Other	40	70

The bank management would like to know the probability that a household head knows the current interest rate, whether or not this knowledge is related to occupation, and, if so, the nature of the relationship.

a. Estimate by means of a 90 percent confidence interval the probability that a household head knows the current interest rate. Interpret your interval estimate.

b. Test whether or not occupation and knowledge of the current interest rate are statistically independent, controlling the α risk at .01. State the alternatives, decision rule, and conclusion. What are the implications of your conclusion for the bank?

c. For each occupation, obtain the estimated conditional probability distribution of knowledge of the current interest rate. What do these distributions show about the nature of the relationship between occupation and knowledge of the current interest rate?

*17.13 Refer to Problem 17.5. The study described there was preceded by another study conducted two years earlier. This earlier study used an independent random sample of 150 regular users and yielded the following preference responses:

Brand most preferred:	A	B	C
Number of women:	53	55	42

a. Test whether or not the preference patterns in the two studies are identical, controlling the α risk at .10. State the alternatives, decision rule, and conclusion.

b. Estimate by means of a 90 percent confidence interval how much change has occurred in the probability that a regular user prefers brand A. (*Hint:* See Section 13.4.)

17.14 Random samples of machine parts selected independently from each of three successive large production runs revealed the following data on the number of parts in acceptable condition:

Quality of Part		Run 1	Run 2	Run 3
Acceptable		80	89	56
Unacceptable		10	31	4
	Total	90	120	60

a. Test whether or not the probability of an acceptable part is the same for the three runs. Control the α risk at .05. State the alternatives, decision rule, and conclusion.

b. Test whether or not the probability of an acceptable part is the same for runs 1 and 3; use $\alpha = .05$. What do the results of this test and that in a suggest about run 2? Discuss.

17.15 A market research organization conducted three surveys at about the same time. All pertained to the population of families in the United States, though the three samples were selected independently. The first survey—the income survey—sought information on the relation between clothing expenditures and family income. The second survey—the brand preference survey—sought information on brand preferences among various packaged foods. The third survey—the family-planning survey—sought information on families' procreative intentions. The respective sizes of the three surveys were 1,000, 1,500, and 500 families. The numbers of families refusing to participate in the surveys were 150, 140, and 130, respectively. Assume that each sample is a simple random one, and that all were carried out in the same manner and with the same care and supervision.

a. Do the results indicate that the subject matter of the survey affects the refusal rate? Perform an appropriate test, controlling the α risk at .01. State the alternatives, decision rule, and conclusion.

b. Test whether or not the refusal rates for the income and brand preference surveys differ, controlling the α risk at .01. State the alternatives, decision rule, and conclusion.

c.   Given the conclusion in **a**, was the test in **b** superfluous in the sense that the conclusion obtained in **a** logically implies the conclusion obtained in **b**? Comment.

d.   Explain why the refusal rate might depend on the nature of the survey.

*17.16   Refer to Problem 17.5. Construct a 95 percent confidence interval for the ratio of the probability that a regular user prefers brand A to the probability that brand B is preferred. Does your interval indicate that more regular users prefer brand A than brand B?

17.17   Refer to Problem 17.6. The manager of the customer services department is concerned about how much more frequently complaints are received about service than about quality of merchandise. Construct an appropriate 99 percent confidence interval, and interpret your interval estimate for the manager.

EXERCISES

17.18   Derive the multinomial probability function (17.2), using (A.28) in Appendix A.

17.19   Refer to Problem 17.1. In a random sample of five components from the lot, two are in excellent condition and three are not in excellent condition. What is the conditional probability that the latter three components consist of one component in each of good, poor, and scrap condition, respectively?

17.20   (Based on Optional Topic Section 11.10.) Refer to Problem 17.6.

a.   Test whether the probability that a complaint deals with service is .24, using a chi-square test procedure. Control the α risk at .01. State the alternatives, decision rule, and conclusion. (*Hint:* Combine the other attribute classes.)

b.   For the test in **a**, obtain the standardized test statistic (12.12) used for testing a population proportion, and show that the square of this test statistic equals the test statistic in **a**.

17.21   (Based on Optional Topic Section 11.11.) Refer to Problem 17.11. A computer package that performs the contingency table test for this problem gives as output a P-value of .006. Does this P-value indicate that H_0, the hypothesis of independence, can be rejected if the α risk is to be controlled at .01? At .005? Explain.

STUDIES

17.22   Refer to Problems 17.5 and 17.7. It is known that 30 percent of all users are regular users. Estimate the preference probabilities for nonregular users. Do there appear to be substantial differences in the preference patterns for regular and nonregular users? Discuss.

17.23   Refer to Problems 17.6 and 17.8.

a.   Combine the categories service and other into a single complaint class and test whether or not the store's complaint pattern is the same as that for the entire chain under the modified classification system; use $\alpha = .05$. What conclusion do you reach?

b.   The conclusion reached in Problem 17.8 is that the store's complaint pattern is not the same as that for the entire chain. How do you explain the difference in conclusions? What does this difference in conclusions suggest about the choice of classes to be used in defining a multinomial population? Discuss.

17.24 In a study of 1,000 randomly selected recent automobile accidents, each accident was classified according to whether it occurred at an intersection (A_1) or elsewhere (A_2), whether it occurred during the day (B_1) or at night (B_2), and whether it involved personal injury (C_1) or no personal injury (C_2). The resulting cross-classification follows.

	B_1		B_2	
	C_1	C_2	C_1	C_2
A_1	52	203	96	34
A_2	47	389	105	74

a. Test whether or not the three variables are independent. Use $\alpha = .01$ for the test. What is your conclusion? [*Hint:* (4.24) extends directly to three variables.]

b. To analyze the nature of the relationship, obtain the marginal bivariate distributions for variables A and B, variables A and C, and variables B and C. For each, test whether the two variables are independent, using $\alpha = .01$ in each case. Are any insights into the nature of the relationships between the three variables provided by these tests? Discuss.

c. Consider the conditional bivariate distribution of variables A and B, given C_1. Test whether variables A and B are independent here, using $\alpha = .01$. Do the same for the conditional bivariate distribution of variables A and B, given C_2. Discuss your findings.

UNIT FIVE

---◆---

Linear
Statistical
Models

18
Simple Linear
Regression

Regression analysis enables us to ascertain and utilize a relation between a variable of interest, called a *dependent* or *response* variable, and one or more *independent* or *predictor* variables.

☐ Examples

1. In a recent regression study, the response variable was family expenditures for food last year and the independent variables were family size and family income last year.
2. In another regression study, the response variable was number of metric tons of lubricating oil consumed regionally during a given period and the independent variables included the number of firms in the region using this oil and a regional index of industrial activity. ☐

Regression analysis is often used to predict the response variable from knowledge of the independent variables. In the lubricating oil study, for instance, it was desired to predict the consumption of lubricating oil in a new marketing region from knowledge of the independent variables for this region. At other times, regression analysis is utilized primarily for examining the nature of the relationship between the independent variables and the response variable. For example, the food expenditures study was designed to determine whether food expenditures account for a declining proportion of family income as family income rises.

In this chapter, we take up some basic concepts of regression analysis. In subsequent chapters, we consider important extensions and applications.

18.1 RELATION BETWEEN TWO VARIABLES

Functional Relation between Two Variables

We begin with the concept of a relation between two variables. It is useful to distinguish between a functional and a statistical relation. We first consider a functional relation:

(18.1) A *functional relation* between two variables X and Y is exact; the value of Y is uniquely determined when the value of X is specified.

☐   Examples

1. The rental fee (Y, in dollars) for an electric motor is related to the number of hours rented (X) as follows:

$$Y = 1.50 + 2.00X$$

Here, 1.50 is the fixed service charge and 2.00 is the hourly charge. Thus, for each number of hours rented, there is a unique rental fee. Figure 18.1a shows the line of relationship $Y = 1.50 + 2.00X$, as well as the observations for three recent rentals—for 1, 2, and 4 hours, respectively. The plotted observations are (1, 3.50), (2, 5.50), and (4, 9.50). Since the value of Y is uniquely determined from X, all observations fall on the line of relationship.

2. The area of a square sheet of metal (Y, in square centimeters) is related to the length of its sides (X, in centimeters) by the functional relation $Y = X^2$. Figure 18.1b shows the curve of relationship, as well as the observations for four sheets whose sides are 10, 25, 30, and 35 centimeters. ☐

Statistical Relation between Two Variables

In most empirical studies where interest centers on a dependent variable Y, the value of Y is not uniquely determined when the level of the independent variable is specified. Thus, in studying the relation between family expenditures for food and family income, we are likely to find that families with the same income level differ in their food expenditures. A main reason is that other factors besides family income also play a role, such as ethnic background, family size, and living style.

Relations where Y is not uniquely determined from knowledge of X are called statistical relations:

FIGURE 18.1   *Examples of functional relations.* All observations fall on the curve of relationship.

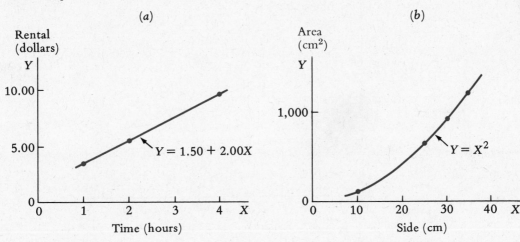

FIGURE 18.2 *Example of a linear statistical relation.* The observations are scattered around the line of statistical relationship.

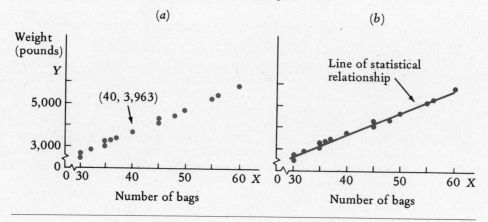

(18.2) A *statistical relation* between two variables X and Y is not exact; the value of Y is not uniquely determined when the value of X is specified.

☐ Examples

1. Figure 18.2a shows observations on weight (Y) and number of bags (X) in 15 recent shipments of a food product in hundred-weight bags from a company's branch plant overseas to its main plant in this country. One of the shipments contained X = 40 bags and weighed Y = 3,963 pounds, so the point for this shipment is plotted at (40, 3,963). The other observations are plotted similarly.

FIGURE 18.3 *Example of a curvilinear statistical relation*

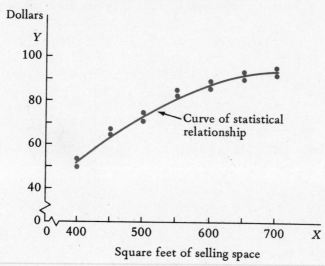

Figure 18.2a shows clearly that weight tends to increase with number of bags. To describe this tendency, we have plotted a line through the concentration of the points, as shown in Figure 18.2b. The relation is not a perfect one, however, and the observations are scattered about the line of relationship. Hence, the relation is a statistical one. Because of the scattering of the points about the line of relationship, Figure 18.2a is called a *scatter diagram or scatter plot*. The statistical relation here is *linear* in that it follows a straight line.

2. Figure 18.3 shows the observations from an experiment involving 14 supermarkets in a chain. Y is dollar sales per square foot of selling space in a specialty foods section and X is number of square feet of selling space in the section. The scattering of points indicates that the relation is a statistical one. In contrast to Figure 18.2b, the relation here is *curvilinear*. With increasing selling space, the growth in sales per square foot is slowed. ☐

18.2 SIMPLE LINEAR REGRESSION MODEL

Development of Model

The two examples of statistical relations just presented portray the two main features of statistical relations:

1. A tendency of the dependent variable Y to vary systematically with the independent variable X, as described by a line or curve of statistical relationship.
2. A scattering of observations around the line or curve of statistical relationship, partly because factors in addition to the independent variable X affect the dependent variable Y, and partly because of inherent variability in Y.

Regression models incorporate these features of a statistical relation by assuming:

1. For each level X of the independent variable, there is a probability distribution of Y.
2. The means of these probability distributions vary in a systematic fashion with X.

☐ Example

Consider again the linear statistical relation between number of bags in shipment and weight of shipment in Figure 18.2b. We portray the corresponding regression model in Figure 18.4. Shown there are the probability distributions of shipment weight Y when the number of bags is X = 40 and X = 50. The regression model assumes that when, say, X = 40 bags are shipped, the observed shipment weight Y will be a random selection from the probability distribution for that level of X.

Note that the means of the probability distributions of Y have a systematic relation to the level of X. This systematic relation is called the *regression function*. In Figure 18.4, the regression function happens to be linear. The fact that the probability distributions of Y are centered around the regression function leads to sample observations that are scattered in a pattern describing the statistical relation, such as the data in Figure 18.2a. ☐

FIGURE 18.4 *Illustration of regression model for shipment example.* The means of the probability distributions fall on the regression function.

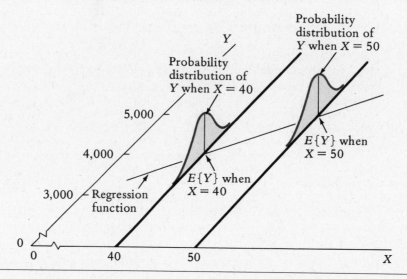

Model of an Observation

The features of a regression model which we have just described can also be expressed by assuming that each observation Y is made up of two components. Let us denote the ith observation on the independent and dependent variables in a regression study by (X_i, Y_i). A regression model then assumes that each observation on the dependent variable, Y_i, is made up of:

1. A component reflecting the line or curve of statistical relationship.
2. A random component reflecting the scatter or deviation about the line or curve of statistical relationship.

For example, if the statistical relation is linear as in Figure 18.2b, the regression model for observation Y_i takes the form:

$$Y_i = \underbrace{\beta_0 + \beta_1 X_i}_{\substack{\text{Line of} \\ \text{Statistical} \\ \text{Relationship} \\ \text{Component}}} + \underbrace{\epsilon_i}_{\substack{\text{Random} \\ \text{Scatter} \\ \text{Component}}}$$

The component reflecting the statistical relation is called the *regression component*, and the component reflecting the random scatter is called the *error term*. Since each observation Y_i contains a random component ϵ_i (Greek epsilon), Y_i is a random variable.

The regression component need not be linear, as just illustrated, but can be curvilinear. Also, the regression component may involve more than one independent variable. Further, various assumptions can be made about the random error term ϵ_i in the regression model.

Simple Linear Regression Model

We now consider a basic regression model where the statistical relation is linear and the following assumptions are made about the error terms ϵ_i:

1. The error terms ϵ_i are normally distributed.
2. The expected value of ϵ_i is $E\{\epsilon_i\} = 0$. (This assumption is necessary so that the random scatter balances around the line or curve of statistical relationship.)
3. The variance of ϵ_i is constant at all levels of X and is denoted by $\sigma^2\{\epsilon_i\} = \sigma^2$.
4. The error terms ϵ_i for different observations are statistically independent.

An equivalent statement of these assumptions is that the error terms ϵ_i are independent $N(0, \sigma^2)$.

We incorporate these conditions into a model for the observations Y_i as follows:

(18.3) $$Y_i = \beta_0 + \beta_1 X_i + \epsilon_i \qquad i = 1, 2, \ldots, n$$

where: Y_i is the response in the ith observation
$\qquad$ X_i is the value of the independent variable in the ith observation, assumed to be a known constant
$\qquad$ β_0 and β_1 are parameters
$\qquad$ ϵ_i's are independent $N(0, \sigma^2)$

Model (18.3) is called the *simple linear regression model*. It has three parameters—β_0 (Greek beta), β_1, and σ^2. We shall now consider some important features of this regression model.

Regression Function

Since the error term ϵ_i is a random variable, so is the response Y_i. We wish to find the expected value of Y_i. Remember that β_0 and β_1 in (18.3) are parameters (hence, constants), and that the value X_i of the independent variable is assumed to be a known constant. Hence, we obtain by (5.6):

$$E\{Y_i\} = E\{\beta_0 + \beta_1 X_i + \epsilon_i\} = \beta_0 + \beta_1 X_i + E\{\epsilon_i\}$$

Since $E\{\epsilon_i\} = 0$, we find:

(18.4) $$E\{Y_i\} = \beta_0 + \beta_1 X_i$$

Thus, when the independent variable has the value X_i, the expected value of Y_i is $E\{Y_i\} = \beta_0 + \beta_1 X_i$.

When (18.4) is considered for any value of X, the relationship between X and $E\{Y\}$ is called the regression function:

(18.5) The *regression function* relates $E\{Y\}$, the mean of Y, to X, the value of the independent variable. For model (18.3), the regression function is:

$$E\{Y\} = \beta_0 + \beta_1 X$$

The regression function thus is the model counterpart to the intuitive curve of statistical relationship. The graph of the regression function is called the *regression line* or *curve*. The parameters β_0 and β_1 are called *regression parameters*. β_0 is the intercept of the regression line, and β_1 is the slope of the line.

☐   Example

UDS (United Data Systems, Incorporated), a data-processing and equipment firm, leases tape drives to customers for extended periods. These drives must be tested periodically. UDS performs this testing as part of the leasing arrangement. An operations analyst for UDS is considering the relation between number of minutes required on a customer service call to test tape drives (Y) and number of tape drives tested (X). She is employing regression model (18.3) since she considers it to be a reasonable model for this application.

Ordinarily, the regression parameters β_0 and β_1 are unknown and must be estimated from sample data. Suppose, however, that the parameters are $\beta_0 = 15$ and $\beta_1 = 45$. The regression function then is:

$$E\{Y\} = 15 + 45X$$

Figure 18.5 illustrates the meaning of the regression parameters. The intercept, $\beta_0 = 15$, indicates that the value of the regression function at $X = 0$ is 15. The slope, $\beta_1 = 45$, signifies that for each additional tape drive tested, the expected time required to service the tape drives increases by 45 minutes.   ☐

Probability Distributions of Y

We have seen that each response Y_i is a random variable with expected value $E\{Y_i\} = \beta_0 + \beta_1 X_i$. Let us next find the variance of Y_i. Since $\beta_0 + \beta_1 X_i$ is a

---

FIGURE 18.5   *Illustration of linear regression function.* β_0 *is the intercept of the regression line and* β_1 *is the slope.*

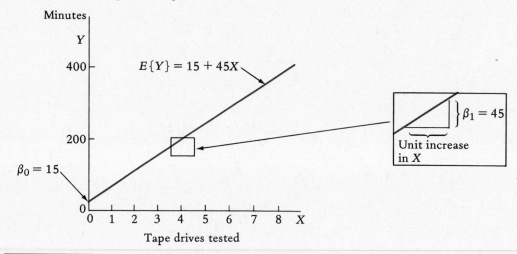

constant, we have by (5.11) that:

$$\sigma^2\{Y_i\} = \sigma^2\{\beta_0 + \beta_1 X_i + \epsilon_i\} = \sigma^2\{\epsilon_i\}$$

But model (18.3) assumes that $\sigma^2\{\epsilon_i\} = \sigma^2$. Hence:

(18.6) $$\sigma^2\{Y_i\} = \sigma^2$$

Thus, the Y_i's have the same variability regardless of the value of X_i.

We further know from theorem (7.8) that each Y_i is normally distributed, because Y_i is a linear function of the normal random variable ϵ_i. Hence, regression model (18.3) implies that for any given value X_i of the independent variable:

1. Y_i is normally distributed.
2. $E\{Y_i\} = \beta_0 + \beta_1 X_i$
3. $\sigma^2\{Y_i\} = \sigma^2$

Finally, since the ϵ_i's are assumed to be independent for the various observations, so are the Y_i's. Hence, an equivalent formulation of regression model (18.3) is:

(18.7) $$Y_i\text{'s are independent } N(\beta_0 + \beta_1 X_i, \sigma^2)$$

In words, the Y_i's are independent normal random variables, with mean $\beta_0 + \beta_1 X_i$ and variance σ^2.

☐ **Example**

Figure 18.6 shows a graphic representation of the simple linear regression model for the UDS example, assuming that $\beta_0 = 15$, $\beta_1 = 45$, and $\sigma = 8$. Two particular distributions of Y are displayed, at $X = 3$ and $X = 4$, respectively. Note that the means of these distributions are located on the regression line. Also note that both distributions have the same amount of variability, and that both are normal.

FIGURE 18.6   *Illustration of simple linear regression model (18.3) for* $\beta_0 = 15$, $\beta_1 = 45$, $\sigma = 8$.

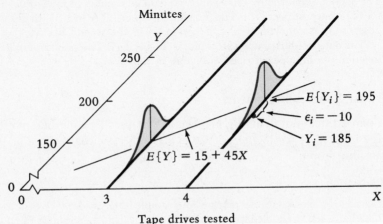

Suppose that for a customer with $X_i = 4$ tape drives, $Y_i = 185$ minutes are required for the service call. This observation is plotted in Figure 18.6. Since the expected value of Y_i is $E\{Y_i\} = 15 + 45(4) = 195$, the error term is $\epsilon_i = Y_i - E\{Y_i\} = 185 - 195 = -10$. Thus, we see again that the error term ϵ_i is simply the deviation of Y_i from its mean, $E\{Y_i\}$, and hence reflects the scatter around the regression line. □

Comments

1. The regression function is also called the *response function,* and the term *response curve* is also used for the regression curve.
2. The normality assumption for ϵ_i is appropriate in many regression applications. A major reason is that there are often many factors influencing Y that are not included among the independent variables of the regression model. Insofar as the effects of these factors are additive and tend to vary with a degree of mutual independence, the composite error term ϵ_i often will tend to comply with the central limit theorem and its distribution will be nearly normal when the number of "missing" factors is large.
3. The designations "response variable" or "dependent variable" on the one hand and "independent variable" on the other carry no connotation that changes in X *cause* changes in Y. No particular cause-effect pattern between the variables is implied necessarily by a regression model. We discuss this point in more detail in Chapter 19.

18.3   POINT ESTIMATION OF β_0 AND β_1

Ordinarily, the regression parameters β_0 and β_1 are unknown and must be estimated from sample data. In this section, we take up point estimation of β_0 and β_1. Point estimates of these parameters are often required in their own right and are also needed when interval estimates of these parameters are desired.

When n sample observations are available, we shall denote the first by (X_1, Y_1), the second by (X_2, Y_2), and the ith by (X_i, Y_i), where $i = 1, 2, \ldots, n$.

□ Example

The analyst employed by UDS collected data on number of tape drives and service times for 12 recent customer calls. The data are shown in Table 18.1. With our notation, the first observation is denoted by $(X_1, Y_1) = (4, 197)$, the second by $(X_2, Y_2) = (6, 272)$, etc.

The data are shown as a scatter plot in Figure 18.7. The assumption of a linear response function would appear to be reasonable in this case. □

Least Squares Estimators

Point estimates of β_0 and β_1 are usually obtained by a method of estimation called the *method of least squares.* We explain the essential nature of this method by an illustration.

□ Illustration

Figure 18.8a shows a scatter plot for four sample observations. We wish to find the straight line that is the "best" fit for these observations. Figure 18.8b shows the horizontal straight line, $\hat{Y} = 1.8 + 0X$, which obviously is a poor fit. $\hat{Y}$ (read "Y hat") here denotes an

TABLE 18.1 *Data for UDS example*

Observation i	Number of Tape Drives Tested X_i	Number of Minutes Required on Call Y_i
1	4	197
2	6	272
3	2	100
4	5	228
5	7	327
6	6	279
7	3	148
8	8	377
9	5	238
10	3	142
11	1	66
12	5	239

ordinate of the fitted line. The distances, or deviations, of the observations from this horizontal line, shown by the vertical rules in Figure 18.8b, are relatively large. For instance, the first deviation is $1 - 1.8 = -.8$.

The method of least squares considers as a measure of fit of a straight line the sum of the squared deviations of the observations from the given straight line. We shall denote such a sum by Q. For the horizontal straight line in Figure 18.8b, the value of Q is:

$$Q = (1 - 1.8)^2 + (2 - 1.8)^2 + (1.5 - 1.8)^2 + (2.5 - 1.8)^2 = 1.260$$

A better fitting straight line is the one shown in Figure 18.8c, for which the sum of the squared deviations is $Q = .566$. A still better fit is given by the line in Figure 18.8d, for

FIGURE 18.7 *Scatter plot for UDS example*

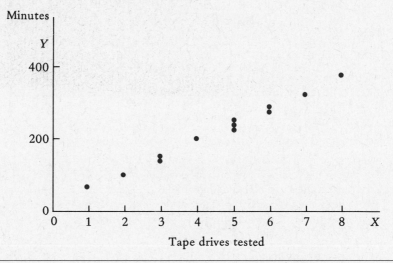

FIGURE 18.8 *Finding the least squares regression line for a scatter plot*

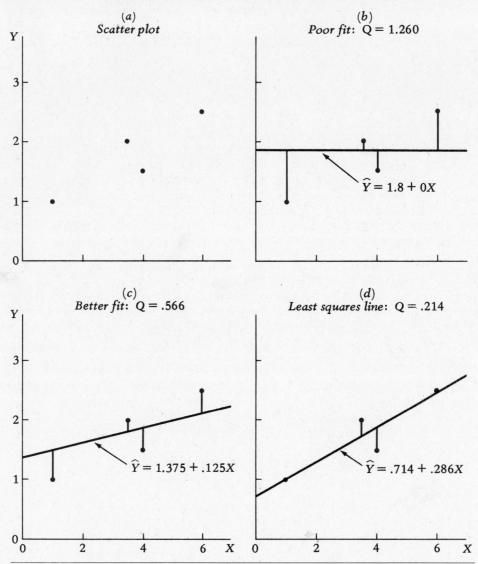

which $Q = .214$. In fact, it can be shown that this latter line, $\widehat{Y} = .714 + .286X$, has a smaller sum of squared deviations than any other straight line. The method of least squares considers that straight line to be the "best" fit for which the criterion Q is a minimum. Thus, $\widehat{Y} = .714 + .286X$ is the least squares regression line, .714 is the least squares estimate of β_0 (the intercept of the regression line), and .286 is the least squares estimate of β_1 (the slope of the line). □

The *least squares line,* as we have seen, is that line for which the sum of the squared vertical distances of the observations from the line is a minimum. We shall denote the *least squares estimators* of β_0 and β_1 by b_0 and b_1, respectively. The least squares criterion requires that b_0 and b_1 minimize Q, the sum of the squared deviations:

$$(18.8) \qquad Q = \sum_{i=1}^{n} [Y_i - (b_0 + b_1 X_i)]^2$$

It can be shown mathematically that the least squares estimators for the simple linear regression model (18.3) are obtained by solving the following two simultaneous equations:

$$(18.9) \qquad \Sigma Y_i = n b_0 + b_1 \Sigma X_i$$
$$\Sigma X_i Y_i = b_0 \Sigma X_i + b_1 \Sigma X_i^2$$

All summations are taken over the sample observations $i = 1, 2, \ldots, n$. We do not show the summation index here and throughout the regression chapters because the nature of the summation is clear.

The equations in (18.9) are called *normal equations*. When they are solved for b_0 and b_1, we obtain the least squares estimators of β_0 and β_1:

(18.10)   For the simple linear regression model (18.3), the least squares estimators of β_1 and β_0 are, respectively:

$$(18.10a) \qquad b_1 = \frac{\Sigma X_i Y_i - \dfrac{(\Sigma X_i)(\Sigma Y_i)}{n}}{\Sigma X_i^2 - \dfrac{(\Sigma X_i)^2}{n}}$$

$$(18.10b) \qquad b_0 = \frac{1}{n}(\Sigma Y_i - b_1 \Sigma X_i)$$

Thus, the least squares estimates b_0 and b_1 are obtained by calculating the quantities ΣX_i, ΣY_i, $\Sigma X_i Y_i$, and ΣX_i^2 from the sample data and substituting into the expressions in (18.10).

Algebraically equivalent formulas for b_1 and b_0 are:

$$(18.11) \qquad b_1 = \frac{\Sigma (X_i - \bar{X})(Y_i - \bar{Y})}{\Sigma (X_i - \bar{X})^2}$$

$$(18.12) \qquad b_0 = \bar{Y} - b_1 \bar{X}$$

☐   Example

The quantities required in (18.10) for our UDS case are calculated from the sample data in Table 18.2. We obtain $\Sigma X_i = 55$, $\Sigma Y_i = 2{,}613$, $\Sigma X_i Y_i = 14{,}060$, $\Sigma X_i^2 = 299$, and $n = 12$. We calculate then by (18.10):

$$b_1 = \frac{14{,}060 - \dfrac{55(2{,}613)}{12}}{299 - \dfrac{(55)^2}{12}} = 44.41385$$

$$b_0 = \frac{1}{12}[2{,}613 - 44.41385(55)] = 14.18652$$

TABLE 18.2 *Basic calculations to obtain b_0 and b_1 in UDS example*

Observation i	Number of Tape Drives Tested X_i	Number of Minutes Required on Call Y_i	X_iY_i	X_i^2
1	4	197	788	16
2	6	272	1,632	36
3	2	100	200	4
4	5	228	1,140	25
5	7	327	2,289	49
6	6	279	1,674	36
7	3	148	444	9
8	8	377	3,016	64
9	5	238	1,190	25
10	3	142	426	9
11	1	66	66	1
12	5	239	1,195	25
Total	$\Sigma\,X_i = 55$	$\Sigma\,Y_i = 2{,}613$	$\Sigma\,X_iY_i = 14{,}060$	$\Sigma\,X_i^2 = 299$

Thus, the point estimates of β_0 and β_1 are, respectively, $b_0 = 14.187$ and $b_1 = 44.414$. ☐

Comments

1. The least squares estimators b_0 and b_1 for regression model (18.3) are unbiased. In addition, we note from (18.10) that both estimators depend linearly on the n values of Y_i. Among all unbiased estimators of β_0 and β_1 which depend linearly on the Y_i's, the least squares estimators are the most efficient. These desirable properties of the least squares estimators do not depend on the normality assumption of model (18.3).

2. (Calculus needed.) The least squares estimators can be derived by calculus. For the given observations (X_i, Y_i), the quantity Q to be minimized in (18.8) is a function of b_0 and b_1. To find the values of b_0 and b_1 that minimize Q, we first take partial derivatives of Q with respect to b_0 and b_1:

$$\frac{\partial Q}{\partial b_0} = \Sigma\,\frac{\partial}{\partial b_0}(Y_i - b_0 - b_1X_i)^2 = -2\,\Sigma\,(Y_i - b_0 - b_1X_i)$$

$$\frac{\partial Q}{\partial b_1} = \Sigma\,\frac{\partial}{\partial b_1}(Y_i - b_0 - b_1X_i)^2 = -2\,\Sigma\,X_i(Y_i - b_0 - b_1X_i)$$

We then set these partial derivatives equal to zero, as follows:

$$-2\,\Sigma\,(Y_i - b_0 - b_1X_i) = 0$$

$$-2\,\Sigma\,X_i(Y_i - b_0 - b_1X_i) = 0$$

After simplifying these equations and rearranging terms, we obtain the normal equations (18.9).

3. When the distribution of the error terms ϵ_i is specified in the regression model, as in model (18.3), estimators of β_0 and β_1 can also be developed by the method of maximum likelihood described in Chapter 10. In applying this method to the normal error regression model (18.3), the same normal equations are obtained as with the method of

least squares. Thus, the least squares estimators b_0 and b_1 for model (18.3) have the combined properties of least squares estimators and maximum likelihood estimators—they are unbiased, consistent, and most efficient among all unbiased estimators.

18.4   POINT ESTIMATION OF MEAN RESPONSE

Estimated Regression Function

The regression function $E\{Y\} = \beta_0 + \beta_1 X$ is estimated by:

(18.13)
$$\widehat{Y} = b_0 + b_1 X$$

The equation in (18.13) is called the *estimated regression function.*

☐   **Example**

Earlier we found for our UDS example that $b_0 = 14.18652$ and $b_1 = 44.41385$. Thus, the estimated regression function is:

$$\widehat{Y} = 14.18652 + 44.41385X$$

This estimated regression function is plotted in Figure 18.9, together with the original observations. Note that the fit of the estimated regression function to the data appears to be good.   ☐

Mean Response

As mentioned earlier, the value of the dependent variable Y is often called a response. The mean response refers to the expected value of Y for a given value of X. Thus, in our UDS example the mean response when $X = 4$ refers to the mean

FIGURE 18.9   *Observations and estimated regression function for UDS example*

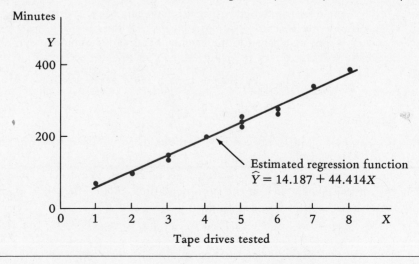

of the probability distribution of service times (Y) for the case where four tape drives are serviced.

(18.14) The *mean response* when $X = X_h$ is denoted by $E\{Y_h\}$ and for regression model (18.3) is:

$$E\{Y_h\} = \beta_0 + \beta_1 X_h$$

Note that X_h denotes any specified level of X, in contrast to X_i which refers to the value of the X variable for the ith sample observation.

Regression analysis is often utilized to estimate mean responses. For instance, management of UDS may be interested in the expected service time, $E\{Y_h\}$, when there are $X_h = 4$ tape drives in order to estimate the direct labor cost of the service calls. An economist may be interested in the expected amount of savings, $E\{Y_h\}$, of families with incomes of $X_h = \$22,000$ to use in a savings model.

Point Estimator of $E\{Y_h\}$

The point estimator of $E\{Y_h\}$ is given by $\widehat{Y}_h$, where $\widehat{Y}_h$ is the value of the estimated regression function when $X = X_h$:

(18.15)
$$\widehat{Y}_h = b_0 + b_1 X_h$$

Since b_0 and b_1 are unbiased estimators of β_0 and β_1, respectively, it follows that $\widehat{Y}_h$ is an unbiased estimator of $E\{Y_h\}$; i.e., $E\{\widehat{Y}_h\} = \beta_0 + \beta_1 X_h = E\{Y_h\}$.

☐ Examples

1. For our UDS case, we wish to find the point estimate of the mean time required on calls to test $X_h = 4$ tape drives. We substitute into our estimated regression function with $X_h = 4$ and obtain:

$$\widehat{Y}_h = 14.18652 + 44.41385(4) = 191.84$$

or 192 minutes.

We interpret this to signify that in a large number of calls to test four tape drives, made under the general conditions prevailing for the 12 calls in the sample, the mean required time on a call is in the neighborhood of 192 minutes.

2. For $X_h = 7$ tape drives, the point estimate of the mean service time is:

$$\widehat{Y}_h = 14.18652 + 44.41385(7) = 325.08 \text{ minutes}$$ ☐

Residuals

For conducting inferences in regression analysis, we need to estimate the magnitude of the random variation in Y_i. The first step is to measure the scatter of the observations around the estimated regression line. We do this by comparing Y_i, the *observed* value of Y in the ith observation, and $\widehat{Y}_i$, the value of the estimated regression function when $X = X_i$:

(18.16)
$$\widehat{Y}_i = b_0 + b_1 X_i$$

$\widehat{Y}_i$ is often called the *fitted* value of Y_i. The difference between the observed value Y_i and the fitted value $\widehat{Y}_i$ is called the residual for the ith observation:

(18.17) The *residual* for the ith observation is denoted by e_i and is defined:

$$e_i = Y_i - \widehat{Y}_i$$

☐   **Example**

The residuals for the UDS example are calculated in Table 18.3. For the first observation, $X_1 = 4$ and $Y_1 = 197$. The corresponding fitted value is $\widehat{Y}_1 = 14.18652 + 44.41385(4) = 191.84$. Thus, the residual for the first observation is:

$$e_1 = 197 - 191.84 = 5.16$$

The observed values, corresponding fitted values, and residuals for our UDS example are portrayed in Figure 18.10.

☐

Distinction between e_i and ϵ_i.   The distinction between the residual e_i and the error term value ϵ_i is an important one. The former measures the deviation of Y_i from the corresponding fitted value $\widehat{Y}_i = b_0 + b_1 X_i$, whereas the latter denotes the deviation of Y_i from the true mean $E\{Y_i\} = \beta_0 + \beta_1 X_i$. Since ϵ_i usually is unknown, it is estimated by e_i.

The residuals e_i are needed not only for estimating the magnitude of the random variation in Y_i for making statistical inferences, but they also are useful for assessing the appropriateness of the regression model being employed. We shall discuss this latter use in the next chapter.

Properties of Least Squares Residuals.   The least squares residuals (18.17) have a number of important properties:

1. The residuals sum to zero:

(18.18)
$$\Sigma\, e_i = 0$$

2. The sum of the squared residuals, $\Sigma\, e_i^2$, is a minimum. This follows because the method of least squares minimizes Q, defined in (18.8).

TABLE 18.3   *Calculation of residuals for UDS example*

Observation i	Number of Tape Drives Tested X_i	Observed Value Y_i	Fitted Value $\widehat{Y}_i$	Residual $e_i = Y_i - \widehat{Y}_i$
1	4	197	191.84	5.16
2	6	272	280.67	−8.67
3	2	100	103.01	−3.01
4	5	228	236.26	−8.26
5	7	327	325.08	1.92
6	6	279	280.67	−1.67
7	3	148	147.43	.57
8	8	377	369.50	7.50
9	5	238	236.26	1.74
10	3	142	147.43	−5.43
11	1	66	58.60	7.40
12	5	239	236.26	2.74
Total	55	2,613	2,613.0	0.0
Mean	4.583	217.75	217.75	0

FIGURE 18.10 *Least squares regression line and residuals for UDS example (observed values and residuals not plotted to scale)*

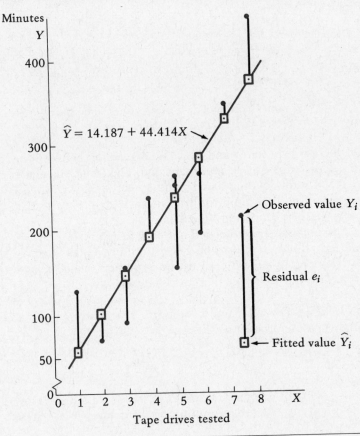

3. The sum of the weighted residuals is zero when the residual in the ith observation is weighted by the level of the independent variable in that observation:

(18.19)
$$\Sigma X_i e_i = 0$$

4. The sum of the weighted residuals is zero when the residual in the ith observation is weighted by the fitted value for that observation:

(18.20)
$$\Sigma \widehat{Y}_i e_i = 0$$

18.5 ANALYSIS OF VARIANCE APPROACH TO SIMPLE LINEAR REGRESSION

Analysis of variance (ANOVA) is a highly useful and flexible mode of analysis for regression models. Here we introduce some basic ANOVA concepts and procedures and use them to obtain a point estimator of σ^2 and to measure the degree of association between X and Y in the observed sample data.

Partitioning of Total Sum of Squares

Basic Concepts.   Consider again the UDS example. If we wish to predict service time without knowing the number of tape drives of the customer, the uncertainty associated with a prediction is related to the variability of the Y observations as measured by the deviations:

$$(18.21) \qquad Y_i - \bar{Y}$$

where $\bar{Y}$ is the mean of the Y observations. These deviations for the UDS example are shown in Figure 18.11a. Clearly, if all Y_i's are equal, there is no variability in the data and all deviations $Y_i - \bar{Y}$ equal zero. The greater the variability in the data, the larger will be the deviations $Y_i - \bar{Y}$, and the greater is the uncertainty associated with a prediction of Y without utilizing knowledge of X.

Conventionally, the measure of the variability of the Y observations is expressed in terms of the sum of squares of the deviations $Y_i - \bar{Y}$ and is denoted by $SSTO$:

$$(18.22) \qquad SSTO = \Sigma \, (Y_i - \bar{Y})^2$$

Here $SSTO$ stands for the *total sum of squares*. If there is no variability in the Y observations, so $Y_i \equiv \bar{Y}$, $SSTO = 0$. The greater is the variability in the Y_i's, the larger is $SSTO$.

We have actually encountered the total sum of squares earlier, but not by this name. Note that the numerator of the sample variance in (8.12b) is $SSTO$ (except that we used X instead of Y to designate the variable).

For the UDS sample data, $\bar{Y} = 217.75$ and we obtain for $SSTO$ (see the data in Table 18.3):

$$SSTO = (197 - 217.75)^2 + (272 - 217.75)^2 + \cdots + (239 - 217.75)^2$$

$$= 92{,}884.25$$

If we use knowledge of the number of tape drives X to predict the service time Y, the uncertainty associated with a prediction is related to the variability of the Y_i's around the fitted regression line as measured by the deviations:

$$(18.23) \qquad Y_i - \hat{Y}_i$$

If all the Y observations fall on the fitted regression line, all deviations $Y_i - \hat{Y}_i$ will equal zero. The larger the deviations $Y_i - \hat{Y}_i$, the greater is the uncertainty associated with a prediction utilizing knowledge of the number of tape drives. The deviations $Y_i - \hat{Y}_i$ for our UDS example are shown in Figure 18.11b.

Again, the conventional measure of the variability around the fitted regression line is the sum of the squared deviations, which is denoted by SSE:

$$(18.24) \qquad SSE = \Sigma \, (Y_i - \hat{Y}_i)^2$$

Here SSE stands for the *error sum of squares*. If all Y observations fall on the fitted regression line, $SSE = 0$. The greater are the deviations $Y_i - \hat{Y}_i$, the larger is SSE.

FIGURE 18.11   *Partitioning of total sum of squares for UDS example (Y values not plotted to scale)*

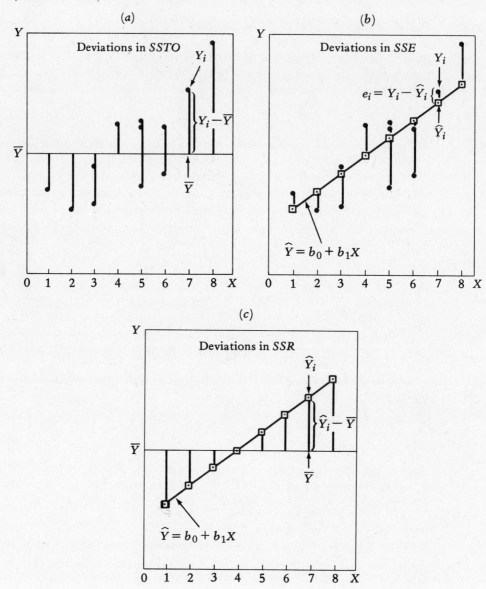

Note that $Y_i - \widehat{Y}_i = e_i$ by (18.17); hence, *SSE* is simply the sum of the squared residuals:

(18.24a)
$$SSE = \Sigma\, e_i^2$$

We calculate *SSE* for our UDS example by using the residuals in Table 18.3:

$$SSE = (5.16)^2 + (-8.67)^2 + \cdots + (2.74)^2 = 336.88$$

We see that the variability of the Y_i's when X is utilized ($SSE = 336.88$) is much smaller than when X is not utilized ($SSTO = 92,884.25$).

This reduction in the variability of the Y_i's is associated with the utilization of knowledge of the number of tape drives for predicting service time. The reduction is, indeed, another sum of squares, denoted by SSR:

(18.25)
$$SSR = SSTO - SSE$$

Here SSR stands for the *regression sum of squares*.

For the UDS example, the reduction in the variability of the Y_i's is:

$$SSR = 92,884.25 - 336.88 = 92,547.37$$

It can be shown that SSR is a sum of squares involving the deviations:

(18.26)
$$\widehat{Y}_i - \bar{Y}$$

Since the mean of the n fitted values $\widehat{Y}_i$ equals $\bar{Y}$ (see Table 18.3 for a confirmation of this), each deviation $\widehat{Y}_i - \bar{Y}$ represents the difference between the fitted value and the mean of the fitted values. The deviations $\widehat{Y}_i - \bar{Y}$ for the UDS example are shown in Figure 18.11c.

The regression sum of squares then is:

(18.27)
$$SSR = \Sigma \, (\widehat{Y}_i - \bar{Y})^2$$

SSR may be viewed as a measure of the effect of the regression relation in reducing the variability of the Y_i's. If $SSR = 0$, the regression relation does not reduce the variability at all. The larger is SSR compared to $SSTO$, the greater is the reduction in the variability of the Y_i's by utilization of knowledge of X through the regression relation.

For our UDS example, the direct calculation of SSR by (18.27) proceeds as follows, using the data in Table 18.3:

$$SSR = (191.84 - 217.75)^2 + (280.67 - 217.75)^2 + \cdots + (236.26 - 217.75)^2$$
$$= 92,547.37$$

This result necessarily agrees with that obtained previously by subtraction.

Formal Statement of Partitioning. Since $SSR = SSTO - SSE$ by (18.25), it follows that $SSTO = SSR + SSE$. We can therefore view the process we have just completed as a partitioning of the total sum of squares $SSTO$ into two additive components: (1) SSR—the portion of the total variability of the Y_i's eliminated when the independent variable X is considered, and (2) SSE—the portion of the total variability of the Y_i's that remains as residual or error variation when X is considered.

(18.28) For simple linear regression, the decomposition of the total sum of squares into two additive components is:

(18.28a)
$$SSTO \quad = \quad SSR \quad + \quad SSE$$

(18.28b)
$$\Sigma \, (Y_i - \bar{Y})^2 = \Sigma \, (\widehat{Y}_i - \bar{Y})^2 + \Sigma \, (Y_i - \widehat{Y}_i)^2$$

☐　Example

In assembling the respective sums of squares for our UDS example, we obtain:

$$92{,}884.25 = 92{,}547.37 + 336.88$$

$$SSTO \quad = \quad SSR \quad + \quad SSE$$　　☐

Computational Formulas.　The preceding formulas for the three sums of squares are the basic definitional ones. The following formulas are algebraically equivalent and tend to be more convenient for manual computation:

(18.29)
$$SSTO = \Sigma\, Y_i^2 - \frac{(\Sigma\, Y_i)^2}{n}$$

(18.30)
$$SSR = \frac{\left(\Sigma\, X_i Y_i - \dfrac{\Sigma\, X_i\, \Sigma\, Y_i}{n}\right)^2}{\Sigma\, X_i^2 - \dfrac{(\Sigma\, X_i)^2}{n}}$$

(18.31)
$$SSE = SSTO - SSR$$

Comment

Formula (18.28b) can be derived by writing:

$$\Sigma\, (Y_i - \bar{Y})^2 = \Sigma\, [(Y_i - \hat{Y}_i) + (\hat{Y}_i - \bar{Y})]^2$$

By expanding the right side, we obtain:

$$\Sigma\, (Y_i - \bar{Y})^2 = \Sigma\, (Y_i - \hat{Y}_i)^2 + \Sigma\, (\hat{Y}_i - \bar{Y})^2 + 2\, \Sigma\, (Y_i - \hat{Y}_i)(\hat{Y}_i - \bar{Y})$$

It can be shown that the last term equals zero.

Partitioning of Degrees of Freedom

A sum of squares has an associated number of degrees of freedom. Recall that the sample variance s^2 in (8.12b) has a denominator $n - 1$. This is the number of degrees of freedom associated with the numerator sum of squares in s^2.

Corresponding to the partitioning of the total sum of squares $SSTO$ into the components SSR and SSE, there is a partitioning of the degrees of freedom. $SSTO$ has $n - 1$ degrees of freedom associated with it. To see why, note that there are n deviations $Y_i - \bar{Y}$ that enter into $SSTO$. However, there is one constraint on these deviations, namely $\Sigma\, (Y_i - \bar{Y}) = 0$, so there are $n - 1$ degrees of freedom in the n deviations.

SSE has $n - 2$ degrees of freedom. There are n residuals $e_i = Y_i - \hat{Y}_i$. However, two degrees of freedom are lost because of two constraints on the e_i's associated with estimating the parameters β_0 and β_1 by the two normal equations.

Finally, SSR has one degree of freedom. There are two parameters in the regression function, but the deviations $\hat{Y}_i - \bar{Y}$ are subject to the constraint $\Sigma\, (\hat{Y}_i - \bar{Y}) = 0$.

We thus see that the degrees of freedom (df) are additive:

(18.32)
$$\underbrace{n - 1}_{\substack{df \text{ for} \\ SSTO}} = \underbrace{1}_{\substack{df \text{ for} \\ SSR}} + \underbrace{n - 2}_{\substack{df \text{ for} \\ SSE}}$$

☐ **Example**

For our UDS example, where $n = 12$, $SSTO$ has associated with it $12 - 1 = 11$ degrees of freedom, SSR has one degree of freedom, and SSE has $12 - 2 = 10$ degrees of freedom. Hence, $11 = 1 + 10$. ☐

Mean Squares

A sum of squares divided by its associated degrees of freedom is called a *mean square*. The sample variance s^2 is an example of a mean square. Two mean squares of importance in regression analysis are the *regression mean square*, denoted by MSR:

(18.33)
$$MSR = \frac{SSR}{1} = SSR$$

and the *error mean square*, denoted by MSE:

(18.34)
$$MSE = \frac{SSE}{n - 2}$$

☐ **Example**

For the UDS data, we have $SSR = 92{,}547.37$, $SSE = 336.88$, and $n = 12$. Hence:

$$MSR = \frac{92{,}547.37}{1} = 92{,}547.37 \qquad MSE = \frac{336.88}{10} = 33.688$$ ☐

Comment

Mean squares are not additive. MSR and MSE do not sum to $SSTO/(n - 1)$. For our example, $SSTO/(n - 1) = 92{,}884.25/11 = 8{,}444.023$, whereas $MSR + MSE = 92{,}547.37 + 33.688 = 92{,}581.06$.

ANOVA Table

It is useful to collect the sums of squares (SS), degrees of freedom (df), and mean squares (MS) in an ANOVA table. Table 18.4a shows a basic ANOVA table for regression analysis, containing the definitional formulas as entries. The corresponding numerical results for the UDS example are displayed in Table 18.4b.

TABLE 18.4   ANOVA *table and results for UDS example*

(a)
General format

Source of Variation	SS	df	MS
Regression	$SSR = \Sigma\ (\widehat{Y}_i - \bar{Y})^2$	1	$MSR = \dfrac{SSR}{1}$
Error	$SSE = \Sigma\ (Y_i - \widehat{Y}_i)^2$	$n - 2$	$MSE = \dfrac{SSE}{n-2}$
Total	$SSTO = \Sigma\ (Y_i - \bar{Y})^2$	$n - 1$	

(b)
UDS example

Source of Variation	SS	df	MS
Regression	92,547.37	1	92,547.37
Error	336.88	10	33.688
Total	92,884.25	11	

Point Estimation of σ^2

For making statistical inferences in regression applications, we require a point estimate of σ^2, the variance of the error terms ϵ_i. Just as we use the mean square:

$$s^2 = \frac{\Sigma\ (X_i - \bar{X})^2}{n - 1}$$

to estimate the variance in a single population where X_i is the ith observation and $\bar{X}$ the estimated mean, we use a mean square to estimate σ^2 in the regression model.

In the regression case, the ith observation is Y_i and its estimated mean is $\widehat{Y}_i$. The relevant deviations therefore are the residuals $e_i = Y_i - \widehat{Y}_i$ and the relevant mean square for estimating σ^2 is the error mean square $MSE = \Sigma\ e_i^2/(n - 2)$.

(18.35) The variance σ^2 of the error terms ϵ_i in regression model (18.3) is estimated by MSE as defined in (18.34).

It can be shown that MSE is an unbiased estimator of σ^2; i.e., $E\{MSE\} = \sigma^2$.

☐   Example

Earlier we found that $MSE = 33.688$ in our UDS example (see Table 18.4b). Thus, our point estimate of σ^2 is 33.688. The corresponding point estimate of σ is $\sqrt{MSE} = \sqrt{33.688} = 5.804$.

Consider again the probability distribution of service times (Y) when the customer has $X_h = 4$ tape drives. We found before that $\widehat{Y}_h = 191.84$ when $X_h = 4$. Hence, the distribution of service times is normal [assuming that model (18.3) is appropriate], with approximate mean 191.84 minutes and approximate standard deviation 5.8 minutes. In the same way, we can describe the probability distribution of Y for any other number of tape drives serviced.   ☐

18.6 COEFFICIENTS OF SIMPLE DETERMINATION AND CORRELATION

We now turn to two measures that are widely used to describe the degree of relationship between the variables X and Y as reflected in the simple linear regression model.

Coefficient of Simple Determination

As we saw, $SSTO$ reflects the variability of the Y_i's when knowledge of X is not utilized. In contrast, SSE reflects the remaining variability of the Y_i's when knowledge of X is utilized by the simple linear regression model. Finally, $SSR = SSTO - SSE$ reflects the reduction in the total variability of the Y_i's associated with use of X to predict Y. To obtain a relative measure of this reduction in variability, SSR is expressed relative to $SSTO$. The measure is denoted by r^2 and is called the coefficient of simple determination:

(18.36) The *coefficient of simple determination* r^2 is defined:

$$r^2 = \frac{SSR}{SSTO} = \frac{SSTO - SSE}{SSTO} = 1 - \frac{SSE}{SSTO}$$

☐ Example

For our UDS example, we have $SSTO = 92{,}884.25$ and $SSR = 92{,}547.37$. Thus:

$$r^2 = \frac{92{,}547.37}{92{,}884.25} = .99637$$

This means that the variability in service times is reduced 99.64 percent when the number of tape drives is considered. This large proportionate reduction in the variability of Y_i's by the introduction of X through the simple linear regression model suggests that the regression relation may be highly useful. For example, it may be possible to anticipate direct labor costs simply on the basis of the number of tape drives serviced. ☐

Limiting Values of r^2. One limiting value of r^2 occurs when the statistical relation is perfect for the sample data, i.e., when all observations fall on the fitted regression line so that all residuals e_i equal zero. Then $SSE = 0$ and $SSR = SSTO - SSE = SSTO$. Hence, $r^2 = SSR/SSTO = 1$. This limiting case is illustrated in Figure 18.12a by the exact fit to the observations of the linear regression function.

The other limiting value of r^2 occurs when the regression relation is of no use in reducing the variability of the Y_i's. Then the remaining variability SSE is the same as $SSTO$. Hence, $SSR = SSTO - SSE = 0$ then, and $r^2 = 0$. Incidentally, the formal condition for SSE to equal $SSTO$ is that the fitted regression line be horizontal, i.e., that $b_1 = 0$. This condition is illustrated in Figure 18.12b. Observe that the sample data exhibit no linear statistical relationship between the two variables X and Y.

In general, we have:

(18.37)
$$0 \le r^2 \le 1$$

FIGURE 18.12 *Scatter plots showing several degrees of linear statistical relationship*

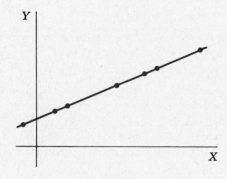

(a)
$$r^2 = 1; r = +1$$

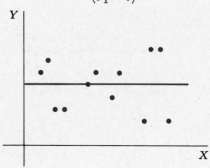

(b)
$$r^2 = 0; r = 0$$
$$(b_1 = 0)$$

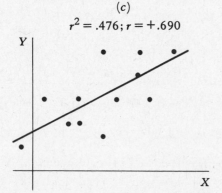

(c)
$$r^2 = .476; r = +.690$$

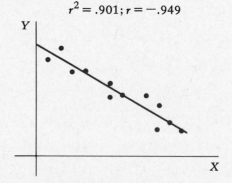

(d)
$$r^2 = .901; r = -.949$$

where the lower bound implies that there is no linear statistical relation exhibited by the observations and the upper bound implies a perfect linear relation. In practice, r^2 usually falls somewhere between 0 and 1; the closer r^2 is to 1, the greater is the degree of linear statistical relation in the observations. The last two panels in Figure 18.12 illustrate two scatter plots with r^2 values that lie between 0 and 1. Note how the scatter plot in Figure 18.12d, for which $r^2 = .901$, is fairly tightly clustered about the regression line, whereas there is quite a bit of scattering in Figure 18.12c, for which $r^2 = .476$.

Coefficient of Simple Correlation

The square root of r^2 is another descriptive measure of the degree of linear relationship between X and Y. It is called the coefficient of simple correlation:

(18.38) The *coefficient of simple correlation r* is defined:

$$r = \pm\sqrt{r^2}$$

where: the sign is that of b_1
r^2 is given by (18.36)

The sign attached to r simply shows whether the slope of the regression line is positive or negative. Since $0 \leq r^2 \leq 1$, we have:

(18.39) $-1 \leq r \leq 1$

The closer the absolute value of r is to 1, the greater is the degree of linear statistical relation in the sample observations.

The coefficient r does not have as clear-cut an operational interpretation as r^2. Even so, there has been a tendency to use r in preference to r^2 in much applied work. It is worth noting that r may give the impression of a closer relation between X and Y than does r^2, since the absolute value of r is closer to 1 than is r^2 (except when r^2 equals 0 or 1).

☐   Examples

1. For the UDS example, $r^2 = .99637$ and b_1 is positive. Hence, $r = +\sqrt{.99637} = +.99818$.
2. For the scatter plot in Figure 18.12d, $r^2 = .901$ and b_1 is negative. Hence, $r = -\sqrt{.901} = -.949$.   ☐

Comment

The coefficient of correlation r in (18.38) is based on data for a regression model but can be shown to be computationally similar to the coefficient of correlation ρ_{XY} in (5.27) for two random variables.

18.7   COMPUTER OUTPUT

Practically all computer software libraries have programs for regression analysis. Indeed, a major reason for the extensive use of regression analysis is the accessibility of computers and software for handling calculations that would otherwise be tedious, time-consuming, and relatively expensive. Many pocket calculators also have regression programs that are either built-in or can be placed in the calculator's memory.

The method of entering data for regression analysis and the output format differ widely from program to program. Figure 18.13 illustrates a typical output format for simple linear regression for our UDS example. Any slight differences between our results and the computer output are due to roundoff effects.

For convenience, we have numbered the blocks of information in the output:

1. Basic information about the independent and dependent variables is given, consisting of the sample mean, sample variance, and sample standard deviation. Variable 1 is X; variable 2 is Y.
2. The estimated regression equation is given.
3. The point estimates b_0 and b_1 are given. The coefficient for variable 0 is b_0; that

FIGURE 18.13 *Computer output for simple linear regression for UDS example*

```
SIMPLE LINEAR REGRESSION

1. PRINT MEANS, VARIANCES AND STANDARD DEVIATIONS?     YES
      VARIABLE          MEAN              VARIANCE        ST. DEV.
         1            4.58333             4.26515         2.06522
         2          217.75000          8444.02273        91.89136


2. ESTIMATED REGRESSION EQUATION:
      YHAT =  14.18650  +44.41385×1

3. VARIABLE    COEFFICIENT     ST. DEV.      T STAT.
      0          14.18650       4.22980        3.35
      1          44.41385       0.84737       52.41

4. PRINT ANOVA TABLE?      YES
      SOURCE                SS            DF            MS
      REGRESSION        92547.3690        1        92547.36901
      RESIDUAL            336.8810       10           33.68810
      TOTAL             92884.2500       11
      F STAT.          = 2747.18

5. RESIDUAL STD DEV.       = 5.80415;   NUMBER OF OBSERVATIONS = 12
   SIMP R SQ (COEFF DETERM) = 0.99637;   SIMP R (COEFF CORREL) = 0.99818

6. ESTIMATE RESPONSE VARIABLE?     YES
   ENTER THE VALUE OF THE INDEPENDENT VARIABLE
   ☐:
         4

   POINT ESTIMATE                               = 191.8419
   STANDARD DEVIATION OF MEAN RESPONSE          = 1.7469
   STANDARD DEVIATION OF INDIVIDUAL RESPONSE    = 6.0613

   ☐:
         STOP

7. PRINT RESIDUALS?      YES
      NO.      OBSERVED        ESTIMATED        RESIDUAL
       1      197.00000       191.84192         5.15808
       2      272.00000       280.66963        ⁻8.66963
       3      100.00000       103.01421        ⁻3.01421
       4      228.00000       236.25577        ⁻8.25577
       5      327.00000       325.08348         1.91652
       6      279.00000       280.66963        ⁻1.66963
       7      148.00000       147.42806         0.57194
       8      377.00000       369.49734         7.50266
       9      238.00000       236.25577         1.74423
      10      142.00000       147.42806        ⁻5.42806
      11       66.00000        58.60036         7.39964
      12      239.00000       236.25577         2.74423
```

for variable 1 is b_1. Information needed for making inferences on the corresponding β's is also provided, which we will discuss in the next chapter.

4. The results for the ANOVA table are displayed. The format is that of Table 18.4a, with the error source of variation labeled the residual source. The item F Stat. immediately below this table will be discussed in the next chapter.

5. The values of $\sqrt{MSE}$, r^2, r, and n are given in the following format:

$$\sqrt{MSE} \qquad n$$

$$r^2 \qquad r$$

6. This block will be discussed in Chapter 19.
7. The observed values Y_i, the estimated (i.e., fitted) values $\widehat{Y}_i$, and the residuals e_i are given by observation number.

PROBLEMS

18.1 A study panel for a professional association suggested the following income guidelines for members in private practice (X is years of experience and Y is target annual income in thousands of dollars):

X:	5	10	15	20	25	30
Y:	39.5	49.0	58.5	68.0	77.5	87.0

a. Examine the relation between experience and income graphically.
b. Do the guidelines represent a statistical or functional relation? Explain.

18.2 For each of the following pairs of variables, explain whether a functional or statistical relation would most likely hold:
a. X = number of beds in hospital; Y = hospital's annual operating budget.
b. X = altitude; Y = atmospheric pressure.
c. X = company's promotional expenditure; Y = company's sales revenue.

18.3 A vending machine company studied the relation between maintenance cost and dollar sales for six machines. The purpose was to identify machines whose costs are "out of line" with their sales volumes.
a. Which variable is the dependent variable here and which is the independent variable? Why?
b. Cost and sales data (in dollars) for the six machines follow.

Machine:	1	2	3	4	5	6
Maintenance cost:	95	110	80	145	125	90
Sales:	900	1,250	550	850	1,500	800

Make a scatter plot of these observations. Does the maintenance cost of any machine seem "out of line"?

18.4 The ages (in years) of the spouses in five married couples follow.

Couple i:	1	2	3	4	5
Age of husband Y_i:	38	24	49	31	55
Age of wife X_i:	35	24	51	25	53

a. Make a scatter plot of these observations.
b. How well does the line $\widehat{Y} = 2 + X$ fit the five observations? Plot the line and comment.

*18.5 Refer to Figure 18.6. For the observation $X_i = 3$, $Y_i = 146$, calculate ϵ_i. What is the implication of the sign of ϵ_i?

18.6 Suppose the regression function relating annual agricultural exports (Y) to agricultural imports (X) for a certain country is $E\{Y\} = 320 + .4X$, where both Y and X are expressed in millions of dollars.
a. What is the expected level of agricultural exports when agricultural imports are $380 million?

       b.   For year i, $X_i = 600$ and $\epsilon_i = 35$. What is the level of agricultural exports for that year? What symbol is used to denote this quantity?

*18.7   Regression model (18.3) applies in a situation with $\beta_0 = -6$, $\beta_1 = 1.3$, and $\sigma^2 = 16$.

       a.   What is the probability that ϵ_i lies between -5 and $+5$? Explain why you can answer without knowing the value of X_i.

       b.   What is the value of Y_i if $X_i = 15$ and $\epsilon_i = 4$?

       c.   (1) Calculate $E\{Y_i\}$ when $X_i = 6$. (2) What is the probability that Y_i will exceed 1 when $X_i = 6$?

18.8   Refer to Problem 18.7. Answer all parts for $\beta_0 = -5$, $\beta_1 = 1.2$, and $\sigma^2 = 9$.

18.9   In an expenditures study, an economist related family expenditures for clothing (Y) to family size (X) using regression model (18.3). Critics of the study commented that numerous factors other than family size affect clothing expenditures. What are some of these other factors and how are they accommodated by model (18.3)?

*18.10   Data gathered in a study of fuel efficiency of eight automobiles follow.

Automobile i:	1	2	3	4	5	6	7	8
Expressway mileage Y_i: (*miles per gallon*)	35	27	31	38	36	40	37	28
Automobile weight X_i: (*hundreds of pounds*)	21	24	23	21	22	18	20	26

It is desired to relate expressway mileage to automobile weight.

       a.   Construct a scatter plot of the data. Is the horizontal line $\widehat{Y} = 34 + 0X$ a good fit for the eight observations? Discuss.

       b.   Obtain the least squares regression line for regressing expressway mileage on automobile weight. Does the least squares line provide a better visual fit to the scatter plot than the horizontal line considered in **a**? Does the least squares line suggest that expressway mileage tends to decrease with greater automobile weight? Explain.

*18.11   Refer to Problem 18.10. Compute Q, the sum of the squared deviations from a fitted line, for the horizontal line in **a** and for the least squares line in **b**. Are the results consistent with theoretical expectations? Explain.

18.12   Refer to Problem 18.4.

       a.   Compute Q, the sum of squared deviations from a fitted line, for the line $\widehat{Y} = 2 + X$.

       b.   Obtain the least squares regression line for regressing husband's age on wife's age. Compute Q for this line and compare it to the measure computed in **a**. Are the results consistent with theoretical expectations? Explain.

*18.13   Refer to Problem 18.10. The estimated regression function is $\widehat{Y} = 72.776 - 1.7726X$.

       a.   Interpret b_1.

       b.   Obtain the value of the estimated regression function when an automobile's weight is 19 hundred pounds. Interpret this number.

       c.   Is it reasonable to estimate expressway mileage for a weight of 19 hundred pounds when none of the eight automobiles in the regression study has this weight? Discuss.

18.14   Thirty homemakers with similar socioeconomic backgrounds served on a consumer panel and were asked the following two questions, among others: (1) "How much did you pay for your present washing machine?" (2) "What is the maximum amount you would be willing to spend to have it repaired?" Model (18.3) was employed on the study results, with purchase price as the independent variable and maximum repair expenditure as the dependent variable (both expressed in dollars). The estimated regression function is $\hat{Y} = 33.297 + .35352X$.

a. Obtain a point estimate of $E\{Y_h\}$ when $X_h = 450$. Interpret this estimate.

b. Give a point estimate of the change in $E\{Y_h\}$ when X_h increases by one dollar.

c. One of the homemakers on the panel paid \$480 for her present machine and would be willing to spend a maximum of \$215 to have it repaired. Obtain the residual for this observation.

18.15 Data for nine large incinerators, showing the cost in thousands of dollars for the most recent reconditioning (Y) and thousands of operating hours since the preceding reconditioning (X), follow.

i:	1	2	3	4	5	6	7	8	9
X_i:	2.2	1.8	2.9	2.5	1.6	2.9	2.7	3.1	1.9
Y_i:	5.0	4.3	6.2	5.1	3.6	5.8	5.2	6.1	4.0

Assume model (18.3) is appropriate.

a. What is meant by *mean response* in the present case?

b. Obtain a point estimate of the change in the mean response when the number of operating hours increases by one thousand.

c. Obtain a point estimate of the mean reconditioning cost when the number of operating hours is 2.0 thousand.

d. Are your point estimates in b and c subject to sampling errors? If so, do the point estimates convey any information about the magnitudes of these errors?

18.16 Data on number of bids prepared weekly by Atlas Structural Corporation (X) and number of manhours required for bid preparation (Y), for the past eight weeks, follow.

i:	1	2	3	4	5	6	7	8
X_i:	6	3	9	6	4	8	2	6
Y_i:	170	78	232	155	107	212	56	162

a. Was the number of manhours required always the same in the weeks in which six bids were prepared? Is this a departure from model (18.3)? Explain.

b. Obtain the estimated regression function and plot it together with the observations. Does the fit of the estimated regression function appear to be satisfactory? Explain.

c. Present the following: (1) a point estimate of the mean response when five bids are prepared, (2) the residual for the third observation, (3) a point estimate of the change in the mean response when the number of bids prepared increases by one.

18.17 A person stated that when model (18.3) is used, "The residual e_i is estimated by the quantity $Y_i - \hat{Y}_i$." Comment.

*18.18 Refer to Problem 18.16. The estimated regression function is $\hat{Y} = 4.325 + 25.85X$.

a. Obtain the residuals.

b. Verify that: (1) $\Sigma e_i = 0$, (2) $\Sigma X_i e_i = 0$.

18.19 Refer to Problem 18.10. The estimated regression function is $\hat{Y} = 72.776 - 1.7726X$.
  a. Obtain the residuals.
  b. Verify that: (1) $\Sigma e_i = 0$, (2) $\Sigma X_i e_i = 0$, (3) $\Sigma e_i^2 = 25.283$, (4) $\Sigma (Y_i - \bar{Y})^2 = 160$.

*18.20 Refer to Problem 18.10. Some calculational results are: $\hat{Y} = 72.776 - 1.7726X$, $SSTO = 160$, and $SSE = 25.283$.
  a. Set up the ANOVA table.
  b. What parameter of regression model (18.3) is estimated by MSE? What is estimated by $\sqrt{MSE}$?
  c. Is SSR a large proportion of $SSTO$ here? What does this fact imply about the extent to which variation in expressway mileage among the eight automobiles is reduced when weight of automobile is considered?
  d. Calculate r^2 and r. What sign did you attach to r and why?

18.21 Refer to Problem 18.14. Some additional calculational results are: $SSR = 25,761$, $SSE = 3,699$.
  a. Set up the ANOVA table.
  b. Obtain point estimates of σ^2 and σ. Explain the meaning of σ in this application.
  c. Does knowledge of the independent variable appear to be helpful here? Comment.
  d. Obtain r^2 and interpret it in this application.

18.22 Refer to Problem 18.16.
  a. Set up the ANOVA table.
  b. Obtain an estimate of σ. According to model (18.3), is this estimate applicable for any probability distribution of required manhours regardless of the number of bids prepared? Explain.
  c. Which, if any, of the entries in your ANOVA table in **a** would change if: (1) each of the Y_i values were increased by 10 manhours, (2) each Y_i value were converted from units of manhours to units of manweeks (using a basis of 40 hours per week)? Explain.
  d. Calculate r^2 and r. Which measure has the clearer operational interpretation? Which measure is closer to 1 in absolute value?

18.23 A discussant stated: "Up to a point, most of the typists we hire improve in speed and accuracy as they gain experience. We have developed two tests to predict performance ratings after improvement levels off. The coefficient of correlation between our speed test score and speed performance rating is $+.85$, while that between the accuracy test score and accuracy performance rating is $+.82$."
  a. Would the predictive usefulness of the tests be reduced if the signs of the coefficients were negative instead of positive? Discuss.
  b. Can we conclude from the correlation coefficients that SSE is smaller for speed than for accuracy? Explain.

EXERCISES

18.24 Demonstrate the algebraic equivalence of the following pairs of formulas: (1) (18.10b) and (18.12), (2) (18.10a) and (18.11).

18.25 Demonstrate that the estimated regression function (18.13) passes through the point $(\bar{X}, \bar{Y})$, where $\bar{X}$ and $\bar{Y}$ denote the means of the X_i and Y_i values, respectively.

18.26 (Calculus needed.) Confirm that the solution to the normal equations (18.9) does indeed lead to a minimum for Q defined in (18.8).

18.27 (Calculus needed.) When β_0 in regression model (18.3) equals zero, the regression function (18.5) becomes $E\{Y\} = \beta_1 X$. This case is called *regression through the origin*. Obtain the least squares estimator of β_1 for this case.

18.28 Refer to Table 18.1 for the UDS example. Suppose a 13th observation were added to the sample, namely: $X = 2$, $Y = 103.0142$.
  a. Verify that this observation will fall directly on the least squares line for the original 12 observations.
  b. If the least squares line were calculated for all 13 observations, would this line differ from the line for the original 12 observations? Explain. (*Hint:* Would the least squares line for the 12 observations still satisfy the least squares criterion as applied to the 13 observations?)
  c. How would the ANOVA table for the 13 observations differ from Table 18.4b? Be specific.

18.29 Prove the following properties of the least squares residuals for model (18.3): (1) $\Sigma\, e_i = 0$, (2) $\Sigma\, X_i e_i = 0$. [*Hint:* Use normal equations (18.9).]

18.30 Prove the following properties of the least squares residuals for model (18.3): (1) $\Sigma\, \hat{Y}_i e_i = 0$, (2) $\Sigma\, Y_i e_i = SSE$.

18.31 Demonstrate that (18.27) and (18.30) are algebraically equivalent formulas.

18.32 In fitting model (18.3) to some data, we obtain: $b_0 = 2.0$, $b_1 = 0$. Explain which of the following will be true and which false: (1) $SSTO = 0$, (2) $r^2 = 0$, (3) $r = 0$, (4) $SSR = SSE$, (5) $SSE = SSTO$, (6) $SSTO = SSR + SSE$.

STUDIES

18.33 Data on workloads for 10 distribution centers maintained in overseas locations by a multinational firm follow (X denotes thousands of work units performed during the reporting period and Y denotes thousands of manhours required).

Center i:	1	2	3	4	5	6	7	8	9	10
X_i:	5.0	3.5	10.0	5.0	6.5	6.0	7.1	2.5	3.0	4.2
Y_i:	27.5	20.1	50.5	26.3	33.5	32.4	36.8	15.5	18.3	22.0

An analyst calculated the number of manhours required per unit of work (Y/X) for each center. He suggested that center 3 appears to be particularly efficient in that its ratio of manhours to work units is lowest.
  a. Calculate the ratio Y/X for each center and present the ratios in the form of a stem-and-leaf display. Does the ratio for center 3 appear to be particularly out of line with respect to those for other centers?
  b. Regress manhours required on work units performed, using model (18.3). Make a scatter plot of the observations and plot your estimated regression function on the same graph.
  c. Contrast your findings with those of the analyst, paying particular attention to the matter of efficiency.

18.34 Refer to Appendix D.

    a. Regress 1981 net sales on 1981 net assets for firms in the electronics industry (companies 249–269 inclusive), using model (18.3).

    b. What is the value of r? Of $\sqrt{MSE}$? Which of these two measures provides more direct information about the variability of net sales of firms with given asset size? Explain.

    c. Examine the residuals. Do any companies seem out of line in terms of the relation of their net sales to net assets? Comment on factors that might explain such "outliers."

    d. Redo the regression analysis in **a**, omitting company 263 from the set of observations. Did this one observation have a substantial effect on the values of b_0 and b_1? On $\sqrt{MSE}$? In general, why do outlying observations have a substantial influence on b_0 and b_1 with the method of least squares?

18.35 Refer to Problem 18.34. Do the problem using the 1980 data for the firms in the electronics industry.

19
Inferences in
Simple Linear Regression

In the previous chapter, we discussed the fundamental concepts of simple linear regression. Here, we consider interval estimation of the parameters of the regression model and of the mean response, as well as a number of other topics of importance in applied work.

The simple linear regression model (18.3) will continue to be employed:

$$(19.1) \qquad Y_i = \beta_0 + \beta_1 X_i + \epsilon_i \qquad i = 1, 2, \ldots, n$$

where: Y_i is the response in the ith observation
   X_i is the value of the independent variable in the ith observation, assumed to be a known constant
   β_0 and β_1 are parameters
   ϵ_i's are independent $N(0, \sigma^2)$

19.1 CONFIDENCE INTERVAL FOR $E\{Y_h\}$

As noted in Chapter 18, regression applications frequently entail estimation of the mean of the distribution of Y at a specified level of X.

☐ Examples

1. A loan officer needs an estimate of the mean repayment period for home improvement loans of the most popular size, namely $3,000.
2. An admissions officer at a university wishes to estimate the mean grade-point average of freshmen students who receive the minimal acceptable score, 450, on an entrance examination. ☐

Point Estimator of $E\{Y_h\}$

Recall that the point estimator of $E\{Y_h\}$, the mean of Y when $X = X_h$, was given in (18.15) and is denoted by $\hat{Y}_h$:

$$(19.2) \qquad \hat{Y}_h = b_0 + b_1 X_h$$

Sampling Distribution of $\widehat{Y}_h$

The estimate $\widehat{Y}_h$ will vary in repeated samples since b_0 and b_1 will vary from sample to sample. The sampling distribution of $\widehat{Y}_h$ describes this variability in $\widehat{Y}_h$ from sample to sample. Specifically, the sampling distribution of $\widehat{Y}_h$ refers to the different possible values of $\widehat{Y}_h$ that would be obtained if repeated samples were selected, each with the same values $X_1, X_2, \ldots, X_n$ for the independent variable.

Properties. Statistical theory provides us with information about the properties of the sampling distribution of $\widehat{Y}_h$:

(19.3) For the simple linear regression model (19.1), the sampling distribution of $\widehat{Y}_h$ has the following properties:

$$\text{Mean: } E\{\widehat{Y}_h\} = \beta_0 + \beta_1 X_h = E\{Y_h\}$$

$$\text{Variance: } \sigma^2\{\widehat{Y}_h\} = \sigma^2\left[\frac{1}{n} + \frac{(X_h - \bar{X})^2}{\Sigma (X_i - \bar{X})^2}\right]$$

Functional form: Normal distribution

We noted in Chapter 18 that $\widehat{Y}_h$ is an unbiased estimator of $E\{Y_h\}$.

Estimated Variance of $\widehat{Y}_h$. Usually $\sigma^2\{\widehat{Y}_h\}$, the variance of the sampling distribution of $\widehat{Y}_h$, is unknown because σ^2, the variance of the error terms ϵ_i, is unknown. However, we know from Chapter 18 that MSE is an unbiased estimator of σ^2. Hence, we can replace σ^2 by MSE to obtain the estimated variance of $\widehat{Y}_h$, denoted by $s^2\{\widehat{Y}_h\}$:

(19.4)
$$s^2\{\widehat{Y}_h\} = MSE\left[\frac{1}{n} + \frac{(X_h - \bar{X})^2}{\Sigma (X_i - \bar{X})^2}\right]$$

□ Examples

1. We continue with our UDS example from Chapter 18. We wish to estimate $\sigma^2\{\widehat{Y}_h\}$ for $X_h = 4$ tape drives. From the earlier calculations in Chapter 18, we have $n = 12$, $\bar{X} = 4.583$, and $MSE = 33.688$. The quantity $\Sigma (X_i - \bar{X})^2$ can be calculated from the data in Table 18.3:

$$\Sigma (X_i - \bar{X})^2 = (4 - 4.583)^2 + (6 - 4.583)^2 + \cdots + (5 - 4.583)^2$$

$$= 46.91667$$

The estimated variance of $\widehat{Y}_h$ for $X_h = 4$ then is:

$$s^2\{\widehat{Y}_h\} = 33.688\left[\frac{1}{12} + \frac{(4 - 4.583)^2}{46.91667}\right] = 3.05139$$

and hence the estimated standard deviation of $\widehat{Y}_h$ is:

$$s\{\widehat{Y}_h\} = \sqrt{3.05139} = 1.747$$

2. We wish to find the estimated variance of $\widehat{Y}_h$ when $X_h = 7$. We obtain:

$$s^2\{\widehat{Y}_h\} = 33.688\left[\frac{1}{12} + \frac{(7 - 4.583)^2}{46.91667}\right] = 7.00204$$

The estimated standard deviation therefore is:

$$s\{\widehat{Y}_h\} = \sqrt{7.00204} = 2.646 \qquad \square$$

Comment

The magnitude of the estimated variance $s^2\{\widehat{Y}_h\}$ in (19.4) is affected by a number of factors:

1. *MSE.*   The larger the variability of the residuals e_i, the larger $s^2\{\widehat{Y}_h\}$ tends to be.
2. *Deviation of X_h from $\bar{X}$.*   The further the specified level X_h is from $\bar{X}$ in either direction, the larger the quantity $(X_h - \bar{X})^2$ and the greater tends to be $s^2\{\widehat{Y}_h\}$. For a given sample, $s^2\{\widehat{Y}_h\}$ is smallest at $X_h = \bar{X}$. This is illustrated in our previous examples where we found $s^2\{\widehat{Y}_h\} = 7.00204$ when $X_h = 7$ and $s^2\{\widehat{Y}_h\} = 3.05139$ when $X_h = 4$. Since $\bar{X} = 4.583$, $X_h = 7$ is further from the mean than $X_h = 4$.
3. *Variability of X_i's.*   The greater is the variability of the X_i's around their mean $\bar{X}$, the larger is $\Sigma\,(X_i - \bar{X})^2$ and the smaller tends to be $s^2\{\widehat{Y}_h\}$.
4. *Sample size n.*   The greater is n, the smaller is $1/n$ and the smaller tends to be $s^2\{\widehat{Y}_h\}$. In addition, the denominator quantity $\Sigma\,(X_i - \bar{X})^2$ often is larger for increasing n, so again the smaller tends to be $s^2\{\widehat{Y}_h\}$.

These properties can be useful when designing a regression study. For example, suppose the major purpose of the regression study is to estimate the mean $E\{Y_h\}$ at $X = X_h$, and the levels of X in the sample may be chosen freely. It would then be desirable to select levels of X with mean $\bar{X} = X_h$.

Confidence Interval for $E\{Y_h\}$

Development of Confidence Interval.   A confidence interval for $E\{Y_h\}$ can be easily constructed on the basis of the following statistical theorem:

(19.5)   For the simple linear regression model (19.1):

$$\frac{\widehat{Y}_h - E\{Y_h\}}{s\{\widehat{Y}_h\}} = t(n - 2)$$

The degrees of freedom for the t random variable are the $n - 2$ degrees of freedom associated with *MSE* (see Table 18.4a). Recall from (19.4) that *MSE* is used to obtain $s^2\{\widehat{Y}_h\}$.

Given theorem (19.5), a confidence interval for $E\{Y_h\}$ is obtained in the usual fashion, utilizing the t distribution:

(19.6)   The confidence limits for $E\{Y_h\}$ with confidence coefficient $1 - \alpha$ for the simple linear regression model (19.1) are:

$$\widehat{Y}_h \pm ts\{\widehat{Y}_h\}$$

where: $t = t(1 - \alpha/2;\ n - 2)$
$s\{\widehat{Y}_h\}$ is given by (19.4)

□ **Examples**

1. The UDS service manager desires a 95 percent confidence interval for the mean time required for service calls involving $X_h = 4$ tape drives. We calculated in Chapter 18 that $\hat{Y}_h = 191.84$ when $X_h = 4$. Earlier in this chapter, we obtained $s\{\hat{Y}_h\} = 1.747$ when $X_h = 4$. Finally, for $n = 12$ and $1 - \alpha = .95$, we require $t(.975; 10) = 2.228$. Hence, the confidence limits are $191.84 \pm 2.228(1.747)$ and the desired confidence interval is:

$$188 \leq E\{Y_h\} \leq 196$$

Thus, with 95 percent confidence, the service manager can conclude that the mean service time for four tape drives is between 188 and 196 minutes.

2. We wish to obtain a 90 percent interval estimate for $E\{Y_h\}$ when $X_h = 7$. Earlier we found for $X_h = 7$ that $\hat{Y}_h = 325.08$ and $s\{\hat{Y}_h\} = 2.646$. We require $t(.95; 10) = 1.812$. Hence, the 90 percent confidence limits are $325.08 \pm 1.812(2.646)$ and the confidence interval is:

$$320 \leq E\{Y_h\} \leq 330$$
□

Comment

The confidence coefficient for confidence interval (19.6) is interpreted in terms of repeated samples in which the set of X levels is the same from sample to sample. Thus, the 95 percent confidence coefficient attached to our interval estimate of $E\{Y_h\}$ when $X_h = 4$ signifies that if many independent samples were taken where the X levels are the same as in the 12 calls in the observed sample and a 95 percent confidence interval for $E\{Y_h\}$ were constructed from each sample, about 95 percent of these intervals would contain the true value of $E\{Y_h\}$.

19.2 PREDICTION OF INDIVIDUAL RESPONSE Y_h

In regression applications, one frequently wishes to predict an individual response:

(19.7) An *individual response*, denoted by Y_h, is the value of Y to be observed in a *new* observation when the level of the independent variable is $X = X_h$.

It is important to distinguish between an individual response Y_h and a mean response $E\{Y_h\}$.

□ **Examples**

1. UDS technicians will test six tape drives on a forthcoming call. The firm wishes to predict the number of minutes required on this particular call. An individual response Y_h is involved here. The required service time will constitute, in effect, a new observation from the distribution of Y when $X_h = 6$. The prediction is not equivalent to estimating $E\{Y_h\}$, the mean of the distribution of Y at $X_h = 6$, since the new observation in all likelihood will deviate from the mean.

2. Anne Smith has just scored 115 on the company's aptitude test. A prediction of Anne Smith's job performance rating based on her test score of 115 entails a prediction of an individual response Y_h when $X_h = 115$. This is in contrast to an estimate of the mean performance rating $E\{Y_h\}$ for all persons who score 115 on the aptitude test.

3. An analyst for a food-processing company has investigated the relation between weight loss in wheels of swiss cheese during storage (Y) and storage temperature (X). He now wishes to estimate the mean weight loss when the storage temperature is $7°C$. Here, an estimate of the mean response $E\{Y_h\}$ is required, and not a prediction for an individual wheel.   ☐

Prediction Interval for Y_h When Parameters Known

To illustrate the basic concept of a prediction interval, we assume first that the parameters β_0, β_1, and σ^2 of the simple linear regression model are known. Ordinarily, of course, they are not known. We then extend the procedure to cover the usual case where the parameters must be estimated.

An operations analyst is studying the mail order operations of a large music store. He believes that the simple linear regression model (19.1) is applicable, where Y is the distance in feet traveled by a clerk in filling the mail order, X is the number of different record albums in the order, $\beta_0 = 20$, $\beta_1 = 15$, and $\sigma = 6.5$.

An order for $X_h = 3$ albums is to be filled next. Figure 19.1 shows the probability distribution of Y when $X_h = 3$. It is normal, with mean $E\{Y_h\} = 20 + 15(3) = 65$ feet and standard deviation $\sigma = 6.5$ feet.

Let us predict the distance to be traveled in filling this order; that is, we wish to predict an individual response Y_h when $X_h = 3$. Of course, a point estimate of Y_h is $E\{Y_h\} = 65$ feet. However, we know that the individual response will deviate around this mean and we wish to set up a prediction interval that recognizes the random deviation of Y_h around $E\{Y_h\}$.

Suppose we use an interval of ± 1.96 standard deviations around the mean. The prediction limits then are:

$$65 \pm 1.96(6.5)$$

FIGURE 19.1   *Prediction of individual response Y_h in music store example—Parameters known*

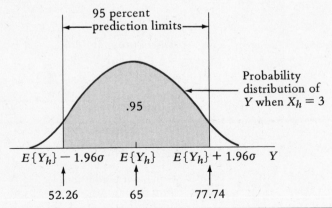

and the prediction interval is:

$$52.26 \leq Y_h \leq 77.74$$

Since 95 percent of the area of a normal distribution is within ± 1.96 standard deviations about the mean, our procedure will give correct predictions 95 percent of the time.

Prediction Interval for Y_h When Parameters Unknown

Development of Prediction Interval.   We now return to the UDS example where β_0, β_1, and σ are not known and must be estimated from the sample. We wish to predict the service time Y_h on the next call with $X_h = 4$ tape drives with confidence coefficient .95. We do not know the exact mean of the distribution of Y when $X_h = 4$. Earlier we did obtain a 95 percent confidence interval for $E\{Y_h\}$—namely, $188 \leq E\{Y_h\} \leq 196$ minutes.

Figure 19.2 shows the distribution of Y_h centered around each of these confidence limits for $E\{Y_h\}$. Of course, any other point between 188 and 196 is another possible location for $E\{Y_h\}$ consistent with our confidence interval.

Figure 19.2 also shows the respective 95 percent prediction limits for Y_h if $E\{Y_h\} = 188$ and if $E\{Y_h\} = 196$. Since we do not know where the distribution of Y_h actually is located when β_0 and β_1 are unknown, the prediction limits we construct must take account of the two sources of variation shown in Figure 19.2:

1. Variation in possible location of the distribution of Y.
2. Variation within the probability distribution of Y.

FIGURE 19.2   *Components of variation in predicting individual response when parameters unknown (not plotted to scale)*

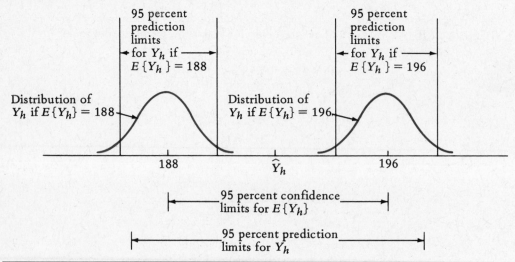

It can be shown that the relevant variance for predicting Y_h is the variance of $Y_h - \widehat{Y}_h$, the difference between the value of the dependent variable to be observed and the estimated expected value of Y_h based on the n sample observations. We denote this difference by d_h:

(19.8)
$$d_h = Y_h - \widehat{Y}_h$$

Since it is assumed that the error term ϵ_h of the new observation is independent of the error terms ϵ_i of the n sample observations, it can be shown that:

(19.9)
$$\sigma^2\{d_h\} = \sigma^2\{\widehat{Y}_h\} + \sigma^2\{Y_h\} = \sigma^2\{\widehat{Y}_h\} + \sigma^2$$

Note that the two components of $\sigma^2\{d_h\}$ relate to the two types of variation shown in Figure 19.2, and are:

1. The variance of the sampling distribution of $\widehat{Y}_h$.
2. The variance of the distribution of Y_h.

An unbiased estimator of $\sigma^2\{d_h\}$ is:

(19.10)
$$s^2\{d_h\} = s^2\{\widehat{Y}_h\} + MSE = MSE\left[1 + \frac{1}{n} + \frac{(X_h - \bar{X})^2}{\Sigma\,(X_i - \bar{X})^2}\right]$$

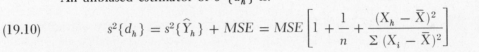

where: $s^2\{\widehat{Y}_h\}$ is given by (19.4)

The prediction interval for Y_h then takes a familiar form:

(19.11) The prediction limits for Y_h with confidence coefficient $1 - \alpha$ for the simple linear regression model (19.1) are:

$$\widehat{Y}_h \pm ts\{d_h\}$$

where: $t = t(1 - \alpha/2; n - 2)$
$s\{d_h\}$ is given by (19.10)

☐ Examples

1. In our UDS example, the customer for the next service call has $X_h = 4$ tape drives. We wish to obtain a 95 percent prediction interval for the service time Y_h. From earlier work, we have for $X_h = 4$:

$$n = 12 \qquad \bar{X} = 4.583 \qquad MSE = 33.688 \qquad \widehat{Y}_h = 191.84 \qquad s^2\{\widehat{Y}_h\} = 3.05139$$

First, we obtain $s^2\{d_h\}$ by (19.10):

$$s^2\{d_h\} = 3.05139 + 33.688 = 36.73939$$

The estimated standard deviation therefore is:

$$s\{d_h\} = \sqrt{36.73939} = 6.0613$$

We also require $t(.975; 10) = 2.228$. The prediction limits therefore are $191.84 \pm 2.228(6.0613)$ and the desired prediction interval is:

$$178 \leq Y_h \leq 205$$

Thus, in the next call in which four tape drives are to be tested, we predict with 95 percent confidence that the time required will be between 178 and 205 minutes.

2. Suppose the customer for the next service call has $X_h = 7$ tape drives. We wish to obtain 99 percent prediction limits for the service time Y_h. From earlier work, we have for $X_h = 7$:

$$n = 12 \qquad \bar{X} = 4.583 \qquad MSE = 33.688 \qquad \widehat{Y}_h = 325.08 \qquad s^2\{\widehat{Y}_h\} = 7.00204$$

Hence, $s^2\{d_h\} = 7.00204 + 33.688 = 40.69004$ and $s\{d_h\} = \sqrt{40.69004} = 6.3789$. We require $t(.995; 10) = 3.169$. The 99 percent prediction limits therefore are $325.08 \pm 3.169(6.3789)$, or 305 and 345 minutes, respectively. ☐

Comments

1. Prediction limits are useful for control purposes. Suppose 230 minutes were required for the new call with four tape drives. Since the 95 percent prediction interval ranges from 178 to 205 minutes, management may wish to ascertain why the service time was so high.
2. Note that the 95 percent prediction interval for Y_h calculated in Example 1 is wider than the corresponding confidence interval for $E\{Y_h\}$. The reason is that the estimated variance of d_h contains not only $s^2\{\widehat{Y}_h\}$ but also an additional component, MSE, which reflects the variability in the probability distribution of Y.
3. The factors affecting the magnitude of $s^2\{\widehat{Y}_h\}$ also affect $s^2\{d_h\}$. For instance, the further X_h is from $\bar{X}$, the larger is $s^2\{d_h\}$ for any given sample, and the wider will be the prediction interval.

19.3 INFERENCES CONCERNING β_1

We turn now to inferences on β_1, the slope of the regression function for model (19.1). In some regression studies, the primary objective is to estimate this slope. Recall that β_1 indicates the change in the mean of the distribution of Y when X increases by one unit. Thus, the UDS service manager may wish to ascertain how much the mean service time increases for each additional tape drive tested on a call.

There are also occasions when a test on β_1 is appropriate. For example, an operations analyst for a large mail order house wishes to ascertain whether a relation exists between weight of incoming mail (X) and dollar value of mail orders (Y). If a relation does exist, the analyst would like to utilize it to predict dollar value of daily mail order sales from the weight of the day's mail.

A test whether a relation between X and Y exists, given that model (19.1) is appropriate, takes the following form:

$$H_0\colon \beta_1 = 0$$

$$H_1\colon \beta_1 \neq 0$$

The reason is that if $\beta_1 = 0$, $E\{Y\} = \beta_0 + \beta_1 X = \beta_0$ for all levels of X. Then all distributions of Y have the same mean. Since model (19.1) requires all distributions of Y to be normal with equal variability, the additional condition of equal means implies that all distributions of Y are identical, regardless of the level of X. Hence, there is no relation between X and Y when $\beta_1 = 0$ for model (19.1).

Sampling Distribution of b_1

We require knowledge of the sampling distribution of b_1 for constructing a confidence interval for β_1, as well as for conducting tests about this parameter:

(19.12) For the simple linear regression model (19.1), the sampling distribution of b_1 has the following properties:

$$\text{Mean: } E\{b_1\} = \beta_1$$

$$\text{Variance: } \sigma^2\{b_1\} = \frac{\sigma^2}{\Sigma\,(X_i - \bar{X})^2}$$

Functional form: Normal distribution

We noted in Chapter 18 that the least squares estimator b_1 is unbiased.

Estimated Variance of b_1. The variance of $\sigma^2\{b_1\}$ usually is unknown and is estimated by replacing σ^2 by MSE. This estimated variance is denoted by $s^2\{b_1\}$:

(19.13)
$$s^2\{b_1\} = \frac{MSE}{\Sigma\,(X_i - \bar{X})^2}$$

☐   Example

In continuing with our UDS example, we know from earlier work that $MSE = 33.688$ and $\Sigma\,(X_i - \bar{X})^2 = 46.91667$. Hence:

$$s^2\{b_1\} = \frac{33.688}{46.91667} = .71804$$

and the estimated standard deviation of the sampling distribution of b_1 is:

$$s\{b_1\} = \sqrt{.71804} = .84737$$ ☐

Confidence Interval for β_1

A confidence interval for β_1 can be constructed readily from the following statistical theorem:

(19.14) For the simple linear regression model (19.1):

$$\frac{b_1 - \beta_1}{s\{b_1\}} = t(n - 2)$$

Again, the degrees of freedom for the t random variable are the $n - 2$ degrees of freedom associated with MSE.

On the basis of theorem (19.14), the confidence interval for β_1 takes a familiar form:

(19.15) The confidence limits for β_1 with confidence coefficient $1 - \alpha$ for the simple linear regression model (19.1) are:

$$b_1 \pm ts\{b_1\}$$

where: $t = t(1 - \alpha/2; n - 2)$
$s\{b_1\}$ is given by (19.13)

☐   **Example**

We wish to obtain a 95 percent confidence interval for β_1 for our UDS case. We have $b_1 = 44.41385$ (from Chapter 18) and $s\{b_1\} = .84737$, and require $t(.975; 10) = 2.228$. The confidence limits therefore are $44.41385 \pm 2.228(.84737)$ and the confidence interval is:

$$42.5 \leq \beta_1 \leq 46.3$$

Thus, we estimate, with 95 percent confidence, that the mean service time increases by somewhere between 42.5 and 46.3 minutes for each additional tape drive tested on a call. ☐

Statistical Tests for β_1

As a result of theorem (19.14), tests on β_1 are constructed in an analogous fashion to those for a population mean μ as outlined in Section 11.9—see Table 11.2 in particular. A test of whether or not $\beta_1 = 0$, using b_1 as the test statistic, takes the following form:

(19.16) When the alternatives are:

$$H_0: \beta_1 = 0$$

$$H_1: \beta_1 \neq 0$$

and the simple linear regression model (19.1) is applicable, the appropriate decision rule to control the α risk is:

If $A_1 \leq b_1 \leq A_2$, conclude H_0

If $b_1 < A_1$ or $b_1 > A_2$, conclude H_1

where: $A_1 = 0 - ts\{b_1\}$
$A_2 = 0 + ts\{b_1\}$
$t = t(1 - \alpha/2; n - 2)$
$s\{b_1\}$ is given by (19.13)

☐   **Example**

We wish to test in our UDS example whether a relation between X and Y exists, given that model (19.1) is appropriate. Thus, we wish to test whether or not $\beta_1 = 0$. The α risk is to be controlled at .05.

We know from a previous calculation that $s\{b_1\} = .84737$ and require $t(.975; 10) = 2.228$. Hence, the two action limits are $0 \pm 2.228(.84737)$ and the decision rule is:

If $-1.888 \leq b_1 \leq 1.888$, conclude H_0

If $b_1 < -1.888$ or $b_1 > 1.888$, conclude H_1

Since $b_1 = 44.414 > 1.888$, we conclude H_1—that $\beta_1 \neq 0$. This implies for model (19.1) that X and Y are related. ☐

The equivalent test can be conducted by means of the standardized test statistic t^*:

$$(19.17) \qquad t^* = \frac{b_1 - 0}{s\{b_1\}} = \frac{b_1}{s\{b_1\}}$$

The decision rule for testing whether or not $\beta_1 = 0$ with the α risk controlled then is:

$$(19.18) \qquad \text{If } |t^*| \leq t, \text{ conclude } H_0$$

$$\text{If } |t^*| > t, \text{ conclude } H_1$$

where: $t = t(1 - \alpha/2; n - 2)$
t^* is given by (19.17)

☐   Example

For our UDS example with $\alpha = .05$, we have $t(.975; 10) = 2.228$. The decision rule thus is:

$$\text{If } |t^*| \leq 2.228, \text{ conclude } H_0$$

$$\text{If } |t^*| > 2.228, \text{ conclude } H_1$$

Since $b_1 = 44.41385$ and $s\{b_1\} = .84737$, we have $t^* = 44.41385/.84737 = 52.41$ and $|t^*| = 52.41 > 2.228$. Hence, we conclude H_1, just as we did with the equivalent test based on the statistic b_1.   ☐

Comments

1. The test just illustrated could also have been conducted via the confidence interval for β_1. Since the 95 percent confidence interval for β_1 obtained earlier does not include $\beta_1 = 0$, we would conclude H_1—that $\beta_1 \neq 0$.
2. Occasionally, one wishes to test whether or not β_1 is some value other than 0, denoted by β_{1I}. For example, an engineer may wish to test whether or not the slope of the regression function for a new chemical plant relating process output to temperature is $\beta_{1I} = 45$ as expected from planning studies. The alternatives in this type of case are:

$$(19.19) \qquad H_0: \beta_1 = \beta_{1I}$$

$$H_1: \beta_1 \neq \beta_{1I}$$

The decision rule to control the α risk is the same as in (19.18) with t^* now defined as follows:

$$(19.20) \qquad t^* = \frac{b_1 - \beta_{1I}}{s\{b_1\}}$$

19.4   INFERENCES CONCERNING β_0

Occasionally, inferences on the intercept β_0 are of interest. For example, sometimes the intercept β_0 can be viewed as a "fixed" component in the relation between X and Y, as when Y is the cost of a production run and X is the number of units in the run. Here, β_0 might represent the fixed setup cost for the production run and β_1 the variable cost per unit.

Inferences concerning β_0 are made in analogous fashion to those for β_1, relying on the following key results:

(19.21) For the simple linear regression model (19.1):

(19.21a)
$$E\{b_0\} = \beta_0$$

(19.21b)
$$\sigma^2\{b_0\} = \sigma^2\left[\frac{1}{n} + \frac{\bar{X}^2}{\Sigma\,(X_i - \bar{X})^2}\right]$$

(19.21c)
$$s^2\{b_0\} = MSE\left[\frac{1}{n} + \frac{\bar{X}^2}{\Sigma(X_i - \bar{X})^2}\right]$$

(19.21d)
$$\frac{b_0 - \beta_0}{s\{b_0\}} = t(n - 2)$$

☐   **Example**

We wish to construct a 95 percent confidence interval for β_0 for our UDS example, where $b_0 = 14.18652$. First, we need to calculate the estimated variance of b_0. Since $MSE = 33.688$, $\bar{X} = 4.583$, and $\Sigma\,(X_i - \bar{X})^2 = 46.91667$, we obtain by (19.21c):

$$s^2\{b_0\} = 33.688\left[\frac{1}{12} + \frac{(4.583)^2}{46.91667}\right] = 17.88895$$

The estimated standard deviation of b_0 therefore is $s\{b_0\} = \sqrt{17.88895} = 4.22953$.

In view of (19.21d), the confidence limits for β_0 are of the usual form: $b_0 \pm ts\{b_0\}$, where $t = t(1 - \alpha/2; n - 2)$. We require $t(.975; 10) = 2.228$. Hence, the confidence limits are $14.18652 \pm 2.228(4.22953)$ and the desired confidence interval is:

$$4.8 \le \beta_0 \le 23.6$$ ☐

Comments

1. The confidence interval that we have just obtained is equivalent to a confidence interval for $E\{Y_h\}$ when $X_h = 0$. This follows because when $X_h = 0$, we have $E\{Y_h\} = \beta_0 + \beta_1 X_h = \beta_0$.

2. If it is known that $\beta_0 = 0$, regression model (19.1) simplifies to one where the regression line goes through the origin. Such a model is considered in Chapter 25 for time series data.

19.5   COVARIANCE OF b_0 AND b_1

Although b_0 and b_1 calculated from a given sample are estimates of different parameters, they usually contain sampling errors that are interrelated. Figure 19.3 illustrates this point. It shows the true linear regression function for a particular application, together with two estimated regression functions obtained from two different samples. The estimated regression function from sample 1 has a steeper slope and lower intercept than the true regression function. Hence, here b_1 overes-

FIGURE 19.3   *Joint error tendencies in b_0 and b_1 when $\sigma\{b_0, b_1\}$ is negative.* The estimates b_0 and b_1 tend to err in opposite directions.

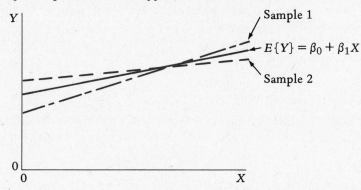

timates β_1 while b_0 underestimates β_0. The estimated regression function from sample 2 shows the reverse pattern, with b_1 underestimating β_1 and b_0 overestimating β_0. Note that in both cases, b_0 and b_1 for a given sample err in opposite directions, so the covariations, as defined in (5.23), are negative. The tendency for b_0 and b_1 to err in opposite directions is not surprising. Intuitively, we sense that in most cases where the slope of an estimated regression line is too high, the intercept will be too low, and vice versa.

The covariance of two random variables was defined in (5.24). It can be shown that the covariance of b_0 and b_1, denoted by $\sigma\{b_0, b_1\}$, is as follows for model (19.1):

(19.22)
$$\sigma\{b_0, b_1\} = -\bar{X}\sigma^2\{b_1\}$$

Consequently when $\bar{X}$ is positive, as in our UDS example, $\sigma\{b_0, b_1\}$ is negative, and b_0 and b_1 tend to err in opposite directions in the same sample. Less commonly, when $\bar{X}$ equals zero, b_0 and b_1 have zero covariance, and when $\bar{X}$ is negative, b_0 and b_1 have positive covariance.

The estimated covariance of b_0 and b_1 is denoted by $s\{b_0, b_1\}$ and is obtained by replacing $\sigma^2\{b_1\}$ by $s^2\{b_1\}$ in (19.22):

(19.23)
$$s\{b_0, b_1\} = -\bar{X}s^2\{b_1\}$$

☐   **Example**

For our UDS example, where $\bar{X} = 4.583$ and $s^2\{b_1\} = .71804$, we calculate:

$$s\{b_0, b_1\} = -4.583(.71804) = -3.29078$$ ☐

19.6   *F* TEST OF $\beta_1 = 0$

In Section 19.3, we discussed a statistical test for ascertaining whether or not $\beta_1 = 0$. This test, which for model (19.1) is equivalent to ascertaining whether or not a relation exists between X and Y, is often utilized as a first step in the analysis of a statistical relation between two variables.

We now take up an equivalent test via the analysis of variance (ANOVA) approach. While this approach does not provide us with anything new here, it will be most useful when we consider multiple regression where there are two or more independent variables.

Expected Mean Squares

We stated earlier that the error mean square MSE, given in (18.34), is an unbiased estimator of σ^2, the variance of the error terms ϵ_i. Thus:

$$(19.24) \qquad E\{MSE\} = \sigma^2$$

It can also be shown that the expected value of MSR, the regression mean square given in (18.33), is as follows for model (19.1):

$$(19.25) \qquad E\{MSR\} = \sigma^2 + \beta_1^2 \Sigma (X_i - \bar{X})^2$$

Thus, when $\beta_1 = 0$, $E\{MSR\} = \sigma^2$, so both MSR and MSE have the same expected value then. This means that when $\beta_1 = 0$, MSR and MSE tend to be of approximately the same order of magnitude. On the other hand, when $\beta_1 \neq 0$, the term $\beta_1^2 \Sigma (X_i - \bar{X})^2$ will be positive and $E\{MSR\} > E\{MSE\}$. Hence, if $\beta_1 \neq 0$, MSR will tend to be larger than MSE.

Development of Statistical Test

The ANOVA test statistic is:

$$(19.26) \qquad F^* = \frac{MSR}{MSE}$$

If F^* is near 1 (MSR and MSE are approximately equal), this would suggest by the earlier reasoning that $\beta_1 = 0$. On the other hand, if F^* is substantially greater than 1 (MSR is much larger than MSE), this would suggest that $\beta_1 \neq 0$. Thus, an upper-tail test is appropriate.

Figure 19.4 portrays the structure of the test. Figure 19.4a shows the appropriate one-sided decision rule, with large values of F^* leading to H_1: $\beta_1 \neq 0$. Figure 19.4b shows the sampling distribution of F^* when $\beta_1 = 0$, the case for which the α risk is to be controlled.

(19.27) For the simple linear regression model (19.1), when $\beta_1 = 0$:

$$F^* = F(1, n - 2)$$

where: F^* is given by (19.26)

Hence, the distribution in Figure 19.4b is an F distribution, where the degrees of freedom are those associated with MSR and MSE. (Readers who need to review the F distribution should turn to Appendix B, Section B.3 before proceeding.)

FIGURE 19.4   *Construction of F test of $\beta_1 = 0$ for model (19.1)*

(a) *Appropriate type of decision rule*

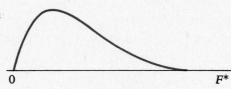

Large values of F^* imply H_1 is correct.

(b) *Sampling distribution of F^* when $\beta_1 = 0$*

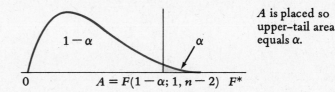

Distribution of F^* when $\beta_1 = 0$ is $F(1, n - 2)$.

(c) *Placement of action limit*

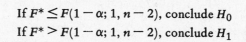

A is placed so upper–tail area equals α.

(d) *Decision rule*

If $F^* \leq F(1 - \alpha; 1, n - 2)$, conclude H_0

If $F^* > F(1 - \alpha; 1, n - 2)$, conclude H_1

Conclude appropriate alternative based on observed F^*.

Finally, Figure 19.4c shows the placement of the action limit so that the upper-tail area equals α, and Figure 19.4d presents the decision rule. Recall that $F(1 - \alpha; 1, n - 2)$ is the $100(1 - \alpha)$ percentile of the F distribution with 1 degree of freedom for the numerator and $n - 2$ for the denominator.

We now present the entire decision procedure:

(19.28) When the alternatives are:

$$H_0: \beta_1 = 0$$

$$H_1: \beta_1 \neq 0$$

and the simple linear regression model (19.1) is applicable, the appropriate decision rule to control the α risk is:

If $F^* \leq F(1 - \alpha; 1, n - 2)$, conclude H_0

If $F^* > F(1 - \alpha; 1, n - 2)$, conclude H_1

where: F^* is given by (19.26)

☐ Example

For the UDS example, we wish to use the ANOVA approach to test:

$$H_0: \beta_1 = 0$$

$$H_1: \beta_1 \neq 0$$

The α risk is to be controlled at .05. We found in Chapter 18 (see Table 18.4b) that $MSR = 92,547.37$ and $MSE = 33.688$. For $\alpha = .05$ and $n - 2 = 12 - 2 = 10$, we require $F(.95; 1, 10) = 4.96$. The decision rule therefore is:

$$\text{If } F^* \leq 4.96, \text{ conclude } H_0$$

$$\text{If } F^* > 4.96, \text{ conclude } H_1$$

The test statistic is $F^* = 92,547.37/33.688 = 2,747.2$. Since $F^* = 2,747.2 > 4.96$, we conclude H_1—that $\beta_1 \neq 0$. This is the same result we obtained by the t test. ☐

Comments

1. The F test in (19.28) is equivalent to the t test in (19.18), the F^* statistic being the square of the standardized t^* statistic. Recall that we had $t^* = 52.41$, and $(52.41)^2 = 2,747 = F^*$. Further, the action limit for the t test was 2.228, and $(2.228)^2 = 4.96$, the action limit for the ANOVA test. This equivalence is based on the second relation in (B.16) of Appendix B.
2. The F test can be used only for testing $\beta_1 = 0$ versus $\beta_1 \neq 0$, and not for testing one-sided alternatives.

19.7 EVALUATION OF APTNESS OF MODEL

When a regression model is applied in practice, one cannot usually be sure in advance that the model employed is appropriate for the situation at hand. Consequently, the model considered needs to be checked for aptness or suitability, using the observed data for guidance. Indeed, frequently several models need to be investigated before one is finally selected.

A basic technique for investigating the aptness of a regression model is based on analyzing the residuals defined in (18.17). If the model is apt, the observed residuals $e_i = Y_i - \hat{Y}_i$ should reflect the properties ascribed to the error terms ϵ_i. For instance, if the model assumes that the ϵ_i's are normal random variables with constant variance, the observed e_i's should show a pattern consistent with these properties.

We now consider how analysis of residuals is used to study the aptness of the simple linear regression model (19.1). Specifically, we shall examine the following departures from model (19.1):

1. The regression function is not linear.
2. The distributions of Y do not have constant variances at all levels of X; or equivalently, the ϵ_i's do not have constant variances.
3. The distributions of Y are not normal; or equivalently, the ϵ_i's are not normal.
4. The error terms ϵ_i are not independent.

Nonlinearity of Regression Function

Whether the regression function is linear or curvilinear can be studied from either a scatter plot or a residual plot.

☐   Examples

1. Figure 19.5a contains a scatter plot for a study of discount card usage by students at 10 community colleges. Here, X is the number of merchants in the discount card plan at a college and Y is the average discount card expenditure per student at the college during the period covered by the study. A linear regression function has been fitted to the data by the method of least squares. The observed values Y_i, the fitted values $\hat{Y}_i$, and the residuals e_i are as follows:

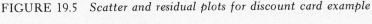

i:	1	2	3	4	5	6	7	8	9	10
Y_i:	75.0	112.0	38.0	120.0	105.0	52.0	116.0	118.0	105.0	110.0
$\hat{Y}_i$:	70.5	94.3	54.6	134.0	86.4	62.6	110.2	126.1	110.2	102.2
e_i:	4.5	17.7	−16.6	−14.0	18.6	−10.6	5.8	−8.1	−5.2	7.8

If the true regression function is linear, the observations should scatter at random around the fitted straight line. This does not appear to be the case here, as may be seen from Figure 19.5a; the points systematically fall below, then above, then again below the fitted line as X increases. It would seem that a linear regression function is not in accord with the observations, and that a curvilinear function is required instead.

An alternative, and frequently more effective, graphic display is a *residual plot*, obtained by plotting the residuals e_i against the fitted values $\hat{Y}_i$. This plot is shown in Figure 19.5b. If the specified regression function is apt, the residuals will tend to scatter at random around the zero line when plotted against $\hat{Y}_i$. Note how this fails to occur in Figure 19.5b. Instead, we see a pattern of deviations which mirrors that in Figure 19.5a. We reach the same conclusion from either graph.

2. Figure 19.6 shows the residual plot for our UDS example. Here, no pattern of system-

FIGURE 19.5   *Scatter and residual plots for discount card example*

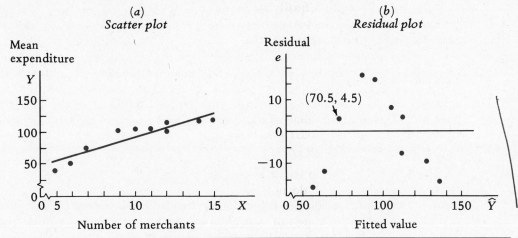

FIGURE 19.6 *Residual plot for UDS example*

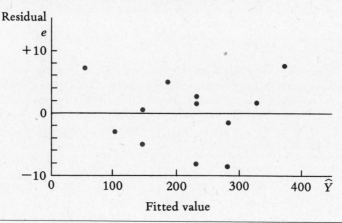

atic departure of the points around the zero line is evident. Thus, the residual plot in Figure 19.6 suggests that the linear regression function specified in model (19.1) is apt in the UDS case. ☐

Nonconstancy of Error Variances

A residual plot also provides information as to whether or not the error terms ϵ_i have constant variance.

☐ Examples

1. Figure 19.7 shows a residual plot for a study of the relation between typing speed (Y) and hours of training (X) for beginning typists using a new keyboard design. Note that the pattern of spread in the residuals becomes greater as $\hat{Y}$ increases. This suggests that the distributions of Y have increasing variances as $E\{Y\}$ becomes larger.
2. The scatter plot in Figure 19.6 for our UDS example illustrates a pattern of residuals when the error variances are constant. Here, the e_i's tend to fall at random in a horizontal band when plotted against $\hat{Y}_i$. ☐

Lack of Normality

A variety of checks for normality of the error terms—and hence of the distributions of Y—are available. One of these employs *standardized residuals*, denoted by e_i':

(19.29)
$$e_i' = \frac{e_i}{\sqrt{MSE}}$$

Thus, a standardized residual e_i' is the residual divided by the estimated standard deviation. If the distributions of Y are normal with constant variance and the linear regression function is appropriate, the standardized residuals will tend to

FIGURE 19.7 *Residual plot illustrating nonconstant error variance*

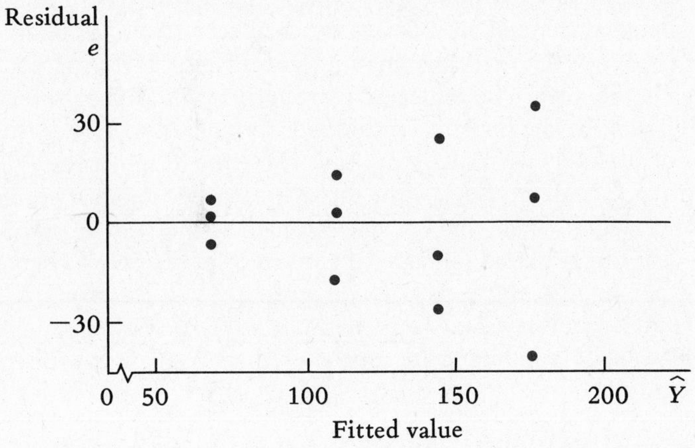

follow the standard normal distribution when n is large. Hence, about half of the standardized residuals should then be positive and half negative, about 68 percent should fall between -1 and $+1$, etc. If n is not large, the standardized residuals should be compared to the $t(n - 2)$ distribution.

Example

For our UDS case, we have $\sqrt{MSE} = 5.804$. The residual for the first observation is $e_1 = 5.16$ (see Table 18.3). Hence, the standardized residual is $e_1' = 5.16/5.804 = .89$. The standardized residuals for the 12 observations are:

i:	1	2	3	4	5	6
e_i':	.89	-1.49	$-.52$	-1.42	.33	$-.29$
i:	7	8	9	10	11	12
e_i':	.10	1.29	.30	$-.94$	1.27	.47

In the t distribution with $12 - 2 = 10$ degrees of freedom, $t(.90; 10) = 1.372$ and $t(.975; 10) = 2.228$. Hence, the central 80 percent of the distribution lies between -1.372 and $+1.372$ and the central 95 percent lies between -2.228 and $+2.228$. Ten of the 12 standardized residuals fall in the first interval, and all fall in the second interval. Also, 5 of the standardized residuals are negative and 7 are positive. Thus, this analysis provides no evidence of any serious departure from normality.

Comments

1. When n is large, the formal tests for normality discussed in Chapter 16 can be employed and a histogram of the residuals can provide a useful visual check.
2. Many regression computer packages print standardized residuals as a routine part of the output. Some packages, however, define standardized residuals somewhat differently from (19.29), using the estimated standard deviation of the residual in the denominator rather than $\sqrt{MSE}$.

Lack of Independence

In regression applications, the observations are often obtained in a time sequence, as when the value of an index of business activity (X) is related to the dollar value of new orders received by a manufacturing corporation (Y) for each of the past 48 months. In such data, there is a strong possibility that the error terms ϵ_i are not independent. For instance, if the error term is positive (negative) in a given month, it is often likely with such data that the error term for the following month is also positive (negative). Such error terms are said to be *autocorrelated*, in contrast to model (19.1), where the ϵ_i's are assumed to be statistically independent.

A plot of the residuals against the time order of the observations helps to assess whether or not autocorrelation is present. We discuss this plot in Chapter 25. When the sample size is large, an approximate test for independence of the error terms can be conducted by means of the runs test described in Chapter 15.

Remedial Actions

When residual analysis discloses that model (19.1) is not apt for the data under study, remedial action may be required. For instance, if residual analysis discloses that the regression function is not linear, a curvilinear regression model might be employed. We discuss curvilinear regression models in the next chapter. Alternatively, a mathematical transformation of one or both variables may be made in order to linearize the regression relation.

☐ Example

Figure 19.8a presents a scatter plot of observations for 11 athletes on an index of physical fitness (X) and the performance time by the athlete in a track event (Y). The regression relation is clearly not linear. If, however, the independent variable is transformed to $1/X$, the reciprocal of the fitness index, the relation becomes linear, as shown in Figure 19.8b. ☐

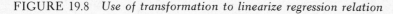

FIGURE 19.8 *Use of transformation to linearize regression relation*

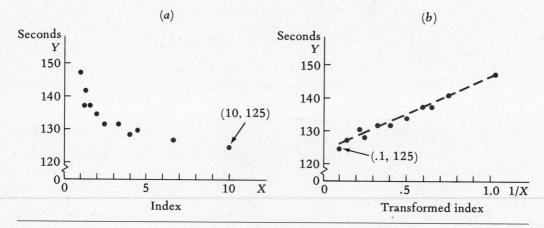

If the distribution of the error terms is far from normal, a mathematical transformation of the Y variable may bring the distribution of the error terms closer to normality. Thus, if the distribution of the ϵ_i's is sharply right-skewed, log Y might be used as the dependent variable instead of Y. Moderate departures from normality are not serious, and even substantial departures usually have little effect on the specified level of confidence or the α risk when the sample size is large.

19.8   PRACTICAL CONSIDERATIONS IN USING REGRESSION ANALYSIS

In practice, there are a number of important considerations in using a regression model.

Scope of Model

Range of Observations.   Caution must be exercised in applying a regression model outside the range of the observations that have been used to estimate the model's parameters and to verify the aptness of the model's assumptions. For instance, in our UDS example we fitted regression model (19.1) to observations where the number of tape drives (X) was between 1 and 8. The fit appeared to be satisfactory for these observations, but in the absence of other information, we do not know whether the model is apt outside the interval $1 \leq X \leq 8$. Thus, it may be dangerous to use the fitted model in our UDS case to estimate the mean service time when, say, $X_h = 25$ because the linear model might not be a good fit when extended to 25 tape drives. A curvilinear model might be needed instead.

Similarly, an estimate of β_0 in model (19.1) should only be used as an indication of the "fixed" component of the relationship when there are observations near $X = 0$ or when theoretical considerations indicate that the regression function is linear for the range including $X = 0$.

Situations frequently arise in practice where inferences are required outside the range of past observations. For example, it may be desired to use the economic indicator per capita disposable personal income (X) to predict the level of business activity (Y) for a firm. In an expanding economy, the level of X required for this prediction will frequently be outside the range of the past data. In such cases, inferences obtained with the regression model must be applied with considerable judgment.

Continuation of Causal Conditions.   Whenever a regression model is utilized to make an inference for the future, the validity of the inference requires that future causal conditions be the same as during the period covered by the observed data. Thus, a prediction from the regression model in our UDS case of the service time on a customer call six months later assumes that the causal conditions affecting the servicing of tape drives has not changed.

When causal conditions have changed, a fitted model may no longer be appropriate. For example, the response of butter consumption to changes in the price of butter was different after the availability of margarine than before.

Causality

As noted earlier, the presence of a regression relation between two variables X and Y does not imply a cause-and-effect relation between them. In some instances, a change in X does force a change in Y as, for example, when an increase in drug dosage X causes a decrease in blood pressure Y. In other instances, both X and Y change in response to changes in one or more other variables without a direct causal linkage between them.

☐ Example

Reading ability (Y) was regressed on shoe size (X) for a sample of elementary school children and a positive regression relation was observed. This relation does not imply, of course, that larger shoes cause children to read better. The age of the child is an intervening variable in this case—older children tend to wear bigger shoes and to read better.   ☐

The omission of an intervening variable can sometimes also hide a relation between two variables.

☐ Example

Figure 19.9a shows a scatter plot of years of education (X) and salary (Y) for 16 corporate middle managers aged 25 to 30 in a multinational firm. The regression line is horizontal, indicating no regression relation between the two variables. In Figures 19.9b and 19.9c, the observations are divided into two groups—sales managers and others. The separate scatter plots show a persistent increase in average salary with more years of education for each group. The fact that managers not in sales activities tend to be paid less than sales manag-

---

FIGURE 19.9 *Effect of an intervening variable on a regression relation.* The relation between salary and education is hidden when the scatter plots for sales managers and others are combined.

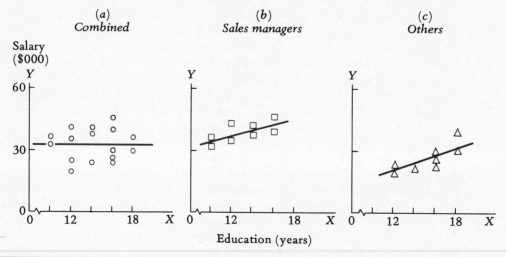

ers for any given number of years of education disguises the relation between years of education and salary when all managers are combined. □

Finally, there are instances where causation acts in the opposite direction to the regression relation in the sense that changes in Y force changes in X.

□ Example

A thermometer is being calibrated by obtaining readings at various known temperatures and regressing the actual temperature (Y) on the thermometer reading (X). Clearly, the thermometer readings X are affected by the actual temperatures Y, and not vice versa. □

Independent Variable Must Be Predicted

Sometimes, we must predict the value to be taken by the independent variable in obtaining a required prediction for Y. This happens quite often in business fore-casting. For example, suppose X_h is the value projected for an economic indicator variable in the next quarter and a prediction interval is desired for company sales (Y_h) in that period. The sales prediction then is a conditional one, dependent on the correctness of the projected X_h.

Independent Variable Is Random

Sometimes in regression applications it is more reasonable to consider both X and Y as random, rather than to take X as fixed in accord with model (19.1). Consider an analyst in an electric company who is studying the regression relation between daily residential electricity consumption in a service area (Y) and mean daily out-door temperature in the area (X). Since the outdoor temperature cannot be con-trolled, X might be considered as random in this application.

When X is random, the distribution of Y at a given level of X is considered to be a conditional distribution with a conditional mean and a conditional variance. It can be shown that all of the results presented for model (19.1) where the X_i's are fixed still apply when the X_i's are random if the following conditions hold: (1) the conditional distributions of Y are normal, with conditional mean $\beta_0 + \beta_1 X$ and conditional variance σ^2; and (2) the X_i's are independent random variables whose probability distribution does not depend on the parameters β_0, β_1, and σ^2.

When the independent variable is random, the interpretation of confidence coefficients and risks of errors differs from before. The interpretation now refers to repeated sampling where *both* the X and Y values change from one sample to the next. Thus, in our electricity consumption example, the confidence coefficient for the interval estimate for, say, β_1 would refer to the proportion of correct interval estimates obtained if repeated random samples of n pairs of (X_i, Y_i) values were taken and the confidence interval for β_1 calculated for each sample. There is no implication that the X levels remain the same from sample to sample.

Comment

When X and Y are both random variables, a population coefficient of simple correlation ρ_{XY} is defined as in (5.27). The coefficient of simple correlation r defined in (18.38) for the n sample values (X_i, Y_i) is then not only a descriptive measure of the degree of relationship between X and Y in the sample but also serves as an estimator of ρ_{XY}.

19.9 COMPUTER OUTPUT

Figure 18.13 (Section 18.7) presents the computer output for our UDS example. We now complete our discussion of this output. Block 3 gives, for b_0 and b_1, calculated results in the following format:

Variable	Coefficient	St. Dev.	T Stat.
0	b_0	$s\{b_0\}$	$t^* = b_0/s\{b_0\}$
1	b_1	$s\{b_1\}$	$t^* = b_1/s\{b_1\}$

The test statistics t^* are for testing whether or not the respective regression coefficient β_0 or β_1 equals zero. Formula (19.17) gives this test statistic for the test concerning β_1.

In block 4, the final item is F^* defined in (19.26) for the F test of $\beta_1 = 0$.

Block 6 gives results for inferences on the response variable Y for values X_h specified by the user. Here, $X_h = 4$. The statistics presented are $\hat{Y}_h$, $s\{\hat{Y}_h\}$, and $s\{d_h\}$.

PROBLEMS

*19.1 Refer to Problem 18.10. Calculational results are: $b_0 = 72.776$, $b_1 = -1.7726$, $MSE = 4.2138$, $\bar{X} = 21.875$, $\Sigma (X_i - \bar{X})^2 = 42.875$.
  a. Obtain $\hat{Y}_h$ and $s\{\hat{Y}_h\}$ when $X_h = 23$.
  b. Obtain a 95 percent confidence interval for $E\{Y_h\}$ when $X_h = 23$. Interpret your confidence interval.

19.2 Refer to Problem 18.16. Calculational results are: $b_0 = 4.325$, $b_1 = 25.85$, $MSE = 29.85$, $\bar{X} = 5.5$, $\Sigma (X_i - \bar{X})^2 = 40$.
  a. Obtain $\hat{Y}_h$ and $s\{\hat{Y}_h\}$ when $X_h = 8$.
  b. Obtain a 90 percent confidence interval for $E\{Y_h\}$ when $X_h = 8$. Interpret your confidence interval. How is the 90 percent confidence coefficient interpreted here?

19.3 Refer to Problems 18.16 and 19.2. The estimated variance of $\hat{Y}_h$ for several values of X_h is:

X_h:	5	6	7
$s^2\{\hat{Y}_h\}$:	3.918	3.918	5.410

A student asks why the third entry for $s^2\{\hat{Y}_h\}$ differs from the first two. Explain why.

19.4 An analyst, when regressing contributions to the pension plan (Y) on years of job seniority (X) for employees in the company, obtained computer output showing the 95 percent confidence interval for $E\{Y_i\}$ for each X_i value in the data set. The analyst expected to find that about 95 percent of the Y_i values would be contained in the corresponding confidence intervals for $E\{Y_i\}$ but discovered that this occurred in a substantially smaller percent of the cases. Explain the analyst's error in thinking.

19.5 For each of the following questions, explain whether a confidence interval for $E\{Y\}$ or a prediction interval for Y is appropriate:

    a. What is the average statistics grade of students who receive 530 on the graduate admissions test?

    b. What will the Cramer family spend on restaurant meals next year if the family's income is $34,000?

    c. What do companies with 500 full-time employees spend each year, on the average, on group life insurance premiums?

*19.6 Refer to Problems 18.10 and 19.1. Another automobile, not included in the regression study, has a weight of 23 hundred pounds.

    a. Construct a 95 percent prediction interval for the expressway mileage of this automobile. Interpret your prediction interval.

    b. Why is the prediction interval in a wider than the confidence interval in Problem 19.1b?

19.7 Refer to Problems 18.16 and 19.2.

    a. Construct a 90 percent prediction interval for Y_h when $X_h = 8$.

    b. Why is the prediction interval in a wider than the confidence interval in Problem 19.2b?

    c. Which interval—the prediction interval or the confidence interval—would be more useful for evaluating the actual number of manhours required for bid preparation next week when eight bids are to be prepared? Explain.

19.8 Refer to Problem 18.15.

    a. Obtain a 95 percent confidence interval for the mean reconditioning cost when the number of operating hours is 2.5 thousand. Interpret your interval estimate.

    b. The reconditioning cost for an incinerator with 2.5 thousand operating hours is to be predicted. Obtain a 99 percent prediction interval. Interpret your prediction interval. Is it precise enough to be useful if a prediction with ±10 percent precision is required? Discuss.

    c. If the reconditioning cost for an incinerator with 5.2 thousand operating hours were to be predicted, would the width of the prediction interval be much wider than that in b? Would there be other problems in this case?

*19.9 Refer to Problems 18.10 and 19.1. Obtain a 95 percent confidence interval for β_1. Interpret your confidence interval.

19.10 Refer to Problems 18.16 and 19.2. Obtain a 99 percent confidence interval for β_1. Does the interval indicate that the means of the probability distributions of Y at different levels of X are not the same? Explain.

19.11 A company has 42 sales districts. When district sales levels (Y, in thousands of dollars) were regressed on district promotional expenditures (X, in thousands of dollars) for the past quarter, the following results were obtained: $b_0 = 389$, $b_1 = -.271$, $MSE = 413$, $\bar{X} = 34$, $\Sigma (X_i - \bar{X})^2 = 2,780$. An alarmed executive, citing the negative value for b_1, stated: "The more we advertise, the less we sell." Obtain a 90 percent confidence interval for β_1 and comment.

*19.12 Refer to Problems 18.10 and 19.1. Test whether or not $\beta_1 = 0$ here, using either (19.16) or (19.18). Control the α risk at .05. State the alternatives, decision rule, and conclusion.

19.13 Refer to Problem 19.11. Test whether or not $\beta_1 = 0$ here, using either (19.16) or (19.18). Control the α risk at .10. State the alternatives, decision rule, and conclusion. What is the implication of your conclusion?

19.14 Refer to Problems 18.16 and 19.2. Test whether or not $\beta_1 = 0$ here, using either (19.16) or (19.18). Control the α risk at .01. State the alternatives, decision rule, and conclusion. What does your conclusion imply about a relationship between X and Y? Discuss.

19.15 Refer to Problem 18.15.
   a. Obtain a 99 percent confidence interval for the change in the mean response when the number of operating hours increases by one thousand. Interpret your confidence interval.
   b. Does a test of whether or not $\beta_1 = 0$, with $\alpha = .01$, provide the same information about β_1 as the confidence interval in **a**? Explain. Conduct such a test. What is your conclusion?

*19.16 Refer to Problems 18.16 and 19.2. Construct a 95 percent confidence interval for β_0.

19.17 Refer to Problem 19.11. Construct a 90 percent confidence interval for β_0. Does β_0 have an operational interpretation here since no sales district had promotional expenditures near zero? Discuss.

*19.18 Refer to Problems 18.16 and 19.2. Estimate the covariance of b_0 and b_1. What does this estimate show about the sampling errors in b_0 and b_1 here?

19.19 Refer to Problem 19.11. Estimate the covariance of b_0 and b_1. If b_1 happens to underestimate β_1 here, what is the likely direction of the sampling error in b_0?

*19.20 Refer to Problem 18.10. Some calculational results are: $MSE = 4.2138$, $MSR = 134.72$. Conduct an F test of whether or not $\beta_1 = 0$, controlling the α risk at .05. State the alternatives, decision rule, and conclusion.

19.21 Refer to Problem 18.16. Some calculational results are: $MSE = 29.85$, $MSR = 26,729$.
   a. Conduct an F test of whether or not $\beta_1 = 0$, controlling the α risk at .01. State the alternatives, decision rule, and conclusion.
   b. What does your conclusion imply about a statistical relation between X and Y? Would the implication still be the same if model (19.1) were not appropriate? Discuss.

19.22 Refer to Problem 18.15.
   a. Use an F test to decide if the expected reconditioning cost for incinerators varies with operating hours. Control the α risk at .01. State the alternatives, decision rule, and conclusion.
   b. Could your conclusion in **a** have been reached from a 99 percent confidence interval for β_1? Explain.
   c. Do the same test as in **a** by means of the test statistic (19.17). Show the equivalence of the test statistics t^* and F^* here.

19.23 An auditor tested the inventory records at a warehouse of a large manufacturer by conducting a physical count of inventory stocks for 25 line items. As part of the analysis, the auditor regressed the actual inventory value of line items (Y, in thousands of dollars) against the value as shown in the inventory records (X, in thousands of dollars). The fitted values and residuals follow.

$\widehat{Y}_i$	e_i	$\widehat{Y}_i$	e_i	$\widehat{Y}_i$	e_i	$\widehat{Y}_i$	e_i	$\widehat{Y}_i$	e_i
3.59	−.70	4.08	.18	5.32	.20	6.09	−.81	13.60	.08
2.24	−.41	4.72	.20	3.23	.87	7.39	.71	8.63	.63
8.42	.69	7.00	−.07	1.60	−.72	7.68	−.48	8.25	−.71
.91	−.07	10.53	−.03	10.18	−.47	4.00	−.98	4.23	−.06
11.50	.06	4.87	.72	6.62	.80	8.93	.25	4.19	.12

a. Construct a residual plot.
b. What does the plot indicate about the aptness of model (19.1) with regard to linearity of the regression function and constancy of the error variance? State your findings.
c. Check for normality of the error terms. What is your conclusion?

*19.24 Refer to Problems 18.16 and 18.18.
a. Construct a residual plot.
b. What does the plot indicate about the aptness of model (19.1) with regard to linearity of the regression function and constancy of the error variance? State your findings.

19.25 Refer to Problems 18.10 and 18.19.
a. Construct a residual plot.
b. Examine the residuals and your residual plot for the four types of departures from model (19.1) listed in Section 19.7. Assume that the automobiles were tested in the order given. State your findings.

19.26 A business psychologist created 16 tasks of varying difficulty, ranging from a difficulty rating of 0 (trivial) to one of 100 (extremely complex). She then organized 16 decision-making groups of roughly equal composition and ability, and randomly assigned one group to each task. The performance of the group in completing its task was rated on a scale from 0 (very low) to 100 (very high). As part of the data analysis, the psychologist regressed group performance rating (Y) on task difficulty (X) using model (19.1), and obtained the following fitted values and residuals:

$\widehat{Y}_i$	e_i	$\widehat{Y}_i$	e_i	$\widehat{Y}_i$	e_i	$\widehat{Y}_i$	e_i
67.4	3.6	74.5	.5	77.1	−9.1	71.4	9.6
71.7	13.3	75.0	−10.0	70.0	13.0	67.6	7.4
64.3	−12.3	69.1	10.9	72.4	.6	65.2	−17.2
65.5	−.5	66.7	−3.7	72.2	6.8	75.7	−12.7

a. Construct a residual plot and examine it for evidence of departure from linearity. Does a linear regression function appear to be appropriate here? Discuss.
b. Can your residual plot be readily examined for constancy of the error variance here? Explain.

19.27 An analyst finds that a statistical relation holds between thickness of silver film deposited in a process and frequency shift of a crystal used as a sensor. She proposes to investigate whether model (19.1) can be used to estimate film thickness (taken as the dependent variable) from knowledge of the magnitude of the frequency shift (taken as the independent variable). An observer states that model (19.1) cannot be used in this manner because frequency shift actually depends on film thickness, not vice versa. Comment.

EXERCISES

19.28 Refer to formula (19.9).
    a. Do both components of $\sigma^2\{d_h\}$ become smaller as the sample size increases? Explain.
    b. What does your answer in **a** imply about our ability to make $\sigma^2\{d_h\}$ small by increasing n?

19.29 A student asks why the F test for deciding between $\beta_1 = 0$ and $\beta_1 \neq 0$ is one-sided, even though the latter alternative implies that either $\beta_1 < 0$ or $\beta_1 > 0$. Explain with specific reference to formulas (19.24) and (19.25).

19.30 In an application of regression model (19.1), $\sigma^2 = 100$ and $\Sigma (X_i - \bar{X})^2 = 50$. Obtain $E\{MSE\}$ and $E\{MSR\}$ when $\beta_1 = 0$, $\beta_1 = 10$, $\beta_1 = -10$, and $\beta_1 = 50$. What do your results imply about the type of decision rule appropriate for the F test of a regression relation?

19.31 Derive formulas (19.21a) and (19.21b) from theorem (19.3) by letting $X_h = 0$.

19.32 Prove (19.25) using the fact that $MSR = b_1^2 \Sigma (X_i - \bar{X})^2$. [*Hint:* Obtain $E\{b_1^2\}$ by using (5.8a).]

19.33 Show that the F^* test statistic (19.26) is equal to the square of the t^* statistic (19.17), using the fact that $MSR = b_1^2 \Sigma (X_i - \bar{X})^2$.

STUDIES

19.34 Refer to Problem 18.15.
    a. Extensive operating data for other large incinerators show that the expected reconditioning cost increases by $1,200 per 1,000 operating hours. Test whether or not $\beta_1 = 1.2$ here, controlling the α risk at .01. State the alternatives, decision rule, and conclusion.
    b. Could the same conclusion be obtained from a 99 percent confidence interval for β_1? Explain.
    c. Why would test statistic (19.17) not be appropriate here?

19.35 Patients in a chronic-care facility are rated on a 100-point scale as to their overall health upon admission (X) and then again two weeks later (Y). When a simple regression analysis was performed for the last 30 patients admitted, the computer output showed: $b_1 = .383$, $s\{b_1\} = .172$. Assume that model (19.1) is appropriate here.
    a. Test whether or not $\beta_1 \leq 0$ using test statistic (19.17), with the α risk controlled at .05 when $\beta_1 = 0$. State the alternatives, decision rule, and conclusion.
    b. The nursing director interpreted the positive value of b_1 to imply that patients tend to improve in overall health after two weeks of care (i.e., Y_i tends to be larger than X_i). Demonstrate that this conclusion does not necessarily follow by constructing an illustrative numerical example for $n = 4$ where X and Y are positively related but in each instance $X_i > Y_i$.

19.36 Refer to Appendix D.
    a. For the firms in the paper industry, regress 1981 net income on 1981 net assets, using model (19.1).
    b. Construct a 95 percent confidence interval for β_1. Interpret your interval.

c. What is the maximum half-width of a 95 percent confidence interval for $E\{Y_h\}$ when X_h is within $\pm\$100$ million of the mean net assets for all 24 firms? Would you consider such a confidence interval to be reasonably precise? Discuss.

d. What is the maximum half-width of a 95 percent prediction interval for Y_h when X_h is within $\pm\$100$ million of the mean net assets for all 24 firms? Would you consider such a prediction interval to be reasonably precise? Discuss.

e. Obtain the standardized residuals and construct a residual plot. Examine the plot for departures from model (19.1) with regard to linearity of the regression function and constancy of the error variance. Check the residuals for normality of the error terms. State your findings.

19.37 Refer to Problem 19.36. Do the problem using the 1980 data for the firms in the paper industry.

19.38 In a finance study, bond yield (Y) was regressed on bond maturity (X) for 400 issues, using model (19.1). About half of the issues were corporate bond issues and the other half were municipal bond issues. The frequency distribution of the standardized residuals obtained from the regression fit follows.

e'_i	Number of Issues	e'_i	Number of Issues
Under -1.5	14	0–under .5	38
-1.5–under -1.0	70	.5–under 1.0	80
-1.0–under $-$ .5	90	1.0–under 1.5	59
$-$ .5–under 0	29	1.5 or more	20

a. What probability distribution should the standardized residuals approximately follow under model (19.1)? Does this model seem to hold in this application? Explain.

b. If the standardized residuals are approximately normally distributed, what is the approximate probability that a standardized residual has a value below -1.5 or above 1.5? Are the tails of the distribution of the 400 standardized residuals beyond ±1.5 consistent with those of a normal distribution?

c. Would the frequency pattern of the residuals differ from that of the standardized residuals? Explain.

20
Multiple
Regression

In many situations, two or more independent variables must be included in a regression model to provide an adequate description of the process under study or to yield sufficiently precise inferences.

☐ Examples

1. A regression model to control the diameter of plastic pellets produced by an extrusion process uses as independent variables the initial temperature of the process, the die temperature, and the extrusion rate.
2. A regression model for predicting the demands for a firm's product in its 25 sales territories uses socioeconomic variables (mean household income, average years of schooling of head of household), demographic variables (average family size, percent of population over 65 years of age), and environmental variables (mean daily temperature, index of atmospheric pollution). ☐

Regression models containing two or more independent variables are called *multiple regression* models. In this chapter, we extend the procedures for simple linear regression to multiple regression and also consider some special topics of importance when multiple regression models are used.

20.1 MULTIPLE REGRESSION MODELS

Model in Two Independent Variables

The simple linear regression model (19.1) can easily be extended to include two independent variables, X_1 and X_2:

(20.1)
$$Y_i = \beta_0 + \beta_1 X_{i1} + \beta_2 X_{i2} + \epsilon_i \qquad i = 1, 2, \dots, n$$

where: Y_i is the response in the ith observation

X_{i1} and X_{i2} are the values of the independent variables in the ith observation, assumed to be known constants

β_0, β_1, and β_2 are parameters

ϵ_i's are independent $N(0, \sigma^2)$

As for the simple linear regression model (19.1), we are assuming here that the error terms ϵ_i are statistically independent normal random variables, with mean zero and constant variance σ^2.

Regression Function. The regression or response function for model (20.1) is:

(20.2)
$$E\{Y\} = \beta_0 + \beta_1 X_1 + \beta_2 X_2$$

This function is frequently called the regression or response *surface*. The surface for response function (20.2) is a plane, as illustrated in Figure 20.1. The parameters of a multiple regression model are interpreted analogously to those in the simple linear case. Thus, in response function (20.2):

1. β_0 is the Y intercept of the plane; it is the mean of the distribution of Y when $X_1 = 0$ and $X_2 = 0$.
2. β_1 indicates the change in $E\{Y\}$ when X_1 increases by one unit while X_2 remains constant.
3. β_2 indicates the change in $E\{Y\}$ when X_2 increases by one unit while X_1 remains constant.

☐ Example

Figure 20.1 shows a response plane for model (20.1) where β_0 and β_1 are positive and β_2 is negative. Just as the regression line for the simple linear regression model (19.1) gives the mean of the distribution of Y for any level of X, the response plane in Figure 20.1 gives the mean of the distribution of Y for any combination of X_1, X_2 values. Shown in Figure 20.1 is a response Y_i when $X_1 = X_{i1}$ and $X_2 = X_{i2}$. The mean of this probability distribution, $E\{Y_i\}$, is a point on the regression plane as indicated in the figure. The error term ϵ_i for

FIGURE 20.1 *Example of a response plane.* β_0 and β_1 are positive and β_2 is negative here. Point $E\{Y_i\}$ on the plane is the mean of the probability distribution corresponding to observation Y_i at coordinates (X_{i1}, X_{i2}). The error term is $\epsilon_i = Y_i - E\{Y_i\}$.

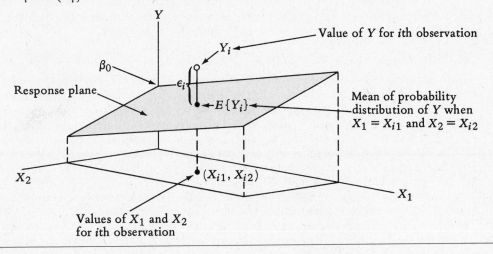

this observation is represented by the vertical distance between Y_i and $E\{Y_i\}$, since $\epsilon_i = Y_i - E\{Y_i\}$ as for simple linear regression. □

Model in $p - 1$ Independent Variables

The regression model (20.1) in two independent variables can be readily extended to include $p - 1$ independent variables $X_1, X_2, \ldots, X_{p-1}$:

(20.3) $$Y_i = \beta_0 + \beta_1 X_{i1} + \beta_2 X_{i2} + \cdots + \beta_{p-1} X_{i,p-1} + \epsilon_i \qquad i = 1, 2, \ldots, n$$

where: Y_i is the response in the ith observation

$X_{i1}, X_{i2}, \ldots, X_{i,p-1}$ are the values of the independent variables in the ith observation, assumed to be known constants

$\beta_0, \beta_1, \ldots, \beta_{p-1}$ are parameters

ϵ_i's are independent $N(0, \sigma^2)$

Note that when $p - 1 = 1$, model (20.3) becomes the simple linear regression model (19.1); and when $p - 1 = 2$, it becomes model (20.1), the multiple regression model in two independent variables.

The parameters in model (20.3) are interpreted in the standard manner:

1. β_0, the Y intercept, indicates the mean of the distribution of Y when $X_1 = X_2 = \cdots = X_{p-1} = 0$.
2. $\beta_k (k = 1, 2, \ldots, p - 1)$ indicates the change in $E\{Y\}$ when X_k increases by one unit while the other independent variables remain constant.
3. σ^2 is the common variance of the distributions of Y.

20.2 BASIC CALCULATIONAL RESULTS FOR MULTIPLE REGRESSION MODELS

Most calculational procedures for multiple regression are direct extensions of those already discussed for simple linear regression. We demonstrate these procedures with an illustration involving two independent variables.

□   Illustration

An experiment was designed by a market researcher to study the effects of two types of promotional expenditures on sales of a line of food products sold in supermarkets. Sixteen localities were selected for the test. They were similar in market potential and were representative of the target market for this line of products. Different combinations of media advertising expenditures (X_1) and point-of-sale expenditures (X_2) were specified for the study, and the localities were assigned at random to one of these X_1, X_2 combinations. Table 20.1 shows the two expenditures levels, together with the dollar sales (Y), for the test period for each of the 16 localities.

Figure 20.2 contains computer output for a regression analysis of the results in Table 20.1, using regression model (20.1). We have numbered the blocks of information in Figure 20.2 for ready identification. The format is similar to that of Figure 18.13 for simple linear regression. For instance, block 1 gives the means, variances, and standard deviations for X_1, X_2, and Y, respectively. □

TABLE 20.1 *Data for promotional expenditures study*

Locality i	Variable 1: Media Expenditures ($ thousands) X_{i1}	Variable 2: Point-of-Sale Expenditures ($ thousands) X_{i2}	Variable 3: Sales Volume ($ ten thousands) Y_i
1	2	2	8.74
2	2	3	10.53
3	2	4	10.99
4	2	5	11.97
5	3	2	12.74
6	3	3	12.83
7	3	4	14.69
8	3	5	15.30
9	4	2	16.11
10	4	3	16.31
11	4	4	16.46
12	4	5	17.69
13	5	2	19.65
14	5	3	18.86
15	5	4	19.93
16	5	5	20.51

Correlation Matrix

The interpretation of multiple regression results often requires information about the coefficients of correlation between pairs of the variables in the study. Let us denote the coefficient of simple correlation between Y and X_1, as defined in (18.38), by r_{Y1}. Similarly, r_{Y2} denotes the coefficient of simple correlation between Y and X_2, and r_{12} the coefficient of simple correlation between X_1 and X_2. This latter coefficient relates to the degree of linear association between the independent variables X_1 and X_2.

Example

The correlation coefficients for the promotional expenditures study are given in block 2 of Figure 20.2 in a matrix called *correlation matrix*. The format and the values of the coefficients for the study are as follows:

$$\begin{array}{c} \\ X_1 \\ X_2 \\ Y \end{array} \begin{array}{ccc} X_1 & X_2 & Y \end{array} \\ \begin{bmatrix} 1 & r_{12} & r_{Y1} \\ & 1 & r_{Y2} \\ & & 1 \end{bmatrix} = \begin{bmatrix} 1 & 0 & .965 \\ & 1 & .225 \\ & & 1 \end{bmatrix}$$

The analyst noted with interest that the coefficient of simple correlation between sales (Y) and media expenditures (X_1) is substantially greater than that between sales (Y) and point-of-sale expenditures (X_2), namely that $r_{Y1} = .965$ while $r_{Y2} = .225$. Also, he knew in advance that the coefficient of simple correlation between X_1 and X_2 would be $r_{12} = 0$, since he had designed this feature into the experiment by setting up the X_1, X_2 combinations in the manner shown in Table 20.1. We will discuss the reason for this design later.

FIGURE 20.2 *Example of computer output for mutiple regression: Promotional expenditures study*

```
MULTIPLE REGRESSION

1. PRINT MEANS, VARIANCES AND STANDARD DEVIATIONS?    YES
   VARIABLE          MEAN              VARIANCE          ST. DEV.
      1            3.50000             1.33333           1.15470
      2            3.50000             1.33333           1.15470
      3           15.20688            13.15002           3.62630

2. PRINT SIMPLE CORRELATION MATRIX?      YES
      1.000     0.000     0.965
                1.000     0.225
                          1.000

   ENTER LIST OF VARIABLES, RESPONSE VAR. FIRST.
   □:
         3  1  2

3. VARIABLE    COEFFICIENT     ST. DEV.      T STAT.
      0          2.13438        0.61036         3.50
      1          3.02925        0.12028        25.18
      2          0.70575        0.12028         5.87
   INDEP. VARS.: 1  2; RESPONSE VAR.: 3

4. PRINT ANOVA TABLE?     YES
      SOURCE                   SS          DF            MS
   REGRESSION              193.4888        2         96.74439
   RESIDUAL                  3.7616       13          0.28935
   TOTAL                   197.2503       15
   F STAT.            = 334.35

5. RESIDUAL STD DEV.      = 0.53791;   NUMBER OF OBSERVATIONS = 16
   MULT R SQ (COEFF DETERM) = 0.98093;   MULT R (COEFF CORREL) = 0.99042

6. PRINT VARIANCE-COVARIANCE MATRIX?     YES
      0.37254          ‾0.05064        ‾0.05064
                        0.01447         0.00000
                                        0.01447

7. ESTIMATE RESPONSE VARIABLE?     YES
   ENTER THE VALUES OF THE INDEPENDENT VARIABLES IN
      THIS ORDER: VARIABLE 1, VARIABLE 2
   □:
         5  2

   POINT ESTIMATE                                  = 18.6921
   STANDARD DEVIATION OF MEAN RESPONSE             = 0.2884
   STANDARD DEVIATION OF INDIVIDUAL RESPONSE       = 0.6104

   □:
      STOP

8. PRINT RESIDUALS?     YES
   NO.      OBSERVED          ESTIMATED          RESIDUAL
    1        8.74000           9.60438          ‾0.86438
    2       10.53000          10.31013           0.21987
    3       10.99000          11.01588          ‾0.02588
    4       11.97000          11.72163           0.24837
    5       12.74000          12.63363           0.10637
```

Estimated Regression Function

For regression model (20.1), the estimated regression function is:

(20.4)
$$\widehat{Y} = b_0 + b_1 X_1 + b_2 X_2$$

where b_0, b_1, and b_2 are the least squares estimators of the corresponding parameters.

The method of least squares leads to a system of normal equations for obtaining b_0, b_1, and b_2, as follows:

$$(20.5) \qquad \Sigma\, Y_i = nb_0 + b_1 \Sigma\, X_{i1} + b_2 \Sigma\, X_{i2}$$

$$\Sigma\, X_{i1}Y_i = b_0 \Sigma\, X_{i1} + b_1 \Sigma\, X_{i1}^2 + b_2 \Sigma\, X_{i1}X_{i2}$$

$$\Sigma\, X_{i2}Y_i = b_0 \Sigma\, X_{i2} + b_1 \Sigma\, X_{i1}X_{i2} + b_2 \Sigma\, X_{i2}^2$$

The least squares estimates are obtained by solving this system of equations simultaneously. Finding the solution manually can be tedious. Hence, computer programs are usually utilized, as here.

☐ **Example**

Block 3 in Figure 20.2 presents information on the estimated regression coefficients for our promotional expenditures study. The format is an extension of that in Figure 18.13 for simple linear regression:

Variable	Coeff.	St. Dev.	T Stat.
0	b_0	$s\{b_0\}$	$t^* = b_0/s\{b_0\}$
1	b_1	$s\{b_1\}$	$t^* = b_1/s\{b_1\}$
2	b_2	$s\{b_2\}$	$t^* = b_2/s\{b_2\}$

The estimated regression function is seen to be:

$$\widehat{Y} = 2.13438 + 3.02925X_1 + .70575X_2$$

☐

Analysis of Variance

Sums of Squares. The sums of squares for the analysis of variance are defined identically in simple and multiple regression. For convenience, we repeat the definitional formulas for the total sum of squares $SSTO$, the error sum of squares SSE, and the regression sum of squares SSR:

$$(20.6) \qquad SSTO = \Sigma\, (Y_i - \bar{Y})^2$$

$$(20.7) \qquad SSE = \Sigma\, e_i^2 = \Sigma\, (Y_i - \widehat{Y}_i)^2$$

$$(20.8) \qquad SSR = \Sigma\, (\widehat{Y}_i - \bar{Y})^2$$

Degrees of Freedom. As usual, the total sum of squares $SSTO$ has $n - 1$ degrees of freedom associated with it.

The error sum of squares SSE has associated with it $n - p$ degrees of freedom. There are n residuals e_i, but p constraints arise on these residuals because p parameters—$\beta_0, \beta_1, \ldots, \beta_{p-1}$—have to be estimated for obtaining the fitted values $\widehat{Y}_i$.

The regression sum of squares SSR has $p - 1$ degrees of freedom associated with it. There are p parameters in the regression function, but the deviations $\widehat{Y}_i - \bar{Y}$ are subject to one constraint, $\Sigma\, (\widehat{Y}_i - \bar{Y}) = 0$.

Note that when $p - 1 = 1$, the simple linear regression case, the degrees of freedom for SSE are $n - p = n - 2$ and those for SSR are $p - 1 = 1$. These, of course, are the degrees of freedom stated in Table 18.4a for simple linear regression.

The degrees of freedom are additive:

$$n - 1 = (p - 1) + (n - p)$$

Table 20.2 shows the general form of the ANOVA table for multiple regression. The mean squares MSR and MSE are defined as for simple linear regression, i.e., as a sum of squares divided by degrees of freedom. Block 4 in Figure 20.2 contains the ANOVA table in our promotional expenditures example.

Point Estimation of σ^2. As in simple linear regression, the error mean square MSE is an unbiased point estimator of σ^2. The ANOVA table in block 4 shows that $MSE = .28935$ for our example. The corresponding point estimate of σ is $\sqrt{MSE} = \sqrt{.28935} = .53791$.

Coefficient of Multiple Determination

The coefficient of multiple determination, denoted by R^2, is defined analogously to r^2, the coefficient of simple determination:

(20.9) The *coefficient of multiple determination R^2* is defined:

$$R^2 = \frac{SSR}{SSTO} = 1 - \frac{SSE}{SSTO}$$

Thus, R^2 measures the proportionate reduction in $SSTO$ associated with the use of the independent variables $X_1, X_2, \ldots, X_{p-1}$ in the multiple regression model. When $p - 1 = 1$, R^2 reduces to r^2. R^2, like r^2, ranges from 0 to 1:

(20.10) $$0 \leq R^2 \leq 1$$

We obtain $R^2 = 0$ when $b_1 = b_2 = \cdots = b_{p-1} = 0$, and $R^2 = 1$ when all the observed Y_i's fall directly on the estimated regression surface.

☐ Example

For our promotional expenditures study, we find from the ANOVA table in Figure 20.2 that $SSR = 193.4888$ and $SSTO = 197.2503$. Hence:

$$R^2 = \frac{193.4888}{197.2503} = .98093$$

Thus, the variability of locality sales is reduced by 98 percent when media and point-of-sale promotional expenditures are considered in the regression model. ☐

TABLE 20.2 *General format of ANOVA table for multiple regression*

Source of Variation	SS	df	MS
Regression	$SSR = \Sigma (\hat{Y}_i - \bar{Y})^2$	$p - 1$	$MSR = \dfrac{SSR}{p - 1}$
Error	$SSE = \Sigma (Y_i - \hat{Y}_i)^2$	$n - p$	$MSE = \dfrac{SSE}{n - p}$
Total	$SSTO = \Sigma (Y_i - \bar{Y})^2$	$n - 1$	

Comment

It can be shown that the coefficient of multiple determination R^2 is identical to the coefficient of simple determination obtained when the observations Y_i are regressed on the fitted values $\hat{Y}_i$ from the multiple regression model.

Coefficient of Multiple Correlation. The positive square root of R^2 is called the coefficient of multiple correlation:

(20.11) The *coefficient of multiple correlation* R is defined:

$$R = +\sqrt{R^2}$$

where: R^2 is given by (20.9)

For our promotional expenditures study, we have:

$$R = \sqrt{R^2} = \sqrt{.98093} = .99042$$

The values of R^2 and R for our example are included in block 5 in Figure 20.2. The format for this block is:

$$\sqrt{MSE} \qquad n$$
$$R^2 \qquad R$$

Adjusted R^2. Some statisticians have suggested that the coefficient of multiple determination defined in (20.9) be modified to recognize the number of independent variables in the model. The reason is that this coefficient can generally be made larger if additional independent variables are added to the model. To see this, note that SSE tends to become smaller (it cannot increase) with each additional independent variable, while $SSTO$ remains fixed. A measure that recognizes the number of independent variables in the model is called the *adjusted coefficient of multiple determination* and is denoted by R_a^2:

(20.12)
$$R_a^2 = 1 - \frac{n-1}{n-p}\left(\frac{SSE}{SSTO}\right)$$

When an independent variable is added to the model, p increases by one and $n - p$ decreases correspondingly. R_a^2 can therefore become smaller if the decrease in $n - p$ is not offset by a sufficient decrease in SSE.

For our example, we have:

$$R_a^2 = 1 - \frac{15}{13}\left(\frac{3.7616}{197.2503}\right) = .97800$$

Here, the adjustment has only a small effect; note that R^2 and R_a^2 are almost equal.

Coefficient of Partial Determination

The coefficient of multiple determination R^2 is a measure of the combined effect of all independent variables $X_1, X_2, \ldots, X_{p-1}$ in the regression model in reducing the total variability *SSTO*. Often, an additional measure is needed when considering whether or not to add another variable to the regression model.

☐ Illustration

In the promotional expenditures study in Table 20.1, a regression of sales volume Y on point-of-sale expenditures X_2 alone yields the estimated regression function $\hat{Y} = 12.73675 + .70575X_2$ (computations not shown). The error sum of squares for this simple regression model is 187.2886, which we shall denote by $SSE(X_2)$. The question of concern is whether it is worth adding the other independent variable X_1, media expenditures, to the regression model. We see from block 4 of Figure 20.2 that the error sum of squares when both X_1 and X_2 are in the regression model, to be denoted by $SSE(X_1, X_2)$, is $SSE(X_1, X_2) = 3.7616$. Thus, the remaining variability in sales volume when X_2 is already in the regression model, 187.2886, is reduced to 3.7616 by adding X_1 to the regression model. The proportionate reduction is 98 percent:

$$\frac{SSE(X_2) - SSE(X_1, X_2)}{SSE(X_2)} = \frac{187.2886 - 3.7616}{187.2886} = .9799$$

This measure indicates the marginal effect of X_1 in reducing the variability in Y when X_2 is already in the model, and is called a *coefficient of partial determination*. ☐

For the case of regression model (20.1), which has two independent variables, the coefficient of partial determination between Y and X_1, given that X_2 is already in the model, is denoted by $r^2_{Y1.2}$ and is defined:

(20.13)
$$r^2_{Y1.2} = \frac{SSE(X_2) - SSE(X_1, X_2)}{SSE(X_2)} = 1 - \frac{SSE(X_1, X_2)}{SSE(X_2)}$$

Comments

1. The square root of the coefficient of partial determination is called the *coefficient of partial correlation*. It is given the same sign as the corresponding regression coefficient in the fitted regression function which contains all the independent variables. For the promotional expenditures study, $r_{Y1.2} = +\sqrt{.9799} = +.9899$ because b_1 in the fitted regression function (see block 3 of Figure 20.2) is positive ($b_1 = +3.02925$).
2. A coefficient of partial determination between Y and X_2, given that X_1 is already in the regression model, is denoted by $r^2_{Y2.1}$ and is defined in an analogous way to (20.13) by reversing the roles of X_1 and X_2.
3. The coefficient of partial determination may also be defined in cases where there are more than two independent variables. For instance, when X_4 enters a regression model already containing the variables X_1, X_2, and X_3, the coefficient of partial determination is given by:

$$r^2_{Y4.123} = \frac{SSE(X_1, X_2, X_3) - SSE(X_1, X_2, X_3, X_4)}{SSE(X_1, X_2, X_3)}$$

Here, $r^2_{Y4.123}$ measures the relative reduction in the error sum of squares by adding X_4 into the model when X_1, X_2, and X_3 are already in the regression model.

20.3 STATISTICAL INFERENCES WITH MULTIPLE REGRESSION MODELS

We turn now to statistical inferences with multiple regression models. We continue to utilize our promotional expenditures study.

Residuals

Usually the first step in the analysis of a multiple regression model is an examination of the aptness of the model. Residuals are studied for this purpose in the same manner described earlier for the simple linear regression model. Thus, residuals are analyzed for randomness, normality, constancy of error variance, and appropriateness of the regression function. Plots of the residuals e_i against the fitted values $\hat{Y}_i$ and against time order are useful. Plots against the independent variables X_1, $X_2, \ldots, X_{p-1}$ one at a time are also helpful in identifying whether the effects of any of the independent variables differ from those postulated in the model. Finally, plots of the residuals e_i against selected new variables not already in the model help to discover if one or more important variables have been omitted from the model.

□ Example

The observed values Y_i, the fitted values $\hat{Y}_i$, and the residuals e_i for our promotional expenditures example are presented in Figure 20.2, block 8. The printout shown is arbitrarily terminated at the fifth observation because of space limitations. We do not present the residual plots, but the market researcher analyzing the regression results felt that they did not challenge the appropriateness of the multiple regression model (20.1). □

F Test for Regression Relation

Frequently, the next step in the analysis of a multiple regression model is to test whether or not there is a relation between the dependent variable Y and the independent variables $X_1, X_2, \ldots, X_{p-1}$. The expectation, of course, is that such a relation exists, because the model is developed in the first place to exploit this relation.

The F test for a multiple regression relation is a direct extension of the F test for $\beta_1 = 0$ versus $\beta_1 \neq 0$ in simple linear regression. It is a test whether or not $\beta_1 = \beta_2 = \cdots = \beta_{p-1} = 0$, i.e., a test whether or not there is a relation between $E\{Y\}$ and the independent variables $X_1, X_2, \ldots, X_{p-1}$. To see what is involved here, let us return to the case of two independent variables and model (20.1). If β_1 and β_2 both equal zero, the response function reduces to $E\{Y\} = \beta_0$. Thus, the means of the distributions of Y are the same for all X_1, X_2 combinations, and there is no regression relation between Y and the independent variables X_1 and X_2. On the other hand, if β_1 and β_2 are not both equal to zero (at least one of the two β's is nonzero), the means of Y are not constant for all X_1, X_2 combinations, and a regression relation does exist between Y and the independent variables.

These ideas extend directly to model (20.3) with $p - 1$ independent variables. Thus, a test of whether or not Y is related to $X_1, X_2, \ldots, X_{p-1}$ involves the following alternatives:

$$H_0: \beta_1 = \beta_2 = \cdots = \beta_{p-1} = 0$$

$$H_1: \text{Not all } \beta_k = 0 \qquad k = 1, 2, \ldots, p - 1$$

The test procedure is similar to that for the F test of $\beta_1 = 0$ versus $\beta_1 \neq 0$ in simple linear regression. The test statistic is the same:

(20.14)
$$F^* = \frac{MSR}{MSE}$$

and the decision rule to control the α risk takes a corresponding form:

(20.15) When the alternatives are:

$$H_0: \beta_1 = \beta_2 = \cdots = \beta_{p-1} = 0$$

$$H_1: \text{Not all } \beta_k = 0 \qquad k = 1, 2, \ldots, p - 1$$

and the multiple regression model (20.3) is applicable, the appropriate decision rule to control the α risk is:

If $F^* \leq F(1 - \alpha; p - 1, n - p)$, conclude H_0

If $F^* > F(1 - \alpha; p - 1, n - p)$, conclude H_1

where: F^* is given by (20.14)

☐ Example

The market researcher conducting the promotional expenditures study, after being satisfied from the residual analysis that model (20.1) is apt, wished next to test whether or not a relation between sales and the two types of promotional expenditures does indeed exist. If so, as he expected, he would then analyze the multiple regression model in some detail. The researcher specified that the α risk be controlled at .05.

The alternatives are:

$$H_0: \beta_1 = \beta_2 = 0$$

$$H_1: \text{Not both } \beta_1 = 0 \text{ and } \beta_2 = 0$$

For $\alpha = .05$, $n = 16$, and $p - 1 = 2$, we require $F(.95; 2, 13) = 3.81$. From the computer output in Figure 20.2 (block 4), we see that $MSR = 96.74439$ and $MSE = .28935$. Thus, $F^* = 96.7444/.28935 = 334 > 3.81$. Hence, we conclude H_1—that there is a regression relation between the two types of promotional expenditures and sales. Note, incidentally, that the value of F^* for this test is given in the computer output in Figure 20.2 immediately below the ANOVA table. ☐

Inferences Concerning Regression Coefficients

Estimates and tests for individual regression coefficients β_k in the multiple regression model (20.3) are conducted in the same manner as in simple linear regression.

The only difference is that the t multiple now involves $n - p$ degrees of freedom, because MSE in multiple regression has associated with it $n - p$ degrees of freedom. The confidence interval for any β_k takes the usual form:

(20.16) The confidence limits for β_k with confidence coefficient $1 - \alpha$ for multiple regression model (20.3) are:

$$b_k \pm ts\{b_k\}$$

where: $t = t(1 - \alpha/2; n - p)$
$s\{b_k\}$ is the estimated standard deviation of b_k

The estimated standard deviation of b_k, $s\{b_k\}$, is given in the computer output of almost all regression packages.

A test of whether or not $\beta_k = 0$ is conducted in the manner explained for simple regression. We present the decision rule based on the standardized test statistic t^* defined in (19.17):

(20.17) When the alternatives are:

$$H_0: \beta_k = 0$$

$$H_1: \beta_k \neq 0$$

and multiple regression model (20.3) is applicable, the appropriate decision rule to control the α risk is:

$$\text{If } |t^*| \leq t(1 - \alpha/2; n - p), \text{ conclude } H_0$$

$$\text{If } |t^*| > t(1 - \alpha/2; n - p), \text{ conclude } H_1$$

where: $t^* = \dfrac{b_k}{s\{b_k\}}$

☐ **Example**

In the promotional expenditures study, the market researcher was interested in estimating β_1 and β_2, each with a 95 percent confidence interval. Since $n - p = 16 - 3 = 13$, we require $t(.975; 13) = 2.160$ for each interval. From the computer output in Figure 20.2 (block 3), we obtain $b_1 = 3.02925$, $s\{b_1\} = .12028$, $b_2 = .70575$, and $s\{b_2\} = .12028$. The confidence limits by (20.16) then are $3.02925 \pm 2.160(.12028)$ for β_1 and $.70575 \pm 2.160(.12028)$ for β_2. Hence, the desired confidence intervals are $2.77 \leq \beta_1 \leq 3.29$ and $.446 \leq \beta_2 \leq .966$.

The confidence intervals indicate that an increase of \$1,000 in media expenditures (with point-of-sale expenditures held fixed) is associated with an increase in expected sales between \$27,700 and \$32,900, while the same increase in point-of-sale expenditures in a locality (with media expenditures held fixed) is associated with an increase in expected sales of only between \$4,460 and \$9,660. Recall from Table 20.1 that the sales volume variable is expressed in units of ten thousand dollars. ☐

Comment

If one concludes that $\beta_k = 0$ for some variable X_k in a multiple regression model, it does not necessarily follow that X_k is not related to Y. It simply means that, when the other independent variables are already in the regression model, the marginal contribution of X_k in further reducing the error sum of squares is negligible. We shall consider this point further in our discussion of multicollinearity.

Confidence Interval for $E\{Y_h\}$

As with simple linear regression, a key objective in many applications of multiple regression is to make an inference on a mean response, $E\{Y_h\}$. Our discussion of such inferences in simple linear regression extends readily to multiple regression.

Let the levels of the independent variables for which the mean response is to be estimated be denoted by $X_{h1}, X_{h2}, \ldots, X_{h,p-1}$. Then $E\{Y_h\}$ denotes the mean response when $X_1 = X_{h1}, X_2 = X_{h2}, \ldots, X_{p-1} = X_{h,p-1}$. The point estimator of $E\{Y_h\}$ is:

(20.18)
$$\widehat{Y}_h = b_0 + b_1 X_{h1} + b_2 X_{h2} + \cdots + b_{p-1} X_{h,p-1}$$

The confidence interval for $E\{Y_h\}$ takes the usual form:

(20.19) The confidence limits for $E\{Y_h\}$ with confidence coefficient $1 - \alpha$ for multiple regression model (20.3) are:

$$\widehat{Y}_h \pm ts\{\widehat{Y}_h\}$$

where: $t = t(1 - \alpha/2; n - p)$
$\widehat{Y}_h$ is given by (20.18)
$s\{\widehat{Y}_h\}$ is the estimated standard deviation of $\widehat{Y}_h$

The estimated standard deviation of $\widehat{Y}_h$, $s\{\widehat{Y}_h\}$, is a complex expression involving the variances and covariances of the regression coefficients and the specified levels of the independent variables. For the simple case of only two independent variables, for instance, $s^2\{\widehat{Y}_h\}$ is defined as follows:

(20.20)
$$s^2\{\widehat{Y}_h\} = s^2\{b_0\} + X_{h1}^2 s^2\{b_1\} + X_{h2}^2 s^2\{b_2\} + 2X_{h1} s\{b_0, b_1\}$$
$$+ 2X_{h2} s\{b_0, b_2\} + 2X_{h1} X_{h2} s\{b_1, b_2\}$$

Computer programs are almost always used to calculate $s\{\widehat{Y}_h\}$.

☐ Example

In our promotional expenditures study, the market researcher wishes to estimate expected sales when $X_{h1} = 5$ and $X_{h2} = 2$. A 95 percent confidence coefficient is to be employed.

We know from Figure 20.2 (block 3) that $b_0 = 2.13438$, $b_1 = 3.02925$, and $b_2 = .70575$. Hence, the point estimate of mean sales in localities where $X_{h1} = 5$ and $X_{h2} = 2$ is:

$$\widehat{Y}_h = 2.13438 + 3.02925(5) + .70575(2) = 18.6921$$

To obtain $s\{\widehat{Y}_h\}$, we refer to the computer printout in Figure 20.2. Many computer programs, like the one here, permit the user to specify the levels of the independent variables of interest and then will calculate $\widehat{Y}_h$ and $s\{\widehat{Y}_h\}$. We see from Figure 20.2, block 7, that the standard deviation of the mean response when $X_{h1} = 5$ and $X_{h2} = 2$ is $s\{\widehat{Y}_h\} = .2884$. $\widehat{Y}_h$ is, of course, the same as calculated previously.

We require $t(.975; 13) = 2.160$. The confidence limits then are $18.6921 \pm 2.160(.2884)$ and the confidence interval is:

$$18.07 \leq E\{Y_h\} \leq 19.32$$

We therefore estimate, with 95 percent confidence, that mean sales are between \$181 thousand and \$193 thousand in localities where media and point-of-sale expenditures are \$5 thousand and \$2 thousand, respectively. □

Comment

The estimated standard deviation of $\widehat{Y}_h$ in our example could have been calculated directly using (20.20). To do so, we would need the estimated variances and covariances of the regression coefficients. These are often provided in computer printouts. For instance, the printout in Figure 20.2 provides them in block 6 in the form of a matrix called *variance-covariance matrix of the regression coefficients*. The format and results for our example are:

$$\begin{bmatrix} s^2\{b_0\} & s\{b_0, b_1\} & s\{b_0, b_2\} \\ & s^2\{b_1\} & s\{b_1, b_2\} \\ & & s^2\{b_2\} \end{bmatrix} = \begin{bmatrix} .37254 & -.05064 & -.05064 \\ & .01447 & 0 \\ & & .01447 \end{bmatrix}$$

Prediction Interval for Y_h

Frequently, we wish to employ a multiple regression model to predict an individual response Y_h when the independent variables are at specified levels $X_{h1}, X_{h2}, \ldots, X_{h, p-1}$. The prediction interval takes the usual form:

(20.21) The prediction limits for Y_h with confidence coefficient $1 - \alpha$ for multiple regression model (20.3) are:

$$\widehat{Y}_h \pm ts\{d_h\}$$

where: $t = t(1 - \alpha/2; n - p)$
$$s^2\{d_h\} = s^2\{\widehat{Y}_h\} + MSE$$

□ ### Example

The market researcher wishes to predict sales in a locality for the next period when $X_{h1} = 5$ and $X_{h2} = 2$, using a 95 percent prediction interval. From earlier work, we know that for this case, $\widehat{Y}_h = 18.6921$ and $s^2\{\widehat{Y}_h\} = (.2884)^2 = .0832$. We also know that $MSE = .28935$. Hence, we calculate by (20.21):

$$s^2\{d_h\} = .0832 + .28935 = .3726$$

so $s\{d_h\} = \sqrt{.3726} = .6104$. This statistic could also have been obtained directly from block 7 in Figure 20.2. Finally, we need $t(.975; 13) = 2.160$. The prediction limits then are $18.6921 \pm 2.160(.6104)$ and the prediction interval is:

$$17.37 \leq Y_h \leq 20.01$$

Hence, we predict, with 95 percent confidence, that sales will be between \$174 thousand and \$200 thousand in the locality when media and point-of-sale expenditures are \$5 thousand and \$2 thousand, respectively. □

20.4 OPTIONAL TOPIC—SPECIAL REGRESSION MODELS

The multiple regression model (20.3):

$$Y_i = \beta_0 + \beta_1 X_{i1} + \cdots + \beta_{p-1} X_{i,p-1} + \epsilon_i$$

is called the *general linear regression model* because it encompasses many special cases. It is called a *linear model* because it is *linear in the parameters* β_0, $\beta_1, \ldots, \beta_{p-1}$. Thus, no parameter appears as an exponent or is multiplied by another parameter. The general linear regression model is not, however, restricted to linear relationships; it encompasses curvilinear relationships, as we shall now see.

Quadratic Model

Frequently, the relationship between the dependent variable Y and the independent variable X is not linear. For example, we would expect increases in sales (Y) with increasing advertising expenditures (X) to slow as saturation is approached.

A widely encountered curvilinear regression model is the *quadratic* regression model:

(20.22) $$Y_i = \beta_0 + \beta_1 X_i + \beta_2 X_i^2 + \epsilon_i$$

Here, the regression function is $E\{Y\} = \beta_0 + \beta_1 X + \beta_2 X^2$. The regression curve shown in Figure 18.3 for the experiment with specialty foods sections in supermarkets is quadratic. β_1 is called the *linear effect coefficient* and β_2 the *curvature effect coefficient*.

To see that model (20.22) is a special case of the general linear regression model (20.3), let $X_{i1} = X_i$ and $X_{i2} = X_i^2$. We can then write model (20.22) in the standard form:

$$Y_i = \beta_0 + \beta_1 X_{i1} + \beta_2 X_{i2} + \epsilon_i$$

Models Containing Interactions

Frequently, regression models contain *cross-product* or *interaction* terms, such as the term $X_{i1}X_{i2}$ in the following model:

(20.23) $$Y_i = \beta_0 + \beta_1 X_{i1} + \beta_2 X_{i2} + \beta_3 X_{i1}X_{i2} + \epsilon_i$$

With this type of model, the effect of X_1 on Y depends on the level of X_2, and similarly the effect of X_2 on Y depends on the level of X_1. Hence, the effects of X_1 and X_2 on Y are no longer additive.

Model (20.23) is a special case of the general linear regression model. To see this, let $X_{i3} = X_{i1}X_{i2}$. We can then write model (20.23) in the form of the general linear regression model (20.3).

Inherently Linear Models

Often models not in the form of model (20.3) can be put into this form by an appropriate mathematical transformation. An example is the following regression function often used in studying the growth of a phenomenon over time, such as sales:

(20.24)
$$E\{Y\} = \gamma_0 \gamma_1^X$$

When we take the logarithms of both sides, we obtain:

$$\log E\{Y\} = \log \gamma_0 + X \log \gamma_1$$

If we let $E\{Y\}' = \log E\{Y\}$, $\beta_0 = \log \gamma_0$, and $\beta_1 = \log \gamma_1$, we can write the transformed regression function as a linear regression function:

$$E\{Y\}' = \beta_0 + \beta_1 X$$

Thus, a linear regression of $\log Y$ on X may be appropriate here.

☐   Example

We now illustrate the analysis of a quadratic regression model, and show how the same multiple regression computer program used for the promotional expenditures study can handle any regression model that is of the general linear model form.

An analyst was asked to investigate whether job proficiency of applicants seeking employment as assemblers could be predicted with satisfactory confidence and precision from the applicants' scores on two tests administered during the interview. Table 20.3 presents a portion of the results for a random sample of 25 applicants who were hired for purposes of the experiment irrespective of their test scores. Here, Y denotes job proficiency score after a learning period, and X_1 and X_2 are the respective scores on manual dexterity and depth perception tests.

From earlier studies of the assembling operation, the analyst knew that manual dexterity and depth perception have linear effects on job proficiency. However, past evidence was inconclusive as to whether or not depth perception also has a curvature effect. Hence, she employed the following model for the study:

$$Y_i = \beta_0 + \beta_1 X_{i1} + \beta_2 X_{i2} + \beta_3 X_{i2}^2 + \epsilon_i$$

with the intent of examining the data to determine whether or not the curvature effect $\beta_3 X_{i2}^2$ is really needed in the model.

TABLE 20.3   *Data for job proficiency example*

i	Variable 1: Manual Dexterity Score X_{i1}	Variable 2: Depth Perception Score X_{i2}	Variable 3: $X_{i3} = X_{i2}^2$	Variable 4: Job Proficiency Score Y_i
1	60	82	6,724	1,102
2	135	163	26,569	2,333
3	101	61	3,721	1,384
⋮	⋮	⋮	⋮	⋮
23	103	102	10,404	1,655
24	156	170	28,900	2,588
25	77	122	14,884	1,513

In entering the data for computation, the analyst employed the transformation $X_{i3} = X_{i2}^2$ to create a third X variable. The observations thus were entered in the following format:

Variable 1:	Variable 2:	Variable 3:	Variable 4:
X_{i1}	X_{i2}	$X_{i3} = X_{i2}^2$	Y_i

as shown in Table 20.3.

Figure 20.3 presents the output obtained by the analyst. As before, we have numbered the blocks for identification. Note that only selected blocks of output were requested in the time-sharing computer program.

The analyst now wished to test for the curvature effect for depth perception. The alternatives are:

$$H_0: \beta_3 = 0$$

$$H_1: \beta_3 \neq 0$$

H_0 implies that a curvature effect is not present, while H_1 implies it is present.

As shown in Figure 20.3 (block 3), the standardized test statistic for variable 3 is $t^* = b_3/s\{b_3\} = -1.36$. The analyst wishes to control the α risk at .05. For $\alpha = .05$, $n = 25$, and $p = 4$, we require $t(.975; 21) = 2.080$. Since $|t^*| = 1.36 \leq 2.080$, we conclude H_0—that the curvature effect is not required in the model.

Consequently, the analyst dropped the quadratic term X_{i2}^2 and fitted the model:

$$Y_i = \beta_0 + \beta_1 X_{i1} + \beta_2 X_{i2} + \epsilon_i$$

FIGURE 20.3　*Computer output for job proficiency example*

```
MULTIPLE REGRESSION

1. PRINT MEANS, VARIANCES AND STANDARD DEVIATIONS? NO
2. PRINT SIMPLE CORRELATION MATRIX?    YES
      1.000    0.796    0.781    0.950
               1.000    0.977    0.945
                        1.000    0.924
                                 1.000

   ENTER LIST OF VARIABLES, RESPONSE VAR. FIRST.
   □:
      4  1  2  3

3. VARIABLE    COEFFICIENT    ST. DEV.      T STAT.
      0          0.08042      4.80373        0.02
      1         10.00907      0.04568      219.14
      2          6.12105      0.08390       72.96
      3         ⁻0.00045      0.00033       ⁻1.36
   INDEP. VARS.: 1  2  3; RESPONSE VAR.: 4

4. PRINT ANOVA TABLE?    YES
      SOURCE            SS            DF           MS
   REGRESSION     10735330.0215       3      3578443.34048
   RESIDUAL            502.5385      21           23.93041
   TOTAL         10735832.5600      24
   F STAT.            = 149535.42

5. RESIDUAL STD DEV.     = 4.89187;   NUMBER OF OBSERVATIONS = 25
   MULT R SQ (COEFF DETERM) = 0.99995;  MULT R (COEFF CORREL) = 0.99998

6. PRINT VARIANCE-COVARIANCE MATRIX? NO

7. ESTIMATE RESPONSE VARIABLE? NO

8. PRINT RESIDUALS? NO
```

She tested this model for aptness and then proceeded to study how well proficiency scores can be predicted from the two aptitude tests. ☐

20.5   OPTIONAL TOPIC—INDICATOR VARIABLES

In our discussion of regression analysis up to this point, we have utilized *quantitative* independent variables, such as test scores or dollar expenditures. Frequently, however, an independent variable of interest is qualitative. Examples of qualitative variables are job location of employee (plant, office, field), sex of respondent (male, female), and nature of financial disclosure (none, footnotes only, line item discussion).

In this section, we show how qualitative variables can be included among the independent variables in regression analysis.

☐   Illustration

A regression analysis was made for a chain of fast-food restaurants of the relation between restaurant dollar sales during a recent period (Y) and number of households in the restaurant's trading area (X_1) and location of restaurant (highway, shopping mall). Since the location variable is qualitative, numerical indicators must be used to denote the qualitative outcomes. The most widely used indicators are variables that take on a value of 0 or 1. For our example, location of restaurant has two qualitative classes, so one indicator variable suffices. It will be coded as follows:

(20.25) $$X_{i2} = \begin{cases} 1 \text{ if restaurant in } i\text{th observation located in shopping mall} \\ 0 \text{ if restaurant in } i\text{th observation located on highway} \end{cases}$$

The data for the analysis appear in Table 20.4.

We employ regression model (20.1) for two independent variables:

(20.26) $$Y_i = \beta_0 + \beta_1 X_{i1} + \beta_2 X_{i2} + \epsilon_i$$

TABLE 20.4   *Data for indicator variable example*

i	Variable 1: Number of Households (*thousands*) X_{i1}	Variable 2: Location (*1 if shopping mall,* *0 if highway*) X_{i2}	Variable 3: Sales Volume (*$ thousands*) Y_i
1	155	0	135.27
2	178	1	179.86
3	215	1	220.14
4	93	0	72.74
5	128	0	114.95
6	114	0	102.93
7	172	1	179.64
8	158	0	131.77
9	197	1	185.92
10	207	1	207.82
11	95	1	113.51
12	183	0	160.91

The response function is:

(20.27)
$$E\{Y\} = \beta_0 + \beta_1 X_1 + \beta_2 X_2$$

Note that for highway restaurants, for which $X_2 = 0$, the response function reduces to:

(20.27a)
$$E\{Y\} = \beta_0 + \beta_1 X_1 + \beta_2(0) = \beta_0 + \beta_1 X_1 \qquad \text{(Highway location)}$$

For mall restaurants, for which $X_2 = 1$, the response function becomes:

(20.27b)
$$E\{Y\} = \beta_0 + \beta_1 X_1 + \beta_2(1) = (\beta_0 + \beta_2) + \beta_1 X_1 \qquad \text{(Mall location)}$$

These two response functions are portrayed in Figure 20.4. Note that both response functions are linear, with the same slope β_1. The intercept for the highway response function is β_0, while that for the mall response function is $\beta_0 + \beta_2$. Consequently, the mean of Y for mall restaurants for any given number of households differs by β_2 from the mean for highway restaurants. Thus, β_2 measures the differential effect on mean sales of mall location compared to highway location. If β_2 is positive, expected mall sales are higher by that amount than expected highway sales for any given number of households; if β_2 is negative, they are lower by that amount. In Figure 20.4, β_2 is portrayed as being positive.

Model (20.26) was fitted to the data using a standard multiple regression computer package. Some selected results are:

$$\hat{Y} = .34022 + .86225 X_1 + 27.90267 X_2 \qquad MSE = 38.43746$$

$$s\{b_1\} = .0508 \qquad s\{b_2\} = 4.0867$$

The analyst was particularly interested in the effect of location on sales. We can estimate this by obtaining a confidence interval for β_2, which reflects the difference in the mean sales, for any number of households, for the two locations. A 90 percent confidence interval is desired.

For $n = 12$ and $p - 1 = 2$, we require $t(.95; 9) = 1.833$. Since $b_2 = 27.90267$ and

FIGURE 20.4 *Illustration of meaning of model parameters for indicator variable example.* β_1 *is the common slope and* β_2 *is the difference in the mean responses for mall and highway locations for any given number of households.*

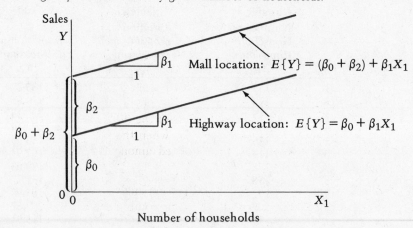

$s\{b_2\} = 4.0867$, we set up the confidence limits $27.90267 \pm 1.833(4.0867)$, and the confidence interval is:

$$20.4 \leq \beta_2 \leq 35.4$$

We therefore estimate, with 90 percent confidence, that for any given number of households in the trading area, the expected sales in a mall location are somewhere between $20.4 thousand and $35.4 thousand greater than in a highway location. □

Comments

1. Indicator variables are sometimes called *dummy variables* or *binary variables*.
2. In general, when the qualitative variable has k classes, $k - 1$ indicator variables are required. Thus, if the fast-food chain had restaurants in three types of locations (highway, shopping mall, city street), two indicator variables (X_{i2} and X_{i3}) would be required. These may be coded as follows:

(20.28)
$$X_{i2} = \begin{cases} 1 & \text{if restaurant in } i\text{th observation located in shopping mall} \\ 0 & \text{otherwise} \end{cases}$$

$$X_{i3} = \begin{cases} 1 & \text{if restaurant in } i\text{th observation located on city street} \\ 0 & \text{otherwise} \end{cases}$$

With this coding, X_2 and X_3 take on the following values for the three locations:

Location	X_2	X_3
Highway	0	0
Mall	1	0
Street	0	1

The regression coefficient β_2 here represents the differential effect on mean sales of mall location compared to highway location, and β_3 represents the differential effect of street location compared to highway location.

20.6 OPTIONAL TOPIC—MULTICOLLINEARITY

In some regression applications, a major objective is to measure the separate effects of the independent variables on the dependent variable. Generally, the regression coefficients are utilized for this purpose because the regression coefficient β_k in the general linear model (20.3) indicates the change in the mean of the distribution of Y when X_k changes by one unit and the other independent variables remain constant.

Unfortunately, the separate effects of the different independent variables cannot usually be measured satisfactorily when the sample observations of the independent variables are highly correlated among themselves. Multicollinearity is then said to exist:

(20.29) *Multicollinearity* is present in a regression analysis when the sample observations of the independent variables, or linear combinations of them, are highly correlated.

When multicollinearity prevails, two major, related problems are encountered in assessing the separate effects of the different independent variables:

1. The estimated regression coefficient b_k for the independent variable X_k may vary substantially, depending on which other independent variables are included in the model. Thus, the value obtained for b_k in any particular fitted model does not indicate the effect of X_k on $E\{Y\}$ in any absolute sense.
2. The estimated regression coefficients tend to have extremely large sampling errors, indicating that they vary widely in repeated samples. As a result, they will give very imprecise information about the regression parameters.

Fortunately, the difficulties just cited for the regression coefficients generally do not carry over to inferences on Y. Providing inferences are made within the region of sample observations on the independent variables, multicollinearity usually causes no special problems in estimating a mean response or predicting an individual response.

We shall illustrate the effects of multicollinearity by two examples. In the first example the independent variables are highly correlated, while in the second they are uncorrelated.

☐ Example 1: Independent Variables Highly Correlated

A machine shop receives rough shafts and smoothes them to specifications. When a shaft is smoothed, it is first carefully mounted on a machine, then excess metal is machined off. The time required to process a shipment of shafts is affected by the time required to mount the shafts for smoothing, and by the time required to machine the shafts once they have been mounted. The total mounting time depends on the number of shafts in the shipment, while the total machining time depends on the weight of the shipment (since the aggregate amount of metal to be machined off varies with the total weight).

Thus, two physically distinct variables affect the processing time for a shipment, namely the number of shafts in the shipment and the total weight of the shipment. Table 20.5 shows, for the 15 most recent shipments, the number of shafts in the shipment (X_1), the total weight of the shipment (X_2), and the processing time for the shipment (Y). The independent variables are highly correlated, as the scatter plot in Figure 20.5 shows. The coefficient of simple correlation between X_1 and X_2 is above .99.

We have fitted three regression models to the data on the 15 shipments, as follows:

Fitted Model	Independent Variables in Model
1	X_1
2	X_1, X_2
3	X_2

Key results are shown in Table 20.6. They illustrate the three points that we made earlier about the effects of multicollinearity:

1. The point estimate b_1 changes substantially when X_2 is added to the model. When the fitted model contains X_1 alone we obtain $b_1 = 34.9$, but when the model contains both X_1 and X_2 we obtain $b_1 = 57.8$. Similarly, the value of b_2 changes when X_1 is added. In fact, the sign of the coefficient changes.

   Thus, b_k does not indicate the effect of X_k on $E\{Y\}$ in any absolute sense when multicollinearity is present.

TABLE 20.5 *Data for example with highly correlated independent variables*

Shipment i	Variable 1: Number of Shafts X_{i1}	Variable 2: Total Weight (*pounds*) X_{i2}	Variable 3: Processing Time (*minutes*) Y_i
1	55	5,563	1,738
2	20	2,041	491
3	35	3,594	999
4	45	4,523	1,370
5	40	4,082	1,150
6	25	2,534	684
7	55	5,556	1,650
8	30	3,044	876
9	60	6,095	1,910
10.	45	4,561	1,380
11	35	3,562	995
12	25	2,546	660
13	45	4,576	1,390
14	35	3,529	1,025
15	30	3,056	821

2. Table 20.6 also shows that the estimated standard deviation of b_k tends to become larger when other variables highly correlated with X_k are added to the fitted model. Thus, when X_1 is the only independent variable in the model we have $s\{b_1\} = .62818$, but when both X_1 and X_2 are included we have $s\{b_1\} = 39.51970$. Similarly, $s\{b_2\}$ increases from .00667 to .39187 when X_1 is added to the model.

3. Multicollinearity causes no special problems when inferences on Y are made within the

FIGURE 20.5 *Region of sample observations for independent variables in machine shop example.* The (X_1, X_2) observations fall in a narrow band here.

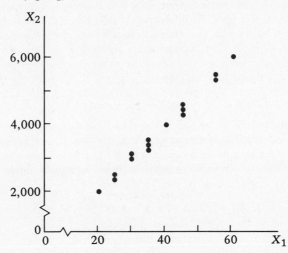

TABLE 20.6 *Results for three fitted regression models for example with highly correlated independent variables*

Statistic	Fitted Model 1: X_1	Fitted Model 2: X_1, X_2	Fitted Model 3: X_2
b_1	34.85684	57.75047	—
$s\{b_1\}$	.62818	39.51970	—
b_2	—	−.22704	.34553
$s\{b_2\}$	—	.39187	.00667
		$X_{h1} = 40, X_{h2} = 4{,}060$	
$\widehat{Y}_h$	1,189.0758	1,188.7534	1,189.5460
$s\{\widehat{Y}_h\}$	7.3437	7.5593	7.8622

region of sample observations on the independent variables, as shown by the estimation of the mean response when $X_{h1} = 40$ and $X_{h2} = 4{,}060$. Note from Figure 20.5 that this pair of values falls within the region of sample observations on the independent variables. As Table 20.6 shows, the estimated means $\widehat{Y}_h$ and the estimated standard deviations $s\{\widehat{Y}_h\}$ are practically the same for the three fitted models. This shows, incidentally, that in the presence of high correlation between X_1 and X_2, either variable alone contains much of the information provided by the other variable. ☐

Comments

1. Note that $s\{\widehat{Y}_h\}$ for the fitted model containing both X_1 and X_2 is of about the same magnitude as when only one of the independent variables is utilized despite the much larger sampling errors for b_1 and b_2. The reason is that these large sampling errors tend to be offsetting here.
2. The outcome of a test of whether or not $\beta_k = 0$ in the presence of multicollinearity depends very much on which other independent variables are included in the regression model. For instance, note in Table 20.6 for the fitted model containing both X_1 and X_2 that one would conclude that $\beta_1 = 0$ or that $\beta_2 = 0$, for any usual level of the α risk. Yet $R^2 = .996$ here. As noted previously, the test for $\beta_k = 0$ is a marginal test and depends on the marginal effect of X_k, given that the other variables are already in the model. When X_1 and X_2 are highly correlated, as here, the marginal effect of either variable may be very small when the other variable is already in the model, yet there may be a strong relation between X_1 and Y and between X_2 and Y.

Example 2: Independent Variables Uncorrelated

For an illustration where the independent variables are uncorrelated, we return to our promotional expenditures example where the two types of promotional expenditures were set up so that X_1 and X_2 are uncorrelated. The basic data were presented in Table 20.1. Table 20.7 shows pertinent results for three regression models fitted to these data, as follows:

Fitted Model	Independent Variables in Model
1	X_1
2	X_1, X_2
3	X_2

TABLE 20.7   *Results for three fitted regression models for example with uncorrelated independent variables*

Statistic	Fitted Model 1: X_1	Fitted Model 2: X_1, X_2	Fitted Model 3: X_2
b_1	3.02925	3.02925	—
$s\{b_1\}$	.22139	.12028	—
b_2	—	.70575	.70575
$s\{b_2\}$	—	.12028	.81786

Table 20.7 illustrates two important points:

1. The values of b_1 and b_2 are unaffected by whether or not the other independent variable is included in the model. This is true in general for any number of independent variables that are mutually uncorrelated.
2. Neither $s\{b_1\}$ nor $s\{b_2\}$ becomes enlarged when the other variable is added to the model. □

The stability of the regression coefficients when the independent variables are mutually uncorrelated makes it desirable to use experimental control to obtain uncorrelated independent variables. Unfortunately, experimental control is often not feasible and one may then have to employ correlated independent variables.

Remedial Measures

Remedial measures for the difficulties caused by multicollinearity can be employed at times. Sometimes it is possible to obtain additional observations which break the pattern of multicollinearity. In other cases, one can estimate some of the regression coefficients from other data. Thus, in our machine shop example, it may be feasible to estimate β_1 by a time-and-motion study. On still other occasions, it may be possible to transform some of the independent variables or to drop some from the model in order to lessen the degree of multicollinearity. In all cases, however, one must be cautious in the interpretation of regression coefficients in the presence of multicollinearity.

PROBLEMS

*20.1 Suppose a firm's earnings (Y, in millions of dollars) are related to its sales revenues (X_1, in millions of dollars) and its total long-term debt (X_2, in millions of dollars) according to model (20.1), and that the parameter values are: $\beta_0 = -5$, $\beta_1 = .15$, $\beta_2 = -.08$, $\sigma^2 = 4$.
   a. State the regression function. What is the value of $E\{Y\}$ if $X_1 = 100$ and $X_2 = 50$?
   b. How is β_1 interpreted here? How is σ interpreted here?

20.2 Refer to Problem 20.1. If $X_{i1} = 100$ and $X_{i2} = 50$: (1) what is the probability that Y_i is less than 3; (2) what is the value of Y_i if $\epsilon_i = -8$?

20.3 Consider regression model (20.1) and suppose that $E\{Y\} = -10 + 2X_1 - 3X_2$, and $\sigma^2 = 1$.

a. Graph the relationship between $E\{Y\}$ and X_2 when $X_1 = 6$. How is $\beta_2 = -3$ interpreted here?

b. Describe the probability distribution of Y when $X_1 = 6$ and $X_2 = 3$.

20.4 A small-business consultant used model (20.1) to study the relation between bad debt losses (Y, in thousands of dollars), business volume (X_1, in thousands of dollars) and credit limit (X_2, in thousands of dollars) for 32 owner-managed building supplies stores. The correlation coefficients for these variables are: $r_{12} = .779$, $r_{Y1} = .985$, $r_{Y2} = .677$.

a. Does the sign of r_{Y2} indicate that bad debt losses tend to increase with the credit limit in these stores? Explain.

b. What does the magnitude of r_{Y1} tell us about the degree of linear association between bad debt losses and business volume? Does it suggest that bad debt losses are substantially higher in high-volume stores as compared to low-volume stores? Explain.

c. The consultant reported: "In owner-managed building supplies stores, there is a very close relation between your credit limit and the amount of business volume you get." Comment.

20.5 Consider regression model (20.1). Describe what is implied if $r_{Y1} = -1$. If $r_{Y2} = 0$.

*20.6 Refer to Problem 20.4. Results of fitting the regression model are: $b_0 = -1.707$, $b_1 = .01681$, $b_2 = -3.507$, $s\{b_1\} = .0003937$, $s\{b_2\} = .4132$, $SSR = 6,444.79$, $SSE = 55.14$, $n = 32$.

a. State the estimated regression equation. Interpret b_1.

b. Obtain a point estimate of mean bad debt losses in stores for which $X_1 = 1,200$ and $X_2 = .7$.

c. Set up the ANOVA table. Calculate R^2 and interpret it.

d. Obtain a point estimate of σ^2.

20.7 An experiment was conducted for two weeks to study the effect of freezer temperature (X_1) and freezer storage density (X_2) on freezer power usage (Y, in kilowatt-hours) for a certain type of commercial freezer. The independent variables are measured in terms of deviations from the levels normally used. Thus, in the first observation, the temperature setting was 10°C below the normal setting and the storage density was 10 percent points less than the normal density. The study results follow.

i:	1	2	3	4	5	6	7	8	9
X_{i1}:	-10	0	$+10$	-10	0	$+10$	-10	0	$+10$
X_{i2}:	-10	-10	-10	0	0	0	$+10$	$+10$	$+10$
Y_i:	235	183	177	261	203	110	284	187	167

Regression model (20.1) is to be employed.

a. Verify, using the normal equations (20.5), that the estimated regression function is $\hat{Y} = 200.8 - 5.433X_1 + .7167X_2$. (Notice the calculational simplifications that arise because of the balanced nature of the X_1 and X_2 values in this experiment's design.) Interpret b_2.

b. Use the facts that $\Sigma (Y_i - \bar{Y})^2 = 22,182$ and $R^2 = .81242$ to set up the ANOVA table.

c. Obtain a point estimate of σ.

20.8 Refer to Problem 20.7.

   a. Obtain point estimates of the estimated change in expected power usage: (1) when freezer temperature is reduced $1°C$ and density is held fixed, (2) when freezer temperature is reduced $10°C$ and density is held fixed.

   b. Obtain a point estimate of the expected power usage when freezer temperature and storage density are both at their normal levels. Is the variability of the probability distribution of power usage in this case large relative to the mean of the distribution? Discuss.

*20.9 **a.** Calculate R^2 and R_a^2 in each of the following cases where $SSTO = 1,000$ and $SSE = 100$: (1) $n = 10$, $p = 7$; (2) $n = 100$, $p = 7$.

   b. If p is small relative to n, is there a large difference between R^2 and R_a^2?

20.10 Refer to Problem 20.7.

   a. What is the meaning of R^2 here?

   b. Calculate R_a^2. Is the difference between R^2 and R_a^2 substantial here? Does n affect this difference? Does p? Comment.

*20.11 Refer to Problems 20.4 and 20.6. When bad debt losses (Y) are regressed only on business volume (X_1) for the 32 stores, the error sum of squares is $SSE(X_1) = 192.1$. Calculate $r_{Y2.1}^2$. Explain its meaning in this setting. Also obtain $r_{Y2.1}$.

20.12 A medical investigator obtained data on the weight (Y), height (X_1), and girth (X_2) for a random sample of 40 middle-aged men. She performed three regression analyses and obtained the following results:

Regression:	Y on X_1	Y on X_2	Y on X_1 and X_2
Error sum of squares:	459	1,104	111

   a. Which of the two variables alone, height or girth, is the better linear predictor of weight? Explain why you can answer without knowing $SSTO$.

   b. Is the variability in weight remaining when the height variable is included in the model reduced substantially when the girth variable is added to the regression model? Comment, calculating the appropriate coefficient of partial determination.

20.13 Refer to Problems 20.4 and 20.6. The fitted values and residuals are:

i:	1	2	3	4	5	6	7	8
$\widehat{Y}_i$:	17.95	15.81	41.70	28.84	38.81	27.97	34.50	4.26
e_i:	2.49	−.16	−.39	1.41	2.19	−.22	.36	.11

i:	9	10	11	12	13	14	15	16
$\widehat{Y}_i$:	6.98	34.65	5.09	4.80	4.91	15.05	40.55	3.19
e_i:	−1.50	.21	2.58	1.63	−.82	−2.21	−.51	1.49

i:	17	18	19	20	21	22	23	24
$\widehat{Y}_i$:	12.17	3.21	25.79	18.68	3.12	3.95	23.02	8.95
e_i:	−1.22	−.87	−1.57	1.11	−1.38	1.21	1.40	−1.81

i:	25	26	27	28	29	30	31	32
$\widehat{Y}_i$:	4.68	15.82	15.30	61.57	13.84	5.99	23.36	20.08
e_i:	.66	−.89	−1.09	−.38	−1.19	1.21	−1.39	−.45

   a. Construct a residual plot and analyze it with regard to the appropriateness of the regression function and the constancy of the error variance. State your findings.

b. Analyze the residuals as to whether or not the error terms are normally distributed. What is your conclusion?

20.14 Refer to Problem 20.7.

a. Obtain the fitted values and residuals. Do the residuals sum to zero as they should?

b. Prepare a plot of the residuals against the fitted values. Analyze it for aptness of the regression function and constancy of the error variance. State your findings.

*20.15 Refer to Problems 20.4 and 20.6. Test whether or not a regression relation holds, controlling the α risk at .01. State the alternatives, decision rule, and conclusion. Does your conclusion imply that both β_1 and β_2 are not equal to zero?

20.16 Refer to Problem 20.7. Test whether or not a regression relation holds, controlling the α risk at .05. State the alternatives, decision rule, and conclusion. Does your conclusion necessarily imply that Y is related to X_1? Explain.

20.17 Refer to Figure 20.2. Are the T Stat. values in block 3 appropriate for testing whether a regression relation exists between the dependent variable and the set of independent variables? Explain.

*20.18 Computer output for a regression study in which $n = 22$ and $p - 1 = 3$ showed:

k:	1	2	3
b_k:	46.878	1.0792	-60.908
$s\{b_k\}$:	10.466	.12546	9.6165

a. Test whether or not $\beta_1 = 0$, controlling the α risk at .01. State the alternatives, decision rule, and conclusion.

b. Construct a 99 percent confidence interval for β_2. Interpret your interval estimate.

20.19 Refer to Problem 20.7. Computer output gives: $s\{b_1\} = 1.075$, $s\{b_2\} = 1.075$.

a. Obtain a 95 percent confidence interval for β_1. Interpret your interval estimate.

b. Test whether or not $\beta_2 = 0$, controlling the α risk at .05. State the alternatives, decision rule, and conclusion. What is the implication of your conclusion?

c. Calculate $r_{Y2.1}^2$. [Hint: When Y is regressed on X_1 alone, $SSE(X_1) = 4,469$.] Is the magnitude of the coefficient of partial determination consistent with your conclusion in b? Discuss.

*20.20 Refer to Problems 20.4 and 20.6. Computer output for the case $X_{h1} = 1,200$ and $X_{h2} = .7$ shows: $\widehat{Y}_h = 16.010$, $s^2\{\widehat{Y}_h\} = .06255$.

a. Estimate the expected bad debt losses for stores for which $X_{h1} = 1,200$ and $X_{h2} = .7$, using a 95 percent confidence interval. Interpret your confidence interval.

b. Construct a 95 percent prediction interval for a particular store with volume $X_{h1} = 1,200$ and credit limit $X_{h2} = .7$. Interpret your prediction interval.

20.21 A government is leasing tracts on the continental shelf to petroleum companies for hydrocarbon development. An energy consultant has employed regression model (20.1) to study data for 36 recent leases. He has regressed the high bid price per acre (Y, in dollars) against the government's reservation price per acre (X_1, in dollars) and the number of acres in the lease (X_2). Computer output for the case $X_{h1} = 359$ and $X_{h2} = 4,670$ shows: $\widehat{Y}_h = 1,842$, $s\{\widehat{Y}_h\} = 103.1$, $MSE = 318,480$.

a. Construct a 90 percent confidence interval for $E\{Y_h\}$ when $X_{h1} = 359$ and $X_{h2} = 4{,}670$. Interpret your interval estimate. [*Hint:* $t(.95; 33) = 1.692$.]

b. Construct a 90 percent prediction interval for the high bid price per acre for a new lease under consideration for which $X_{h1} = 359$ and $X_{h2} = 4{,}670$. Assume that conditions governing bids for this lease are the same as for the 36 recent leases. Interpret your prediction interval.

c. Would you expect your prediction interval in **b** to be much narrower if the regression function had been estimated from a very large number of leases rather than from only 36? Discuss.

20.22 Refer to Problem 20.7. The estimated variance-covariance matrix of the regression coefficients is as follows:

$$
\begin{array}{ccc}
77.050 & 0 & 0 \\
& 1.1558 & 0 \\
& & 1.1558
\end{array}
$$

Also, $MSE = 693.5$.

a. Obtain a 90 percent confidence interval for β_0.

b. Obtain a 95 percent confidence interval for the mean power usage of freezers when $X_{h1} = -10$ and $X_{h2} = -10$, using (20.20). Interpret your interval estimate.

c. Predict the power usage of a particular freezer when $X_{h1} = -10$ and $X_{h2} = -10$, using a 95 percent prediction interval.

d. Why is the prediction interval in **c** wider than the confidence interval in **b**?

20.23 Refer to Problems 20.7 and 20.22.

a. Is β_0 a meaningful parameter in this regression application? Explain.

b. The off-diagonal elements of the estimated variance-covariance matrix of the regression coefficients in Problem 20.22 are zeros. What does this tell us about the relations between the sampling errors in the estimated regression coefficients? Explain.

*20.24 The following data and fitted values were obtained when regressing number of defective pipes in shipment (Y) against total number of pipes in shipment $(X,$ in hundreds) for 12 recent shipments with the simple linear regression model (18.3):

i:	1	2	3	4	5	6	7	8	9	10	11	12
X_i:	5	10	4	10	7	8	8	5	10	5	12	6
Y_i:	30	51	26	52	40	43	45	31	52	30	59	36
$\hat{Y}_i$:	30.8	51.7	26.7	51.7	39.2	43.3	43.3	30.8	51.7	30.8	60.0	35.0

An analyst reviewing this regression application was concerned whether the regression function actually is quadratic. Some results of fitting the quadratic model (20.22) to the data follow.

k:	0	1	2	
b_k:	4.084	5.8478	$-.10736$	$MSE = .481$
$s\{b_k\}$:	—	.6036	.03830	

a. For the fitted simple linear regression model, obtain the residuals and construct a residual plot. Does the plot suggest that the regression function is curvilinear? Explain.

b. State the estimated quadratic regression function. Show the data setup for computer input, assuming the analyst followed the format of Table 20.3.

    c. Test whether or not the curvature effect can be dropped from the quadratic regression model, controlling the α risk at .05. State the alternatives, decision rule, and conclusion. What is the implication of your conclusion?

20.25 A sports statistician studied the relation between the time (Y, in seconds) for a particular competitive swimming event and the swimmer's age (X, in years) for 20 swimmers with ages ranging from 8 to 18. She employed model (20.22) and obtained the following results: $\hat{Y} = 147.3 - 11.11X + .2730X^2$, $s\{b_2\} = .1157$.

    a. Plot the estimated regression function. Would it be reasonable to use this regression function when the swimmer's age is 40? Comment.

    b. Construct a 95 percent confidence interval for the curvature effect coefficient. Interpret your interval estimate.

20.26 The personnel department of a company studied salaries of employees who did not complete high school. In the study, employee's salary (Y, in thousands of dollars) was regressed on number of years of work experience (X_1) and number of years of high school completed (X_2). The interaction regression model (20.23) was employed and the estimated regression function was $\hat{Y} = 8.90 + 1.10X_1 + 3.05X_2 - .20X_1X_2$.

    a. Obtain a point estimate of the mean salary of employees with 5 years of work experience and 1 year of high school completed.

    b. Plot $\hat{Y}$ against X_1 when $X_2 = 0$. On the same graph, plot $\hat{Y}$ against X_1 when $X_2 = 3$. Describe how the effect of X_1 on $\hat{Y}$ depends on the level of X_2 for this interaction model.

*20.27 Refer to Problem 18.10. Expressway mileage was measured for regular gasoline except in the cases of automobiles 4, 5, and 8, for which premium gasoline was inadvertently used. Let X_1 represent an automobile's weight (in hundred pounds) and X_2 be an indicator variable coded so that $X_2 = 1$ designates premium gasoline and $X_2 = 0$ designates regular gasoline. When expressway mileage was regressed on these two variables using model (20.1), the results were: $b_0 = 77.78$, $b_1 = -2.065$, $b_2 = 3.717$, $s\{b_1\} = .1286$, $s\{b_2\} = .6150$.

    a. Obtain a point estimate of the mean expressway mileage for an automobile weighing 19 hundred pounds using: (1) premium gasoline, (2) regular gasoline.

    b. Construct a 95 percent confidence interval for β_2. Interpret your interval estimate. What effect does β_2 represent? Does the regression model employed here assume that this effect is the same irrespective of automobile weight?

20.28 A financial analyst is studying the relation between bond yield (Y, in percent), bond maturity (X_1, in years), and whether the bond is from a corporate or municipal issue (X_2). Here, $X_2 = 1$ denotes a corporate bond issue and $X_2 = 0$ a municipal bond issue. The analyst has performed a regression analysis using model (20.1) for a random sample of 300 bond issues. Some regression results are: $b_0 = 8.223$, $b_1 = .1421$, $b_2 = .3381$, $s\{b_1\} = .006270$, $s\{b_2\} = .04816$.

    a. Conduct an appropriate test to determine whether or not there is a difference in the mean yields of corporate and municipal bonds of the same maturity. Control the α risk at .05. State the alternatives, decision rule, and conclusion. What is the implication of your conclusion?

    b. Plot $\hat{Y}$ against X_1 when $X_2 = 0$. On the same graph, plot $\hat{Y}$ against X_1 when $X_2 = 1$. Let X_1 range between 0 and 15 years in your plots. How is b_1 interpreted here? How is b_2 interpreted here?

20.29 Consider the indicator variables X_2 and X_3 for the Text restaurant example as coded in (20.28).

a. State the regression model for this case when there are three independent variables.

b. An observer has stated that when $\beta_2 > 0$ and $\beta_3 = 0$, this means that location does have an impact on sales if the restaurant is in a shopping mall or on a highway, but has no impact if the restaurant is on a city street. Comment.

c. What interpretation can be given to the quantity $\beta_3 - \beta_2$ here?

20.30 In an analysis of the relation between sales (Y) and target population (X_1) and advertising expenditures (X_2) for 10 sales territories of a large corporation, the following correlation matrix was obtained (format is that of block 2 in Figure 20.2, p. 516):

$$
\begin{array}{ccc}
1.000 & .999 & .992 \\
& 1.000 & .992 \\
& & 1.000
\end{array}
$$

a. What difficulties will arise here in attempting to measure the separate effects of population and advertising expenditures on sales? Discuss.

b. Will the difficulties cited in **a** necessarily carry over to the making of inferences on Y? Explain.

EXERCISES

20.31 (Calculus needed.) Derive the normal equations (20.5) for regression model (20.1).

20.32 Show that the coefficient of multiple determination R^2 is identical to the coefficient of simple determination obtained when the observations Y_i are regressed on the fitted values $\widehat{Y}_i$ from the multiple regression model.

20.33 a. Show that $r_{Y1.2}^2 = (R^2 - r_{Y2}^2)/(1 - r_{Y2}^2)$, where R^2 is the coefficient of multiple determination for the regression of Y on X_1 and X_2.

b. Prove that $F^* = (n - p)R^2/(p - 1)(1 - R^2)$ for regression model (20.3).

20.34 Consider the regression function $E\{Y\} = \beta_0 X_1^{\beta_1} X_2^{\beta_2}$, where β_0, β_1, and β_2 denote regression coefficients and X_1 and X_2 denote independent variables. Show how the transformation $\log E\{Y\}$ converts this function into a linear regression function, and explain the relation of the regression coefficients in the resulting linear regression function to the original regression coefficients.

20.35 Refer to Problem 20.27. What would have been the values of b_0, b_1, and b_2 if X_2 had been coded so that $X_2 = 1$ designates regular gasoline and $X_2 = 0$ designates premium gasoline? (*Hint:* The original indicator variable X_2 and the new one X_2' are related by $X_2' = 1 - X_2$.) Would the fitted values $\widehat{Y}_i$ be affected by this change in the coding scheme? Explain.

STUDIES

20.36 (Computer needed.) The manager of a university computer system wishes to employ model (20.1) in a study of the relation between the average response time of the system (Y, in seconds) and the number of university terminal users (X_1) and commercial terminal users (X_2) signed on the system. The data for 24 observations follow.

X_1	X_2	Y	X_1	X_2	Y	X_1	X_2	Y	X_1	X_2	Y
10	0	.08	48	4	.52	23	3	.13	69	12	1.06
36	8	.59	21	1	.13	66	7	.81	58	10	.74
75	5	.77	66	10	.88	10	2	.15	26	2	.23
16	4	.21	30	3	.28	70	10	1.00	14	4	.18
35	5	.45	55	9	.70	44	0	.28	62	9	.72
50	8	.56	49	6	.63	42	4	.43	63	3	.55

a. Obtain: (1) the estimated regression equation; (2) $s\{b_1\}$, $s\{b_2\}$; (3) the fitted values and residuals; (4) the correlation matrix.

b. Obtain 95 percent confidence intervals to estimate the change in the mean response time of the system when: (1) an additional university terminal user signs on the system and the number of commercial users is held constant, (2) an additional commercial terminal user signs on the system and the number of university users is held constant.

c. Is multicollinearity a problem in this regression study? Comment, referring to appropriate results in **a** to support your answer.

d. Use the residuals and selected residual plots to check the aptness of model (20.1) with respect to: (1) the need for a curvature effect in the regression model for variable X_2, (2) the constancy of the error variance, (3) the normality of the error terms.

20.37 (Computer needed.) The following tabulation gives the region, number of beds (X_1), and number of admissions (Y) last year for each of 24 small acute-care hospitals:

Region A

Number of beds X_1:	19	120	49	100	33	22
Admissions Y:	460	3,374	2,244	3,606	950	703

Region B

Number of beds X_1:	96	48	148	101	66	138	25	193	44
Admissions Y:	2,958	1,487	4,700	3,308	2,696	4,845	1,159	5,692	1,576

Region C

Number of beds X_1:	76	75	84	13	40	69	125	13	32
Admissions Y:	2,648	2,757	2,881	402	1,600	1,646	4,825	370	987

Use X_2 and X_3 to define the regions, as follows:

Region	X_2	X_3
A	1	0
B	0	1
C	0	0

Assume model (20.3) is appropriate.

a. Obtain the estimated regression function. Interpret the meaning of b_1, b_2, and b_3. Give a point estimate of the mean number of admissions for 100-bed hospitals in region B.

b. Obtain the ANOVA table. Test if a regression relation holds, controlling the α risk at .01. State the alternatives, decision rule, and conclusion. Does your conclusion necessarily imply that mean admissions differ among the three regions for hospitals with a given number of beds? Comment.

c. Calculate $\sqrt{MSE}$. What does this number measure in this study?

d. Calculate the residuals and make appropriate residual plots to see if the fitted regression function is apt here. State your findings.

20.38 (Computer needed.) A student, after studying the Text discussion for the example in Table 20.5, stated that the multicollinearity in the regression of Y on X_1 and X_2 would largely disappear if the total weight in excess of 100 pounds per shaft for each shipment were used in place of the total weight of shafts in the shipment, i.e., if the variable $X_2' = X_2 - 100X_1$ were to replace X_2 in the regression model.

a. Calculate the coefficient of correlation between X_1 and X_2'. Do you agree with the student's statement? Explain.

b. For the new regression of Y on X_1 and X_2', obtain the regression coefficients denoted by b_1' and b_2', and also obtain their estimated standard deviations $s\{b_1'\}$ and $s\{b_2'\}$. How do these quantities compare with the corresponding values in Table 20.6 when Y is regressed on X_1 and X_2? Comment.

c. Obtain R^2 for the regression of Y on X_1 and X_2'. Is it the same, larger, or smaller than R^2 for the regression of Y on X_1 and X_2? (Recall from the Text that in the latter case, $R^2 = .996$.) Discuss.

21
Analysis of
Variance

In this chapter, we consider analysis of variance (ANOVA) models, which are useful for studying the statistical relation between a dependent variable and one or more independent variables. Although these models may be viewed as a special case of the general linear regression model, they allow a study of statistical relations from a different perspective and are widely used. The nature of the ANOVA approach was partly explained in the preceding chapters on regression; we now undertake a fuller study of this approach, with particular emphasis on ANOVA models.

21.1 BASIC CONCEPTS

We illustrate the nature of ANOVA models by two examples.

☐   Examples

1. A firm developing a new citrus-flavored soft drink conducted an experiment to study consumer preferences for the color of the drink. Four colors were under consideration: colorless, pink, orange, and lime green. Twenty test localities were selected which were similar in sales potential and representative of the target market for this product. Each color was then randomly assigned to five of these localities for test marketing. The dependent variable was number of cases sold during the test period per 1,000 population, and the independent variable was color. Other factors, such as price, flavor, degree of carbonation, sweetness, and calorie content, were held fixed in all the localities.

   An ANOVA model was employed to study the comparative effects of color on sales. Figure 21.1a portrays this model. For each color, there is a probability distribution of locality sales per 1,000 population. Each of the distributions is normal, with the same variability, but the means of the distributions may differ. In Figure 21.1a, the illustration shows that the green color has the largest mean, and thus tends to lead to highest sales. Since all probability distributions are normal with constant variance, the means μ_1, μ_2, μ_3, and μ_4 of the four probability distributions convey the information about the effects of color on sales. Thus, Figure 21.1a illustrates the situation where sales of

FIGURE 21.1   *Representations of single-factor ANOVA model*

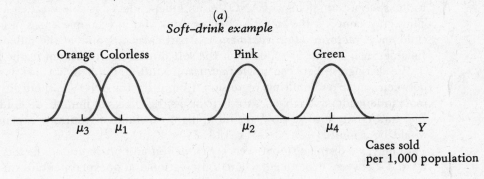

(a)
Soft–drink example

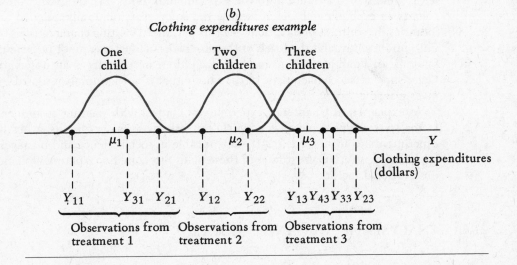

(b)
Clothing expenditures example

orange and colorless drinks tend to be substantially smaller than those of pink and green drinks.

2. Families with one, two, or three children were selected in a middle-income school district to study expenditures on children's clothing during the past year. Here the dependent or response variable is annual family expenditures for children's clothing (in dollars). The independent variable is number of children, and three levels of the independent variable are included in the study, namely one, two, or three children.

   Figure 21.1b portrays the ANOVA model for this case. Note that there is a probability distribution of clothing expenditures for each level of the independent variable, i.e., for each number of children, and that these probability distributions are normal with the same variance. Thus, μ_1, μ_2, and μ_3 convey information about the effect of number of children on annual children's clothing expenditures. As might be expected, the positions of the means μ_1, μ_2, and μ_3 in Figure 21.1b indicate that a family's expenditures for children's clothing increase with number of children. □

Factors and Factor Levels

In the special terminology of ANOVA, an independent variable is called a *factor*. Thus, in Example 1 the factor is color of drink and in Example 2 it is number of children. A *factor level* or *treatment* is a particular outcome of the independent variable, such as lime green color of the soft drink or two children in the family.

We distinguish between *single-factor* and *multifactor* ANOVA just as we did between simple and multiple regression. In multifactor ANOVA, there are two or more independent variables. A study to assess the effects of number of children (1, 2, 3) and family income (low, medium, high) on children's clothing expenditures would be a multifactor ANOVA study.

Finally, we distinguish between *experimental* and *observational* factors just as we did between experimental and observational (nonexperimental) studies in Chapter 1. Experimental factors are under the control of the investigator and the levels are assigned at random to the experimental units. In Example 1, soft-drink color is an experimental factor because the colors were randomly assigned to test localities. In contrast, observational factors pertain to existing characteristics of the units under study, and the levels of the observational factors cannot be assigned at random. In Example 2, the number of children in a family is an observational factor since it is a characteristic of the family that is not randomly assigned by the investigator.

As explained in Chapter 1, experimental control with random assignments of factor levels helps avoid biases which can otherwise arise. On the other hand, experimental control is frequently not possible in social science and management applications, so ANOVA studies in these fields often are nonexperimental and use observational factors.

21.2 ANALYSIS OF VARIANCE MODEL

Development of Model

The ANOVA model, which we illustrated graphically for two examples, has the following features:

1. There are r factor levels or treatments under study.
2. For each treatment, the probability distribution of Y is normal.
3. All probability distributions of Y have constant variance, denoted by σ^2.
4. The mean of the probability distribution of Y for the jth treatment ($j = 1, 2, \ldots, r$) is denoted by μ_j. The means μ_j may differ, reflecting the varying effects of the treatments.

In an ANOVA study, n_j observations are selected for the jth treatment. The ith observation ($i = 1, 2, \ldots, n_j$) for the jth treatment is denoted by Y_{ij}. This observation is assumed to represent a random selection from the probability distribution of Y for the jth treatment. The sampling process is illustrated in Figure 21.1b for the clothing expenditures example. Note that the sample sizes n_j need not be the same for all treatments.

As with regression models, we state the ANOVA model formally by indicating the components that make up an observation Y_{ij}:

(21.1)
$$Y_{ij} = \mu_j + \epsilon_{ij} \qquad i = 1, 2, \ldots, n_j; \; j = 1, 2, \ldots, r$$

where: Y_{ij} is the response in the ith observation for the jth treatment

μ_j's are parameters

ϵ_{ij}'s are independent $N(0, \sigma^2)$

Model (21.1) is called a *single-factor ANOVA model.*

Important Features of Model

1. Model (21.1) assumes that the observed value Y_{ij} is the sum of two components—a constant term μ_j and a random deviation or error term ϵ_{ij}. Thus, Y_{ij} is a random variable.
2. Since the error terms ϵ_{ij} are independent and normally distributed, so are the Y_{ij}'s.
3. Since $E\{\epsilon_{ij}\} = 0$, it follows that $E\{Y_{ij}\} = \mu_j$. The parameter μ_j thus is the mean response for the jth treatment.
4. The parameter σ^2 is the common variance of the distributions of Y because $\sigma^2\{Y_{ij}\} = \sigma^2\{\epsilon_{ij}\} = \sigma^2$ by (5.11a).
5. In view of these features, model (21.1) can be expressed alternatively as follows:

(21.2)
$$Y_{ij}\text{'s are independent } N(\mu_j, \sigma^2) \qquad i = 1, 2, \ldots, n_j; \; j = 1, 2, \ldots, r$$

6. Model (21.1) is a linear statistical model because it is linear in the parameters $\mu_1, \mu_2, \ldots, \mu_r$. Thus, no parameter μ_j appears as an exponent or is multiplied by another parameter.

Steps in Analysis

Since the probability distributions of Y corresponding to the different treatments are each normal with constant variability, differences in treatment effects are associated with differences between the treatment means μ_j. For this reason, ANOVA studies concentrate on making inferences about the μ_j's and usually proceed in two stages:

1. Determine whether or not the μ_j's are equal. If they are equal, we say there is no factor effect, and no further analysis is required.
2. If the μ_j's are not equal, study the nature of the treatment effects.

We now consider each of these stages of analysis in turn.

21.3 TEST FOR EQUALITY OF TREATMENT MEANS

☐ Illustration

We shall utilize the soft-drink example mentioned earlier as an illustration. Table 21.1 contains the sales data for the study. Each flavor was assigned to five localities for test marketing, and number of cases sold per 1,000 population during the study period was recorded for each locality. Thus, we see from Table 21.1 that in the first locality assigned to the colorless drink, 26.5 cases per 1,000 population were sold during the study period, or $Y_{11} = 26.5$. Similarly, $Y_{12} = 31.2$. ☐

Point Estimation of μ_j

It can be shown that the least squares estimator of the treatment mean μ_j is the sample mean of the observations for the jth treatment, denoted by $\bar{Y}_j$:

(21.3)
$$\bar{Y}_j = \frac{\sum\limits_{i=1}^{n_j} Y_{ij}}{n_j}$$

$\bar{Y}_j$ is an unbiased estimator of μ_j.

☐ Example

We see from Table 21.1 that the point estimate of μ_1, the mean sales for the colorless drink, is $\bar{Y}_1 = 27.32$ cases per 1,000 population. Note from Table 21.1 that mean sales were highest for the green color. ☐

TABLE 21.1 *Data for soft-drink experiment (cases sold per 1,000 population)*

Observation i	Treatment j				Total
	1 Colorless	2 Pink	3 Orange	4 Green	
1	26.5	31.2	27.9	30.8	
2	28.7	28.3	25.1	29.6	
3	25.1	30.8	28.5	32.4	
4	29.1	27.9	24.2	31.7	
5	27.2	29.6	26.5	32.8	
Total	136.6	147.8	132.2	157.3	573.9
Mean:	$\bar{Y}_1 = 27.32$	$\bar{Y}_2 = 29.56$	$\bar{Y}_3 = 26.44$	$\bar{Y}_4 = 31.46$	
Number of observations:	$n_1 = 5$	$n_2 = 5$	$n_3 = 5$	$n_4 = 5$	

Total number of observations: $n_T = 5 + 5 + 5 + 5 = 20$

Overall mean: $\bar{\bar{Y}} = \dfrac{573.9}{20} = 28.695$

Partitioning of Total Sum of Squares

Basic Concepts. As the term *analysis of variance* suggests and as was illustrated by the use of ANOVA in regression analysis, ANOVA procedures utilize partitions of the total sum of squares. We denote the overall mean for all the sample data by $\bar{\bar{Y}}$:

$$(21.4) \qquad \bar{\bar{Y}} = \frac{\Sigma \Sigma Y_{ij}}{n_T}$$

where:

$$(21.5) \qquad n_T = \Sigma n_j$$

Here, n_T represents the total number of sample observations in the entire study. Again we omit the indexes of summation when the nature of the summation is clear.

Recall from regression analysis that *SSTO*, the *total sum of squares*, is based on the deviations of the observations around their mean. In our current notation, these deviations are:

$$(21.6) \qquad Y_{ij} - \bar{\bar{Y}}$$

SSTO is the sum of the squares of these deviations:

$$(21.7) \qquad SSTO = \Sigma \Sigma (Y_{ij} - \bar{\bar{Y}})^2$$

☐ **Example**

In our soft-drink example, the total sample size is $n_T = 20$ (see Table 21.1) and the overall mean is:

$$\bar{\bar{Y}} = \frac{26.5 + 28.7 + \cdots + 32.8}{20} = \frac{573.9}{20} = 28.695$$

Hence, we find *SSTO* as follows:

$$SSTO = (26.5 - 28.695)^2 + (28.7 - 28.695)^2 + \cdots + (32.8 - 28.695)^2$$

$$= 115.92950$$

☐

The *error sum of squares SSE* measures the variability of the observations Y_{ij} when information about the treatments is utilized. Hence, the deviations are taken around the sample treatment means $\bar{Y}_j$, as follows:

$$(21.8) \qquad Y_{ij} - \bar{Y}_j$$

The error sum of squares is then simply the sum of squares of these deviations, first summed within a treatment and then summed over all treatments:

$$(21.9) \qquad SSE = \sum_j \left[\sum_i (Y_{ij} - \bar{Y}_j)^2 \right]$$

Note that if for each treatment all observations within the treatment are the same, we have:

$$\sum_i (Y_{ij} - \bar{Y}_j)^2 = 0$$

for all treatments, and $SSE = 0$. At the other extreme, if all treatment means $\bar{Y}_j$ are equal, so that $\bar{Y}_j \equiv \bar{\bar{Y}}$, a comparison of (21.7) and (21.9) shows that $SSE = SSTO$ then. Thus, the closer the treatment means $\bar{Y}_j$ are to each other, the closer will SSE be to $SSTO$.

☐   **Example**

For our soft-drink example, we obtain SSE by first calculating the sums of squares within each treatment. Thus, for treatment 1, we have:

$$\Sigma\, (Y_{i1} - \bar{Y}_1)^2 = (26.5 - 27.32)^2 + (28.7 - 27.32)^2 + \cdots + (27.2 - 27.32)^2$$

$$= 10.68800$$

The corresponding calculations for the other three treatments are:

$$(31.2 - 29.56)^2 + \cdots + (29.6 - 29.56)^2 = 8.57200$$

$$(27.9 - 26.44)^2 + \cdots + (26.5 - 26.44)^2 = 13.19200$$

$$(30.8 - 31.46)^2 + \cdots + (32.8 - 31.46)^2 = 6.63200$$

SSE is then the sum of these within-treatments sums of squares:

$$SSE = 10.68800 + 8.57200 + 13.19200 + 6.63200 = 39.08400 \qquad ☐$$

The difference between $SSTO$ and SSE is called the *treatment sum of squares* and is denoted by $SSTR$:

(21.10)
$$SSTR = SSTO - SSE$$

As we noted earlier, if the $\bar{Y}_j$'s are close to each other, SSE will be close to $SSTO$ and $SSTR$ will then be small. If, however, the sample treatment means $\bar{Y}_j$ are not close to each other, $SSTR$ will tend to be larger.

The treatment sum of squares is actually a sum of squares made up of deviations of the sample treatment means $\bar{Y}_j$ around the overall mean $\bar{\bar{Y}}$:

(21.11)
$$\bar{Y}_j - \bar{\bar{Y}}$$

and can be shown to equal:

(21.12)
$$SSTR = \Sigma\, n_j(\bar{Y}_j - \bar{\bar{Y}})^2$$

Thus, $SSTR$ is a sum of the squared deviations of $\bar{Y}_j$ around the overall mean $\bar{\bar{Y}}$, weighted by the number of observations n_j. Formula (21.12) makes it very clear that $SSTR$ reflects the variability of the sample treatment means $\bar{Y}_j$. The further apart they are, the larger will be $SSTR$. If all $\bar{Y}_j$'s are equal, then $SSTR = 0$.

☐   Example

In our soft-drink example, we obtain by (21.10):

$$SSTR = SSTO - SSE = 115.92950 - 39.08400 = 76.84550$$

Here, SSE is not close to $SSTO$, so $SSTR$ is comparatively large.
   The direct calculation of $SSTR$ by (21.12) is as follows:

$$SSTR = 5(27.32 - 28.695)^2 + 5(29.56 - 28.695)^2$$
$$+ 5(26.44 - 28.695)^2 + 5(31.46 - 28.695)^2$$
$$= 76.84550$$

This result must be the same as that obtained by subtraction.   ☐

Formal Statement of Partitioning.   As in regression analysis, we have now obtained a partitioning of the total sum of squares into two components:

(21.13)   For single-factor ANOVA, the decomposition of the total sum of squares into two additive components is:

(21.13a) $$SSTO \quad = \quad SSTR \quad + \quad SSE$$
(21.13b) $$\Sigma\,\Sigma\,(Y_{ij} - \bar{\bar{Y}})^2 = \Sigma\,n_j(\bar{Y}_j - \bar{\bar{Y}})^2 + \Sigma\,\Sigma\,(Y_{ij} - \bar{Y}_j)^2$$

☐   Example

We obtain the following decomposition for our soft-drink example:

$$115.92950 = 76.84550 + 39.08400$$
$$SSTO \quad = \quad SSTR \quad + \quad SSE$$ ☐

Calculational Formulas.   We have used the basic definitional formulas for $SSTO$, $SSTR$, and SSE to bring out the meaning of these different sums of squares. The following formulas are algebraically equivalent and tend to be more convenient for hand computation:

(21.14)
$$SSTO = \Sigma\,\Sigma\,Y_{ij}^2 - \frac{(\Sigma\,\Sigma\,Y_{ij})^2}{n_T}$$

(21.15)
$$SSTR = \sum_j \frac{\left(\sum_i Y_{ij}\right)^2}{n_j} - \frac{(\Sigma\,\Sigma\,Y_{ij})^2}{n_T}$$

(21.16)
$$SSE = SSTO - SSTR$$

Comments

1. $SSTR$ and SSE are sometimes called, respectively, the *between-treatments* and *within-treatments* sums of squares.
2. Formula (21.13b) can be derived by writing:

$$\Sigma\,\Sigma\,(Y_{ij} - \bar{\bar{Y}})^2 = \Sigma\,\Sigma\,[(Y_{ij} - \bar{Y}_j) + (\bar{Y}_j - \bar{\bar{Y}})]^2$$

Upon expanding the right side, we obtain:

$$\Sigma \Sigma (Y_{ij} - \bar{\bar{Y}})^2 = \Sigma \Sigma (Y_{ij} - \bar{Y}_j)^2 + \Sigma \Sigma (\bar{Y}_j - \bar{\bar{Y}})^2 + 2 \Sigma \Sigma (Y_{ij} - \bar{Y}_j)(\bar{Y}_j - \bar{\bar{Y}})$$

The second sum on the right can be written $\Sigma n_j(\bar{Y}_j - \bar{\bar{Y}})^2$ since the term $(\bar{Y}_j - \bar{\bar{Y}})^2$ is constant when summing over i and is picked up n_j times ($i = 1, 2, \ldots, n_j$). The last sum on the right can be shown to equal zero.

Partitioning of Degrees of Freedom

As in regression analysis, each sum of squares for the single-factor ANOVA model has associated with it a number of degrees of freedom. SSTO has $n_T - 1$ degrees of freedom associated with it. There are n_T deviations $Y_{ij} - \bar{\bar{Y}}$ but one constraint exists, namely $\Sigma \Sigma (Y_{ij} - \bar{\bar{Y}}) = 0$.

SSTR has $r - 1$ degrees of freedom associated with it. There are r deviations $\bar{Y}_j - \bar{\bar{Y}}$ but one constraint exists, namely $\Sigma n_j(\bar{Y}_j - \bar{\bar{Y}}) = 0$.

Finally, SSE has $n_T - r$ degrees of freedom associated with it. Each of the r levels of the factor contributes a component sum of squares $\underset{i}{\Sigma} (Y_{ij} - \bar{Y}_j)^2$ to SSE. The component sum of squares for the jth treatment is equivalent in form to a total sum of squares and therefore has $n_j - 1$ degrees of freedom. The degrees of freedom associated with SSE are then the sum of the degrees of freedom for each of the r components:

$$\sum_{j=1}^{r} (n_j - 1) = n_T - r$$

We see again that the degrees of freedom are additive:

(21.17)
$$\underbrace{n_T - 1}_{\substack{df \text{ for} \\ SSTO}} = \underbrace{r - 1}_{\substack{df \text{ for} \\ SSTR}} + \underbrace{n_T - r}_{\substack{df \text{ for} \\ SSE}}$$

☐ **Example**

For our soft-drink example, where $n_T = 20$ and $r = 4$, the degrees of freedom associated with the sums of squares are as follows:

$$
\begin{aligned}
SSTR: \quad & 4 - 1 = 3 \\
SSE: \quad & 20 - 4 = 16 \\
\hline
SSTO: \quad & 20 - 1 = 19
\end{aligned}
$$

☐

Mean Squares

Our interest is in the *treatment mean square*, denoted by MSTR, and in the *error mean square*, denoted by MSE. These are obtained in the usual way by dividing the respective sums of squares by their associated degrees of freedom:

(21.18)
$$MSTR = \frac{SSTR}{r - 1}$$

(21.19)
$$MSE = \frac{SSE}{n_T - r}$$

☐   Example

For our soft-drink example, we found earlier that $SSTR = 76.84550$ and $SSE = 39.08400$. Also, $r - 1 = 3$ and $n_T - r = 16$. Thus, we calculate:

$$MSTR = \frac{76.84550}{3} = 25.61517$$

$$MSE = \frac{39.08400}{16} = 2.44275$$ ☐

ANOVA Table

The ANOVA table for single-factor analysis of variance is shown in Table 21.2. Table 21.2a displays the general format of the table, and Table 21.2b shows the results for our soft-drink example.

F Test for Equality of Treatment Means

We are now in a position to conduct an F test for equality of treatment means. The alternative conclusions are:

(21.20)
$$H_0: \mu_1 = \mu_2 = \cdots = \mu_r$$
$$H_1: \text{Not all } \mu_j\text{'s are equal}$$

H_0 implies that all distributions of Y have the same mean, and hence for model (21.1) are identical, while H_1 implies they do not all have the same mean. Note

TABLE 21.2   *ANOVA table for single-factor study and results for soft-drink example*

(a) General format

Source of Variation	SS	df	MS
Treatments	$SSTR = \Sigma\, n_j (\overline{Y}_j - \overline{\overline{Y}})^2$	$r - 1$	$MSTR = \dfrac{SSTR}{r - 1}$
Error	$SSE = \Sigma \Sigma\, (Y_{ij} - \overline{Y}_j)^2$	$n_T - r$	$MSE = \dfrac{SSE}{n_T - r}$
Total	$SSTO = \Sigma \Sigma\, (Y_{ij} - \overline{\overline{Y}})^2$	$n_T - 1$	

(b) Soft-drink example

Source of Variation	SS	df	MS
Treatments	76.84550	3	25.61517
Error	39.08400	16	2.44275
Total	115.92950	19	

that Figure 21.1 portrays two situations where H_1 is true—one for the soft-drink example and another for the clothing expenditures example.

The test statistic is analogous to that for regression analysis:

(21.21)
$$F^* = \frac{MSTR}{MSE}$$

Large values of F^* lead to conclusion H_1 because $MSTR$ tends to be larger than MSE when H_1 holds, whereas the two mean squares tend to be of the same magnitude when H_0 holds. This can be seen from the expected values of the mean squares for model (21.1), which are as follows:

(21.22)
$$E\{MSE\} = \sigma^2$$

(21.23)
$$E\{MSTR\} = \sigma^2 + \frac{\Sigma\, n_j(\mu_j - \mu)^2}{r - 1}$$

where: $\mu = \dfrac{\Sigma\, n_j\mu_j}{n_T}$

Note that MSE is always an unbiased estimator of the error variance σ^2. Also note that $E\{MSTR\} = \sigma^2$ when all μ_j's are equal because then $\mu_j \equiv \mu$, where μ is a weighted average of the μ_j's. On the other hand, when the μ_j's are not all equal, the second term on the right in (21.23) is positive and $E\{MSTR\} > E\{MSE\} = \sigma^2$.

The development of the test is illustrated in Figure 21.2. The test is an upper-tail one, as shown in Figure 21.2a. It can be shown that F^* follows the F distribution when H_0 holds:

(21.24) For the single-factor ANOVA model (21.1), when $\mu_1 = \mu_2 = \cdots = \mu_r$:

$$F^* = F(r - 1, n_T - r)$$

where: F^* is given by (21.21)

Figure 21.2b shows this sampling distribution of F^* when H_0 holds, the case for which the α risk is to be controlled.

Finally, Figure 21.2c shows the placement of the action limit so that the upper-tail area equals α, and Figure 21.2d presents the decision rule.

We now summarize the entire decision procedure:

(21.25) When the alternatives are:

$$H_0: \mu_1 = \mu_2 = \cdots = \mu_r$$

$$H_1: \text{Not all } \mu_j\text{'s are equal}$$

and the single-factor ANOVA model (21.1) is applicable, the appropriate decision rule to control the α risk is:

$$\text{If } F^* \leq F(1 - \alpha;\, r - 1, n_T - r), \text{ conclude } H_0$$

$$\text{If } F^* > F(1 - \alpha;\, r - 1, n_T - r), \text{ conclude } H_1$$

where: F^* is given by (21.21)

FIGURE 21.2 *Construction of decision rule for test of equality of treatment means in single-factor ANOVA study*

(a) *Appropriate type of decision rule*

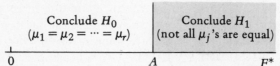

Large values of F^* imply H_1 is correct.

(b) *Sampling distribution of F^* when H_0 holds*

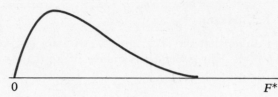

Distribution of F^* when H_0 holds is $F(r-1, n_T-r)$.

(c) *Placement of action limit*

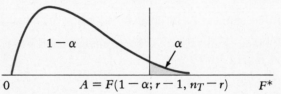

A is placed so upper–tail area equals α.

(d) *Decision rule*

If $F^* \leq F(1-\alpha; r-1, n_T-r)$, conclude H_0
If $F^* > F(1-\alpha; r-1, n_T-r)$, conclude H_1

Conclude appropriate alternative based on observed F^*.

Example

For our soft-drink example, the alternative conclusions are:

$$H_0: \mu_1 = \mu_2 = \mu_3 = \mu_4$$

$$H_1: \text{Not all } \mu_j\text{'s are equal}$$

Suppose it is desired to control the α risk at .05. Since $\alpha = .05$, $r - 1 = 3$, and $n_T - r = 16$, we require $F(.95; 3, 16) = 3.24$. The decision rule is:

$$\text{If } F^* \leq 3.24, \text{ conclude } H_0$$

$$\text{If } F^* > 3.24, \text{ conclude } H_1$$

From Table 21.2b, we calculate $F^* = 25.61517/2.44275 = 10.49 > 3.24$. Hence, we conclude H_1—that mean sales are not the same for the different colors. The next step in the analysis is then to study the nature of the color effects in some detail. ☐

Comment

In Chapter 13, we discussed the comparison of two population means when the populations are normal with equal variance and the samples are independent. The test whether or not $\mu_1 = \mu_2$ based on statistic (13.6) is a special case of the ANOVA F test.

21.4 ANALYSIS OF TREATMENT EFFECTS

We now take up some procedures for investigating treatment effects when it is concluded that the treatment means μ_j are not all equal. Important avenues of investigation include interval estimation of the mean response for a given treatment and comparison of mean responses for different treatments. We consider each of these in turn.

Interval Estimation of μ_j

We noted earlier that the least squares estimator of the mean response μ_j is the sample mean $\bar{Y}_j$. The estimated variance of $\bar{Y}_j$, denoted by $s^2\{\bar{Y}_j\}$, is given by:

(21.26)
$$s^2\{\bar{Y}_j\} = \frac{MSE}{n_j}$$

A two-sided confidence interval for μ_j takes the usual form:

(21.27) The confidence limits for μ_j with confidence coefficient $1 - \alpha$ for the single-factor ANOVA model (21.1) are:

$$\bar{Y}_j \pm ts\{\bar{Y}_j\}$$

where: $t = t(1 - \alpha/2; n_T - r)$
$s\{\bar{Y}_j\}$ is given by (21.26)

The t multiple involves $n_T - r$ degrees of freedom because these are the degrees of freedom associated with MSE.

☐ Example

It is desired to estimate mean sales for the colorless version of the drink (treatment 1) with a 90 percent confidence interval. We know from Tables 21.1 and 21.2 that $\bar{Y}_1 = 27.32$, $MSE = 2.44275$, $n_T - r = 16$, and $n_1 = 5$. We calculate by (21.26):

$$s^2\{\bar{Y}_1\} = \frac{2.44275}{5} = .48855$$

Hence, $s\{\bar{Y}_1\} = \sqrt{.48855} = .69896$. We require $t(.95; 16) = 1.746$. The confidence limits therefore are $27.32 \pm 1.746(.69896)$ and the confidence interval is:

$$26.1 \leq \mu_1 \leq 28.5$$

Thus, it can be reported with 90 percent confidence that mean sales for the colorless version are between 26.1 and 28.5 cases per 1,000 population. ☐

Comparison of Two Treatment Means

Frequently, we wish to compare the mean responses for two treatments. Let the two treatments of interest be denoted by j and j'. The difference of interest then is:

(21.28)
$$\mu_j - \mu_{j'}$$

Such a difference between a pair of treatment means is called a *pairwise comparison*. An unbiased point estimator of $\mu_j - \mu_{j'}$ is:

$$(21.29) \qquad \bar{Y}_j - \bar{Y}_{j'}$$

Because all error terms are independent, $\bar{Y}_j$ and $\bar{Y}_{j'}$ are independent and, by (5.14b), the variance of $\bar{Y}_j - \bar{Y}_{j'}$ is simply the sum of the variances of $\bar{Y}_j$ and $\bar{Y}_{j'}$. Thus, the estimated variance of $\bar{Y}_j - \bar{Y}_{j'}$, which is denoted by $s^2\{\bar{Y}_j - \bar{Y}_{j'}\}$, is:

$$(21.30) \qquad s^2\{\bar{Y}_j - \bar{Y}_{j'}\} = \frac{MSE}{n_j} + \frac{MSE}{n_{j'}}$$

A two-sided confidence interval for $\mu_j - \mu_{j'}$ is of the usual form:

(21.31) The confidence limits for $\mu_j - \mu_{j'}$ with confidence coefficient $1 - \alpha$ for the single-factor ANOVA model (21.1) are:

$$(\bar{Y}_j - \bar{Y}_{j'}) \pm t s\{\bar{Y}_j - \bar{Y}_{j'}\}$$

where: $t = t(1 - \alpha/2; n_T - r)$
$s\{\bar{Y}_j - \bar{Y}_{j'}\}$ is given by (21.30)

☐ Example

In our soft-drink example, the firm's advertising agency had initially recommended, on the basis of interviews with a panel of consumers, that either lime green or pink be used in launching the product nationally. Hence, a major purpose of the experiment was to compare the effects of these two colors in an actual market setting.

The pairwise comparison of interest is $\mu_4 - \mu_2$, and it is to be estimated with a 90 percent confidence interval. The point estimator is $\bar{Y}_4 - \bar{Y}_2$. From Table 21.1, we see that $\bar{Y}_4 - \bar{Y}_2 = 31.46 - 29.56 = 1.90$. We obtain $t(.95; 16) = 1.746$, and require $s\{\bar{Y}_4 - \bar{Y}_2\}$. Since $MSE = 2.44275$, $n_4 = 5$, and $n_2 = 5$, we calculate by (21.30):

$$s^2\{\bar{Y}_4 - \bar{Y}_2\} = \frac{2.44275}{5} + \frac{2.44275}{5} = .97710$$

and hence $s\{\bar{Y}_4 - \bar{Y}_2\} = \sqrt{.97710} = .98848$. The confidence limits therefore are $1.90 \pm 1.746 (.98848)$ and the desired confidence interval is:

$$.2 \leq \mu_4 - \mu_2 \leq 3.6$$

Thus, it is estimated, with 90 percent confidence, that mean sales with green color are somewhere between .2 and 3.6 cases per 1,000 population greater than those with pink color. ☐

Simultaneous Comparisons

In studies with three or more treatments, interest often centers on several pairwise comparisons. In the soft-drink example, for instance, the sales manager wishes to compare mean sales for both green and pink colors with those for the colorless mixture. The set of comparisons here is:

Pairwise Comparison	Colors	Treatment Mean Difference
1	Green − colorless	$\mu_4 - \mu_1$
2	Pink − colorless	$\mu_2 - \mu_1$

Since the results of both comparisons will affect management's final marketing decision, the sales manager would like to have known assurance that both confidence intervals in the set will be correct simultaneously. This situation calls for *simultaneous confidence intervals* where the *joint confidence coefficient* indicates the probability that the procedure will lead to all confidence intervals in the set being simultaneously correct. There are a variety of procedures available for obtaining simultaneous confidence intervals. We shall discuss a simple procedure that is highly useful when the number of confidence intervals in the set is not too large.

Simultaneous Confidence Intervals. If a $1 - \alpha$ confidence interval is constructed for each of, say, m pairwise comparisons using the usual procedure of the preceding section, the level of confidence that *all* m confidence intervals are simultaneously correct will be less than $1 - \alpha$. The reason is that each confidence statement involves a risk α of being incorrect, and these risks compound when several such statements are made. Hence, if one desires a joint confidence coefficient $1 - \alpha$ that all comparisons in the set are simultaneously correct, each individual confidence interval will need to be constructed with a confidence coefficient that involves a risk smaller than α.

The following theorem states how much smaller than α the risk of an incorrect confidence interval must be so that the overall risk for all m confidence intervals in the set does not exceed α:

(21.32) If each of m confidence intervals is constructed with confidence coefficient $1 - (\alpha/m)$, the joint confidence coefficient for the set of m confidence intervals is at least $1 - \alpha$.

In other words, if the confidence interval for each pairwise comparison is constructed with confidence coefficient $1 - (\alpha/m)$, the probability that all m confidence intervals are simultaneously correct will be at least $1 - \alpha$.

Theorem (21.32) thus provides a simple procedure for controlling the joint confidence level at $1 - \alpha$ when m pairwise comparisons are to be made:

(21.33) The confidence limits for m simultaneous pairwise comparisons $\mu_j - \mu_{j'}$ with joint confidence coefficient of at least $1 - \alpha$ for the single-factor ANOVA model (21.1) are:

$$(\bar{Y}_j - \bar{Y}_{j'}) \pm ts\{\bar{Y}_j - \bar{Y}_{j'}\}$$

where: $t = t(1 - \alpha/2m; n_T - r)$
   $s\{\bar{Y}_j - \bar{Y}_{j'}\}$ is given by (21.30)

Note that the percentile of t that must be used is $100(1 - \alpha/2m)$. By placing risk

$\alpha/2m$ in each tail, the combined risk in both tails is α/m and the confidence coefficient is $1 - (\alpha/m)$ for each confidence interval.

☐ **Example**

For the two pairwise comparisons given earlier, it is desired to control the joint confidence coefficient at 90 percent. Hence, for $\alpha = .10$ and $m = 2$, each pairwise comparison requires $t(.975; 16) = 2.120$ because $1 - (\alpha/2m) = .975$ and $n_T - r = 16$. The estimated standard deviation $s\{\bar{Y}_4 - \bar{Y}_2\}$ was calculated earlier from (21.30) to be .98848; it is applicable for all pairwise comparisons in this study because each treatment has the same number of observations. The remaining calculational details are:

Colors	$\mu_j - \mu_{j'}$	$\bar{Y}_j - \bar{Y}_{j'}$	$\pm ts\{\bar{Y}_j - \bar{Y}_{j'}\}$
Green — colorless	$\mu_4 - \mu_1$	$31.46 - 27.32 = 4.14$	$\pm 2.120(.98848) = \pm 2.096$
Pink — colorless	$\mu_2 - \mu_1$	$29.56 - 27.32 = 2.24$	$\pm 2.120(.98848) = \pm 2.096$

Hence, the two confidence intervals are $2.04 \le \mu_4 - \mu_1 \le 6.24$ and $.14 \le \mu_2 - \mu_1 \le 4.34$. The joint confidence coefficient for the two intervals is at least .90. Hence, with confidence of at least 90 percent, it can be concluded that both the green and pink drinks have greater mean sales than the colorless drink. ☐

21.5 COMPUTER OUTPUT

Computer program packages for ANOVA are widely available, and a number of pocket calculators also can perform ANOVA calculations. Figure 21.3 shows output for our soft-drink example from a computer package.

FIGURE 21.3 *Computer output for soft-drink example*

```
SINGLE FACTOR ANOVA

1. DATA
        26.5          31.2          27.9          30.8
        28.7          28.3          25.1          29.6
        25.1          30.8          28.5          32.4
        29.1          27.9          24.2          31.7
        27.2          29.6          26.5          32.8

2. TREATMENT        N              MEAN          STD.DEV.
        1           5              27.32          1.6346
        2           5              29.56          1.4639
        3           5              26.44          1.816
        4           5              31.46          1.2876
        0           20             28.695         2.4701

        TREATMENT '0' IS ENTIRE DATA SET

3. ANOVA TABLE
        SOURCE          SS           DF          MS
        TREATMENTS      76.8455       3          25.61517
        ERROR           39.0840      16           2.44275
        TOTAL          115.9295      19
```

Block 1 contains the data input. The observations here are entered in columns, with each column representing a treatment.

Block 2 shows, for each treatment, the sample size n_j, the sample mean $\overline{Y}_j$, and the sample standard deviation. It also shows, for treatment 0, the corresponding statistics for the entire sample.

Block 3 contains the ANOVA table.

21.6   RESIDUAL ANALYSIS

Residual

The aptness of ANOVA model (21.1) needs to be evaluated, just as evaluation of the aptness of regression models is required. Residual plots again are most helpful. The residual e_{ij} for the ANOVA model (21.1) is defined as follows:

$$(21.34) \qquad e_{ij} = Y_{ij} - \overline{Y}_j$$

Thus, e_{ij} represents the deviation of the observation Y_{ij} from the sample mean for the jth treatment. Table 21.3 contains the residuals for our soft-drink example, calculated from the data in Table 21.1. For example, $e_{11} = Y_{11} - \overline{Y}_1 = 26.5 - 27.32 = -.82$.

Comment

Note from (21.9) and (21.34) that $SSE = \Sigma \Sigma e_{ij}^2$. Also note from Table 21.3 that the residuals sum to zero for each treatment:

$$(21.35) \qquad \sum_i e_{ij} = 0 \qquad j = 1, 2, \ldots, r$$

Since there are r treatments, the n_T residuals are subject to r constraints of the form (21.35). This is consistent with our earlier discussion where we noted that SSE is associated with $n_T - r$ degrees of freedom.

Residual Plots

As in regression, the residuals e_{ij} should reflect the properties ascribed to the error terms ϵ_{ij} if the model is apt. ANOVA model (21.1), as we know, requires the error

TABLE 21.3 *Residuals for soft-drink experiment*

Observation i	Treatment j 1	2	3	4
1	−.82	1.64	1.46	−.66
2	1.38	−1.26	−1.34	−1.86
3	−2.22	1.24	2.06	.94
4	1.78	−1.66	−2.24	.24
5	−.12	.04	.06	1.34
Total	0.00	0.00	0.00	0.00

terms ϵ_{ij} to be independent and normally distributed with constant variance σ^2. The construction and use of residual plots for ANOVA proceeds in a manner similar to that for regression. We will illustrate the use of residual plots for two departures from the ANOVA model (21.1).

Lack of Independence. In ANOVA studies, the observations frequently are taken in a time sequence. The residuals should then be plotted against time order and checked for independence over time.

☐ Example

Figure 21.4 contains residual plots for a single-factor ANOVA study based on model (21.1), in which respondents were asked to estimate the price of a rug. The three treatments were different pile lengths of the rug. The residuals are plotted against observation number in Figure 21.4 and the plots show tracklike patterns instead of random scatter. Subsequent investigation found that many respondents had been able to overhear the price estimate of the preceding respondent for the same rug.   ☐

Lack of independence in the error terms ϵ_{ij} can have serious effects on inferences in ANOVA, causing the true risks of errors and confidence coefficients to differ materially from the stated ones. Thus, it is important to prevent this difficulty whenever possible.

Nonconstancy of Error Variances. A convenient graphic check for constancy of the error variance from treatment to treatment can be made by comparing the scatter in the residual plots for the several treatments.

☐ Example

Figure 21.5 contains the residuals taken from a study of productivity of salespeople under four different compensation plans. The plots suggest that the error variances are unequal, those for plans 1 and 4 being greater than those for plans 2 and 3.   ☐

FIGURE 21.4   *Residual plots suggesting lack of independence*

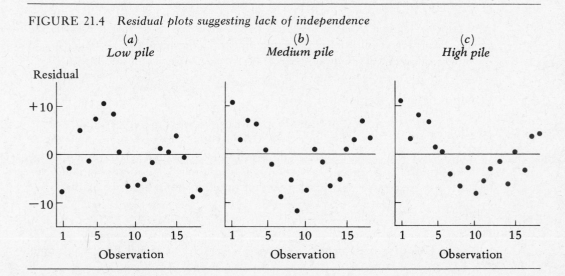

FIGURE 21.5 *Residual plots suggesting unequal treatment variances*

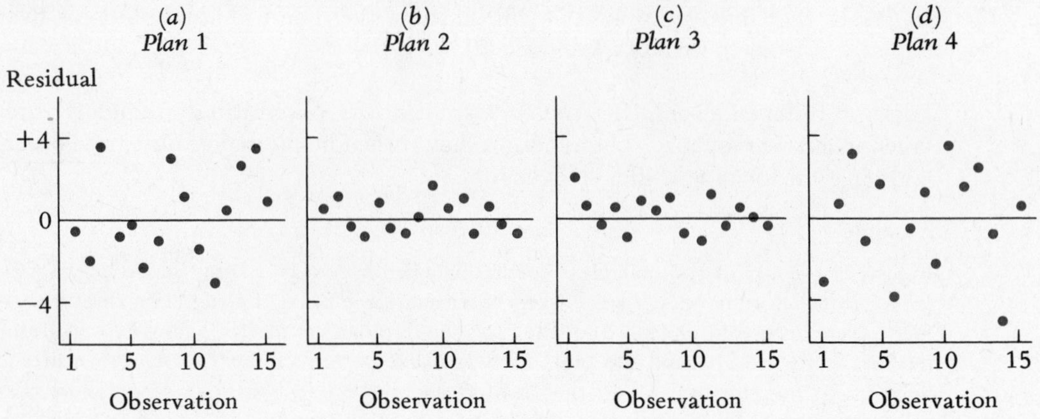

Unequal error variances can also arise as a result of time-related effects, as when random measurement errors become smaller in the course of an experiment because the laboratory technicians become increasingly proficient with the instruments.

Inferences for model (21.1) involving the F test are not seriously affected by unequal error variances if the sample sizes n_j are equal. However, pairwise comparisons can be seriously affected so that the actual and specified confidence coefficients may differ markedly. Frequently, it is possible to find a mathematical transformation of Y that will produce approximately equal error variances.

21.7 OPTIONAL TOPIC—CORRESPONDENCE BETWEEN ANOVA AND REGRESSION

As was noted earlier, ANOVA models may be viewed as a special case of the general linear regression model. We now illustrate the regression formulation of the single-factor ANOVA model.

Development of Regression Model

In Chapter 20, we saw how indicator variables can be used in a regression model to represent a qualitative variable. Since the factor of interest in a single-factor ANOVA study (e.g., color of soft drink) is equivalent to a qualitative variable, with each factor level (e.g., green) corresponding to one outcome of the variable, the factor levels are represented in a regression function by indicator variables.

☐   Example

For our soft-drink example, the factor is color of soft drink, and the factor levels are the four colors: colorless, pink, orange, and green. The ANOVA model for the soft-drink example is:

$$Y_{ij} = \mu_j + \epsilon_{ij} \qquad j = 1, 2, 3, 4$$

To develop the equivalent regression model, we require three indicator variables, which we shall define as follows:

(21.36)
$$X_1 = \begin{cases} 1 \text{ if color is pink} \\ 0 \text{ otherwise} \end{cases}$$

$$X_2 = \begin{cases} 1 \text{ if color is orange} \\ 0 \text{ otherwise} \end{cases}$$

$$X_3 = \begin{cases} 1 \text{ if color is green} \\ 0 \text{ otherwise} \end{cases}$$

The equivalent regression model then is:

(21.37)
$$Y_i = \beta_0 + \beta_1 X_{i1} + \beta_2 X_{i2} + \beta_3 X_{i3} + \epsilon_i \qquad i = 1, 2, \ldots, n_T$$

where the index i now ranges over all $n_T = 20$ observations in the experiment.

Table 21.4 contains the soft-drink data with the indicator variables coded in accord with (21.36). For example, for each observation for treatment 1 (colorless), the associated values of the independent variables are $X_1 = 0$, $X_2 = 0$, and $X_3 = 0$. Similarly, for each observation for treatment 2 (pink), the values of the independent variables are $X_1 = 1$, $X_2 = 0$, and $X_3 = 0$. ☐

TABLE 21.4 *Data for regression analysis in soft-drink example*

Observation i	Color	Indicator Variables			Y_i
		X_{i1}	X_{i2}	X_{i3}	
1	Colorless	0	0	0	26.5
2	Colorless	0	0	0	28.7
3	Colorless	0	0	0	25.1
4	Colorless	0	0	0	29.1
5	Colorless	0	0	0	27.2
6	Pink	1	0	0	31.2
7	Pink	1	0	0	28.3
8	Pink	1	0	0	30.8
9	Pink	1	0	0	27.9
10	Pink	1	0	0	29.6
11	Orange	0	1	0	27.9
12	Orange	0	1	0	25.1
13	Orange	0	1	0	28.5
14	Orange	0	1	0	24.2
15	Orange	0	1	0	26.5
16	Green	0	0	1	30.8
17	Green	0	0	1	29.6
18	Green	0	0	1	32.4
19	Green	0	0	1	31.7
20	Green	0	0	1	32.8

Correspondence of Parameters

We now develop the correspondence between the ANOVA parameters μ_j and the regression parameters β_k for the soft-drink example which, from (21.37), has the response function:

$$(21.38) \qquad E\{Y\} = \beta_0 + \beta_1 X_1 + \beta_2 X_2 + \beta_3 X_3$$

For an observation from treatment 1 (colorless), for which $X_1 = X_2 = X_3 = 0$, the response function simplifies to:

$$E\{Y\} = \beta_0 + \beta_1(0) + \beta_2(0) + \beta_3(0)$$
$$= \beta_0$$

But, according to the ANOVA model, the mean for treatment 1 is μ_1, so we have $\beta_0 = \mu_1$.

For an observation from treatment 2 (pink), for which $X_1 = 1$, $X_2 = 0$, and $X_3 = 0$, the response function (21.38) reduces to:

$$E\{Y\} = \beta_0 + \beta_1(1) + \beta_2(0) + \beta_3(0)$$
$$= \beta_0 + \beta_1$$

Since the mean for treatment 2 in the ANOVA model is μ_2, it follows that $\beta_0 + \beta_1 = \mu_2$. Moreover, we have just shown that $\beta_0 = \mu_1$. Hence, $\beta_1 = \mu_2 - \mu_1$.

Continuing in this fashion, we establish the following correspondences between the parameters of the single-factor ANOVA model and the equivalent regression model:

$$(21.39a) \qquad \beta_0 = \mu_1 \qquad\qquad \beta_2 = \mu_3 - \mu_1$$
$$\beta_1 = \mu_2 - \mu_1 \qquad\qquad \beta_3 = \mu_4 - \mu_1$$

$$(21.39b) \qquad \mu_1 = \beta_0 \qquad\qquad \mu_3 = \beta_0 + \beta_2$$
$$\mu_2 = \beta_0 + \beta_1 \qquad\qquad \mu_4 = \beta_0 + \beta_3$$

Note in (21.39a) that β_1, β_2, and β_3 represent the *differential effects* of treatments 2, 3, and 4, respectively, compared with treatment 1 (the treatment for which all indicator variables are zero).

Inferences with Regression Model

Estimates and tests for the treatment means in the ANOVA model can be carried out equivalently in terms of inferences about the regression coefficients. We demonstrate two of these equivalent procedures for the soft-drink example.

☐   **Example 1: Estimating Treatment Means**

Key results of fitting regression model (21.37) to the soft-drink data in Table 21.4 by a computer package are shown in Table 21.5. Observe that b_0 is identical to the sample

TABLE 21.5 *Key results in regression analysis of soft-drink data using indicator variables*

Estimated regression coefficients

$$b_0 = 27.32 \qquad b_2 = -.88$$
$$b_1 = 2.24 \qquad b_3 = 4.14$$

ANOVA table

Source of Variation	SS	df	MS
Regression	76.84550	3	25.61517
Error	39.08400	16	2.44275
Total	115.92950	19	

mean for treatment 1 as given in Table 21.1; i.e., $\bar{Y}_1 = b_0 = 27.32$. Likewise, observe that:

$$\bar{Y}_2 = b_0 + b_1 = 27.32 + 2.24 = 29.56$$

$$\bar{Y}_3 = b_0 + b_2 = 27.32 - .88 = 26.44$$

$$\bar{Y}_4 = b_0 + b_3 = 27.32 + 4.14 = 31.46$$

These correspondences follow from the relations between the μ_j ($j = 1, 2, 3, 4$) and β_k ($k = 0, 1, 2, 3$) given in (21.39b). □

Example 2: Testing the Equality of Treatment Means

The ANOVA test for equality of means in our soft-drink example involves the alternatives:

$$H_0: \mu_1 = \mu_2 = \mu_3 = \mu_4$$

$$H_1: \text{Not all } \mu_j\text{'s are equal}$$

We see from (21.39a) that when $\mu_1 = \mu_2 = \mu_3 = \mu_4$, the regression parameters β_1, β_2, and β_3 equal zero. Hence, the equivalent test in the regression framework is:

$$H_0: \beta_1 = \beta_2 = \beta_3 = 0$$

$$H_1: \text{Not all } \beta_k = 0 \qquad k = 1, 2, 3$$

These are precisely the alternatives for testing the existence of a regression relation. The appropriate test statistic, as given in (20.14), is $F^* = MSR/MSE$, and the test procedure is given in (20.15).

   Referring to the regression output in Table 21.5, we see that the entries in the regression ANOVA table are identical to those in Table 21.2b for ANOVA model (21.1), with SSR corresponding to SSTR. Thus, the F test for the equality of treatment means is identical to the F test for the existence of a regression relation when indicator variables are used in the regression model. □

Comment

When the factor under study is qualitative, such as color of soft drink, the ANOVA model and the regression model with indicator variables are equivalent models, and there is no choice of any other type of regression model. When the factor is quantitative, however, the analyst can choose whether to use: (1) an ANOVA model or the equivalent regression model with indicator variables, or (2) a regression model with the factor treated as a continuous variable. Consider a study of the effect of size of crew (4, 12, 20 persons) on volume of output. If interest centers solely on the three crew sizes under study, the factor can be treated as a qualitative one by means of the ANOVA model or the equivalent regression model with indicator variables, requiring no specification of a functional relation between crew size and volume of output. On the other hand, if the factor is treated as a quantitative variable in an ordinary regression model, a specification of the nature of the relation between crew size and output volume (e.g., linear, quadratic) must be made. Making this specification is more restrictive than treating the independent variable as qualitative, but it does have the advantage of permitting inferences about crew sizes other than those in the study (e.g., crews of 5 or 17 persons).

PROBLEMS

21.1 In each of the following single-factor ANOVA studies, identify: (1) the factor, stating whether it is experimental or observational; (2) the factor levels; (3) the dependent variable.

    a. An automobile service chain operates three kinds of service centers—namely, tire centers, brake system centers, and exhaust system centers. An analyst wishes to investigate quarterly profit rates for the different kinds of centers.

    b. Three interviewers have been hired for a household survey. Households to be surveyed are randomly assigned to the interviewers in order to compare the effects of the interviewers on responses about household income.

21.2 In each of the following single-factor ANOVA studies, identify: (1) the factor, stating whether it is experimental or observational; (2) the factor levels; (3) the dependent variable.

    a. In a study of heat resistance of a certain type of wood paint, identical painted wood panels are exposed to one of three different temperatures in laboratory test rooms. Temperature levels are randomly assigned to the panels and resistance is measured by the time until heat blistering occurs.

    b. Financial reports of companies from five different extractive industries are examined to compare the effects of a depletion tax allowance on reported annual earnings in these industries.

21.3 Each of 12 persons in a consumer panel was shown one of three versions of a television documentary and was asked to terminate viewing it when interest waned. Four persons were assigned at random to each of the three versions. Assume that model (21.1) is applicable, and that the parameter values (in minutes) are: $\mu_1 = 8.5$, $\mu_2 = 5.3$, $\mu_3 = 7.6$, $\sigma = 1.0$.

    a. Portray the model for this case graphically, as in Figure 21.1a.

    b. Explain the meaning of the following symbols: (1) μ_1, (2) Y_{32}, (3) ϵ_{41}, (4) σ.

    c. Will Y_{23} necessarily be smaller than Y_{41}? Are Y_{23} and Y_{41} statistically independent? Explain.

21.4 In a study of bonus pay earned under four bonus plans, five salespeople were assigned at random to each of these plans. Assume that model (21.1) is applicable, and that the parameter values (in dollars) are: $\mu_1 = 370$, $\mu_2 = 850$, $\mu_3 = 450$, $\mu_4 = 590$, $\sigma = 40$.

    a. Portray the model for this case graphically, as in Figure 21.1a.

    b. Explain the meaning of the following symbols: (1) μ_3, (2) Y_{14}, (3) ϵ_{32}, (4) σ.

c. If $Y_{52} = 800$, what is ϵ_{52}?

*21.5 For each of the following cases, state the degrees of freedom associated with: (1) *SSTR*, (2) *SSE*, (3) *SSTO*:

a. Four treatments, seven observations for each treatment.

b. Four treatments, 20 observations for each treatment.

c. Five treatments, three observations for each of treatments 1, 2, 3, and 4, and four observations for treatment 5.

21.6 For each of the following cases, state the degrees of freedom associated with: (1) *SSTR*, (2) *SSE*, (3) *SSTO*:

a. Three treatments, four observations for each treatment.

b. Five treatments, 16 observations for each treatment.

c. Four treatments, three observations for each of treatments 1 and 2, and eight observations for each of treatments 3 and 4.

*21.7 Fifteen students enrolled in a mathematics course were randomly divided into three groups of 5 students each. Each group was then randomly assigned to one of three instructional modes, augmenting the traditional course materials: (1) programmed text, (2) video tapes, (3) conversational computer programs. At the end of the course, each student was given the same achievement test. The test scores follow.

Observation i	Treatment i		
	1 Programmed Text	2 Video Tapes	3 Computer Programs
1	86	80	88
2	82	69	80
3	94	78	75
4	77	77	84
5	86	86	73

Assume that model (21.1) is applicable.

a. State the values of: (1) n_T, (2) n_1, (3) Y_{32}, (4) $\bar{Y}_2$, (5) $\bar{\bar{Y}}$.

b. For which instructional mode is the sample mean $\bar{Y}_j$ largest? Is the treatment mean μ_j for this instructional mode necessarily the largest of the three treatment means? Explain.

c. Obtain the ANOVA table.

21.8 A product rating organization tested battery life for four makes of multiband portable radios. Six sets of each make were purchased "off the shelf" from local retail outlets, and fresh batteries were inserted on a random basis. The observations on battery playing hours at high volume follow.

Observation i	Treatment i			
	1	2	3	4
1	5.5	4.7	6.1	4.5
2	5.0	3.9	5.7	5.1
3	5.2	4.3	5.0	4.3
4	5.3	4.5	5.3	4.1
5	4.8	4.1	5.2	4.5
6	4.8	4.3	6.3	5.1

Assume that model (21.1) is applicable.

a. State the values of: (1) n_T, (2) n_3, (3) Y_{24}, (4) $\bar{Y}_3$, (5) $\bar{\bar{Y}}$.

b. Use (21.7), (21.9), and (21.12) to compute $SSTO$, SSE, and $SSTR$, respectively. Are your results consistent with identity (21.13a)?

c. Obtain the ANOVA table.

*21.9 Refer to Problem 21.7. It is desired to determine whether or not mean achievement (as measured by the test) is the same for the three instructional modes.

a. State the alternatives for the test of equality of treatment means here. Specify the distribution of F^* if H_0 is correct.

b. Test for the equality of the treatment means, controlling the α risk at .05. State your decision rule and conclusion.

c. Given your conclusion in b, should the next step involve an examination of the differences among the individual treatment means? Comment.

21.10 Refer to Problem 21.8.

a. Test for the equality of the treatment means, controlling the α risk at .01. State the alternatives, decision rule, and conclusion.

b. Does your conclusion in a imply that radio makes 2 and 4 have different mean playing hours at high volume? Comment.

c. If playing volume had not been controlled in the tests, would the interpretation of the results be affected? Discuss.

d. Is it likely that the differences between makes are actually due to differences in the test batteries? Explain.

*21.11 Random samples of families with one, two, and three children were selected in a middle-income school district to study expenditures on children's clothing (in dollars) during the past year. Some results of the study follow.

	Number of Children		
	One ($j = 1$)	Two ($j = 2$)	Three ($j = 3$)	
$\bar{Y}_j$:	230	410	580	
n_j:	41	41	41	$MSE = 1{,}601$

Assume that model (21.1) is applicable.

a. Obtain a 95 percent confidence interval for μ_2.

b. Obtain a 95 percent confidence interval for the pairwise comparison $\mu_3 - \mu_2$. Does this interval indicate that families with three children spend, on the average, more on children's clothing than families with two children? Comment.

21.12 Refer to Problem 21.11. Answer both parts, assuming the sample sizes are: $n_1 = 11$, $n_2 = 21$, $n_3 = 31$.

21.13 Refer to Problem 21.8.

a. Obtain a 99 percent confidence interval for the mean playing hours at high volume for radio make 3. (*Hint: MSE = .151.*)

b. Obtain a 99 percent confidence interval for the difference in mean playing hours at high volume between radio makes 3 and 1. Interpret your confidence interval.

c. If another radio of make 3 were to be tested under the same conditions, would your confidence interval in a provide appropriate limits for the number of playing hours for this radio? Explain.

*21.14 Refer to Problem 21.8. It is desired to construct simultaneous confidence intervals for the difference in mean playing hours for the two least expensive makes of radio (makes 2 and 4) and for the difference between the two makes with best tone (makes 1 and 4). Obtain these confidence intervals with a joint confidence coefficient of at least 90 percent. Interpret your confidence intervals. (*Hint*: $MSE = .151$.)

21.15 Refer to Problem 21.11. Construct simultaneous confidence intervals for the pairwise comparisons $\mu_3 - \mu_2$, $\mu_3 - \mu_1$, and $\mu_2 - \mu_1$ such that the joint confidence coefficient is at least 97 percent. Do your intervals indicate that all three treatment means differ from each other and, if so, how do they differ? Discuss.

21.16 A kitchen utensil manufacturer selected 18 similar stores to try out three different promotional displays for a new low-energy cooking pot. The display that generates the highest sales in this study is to be used in the manufacturer's national promotion program. Each display was assigned at random to six stores. One store carrying display 3 had a fire during the period and had to be dropped from the study. Sales (in dollars) for the stores after a two-week observation period follow (note that the treatments are placed in rows).

Display	Store					
	1	2	3	4	5	6
1	2,161	1,369	2,748	1,782	3,530	3,983
2	2,379	1,913	1,119	1,208	1,962	1,689
3	1,479	1,024	598	963	1,913	

a. Test whether or not the displays are equally effective in generating sales, controlling the α risk at .05. State the alternatives, decision rule, and conclusion.
b. The marketing manager had surmised in advance of seeing the study's sales figures that display 1 would be the most effective. Estimate the treatment mean for this display, using a 95 percent confidence interval. Interpret your interval.
c. Construct simultaneous confidence intervals for the differences in mean sales per store between displays 1 and 2 and between displays 1 and 3. Use a joint confidence coefficient of at least .96. Do your intervals indicate that display 1 is clearly superior to both of the others? Comment.

*21.17 Refer to Problem 21.7.
a. Obtain the residuals. Do they sum to zero for each treatment? Is this necessarily the case?
b. Construct a residual plot, as in Figure 21.5, and analyze it for constancy of the treatment error variances. State your findings.

21.18 Refer to Problem 21.16. Obtain the residuals and make a residual plot, as in Figure 21.5. Analyze this plot for unequal treatment error variances. State your findings.

*21.19 Refer to Problem 21.7. Let two indicator variables be defined as follows:

$X_1 = 1$ if instructional mode is video tapes, 0 otherwise

$X_2 = 1$ if instructional mode is computer programs, 0 otherwise

a. State the regression model equivalent to ANOVA model (21.1) for this study.
b. Arrange the data for computer input in the format of Table 21.4.
c. The following results were obtained in fitting the regression model in a:

$b_0 = 85.0$, $s\{b_0\} = 2.769$, $b_1 = -7.0$, $s\{b_1\} = 3.916$, $b_2 = -5.0$, $s\{b_2\} = 3.916$, $SSR = 130.0$, $SSE = 460.0$. Test for equality of the treatment means via the regression approach, controlling the α risk at .05. State the alternatives, decision rule, and conclusion. Show the correspondence to the ANOVA approach in Problem 21.9b.

d. Construct a 95 percent confidence interval for β_0. What is being estimated here?

e. Construct a 95 percent confidence interval for β_1. Interpret your interval estimate.

21.20 Refer to Problem 21.8. Let three indicator variables be defined as follows:

$$X_1 = 1 \text{ if radio is make 2, 0 otherwise}$$

$$X_2 = 1 \text{ if radio is make 3, 0 otherwise}$$

$$X_3 = 1 \text{ if radio is make 4, 0 otherwise}$$

a. State the regression model equivalent to ANOVA model (21.1) for this experiment.

b. The following results were obtained in fitting the regression model in a: $b_0 = 5.10$, $s\{b_0\} = .1586$, $b_1 = -.80$, $s\{b_1\} = .2244$, $b_2 = .50$, $s\{b_2\} = .2244$, $b_3 = -.50$, $s\{b_3\} = .2244$, $SSR = 5.8800$, $SSE = 3.0200$. Test for equality of the treatment means via the regression approach, controlling the α risk at .01. State the alternatives, decision rule, and conclusion. Show the correspondence to the ANOVA approach in Problem 21.10a.

c. Construct a 99 percent confidence interval for β_2. What differential effect is being estimated here? Interpret your interval estimate.

EXERCISES

21.21 If $\overline{Y}_j = \overline{\overline{Y}}$ for all r treatments, which of the following will be true? Which false? Why? (1) $SSTR = 0$, (2) $SSTO = SSTR + SSE$, (3) $SSE = 0$.

21.22 In an experiment, three treatments were investigated and five subjects were utilized for each treatment. Assume that model (21.1) is applicable and that $\sigma^2 = 9$.
a. Obtain $E\{MSTR\}$ and $E\{MSE\}$ when: (1) $\mu_1 = \mu_2 = \mu_3 = 10$; (2) $\mu_1 = 6$, $\mu_2 = 10$, $\mu_3 = 14$; (3) $\mu_1 = 0$, $\mu_2 = 10$, $\mu_3 = 20$.
b. What do your results in a imply about the type of decision rule appropriate for the F test of the equality of treatment means? Explain.
c. Redo a assuming that 10 subjects were utilized for each treatment. What do your results imply about the power of the test when the sample sizes are increased? Discuss.

21.23 Refer to Comment 2 on p. 551. Show that $\Sigma \Sigma (Y_{ij} - \overline{Y}_j)(\overline{Y}_j - \overline{\overline{Y}}) = 0$.

21.24 Use the properties of model (21.1) to prove (21.22).

$$\left[\text{Hint: } E\left\{ \sum_{i=1}^{n_j} (Y_{ij} - \overline{Y}_j)^2/(n_j - 1) \right\} = \sigma^2. \right]$$

21.25 Use the properties of model (21.1) to prove that $MSE/\sigma^2 = \chi^2(n_T - r)/(n_T - r)$. [Hint: Use (13.29) and Exercise B.14b in Appendix B.]

21.26 Two confidence intervals are being constructed from the same sample data, so they are not statistically independent. Let α_1 and α_2 denote the probabilities for the two intervals, respectively, that the procedure will lead to an *incorrect* interval. Use the addition theorem (4.16) to prove that the probability both intervals will be correct is at least $1 - \alpha_1 - \alpha_2$. [*Hint:* $P(E_1 \cap E_2) = 1 - P(E_1^* \cup E_2^*)$.]

STUDIES

21.27 Refer to Appendix D. Consider all companies in the electronic computer equipment industry with 1981 net assets of $100 million or less (i.e., exclude companies 217, 219, 220, 221, 224, and 227). Divide these companies into two groups—those that made a profit in 1980 and those that did not. View the 1981 income figures for the two groups as independent samples from normal populations with equal variances.

    a. Investigate whether mean net income in 1981 differs between companies that made a profit in 1980 and those that did not. In your investigation, perform an ANOVA test for the equality of means using $\alpha = .05$. Also estimate the difference between the two population means by a 95 percent confidence interval.

    b. (Based on Optional Topic Section 11.10.) For the test in **a**, obtain test statistic (13.6) and show the equivalence of this test statistic to the test statistic F^* in **a**.

    c. (Based on Optional Topic Section 13.5.) Test whether or not the two populations of 1981 incomes have equal variances, using $\alpha = .02$. [*Hint:* $F(.99; 18, 11) = 4.15$, $F(.99; 11, 18) = 3.43$.]

    d. Using the assumption that the populations of net incomes are normal, obtain a point estimate of the probability that a profitable company in 1980 is unprofitable in 1981. Compare this estimate with a direct estimate of this probability based on the sample proportion. Does the direct estimate depend on any assumptions about the population distribution?

21.28 A maintenance report for five aircraft of the same type in an airline's fleet contains data on the time intervals (in operating hours) between successive breakdowns in each aircraft's airconditioning equipment. The numbers of operating hours between breakdowns, shown in chronological order for each aircraft, follow.

Breakdown	Aircraft				
	1	2	3	4	5
1	17	100	54	6	13
2	31	98	15	36	270
3	179	87	33	281	91
4	45	230	68	54	38
5	27	81	132	254	603
6	198	98	67		118
7		140	60		450
8		488	229		
9		48	209		
10		89	102		

Assume that model (21.1) is applicable when the logarithms of the time intervals are employed.

a. Test whether the aircraft differ in terms of the mean log-time between successive breakdowns, using the log-time data. Control the α risk at .01. Use logarithms to base 10. [Hint: $F(.99; 4, 33) = 3.95$.]

b. Given your conclusion in a, would it be appropriate to combine all 38 sample observations and treat them as a random sample from the same population? Discuss.

c. Obtain the residuals for the log-time data and construct a residual plot, as in Figures 21.4 and 21.5. Examine the plot for constancy of the aircraft error variances and for independence of the error terms for each aircraft. State your findings.

d. Combine the residuals into one group and examine them for normality of the error terms. State your findings, taking into account your conclusions in c.

UNIT SIX

Bayesian
Decision
Making

22
Bayesian Decision
Making—I

Decision making has been studied intensively by researchers in the social and behavioral sciences. Basic principles and methods of decision making have been developed which are widely applied in administration, economics, and other traditional fields as well as in special fields, such as genetic counseling and environmental conservation. In this chapter and the next, we extend our discussion of decision making begun in earlier chapters. To understand the nature of the extensions, it is important to recognize three pertinent points about statistical decision making presented in earlier chapters:

1. Decision rules were constructed and evaluated in terms of risks of errors at different possible values of a population parameter, such as the population mean or the population proportion.
2. Monetary and other consequences of a decision were considered in formulating the alternatives (e.g., one-sided versus two-sided alternatives) and in specifying the α and β risks, but these consequences were not measured explicitly and were not incorporated into the calculations.
3. The only information used in reaching a conclusion about the population parameter was the information in the sample itself.

We now broaden our approach to decision making by incorporating monetary and other consequences of a decision into our calculations, and by using information about the population that is available prior to sampling. In this chapter, we discuss decision making when no sample information is available, and in the next chapter we take up decision making when sample information is available.

First in this chapter, we consider the common structure of decision problems. Then we take up decision making under certainty and decision making under uncertainty, in both of which the consequences of decisions are explicitly taken into account. Finally, we consider Bayesian decision making, where not only the consequences of decisions are explicitly taken into account but prior information about the population is also utilized.

22.1   DECISION PROBLEMS

Decision problems, whether in business, government, or any other sphere, have a common structure. To illustrate this common structure, we consider two examples.

☐   Examples

1. Firm A has developed a new engine that will give greater fuel economy than conventional engines. Tooling up for mass production can be initiated now or can be delayed for another year in the hope that further research and development will yield additional major refinements. The engineering group assesses the probability of major additional refinements in the next year to be .6. However, a rival—firm B—has also developed a new engine and has just begun to tool up for mass production. This engine is competitive with the present version of firm A's new engine. Thus, firm A will be at a long-term competitive disadvantage if it delays mass production of its new engine and does not discover major additional refinements in the next year. However, if additional major refinements are discovered within the next year, firm A will more than recoup the losses from the delay in getting into mass production. The executive committee of firm A must now decide whether to initiate immediate tooling up for mass production or undertake further research and development for another year.

2. A small oil tanker has just discharged its cargo and is about to begin another round trip. The tanker's operator must decide whether to air-blow the ship's wing tanks to clear the gas residue or to fill the tanks with water ballast. The former procedure costs $200, while pumping the water ballast costs $2,000. Air-blowing will suffice unless there is a collision affecting a wing tank. If this occurs, a wing tank that is merely air-blown will explode, with damage to the ship and possible risk to the crew—about a $10 million estimated loss altogether—whereas a wing tank filled with water ballast cannot explode. The tanker operator is not able to assess the probability of a wing-tank collision, except that he knows the possibility cannot be ignored since two narrow and busy channels must be negotiated on the return trip for another cargo of oil.   ☐

Structural Elements of Decision Problems

The two examples just presented illustrate the common structural elements of decision problems:

(22.1)   The structural elements of decision problems are:

1. *Decision maker*   The decision maker is either a single individual (the tanker operator) or a group of persons acting as a single individual (the executive committee).

2. *Acts or strategies* A_j These are the alternative courses of action open to the decision maker (tool up now, delay; air-blow, use water ballast). They reflect the *controllable variables* in the decision problem—i.e., the variables under the decision maker's control.

3. *Outcome states* S_i These are the different situations that may prevail and affect the consequences of the acts (further refinements in engine will be discovered, no further refinements in engine will be discovered; wing-tank collision, no

wing-tank collision). The outcome states reflect the *uncontrollable variables* in the decision problem—i.e., the variables not under the decision maker's control.

4. *Consequences or payoffs C_{ij}* These are the different gains, rewards, etc., that may be experienced by the decision maker, measured in monetary or other units. C_{ij} denotes the gain or payoff experienced if the *j*th course of action is chosen and the *i*th outcome state prevails (if the operator uses air-blowing and a wing-tank collision occurs).

5. *Outcome state probabilities $P(S_i)$* These are probabilities that the decision maker may assign to the outcome states. Outcome state probabilities are not present in all decision problems (they were present in Example 1 but not in Example 2).

6. *Criterion* This is a basis for identifying the act that is "best" among the courses of action available to the decision maker (the examples did not illustrate any criterion).

Payoff Tables and Decision Trees

Two convenient devices for displaying the acts, outcome states, consequences, and probabilities in a decision problem are the *payoff table* and the *decision tree*. They present the same information; hence, either one can be used depending on whether a tabular or graphic display is more effective in the given situation.

Payoff Table. Table 22.1a shows the format and notation we shall employ for payoff tables, when there are *r* outcome states and *c* acts. Table 22.1b presents the entries for the tanker example, where $r = c = 2$. Since the tanker operator was unable to assign probabilities $P(S_i)$ to the different possible outcome states S_i, "n.a." (not applicable) is entered in place of the $P(S_i)$. Ordinarily, we shall omit

TABLE 22.1 *Illustration of payoff table*

(a)
General format

Outcome State	Probability	Act			
		A_1	A_2	$\cdots$	A_c
S_1	$P(S_1)$	C_{11}	C_{12}	$\cdots$	C_{1c}
S_2	$P(S_2)$	C_{21}	C_{22}	$\cdots$	C_{2c}
$\vdots$	$\vdots$	$\vdots$	$\vdots$	$\vdots$	$\vdots$
S_r	$P(S_r)$	C_{r1}	C_{r2}	$\cdots$	C_{rc}

(b)
Tanker example

Outcome State	Probability	Act	
		Air-Blow	Water Ballast
No wing-tank collision	n.a.	−200	−2,000
Wing-tank collision	n.a.	−10 million	−2,000

this column when probabilities are not used. The entries in Table 22.1b are negative because they entail losses to the operator. Thus, $C_{11} = -200$, since the cost of air-blowing is $200 and no other losses are encountered when no wing-tank collision occurs.

Note how effectively Table 22.1b consolidates the information in the tanker decision problem. We see quickly that water ballast entails a loss of $2,000 regardless of whether or not there is a collision. Air-blowing, on the other hand, entails a loss of only $200 if there is no wing-tank collision but a loss of $10 million if such a collision occurs.

Decision Tree. Figure 22.1a shows the format of a decision tree for a decision problem with two acts and two outcome states, and Figure 22.1b illustrates the decision tree for the tanker example. A decision tree consists of a sequence of nodes and branches. We enter from the left. The branches emanating from the first node represent the alternative courses of action A_j. Since this node entails a

FIGURE 22.1   *Illustration of a decision tree*

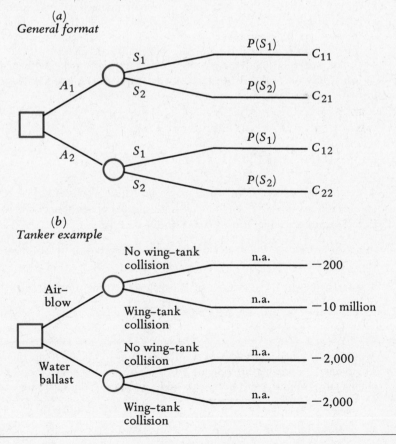

(a)
General format

(b)
Tanker example

controllable variable, it is called a *decision node*. The next set of branches represents the possible outcome states S_i for each course of action. The nodes from which these branches emanate involve uncontrollable variables and are called *chance nodes*. The outcome state probabilities $P(S_i)$—when applicable—are shown on the branches emanating from each chance node. Finally, the payoff for each sequence of A_j and S_i is shown at the right. We follow the convention of indicating decision nodes by squares, and chance nodes by circles.

Comments

1. Two or more courses of action must be available to the decision maker before a decision problem exists. If only one course of action is available, there is no decision problem.
2. The identification of feasible acts and relevant outcome states, and the measurement of payoffs, can be highly demanding in actual decision problems. For instance, in our tanker example it is very difficult to assess the loss resulting from a wing-tank explosion. In addition, it is entirely possible that some desirable courses of action are overlooked when the available acts are formulated. One may need to examine the causal basis of a problem in great detail to identify all available acts for solving it.
3. Note that the losses considered in the tanker example are those directly produced by the course of action. Losses that would run on unaffected by the action or that occurred in the past are excluded. Thus, in our tanker example, there is no need to consider the crew's wages, fuel costs, and other costs of the next trip which will occur regardless whether water ballast is used or the wing tanks are air-blown.
4. In the tanker example, the payoffs are expressed as losses, and hence are negative. In many decision problems, the payoffs are expressed as gains, such as profits, and are therefore positive.

22.2 DECISION MAKING UNDER CERTAINTY

The first class of decision problems where the consequences of acts are explicitly considered is one where each act has a unique consequence. Decision problems of this type involve decision making under certainty:

(22.2) In *decision making under certainty*, each act A_j has a unique consequence that is known in advance.

A key feature of decision making under certainty is that there are no uncontrollable variables present.

☐ Example

A municipality had to choose an investment firm to handle its forthcoming bond issue. Three highly reputable firms made written bids guaranteeing the net interest rate the municipality would pay. The payoff table, indicating the guaranteed interest rates, is:

	Firm A	Firm B	Firm C
Guaranteed interest rate	9.92%	9.80%	9.88%

Note there is only a single outcome state here, or equivalently there is no uncontrollable variable present.

The municipality decided to employ the criterion: Minimize the total interest to be paid. Hence, firm B was selected. □

Comment

Decision problems involving certainty sometimes are referred to as *deterministic*. These decision problems are not necessarily simple. Sophisticated mathematical methods may be required to find the act that is best according to the criterion. But the characteristic feature of all such problems, whether simple or complex, is that for each act there is a unique payoff which is known in advance.

22.3 DECISION MAKING UNDER UNCERTAINTY

We turn next to decision problems where uncontrollable variables are present. Here, there are several outcome states and the payoff for act A_j depends on the outcome state S_i that prevails. Since the outcome state S_i is not known with certainty, two different approaches may be taken:

1. The decision maker is unwilling or unable to assign probabilities to the outcome states S_i and regards them simply as uncertain.
2. The decision maker assigns probabilities $P(S_i)$ to the outcome states.

We consider the first approach in this section. It is called decision making under uncertainty:

(22.3) In *decision making under uncertainty*, there are two or more outcome states S_i that affect the payoff, and there is no assignment of probabilities $P(S_i)$ to these states.

The uncertainty approach tends to be appealing when there is no relevant past experience or other basis for assessing the probabilities $P(S_i)$. We now describe an example where this approach was utilized.

□ Illustration

A firm planning to produce a special camera and film for making photographs with a distinct three-dimensional effect is being sued for patent infringements in separate lawsuits by a camera manufacturer and a film manufacturer. The alternative courses of action being considered by management are shown in Table 22.2. Note that they involve different degrees of tooling up to produce the camera and film. For example, act A_1 entails tooling up for both camera and film, while A_2 entails tooling up for the camera but delaying tooling up for the film until the outcomes of the lawsuits are known. The firm has decided to oppose both suits and obtain definitive legal decisions rather than seek any compromise settlements. Hence, the outcome states are the four possible combinations of legal decisions in the two suits, as shown in Table 22.2. The firm's legal counsel is unwilling to assess probabilities for the outcome states because legal issues without precedent are involved. Consequently, management has decided to regard the outcome states simply as uncertain. Thus, the problem is one of decision making under uncertainty.

TABLE 22.2 *Payoff table for camera firm case ($ million)*

| | Act | | | |
Outcome State	A_1 Tool up for Both Camera and Film	A_2 Tool up for Camera Only	A_3 Tool up for Film Only	A_4 Do Not Tool up for Either
S_1: Win both suits	1,450	1,400	1,100	1,100
S_2: Win camera suit only	1,300	1,425	1,000	1,000
S_3: Win film suit only	700	825	850	850
S_4: Lose both suits	400	775	800	810
Minimum payoff	400	775		810

The payoffs C_{ij} are shown in the payoff table in Table 22.2. They represent the estimated present value of future net income in millions of dollars. These payoffs are affected by two factors: (1) If the firm loses a given lawsuit, it can still manufacture the item, but at a higher production cost because of modifications to avoid patent infringement. (2) Competing firms will develop their own versions of the camera and film, so it is advantageous for the firm to place its product on the market quickly if the legal cases are won. □

Admissible Acts

Sometimes not all available acts need be considered. An act may be so poor that it should be dropped immediately. To determine whether such poor acts are present, we compare the payoffs for each pair of acts. In our camera firm case, suppose we compare acts A_3 and A_4 in Table 22.2. We note that the payoffs are equally good under outcome states S_1, S_2, and S_3. However, under S_4 the payoff for A_3 ($C_{43} = 800$) is less than that for A_4 ($C_{44} = 810$). Hence, there is no need to consider A_3 any further because the payoff with A_3 cannot be higher than with A_4 but can be lower. We say that A_4 dominates A_3 here.

(22.4) Act A_j *dominates* act $A_{j'}$ if A_j offers as good a payoff as $A_{j'}$ in every outcome state and a better payoff in at least one outcome state, i.e., if $C_{ij} \geq C_{ij'}$ for all outcome states i, and $C_{ij} > C_{ij'}$ for at least one outcome state.

Act A_j is *admissible* if it is not dominated by any other available act, and is *inadmissible* if it is dominated by another act.

Note from Table 22.2 that acts A_1, A_2, and A_4 are not dominated by any other act and hence are each admissible. Thus, in the camera firm case we shall only need to consider A_1, A_2, and A_4 as candidates for best act.

Criteria for Selecting Best Act

To choose the "best" act, a criterion is required. We now discuss two criteria that have been proposed by decision theorists for decision making under uncertainty.

Maximin Criterion. A decision maker who tends to be cautious might emphasize the guaranteed payoff for each act—that is, the level below which the actual payoff cannot go—and select the act that offers the best guarantee. To put this another way, the decision maker would select the act that maximizes the minimum possible payoff:

(22.5)   The *maximin criterion* considers the minimum payoff that can be obtained with each act A_j—namely, $\min_i C_{ij}$—and selects that act for which the minimum payoff is maximized: $\max_j \left(\min_i C_{ij} \right)$.

☐   **Example**

In our camera firm case, the minimum payoffs for the three admissible acts A_1, A_2, and A_4 are the respective column minimums shown in Table 22.2. For example, for act A_1 the minimum payoff is 400. We see from Table 22.2 that act A_4 maximizes the minimum possible payoff and would be chosen under the maximin criterion. Of course, the actual payoff with A_4 can be higher than the minimum 810 (if the outcome state is S_1, S_2, or S_3), but it cannot be lower.   ☐

Comment

In some decision problems involving uncertainty, the outcome states are under the control of an adversary (e.g., a business competitor) whose objective is to hold the decision maker to as poor a payoff as possible. The maximin criterion may have particular appeal to the decision maker here since it guarantees that he will receive the best of the worst possible payoffs that can be forced on him.

Minimax Regret Criterion. A second criterion for decision making under uncertainty appeals to a decision maker who reasons as follows: "I must commit myself now to a course of action before I know the outcome state. But once the outcome state is known, the 'Monday morning quarterbacks' will compare my result against the best that could have been obtained in light of the actual outcome state which prevailed. I would like to protect myself against falling too far short!"

A decision maker taking this position is concerned about regrets or opportunity losses, which we denote by O_{ij} for the ith outcome state and jth act:

(22.6)   The *regret* or *opportunity loss* O_{ij} is the difference between the best payoff attainable under the ith outcome state—namely, $\max_j C_{ij}$—and the payoff attained with the jth act under that state; i.e.:

$$O_{ij} = \left(\max_j C_{ij} \right) - C_{ij}$$

The minimax regret criterion considers the maximum regret for each act and selects the act with the smallest maximum regret as best:

(22.7)   The *minimax regret criterion* considers the maximum regret that can be obtained with each act A_j—namely, $\max_i O_{ij}$—and selects that act for which the maximum regret is minimized: $\min_j \left(\max_i O_{ij} \right)$.

◻ Example

The regrets for our camera firm case are obtained in Table 22.3. The payoffs C_{ij} for the admissible acts are repeated in Table 22.3a. The maximum payoffs attainable under the four outcome states are the respective row maximums. These are starred in the table. Thus, $C_{11} = 1,450$ is starred since it is the maximum payoff attainable under S_1.

The regrets O_{ij} are shown in Table 22.3b. Note that $O_{11} = 1,450 - 1,450 = 0$, because A_1 is the best act when S_1 prevails. Similarly $O_{12} = 1,450 - 1,400 = 50$, which indicates that the payoff under S_1 with act A_2 is $50 million worse than with the best act for this outcome state. The other regrets are obtained in the same manner.

To obtain the maximum regret for each act, we find the largest regret within each column. These are shown in Table 22.3b; for example, for act A_1 the maximum regret is 410. The smallest maximum regret is for act A_2. The decision maker, in selecting A_2, is assured that the payoff actually achieved when the outcome state materializes will fall short of the best attainable under that outcome state by no more than $50 million. ◻

Choice of Criterion. We have considered two criteria for selecting a course of action in decision making under uncertainty. Still other criteria might have been used. Unfortunately, different criteria frequently select different acts in the same problem. Thus, in our camera firm case the maximin criterion selected A_4 while the minimax regret criterion selected A_2. Since the theory of decision making under uncertainty does not identify any criterion as objectively superior to the others, the choice of a criterion remains subjective.

TABLE 22.3 *Payoffs and regrets for camera firm case ($ million)*

(a)
Payoffs (C_{ij})

Outcome State	Act		
	A_1	A_2	A_4
S_1	1,450*	1,400	1,100
S_2	1,300	1,425*	1,000
S_3	700	825	850*
S_4	400	775	810*

*Maximum payoff for outcome state.

(b)
Regrets or opportunity losses (O_{ij})

Outcome State	Act		
	A_1	A_2	A_4
S_1	0	50	350
S_2	125	0	425
S_3	150	25	0
S_4	410	35	0
Maximum regret	410	50	425

22.4   BAYESIAN DECISION MAKING WITH PRIOR INFORMATION ONLY

We now consider the second of the two approaches listed earlier for decision problems in which uncontrollable variables are present—namely, the approach where the decision maker assigns probabilities $P(S_i)$ to the outcome states. This approach is called Bayesian decision making:

(22.8)   In *Bayesian decision making*, there are two or more outcome states S_i that affect the payoff, and probabilities $P(S_i)$ are assigned to these outcome states.

Bayesian decision making plays an important role in both the theory and practice of decision making, and we shall study it in the remainder of this chapter and in the following one. In this chapter, we take up Bayesian decision making using only prior information about the outcome states in the form of probabilities $P(S_i)$, but no sample information. In the next chapter, we consider Bayesian decision making when both prior information about the outcome states and sample information are available. Bayesian decision making with prior information only is also called *decision making under risk*.

☐   Illustration

Mr. Adams operates a tourist service on an island noted for its historic sites. Tourists on incoming packaged tours to a nearby mainland area may make optional side trips by air to this island. Mr. Adams handles the travel and lodging reservations for these side trips. He must make reservations for lodging and travel before he knows how many tourists will make the side trip. He gains $50 for each filled reservation up to four reservations, but loses a $30 advance deposit for each unfilled reservation. Mr. Adams neither gains nor loses for any tourists in excess of the number of reservations made. Finally, if five or more persons in an incoming tour take the side trip, the travel agency handling the tour makes the side trip arrangements, reimbursing Mr. Adams for any deposits made and giving him a fixed fee of $150 for servicing the group's visit.

The alternative courses of action available to Mr. Adams for a tour are the number of reservations to be made, as shown in Table 22.4. The outcome states are the actual number of tourists making the side trip, and these are also shown in Table 22.4. We illustrate the entries in the payoff table by considering the payoffs for act A_3 (two reservations). If no tourists make the side trip, Mr. Adams forfeits two $30 deposits for a total loss of $60.

TABLE 22.4   *Payoff table for island tour example*

Number of Tourists Making Side Trip	Probability $P(S_i)$	Number of Reservations				
		A_1 None	A_2 One	A_3 Two	A_4 Three	A_5 Four
S_1: None	.25	0	−30	−60	−90	−120
S_2: One	.10	0	50	20	−10	−40
S_3: Two	.20	0	50	100	70	40
S_4: Three	.20	0	50	100	150	120
S_5: Four	.15	0	50	100	150	200
S_6: Five or more	.10	150	150	150	150	150

If one tourist makes the side trip, he gains $50 for that one and loses $30 on the one unfilled reservation for a net gain of $20. If two tourists make the side trip, he gains $50 for each for a total gain of $100. If three or four tourists make the side trip, Mr. Adams's gain still is $100 since he neither gains nor loses on tourists in excess of the number of reservations. Finally, if five or more tourists make the side trip, the gain is the fixed fee of $150 paid to him by the travel agency.

No more than four reservations need be considered as available acts in the decision problem. Any act entailing more than four reservations is dominated by act A_5 (four reservations).

Mr. Adams was able to assign probabilities to each of the outcome states based on past experience. For example, in about 25 percent of past tours, no person elected the side trip; hence, he assessed the probability that no person will elect the side trip on the next tour as $P(S_1) = .25$. The outcome state probabilities are also shown in Table 22.4. ☐

Comment

An objective interpretation can be given to Mr. Adams's outcome state probabilities since they are based on observed relative frequencies for past tours conducted under conditions similar to those prevailing currently. In many cases, such relative frequencies are not available; for instance, there may be no relevant past experience. In these cases, subjective or personal probability assessments may be employed. We discussed such assessments in Chapter 4. Bayesian decision making proceeds in the same fashion regardless whether the probabilities are objective or subjective.

Expected Payoff Criterion

When probabilities $P(S_i)$ are assigned to the outcome states, the payoff for an act is considered to be a random variable with an associated probability distribution. Thus, we note from Table 22.4 that if Mr. Adams selects act A_5 (four reservations), the probability distribution of payoffs is as follows:

Payoff C_{i5}	Probability $P(S_i)$
−120	.25
− 40	.10
40	.20
120	.20
200	.15
150	.10
Total	1.00

For brevity we use the term *lottery* to denote a probability distribution of payoffs. When a decision maker selects a given act in Bayesian decision making, he or she chooses, in effect, a given lottery in preference to the other lotteries available.

Which of the available lotteries should Mr. Adams choose? Since his handling of the side trips is a recurrent activity, Mr. Adams may well prefer the lottery that yields the largest expected payoff per tour. This is equivalent to maximizing his total gain over the long run. The act with the largest expected payoff is said to satisfy the expected payoff criterion:

(22.9) The *expected payoff criterion* selects that act for which the expected payoff is maximized.

We calculate the expected payoff in the manner of any other expected value. In adapting formula (5.5) to our present notation, we obtain:

(22.10)
$$EP(A_j) = \sum_i C_{ij} P(S_i)$$

where $EP(A_j)$ denotes the expected payoff for act A_j. The act with the largest expected payoff is called the Bayes act:

(22.11) The act selected by the expected payoff criterion is called the *Bayes act*.

The expected payoff for the Bayes act is denoted by BEP, where:

$$BEP = \max_j EP(A_j)$$

☐   **Example**

To find the Bayes act for the island tour example, we need to obtain the expected payoff for each act. We illustrate the calculations for act A_5:

$$EP(A_5) = \sum_{i=1}^{6} C_{i5} P(S_i) = (-120)(.25) + (-40)(.10) + \cdots + 150(.10) = 43$$

Thus, if Mr. Adams makes four reservations, his expected gain per tour is $43. The other expected payoffs are obtained in similar fashion. We find:

Act:	A_1	A_2	A_3	A_4	A_5
$EP(A_j)$:	15	40	57	58	43

We see that the expected payoff is largest for A_4, so act A_4 (three reservations) is the Bayes act and the expected payoff for the Bayes act is $BEP = EP(A_4) = \$58$ per tour.   ☐

Comments

1. Note that $EP(A_j)$ in our example becomes larger and then declines as the number of reservations increases. The act of making three reservations strikes the best balance here between losses arising from unused reservations and gains from reservations that are utilized.
2. In our example, as in most applications of Bayesian decision making, the outcome state probabilities do not depend on the acts. Hence, the same set of probabilities $P(S_i)$ is used to calculate the expected payoff for each act. In some decision problems, however, the outcome state probabilities are affected by the acts. For example, a firm's bid price for a construction contract will affect its probability of winning the contract. In this type of decision problem, different sets of outcome state probabilities must be used to calculate the expected payoffs for the different acts.

Decision Tree Representation.   Bayesian decision problems can be represented by decision trees, though this representation becomes impractical for large problems.

☐ Example

Figure 22.2 shows the decision tree for the island tour example. We have truncated most of the branches for convenience. The act branches are shown emanating from the decision node. Let us follow the branch for act A_4. This branch leads to a chance node with six outcome state branches. The probabilities $P(S_i)$ and payoffs C_{i4} are shown on the tree and the expected payoff $EP(A_4) = 58$ is recorded in the chance node. The other portions of the tree are completed in a similar manner. The decision maker, standing as it were at the decision node, now selects the act branch that leads to the highest expected payoff, namely branch A_4. The other act branches are marked by blocking signs to indicate they are not optimal.   ☐

Value of Information

Frequently, a decision maker has the option of obtaining information about the outcome state before making a decision. Of course, this information normally entails a cost. We now consider how much a decision maker should be willing to pay to learn with certainty which outcome state prevails. We continue with our island tour example.

Expected Payoff with Perfect Information.   Suppose it were possible for Mr. Adams to ascertain in advance how many tourists will make the side trip (e.g., by

FIGURE 22.2   *Decision tree for island tour case*

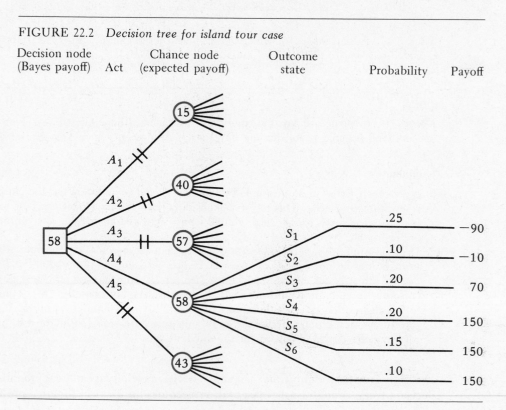

Decision node (Bayes payoff)	Act	Chance node (expected payoff)	Outcome state	Probability	Payoff

radio-telephone to the mainland). In that case, Mr. Adams would know the outcome state for each tour—i.e., he would have *perfect information* about the outcome state. He would then be in the position of choosing the best act for the given outcome state. Thus, we see from Table 22.4 that if four tourists will make the side trip, he should select act A_5 (make four reservations), which will give him a payoff of $200. The best act and its payoff for each outcome state, and the probability of each outcome state, are as follows:

Outcome State:	S_1	S_2	S_3	S_4	S_5	S_6
Best act:	A_1	A_2	A_3	A_4	A_5	A_5
$\max_j C_{ij}$:	0	50	100	150	200	150
$P(S_i)$:	.25	.10	.20	.20	.15	.10

Note that we have another probability distribution of payoffs here, this time for the case where there is perfect information about the outcome state and the best act can be chosen for the outcome state. We can now obtain the expected payoff with perfect information, denoted by *EPPI*:

$$EPPI = 0(.25) + 50(.10) + 100(.20) + 150(.20) + 200(.15) + 150(.10)$$

$$= 100$$

Thus, if Mr. Adams knew the outcome state in advance for each tour and could select the best act for that outcome state, his expected payoff would be $EPPI = \$100$ per tour.

We now define the expected payoff with perfect information formally:

(22.12) The *expected payoff with perfect information*, denoted by *EPPI*, is the expected payoff when the act is chosen based on information that indicates the exact outcome state, and is defined:

$$EPPI = \sum_i \left(\max_j C_{ij} \right) P(S_i)$$

Note that any cost incurred in obtaining the perfect information is not considered in the definition of *EPPI*.

Expected Value of Perfect Information. In our island tour example, the expected payoff of the Bayes act, which utilizes only the probabilistic information about the outcome states, is $BEP = 58$. Since the expected payoff with perfect information is $EPPI = 100$, the difference $EPPI - BEP = 100 - 58 = 42$ is the incremental gain due to perfect information about the outcome state. This incremental gain is called the expected value of perfect information. Assuming that the payoffs measured in monetary terms are meaningful to Mr. Adams and that the expected payoff criterion is relevant, he should be willing to pay up to $42 per tour for perfect information about the number of tourists who will make the side trip.

We now define the expected value of perfect information formally:

(22.13) The *expected value of perfect information*, denoted by *EVPI*, is the difference between the expected payoff with perfect information (*EPPI*) and the expected payoff of the Bayes act (*BEP*); i.e.:

$$EVPI = EPPI - BEP$$

Comments

1. Assuming that payoffs expressed in monetary units are meaningful and that the expected payoff criterion is relevant, *EVPI* is the maximum amount that should be spent for any type of information about the outcome state, whether the information is perfect or not. As we shall demonstrate in the next chapter, sample information is imperfect and so is worth less than perfect information.
2. Regrets or opportunity losses can be used in Bayesian decision making. The Bayes act is then the one that minimizes the expected regret. Two results can be shown to hold: (a) The same act that maximizes expected payoff minimizes expected regret. (b) The magnitude of the minimum expected regret is identical to *EVPI*.

Sensitivity Analysis

Frequently, the data on outcome state probabilities and payoffs in decision problems are unreliable to some degree. Consequently, it is advisable that one study whether the choice of the best act is sensitive to variations in the data. This analysis is called *sensitivity analysis*. Sensitivity analysis in large-scale decision problems is conducted by mathematical methods or computer simulation. In small decision problems, one can simply insert new values for the probabilities or payoffs in the problem to see if the optimal act remains the same and, if not, whether the expected payoffs of the different optimal acts vary greatly.

☐ Example

In our island tour example, if the probabilities:

Outcome State:	S_1	S_2	S_3	S_4	S_5	S_6
$P(S_i)$:	.20	.20	.20	.15	.15	.10

had been used instead of those given in Table 22.4, the expected payoffs of the different acts would have been:

Act:	A_1	A_2	A_3	A_4	A_5
$EP(A_j)$:	15	44	57	54	39

Note that the Bayes act would change from A_4 to A_3, although the expected payoff with act A_4 would only be $3 less than that of the Bayes act for the second set of outcome state probabilities.   ☐

22.5 MULTISTAGE DECISION PROBLEMS

The decision problems discussed so far in this chapter are called *single-stage* problems since a single decision is involved. Other decision problems entail a sequence of two or more decisions and are called *multistage decision* problems. Decision trees are useful for studying simple multistage problems.

☐ Illustration

A manufacturer of minicomputers has determined that the firm's product line must be broadened to include a line of office machines utilizing the latest technology. Investigation shows that two alternative courses of action are worth considering:

A_1: Undertake research and development to achieve production of the desired line

A_2: Acquire a company already producing a satisfactory line and undertake manufacturing integration

These alternative acts are represented in Figure 22.3 by the branches emanating from the first-stage decision node. (Ignore for now the blocking marks on some branches and the numerical entries in the nodes.) Note that if the research and development (R & D) strategy is followed (A_1), the possible outcomes are "unsuccessful" and "successful" and the respective probabilities are assessed at .4 and .6. The payoff is the estimated net income position (in million dollars) of the company five years hence.

In turning to the path for acquisition and manufacturing integration (A_2), we note that the attempt to integrate manufacturing could be successful (with probability .7) or unsuccessful (with probability .3). If success is achieved, an attempt to consolidate the overall management structure will be made. The alternative courses of action in this second-stage

FIGURE 22.3 *Example of two-stage decision problem*

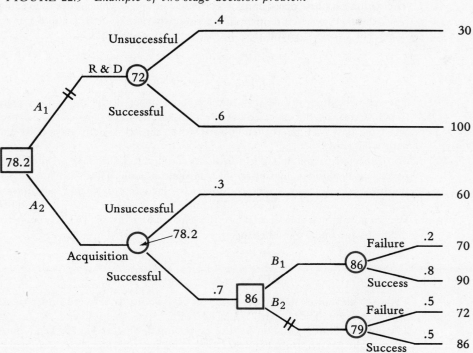

decision are:

B_1: Attempt consolidation on a product basis

B_2: Attempt consolidation on a geographic basis

Thus, a second-stage decision node is indicated in the tree at this point. The branches emanating from this decision node lead to respective chance nodes and the possible outcomes failure and success, with probabilities as shown. Finally, the payoffs are given for each of the branches.   □

Backward Induction

Bayesian decision making for multistage problems can be based on the principle of *backward induction*. This principle requires that the optimal acts at the last decision nodes be identified first, that the optimal acts at the preceding set of decision nodes then be identified using the expected payoffs of the previously selected optimal acts, etc. In terms of a decision tree, the principle of backward induction requires us to work from right to left, using the expected payoffs of the optimal acts to the right in succeeding calculations.

□   Example

We illustrate the principle of backward induction for our computer manufacturer case represented in Figure 22.3. The steps are as follows:

1. There is only a single decision node at the second stage, which arises if first-stage act A_2 is successful. We therefore need to ascertain whether B_1 or B_2 is the Bayes act at this stage. The expected payoffs for the two acts are as follows:

$$EP(B_1) = 70(.2) + 90(.8) = 86$$
$$EP(B_2) = 72(.5) + 86(.5) = 79$$

These expected payoffs are recorded for convenience in the respective chance nodes. Act B_1 maximizes the expected payoff in the second-stage decision. The expected payoff of the Bayes act, 86, is then entered in the second-stage decision node and the B_2 path is blocked.

2. The expected payoff in a decision node consolidates the data on probabilities and payoffs on all paths emanating rightward from that node. Thus, in Figure 22.3, act A_2 in the first-stage decision is now considered to lead to two chance outcomes offering, respectively, a payoff of 60 with probability .3 and a payoff of 86 with probability .7.

3. We are now in position to calculate the expected payoffs for the acts emanating from the first-stage node. We obtain:

$$EP(A_1) = 30(.4) + 100(.6) = 72$$
$$EP(A_2) = 60(.3) + 86(.7) = 78.2$$

Thus, A_2 maximizes the expected payoff and would be selected under the expected payoff criterion. The A_1 path is blocked off.

4. The optimal strategy can now be seen by tracing through the unblocked branches from left to right. We see from Figure 22.3 that the firm should acquire a company already producing a satisfactory line and undertake manufacturing integration (A_2). If this is

successful, managerial consolidation on a product basis should be attempted (B_1). The possible payoffs for the optimal strategy, with attendant probabilities, are as follows by forward induction:

Payoff	Probability
60	.30
70	.7(.2) = .14
90	.7(.8) = .56
Total	1.00

The expected payoff for the optimal strategy by forward induction is $60(.30) + 70(.14) + 90(.56) = 78.2$, the same as calculated before by backward induction. □

Comments

1. In multistage decision problems, second- or later-stage decisions are not necessarily acted on. If events intervene and new conditions arise, later-stage decisions must be determined anew in the light of the then-existing situation.
2. The probabilities in the branches emanating from any chance node always sum to 1. Thus, for any chance node after the first, these probabilities are conditional ones, the conditioning events being the relevant outcomes earlier in the tree.
3. Multistage decision problems, like single-stage ones, can be handled in terms of uncertainty—i.e., without assigning probabilities $P(S_i)$ to the branches emanating from the chance nodes. For example, if the computer manufacturer problem in Figure 22.3 were approached in terms of uncertainty and the maximin criterion employed, the acquisition strategy would be implemented; the minimum possible payoff for this strategy is 60 while that for the R & D strategy is 30.

22.6 OPTIONAL TOPIC—UTILITY

Problem with Monetary Payoffs

Our discussion of Bayesian decision making so far has utilized the expected payoff criterion, where the payoffs are measured in monetary units. However, payoffs in monetary units may not fully reflect the decision maker's attitude toward risk in a given situation. Hence, the comparative expected payoffs may not be in accord with the decision maker's actual preferences. Consider the owner of a small construction company who must decide whether to be the general contractor on a large project (A_1) or simply a subcontractor (A_2). The payoff table is:

		Act	
Outcome State	Probability	A_1	A_2
S_1: No delays	.8	$80,000	$30,000
S_2: Delays	.2	−$40,000	−$ 5,000

The contractor knows that a loss of $40,000 with act A_1 if delays occur would place him in extreme financial difficulties, whereas a loss of $5,000 with act A_2 could be absorbed. In other words, a $40,000 loss would be much more than eight times as severe for him as a \$5,000 loss. Although the expected payoffs are $EP(A_1) = \$56,000$ and $EP(A_2) = \$23,000$ (calculations not shown), the contractor finds A_2 preferable.

Basic Concepts of Utility Theory

Expected Utility Axioms. Utility theory has been developed to take into account the decision maker's attitude toward risk. Recall that in Bayesian decision making, each act entails a probability distribution of payoffs, which is called a lottery. Utility theory makes some basic assumptions about the consistency of the decision maker's preferences among lotteries. One of these is the *transitivity* assumption or axiom, which states that if lottery A_1 is preferred to lottery A_2 and lottery A_2 is preferred to lottery A_3, then the decision maker will prefer A_1 to A_3. The transitivity axiom, together with several other axioms, make up a set of axioms called the *expected utility axioms*. If a decision maker's preferences are consistent with this set of axioms, then it can be shown that a utility function $u(C)$ exists which properly reflects the decision maker's attitude toward risk. Here, C is the monetary consequence and $u(C)$ is the *utility number* associated with the money amount C. We call the function $u(C)$ the decision maker's *utility function for money*. Roughly interpreted, $u(C)$ is the subjective worth to the decision maker of a dollar payoff of amount C.

Expected Utility Criterion. Utility theory shows that when utility numbers rather than monetary payoffs are used in Bayesian decision making, the preferred act will be the one with the largest expected value of the outcomes expressed in utility numbers. Consider a Bayesian decision problem, where C_{ij} is the monetary payoff when the jth act is selected and the ith outcome state prevails. The expected payoff for the jth act, according to (22.10), is:

$$EP(A_j) = \sum_i C_{ij} P(S_i)$$

If we now express each monetary payoff C_{ij} in terms of its corresponding utility number $u(C_{ij})$, we can calculate for the jth act the expected value of the outcomes in utility numbers. We denote this expected value by $EU(A_j)$ and have:

(22.14)
$$EU(A_j) = \sum_i u(C_{ij}) P(S_i)$$

Here, $EU(A_j)$ stands for the expected utility of act A_j.
 As mentioned before, utility theory shows that the decision maker's preferred act is the one with the greatest expected utility:

(22.15) The *expected utility criterion* selects that act for which the expected utility, defined in (22.14), is maximized.
 The act selected by the expected utility criterion is called the *Bayes act*.

Assessing the Utility Function

The major problem in utilizing utility theory is the assessment of the utility function for the decision maker. Once the utility function $u(C)$ has been ascertained, the finding of the Bayes act proceeds as before, except that utility numbers rather than monetary payoffs are used. We shall use an illustration to explain how the utility function for a decision maker may be assessed.

Illustration

An investment counselor is offering a speculative venture to a client. If the venture is successful the client will gain $10 thousand, but if unsuccessful she will lose $6 thousand. Alternatively, the client can decline the venture, in which case her payoff is $0—she neither gains nor loses. Thus, there are three possible dollar outcomes, namely $10 thousand, $0, and −$6 thousand. We wish to assign utility numbers to these dollar amounts that will reflect the client's attitude toward risk.

Specifying the Utility Scale. Since utility is synonymous with subjective worth, the more desirable an outcome, the larger is the utility number assigned to it. Thus, utility functions for money assign larger utility numbers $u(C)$ as the monetary amount C increases.

There is no unique scale for utility, and two points on the utility function can be defined arbitrarily. For convenience, we shall usually follow the convention of assigning to the smallest monetary outcome in the decision problem the utility number 0 and to the largest monetary outcome the utility number 1. Thus, for our investment example, we shall let $u(-6) = 0$ and $u(10) = 1$, where the monetary outcomes are expressed in thousands of dollars.

We have all encountered other scales where two points can be assigned arbitrary values. The temperature scale is one such scale. The Celsius scale assigns 0 to freezing temperature and 100 to boiling, while the Fahrenheit scale assigns 32 and 212 to these temperatures, respectively.

Reference Lottery Method. A variety of methods are available for assessing utility functions. We shall explain one of these, the *reference lottery method*. In our investment illustration, we have defined $u(-6) = 0$ and $u(10) = 1$. We still require the utility of $0, $u(0)$, as well as some other utility numbers so that we can assess the shape of the decision maker's utility function.

To find $u(0)$, we give the decision maker a choice between $0 for certain and the following reference lottery (made up of equal probabilities for the two extreme outcomes):

Outcome ($000)	Probability
10	.5
−6	.5

Suppose the decision maker prefers $0 for certain to this reference lottery. This means that $u(0)$ exceeds the expected utility of the reference lottery. Since

$u(10) = 1$ and $u(-6) = 0$, the expected utility of the reference lottery is:

$$u(10)(.5) + u(-6)(.5) = 1(.5) + 0(.5) = .5$$

Since we wish to find the indifference point between $0 for certain and a reference lottery, we need to increase the probability of gaining $10 thousand in the reference lottery.

Suppose that after several steps, we have reached the reference lottery:

Outcome ($000)	Probability
10	.7
−6	.3

Now, the client indicates that she is indifferent between $0 for certain and this reference lottery. Hence, $u(0)$ is equal to the expected utility of this reference lottery:

$$u(0) = u(10)(.7) + u(-6)(.3) = 1(.7) + 0(.3) = .7$$

To ascertain some other points for the utility function, suppose next we wish to find $u(2)$. We utilize the same approach, giving the client a choice each time between a $2 thousand gain for certain and a reference lottery. If we eventually reach the indifference point for the following reference lottery:

Outcome ($000)	Probability
10	.8
−6	.2

we have:

$$u(2) = 1(.8) + 0(.2) = .8$$

In continuing in this manner, suppose we obtain the following results for the investment client:

C:	−6	−3	−1	0	+2	+5	+8	+10
$u(C)$:	0	.45	.65	.70	.80	.90	.98	1.00

These points are plotted in Figure 22.4. When enough points are obtained, a curve can be drawn through them as shown in the figure. This curve is the graph of the client's utility function for money.

Often, a mathematical function is fitted to the data, such as a *quadratic utility function:*

(22.16) $$u(C) = b_0 + b_1C + b_2C^2$$

or an *exponential utility function:*

(22.17) $$u(C) = b_0 + b_1\exp(b_2C)$$

where b_0, b_1, and b_2 are constants.

FIGURE 22.4 *Investment client's utility function for money*

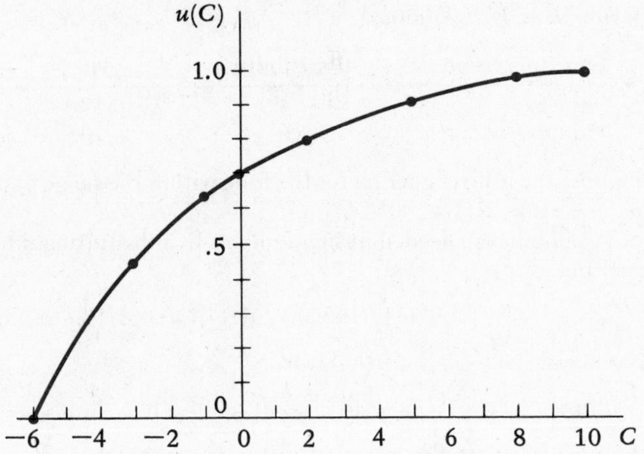

Comment

Since persons may give inconsistent responses when utility functions are assessed, it is good practice to check the responses for internal consistency with respect to the expected utility rule. In our investment client case, for example, the client might be given an additional choice between $2 thousand for certain and a reference lottery offering a 50-50 chance at $0 and $5 thousand. Recall from earlier questioning that $u(2) = .8$ and note that the expected utility of the reference lottery is also .8, i.e., $.9(.5) + .7(.5)$. If the client is not indifferent here, additional questioning and possible revisions of earlier responses will be required.

Application of Utility Function

Once the decision maker's utility function has been assessed, the finding of the Bayes act is routine, with the utility numbers simply replacing the monetary outcomes.

☐ Examples

1. For the investment client case, suppose the speculative venture probabilities are as follows:

Outcome ($000)	Probability
10	.5
−6	.5

We have $u(10) = 1$ and $u(-6) = 0$. If A_1 denotes pursuing the speculative venture, we therefore have $EU(A_1) = 1(.5) + 0(.5) = .5$. Let A_2 denote the act not to pursue the speculative venture. Then $EU(A_2) = u(0) = .7$. Hence, act A_2 has the higher expected utility and is the preferred act.

2. Management of a firm has the choice of conducting mineral exploration alone (A_1), as a joint venture with another firm (A_2), or not conducting the exploration (A_3). The payoff table is as follows:

Outcome State	Probability	A_1	A_2	A_3
Success	.4	17	8.5	0
Failure	.6	−10	−5.0	0

Suppose the utility function for the firm, within the range of payoffs for this problem, is $u(C) = .2 + .03C + .001C^2$.

We obtain the needed utility numbers by substituting into the utility function. For example:

$$u(17) = .2 + .03(17) + .001(17)^2 = 1.0$$

We obtain:

$$u(0) = .2 \quad u(-5) = .075 \quad u(-10) = 0 \quad u(8.5) = .53 \quad u(17) = 1.0$$

We now calculate the expected utility for each act:

$$EU(A_1) = 1.0(.4) + 0(.6) = .4$$
$$EU(A_2) = .53(.4) + .075(.6) = .257$$
$$EU(A_3) = u(0) = .2$$

Hence, act A_1 is the preferred act. ☐

Types of Utility Functions for Money

Researchers have investigated utility functions for money in some detail. Three prototype utility functions are graphed in Figure 22.5. The *concave utility function* in Figure 22.5a, labeled *risk-averse*, is the type exhibited by the investment

FIGURE 22.5 *Examples of utility functions for money showing different attitudes toward risk*

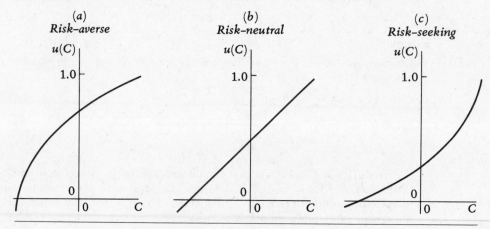

client in our earlier case. It can be shown that a decision maker with a concave utility function will prefer a given dollar payoff for certain to a lottery offering an equal expected dollar payoff. For instance, such a person would prefer a sure gain of $5 thousand to a 50-50 chance of obtaining $0 or $10 thousand.

Figure 22.5b shows a *linear utility function*:

$$u(C) = b_0 + b_1 C \qquad b_1 > 0 \tag{22.18}$$

A decision maker with this type of utility function for money is said to be *risk-neutral*. It is easy to show that the expected utility of an act A_j for a risk-neutral person can be expressed in terms of the expected payoff of the act as follows:

$$EU(A_j) = b_0 + b_1[EP(A_j)] \qquad b_1 > 0 \tag{22.19}$$

Hence, the act that maximizes expected payoff also maximizes expected utility. Thus, there is no need to use utility numbers for risk-neutral persons; the expected payoff criterion will correctly indicate the preferred act.

The *convex utility function* in Figure 22.5c pertains to decision makers said to be *risk-seeking*. The attitude toward risk here is the opposite of risk aversion. For instance, a decision maker with a convex utility function for money will prefer a lottery offering a 50-50 chance at $0 or $10 thousand to $5 thousand for certain. Thus, an individual who buys a ticket in a state lottery is exhibiting risk-seeking behavior because the cost of the ticket exceeds the expected amount of winning. The quadratic utility function in the mineral exploration case is convex, indicating that the management of the firm is risk-seeking.

Comments

1. Our discussion of utility focused on utility functions for money. Utility functions can also be developed for other types of outcomes, such as rate of return on investment, health status, and location of next job.
2. Utility is subjective. We cannot say that an individual is correct or incorrect in his or her attitude toward risk. Morever, a decision maker's utility function may change over time as new circumstances arise. Finally, because utility is a subjective measure and arbitrarily scaled, it is meaningless to compare the utility numbers for different individuals.
3. In many practical situations, the decision maker's utility function for money is approximately linear within the range of the monetary outcomes encountered. Hence, as we have seen, in these situations utility assessments are not required since the act with the largest expected payoff (in monetary terms) will be the decision maker's preferred act.

PROBLEMS

22.1 An executive of a multinational corporation told a group of graduate students: "In inflation-prone countries, we employ a flexible credit policy adjusted to the short-run outlook for inflation in the local currency. If inflation will be slight, we allow retailers a six-week credit limit, since experience shows this limit will generate the highest remitted profits in dollar terms. If inflation will be moderate, we cut this limit to four weeks, since a four-week limit then will result in the highest remitted dollar profits, and so on down to a limit of one week if inflation is severe. In one of the countries, the government has told us we must announce a credit policy that will be fixed for the next two years. We no longer can shift our limit back and forth as tactical considerations dictate. In the next few days I must decide what our two-year credit policy will be in that country." Explain what each of the structural elements of a decision problem in (22.1) corresponds to here.

*22.2 Substitute teachers for a school can be hired on a contract or noncontract basis. Contract substitutes receive $15 each school day for the whole year simply for being available to teach, and also receive an additional $75 on any day they actually teach. When the number of contract substitutes is inadequate on any school day, noncontract substitutes must be called in. Noncontract substitutes receive $110 for each day they teach but nothing when they do not teach. The school administrator must now decide how many substitutes should be given contracts for the coming school year. Records show that the following probability distribution describes the number of substitutes required on any school day:

S_i:	0	1	2
$P(S_i)$:	.4	.5	.1

Construct the payoff table for the administrator's decision problem, showing the acts, outcome states, and payoffs for any given school day. Display the costs as negative payoffs.

22.3 Consider playing the following game with a friend. Your friend writes either "0¢" or "$1.01" on a slip of paper. Concurrently, you write either "50¢" or "51¢" on your slip of paper. The two amounts are then added. If the sum is even, you pay your friend that amount; if the sum is odd, your friend pays you that amount.
a. Construct your payoff table for a single trial of this game.
b. If you knew the number your friend will write on the next play, would you have a decision problem on this play? If so, would it involve certainty or uncertainty?

22.4 Refer to Problem 22.2. Construct the decision tree corresponding to the payoff table.

22.5 Refer to Problem 22.3.
a. Construct the decision tree corresponding to the payoff table. Does your decision tree contain the same information as your payoff table? Explain.
b. Suppose you have already played this game five times and are 50¢ behind. Should this loss be included in the payoffs in your decision tree for the sixth play? Explain.

22.6 Paul McQue helps finance his education at Bushnell University by undertaking small projects. McQue's next project requires a decision concerning the number of copies of a brochure to be printed for homecoming weekend. The printing order must be a multiple of 500. The size of the homecoming crowd is uncertain because it depends on weather conditions. The payoff table follows (payoffs are net profits in dollars).

Size of Crowd	Printing Order				
	A_1 1,000	A_2 1,500	A_3 2,000	A_4 2,500	A_5 3,000
S_1: Above average	200	300	400	500	450
S_2: Average	200	300	250	200	150
S_3: Below average	200	150	100	50	0

a. Are all acts admissible? What is the implication of this?

b. Does it appear that McQue has missed an opportunity by not including printing sizes above 3,000 in his possible courses of action? Comment.

*22.7 Refer to Problem 22.6.

a. Which act satisfies the maximin criterion? What is the guaranteed payoff with this act? Can you tell what will be the actual payoff here with the maximin act? Can you always tell?

b. Convert the payoffs to regrets and ascertain which act satisfies the minimax regret criterion. Is this act the same one selected by the maximin criterion in **a**? Is this necessarily the case?

22.8 Union Appliance, Ltd., is a market leader in cast-iron cooking utensils. Last year, a firm specializing in advanced metal technology introduced a line of competitive lightweight utensils containing a new design. These utensils are made of a new alloy that exactly duplicates the properties of cast iron for cooking. The competitive line has cut sharply into sales of Union's line. Union can incorporate the new design into its own line, but not the lighter weight. Changing its lines to the new design will be expensive for Union, so management does not wish to undertake it unless the new design is the major factor in consumer acceptance of the competitive line. The payoff table follows (payoffs are discounted profits in millions of dollars).

Importance of New Design	Act	
	A_1 Incorporate New Design	A_2 Do Not Incorporate New Design
S_1: Major factor	3.5	.8
S_2: Not major factor	1.9	3.9

a. Which act satisfies the maximin criterion? What minimum payoff does this act guarantee? Could the actual payoff exceed the guaranteed minimum? Explain.

b. Convert the payoffs to opportunity losses and ascertain which act satisfies the minimax regret criterion. Is this act the same one selected by the maximin criterion in **a**? Is this always the case?

*22.9 Refer to Problem 22.6. Suppose McQue is able to assign probabilities to each outcome state, and that these are: $P(S_1) = .3$, $P(S_2) = .5$, $P(S_3) = .2$.

a. Which act is the Bayes act? What is the expected payoff with this act?

b. Under some other set of probabilities, could A_1 be the Bayes act here? Could A_5 ever be the Bayes act? Explain.

22.10 A small manufacturer of computer graphics equipment has accumulated substantial cash reserves. The executive committee will now decide whether to expand vertically by buying another computer graphics company, or expand horizontally by buying a company that manufactures medical monitoring equipment. The outcome states correspond to different possible degrees of elimination of financially weaker companies in the computer graphics field in the next several years. The payoff table follows (payoffs are present values of future earnings in millions of dollars).

Degree of Elimination	Act	
	A_1 Vertical	A_2 Horizontal
S_1: Moderate	8	15
S_2: Severe	16	10

The executive committee assesses the outcome state probabilities to be: $P(S_1) = .4$, $P(S_2) = .6$.

a. Show the probability distributions of payoffs associated with acts A_1 and A_2, respectively.

b. Find the Bayes act. What is its expected payoff?

22.11   Refer to Problem 22.2. The payoff table for the school administrator's decision problem follows.

Substitutes Required for School Day	Probability $P(S_i)$	Contract Substitutes to Be Hired		
		A_1 0	A_2 1	A_3 2
S_1: 0	.4	0	−15	−30
S_2: 1	.5	−110	−90	−105
S_3: 2	.1	−220	−200	−180

a. Find the Bayes act. What is its expected payoff?

b. What will be the expected earnings per school day of a contract substitute if the administrator adopts the Bayes act in a?

*22.12 Refer to Problem 22.10. Obtain EPPI and EVPI. Interpret their meanings here.

22.13   Refer to Problems 22.6 and 22.9.

a. Obtain EPPI and EVPI. Interpret their meanings here.

b. By paying a $65 premium for overtime work, McQue can delay the printing order until just prior to the weekend, when the size of the crowd is known. Do you recommend that he follow this option? Explain.

22.14   Refer to Problem 22.8. Suppose management is able to assess the probabilities of the outcome states as follows: $P(S_1) = .6$ and $P(S_2) = .4$, and wishes to maximize expected payoff. A consumer research study will be undertaken to obtain information on which outcome state prevails.

a. What is the maximum amount management should be willing to spend on such a study? Explain.

b. Since the study involves a sample survey, information obtained from it about the prevailing outcome state will be subject to sampling error and perhaps also other errors. What effect does this have on the amount management should be willing to spend on the study?

22.15   If for a decision problem it is known that EPPI = 160 and EVPI = 30, what is the expected payoff of the Bayes act? What is the expected regret or opportunity loss of the Bayes act?

22.16 A respondent, when asked if he uses Bayesian decision analysis, replied: "Yes. I find the outcome state that has the highest probability and pick the act with the best payoff under this outcome state." Comment.

*22.17 Refer to Figure 22.3. Suppose that if management follows the R & D route and the outcome is unsuccessful, there is still an opportunity to broaden the product line by undertaking negotiations with another firm on either a merger or a license arrangement. (It would be extremely disadvantageous to attempt to negotiate in both areas, so one or the other must be selected for negotiation.) The acts, outcomes, payoffs, and probabilities of success for the negotiations are:

Merger: success, 65; failure, 20; probability of success = .70

License: success, 60; failure, 30; probability of success = .60

    a. Draw a decision tree to reflect the modified situation.
    b. Using backward induction, ascertain the optimal strategy. What is the expected payoff for the optimal strategy?
    c. Confirm the expected payoff for the optimal strategy by forward induction.

22.18 A firm making industrial products developed a marketable consumer item as a by-product of its research and development. A decision is to be made on whether to sell this item through the company's own sales and distribution channels (A_1) or to use independent sales agents (A_2). A second-stage decision problem will arise if the firm selects A_1 and is relatively unsuccessful, since management must then decide between switching to sales agents immediately (B_1) or first attempting to dispose of the item by selling the key patent at favorable terms to another firm and switching only if the attempt is unsuccessful (B_2). The decision tree for this problem is shown in Figure A (payoffs are present values of estimated future earnings in millions of dollars).

    a. Using backward induction, determine the optimal strategy. What is the expected payoff for the optimal strategy?
    b. Confirm the expected payoff for the optimal strategy by forward induction.

*22.19 A decision maker's utility function for money is $u(C) = .02C - .0001C^2$, where $0 \leq C \leq 100$ and C is in thousand dollars.

    a. Plot $u(C)$. Is the decision maker risk-averse, risk-neutral, or risk-seeking?
    b. Which will the decision maker prefer: (1) a lottery offering an even chance at $0 or $75 thousand, or (2) $30 thousand for certain?

22.20 Refer to Problem 22.19. Answer both parts assuming the utility function is $u(C) = \sqrt{(C + 40)}/120$, where $-40 \leq C \leq 80$ and C is in thousand dollars.

FIGURE A

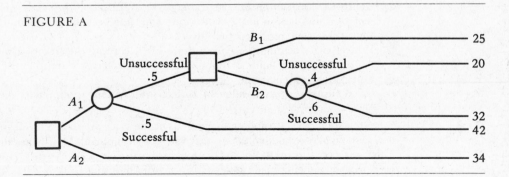

22.21 Refer to Figure 22.4, containing the utility function for the investment client in the Text illustration. Suppose this client is offered a lottery in which there are 2 chances in 3 of losing $3 thousand and 1 chance in 3 of winning $8 thousand, and the client indicates indifference between playing this lottery and not playing it. Is this response consistent with the client's utility function in Figure 22.4? Explain.

EXERCISES

22.22 Show that the act selected by the maximin criterion is necessarily admissible.

22.23 **a.** Show that the act that maximizes expected payoff minimizes expected regret.
b. Show that the expected regret for the Bayes act equals *EVPI*.

22.24 Refer to Problem 22.8. For what value of $P(S_1)$ do the two acts have equal expected payoffs? How is this value useful if management is unsure about the exact outcome state probabilities but does believe that $P(S_1)$ is somewhere between .15 and .35? Between .50 and .80?

22.25 Refer to Problem 22.10. The executive committee is somewhat tentative about its assessed value for $P(S_1)$. How far off could this assessment be before it would affect the optimal act?

22.26 Can the optimal strategy in a multistage decision problem be ascertained by either forward or backward induction? If so, what are the advantages of the backward induction approach? Explain.

22.27 Refer to Figure 22.3. The manufacturer conjectures that the payoff of 86 found on the bottom branch of the tree might be an understatement. How much larger could this payoff be before it would alter the manufacturer's optimal strategy, assuming that the other payoffs and the probabilities remained unchanged?

STUDIES

22.28 Someone has given you a ticket in a state lottery. You now find that your ticket number is one of 10 in a final runoff. Three numbers will be drawn from the 10 at random without replacement. If your number is drawn first, second, or third, you will win $250,000, $100,000, or $50,000, respectively. If your number remains undrawn, you will win $1,000. A consortium has offered you $25,000 for your ticket before the runoff. Assume that this is the only offer you will receive.
a. Construct the decision tree associated with your decision problem. Also present this information in a payoff table. State the probability distribution of your winnings if you keep the lottery ticket.
b. Which act (accepting the consortium's offer, keeping the ticket) is optimal under: (1) the maximin criterion, (2) the minimax regret criterion, (3) the expected payoff criterion?
c. Calculate the expected value of perfect information. Interpret its meaning here.

22.29 Refer to Problem 22.3. Suppose many trials of the game are to be played and you decide to employ a randomized strategy whereby with probability .5 you will write "50¢" on your slip of paper and with probability .5 you will write "51¢," the random choices being made independently from one trial to the next. Suppose your friend uses the same probabilities to choose between "0¢" and "$1.01," the random choices being made independently of yours. What is your expected payoff? Is the game played in this fashion a fair one in the sense that neither you nor your friend can expect to gain over many trials of the game?

22.30 Refer to Problem 22.28. Use the reference lottery method to assess your own utility function for the payoffs in this decision problem. Test your utility function for consistency and revise it if necessary, and then apply it to ascertain your optimal act. Is the result obtained from your derived utility function consistent with your actual preference? If not, what are the implications of this?

22.31 Mr. Gatling has an opportunity to join in two oil exploration ventures. In each, he will gain $40 thousand if the venture is successful or lose $10 thousand if the venture is unsuccessful. The probability of success in a venture is .25. The outcomes in the two ventures are statistically independent. The choices open to Gatling are:

$$A_1: \text{Participate in both ventures}$$

$$A_2: \text{Participate in venture 1 only}$$

$$A_3: \text{Don't participate in either venture}$$

Gatling's utility function for money is the one given in Problem 22.20.
a. Which strategy is optimal for Gatling? Give your calculations.
b. Would the expected payoff criterion have led to the same act? Give your calculations.
c. What amount of money offered with certainty would be equally attractive to Gatling as the optimal choice in a?

22.32 Refer to Problem 22.18 and the corresponding decision tree in Figure A. Suppose management's utility function is $u(C) = 1 - \exp(-.05C)$, where $0 \leq C < \infty$ and C is in million dollars.
a. Plot the utility function. What attitude toward risk is implicit in this function?
b. What is management's optimal strategy in terms of maximizing expected utility? Is this the same strategy as the one that maximizes expected payoff? Discuss.
c. Suppose probabilities could not be assigned to the outcome states in the decision tree. Which strategy would be selected with the maximin criterion? Does it matter here whether management considers payoffs or utility values? Explain.

23
Bayesian Decision Making—II

In this chapter, we discuss the use of Bayesian decision making when: (1) prior information about the outcome states in the form of probabilities $P(S_i)$ is available, and (2) it is possible to obtain information about the prevailing outcome state by sampling. This type of decision situation is encountered frequently.

☐ Examples

1. A decision must be made whether or not to accept a shipment from a supplier. The payoff is affected by the proportion of defective items in the shipment. The outcome states here are the different possible proportions defective. From past experience with many shipments from the same supplier, probabilities can be assigned to the outcome states. In addition, a sample can be selected from the particular shipment under consideration to obtain information on the proportion defective in this shipment.
2. A decision must be made whether or not to market a product currently sold by competing firms. The payoff is affected by the number of retailers who now carry this product. The outcome states here are the different possible number of retailers who now carry the product. No firm information is available on this, but management is willing to assign subjective probabilities to the different outcome states. In addition, it is possible to sample the retailers in the industry to obtain information about the number currently carrying the product. ☐

When it is possible to obtain sample information about the prevailing outcome state to supplement prior information in the form of outcome state probabilities, two major questions arise:

1. For given sample size n, how does one select the best act based on both the prior probabilities and the sample information?
2. If the sample size is not predetermined, how does one ascertain the best sample size?

We shall discuss each of these questions in turn, employing a detailed case illustration. Throughout this chapter, we assume that the expected payoff criterion is the appropriate guide for finding the best act.

In the context of decision making with sample information, the outcome state probabilities $P(S_i)$ are called *prior probabilities* because the sample information subsequently provides further information about the prevailing outcome state.

Comment

If the decision maker's utility function for money is not linear (see Section 22.6), utility numbers will need to replace the monetary outcomes in the calculations for identifying the best act.

23.1 ILLUSTRATION

A firm manufactures an air pollution detector. A key component is a large membrane produced periodically by the firm in batches of several hundred and inventoried to meet production and replacement needs. A chemical purchased from a supplier and used in producing the membrane must be pure, or else some membranes will be inert and have to be discarded. From experience it is known that if there are slight trace impurities, about 10 percent of membranes will be inert. With higher levels of impurity, the proportion of inert membranes will be higher.

To keep the problem relatively simple, we shall assume that the proportion of inert membranes as a result of the impurities in the chemical will be one of only five values—these are the outcome states for the decision problem:

Outcome State	Proportion of Inert Membranes
S_1	0
S_2	.1
S_3	.2
S_4	.3
S_5	.4

The chemical tends to deteriorate in storage, so the firm receives a fresh shipment from the supplier just before making a batch of membranes. The decision problem then arising involves the following alternative acts:

A_1: Use the shipment as is

A_2: Put the shipment through a purification process

The purification process costs \$1,700 but removes all impurities.

The outcome states S_i, prior probabilities $P(S_i)$, and payoffs C_{ij} for this decision problem are shown in Table 23.1. The prior probabilities are based on experience with previous shipments. The payoffs represent the anticipated dollar gain to the firm in producing a batch of the membranes. For act A_1, the payoff varies with the purity of the shipment. For A_2, the payoff is fixed at \$500—the gain when the chemical is pure (2,200) less the cost of purification (1,700).

Analysis without Sample Information

If a decision had to be made without sample information, we would follow the procedures presented in Chapter 22. For review, we shall apply these to our pollu-

TABLE 23.1 *Outcome states, prior probabilities, and payoffs for pollution detector case*

			Act	
i	Proportion of Inert Membranes S_i	Prior Probability $P(S_i)$	A_1 Do Not Purify	A_2 Purify
1	0	.2	2,200	500
2	.1	.2	1,600	500
3	.2	.1	1,000	500
4	.3	.2	400	500
5	.4	.3	−200	500

tion detector example. The expected payoffs for the two acts are, by (22.10):

$$EP(A_1) = 2,200(.2) + 1,600(.2) + 1,000(.1) + 400(.2) + (-200)(.3) = 880$$

$$EP(A_2) = 500(.2) + 500(.2) + 500(.1) + 500(.2) + 500(.3) = 500$$

Hence, if no information about the purity of a particular shipment is available, the best act is A_1—to use the shipment as is. A_1 therefore is the Bayes act, and the Bayes expected payoff is $BEP(0) = 880$. We now use $BEP(0)$ to denote the Bayes expected payoff, indicating that a sample of size $n = 0$ is utilized.

If perfect information about the purity of shipments were available, the best act would be A_1 for outcome states S_1, S_2, and S_3, and A_2 for outcome states S_4 and S_5 (see Table 23.1). Hence, the expected payoff with perfect information is, by (22.12):

$$EPPI = 2,200(.2) + 1,600(.2) + 1,000(.1) + 500(.2) + 500(.3) = 1,110$$

Finally, the expected value of perfect information is, by (22.13):

$$EVPI = EPPI - BEP(0) = 1,110 - 880 = 230$$

Thus, the decision maker should pay no more than $230 per shipment for information about the purity of the shipment, whether the information is perfect or imperfect.

The results of the Bayesian analysis without sample information are summarized in Table 23.2.

TABLE 23.2 *Bayesian decision analysis without sample information in pollution detector case*

Expected Payoffs	Expected Payoff with Perfect Information
$EP(A_1) = 880$ $EP(A_2) = 500$	$EPPI = 1,110$
Bayes Act	Expected Value of Perfect Information
A_1 $BEP(0) = 880$	$EVPI = 230$

Sample Information

Management need not make a decision here without sample information about the purity of a particular shipment of chemical, because an engineer has developed a process for making test membranes by hand. While a test membrane is not sensitive enough to be used in the monitoring equipment, it can be checked for inertness. Thus, it is possible for the company to obtain sample information on the outcome state for an incoming shipment by making a number of test membranes and ascertaining how many of these are inert.

23.2 OPTIMAL DECISION RULE FOR GIVEN SAMPLE SIZE

We now consider the first of the two questions posed earlier: For given sample size n, how does one select the best act based on both the prior probabilities and the sample information? In practice, this case arises when a sample has already been taken and the results are now available, or when the sample size is predetermined by policy or by constraints involving time, money, or availability of personnel.

We shall continue to utilize the pollution detector case and suppose that management has decided to make $n = 6$ test membranes for each shipment of chemical before choosing its course of action. We let X denote the number of test membranes in the sample that are inert.

Decision Rule

In Bayesian decision making with sample information, we use the same type of statistical decision rule as discussed in earlier chapters for statistical decision making. Thus, since A_1 yields a higher payoff than A_2 when the chemical is relatively pure and a lower payoff when the chemical is relatively impure, small values of X (number of inert membranes in sample) will lead to act A_1 while large values will lead to act A_2. Hence, the appropriate form of the decision rule is (where K denotes the action limit):

(23.1) For a sample of size n:

$$\text{If } X \leq K, \text{ select } A_1 \quad \text{(Do not purify)}$$
$$\text{If } X > K, \text{ select } A_2 \quad \text{(Purify)}$$

This decision rule is denoted by (n, K).

In contrast to our earlier approach to constructing statistical decision rules, in Bayesian decision analysis the action limit K is determined by explicitly utilizing information about the outcome state probabilities and payoffs.

Set of Decision Rules under Consideration

The choice of the optimal action limit K when $n = 6$ test membranes are utilized must be made from among eight possibilities, as shown in Table 23.3. For com-

TABLE 23.3 *Set of decision rules under consideration for pollution detector case*

Decision Rule	Select A_1 if	Select A_2 if
(6, 0)	$X = 0$	$X > 0$
(6, 1)	$X \leq 1$	$X > 1$
(6, 2)	$X \leq 2$	$X > 2$
(6, 3)	$X \leq 3$	$X > 3$
(6, 4)	$X \leq 4$	$X > 4$
(6, 5)	$X \leq 5$	$X > 5$
(6, 6)	Always	Never
(6, −1)	Never	Always

pactness, we express decision rule (23.1) by (n, K). Note that decision rule (6, 6) leads to act A_1 whenever $X \leq 6$. In a sample of size $n = 6$, $X \leq 6$ always. Hence, rule (6, 6) leads to act A_1 regardless of the sample outcome. Similarly, rule (6, −1) always leads to act A_2 since X can never be less than zero.

Expected Payoff for Decision Rule

We now seek the decision rule among those in Table 23.3 that has the greatest expected payoff. Consider rule (6, 0). This rule directs that six test membranes be made and that A_1 be selected if none of the six is inert; otherwise, A_2 is to be selected. The procedure for calculating the expected payoff for this rule is shown in Table 23.4. The basic idea is: (1) to determine the probabilities of the sample outcome leading to acts A_1 and A_2 for a given outcome state, (2) to obtain the expected payoff for that outcome state, and (3) to determine the expected payoff over all outcome states.

In columns 1 through 3 of Table 23.4, we repeat from Table 23.1 the outcome states and payoffs for the pollution detector case.

TABLE 23.4 *Calculation of expected payoff for decision rule (6, 0)*

(1)	(2)	(3)	(4)	(5)	(6)	(7)	(8)
Proportion of Inert Membranes S_i	Payoffs		Act Probabilities		Conditional Expected Payoff $(2) \times (4)$ $+ (3) \times (5)$	Prior Probability $P(S_i)$	
	C_{i1}	C_{i2}	$P(A_1|S_i)$	$P(A_2|S_i)$			$(6) \times (7)$
0	2,200	500	1.00000	0	2,200.00	.2	440.000
.1	1,600	500	.5314	.4686	1,084.54	.2	216.908
.2	1,000	500	.2621	.7379	631.05	.1	63.105
.3	400	500	.1176	.8824	488.24	.2	97.648
.4	−200	500	.0467	.9533	467.31	.3	140.193
							$EP(6, 0) = 957.85$

Act Probabilities.   The next two columns of Table 23.4 show the respective probabilities of selecting A_1 and A_2 for rule (6, 0), conditional on the outcome state S_i. We designate these probabilities as act probabilities:

(23.2)   An *act probability*, denoted by $P(A_j|S_i)$, is the conditional probability that act A_j will be selected under the given decision rule when the outcome state is S_i.

The act probabilities $P(A_j|S_i)$ are analogous to the probabilities of selecting H_0 and H_1 in a statistical test, discussed in earlier chapters. These latter probabilities are conditional on the value of a parameter, such as the population mean μ.

Sampling Distribution of X.   To ascertain the act probabilities for our pollution detector case, we must obtain the sampling distribution of X, the number of inert membranes in the sample of six. It is reasonable to assume here that X is a binomial random variable, whose probability function is given by (6.3). The parameters of this probability function are $n = 6$ and the proportion of inert membranes for the batch, denoted by p, which is given for each outcome state.

Calculating Act Probabilities.   Consider first outcome state S_1. Here, the proportion of inert membranes for the batch is $p = 0$; thus, the probability of an inert membrane is $p = 0$. Hence, no test membrane will be inert under this outcome state, and rule (6, 0) will always lead to A_1. The respective act probabilities therefore are $P(A_1|S_1) = 1.0$ and $P(A_2|S_1) = 0$. These are entered in columns 4 and 5, respectively, for outcome state S_1.

When the outcome state is S_2, the proportion of inert membranes for the batch is $p = .1$. Since decision rule (6, 0) leads to A_1 if $X = 0$, we need to find $P(X = 0)$ when $n = 6$ and $p = .1$. We use Table C–5 and find $P(0) = .5314$. Hence, $P(A_1|S_2) = .5314$, and by complementation theorem (4.19) we obtain $P(A_2|S_2) = 1 - .5314 = .4686$. These act probabilities are entered in Table 23.4 for outcome state S_2.

For outcome state S_3, when $p = .2$, we find from Table C–5 that $P(0) = .2621$. Hence, $P(A_1|S_3) = .2621$ and $P(A_2|S_3) = 1 - .2621 = .7379$.

The act probabilities for the other outcome states are obtained in similar fashion.

Conditional Expected Payoffs.   We can now obtain the expected payoff for rule (6, 0) for each outcome state. Such an expected payoff is called a conditional expected payoff, because it is conditional on a particular outcome state:

(23.3)   The *conditional expected payoff* for decision rule (n, K) and outcome state S_i, denoted by $EP(n, K|S_i)$, is the expected payoff for this rule conditional on outcome state S_i; i.e.:

$$EP(n, K|S_i) = \sum_j C_{ij} P(A_j|S_i)$$

For outcome state S_1, we see from Table 23.4 that $C_{11} = 2,200$, $C_{12} = 500$, $P(A_1|S_1) = 1.0$, and $P(A_2|S_1) = 0$. Hence, the conditional expected payoff for rule $(6, 0)$ when the outcome state is S_1 is:

$$EP(6, 0|S_1) = 2,200(1.0) + 500(0) = 2,200$$

We interpret this to mean that if rule $(6, 0)$ is used repeatedly with many incoming shipments for which the proportion of inert membranes is $p = 0$, the average payoff is \$2,200 per shipment.

For outcome state S_2, the conditional expected payoff for rule $(6, 0)$ is:

$$EP(6, 0|S_2) = 1,600(.5314) + 500(.4686) = 1,084.54$$

Thus, if rule $(6, 0)$ is used repeatedly with many incoming shipments for which the proportion of inert membranes is $p = .1$, the average payoff is \$1,084.54 per shipment.

The conditional expected payoffs under S_3, S_4, and S_5 are calculated and interpreted in a similar manner. Column 6 of Table 23.4 shows all of the conditional expected payoffs for rule $(6, 0)$.

Expected Payoff. We can now calculate the expected payoff for decision rule $(6, 0)$ over all outcome states. The prior probabilities $P(S_i)$ assigned on the basis of past experience to the outcome states S_i are repeated in column 7 of Table 23.4. For instance, $P(S_1) = .2$. Thus, the probability of conditional expected payoff $EP(6, 0|S_1) = 2,200$ is .2 since S_1 will be the outcome in 20 percent of the shipments. Similarly, the probability of conditional expected payoff $EP(6, 0|S_2) = 1,084.54$ is $P(S_2) = .2$. In taking the expected value of the conditional expected payoffs, we obtain the expected payoff for rule $(6, 0)$.

(23.4) The *expected payoff* for decision rule (n, K), denoted by $EP(n, K)$, is the expected value of the conditional expected payoffs under the prior probabilities $P(S_i)$; i.e.:

$$EP(n, K) = \sum_i EP(n, K|S_i) P(S_i)$$

For our example, we calculate (as shown in column 8 of Table 23.4):

$$EP(6, 0) = 2,200(.2) + 1,084.54(.2) + 631.05(.1) + 488.24(.2) + 467.31(.3)$$

$$= 957.85$$

We interpret this to mean that if decision rule $(6, 0)$ is used repeatedly with many incoming shipments, the average payoff per shipment in the long run will be \$957.85.

Optimal Decision Rule

Now that we have obtained the expected payoff for decision rule $(6, 0)$, we need to find the expected payoffs for the other decision rules in Table 23.3.

In Table 23.5, we show the calculations for decision rule $(6, 1)$. We illustrate how the act probabilities for outcome state S_2 are obtained. Rule $(6, 1)$ states that

TABLE 23.5 *Calculation of expected payoff for decision rule (6, 1)*

(1)	(2)	(3)	(4)	(5)	(6)	(7)	(8)
					Conditional Expected Payoff	Prior	
Proportion of Inert Membranes	Payoffs		Act Probabilities		$(2) \times (4)$	Probability	
S_i	C_{i1}	C_{i2}	$P(A_1|S_i)$	$P(A_2|S_i)$	$+ (3) \times (5)$	$P(S_i)$	$(6) \times (7)$
0	2,200	500	1.0000	0	2,200.00	.2	440.000
.1	1,600	500	.8857	.1143	1,474.27	.2	294.854
.2	1,000	500	.6553	.3447	827.65	.1	82.765
.3	400	500	.4201	.5799	457.99	.2	91.598
.4	−200	500	.2333	.7667	336.69	.3	101.007
							$EP(6, 1) = 1,010.22$

act A_1 should be selected when $X \leq 1$; otherwise, act A_2 should be selected. Since outcome state S_2 pertains to a proportion of inert membranes of $p = .1$, we need to find $P(X \leq 1)$ when $n = 6$ and $p = .1$. From Table C-5, we find $P(X \leq 1) = P(0) + P(1) = .5314 + .3543 = .8857$. Hence, $P(A_1|S_2) = .8857$ and $P(A_2|S_2) = 1 - .8857 = .1143$.

The expected payoffs for all of the decision rules in Table 23.3 are as follows (further calculations are not shown):

K:	0	1	2	3	4	5	6	−1
$EP(6, K)$:	958	1,010	972	918	889	881	880	500

Bayes Decision Rule. The expected payoff criterion leads to the selection of the decision rule with the largest expected payoff. In our pollution detector case, this rule is (6, 1). Rule (6, 1) is called the Bayes decision rule for sample size $n = 6$.

(23.5) The *Bayes decision rule* for a sample of size n is the one that maximizes the expected payoff among all possible decision rules for that sample size.

The expected payoff for the Bayes decision rule for sample size n is denoted by $BEP(n)$, where:

$$BEP(n) = \max_K EP(n, K)$$

Thus, in our pollution detector case the expected payoff for the Bayes rule (6, 1) is $BEP(6) = EP(6, 1) = 1,010$.

Bayes Acts. The acts to be selected according to the Bayes decision rule for each sample outcome are called the *Bayes acts*. In our pollution detector case where the Bayes rule for $n = 6$ is (6, 1), the Bayes acts are as follows for the different sample outcomes X:

X:	0	1	2	3	4	5	6
Bayes act:	A_1	A_1	A_2	A_2	A_2	A_2	A_2

Summary of Procedure

We now summarize the procedure for finding the Bayes decision rule when the sample size is given:

(23.6) The steps for finding the Bayes decision rule for a given sample size n are:

1. Identify all possible decision rules (n, K) for the given sample size n.
2. For each possible decision rule:
   a. Calculate the act probabilities $P(A_j|S_i)$.
   b. Obtain the conditional expected payoff $EP(n, K|S_i)$ for each outcome state S_i.
   c. Use the prior probabilities $P(S_i)$ to obtain the expected payoff $EP(n, K)$ for the decision rule (n, K).
3. Select the decision rule for which the expected payoff $EP(n, K)$ is largest.

In actual applications, Bayesian decision rules are often calculated by using computer packages.

Comments

1. The action limit $K = 1$ is optimal in our pollution detector case for the given sample size, payoffs, and prior probabilities. If any of these elements change, some other action limit might be optimal.
2. The expected payoffs for rules $(6, 6)$ and $(6, -1)$, which do not utilize the sample information, are those for decision making with prior information only, as may be seen from Table 23.2. Decision rule $(6, 6)$ has expected payoff $EP(6, 6) = EP(A_1) = 880$, while rule $(6, -1)$ has expected payoff $EP(6, -1) = EP(A_2) = 500$.
3. If a decision maker is unable or unwilling to assign prior probabilities $P(S_i)$, he or she can follow the uncertainty approach described in the preceding chapter. The chosen criterion for decision making under uncertainty is then simply applied to the conditional expected payoffs $EP(n, K|S_i)$.

23.3 DETERMINATION OF OPTIMAL SAMPLE SIZE

We now come to the second question raised at the beginning of this chapter: If the sample size is not predetermined, how does one ascertain the best sample size? In practice, this case arises when no firm decision on sample size has been made as yet.

The procedure for finding the optimal sample size is conceptually very simple, though extensive computations may be required. We determine for each possible sample size the expected payoff of the Bayes rule net of sampling costs and select that sample size for which this net expected payoff is a maximum. Use of available computer packages can be very helpful. We now consider this procedure in more detail.

Cost of Sampling

We denote the cost of a sample of size n by $s(n)$. We shall assume here that the total sample cost is a linear function of sample size:

(23.7)
$$s(n) = \begin{cases} a + bn & n > 0 \\ 0 & n = 0 \end{cases}$$

where: $a \geq 0$
$b > 0$

Here, a is the overhead cost of setting up the sample and b is the variable cost of including one additional observation in the sample. Note that $s(n) = 0$ when no sample is employed. This simple cost model is often an adequate representation.

Upper Bound on Optimal Sample Size

If many possible sample sizes had to be examined in order to find the optimal n, the computational task would be very large. Fortunately, it is possible to obtain an upper bound for the optimal sample size, beyond which no sample size can be optimal. Recall that the expected value of perfect information, $EVPI$, represents the maximum amount that a decision maker should spend for any information about the outcome state, whether it be perfect or imperfect. Hence, the sample cost must not exceed $EVPI$; i.e.:

$$s(n) \leq EVPI$$

For cost model (23.7), the inequality becomes:

$$a + bn \leq EVPI$$

or:

$$n \leq \frac{EVPI - a}{b}$$

Therefore:

(23.8) For cost model (23.7), an *upper bound* on the optimal sample size, denoted by $\max(n)$, is:

$$\max(n) = \begin{cases} \dfrac{EVPI - a}{b} & \text{if } a < EVPI \\ 0 & \text{if } a \geq EVPI \end{cases}$$

Optimal Sample Size

We can now state the procedure for finding the optimal sample size n more formally:

(23.9) The steps for finding the optimal sample size n are:

1. For each sample size in the interval $0 \leq n \leq \max(n)$, ascertain the Bayes decision rule.
2. For the Bayes rule for given n, determine the expected payoff $BEP(n)$ and the *net expected payoff*, denoted by $BNEP(n)$, where:

$$BNEP(n) = BEP(n) - s(n)$$

3. Select that sample size and associated Bayes decision rule which maximizes the net expected payoff $BNEP(n)$.

☐ **Example**

Suppose that the sample size in our pollution detector case has not yet been determined. An analyst has been asked to obtain the optimal sample size. She has determined that cost model (23.7) is applicable, and that there are no overhead costs, i.e., $a = 0$. The variable cost per test membrane is $b = \$7.50$, so the cost function is:

$$s(n) = 7.5n$$

Since $EVPI = 230$ (see Table 23.2), we obtain by (23.8):

$$\max(n) = \frac{230}{7.5} = 30.7$$

The upper bound on the optimal sample size therefore is 30 test membranes.

Consequently, the analyst has to obtain the Bayes decision rule for each sample size from 0 to 30, and then determine the net expected payoff for each. Recall that we found earlier for $n = 6$ that the Bayes rule is (6, 1) and its expected payoff is $BEP(6) = 1,010$. Since the cost of six test membranes is $7.50(6) = 45$, the net expected payoff for the Bayes rule is:

$$BNEP(6) = BEP(6) - s(6) = 1,010 - 45 = 965$$

The analyst used a computer package to perform the needed calculations. Figure 23.1 shows the computer printout she obtained. For each $n > 0$, the printout shows the action limit K for the Bayes rule for that sample size and the net expected payoff $BNEP(n)$. Note that the net expected payoff for the Bayes decision rule for $n = 6$ is shown to be 965.24, the same as our result from hand computation except for the cents that we had dropped.

The analyst could readily see from the printout in Figure 23.1 that the optimal sample size is $n = 6$ and that the Bayes rule for this sample size is (6, 1). ☐

Comments

1. The printout in Figure 23.1 for $n = 0$ provides the expected payoff for the Bayes act with prior information only, i.e., $BEP(0) = 880$ (see Table 23.2).
2. The output in Figure 23.1 can also be used to ascertain the optimal action limit when the sample size is fixed in advance. For instance, if only $n = 3$ test membranes can be made in the time available, it is seen that the Bayes decision rule is (3, 0)—that is, act A_1 should be selected if there are no inert membranes in the sample of three; otherwise, A_2 should be selected. The net expected payoff of this rule is 951.26, not very much less than the net expected payoff for the optimal sample size $n = 6$.
3. If the upper bound (23.8) on the optimal sample size is zero, the optimal approach is to make the decision using prior information only.
4. The optimal sample size and decision rule depend on the interrelations between the payoffs, prior probabilities, and sample cost function. Computer-assisted calculations allow the decision maker to check easily whether the optimal sample size and decision rule are sensitive to variations in the entries of the payoff table. For example, suppose management is considering modifying the prior probabilities for our pollution detector case as follows because of a suspected recent decline in the quality of shipments:

	Prior Probabilities	
Outcome State	Original	Modified
S_1	.2	.1
S_2	.2	.2
S_3	.1	.2
S_4	.2	.2
S_5	.3	.3

A computer run employing the modified prior probabilities readily shows that the Bayes decision rule still is (6, 1), but with a net expected payoff of $828. Thus, modification of the prior probabilities is not a pressing matter here since the optimal decision rule is unaffected. The difference in the net expected payoffs, $965 - 828 = \$137$, measures the expected cost per shipment to the firm if the suspected quality decline is real.

Value of Sample Information

We now present two measures that deal with the value of sample information in Bayesian decision making, and shall illustrate them in terms of our pollution detector case.

FIGURE 23.1 *Computer output showing Bayes decision rules for sample sizes 0 to 30 in pollution detector case*

BAYES RULE N	BAYES RULE K	BAYES RULE NEP
0	–	880.000
1	0	930.500
2	0	949.800
3	0	951.260
4	1	946.628
5	1	960.117
6	1	965.237
7	1	963.502
8	1	956.604
9	2	959.824
10	2	960.661
11	2	957.439
12	2	950.946
13	3	948.537
14	3	946.992
15	3	942.743
16	3	936.259
17	4	931.306
18	4	928.187
19	4	923.158
20	4	916.534
21	5	910.427
22	5	906.189
23	5	900.552
24	5	893.740
25	6	887.155
26	6	882.091
27	6	875.977
28	6	868.974
29	7	862.244
30	7	856.560

OPT N = 6   NEP = 965.237

Expected Value of Sample Information. For a sample of size n, the *expected value of sample information*, denoted by $EVSI(n)$, is the difference between the expected payoff of the Bayes decision rule for that n and the expected payoff for the Bayes act with prior information only; i.e.:

(23.10)
$$EVSI(n) = BEP(n) - BEP(0)$$

In our pollution detector case, we calculate from results for $n = 6$ obtained earlier:

$$EVSI(6) = 1{,}010 - 880 = 130$$

In a choice between making a decision using prior information only or taking a sample of size $n = 6$, the decision maker should not pay more than \$130 for the sample information.

Expected Net Gain from Sampling. For a sample of size n, the *expected net gain from sampling*, denoted by $ENGS(n)$, is the difference between the net expected payoff for the Bayes rule for that sample size and the expected payoff of the Bayes act with prior information only; i.e.:

(23.11)
$$ENGS(n) = BNEP(n) - BEP(0)$$

For our pollution detector case for $n = 6$, we have (see Figure 23.1):

$$ENGS(6) = 965 - 880 = 85$$

Note this is equivalent to $EVSI(6) - s(6) = 130 - 45 = 85$. In using the optimal sample size, we automatically maximize $ENGS(n)$.

23.4 DETERMINING BAYES ACT BY REVISION OF PRIOR PROBABILITIES

Once a sample has been selected and the result observed, it is possible to ascertain directly the Bayes act for that given sample result without developing the decision rule covering all possible sample results. The procedure employs Bayes' theorem. (A discussion of Bayes' theorem is presented in Section 4.7.)

We shall first restate Bayes' theorem (4.25) in terms of our present notation, using the pollution detector case for illustration. The variable of interest here is the outcome state S_i, denoting the proportion of inert membranes. The prior probabilities of the outcome states are denoted by $P(S_i)$. The additional information variable is the number of inert membranes among the n test membranes, denoted by X_j. We shall use the notation X_0 to represent no inert membrane in the sample, X_1 to denote one inert membrane in the sample, X_j to denote j inert membranes in the sample, etc. With this notation, Bayes' theorem is as follows:

(23.12)
$$P(S_i|X_j) = \frac{P(S_i)P(X_j|S_i)}{\sum_i P(S_i)P(X_j|S_i)}$$

The posterior probability $P(S_i|X_j)$ is the probability that outcome state S_i is the prevailing one, given the sample outcome X_j.

To find the Bayes act for a given sample result X_j by means of Bayes' theorem, we employ the following procedure:

(23.13) The steps for finding the Bayes act for a given sample result by utilizing Bayes' theorem are:

1. For the prior probabilities $P(S_i)$ and the observed sample result X_j, obtain the posterior probabilities $P(S_i|X_j)$ using Bayes' theorem (23.12).
2. Calculate for each act the expected payoff under the posterior probabilities $P(S_i|X_j)$.
3. Select the act for which the expected payoff is largest.

We now illustrate this procedure for the pollution detector case where $n = 6$ test membranes were examined and two were found inert, i.e., the sample outcome was X_2.

Obtaining Posterior Probabilities

The steps for revising the prior probabilities by Bayes' theorem (23.12) are shown in Table 23.6. Columns 1 and 2 repeat the outcome states and prior probabilities given initially in Table 23.1. Column 3 shows the conditional probabilities $P(X_2|S_i)$ of obtaining the observed sample result X_2 for the different possible outcome states S_i. For instance, if the state is S_2—i.e., the proportion of inert membranes is $p = .1$—we find from Table C–5 for $n = 6$ and $p = .1$ that $P(2) = .0984$. Thus, $P(X_2|S_2) = .0984$. For state S_3, the proportion of inert membranes is $p = .2$, so for $n = 6$ we find $P(2) = .2458$; hence, $P(X_2|S_3) = .2458$. The other conditional probabilities are found in a similar manner.

The joint probabilities $P(S_i \cap X_2)$ are calculated in column 4 by multiplying $P(X_2|S_i)$ by $P(S_i)$ for each outcome state. The sum of this column is the denominator of Bayes' theorem (23.12). The posterior probabilities $P(S_i|X_2)$ are obtained in column 5 by dividing each entry in column 4 by the sum of column 4. We see,

TABLE 23.6   *Revision of prior probabilities for pollution detector case—$n = 6$ membranes tested and X_2 (i.e., two membranes) found inert*

i	(1) Proportion of Inert Membranes S_i	(2) Prior Probability $P(S_i)$	(3) $P(X_2|S_i)$	(4) Joint Probability $P(S_i \cap X_2)$ (2) × (3)	(5) Posterior Probability $P(S_i|X_2)$ (4) ÷ .20238
1	0	.2	0	0	0
2	.1	.2	.0984	.01968	.0972
3	.2	.1	.2458	.02458	.1215
4	.3	.2	.3241	.06482	.3203
5	.4	.3	.3110	.09330	.4610
	Total	1.0		.20238	1.0000

for instance, that $P(S_3|X_2) = .02458/.20238 = .1215$. Thus, given that two membranes were inert among the six test membranes, the conditional probability that the shipment will produce proportion $p = .2$ of inert membranes (i.e., that the outcome state is S_3) is .1215. The other posterior probabilities are similarly interpreted.

It is interesting to note how the prior probabilities have been revised in light of the sample information of two inert membranes among the six test membranes. The prior probabilities $P(S_1)$ and $P(S_2)$ have been revised downward substantially and the prior probabilities $P(S_4)$ and $P(S_5)$ have been revised upward substantially. Thus, finding two test membranes inert makes it much more likely that the proportion p of inert membranes in the shipment will be high.

Determining Expected Payoffs

Proceeding now to step two of (23.13), we need to find the expected payoffs under the posterior probabilities for each act. We utilize the payoffs in Table 23.1 and the posterior probabilities in Table 23.6. The expected payoffs are calculated in the usual way, and are denoted by $EP(A_1|X_2)$ and $EP(A_2|X_2)$ to indicate that they are conditional on the given sample outcome. We obtain:

$$EP(A_1|X_2) = 2{,}200(0) + 1{,}600(.0972) + 1{,}000(.1215) + 400(.3203)$$
$$+ (-200)(.4610)$$
$$= 312.94$$

$$EP(A_2|X_2) = 500(0) + 500(.0972) + 500(.1215) + 500(.3203)$$
$$+ 500(.4610)$$
$$= 500$$

Determining Bayes Act

It is evident from the preceding calculations that A_2 is the Bayes act here since it has the larger expected payoff. Thus, when two test membranes among six are inert, the company should purify the shipment. Of course, this result is in accord with the Bayes decision rule (6, 1) obtained earlier.

Extensive and Normal Forms of Bayesian Analysis

The procedure described in this section for ascertaining the Bayes act for a given sample result can be extended to determine the Bayes decision rule for a given sample size, and to determine the optimal sample size. It is called the *extensive form* of Bayesian decision analysis. It contrasts with the form employed in the preceding sections of this chapter, called the *normal form*. As we just saw, the extensive form introduces the prior probabilities in the first stage of the analysis and revises them via Bayes' theorem. The normal form, on the other hand, introduces the prior probabilities as weights for the conditional expected payoffs in the final stage of the calculations and does not employ Bayes' theorem.

Each form of Bayesian analysis has its respective advantages. As noted earlier, if we wish to ascertain the Bayes act for a specific sample result, the extensive form will identify this act directly. The normal form, in contrast, develops the complete Bayes decision rule for that sample size. Another advantage of the extensive form is its adaptability to shortcut mathematical procedures.

On the other hand, the normal form of Bayesian analysis has the advantage that the prior probabilities are introduced last. The decision maker then has the option of treating the outcome states S_i as uncertain or of assigning prior probabilities $P(S_i)$ to them. Also, many users find the normal form to be more suitable for sensitivity analysis. In part, this is because the more objective elements of the decision problem (payoffs and sample cost function) are introduced first. Finally, many statisticians and management scientists find the normal form easier to explain to managers because it proceeds in straightforward steps. The extensive form, on the other hand, requires Bayes' theorem at the onset and its rationale is grasped somewhat less easily.

PROBLEMS

*23.1 Refer to the pollution detector case in the Text. Assume that the payoff table remains as in Table 23.1 except that new prior probabilities have been determined, as follows:

S_i:	S_1	S_2	S_3	S_4	S_5
$P(S_i)$:	4	.2	.1	.1	.2

   a. Determine the Bayes act based on prior information alone. What is the expected payoff of the Bayes act?

   b. Obtain EPPI and EVPI. Since perfect information is assumed, why are these quantities expected values?

23.2 A home products firm has developed a new solar energy heater and now must decide whether to market it or not. The success of the product, as measured by its profit contribution, depends on the proportion of households in the target market that will purchase it. Research by the marketing department has produced the following payoff table and prior probabilities (payoffs are in thousands of dollars):

	Proportion Purchasing Product S_i	Prior Probability $P(S_i)$	Act A_1 Don't Market Product	Act A_2 Market Product
i				
1	.05	.5	0	−500
2	.10	.1	0	0
3	.15	.1	0	500
4	.20	.3	0	1,000

Assume that expected payoff is to be maximized.

   a. Determine the Bayes act if a decision must be made without any sample information. What is the expected payoff of the Bayes act? What is the probability of a negative payoff with the Bayes act?

   b. Obtain EVPI and interpret your result.

*23.3 Refer to Problem 23.1. For the new prior probabilities, the expected payoffs for the decision rules that need to be considered when six test membranes are made follow.

(6, K):	(6, 0)	(6, 1)	(6, 2)	(6, 3)	(6, 4)	(6, 5)	(6, 6)	(6, −1)
EP(6, K):	_____	1,371	1,358	_____	1,306	1,301	_____	_____

   a. Obtain the missing expected payoffs.
   b. For the new prior probabilities, which is the Bayes decision rule for sample size $n = 6$? What is the value of $BEP(6)$? Interpret the meaning of this value.
   c. If six test membranes are made for a shipment and one of these is inert, which act is best?

23.4 Refer to Problem 23.2. Suppose a random sample of $n = 15$ households is to be selected from the target market to obtain information about the proportion of households that will purchase the heater. The decision rule to be employed is of the form: If $X \leq K$, select A_1; if $X > K$, select A_2. Here, X is the number of households in the sample that will purchase the heater, and it can be treated as a binomial random variable.
   a. For outcome state S_1 and the decision rule employing $K = 2$, obtain: (1) the act probabilities, (2) the conditional expected payoff. Interpret the latter result.
   b. Obtain the expected payoff for decision rule (15, 2).
   c. The expected payoffs for selected decision rules based on sample size 15 follow.

(15, K):	(15, 0)	(15, 1)	(15, 2)	(15, 3)	(15, 4)	(15, 15)	(15, −1)
EP(15, K):	201	241	_____	113	52	_____	_____

   Obtain the missing expected payoffs. What appear to be the Bayes decision rule for $n = 15$ and the value of $BEP(15)$? Comment.

23.5 Refer to Problems 23.2 and 23.4. The Bayes rule is (15, 1).
   a. What is the probability of a negative payoff with the decision rule (15, 1)? How does this probability compare with the corresponding one for the Bayes rule based on no sampling?
   b. What other factors should be considered in deciding whether it is worthwhile to select a sample of 15 households?

*23.6 Refer to Problem 23.1. If the cost function is $s(n) = 10n$, what is the upper bound on the optimal sample size given by (23.8)? Interpret this number.

23.7 Refer to Problem 23.2. The setup cost of a survey is $8 thousand and, because of the in-depth interviewing that must be conducted, the cost per respondent is $.05 thousand—i.e., $50.
   a. State the cost function for sampling.
   b. Obtain the upper bound on the optimal sample size given by (23.8).

*23.8 A government auditor is to check the accuracy of a company's entertainment expense vouchers by examining a sample of n vouchers. The expected payoffs of the Bayes decision rules for several sample sizes follow. The payoffs represent recaptured tax revenues from ineligible entertainment expenses in thousand dollars.

n:	0	100	200	500	1,000
$BEP(n)$:	0	63	86	98	99

Assume the cost function for sampling is $s(n) = 6 + .01n$.

a. Find $BNEP$ for each of the values of n considered here. Which of these sample sizes leads to the largest expected payoff net of sampling costs?

b. Find $EVSI$ and $ENGS$ for each of the values of n considered. Interpret the two measures for sample size 100.

23.9 Refer to Problem 23.2. Assume that the cost function for sampling is $s(n) = 8 + .05n$. The values of $BEP(n)$ for selected values of n follow.

n:	0	20	50	100	300
$BEP(n)$:	100.0	254.4	315.2	340.4	349.8

a. Find $BNEP$, $EVSI$, and $ENGS$ for each of the values of n considered here. Which of these sample sizes leads to the largest expected payoff net of sampling costs?

b. What is the meaning of: (1) $BNEP(50)$, (2) $ENGS(50)$?

c. For the sample sizes considered here, how large would the setup cost of the survey have to be before it would not pay to conduct a survey? (Assume the cost per respondent is still .05.)

23.10 Refer to Table 23.1 and Figure 23.1, pertaining to the pollution detector case.

a. Obtain $BEP(n)$ for $n = 0, 1, 2, 3$, and 4. Recall that the cost function for sampling is $s(n) = 7.5n$.

b. What would be the optimal sample size and decision rule if the sampling cost function were: (1) $s(n) = 100 + 7.5n$, (2) $s(n) = 30 + 20n$? Give the net expected payoff for the optimal decision rule in each case.

23.11 In a marketing seminar on Bayesian decision making, one participant stated: "We know that risks of errors can be reduced by increasing the sample size. Therefore, the best sample size is the largest one having a cost less than the expected value of perfect information." Comment.

*23.12 Refer to Table 23.1, pertaining to the pollution detector case. Suppose eight test membranes are made from a shipment of chemical and none of them is found to be inert.

a. Revise the prior probabilities to obtain the posterior probabilities for the outcome states. Interpret the posterior probability for outcome state S_2.

b. Determine the Bayes act for this sample result. What is the expected payoff of the Bayes act, given the sample result?

23.13 Refer to Problem 23.2. Suppose that two respondents in a random sample of 20 indicate they will purchase the heater.

a. Revise the prior probabilities to obtain the posterior probabilities for the outcome states.

b. Determine the Bayes act for this sample result. What is the expected payoff of the Bayes act, given the sample result?

c. Explain how the procedure in **a** and **b** could be extended to determine the Bayes decision rule for the given sample size.

d. In advance of selecting the sample, what is the probability that the sample will contain exactly two respondents who indicate they will purchase the heater?

23.14 Refer to Problem 23.13. Answer all parts assuming that two respondents in a random sample of 12 indicate they will purchase the heater.

EXERCISES

23.15 a. Prove that the expected opportunity loss associated with decision rule (n, K) equals $EPPI - EP(n, K)$.
   b. Explain, for a given sample size, why maximizing the expected payoff and minimizing the expected opportunity loss yield the same optimal decision rule.

23.16 (Based on Optional Topic Section 22.6.) Can the upper bound for the optimal sample size in (23.8) be employed when the decision maker's utility function is not linear? Discuss.

23.17 Show that revision of prior probabilities by (23.12) must yield posterior probabilities that are numbers between 0 and 1 and sum to 1.

STUDIES

23.18 A quality control official for a government agency inspects large shipments of shirts that are received periodically from a supplier. A shirt is classified as defective if workmanship, fabric quality, or pattern does not meet specifications. Rejecting a shipment costs $150 because it entails 100 percent inspection and administrative costs of obtaining replacements for defective shirts. The cost of accepting a shipment is $15,000p$, where p is the proportion of shirts in the shipment that are defective. Extensive past experience with the supplier indicates that shipment quality has the following probability distribution:

Proportion defective:	0	.01	.04	.08
Probability:	.30	.50	.15	.05

   a. Set up the payoff table for the official's decision problem. Show costs as negative payoffs. Denote the acts of accepting and rejecting a shipment by A_1 and A_2, respectively.
   b. A random sample of 10 shirts contains one defective. Assume the number of defective shirts in the sample can be taken as a binomial random variable. Revise the prior probabilities and determine the Bayes act.
   c. Calculate the values of $EP(n, K)$ for $n = 10$ and $K = -1, 0, 1,$ and 2. What appears to be the Bayes decision rule for this sample size? Is the Bayes act identified in **b** consistent with this rule?
   d. Calculate the values of $EP(n, K)$ for $n = 50$ and $K = -1, 0, 1,$ and 2. Use a Poisson distribution with $\lambda = np$ to approximate the required binomial probabilities (see p. 298). What appear to be the Bayes decision rule and the value of $BEP(50)$?

23.19 A food company imports nuts in the shell in large lots. Since consumer reaction to purchasing a package of nuts in the shell containing a high proportion of spoiled nuts is quite adverse, each lot is sampled to decide whether: (1) to package the nuts in the shell for sale; or (2) to shell all nuts first, remove the spoiled ones manually by sorters, and then package the remaining shelled nuts for sale. The company currently follows a zero-defects inspection plan whereby a random sample of 20 nuts is selected from each lot and shelled. If no nuts in the sample are spoiled, the lot is judged acceptable and the nuts are packaged in the shell; otherwise, the lot is judged unacceptable and the nuts are shelled and sorted before being packaged.

The payoffs and prior probabilities for the quality of a lot (as measured by the proportion spoiled) follow. The payoffs (in dollars) represent profit contributions.

	Proportion Spoiled	Prior Probability	Act	
			---	---
i	S_i	$P(S_i)$	A_1 Acceptable	A_2 Unacceptable
1	0	.6	700	600
2	.01	.1	650	580
3	.10	.3	−300	400

Assume that expected payoff is to be maximized.

a. What is the expected payoff of the company's zero-defects inspection plan?

b. Construct the probability distribution of payoffs for a lot randomly received under the company's zero-defects inspection plan.

c. Does the company's zero-defects inspection plan correspond to the Bayes decision rule for a sample size of $n = 20$? [*Hint:* Consider the decision rules (20, 0), (20, 1), (20, 2), (20, 20), and (20, −1).]

23.20 (Based on Optional Topic Section 22.6.) Refer to Problem 23.2. Suppose that the utility function for money of the decision maker in this firm is $u(C) = 1 - \exp[-.001(C + 500)]$, where $-500 \leq C < \infty$ and C is in thousand dollars, and that the expected utility criterion is to be employed.

a. Which act is better, based on the prior probabilities alone?

b. Which act is better if two respondents in a random sample of 20 indicate they will purchase the heater?

c. What is the maximum amount that the firm should be willing to pay for a market survey even if it were to provide perfect information about the proportion of households in the target market that will purchase the heater?

UNIT SEVEN

—◆—

Time Series Analysis
and Index Numbers

24
Time Series—I

Managers and social scientists often deal with processes that vary as time passes. Observations on such a process, when arranged in time sequence, are called a time series. Examples of time series are monthly sales of a product for the past 240 months and annual gross national product in the United States since 1900.

Time series are analyzed to better understand, describe, control, and predict the underlying process. The analysis usually involves a study of components of the time series—such as trend, cyclical, and seasonal components—the particular components of interest tending to vary from one problem to another. Thus, an economist studying generating capacity needed by a power company 20 years hence will focus on the trend component of electric sales which reflects the effects of long-term factors on sales (e.g., population growth and technological changes). In contrast, the sales manager of a carpet company who wishes to incorporate seasonal effects into quarterly sales forecasts for next year will be concerned with the seasonal component of the company's sales.

In this chapter, we consider basic aspects of time series analysis and discuss some descriptive methods that are widely used. In the next chapter, we take up some statistical models for time series analysis.

24.1 TIME SERIES

(24.1) A *time series* is a sequence of n observations $Y_1, Y_2, \ldots, Y_t, \ldots, Y_n$ on a process at equally spaced points in time.

The equally spaced points in time might be months (as in a time series of monthly sales), years (as in annual population of Canada), or hours (as in hourly electricity consumption).

Time series arise in two different ways:

1. Readings are taken on a variable at consecutive points in time. Examples are: (a) The raw materials inventory of a manufacturer is recorded at the close of each fiscal quarter. (b) The level of atmospheric pollution at a downtown location is recorded at 5 P.M. each day.

2. A variable is aggregated or accumulated over an interval of time. Examples are: (a) Sales are aggregated to obtain monthly sales data. (b) Snowfalls are accumulated to obtain annual data on total snowfall.

Comment

Time series as defined in (24.1) are *discrete* time series, because the observations pertain to separated points in time. There are also *continuous* time series, where the variable Y is measured continuously over time. An electrocardiogram of heart action is a graph of a continuous time series.

24.2   SMOOTHING

Smoothing is a statistical procedure which is used to dampen or average out fluctuations in a time series to obtain a smooth component that reflects the systematic movement of the series. We explain smoothing by utilizing the following illustration.

☐ Illustration

A department store extended its evening openings to five nights, Monday through Friday. The store was staffed during these extended hours primarily by new, part-time sales clerks. Shortly after the new hours went into effect, customers began to complain increasingly about incidents of rudeness and inefficiency. Management quickly initiated a remedial program. Data on the daily number of complaints on Mondays through Fridays for the two weeks prior to the initiation of the remedial program and for four weeks thereafter are shown in Figure 24.1a.

An operations analyst perceived two characteristics of relevance in the time series data:

1. A downward shift occurred in the underlying direction of the series coinciding with the initiation of the remedial program.
2. Marked fluctuations within each 5-day period are evident both before and after the initiation of the remedial program.

To present these points effectively at a management review meeting, the analyst wished to average out the fluctuations to obtain a clearer picture of the underlying movement of the series. ☐

Moving Average

A moving average dampens fluctuations in a time series by taking successive averages of groups of observations:

(24.2) A *moving average* of a time series is obtained by replacing each successive, overlapping sequence of k observations in the series by the mean of that sequence. The first sequence contains observations $Y_1, Y_2, \ldots, Y_k$; the second sequence contains observations $Y_2, Y_3, \ldots, Y_{k+1}$; etc. k denotes the *term* of the moving average.

The term k of the moving average should be the same as the period of fluctuations in the time series, or a multiple of this period.

FIGURE 24.1 *Complaints series and components for department store example*

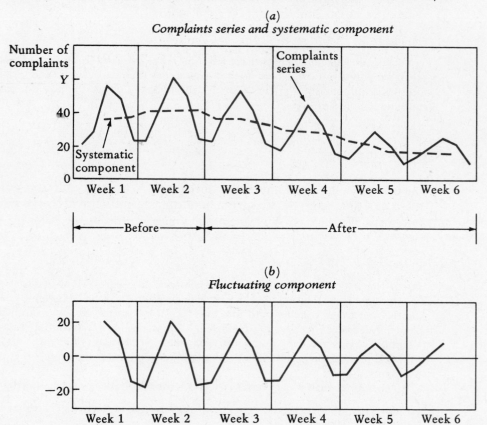

(a)
Complaints series and systematic component

(b)
Fluctuating component

□ Example

In our department store case, Figure 24.1a shows clearly that there is a 5-day cycle of fluctuations which is repeated week after week. Therefore, the analyst decided to use a 5-term moving average. Note that with this term, each sequence $(Y_1, Y_2, \ldots, Y_5)$, $(Y_2, Y_3, \ldots, Y_6)$, etc., contains an observation for each of the five days of the week the store is open in the evening. Thus, each average contains a Wednesday, when the volume of complaints reaches a peak, and a Monday and a Friday, when the volume is lower.

Table 24.1 shows the procedure for taking a 5-term moving average. The observations are given in column 1. Column 2 contains the totals of the overlapping sequences of five observations. For example, the total for the first sequence is $22 + 30 + 57 + 51 + 24 = 184$. This total is placed at the middle of its sequence at $t = 3$. Succeeding totals can be obtained readily by dropping the earliest observation in the preceding total and adding the next observation. Thus, for the total corresponding to $t = 4$, we have $184 - 22 + 24 = 186$, and the total corresponding to $t = 5$ is $186 - 30 + 41 = 197$. The totals in column 2 are called *moving totals*.

When the totals in column 2 are divided by 5, we obtain the moving averages. Thus,

TABLE 24.1 *Calculation of 5-term moving average for department store example*

Week and Day		t	(1) Number of Complaints Y_t	(2) 5-Term Moving Total	(3) 5-Term Moving Average	(4) Fluctuating Component (1) − (3)
1	M	1	22			
	T	2	30			
	W	3	57	184	36.8	20.2
	T	4	51	186	37.2	13.8
	F	5	24	197	39.4	−15.4
2	M	6	24	203	40.6	−16.6
	T	7	41	204	40.8	0.2
	W	8	63	205	41.0	22.0
	T	9	52	205	41.0	11.0
	F	10	25	205	41.0	−16.0
3	M	11	24	198	39.6	−15.6
	T	12	41	190	38.0	3.0
	W	13	56	186	37.2	18.8
	T	14	44	181	36.2	7.8
	F	15	21	170	34.0	−13.0
4	M	16	19	159	31.8	−12.8
	T	17	30	150	30.0	0.0
	W	18	45	146	29.2	15.8
	T	19	35	142	28.4	6.6
	F	20	17	135	27.0	−10.0
5	M	21	15	120	24.0	− 9.0
	T	22	23	108	21.6	1.4
	W	23	30	101	20.2	9.8
	T	24	23	99	19.8	3.2
	F	25	10	96	19.2	− 9.2
6	M	26	13	93	18.6	− 5.6
	T	27	20	92	18.4	1.6
	W	28	27	92	18.4	8.6
	T	29	22			
	F	30	10			

the first moving total, 184 at $t = 3$, is divided by 5 and the average, 36.8, is placed in column 3 at $t = 3$. The averages in column 3 constitute the moving average of the complaints series. Note that there are no moving average values for $t = 1$ and 2 and for $t = 29$ and 30, the first two and last two days of the time series. In general, for a k-term moving average, there will be no moving average values for the first and last $(k − 1)/2$ time periods when k is odd. The moving average of the complaints series, representing the systematic component, is plotted in Figure 24.1a, together with the time series of number of complaints.

Finally, the analyst subtracted the moving averages from the original observations to obtain the series in column 4. This series, plotted in Figure 24.1b, is the *fluctuating component* of the series—in the present case, the pattern of day-to-day fluctuations in complaints.

The analyst presented Figure 24.1 at the management review meeting with the following interpretation:

1. The systematic movement of complaints, portrayed by the moving average, turned downward shortly after the remedial program was introduced. However, the cause of the downturn cannot be established from the smoothed series alone. The downturn could have resulted from the remedial program, from experience gained by the clerks, or from some combination of these or other factors. Further information is required to establish the cause of the downturn.
2. The systematic movement appears to be leveling off in the last week or so. There may be no further improvement unless the situation receives additional attention from management.
3. The fluctuating component exhibits a midweek peak. Further investigation is required to ascertain whether this peak reflects a corresponding peak in the number of transactions handled or whether some other factors are responsible. □

Centered Moving Average. When an even number of terms is required in a moving average, e.g., a 4-term or 12-term moving average, a special problem arises. Consider the following case where a 4-term moving average is taken:

t:	1	2	3	4	5	6	7
Y_t:	2	6	4	8	6	7	2
Moving total ($k = 4$):			20	24	25	23	
Moving average ($k = 4$):			5.0	6.0	6.25	5.75	

The 4-term totals and averages, placed at the center of their sequences, fall at $t = 2.5$, $t = 3.5$, etc. This is unsatisfactory when original observations and moving averages are to be compared. The remedy is to *center* the moving averages as follows:

t:	1	2	3	4	5	6	7
Y_t:	2	6	4	8	6	7	2
Moving total ($k = 4$):			20	24	25	23	
Moving average ($k = 4$):			5.0	6.0	6.25	5.75	
Centered moving average:			5.5	6.13	6.0		

Note that centering simply involves taking a 2-term moving average of the 4-term moving average. Thus, the first average is $(5.0 + 6.0)/2 = 5.5$, corresponding to $t = 3$. Similarly, the second average is $(6.0 + 6.25)/2 = 6.13$ at $t = 4$.

With centered moving averages, we do not obtain moving average values for the first and last $k/2$ time periods. Here, $k/2 = 2$, so there are no moving average values for the first two and last two periods.

Comment

A centered moving average may be viewed as a *weighted moving average*, where the observations in a sequence do not receive equal weights. Specifically, the centered moving average in our previous illustration is a weighted 5-term moving average where the observations in a sequence receive weights (1, 2, 2, 2, 1). Thus, the first centered moving average can be obtained as follows:

$$\frac{1}{8}[2(1) + 6(2) + 4(2) + 8(2) + 6(1)] = \frac{44}{8} = 5.5$$

where 8 is the sum of the weights.

Effect of Term of Moving Average

The term k of the moving average can materially affect the moving average series. We illustrate this for our department store case. Since the fluctuating component had a period of five days there, it was natural to use a 5-term moving average to obtain a smoothed series that portrays effectively the systematic movement of the series. If 3-term or 7-term moving averages had been calculated, they would not have been effective in portraying the systematic component.

☐ Example

Figures 24.2a and 24.2b show the 3-term and 7-term moving averages, respectively, for the complaints data. The moving average series are anything but smooth. Since $k = 3$ and $k = 7$ are not multiples of the period of the fluctuations in the series, each moving average contains a mixture of the systematic and fluctuating components. Further, in the 7-term moving average the pattern of fluctuation is the reverse of that in the original series.

FIGURE 24.2 *3-term, 7-term, and 10-term moving averages for department store example*

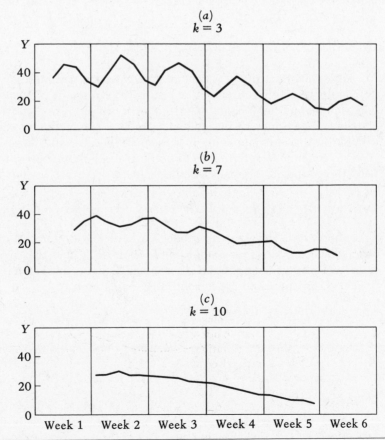

Figure 24.2c shows the centered 10-term moving average for the complaints data. Here, the moving average is smooth because 10 is a multiple of 5, the period of the fluctuating component. However, since moving averages are not obtained for the first and last five time periods when $k = 10$, important information is lost on the possible leveling off of the systematic component in the last week. □

This example demonstrates that the term k of a moving average used for smoothing should equal the period of fluctuation in the time series. Unfortunately, many time series in management, economics, and the social sciences oscillate with considerable irregularity. In these cases, the best compromise is to select the smallest value of k that will dampen the fluctuations. Computer programs for time series smoothing are available, and these enable one to experiment with different values of k.

Comment

Since a time series observation Y_t enters k successive moving averages (except at the beginning and end of the series), an unusually small or large observation can produce oscillations in the moving average series even though the original time series contains no periodicity. For this and other reasons, care must be used in interpreting oscillations in a moving average series.

24.3 CLASSICAL TIME SERIES MODEL

We have seen from the department store case how a time series can be decomposed into a smooth component, reflecting the systematic movement of the series, and a fluctuating component, reflecting the shorter-term and erratic movements in the time series. For many business and economic time series, however, this decomposition into two components is inadequate because the moving average series tends to combine long-term growth patterns and business cycle movements which need to be studied as separate components.

Multiplicative Model

Economists have developed a model for studying economic and business time series which contains four components. These are called the trend, cyclical, seasonal, and irregular components. In the model that is most widely used, the components are assumed to act in multiplicative fashion.

(24.3) The *classical multiplicative time series model* is:

$$Y = T \cdot C \cdot S \cdot I$$

where: $T, C, S,$ and I denote, respectively, the *trend, cyclical, seasonal,* and *irregular* components of the time series

□   Illustration

A company manufactures a large outdoor lighting fixture. Quarterly shipments of this fixture between 1972 and 1981 are shown in Figure 24.3a. The four components of this

FIGURE 24.3   *Shipments of outdoor lighting fixtures and time series components according to classical multiplicative model*

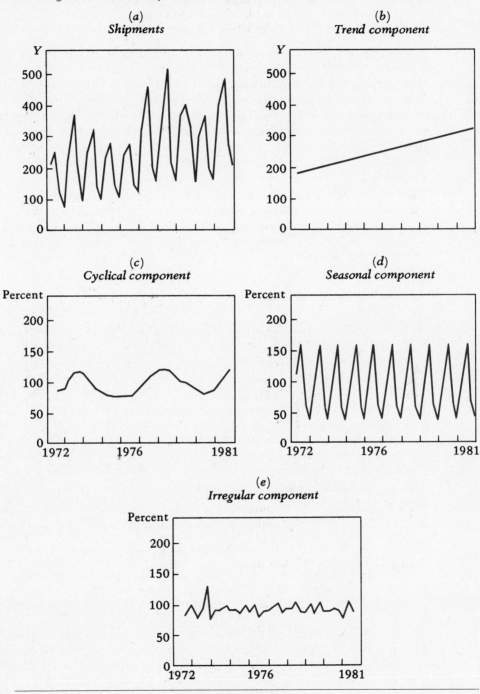

time series according to time series model (24.3) are shown in panels b through e of this same figure. We shall use this figure to illustrate the nature and meaning of the components in the classical model (24.3). ☐

Trend Component

The trend component of the time series is shown in Figure 24.3b. This component describes the net influence of long-term factors whose effects on the shipments tend to change gradually. Generally, these factors include: (1) changes in size, demographic characteristics, and geographic distribution of the population; (2) technological improvements; (3) economic development; and (4) gradual shifts in habits and attitudes. Since these effects tend to operate fairly gradually and in one direction or the other over long periods of time, the trend component usually is modeled by a smooth, continuous curve spanning the entire time series. The curve employed is called the *trend curve*. The trend curve in Figure 24.3b is linear, but in another case a different type of trend curve might be appropriate.

A major application of trend analysis is found in long-term forecasting.

Cyclical Component

The cyclical component of the time series on shipments of outdoor lighting fixtures is shown in Figure 24.3c. Generally, business and economic activities do not grow (or decline) at a steady rate. Instead, there are alternating periods of relative expansion and contraction. The cyclical component of a time series describes this characteristic. It measures the net effect of a variety of interrelated factors that tend to shift in direction from time to time and to vary in intensity and impact. As a result, the cyclical component shows cycles of expansion and contraction of uneven duration and amplitude. Factors leading to cyclical movements in business and economic time series include buildups and depletions of inventories, shifts in rates of capital expenditures by businesses, year-to-year variations in harvests, and changes in governmental monetary and fiscal policy. This list could be continued at length. The cyclical component in the shipments of outdoor lighting fixtures in Figure 24.3c reflects changes in sales arising from variations in the rate of development of new shopping centers and outdoor recreational facilities, which in turn is influenced by variations in underlying economic conditions.

Cyclical movements are studied for information on changes in rates of current activity. Such information is useful in assessing current conditions and in making short-term forecasts.

Seasonal Component

The seasonal component of the time series on shipments of outdoor lighting fixtures is shown in Figure 24.3d. This component describes effects that occur regularly over a period of a year, quarter, month, week, or day. Seasonal effects generally are associated with the calendar or the clock. For example, electricity consumption in the southern United States reaches a peak in summer, year after year, because of airconditioning demand. Department store sales reach a peak every December, because of the Christmas holidays. Shipments of outdoor lighting fixtures in our example reach a peak in each second quarter, because the height of construction activity is in the summer. Subway traffic reaches a peak each weekday morning and late afternoon, because of passengers going to and from work.

Seasonal effects tend to recur fairly systematically. Consequently, the pattern of movement in the seasonal component tends to be more regular than the cyclical pattern and therefore is more predictable. Note that the seasonal pattern in Figure 24.3d is stable over the entire period. In another series, however, we might find that the seasonal pattern is undergoing gradual modification.

Seasonal movements are measured so that seasonal effects can be taken into account in evaluating past and current activity, and so that they can be incorporated into forecasts of future activity.

Irregular Component

The irregular component describes residual movements that remain after the other components have been taken into account. Irregular movements reflect effects of unique and nonrecurring factors, such as strikes, unusual weather conditions, and international crises.

In some business and economic time series, the cyclical component is itself so irregular that any breakdown into separate cyclical and irregular components would be arbitrary. In such cases, a combined cyclical-irregular component is often developed.

Figure 24.3e contains the irregular component in the shipments series. The unusual peak in the third quarter of 1973 was due to the fact that a major competitor had a strike and could make no deliveries.

Definitions

We now define the time series components in the classical time series model (24.3) more formally:

(24.4) The components of a business or economic time series according to the classical time series model (24.3) are:

(24.4a) *Trend* component, which describes the long-term sweep of the series; it is usually modeled by a smooth curve.

(24.4b) *Cyclical* component, which describes the alternating periods of relative expansion and contraction of more than one year's duration in the series; it consists of cycles that vary in amplitude and duration.

(24.4c) *Seasonal* component, which describes the pattern of change recurring within periods of a year or less; it consists of a sequence of relatively repetitious cycles whose duration is a year or less.

(24.4d) *Irregular* component, which describes the effects of all other factors; it tends to have an irregular, saw-toothed pattern.

24.4 LINEAR TREND

Trend Function

As we noted earlier, the trend component is usually described by a smooth curve. We first consider a linear trend, such as the one for our shipments example in Figure 24.3b:

(24.5) The *linear trend* function is:

$$T_t = b_0 + b_1 X_t$$

Here, T_t is the trend value for period $t(t = 1, 2, \ldots, n)$, X_t represents a numerical code denoting period t, and b_0 and b_1 are the intercept and slope of the trend line, respectively.

We shall follow the practice of denoting time periods by $X_1 = 1$, $X_2 = 2, \ldots, X_t = t, \ldots, X_n = n$. Thus, for annual data from 1960 to 1981 ($n = 22$), 1960 is represented by $X_1 = 1$, 1970 by $X_{11} = 11$, and 1981 by $X_{22} = 22$.

Fitting Trend Line

Ordinarily, b_0 and b_1 are calculated by the method of least squares as described in Chapter 18. Recall that this method does not require an explicit specification of the distribution of the Y's; hence, it is useful for time series model (24.3) which does not make any such specification. Our discussion of fitting a linear regression model in Chapter 18 applies entirely to fitting a linear trend function. For convenience, we repeat the least squares formulas for b_0 and b_1, using the subscript t now instead of the earlier subscript i for the X and Y observations:

(24.6)
$$b_1 = \frac{\sum X_t Y_t - \dfrac{(\sum X_t)(\sum Y_t)}{n}}{\sum X_t^2 - \dfrac{(\sum X_t)^2}{n}}$$

(24.7)
$$b_0 = \frac{1}{n}(\sum Y_t - b_1 \sum X_t)$$

Since all summations are over the time series data ($t = 1, 2, \ldots, n$), we do not show the summation index in the formulas.

We now illustrate the calculation of a trend line and its uses by an example.

☐ Example

Figure 24.4a and Table 24.2 present data obtained by an investment analyst on the number of twin-engine executive jets sold annually between 1972 and 1981 by an aircraft manufacturer. The analyst was preparing a report on sales trend patterns for selected companies and now wished to isolate the trend component for this company. He noted that the annual sales observations tend to fall along a straight line when plotted against time (see Figure 24.4a). This finding, together with an analysis of the factors underlying the long-term sales pattern, led him to conclude that the trend of sales could be modeled

FIGURE 24.4   *Annual sales and linear trend and cyclical-irregular components for executive jet example*

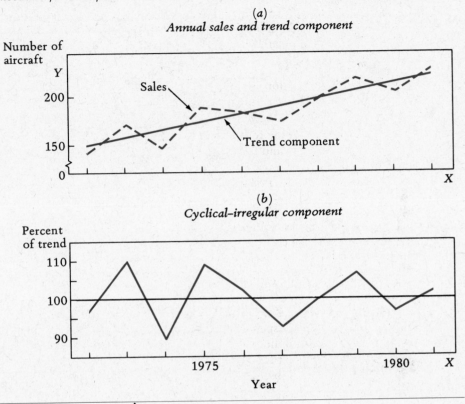

(a)
Annual sales and trend component

(b)
Cyclical–irregular component

TABLE 24.2   *Computations in fitting linear trend for executive jet example*

Year	(1) Year, in Coded Units X_t	(2) Number of Aircraft Sold Y_t	(3) $X_t Y_t$	(4) X_t^2	(5) Trend Value T_t	(6) Percent of Trend
1972	1	147	147	1	152.09	96.7
1973	2	175	350	4	160.25	109.2
1974	3	150	450	9	168.41	89.1
1975	4	191	764	16	176.56	108.2
1976	5	188	940	25	184.72	101.8
1977	6	179	1,074	36	192.88	92.8
1978	7	200	1,400	49	201.04	99.5
1979	8	220	1,760	64	209.19	105.2
1980	9	208	1,872	81	217.35	95.7
1981	10	230	2,300	100	225.51	102.0
Total	55	1,888	11,057	385		

appropriately by a straight line. We shall show the calculational steps, although the analyst obtained the linear trend equation directly from a hand-held calculator with a routine for trend fitting.

The preliminary calculations required for obtaining b_0 and b_1 are given in Table 24.2, columns 1 through 4. Note that $n = 10$ here. We obtain:

$$b_1 = \frac{11{,}057 - \dfrac{55(1{,}888)}{10}}{385 - \dfrac{(55)^2}{10}} = 8.15758$$

$$b_0 = \frac{1}{10}[1{,}888 - 8.15758(55)] = 143.93331$$

The fitted trend equation therefore is:

$$T_t = 143.93331 + 8.15758X_t$$

$$X_t = 1 \text{ at } 1972; \ X \text{ in one-year units}$$

Note that we indicate the coding scheme for X_t when we report the trend equation. The slope $b_1 = 8.2$ indicates that the trend value increases by 8.2 aircraft from one year to the next.

The trend line is plotted in Figure 24.4a. Trend values are calculated in the usual way. For example, the trend value for 1974 ($X_3 = 3$) is:

$$T_3 = 143.93331 + 8.15758(3) = 168.41$$

The trend values are presented in Table 24.2, column 5. ☐

Projection of Trend Values

Projection or extrapolation of trend values is performed by simply substituting the X_t value of interest into the trend equation.

☐ Example

In our executive jet example, it was desired to project the trend to 1986 ($X_{15} = 15$):

$$T_{15} = 143.93331 + 8.15758(15) = 266.30$$

Thus, if the long-term sweep of jet sales continues as in the past, the trend level of sales in 1986 will be 266 aircraft. ☐

Percents of Trend

Additional components of the time series can now be isolated in accord with model (24.3). Since the seasonal component pertains to cyclical movements with a period of one year or less, this component is omitted from the model when annual data are used.

(24.8) The classical multiplicative time series model (24.3) for *annual* data reduces to:

$$Y = T \cdot C \cdot I$$

Thus if we divide each observation Y_t in our annual time series by its trend value T_t, we obtain the combined cyclical-irregular component, denoted by $C \cdot I$:

$$(24.9) \qquad C \cdot I = \frac{Y}{T} = \frac{T \cdot C \cdot I}{T} \qquad \text{(Annual data)}$$

The $C \cdot I$ values are conventionally expressed as percents, and are called *percents of trend.*

☐ **Example**

In our jet aircraft example, for 1972 we have $Y_1 = 147$ and $T_1 = 152.09$. Hence, the percent of trend is $100(147/152.09) = 96.7$. The percents of trend for this example are given in Table 24.2, column 6, and are plotted in Figure 24.4b. Thus, Figure 24.4 shows the decomposition of the jet aircraft series into the trend and cyclical-irregular components. ☐

Comment

The percents of trend $C \cdot I$ may be smoothed by calculating a moving average, and the resulting moving average series is then taken as the cyclical component C. In accord with model (24.8), we would then divide each moving average value C into the corresponding percent of trend $C \cdot I$ to isolate the irregular component I.

24.5 EXPONENTIAL TREND

Nature of Exponential Trend

An exponential trend is one where the trend is changing at a constant *rate* from one period to another. This is in contrast to a linear trend, where the trend is changing by a constant amount. Exponential trends are encountered quite often.

☐ **Illustration**

Figures 24.5a and 24.5b present the time series for manufacturing production in the United States during 1950–1979. The series, which is expressed in index form, is plotted against an ordinary arithmetic Y-scale in Figure 24.5a and against a *ratio* or *logarithmic* scale in Figure 24.5b. Whereas equal distances on an arithmetic scale represent equal *amounts* of increase or decrease, equal distances on a ratio or logarithmic scale represent equal *rates* or *percents* of increase or decrease. For instance, the distances from 40 to 80, 60 to 120, and 50 to 100 are all equal on the ratio scale in Figure 24.5b, since in each case the increase is 100 percent. The grid in Figure 24.5b is often called a *semilogarithmic grid,* in contrast to the *arithmetic grid* in Figure 24.5a.

As seen from Figure 24.5a, the long-term pattern in the series bends upward, indicating that the trend component is increasing by larger increments each year. The same series in Figure 24.5b shows a long-term pattern that is approximately a straight line, indicating that the trend component is increasing by a constant rate from year to year. An exponential trend is appropriate for such a time series. ☐

FIGURE 24.5 *Index of manufacturing production for the United States, 1950–1979, plotted on arithmetic and semilogarithmic grids*

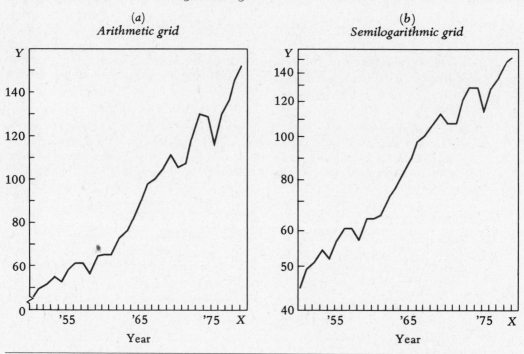

SOURCE: Federal Reserve Board.

Fitting Exponential Trend

If the long-term sweep of a series shows an increase (or decrease) from one period to the next at a constant rate, the logarithms of the series will show a linear trend. (A review of logarithms is found in Appendix A, Section A.2.) Hence, we obtain an exponential trend by first taking logarithms of the observations Y_t and then fitting a linear trend to the logarithms of the observations:

(24.10) The *exponential trend* function, when fitted to the logarithms of the observations, is:

$$T'_t = b_0 + b_1 X_t$$

where: $T'_t = \log T_t$

To obtain the trend value T_t for the original series, we simply take the antilog of T'_t. Thus, there is nothing really new in fitting an exponential trend once we have taken the logarithms of Y_t.

☐   Example

Column 2 of Table 24.3 contains a portion of the data Y_t for our production index example. Column 3 contains log Y_t, denoted by Y'_t. The logarithms as usual are to base 10.

TABLE 24.3 *Basic calculations for exponential trend example*

Year	(1) Year, in Coded Units X_t	(2) Production Index Y_t	(3) $Y'_t = \log Y_t$	(4) $X_t Y'_t$	(5) X_t^2
1950	1	45	1.6532	1.6532	1
1951	2	49	1.6902	3.3804	4
⋮	⋮	⋮	⋮	⋮	⋮
1978	29	147	2.1673	62.8522	841
1979	30	153	2.1847	65.5407	900
Total	465		57.7278	935.8752	9,455

$$b_1 = \frac{\sum X_t Y'_t - \frac{(\sum X_t)(\sum Y'_t)}{n}}{\sum X_t^2 - \frac{(\sum X_t)^2}{n}} = \frac{935.8752 - \frac{465(57.7278)}{30}}{9,455 - \frac{(465)^2}{30}} = .0182844$$

$$b_0 = \frac{1}{n}(\sum Y'_t - b_1 \sum X_t) = \frac{1}{30}[57.7278 - .0182844(465)] = 1.64085$$

Columns 4 and 5 contain $X_t Y'_t$ and X_t^2. The calculations of b_0 and b_1 are also shown in the table. The least squares formulas are the same as before, except Y'_t is used instead of Y_t.

The fitted trend equation is:

$$T'_t = 1.64085 + .0182844 X_t$$

$$X_t = 1 \text{ at } 1950; \text{ X in one-year units}$$

Trend values for the fitted exponential trend curve are obtained in the standard manner. For instance, the logarithmic trend value for 1950 ($X_1 = 1$) is:

$$T'_1 = 1.64085 + .0182844(1) = 1.65913$$

We then take the antilog to obtain the trend value in the original units:

$$T_1 = \text{antilog } 1.65913 = 45.6$$

☐

Rate of Growth

The constant rate of growth implicit in a fitted exponential trend curve, denoted by g, is given by:

(24.11)
$$g = (\text{antilog } b_1) - 1$$

☐ Example

In our production index example, $b_1 = .0182844$, so we have:

$$g = (\text{antilog } .0182844) - 1 = 1.0430 - 1$$

$$= .0430 \text{ or } 4.30 \text{ percent per year}$$

☐

Comments

1. The exponential trend sometimes is called *log-linear* trend because it is a linear trend for the logarithms of the observations Y_t.
2. Once the trend values in the original units are obtained by taking antilogs of the trend values T'_t, the usual decomposition of the time series can be developed by calculating percents of trend and proceeding from there in accordance with model (24.3) or (24.8), depending upon whether or not a seasonal component is present.
3. The observations Y_t must be positive for $\log Y_t$ to be defined, but such is often the case in business and economic series.
4. Graph paper with a semilogarithmic grid, such as the one in Figure 24.5b, is readily available. It can be useful to plot a time series on both arithmetic and semilogarithmic grids, as in Figure 24.5, to study the nature of the trend. If semilogarithmic graph paper is unavailable, an equivalent graph can be obtained by plotting $\log Y_t$ against time on an arithmetic grid.

24.6   GOMPERTZ TREND

Many business and economic activities tend to undergo a period of rapid expansion once they are established, but as the size of the activity increases the rate of expansion slackens and a saturation point is eventually approached. A representative trend pattern for such an activity is shown in Figure 24.6, where the trend values are plotted on both arithmetic and semilogarithmic grids. On the arithmetic grid the pattern forms an elongated S, indicating a slow increase first, then a rapid one, and finally a retardation in absolute growth. The semilogarithmic grid shows a pattern with a declining slope throughout, indicating that the rate of growth is declining during the entire course of the series.

Curves that fit the general pattern of Figure 24.6 are called *growth curves*. One

FIGURE 24.6   *Gompertz growth curve plotted on arithmetic and semilogarithmic grids*

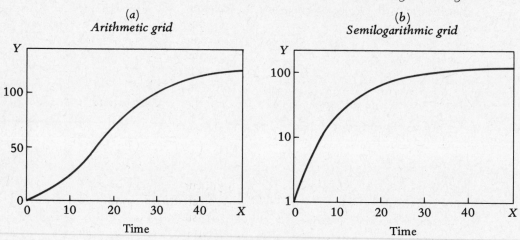

important example of a growth curve is the *Gompertz trend function*. The Gompertz trend function involves an upper asymptote and is usually fitted by special methods. The upper asymptote must be interpreted with caution because it is very sensitive to the observations and may change materially with one or two new observations. Also, there are always serious risks in extrapolating far beyond the range of observations. Finally, while a growth curve may provide a good fit to past data, the assumption of a fixed upper limit may not be reasonable for many business and economic activities in an open and expanding economy.

Comment

The Gompertz trend function is:

$$(24.12) \qquad\qquad T_t = b_0 b_1{}^{b_2{}^{X_t}}$$

The growth curve plotted in Figure 24.6 is a Gompertz trend function with $b_0 = 125$, $b_1 = .008$, and $b_2 = .90$:

$$T_t = 125(.008)^{.90^{X_t}}$$

The coefficient b_0 represents the upper asymptote of the growth curve when $0 < b_1 < 1$ and $0 < b_2 < 1$, and b_2 reflects the degree of retardation in the rate of growth from one period to the next. In Figure 24.6, the upper asymptote of the Gompertz curve is $b_0 = 125$ and the retardation coefficient is $b_2 = .90$.

24.7 FORECASTING BY TREND PROJECTIONS

Trend fitting is used often as an aid in forecasting. This usage is inherently subjective since the fitted model is extrapolated into a future that never is seen clearly. We now take up some considerations in using trend fitting for forecasting.

Short-Term versus Long-Term Forecasting

Short-term operating forecasts must be relatively precise. If the cyclical-irregular component in the activity to be forecast has been negligible and is likely to continue so in the future, it may be feasible to project the trend curve a year or two ahead for short-term forecasting. In most cases, however, the cyclical-irregular component is potentially too large to permit a short-term forecast to be based on trend alone. On the other hand, trend projections often are of assistance in longer-term forecasting, where long-term effects are important and shorter-term fluctuations are of secondary consideration.

Time Period

Ideally, the observed data to which the trend curve is fitted should be relevant to the period into which the curve will be projected. If important new conditions affecting trend have arisen in the recent past (such as a new technological breakthrough), data prior to the change of conditions may have to be discarded or given less weight. If important new conditions affecting trend are expected in the future (such as new environmental protection legislation), the trend projection may have to be modified to take into account the effects of the expected new conditions.

Choice of Trend Curve

The choice of a trend curve cannot be made solely on the basis of how good the fit is to the past data. Usually, there will be a number of trend curves that fit the data well. However, when the trend curves are projected beyond the range of the data, the different trend curves may diverge sharply. Figure 24.7 illustrates this point. Exponential and Gompertz curves both fit the sales data closely but diverge increasingly as the curves are projected beyond the range of the data.

Thus, the choice of the trend curve for forecasting must not only be based on a good fit to the past data but must also reflect an expert's assessment of the long-term factors determining the future course of the activity. If an exponential trend appears to be apt for the past data and it is expected that the constant rate of growth in the past will continue in the future, an exponential trend equation should be used. If the past rate of growth is expected to taper off, on the other hand, a Gompertz growth curve might be used instead.

The choice of the trend curve is also affected by the length of the projection. An exponential trend might be reasonable for a projection of 5 years, but not for a projection of 20 years.

FIGURE 24.7 *Diverging trend projections from exponential and Gompertz trend curves fitted to same series—Semilogarithmic grid*

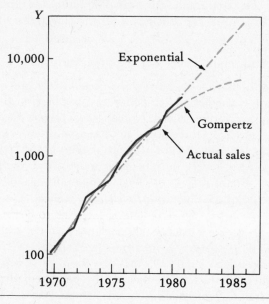

Computer-Assisted Screening

Computer programs are available for assistance in screening alternative trend functions. In a typical program, a number of different types of trend curves are available for fitting to the time series. For each type of curve, the fitted equation can be obtained together with a descriptive measure, such as r^2, showing the degree of relationship between the observations Y_t and the fitted trend values T_t. Of course, as we noted previously, the degree to which a trend curve fits the past observations is only one of the factors to be considered in selecting a trend curve for forecasting.

24.8   CYCLICAL ANALYSIS AND SHORT-TERM FORECASTING

Cyclical movements in business and economic data are studied to assess the direction in which an economic sector is moving, to evaluate the short-term outlook for an industry or company, and to prepare intermediate and short-term forecasts used in planning and controlling day-to-day operations. While many time series in engineering and the physical sciences have cyclical components that are quite regular and can be described by periodic functions, cyclical components in business and economic series tend to vary widely in both duration and amplitude from one cycle to the next, and hence do not lend themselves to fitting of mathematical periodic functions.

Identification of Cyclical Turning Points

A major objective of cyclical analysis of business and economic time series is to predict cyclical downturns and upturns in the activity. Unfortunately, this is a difficult task. Analysts are often satisfied simply to be able to identify cyclical turning points shortly after they have occurred.

(24.13) A *cyclical turning point* is the time point at which the cyclical component of a time series changes direction.

Even the task of identifying a turning point retrospectively is difficult; often it takes months before a cyclical turn can be recognized.

The reasons for the difficulties in ascertaining current and future cyclical developments are illustrated by the deseasonalized monthly series on value of manufacturers' new orders received in the durable goods industries between 1956 and 1979, presented in Figure 24.8. We can visually trace the cyclical pattern in this series and, with some arbitrariness, identify the cyclical turning points. Figure 24.8 has been divided into clear and shaded intervals to correspond, respectively, to the periods of cyclical expansion and contraction in the series. Note in Figure 24.8 the variability of the durations and amplitudes of the cycles, typical of most business and economic series.

In addition, substantial irregular movements are present in the series in Figure 24.8. These irregular movements, together with the variability in the cyclical

FIGURE 24.8 *Deseasonalized value of manufacturers' new orders received monthly in durable goods industries, 1956–1979—Semilogarithmic grid*

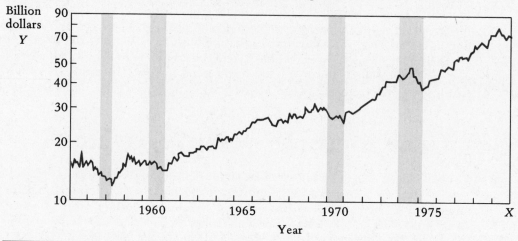

SOURCE: U.S. Department of Commerce.

swings, make it virtually impossible to identify current or imminent cyclical behavior from inspection of the data alone.

Cyclical Indicators

Cyclical Indicators Approach.   A useful framework for cyclical analysis of business conditions is based on the cyclical indicators approach originally developed by the National Bureau of Economic Research (NBER). In this approach, economic time series are sought which are *leading, coinciding,* and *lagging* in comparison with cyclical movements in aggregate economic activity. A leading series, for instance, is one whose cyclical upturns and downturns tend to occur before the turns in aggregate activity.

Although most series are not sufficiently consistent to serve as reliable indicators, there are exceptions. The Bureau of Economic Analysis (BEA) of the Department of Commerce, working with NBER, has examined many hundreds of series and identified and listed the consistent ones. The classification process is ongoing, and the list is subject to change; it currently contains about 150 series. The BEA has identified a short list of cyclical indicator series, culled from the larger list, which represent key activities in the cyclical process, perform particularly well as indicators, and are collected and published on a timely basis. Table 24.4 shows the short list current in the early 1980s.

Use of Cyclical Indicators.   The leading series are studied primarily to help in anticipating cyclical turning points. The coincident and lagging series are studied mainly for assistance in recognizing a cyclical turn once it has occurred. Unfortu-

TABLE 24.4 *Cyclical indicators in BEA composite indexes*

Leading Indicators

Average workweek, production workers, manufacturing
Layoff rate per 100 employees, manufacturing
New orders for consumer goods and materials, 1972 dollars
Vendor performance, percent of companies receiving slower deliveries
Net business formation, index: 1967 = 100
Contracts and orders for plant and equipment, 1972 dollars
New building permits, private housing units, index: 1967 = 100
Change in inventories on hand and on order, smoothed series, 1972 dollars
Percent change in sensitive crude materials prices, smoothed series
Stock prices, 500 common stocks, index: 1941–43 = 10
Percent change in total liquid assets, smoothed series
Money supply (M2), 1972 dollars

Coincident Indicators

Employees on nonagricultural payrolls
Personal income less transfer payments, 1972 dollars
Industrial production, total, index: 1967 = 100
Manufacturing and trade sales, 1972 dollars

Lagging Indicators

Average duration of unemployment
Manufacturing and trade inventories, 1972 dollars
Labor cost per unit of output, manufacturing, index: 1967 = 100
Average prime rate charged by banks
Commercial and industrial loans outstanding
Ratio of consumer installment credit to personal income

Source: U.S. Department of Commerce.

nately, the situation is not always clear-cut. At a given time, 8 of the 12 leading indicators on the short list may be moving in one direction while the others are indeterminate or are moving in the opposite direction. Similar mixed pictures can exist in the coincident and lagging series.

To assist in the monitoring of the indicators, various indexes have been developed, each of which summarizes some aspect of the movements in a group of indicators. For example, the BEA composite index of leading indicators is based on the 12 series in Table 24.4, each of which has been standardized. Similarly, there are composite indexes of lagging and coincident indicators based on the respective indicator series in Table 24.4.

The cyclical indicators approach, despite some difficulties because of inconsistent behavior among the series, has proven to be highly useful, and a broad apparatus of business cycle analysis has developed around it. A key source that brings together the series used in the cyclical indicators approach is *Business Conditions Digest*, published monthly by the Bureau of Economic Analysis, U.S. Department of Commerce. Figure 24.8 is derived from this publication.

Canadian Experience. The cyclical indicators concept also has been evaluated for Canada. The Canadian series perform almost identically, in relation to the Canadian cycle, as their U.S. counterparts perform for the U.S. cycle.

24.9 CALCULATION OF SEASONAL INDEXES

The seasonal component frequently is the chief source of pronounced short-term fluctuations in business and economic time series, as Figures 24.1 and 24.3 demonstrate. In each of these cases, the seasonal component dominates the short-term movements. In this section, we discuss how to measure the seasonal component in the form of seasonal indexes, and in Section 24.10 we show how seasonal indexes are used in analysis of past and current activity and for short-term forecasting.

Method of Ratio to Moving Average

The method most commonly used for measuring the seasonal component is the *method of ratio to moving average*. This method is appropriate for the multiplicative model (24.3). We illustrate it for a quarterly time series but it can be readily adapted to monthly or weekly series.

☐ Illustration

Figure 24.9 presents the number of revenue passenger miles flown quarterly between 1976 and 1981 by Taxair, Inc., an air taxi service. We note a recurrent seasonal pattern marked by relatively low mileage in the first and fourth quarters and peak mileage in the third

FIGURE 24.9 *Revenue passenger miles flown quarterly, Taxair, Inc., 1976–1981*

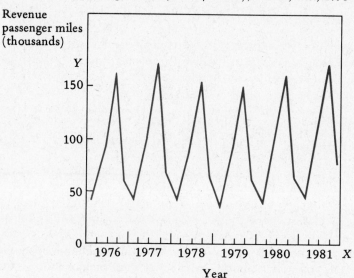

quarter. Of course, the fluctuations are not perfectly repetitive, in large part because irregular effects are present in the series. ☐

Basic Procedure. The procedure for measuring the seasonal component by the method of ratio to moving average involves three steps:

(24.14) The procedure for measuring the seasonal component by the *method of ratio to moving average* is:

1. Smooth the series; the smoothed values reflect $T \cdot C$, the trend-cyclical component.
2. Obtain the seasonal-irregular component, denoted by $S \cdot I$, by dividing the original series by the smoothed values:

$$\frac{Y}{T \cdot C} = \frac{T \cdot C \cdot S \cdot I}{T \cdot C} = S \cdot I$$

3. Isolate the seasonal component S by averaging out the irregular effects in $S \cdot I$.

Smoothing to Obtain $T \cdot C$ Component. Table 24.5, column 1, contains the quarterly revenue passenger miles for the Taxair example. Since the seasonal pattern repeats every four quarters, a centered 4-quarter moving average is appropriate here, as discussed earlier. We do not show the detailed computational steps and present the centered 4-quarter moving average in column 2.

As an illustration, the first two 4-quarter moving averages are:

$$\frac{43.20 + 90.00 + 162.00 + 64.80}{4} = 90.0000$$

$$\frac{90.00 + 162.00 + 64.80 + 42.35}{4} = 89.7875$$

and their average is $(90.0000 + 89.7875)/2 = 89.894$. This is the first entry in column 2, for quarter 3, 1976.

The centered 4-quarter moving average in column 2 represents $T \cdot C$, the trend-cyclical component.

Obtaining $S \cdot I$ Component. The seasonal-irregular component $S \cdot I$ is isolated in column 3 by dividing the observations in column 1 by the corresponding $T \cdot C$ values in column 2 and expressing the result as a percent. For example, $S \cdot I$ for quarter 3, 1976 is $100(162.00/89.894) = 180.21$.

The entries in column 3, representing the seasonal-irregular component $S \cdot I$, are called *specific seasonal relatives* because they measure seasonal and irregular effects specific to each quarter.

Isolation of Stable Seasonal Component S. The procedure for carrying out the third step—i.e., isolating the seasonal component—depends on whether this component is stable from year to year or is undergoing gradual modification. A conven-

TABLE 24.5 *Computation of quarterly specific seasonal relatives for Taxair example by method of ratio to moving average*

Year and Quarter	(1) Revenue Passenger Miles (*thousands*) Y	(2) Centered 4-Quarter Moving Average $T \cdot C$	(3) Specific Seasonal Relative $S \cdot I$
1976　1	43.20		
2	90.00		
3	162.00	89.894	180.21
4	64.80	91.050	71.17
1977　1	42.35	93.719	45.19
2	100.10	95.688	104.61
3	173.25	96.326	179.86
4	69.30	95.078	72.89
1978　1	42.96	91.786	46.80
2	89.50	89.660	99.82
3	157.52	88.874	177.24
4	68.02	88.272	77.06
1979　1	37.95	87.582	43.33
2	89.70	86.559	103.63
3	151.80	86.525	175.44
4	65.55	87.450	74.96
1980　1	40.15	89.656	44.78
2	94.90	91.231	104.02
3	164.25	92.186	178.17
4	65.70	93.667	70.14
1981　1	47.64	95.514	49.88
2	99.25	98.034	101.24
3	174.68		
4	75.43		

ient way to study this question is to assemble the specific seasonal relatives $S \cdot I$ by quarter and to plot them in time sequence. This is done in Figure 24.10.

We examine Figure 24.10 to determine whether any systematic modification is taking place in the seasonal component. For instance, if the seasonal pattern were flattening in recent years, the line graphs for the first and fourth quarters would tend upward while that for the third quarter would tend downward. In Figure 24.10, the line graphs do not appear to have any systematic upward or downward tendencies, though they do show substantial irregular effects. Hence, we conclude that the seasonal pattern in the Taxair series is stable from year to year. (In practice, we would supplement our examination of the data with an analysis of the factors causing the seasonal effects.)

When the seasonal pattern is stable, as here, the seasonal component S is isolated by averaging the specific seasonal relatives for each quarter. The idea is that irregular effects will tend to cancel one another in the average and, as a result, the average will reflect primarily the seasonal effect for that quarter. To avoid the possibility that extreme irregular effects will distort the results, an average that is

FIGURE 24.10 *Specific seasonal relatives by quarter for Taxair example.* No systematic upward or downward tendencies are evident, which suggests that the seasonal pattern has been stable.

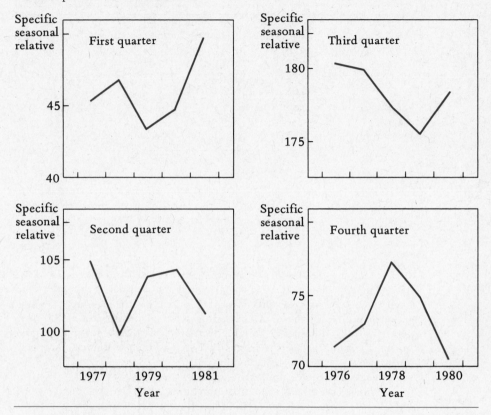

not affected by extreme values, such as a trimmed mean or the median, is usually employed. We shall employ the median here.

The specific seasonal relatives $S \cdot I$ from column 3 of Table 24.5 are assembled by quarter in Table 24.6. We find the median of the specific seasonal relatives for the first quarter to be 45.19. The medians for each of the four quarters are shown in Table 24.6.

One last step remains to obtain seasonal indexes. Since a seasonal pattern must balance out over a year, the multiplicative model (24.3) requires that the seasonal indexes average out to 100 percent. The method of ratio to moving average does not contain a built-in procedure to constrain the medians to average out to 100 percent; hence, a final adjustment is usually required.

In our Taxair example, the medians sum to 399.88. They should sum to 400.00 to average out to 100. Hence, we simply multiply each median by the factor $400.00/399.88 = 1.00030$. The adjusted medians are the *seasonal indexes;* they are shown in Table 24.6.

TABLE 24.6 *Specific seasonal relatives and seasonal indexes for Taxair example*

	Quarter				
Year	1	2	3	4	Total
1976			180.21	71.17	
1977	45.19	104.61	179.86	72.89	
1978	46.80	99.82	177.24	77.06	
1979	43.33	103.63	175.44	74.96	
1980	44.78	104.02	178.17	70.14	
1981	49.88	101.24			
Median	45.19	103.63	178.17	72.89	399.88
Seasonal					
index	45.2	103.7	178.2	72.9	400.0

We thus see that for Taxair the seasonal component ranges from a low of 45.2 in the first quarter to a high of 178.2 in the third quarter, indicating large seasonal effects.

Comments

1. At times the seasonal pattern shifts gradually because of technological developments, changes in customs or habits, and managerial actions such as promotional campaigns timed to dampen seasonal effects. In these cases, we can measure the degree of shift and then project it ahead in calculating seasonal indexes for future periods. One method of doing this is to fit "trend" functions to the specific seasonal relatives for each time unit (e.g., to the specific seasonal relatives in each of the four quarters) and then project each of these "trend" functions.
2. A number of standard computer programs are available for compiling seasonal indexes. One of these is the widely used program developed by the U.S. Bureau of the Census. It employs a refined version of the ratio to moving average method described here.

24.10 APPLICATIONS OF SEASONAL INDEXES

In working with time series data, questions such as the following arise:

1. Is the recent increase in unemployment normal for this time of year?
2. Has a cyclical contraction of retail sales begun which currently is masked by the pre-Christmas buying rush?
3. A sales target of 400,000 units is reasonable for the next 12 months. In breaking down this figure into monthly sales targets, how do we allow for seasonal effects?

Obtaining answers to these questions illustrates two important uses of seasonal indexes:

1. To adjust for seasonal effects in past and current data in order to reveal movements in the other time series components more clearly (Questions 1 and 2).

2. To incorporate seasonal effects into forecasts (Question 3).

We now take up each of these uses in turn.

Adjusting for Seasonal Effects

For the classical multiplicative model (24.3), adjusting for seasonal effects can be represented as follows:

(24.15)
$$\frac{Y}{S} = \frac{T \cdot C \cdot S \cdot I}{S} = T \cdot C \cdot I$$

Thus, the observed value for a period is simply divided by the seasonal index for that period (in decimal form) to obtain the trend-cyclical-irregular component, denoted by $T \cdot C \cdot I$. The $T \cdot C \cdot I$ value for a period is called the *seasonally adjusted value* or *deseasonalized value* of the series for that period.

☐ Example

Table 24.7, column 1, presents the number of transactions of a department store by quarter for the year just ended. The seasonal indexes for the volume of transactions are given in column 2. We first wish to obtain the deseasonalized number of transactions for the first quarter; the actual number is 17.36 thousand. The first-quarter seasonal index of 70 indicates that the actual number of transactions in the first quarter is 70 percent as great as if no seasonal effects were present. Thus, to find the $T \cdot C \cdot I$ component (i.e., the rate of transactions if seasonal effects were not present), we employ (24.15) as follows:

$$T \cdot C \cdot I = \frac{Y}{S} = \frac{17.36}{.70} = 24.80 \text{ thousand}$$

Thus, if no seasonal effects had been present, the trend-cyclical-irregular factors for the first quarter would have led to a quarterly rate of 24.80 thousand transactions.

The rate 24.80 thousand transactions is the deseasonalized or seasonally adjusted rate of transactions for the first quarter. The seasonally adjusted transactions for the other quarters are obtained in like manner, by dividing the number of transactions by the seasonal index for that quarter (in decimal form). The results are shown in column 3 of Table 24.7. Note that the seasonally adjusted transactions declined in the third and fourth

TABLE 24.7 *Calculation of deseasonalized number of transactions and seasonally adjusted annual rate of transactions*

Quarter	(1) Thousands of Transactions Y	(2) Seasonal Index S	(3) Seasonally Adjusted Transactions $T \cdot C \cdot I$	(4) Seasonally Adjusted Annual Rate of Transactions $4(T \cdot C \cdot I)$
1	17.36	70	24.80	99.20
2	21.11	85	24.84	99.36
3	21.68	92	23.57	94.28
4	32.25	153	21.08	84.32

quarters. Although the actual transactions increased during these quarters, this increase was less than would be expected on a seasonal basis alone. Hence, a decline has occurred in the trend-cyclical-irregular component which is possibly cyclical in nature and warrants attention. □

Seasonally Adjusted Annual Rates. Often, seasonally adjusted data are expressed in terms of annual rates for comparison with annual data for past years. The *seasonally adjusted annual rates* for our department store example are given in column 4 of Table 24.7. They are obtained by multiplying the entries in column 3 by four. The seasonally adjusted annual rate of 99.20 thousand transactions for the first quarter, for instance, indicates that if the trend-cyclical-irregular factors for the first quarter prevailed for an entire year, the annual number of transactions would be 99.20 thousand.

Deseasonalized Percents of Trend. If interest is in the cyclical-irregular component $C \cdot I$, rather than in the trend-cyclical-irregular component $T \cdot C \cdot I$, we would first obtain percents of trend:

$$(24.16) \qquad \frac{Y}{T} = C \cdot S \cdot I$$

and then deseasonalize these percents of trend:

$$(24.17) \qquad \frac{C \cdot S \cdot I}{S} = C \cdot I$$

If the irregular effects should also be eliminated so that the cyclical effects stand out, we would then smooth the cyclical-irregular component.

The procedure for obtaining deseasonalized percents of trend is illustrated in Table 24.8 for a fragment of a quarterly series.

Incorporation of Seasonal Effects into Forecasts

In short-term forecasting, it is usually essential to incorporate the seasonal effects into the forecast. The basic procedure is to start with an annual forecast. If quar-

TABLE 24.8 *Procedure for calculating deseasonalized percents of trend*

Year and Quarter		(1) Number of Units Sold Y	(2) Trend Value T	(3) Percent of Trend $C \cdot S \cdot I$ $100[(1) \div (2)]$	(4) Seasonal Index S	(5) Deseasonalized Percent of Trend $C \cdot I$ $100[(3) \div (4)]$
⋮	⋮	⋮	⋮	⋮	⋮	⋮
1980	1	100,000	90,000	111	97	114
	2	150,000	95,000	158	135	117
	3	120,000	100,000	120	106	113
	4	67,000	105,000	64	62	103
⋮	⋮	⋮	⋮	⋮	⋮	⋮

terly forecasts are desired, the annual forecast is usually then divided by four to obtain equal quarterly rates in which no seasonal effects are present. These quarterly rates are then multiplied by the seasonal indexes (in decimal form).

☐ Example

A sales manager has received an annual sales forecast of 200 thousand units for a product. She desires sales forecasts for each of the four quarters. The procedure is illustrated in Table 24.9. First, we break down the annual forecast into the quarterly rates without seasonal effects, shown in column 1. This breakdown into equal quarterly rates assumes that quarter-to-quarter changes in the trend-cyclical component will be negligible. (A different assumption could be made if warranted.) The quarterly seasonal indexes for sales of this product are given in column 2. For the multiplicative model (24.3), the seasonal component now is incorporated by multiplying each entry in column 1 by the corresponding seasonal index (in decimal form). The results, entered in column 3, are the desired quarterly sales forecasts. ☐

24.11 OPTIONAL TOPIC—EXPONENTIAL SMOOTHING

Many forecasting applications require the analysis of a large number of time series on a frequent, periodic basis.

☐ Examples

1. A manufacturer stocks thousands of different types of spare parts, raw materials, components, and finished products. Demand forecasts for these items are required monthly so that inventory carrying costs on the one hand and out-of-stock costs on the other can be controlled at reasonably low levels.
2. A manager of many investment portfolios requires that rates of return for a large number of securities be monitored weekly. ☐

The methods of time series analysis and forecasting discussed earlier in this chapter require much attention by skilled professionals and large amounts of data handling and storage, and would be very expensive to apply to hundreds or thousands of time series on a frequent, periodic basis. In this latter type of situation,

TABLE 24.9 *Procedure for incorporating seasonal effects into forecasts*

Quarter		(1) Forecast without Seasonal Effects	(2) Seasonal Index	(3) Forecast with Seasonal Effects $(1) \times \frac{(2)}{100}$
1		50	123	61.5
2		50	108	54.0
3		50	79	39.5
4		50	90	45.0
	Total	200		200.0

other forecasting methods are frequently employed. *Exponential smoothing* is such a method, being routine and computer-oriented, and requiring a minimum of professional attention and data storage. The basic exponential smoothing model, which is the one considered here, employs a weighted average to make projections for time series where trend and seasonal components are not present.

Basic Model for Exponential Smoothing

The basic model for exponential smoothing, expressed in terms of forecasting demand, is:

(24.18)
$$F_{t+1} = F_t + w(D_t - F_t)$$

where: F_t is the forecast of demand for the current period t, made in period $t - 1$
F_{t+1} is the forecast of demand for period $t + 1$, made in the current period t
D_t is observed demand in the current period t
w is a weight such that $0 < w < 1$

Note that the error in the forecast for the current period t is the quantity $D_t - F_t$. Hence, model (24.18) provides that the forecast for the period ahead is given by the forecast for the current period plus a proportion w of the error in the forecast for the current period. If the forecast for the current period is, say, too low, the quantity $D_t - F_t$ is positive and the forecast for the next period is increased.

□ Example

Weekly demand for a certain type of industrial oil carried in inventory at a distribution center is to be forecast by formula (24.18) with $w = .10$. The forecast of demand for the current week t was $F_t = 363.0$ barrels. Actual demand for the current week is found to be $D_t = 383$ barrels. The forecast for week $t + 1$ then is:

$$F_{t+1} = 363.0 + .10(383 - 363.0) = 365.0 \text{ barrels}$$

Actual demand for week $t + 1$ was $D_{t+1} = 364$ barrels. The forecast for week $t + 2$ then is:

$$F_{t+2} = 365.0 + .10(364 - 365.0) = 364.9 \text{ barrels} \qquad □$$

Properties

The exponential smoothing procedure (24.18) has a number of important properties:

1. The procedure is "adaptive" in that it automatically responds to observed errors in the current forecast by taking corrective action intended to reduce the error in the next forecast.
2. The only information required to calculate the new forecast F_{t+1} is the observed current demand D_t and the forecast for the current period F_t. The only information that must be carried over to the next period is F_{t+1}. Thus, only one

number need be stored for each item in the forecasting system from one period to the next. The small data storage requirement is a key advantage of exponential smoothing over alternative computer-based forecasting systems.

3. Exponential smoothing is a weighted average procedure. Consider the forecast for period $t + 1$, which may be written:

$$F_{t+1} = F_t + w(D_t - F_t) = wD_t + (1 - w)F_t$$

Now forecast F_t, in turn, is given by:

$$F_t = F_{t-1} + w(D_{t-1} - F_{t-1}) = wD_{t-1} + (1 - w)F_{t-1}$$

Substituting this result for F_t in the forecast equation for F_{t+1} gives:

$$F_{t+1} = wD_t + w(1 - w)D_{t-1} + (1 - w)^2 F_{t-1}$$

Repeating this process by substituting for F_{t-1}, we obtain:

$$F_{t+1} = wD_t + w(1 - w)D_{t-1} + w(1 - w)^2 D_{t-2} + (1 - w)^3 F_{t-2}$$

Repeating this process further, one sees that the forecast for period F_{t+1} is a weighted average of the demands $D_t, D_{t-1}, D_{t-2}, \ldots$ for all earlier periods. Notice that the earlier the period, the less weight that is assigned to its demand. This follows, because the weight w is such that $0 < w < 1$. Hence, the weight $w(1 - w)^k$ shrinks as k increases. Indeed, the term *exponential smoothing* refers to the fact that the weight $w(1 - w)^k$ declines geometrically the further the period is in the past (i.e., the larger is k). For $w = .1$, for instance, the weight for D_{t-10} in forecast F_{t+1} is $w(1 - w)^{10} = .03$, while the weight for D_{t-20} is .01. Thus, demands in periods far in the past receive very little weight.

4. The exponential smoothing forecast is sensitive to the choice of w, sometimes called the *smoothing constant*. The closer w is to 1, the faster the weight $w(1 - w)^k$ declines with k. In consequence, the larger is w, the less weight is given to each past demand. For example, if $w = .5$, the weight for D_{t-10} is .0005. (Recall the weight was .03 for $w = .1$.) One of the problems in using exponential smoothing is to determine the best value of w for each time series. As just noted, if w is close to 1, relatively heavy weight is placed on the demands in the current and recent periods in developing the forecast for the next period. The forecasting scheme then becomes overly responsive to current and recent irregular movements in demand. If w is too low, on the other hand, the system responds too slowly to basic changes in the underlying pattern of demand.

If information on past demands is available, one can experiment with various values of w to see which value gives the best agreement between forecasts and actual demands.

Comment

Formula (24.18) for exponential smoothing can be extended to include adjustments for trend and/or seasonal effects, though at an increase in data storage and calculational requirements. Since items carried in exponential smoothing forecasting systems often are subject to trend or seasonal effects or both, extensions of the basic formula frequently are worth the extra cost involved.

PROBLEMS

*24.1 Data on total production of pulp and paper (in millions of metric tons) in one overseas division of a multinational company during 1962–1981 follow.

Year:	1962	'63	'64	'65	'66	'67	'68	'69	'70	'71
Production:	3.3	3.7	3.9	3.5	2.9	3.1	3.5	3.6	3.2	3.5
Year:	'72	'73	'74	'75	'76	'77	'78	'79	'80	'81
Production:	4.0	4.6	4.1	4.8	4.7	4.3	4.5	4.0	3.6	3.8

a. Calculate a 3-term moving average of the production series.
b. Plot the original series and the moving average series on one graph. What information is provided by the moving average series that is not obtained readily from the original series?

24.2 Data on daily city bus traffic (in thousands of passengers) for a recent three-week winter period follow.

	Week 1		Week 2		Week 3	
Day	Number of Passengers	Moving Average	Number of Passengers	Moving Average	Number of Passengers	Moving Average
M	211		163	136.3	212	166.9
T	182		143	136.4	187	_____
W	199		131	142.1	193	_____
T	170	158.0	173	144.7	173	_____
F	206	151.1	207	151.7	204	
S	109	145.6	149	158.0	147	
S	29	135.9	47	166.9	41	

a. A 7-term moving average is to be computed for the series. Calculate the moving average values corresponding to the three blanks.
b. Plot the original series and the moving average series on one graph. What information is provided by the moving average series? Is a 7-term moving average a good choice here? Explain.

*24.3 Refer to Problem 24.1. Calculate the first two values of a centered 6-term moving average for the production series. Would a 3-term moving average or a centered 6-term moving average smooth the series more? Explain.

24.4 Refer to Problem 24.2. Calculate the first two values of a centered 4-term moving average. How many observations enter the first centered moving average value? What are their weights in the centered moving average?

24.5 Refer to Problem 24.2.
a. Calculate the difference between the original observation and the corresponding moving average value—i.e., obtain the residual—for each day of the series. Plot these residuals on a graph against day of the week (M, T, W, T, F, S, S) so that the two Monday residuals are plotted against M, the two Tuesday residuals are plotted against T, etc.
b. How do the following two factors influence the pattern of residuals plotted in **a**? (1) the recurring variation in daily bus traffic within a week, (2) the five-day snow storm that hit the city beginning on Saturday of week 1.

24.6 Suppose that quarterly production of a processed feed in each year during a three-year period were 9, 2, 6, and 7 million pounds, respectively.
   a. Calculate: (1) a centered 4-term moving average series, (2) a 3-term moving average series, (3) a 5-term moving average series.
   b. Why is the centered 4-term moving average series the smoothest one?

*24.7 Refer to Problem 24.1.
   a. Fit a linear trend function to the series. Set $X_t = 1$ at 1962.
   b. Plot the trend line and the original series on one graph. Does the linear trend appear to provide a good description of the trend of the series?
   c. What is the year-to-year increase in the trend line obtained in **a**? Does the intercept of the trend line have any meaning here?
   d. Obtain the projected trend level for production in 1988. Assuming that the linear trend fitted in **a** is appropriate through 1988 and that the best estimate of the cyclical component in 1988 is 115 percent, obtain a forecast of actual production in 1988.

*24.8 Refer to Problems 24.1 and 24.7. Calculate the percents of trend and plot them. If one were to calculate a 5-term moving average series of the percents of trend, which component of model (24.3) would this series represent?

24.9 Data on the number of industrial customers (in thousands, at midyear) of an electric company during the period 1964–1981 follow.

Year:	1964	'65	'66	'67	'68	'69	'70	'71	'72
Customers:	3.92	4.12	4.13	4.31	4.23	4.22	4.19	4.45	4.59
Year:	'73	'74	'75	'76	'77	'78	'79	'80	'81
Customers:	4.51	4.50	4.67	4.64	4.67	4.63	4.60	4.79	4.86

   a. Does this time series contain a seasonal pattern? Explain.
   b. Fit a linear trend function to this series. Set $X_t = 1$ at 1964. Plot the trend line and the original series on one graph. Does the trend line provide a good description of the trend of the series? Comment.
   c. Project the trend to midyear 1988. Is this projection a forecast of the number of industrial customers at midyear 1988? Explain.

24.10 Refer to Problem 24.9. Calculate the percents of trend and plot them. Which components of model (24.3) are represented by these percents of trend?

*24.11 Data on the number of sailboats sold annually by a manufacturer during 1967–1981 follow.

Year:	1967	'68	'69	'70	'71	'72	'73	'74
Units:	2,089	2,138	2,317	2,305	2,473	2,548	2,530	2,651
Year:	'75	'76	'77	'78	'79	'80	'81	
Units:	2,750	2,906	2,996	3,234	3,218	3,366	3,434	

   a. Fit an exponential trend function. Set $X_t = 1$ at 1967.
   b. Plot the original series and the exponential trend curve (expressed in the original units) on a graph with arithmetic grid. Does the trend curve provide a good description of the trend in the series? Explain.

*24.12 Refer to Problem 24.11. The exponential trend function for the logarithms of the observations is $T_t' = 3.305272 + .01569859X_t$.

a. Obtain the annual rate of growth in the trend function.

b. Project the trend curve to 1990 and also to 2030 and express the projected trend values in the original units. Do you believe that the trend projection for 2030 is reasonable? Discuss.

24.13 Data on the amounts of life insurance (in billions of dollars) underwritten in Canada by U.S. companies during 1960–1979 follow.

Year:	1960	'61	'62	'63	'64	'65	'66	'67	'68	'69
Amount:	12.6	13.2	14.1	15.1	16.4	18.3	20.1	21.8	23.8	26.4
Year:	'70	'71	'72	'73	'74	'75	'76	'77	'78	'79
Amount:	28.0	29.8	32.3	35.4	39.0	43.7	49.8	51.7	56.6	64.4

SOURCE: *Canadian Life Insurance Facts*, published by the Canadian Life Insurance Association.

a. Fit an exponential trend to the series. Set $X_t = 1$ at 1960. Plot the original series and the trend function (expressed in the original units) on a graph with arithmetic grid.

b. The linear trend function for this series is $T_t = 3.76000 + 2.55857X_t$ ($X_t = 1$ in 1960). Plot the linear trend on the same graph prepared in a. Which of the two trend functions provides a better fit to the trend of the series? Comment.

24.14 Refer to Problem 24.13.

a. Obtain the annual rate of growth in the exponential trend function.

b. Use the exponential trend function to project the trend level for 1995 and express the projected trend level in the original units. Would the linear trend function fitted to the same series give a similar projected trend level for 1995? Explain.

*24.15 Refer to Problem 24.1. The linear trend values and percents of trend for this time series follow.

Year:	1962	'63	'64	'65	'66	'67	'68	'69	'70	'71
Trend:	3.341	3.393	3.444	3.496	3.547	3.599	3.650	3.701	3.753	3.804
Percent:	98.8	109.1	113.2	100.1	81.8	86.1	95.9	97.3	85.3	92.0
Year:	'72	'73	'74	'75	'76	'77	'78	'79	'80	'81
Trend:	3.856	3.907	3.959	4.010	4.061	4.113	4.164	4.216	4.267	4.319
Percent:	103.7	117.7	103.6	119.7	115.7	104.6	108.1	94.9	84.4	88.0

a. Calculate a 5-term moving average series of the percents of trend to reflect the cyclical component of the time series. Then divide each moving average value into the corresponding percent of trend value to isolate the irregular component.

b. Plot the original series, trend, cyclical, and irregular components in separate graphs, as in Figure 24.3. Has the fluctuation of the series about the trend curve been due predominantly to cyclical or irregular forces? Does the cyclical component appear to be sufficiently regular in period or amplitude that you could confidently project it forward?

24.16 A speaker stated: "We have developed a computer program to make short-term and long-term sales forecasts for our products. For each product, the computer will access annual sales data for the past 12 years and will fit five different trend functions—linear, exponential, and three others. Coefficients of correlation between sales and trend will be calculated for each trend function. The trend function with the highest r will be projected 1 year and 10 years ahead for the short-term and long-term sales forecasts for this product." Discuss the problems involved in this approach to sales forecasting.

24.17 Refer to Problems 24.1 and 24.15. The cyclical component for this time series (calculated as a 5-term moving average of the percents of trend) follows.

Year:	1964	'65	'66	'67	'68	'69	'70	'71
Cyclical component:	100.6	98.1	95.4	92.2	89.3	91.3	94.8	99.2

Year:	'72	'73	'74	'75	'76	'77	'78	'79
Cyclical component:	100.5	107.3	112.1	112.3	110.3	108.6	101.5	96.0

Identify the turning points of the cyclical component. From the information given here, can you now confidently identify the year in which the next cyclical turning point will occur? Discuss.

24.18 Data on the cyclical components of number of building permits issued for private housing units (series A) and value of home mortgage loans made by major lending institutions (series B) during a recent four-year period follow.

Year	Quarter	Series A	Series B	Year	Quarter	Series A	Series B
1	1	115	105	3	1	99	96
	2	106	110		2	105	99
	3	101	103		3	103	102
	4	99	100		4	95	108
2	1	90	95	4	1	92	101
	2	86	91		2	96	97
	3	91	90		3	105	93
	4	95	88		4	114	97

a. Plot the cyclical component of series B on a graph. Identify its turning points.
b. Plot the cyclical component of series A on the same graph. Does the cyclical behavior of series A lead, lag, or coincide with that of series B? Explain. Is this behavior consistent?

24.19 Explain why it is reasonable that average duration of unemployment is a lagging cyclical indicator while contracts and orders for plant and equipment is a leading cyclical indicator according to the BEA list in Table 24.4.

*24.20 Data on quarterly sales of a brewing company (in million liters) during 1976–1981 follow.

Quarter	1976	1977	1978	1979	1980	1981
1	172	169	182	169	179	170
2	227	218	218	245	235	241
3	310	309	313	299	292	307
4	222	209	224	221	213	217

a. Plot the time series. Does a seasonal component appear to be present? Explain.
b. Calculate specific seasonal relatives for the series.
c. Assume the seasonal pattern is stable and obtain the quarterly seasonal indexes for the series. Describe the seasonal pattern in the series.

24.21 The enrollments at a university operating on a trimester system during 1974–1981 follow.

Trimester	1974	'75	'76	'77	'78	'79	'80	'81
I	10,284	10,724	11,052	11,301	12,240	12,958	12,751	13,253
II	9,445	9,828	10,636	10,757	10,946	11,617	12,177	12,880
III	6,307	6,015	6,922	6,883	7,300	7,153	8,170	7,617

   a. Plot the time series. Does the series have a noticeable seasonal component? A noticeable trend-cyclical component? Comment.

   b. Calculate specific seasonal relatives for the series. Was there any need to center the moving average in your calculations? Explain.

   c. Assume that the seasonal pattern is stable and obtain the seasonal indexes. Describe the pattern of seasonal variation.

24.22 Refer to Problem 24.20. Plot the specific seasonal relatives in a graph, as in Figure 24.10. Is the assumption of a stable seasonal pattern reasonable here? Comment.

24.23 Refer to Problem 24.21.

   a. Plot the specific seasonal relatives in a graph, as in Figure 24.10. Is the assumption of a stable seasonal pattern reasonable here? Comment.

   b. Why are there seven specific seasonal relatives for the first and third trimesters but eight for the second trimester?

*24.24 Data on quarterly sales (in millions of dollars) of Nettles, Ltd., during 1981, together with the seasonal indexes for these sales, follow.

Quarter:	1	2	3	4
Sales:	3.40	3.01	4.21	5.33
Seasonal index:	83	75	102	140

   a. Obtain the deseasonalized sales volume for each quarter of 1981. Should management be pleased with the relatively large sales volume experienced in the fourth quarter? Comment.

   b. Suppose that only seasonal changes in sales were expected between the fourth quarter of 1981 and the first quarter of 1982. What would be the expected sales volume for the first quarter of 1982, assuming the seasonal pattern is stable? If actual sales in the first quarter of 1982 were $3.06 million, what would this suggest? Discuss.

24.25 Refer to Problem 24.21. The trimester seasonal indexes are: 119.4, 110.5, 70.1. Obtain the deseasonalized enrollment levels for 1981. Interpret your results.

*24.26 Refer to Problem 24.20. The quarterly seasonal indexes for company sales are: 73.1, 101.1, 132.4, 93.4.

   a. In 1981, company executives forecast that 1982 annual sales would be 930 million liters and that trend-cyclical movements during this year would be negligible. Convert this forecast into quarterly forecasts for 1982, assuming the seasonal pattern is stable.

   b. Actual sales in the first quarter of 1982 were 168 million liters. Obtain the deseasonalized level of sales for this quarter and the seasonally adjusted annual rate of sales for this quarter, and interpret each.

24.27 The quarterly seasonal indexes for the total loans made by a small credit company are: 75, 91, 102, 132. Late in 1981 a forecast of total loans of $15.7 million for 1982 was made and it was expected that trend-cyclical movements during the year would be very small. Convert the annual forecast into quarterly forecasts for 1982. Why do the quarterly forecasts sum to the annual forecast of $15.7 million here? Explain.

*24.28 Data on the number of units of a spare part demanded in the past seven time periods follow.

t:	1	2	3	4	5	6	7
D_t:	638	632	658	653	655	650	657

Note that demand for this part shifted upward in period 3, a result of a sudden increase in orders from overseas users.

a. Obtain the forecasts of demand for periods 2 through 8 inclusive by exponential smoothing formula (24.18) with smoothing constant $w = .4$. Assume the forecast for period 1 is $F_1 = 635.0$.

b. Suppose the smoothing constant were $w = .1$. Obtain the forecasts with this smoothing constant and compare the results with those in **a**. Does the decrease in w make the forecasting system more responsive or less responsive to the basic shift in demand? Explain.

24.29 Refer to Problem 24.28. Answer both parts assuming that demand shifted downward, as follows:

t:	1	2	3	4	5	6	7
D_t:	632	635	601	611	607	609	605

24.30 An inventory manager uses formula (24.18) with $w = .2$ to forecast monthly demand for several hundred types of items stocked in inventory. Demand for one such item was 1,030 units last March, and the corresponding forecast of demand for this item was 1,090.0 units.

a. Forecast April demand for this item.

b. Actual demand in April was 1,075 units. Prepare the forecast for May.

c. What information did you carry over into April from the preceding months in order to prepare the forecast for May?

d. If the inventory manager used a monthly forecasting procedure based on a 6-term moving average, would more or less data have to be carried over from previous months than with exponential smoothing? Explain the practical significance of this.

EXERCISES

24.31 An analyst was studying the trend of a company's monthly production volume. Before fitting any trend curves, however, she divided each month's production by the number of working days in that month.

a. Why did she take this preliminary step before fitting any trend curves? Is it always necessary to divide production data by the number of working days before studying trend? Explain.

b. The analyst obtained the equation $T_t = 891 + 1.23X_t$ when fitting a linear trend to the adjusted monthly production data. Project the trend line to month $X_t = 40$. In this month, there will be 22 working days. What is the projected trend level of total production for this month?

24.32 Construct a logarithmic scale with gradation points corresponding to 1, 2, . . . , 9, 10, 20, . . . , 90, 100. Confirm that the distances on your scale between 2 and 6 and between 30 and 90 are the same, as theoretically they should be.

24.33 Refer to Problems 24.11 and 24.12. Plot the logarithms of the original series and the exponential trend function fitted to the log-observations on graph paper with arithmetic grid. Does the trend function plot as a straight line as expected? Is it a good description of the trend of the log-series?

24.34 A demographer fitted a Gompertz trend function to annual population data for a country and obtained: $b_0 = 50$, $b_1 = .19$, $b_2 = .99$ (T_t is in millions of persons and $X_t = 0$ at 1900).

    a. Identify the upper asymptote of the Gompertz trend curve. What is the coefficient of retardation here? How is each of these values interpreted here?

    b. Compute the trend values for 1900, 1950, and 2000. Using these points and your answers in **a**, sketch the Gompertz trend curve on graph paper with arithmetic grid.

24.35 Refer to Problems 24.21 and 24.25. The linear trend fitted to the enrollment series is $T_t = 8{,}791.99 + 94.0278X_t$, where $X_t = 1$ in the first trimester of 1974.

    a. Assuming the cyclical component is negligible, forecast trimester enrollments in 1982.

    b. The actual enrollment for the first trimester of 1982 was 13,690. Obtain the seasonally adjusted enrollment for this trimester. Interpret this number.

24.36 An analyst stated: "The best indication of the number of pieces of an inventory item demanded next week is given by the numbers of pieces demanded in the past few weeks. Numbers of pieces demanded in weeks further in the past are relatively unimportant." In forecasting demand for this inventory item a week ahead by exponential smoothing, should the smoothing constant be relatively close to 0 or relatively close to 1? Explain.

STUDIES

24.37 Refer to Problem 24.20. The linear trend equation is $T_t = 223.68 + .6422X_t$, where $X_t = 1$ in the first quarter of 1976. The quarterly seasonal indexes are: 73.1, 101.1, 132.4, 93.4.

    a. Obtain deseasonalized percents of trend for the sales data. Which components of the classical time series model are represented by these percents? Explain.

    b. Smooth the percents obtained in **a** using a 5-term moving average. Which component is represented by your moving average series? Plot this series.

    c. Obtain the irregular component and plot it.

    d. The company had a strike in its distribution division in the second quarter of 1978. Estimate the effect of the strike on sales.

    e. What other factors besides strikes might produce irregular effects in the sales series? Discuss.

24.38 (Computer needed.) Refer to the monthly data on the number of marriages (in thousands) in the United States during the period July 1976—June 1980 in Problem 15.28.

    a. Decompose the marriage series into its trend, cyclical, seasonal, and irregular components using the following procedures: (1) Obtain the monthly seasonal indexes by the method of ratio to moving average, using a centered 12-term moving average. (2) Deseasonalize the series. (3) Fit a linear trend function to the deseasonalized series, setting $X_t = 1$ at July 1976. (4) Obtain the deseasonalized percents of trend and smooth them, using a 5-term moving average.

b. Plot the original series and its four components on separate graphs, as in Figure 24.3, and briefly describe the nature of the components.

c. Project the trend and seasonal components to July and August 1980. Assuming that the cyclical component will be 100 in each of the two months, combine the projections to obtain forecasts of the numbers of marriages for July and August 1980. The actual numbers of marriages for these months were 235 thousand and 255 thousand, respectively. How close are your forecasts?

24.39 The following specific seasonal relatives for quarterly sales of large grade A eggs by a supermarket chain were obtained for 1976–1981:

		Quarter		
Year	1	2	3	4
1976			98.9	97.1
1977	102.9	100.4	97.9	99.5
1978	101.5	99.5	99.2	107.7
1979	100.9	95.1	100.0	105.3
1980	99.5	95.1	100.4	106.7
1981	98.0	92.7		

a. Describe the shift in the seasonal pattern of egg sales that has occurred during this period. What factors might have produced this shift?

b. Linear "trend" equations fitted to the specific seasonal relatives for the first three quarters are:

First quarter: $S = 104.10 - 1.18X$; $X = 1$ in 1977
Second quarter: $S = 102.50 - 1.98X$; $X = 1$ in 1977
Third quarter: $S = 97.75 + .510X$; $X = 1$ in 1976

Obtain the "trend" equation for the fourth quarter, letting $X = 1$ in 1976.

c. Calculate the seasonal indexes for 1982 by projecting each "trend" equation to 1982 and adjusting the indexes so that they add to 400.0.

24.40 Consult the latest issue of *Business Conditions Digest* and examine the following three business indicator series: (1) contracts and orders for plant and equipment, (2) manufacturing and trade inventories, (3) manufacturing and trade sales.

a. For each of these series, indicate whether it is a leading, lagging, or coincident indicator series.

b. Find the most recent value for each of these series. Are these values seasonally adjusted?

c. What is the current cyclical phase—expansion or contraction—of each of the series? How have you established the cyclical phase for each series? Is the phase of each series consistent with its classification as a leading, lagging, or coincident indicator? Comment.

25
Time Series—II

The classical time series model (24.3), discussed in the previous chapter, is a descriptive model. No attempt is made to define the irregular component in probabilistic terms, and no parameters are specified about which formal statistical inferences can be made. Uncertainties are handled by judgment, not by statistical theory. With statistical time series models, on the other hand, formal statistical inferences and predictions can be made. In this chapter, we consider the use of a variety of regression models for time series data.

25.1 REGRESSION MODELS WITH INDEPENDENT ERROR TERMS

We take up first a number of regression models in which the error terms are assumed to be independent. The models are special cases of multiple regression model (20.3), which we repeat here for convenience:

$$(25.1) \qquad Y_t = \beta_0 + \beta_1 X_{t1} + \beta_2 X_{t2} + \cdots + \beta_{p-1} X_{t,p-1} + \epsilon_t$$

where: ϵ_t's are independent $N(0, \sigma^2)$

We now use the subscript t to denote the time period $(t = 1, 2, \ldots, n)$. With this type of model, time-related effects are incorporated into the model entirely by means of the independent variables. The random error terms are assumed to be time-independent.

Trend Model

If a time series, such as annual company sales for the past 10 years, contains a linear trend and a random component that is independent from one time period to another, model (25.1) takes the form:

$$(25.2) \qquad Y_t = \beta_0 + \beta_1 X_t + \epsilon_t$$

where: $X_t = t$

In other words, X_t denotes the coded time period; we have $X_1 = 1$, $X_2 = 2$, etc.

Additive Trend and Seasonal Components

When a time series, such as quarterly shipments of television sets during the past 48 quarters, contains both a trend component and a seasonal component and they act additively, regression model (25.1) again can be used if the random effect is time-independent. For a quarterly time series and a linear trend, the model takes the following form:

(25.3)
$$Y_t = \beta_0 + \beta_1 X_{t1} + \beta_2 X_{t2} + \beta_3 X_{t3} + \beta_4 X_{t4} + \epsilon_t$$

where: $X_{t1} = t$

$X_{t2} = \begin{cases} 1 \text{ if period } t \text{ is second quarter} \\ 0 \text{ otherwise} \end{cases}$

$X_{t3} = \begin{cases} 1 \text{ if period } t \text{ is third quarter} \\ 0 \text{ otherwise} \end{cases}$

$X_{t4} = \begin{cases} 1 \text{ if period } t \text{ is fourth quarter} \\ 0 \text{ otherwise} \end{cases}$

Here, $\beta_1 X_1$ is the linear trend effect. X_2, X_3, and X_4 are three *indicator variables* (discussed in Section 20.5) which denote the four quarters of the year. These indicator variables take on the following values for the four quarters:

Quarter	X_2	X_3	X_4
First	0	0	0
Second	1	0	0
Third	0	1	0
Fourth	0	0	1

The regression coefficient β_2 in model (25.3) represents the differential effect on the dependent variable of the second quarter as compared with the first quarter, and the regression coefficients for the other indicator variables represent corresponding differential effects for the other quarters.

Business Indicators and Other Predictors as Independent Variables

In many cases where the dependent variable is a time series, the independent variables in the regression model include business indicators or other predictor series. Business indicators are time series measuring activities in key sectors of the economy.

☐ Examples

1. Annual industry production (Y) is related to gross national product (X_1) and the Consumer Price Index (X_2), for the past 16 years.
2. Annual kindergarten enrollment (Y) is related to number of births five years earlier (X), for the past 20 years.
3. Monthly sales of a firm (Y) are related to the firm's promotional expenditures in the previous month (X), for the past 36 months.
4. Quarterly shipments of cement (Y) are related to two independent variables—the volume of construction contracts awarded (X) during the preceding quarter and the volume during the quarter before that—for the past 20 quarters. ☐

Coincident Independent Variables. Regression model (25.1) for Example 1 takes the following form, assuming the effects of the independent variables are linear and additive:

(25.4)
$$Y_t = \beta_0 + \beta_1 X_{t1} + \beta_2 X_{t2} + \epsilon_t$$

Here, Y_t is industry production, X_{t1} gross national product, and X_{t2} the Consumer Price Index, each for year t.

Y_t, X_{t1}, and X_{t2} are *coincident time series*; i.e., they refer to the same time period. With coincident independent variables, a forecast of the dependent variable requires forecasts of the independent variables. Suppose the data in Example 1 to which the regression model is fitted pertain to the years 1966–1981 inclusive. If industry production for 1982 were to have been forecast, forecasts for GNP and the Consumer Price Index for 1982 would have been needed. Forecasting GNP and the Consumer Price Index in order to obtain a forecast of industry production may seem to be a circuitous procedure. If, however, accurate forecasts of the independent variables are available and there is a close relation between them and industry production, an accurate forecast of the dependent variable can result.

Lagged Independent Variables. Sometimes it is possible to construct a regression model where some or all of the independent variables are lagged. Example 2 is an illustration. A simple linear regression model would be:

(25.5)
$$Y_t = \beta_0 + \beta_1 X_{t-5} + \epsilon_t$$

Here, Y_t is kindergarten enrollment in year t and X_{t-5} is the number of births five years earlier. X_{t-5} is a *lagged time series* relative to Y_t, the lag being five years.

The advantage of a regression model with lagged independent variables is that the independent variables need not be predicted in order to forecast the dependent variable. Suppose that the data for Example 2 span the years 1962–1981 inclusive. If a forecast of kindergarten enrollment for 1985 had been desired in 1982, the independent variable would have been number of births in 1980. This would have been known in 1982, so there would have been no need to forecast the independent variable.

The regression model may contain more than one lagged term of an independent variable. In Example 4, cement shipments in quarter t are related to the volume of construction contracts awarded in quarter $t - 1$ and to the volume in quarter $t - 2$. The regression model here, with linear and additive effects, would be:

(25.6)
$$Y_t = \beta_0 + \beta_1 X_{t-1} + \beta_2 X_{t-2} + \epsilon_t$$

This model is said to contain a *distributed time lag*.

More Complex Models

Regression models for time series data often combine several of the features explained previously. Also, they may involve a transformed dependent variable, such as log Y_t. An example of a more complex regression model is the following for a quarterly time series:

(25.7)
$$Y'_t = \beta_0 + \beta_1 X_{t1} + \beta_2 X_{t2} + \beta_3 X_{t3} + \beta_4 X_{t4} + \beta_5 X_{t-1,5} + \epsilon_t$$

where: $Y'_t = \log Y_t$
$X_{t1} = t$
X_{t2}, X_{t3}, X_{t4} are indicator variables for quarterly seasonal effects
$X_{t-1,5}$ is a business indicator lagged one quarter

Note that model (25.7) involves an exponential trend effect $\beta_1 X_{t1}$ because the dependent variable is $\log Y_t$ (see Section 24.5).

☐   Example

We illustrate the use of a regression model with independent error terms by considering an annual time series where a linear trend component and a lagged business indicator are employed in the model.

A consultant, developing forecasting procedures for a firm specializing in commercial and industrial roofing, found that the dollar volume of contracts received annually for reroofing followed a linear trend over the course of the observations. In any year, the volume also reflected the state of general business activity in the firm's service region during the preceding year. The consultant decided tentatively to use the following model to forecast the dollar volume of reroofing contracts one year ahead:

$$Y_t = \beta_0 + \beta_1 X_{t1} + \beta_2 X_{t-1,2} + \epsilon_t$$

where: Y_t is the dollar volume of reroofing contracts in year t
$X_{t1} = t$ is the linear trend variable
$X_{t-1,2}$ is the index of general business activity in year $t - 1$

The consultant's data are presented in Table 25.1. The fitted model is (calculations not shown):

$$\hat{Y}_t = -188.62891 + 8.77231 X_{t1} + 4.35074 X_{t-1,2}$$

The consultant was particularly interested in whether or not the error terms could be assumed to be independent. The residuals are shown in Table 25.1 and the residual plot by time order is presented in Figure 25.1. On the basis of this residual plot, the consultant concluded that a regression model with independent error terms is not unreasonable here. A residual plot against the fitted values $\hat{Y}$ was also prepared by the consultant. He concluded from it that the fitted regression function is reasonable and that there is no indication that the error variance is not constant.

A forecast of Y_t for year $t = 12$ is desired, by means of a 90 percent prediction interval. The consultant ascertained that the index of business activity for the current year $t = 11$ is $X_{11,2} = 105.8$. Also, $X_{12,1} = 12$ for next year. Hence:

$$\hat{Y}_{12} = -188.62891 + 8.77231(12) + 4.35074(105.8) = 376.9$$

For $n = 11$ and $1 - \alpha = .90$, we require $t(.95; 8) = 1.860$. A time-sharing program provided the consultant with the estimated standard deviation for the prediction interval, as given in (20.21); it is $s\{d_{12}\} = 5.6551$. The prediction limits, by (20.21), therefore are $376.9 \pm 1.860(5.6551)$ and the prediction interval is:

$$366.4 \leq Y_{12} \leq 387.4$$

TABLE 25.1 *Reroofing contracts example*

Year t	Linear Trend Variable X_{t1}	Index of General Business Activity (*lagged*) $X_{t-1,2}$	Dollar Volume of Reroofing Contracts (*$000*) Y_t	Residual e_t
1	1	79.7	173.7	6.8
2	2	83.1	187.0	−3.5
3	3	90.9	232.1	−1.1
4	4	99.1	280.0	2.4
5	5	100.0	287.0	−3.3
6	6	97.5	282.8	−5.4
7	7	91.3	273.9	3.9
8	8	86.6	254.7	−3.6
9	9	94.4	300.8	−.2
10	10	105.5	355.7	−2.4
11	11	108.7	387.2	6.4

Hence, with 90 percent confidence, the consultant predicts that the volume of reroofing contracts in year 12 will be between $366.4 thousand and $387.4 thousand. □

25.2 AUTOCORRELATED ERROR TERMS

The regression models presented in the previous section all assume that the error terms ϵ_t are independent from period to period. In many applications, however, the error terms in the different periods are correlated. When this is the case, the error terms are said to be *autocorrelated* or *serially correlated*.

FIGURE 25.1 *Residual plot against time order for reroofing contracts example*

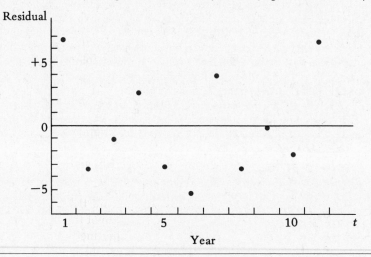

Autocorrelated error terms arise for a variety of reasons. One major cause is the omission of one or more key variables from the fitted regression model, such as the omission of population size in a model predicting sales where population size has a time-ordered effect on sales. Another important reason is that major random effects often tend to persist for several periods, as when the effects of a very poor harvest of an agricultural commodity are felt for several years thereafter.

First-Order Autoregressive Error Model

When the error terms are related over time, a model for the error terms frequently employed is the first-order autoregressive error model:

(25.8)   The *first-order autoregressive error model* is:

$$\epsilon_t = \rho \epsilon_{t-1} + u_t$$

where: ρ (Greek rho) is a parameter such that $-1 < \rho < 1$
u_t's are independent $N(0, \sigma^2)$

This model assumes that the error term ϵ_t for period t contains a component resulting from the error term ϵ_{t-1} for the preceding period (when $\rho \neq 0$) and a random disturbance term u_t which is independent of earlier time periods.

The parameter ρ is called the *autocorrelation parameter* and, in fact, represents the coefficient of correlation between ϵ_t and ϵ_{t-1} as defined in (5.27). When ρ is positive, the autocorrelation is positive; when ρ is negative, the autocorrelation is negative. In most business and economic series with time-dependent error terms, the autocorrelation is positive. For example, large sales in one quarter because of good cyclical conditions are likely to persist into the next quarter.

When $\rho = 0$, the error terms ϵ_t and ϵ_{t-1} are uncorrelated and error model (25.8) simplifies to $\epsilon_t = u_t$. The ϵ_t's are then independent $N(0, \sigma^2)$—the case assumed for all regression models up to now.

Effects of Autocorrelation.   If the method of least squares is employed in the usual fashion for fitting a regression model when the error terms are autocorrelated according to (25.8), it can be shown that the following consequences arise:

1. The least squares regression coefficients are still unbiased estimators, but they no longer have the minimum variance property; they tend to be relatively inefficient.
2. The error mean square *MSE* can seriously underestimate the true variance of the error terms.
3. Standard procedures for confidence intervals and tests using the t and F distributions are no longer strictly applicable.

There are several methods available to avoid these difficulties caused by autocorrelated error terms. One involves estimation of the autocorrelation parameter ρ, and this can be done in several ways. Another method involves a simple transformation of the variables, and we shall discuss this now.

Method of First Differences

We shall explain the method of first differences in terms of a regression model with one independent variable. The method extends readily to multiple regression applications. The basic idea of the method of first differences is to use a simple transformation of the variables which will make the error terms approximately independent so that ordinary regression methods can be used to their full advantage.

Development of Method. The simple linear regression model with first-order autoregressive error terms is as follows:

$$(25.9) \qquad Y_t = \beta_0 + \beta_1 X_t + \epsilon_t$$

where: $\epsilon_t = \rho \epsilon_{t-1} + u_t$
u_t's are independent $N(0, \sigma^2)$

As noted before, the autocorrelation parameter in business and economic applications when the error terms are serially correlated is usually positive and frequently very large. Let us see what happens when $\rho = 1$. The error term ϵ_t in (25.9) then becomes:

$$(25.10) \qquad \epsilon_t = \epsilon_{t-1} + u_t$$

and the *first difference* $Y_t - Y_{t-1}$ becomes:

$$\begin{aligned} Y_t - Y_{t-1} &= (\beta_0 + \beta_1 X_t + \epsilon_t) - (\beta_0 + \beta_1 X_{t-1} + \epsilon_{t-1}) \\ &= (\beta_0 + \beta_1 X_t + \epsilon_{t-1} + u_t) - (\beta_0 + \beta_1 X_{t-1} + \epsilon_{t-1}) \qquad \text{by (25.10)} \end{aligned}$$

or:

$$(25.11) \qquad Y_t - Y_{t-1} = \beta_1(X_t - X_{t-1}) + u_t$$

We shall use the following notation:

$$(25.12) \qquad Y'_t = Y_t - Y_{t-1} \qquad X'_t = X_t - X_{t-1}$$

where Y'_t and X'_t are first differences. Then we can express (25.11) as follows:

$$(25.13) \qquad Y'_t = \beta_1 X'_t + u_t$$

where: u_t's are independent $N(0, \sigma^2)$
X'_t and Y'_t are given by (25.12)

Note that (25.13) is a simple linear regression model with the intercept $\beta_0 = 0$ and *independent* error terms u_t.

This suggests that when the error terms are positively correlated and the autocorrelation parameter ρ is large, the ordinary regression procedures become approximately applicable if:

1. *First differences* $Y_t - Y_{t-1}$ and $X_t - X_{t-1}$ are used for the dependent and independent variables.
2. A regression model with *regression through the origin* ($\beta_0 = 0$) is employed.

TABLE 25.2 *Key formulas in regression through origin:* $Y_i = \beta_1 X_i + \epsilon_i$
($i = 1, 2, ..., n$)

Least squares estimator:	$b_1 = \dfrac{\Sigma \, X_i Y_i}{\Sigma \, X_i^2}$
Error mean square:	$MSE = \dfrac{\Sigma \, (Y_i - b_1 X_i)^2}{n - 1}$
Estimated variance of b_1:	$s^2\{b_1\} = \dfrac{MSE}{\Sigma \, X_i^2}$
Estimated variance of $\widehat{Y}_h$:	$s^2\{\widehat{Y}_h\} = \dfrac{X_h^2 MSE}{\Sigma \, X_i^2}$
Estimated variance of d_h:	$s^2\{d_h\} = MSE\left(1 + \dfrac{X_h^2}{\Sigma \, X_i^2}\right)$

Note that the parameter β_1 in the first-differences model (25.13) is the same slope β_1 as in the original regression model (25.9). Thus, we are still estimating the same slope parameter as in our original regression model.

Regression through Origin. Procedures for regression through the origin follow those already discussed for the standard regression model. Key formulas are given in Table 25.2 for applications with one independent variable. Confidence intervals and tests are set up in the usual way by means of the t distribution. Now, however, MSE has associated with it $n - 1$ degrees of freedom for n observations. In particular for first differences, where there are $n - 1$ first differences for n original observations, the relevant t distribution is that for $(n - 1) - 1 = n - 2$ degrees of freedom.

☐ Example

A securities analyst was studying the relation between sales of a company and of the industry. She obtained deseasonalized data for the past 20 quarters. These are presented (in part) in Table 25.3, columns 1 and 2. The analyst expected high positive autocorrelation.

TABLE 25.3 *First-differences example (sales in millions of dollars)*

Quarter t	(1) Company Sales Y_t	(2) Industry Sales X_t	(3) First Differences Y_t'	(4) First Differences X_t'	(5) $X_t' Y_t'$	(6) $(X_t')^2$
1	77.044	746.512	—	—	—	—
2	78.613	762.345	1.569	15.833	24.842	250.684
3	80.124	778.179	1.511	15.834	23.925	250.716
⋮	⋮	⋮	⋮	⋮	⋮	⋮
20	102.481	1,006.882	1.745	17.592	30.698	309.478
				Total	667.250	6,612.947

Her first step was to fit the standard simple linear regression model $Y_t = \beta_0 + \beta_1 X_t + \epsilon_t$ by the method of least squares in order to study the residuals. These are plotted in time order in Figure 25.2a. Her expectation of high positive autocorrelation was borne out; note the succession of positive residuals, then negative residuals and finally positive residuals, with generally gradual changes in the residuals from one period to the next.

She therefore decided to fit the first-differences model (25.13):

$$Y_t' = \beta_1 X_t' + u_t$$

The first differences for the dependent and independent variables are shown (in part) in columns 3 and 4 of Table 25.3. Note that the first-differences transformation yields one fewer observation than the original number. The necessary calculations with the first differences are shown in columns 5 and 6 of Table 25.3. The estimated slope, using the formula in Table 25.2 and Y_t' and X_t' as the variables, is:

$$b_1 = \frac{\Sigma X_t' Y_t'}{\Sigma (X_t')^2} = \frac{667.250}{6,612.947} = .10090$$

and the fitted first-differences model is:

$$\widehat{Y}_t' = .10090 X_t'$$

The residuals for this model are plotted in time order in Figure 25.2b. The scatter now appears reasonably consistent with time-independent error terms, and the analyst was satisfied by the first-differences model in this regard.

The analyst desired to forecast company sales for quarter $t = 21$. She obtained a reliable industry forecast for that quarter, $X_{21} = 1,015.9$. Since industry sales in quarter 20 were $X_{20} = 1,006.9$, the projected first difference is $X_{21}' = 1,015.9 - 1,006.9 = 9.0$. Hence, a point estimate of Y_{21}' is:

$$\widehat{Y}_{21}' = .10090(9.0) = .91$$

This estimate indicates that deseasonalized company sales are expected to increase by \$.91 million between quarters 20 and 21.

FIGURE 25.2 *Residual plots against time order for company sales example*

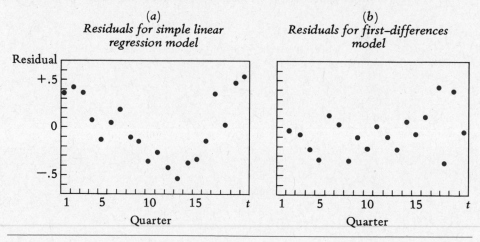

The final step in developing the forecast was to take the deseasonalized company sales in quarter 20, $Y_{20} = 102.5$, and adjust them by the expected increase:

$$\text{Forecast for quarter } 21 = 102.5 + .91 = 103.4$$

The analyst thus predicted that deseasonalized company sales in quarter 21 will be $103.4 million. ☐

Comment

Formal tests of randomness of the error terms designed particularly for regression models are available; these are discussed in specialized texts. The test of runs up and down discussed in Section 15.4 is not strictly applicable to regression problems but can be used on the residuals as an approximate test when the number of observations is large.

PROBLEMS

*25.1 Refer to the quarterly time series in Problem 24.20. When model (25.3) is fitted to the quarterly sales of the brewing company (with $X_{t1} = 1$ in the first quarter of 1976), the estimated regression coefficients are: $b_0 = 172.34$, $b_1 = .10536$, $b_2 = 57.06$, $b_3 = 131.29$, $b_4 = 43.85$.

    a. Obtain the fitted regression value for the third quarter of 1980. Calculate the value of the residual for this quarter. Is it small or large in relation to the fitted value?

    b. (1) By how much do expected sales in any third quarter differ from expected sales in the preceding first quarter because of differential seasonal effects? Because of linear trend? (2) By how much do expected sales in any third quarter differ from expected sales in the following fourth quarter because of differential seasonal effects? Because of linear trend? Give a point estimate in each case.

    c. Obtain a point prediction of sales in the second quarter of 1982.

    d. Give a point estimate of the change in expected sales between the second quarter of one year and the second quarter of the following year. Is the expected year-to-year change the same for the other quarters? Explain.

25.2 The following regression model was employed to forecast annual sales of dishwashers by a firm:

$$Y_t = \beta_0 + \beta_1 X_{t1} + \beta_2 X_{t-1,2} + \epsilon_t$$

where: ϵ_t's are independent $N(0, \sigma^2)$

Here, Y is thousands of units sold, X_1 is a variable for linear trend, and X_2 is thousands of utility connections for new housing. Data for 1972–1981 follow.

Year:	1972	'73	'74	'75	'76	'77	'78	'79	'80	'81
t:	1	2	3	4	5	6	7	8	9	10
X_{t1}:	1	2	3	4	5	6	7	8	9	10
$X_{t-1,2}$:	22.3	28.1	19.6	24.7	26.9	27.3	19.9	20.4	26.7	20.7
Y_t:	6.06	7.55	6.40	7.78	8.53	9.06	8.01	8.42	10.14	9.37

The estimated regression coefficients are: $b_0 = 1.38041$, $b_1 = .398955$, $b_2 = .192618$.

   a. Give a point estimate of the expected year-to-year increase in dishwasher sales when the number of utility connections for new housing is kept constant.

   b. What was the number of utility connections for new housing in 1978? Calculate the fitted regression value for 1979.

   c. In 1981, there were 24.3 thousand utility connections for new housing. Obtain a point estimate of expected sales of dishwashers by the firm in 1982.

*25.3 Refer to Problem 25.1. Additional regression results are: $s\{b_0\} = 4.322$, $s\{b_1\} = .2464$, $s\{b_2\} = 4.767$, $s\{b_3\} = 4.786$, $s\{b_4\} = 4.818$, $SSR = 53,761.2$, $SSE = 1,291.7$.

   a. Construct a 90 percent prediction interval for sales in the second quarter of 1982. A computer printout shows that $s\{d_h\} = 9.551$. Interpret the prediction interval.

   b. Conduct a test of whether or not $\beta_4 = 0$, controlling the α risk at .05. State the alternatives, decision rule, and conclusion. What is the implication of your conclusion? Explain.

25.4 Refer to Problem 25.2. Additional regression results are: $s\{b_0\} = .11878$, $s\{b_1\} = .0052215$, $s\{b_2\} = .0046142$, $SSR = 14.4225$, $SSE = .015217$.

   a. Construct a 99 percent prediction interval for sales in 1982. A computer printout shows that $s\{d_h\} = .05706$. Does it seem likely that 1982 sales will be higher than sales in 1981?

   b. Construct a 99 percent confidence interval for β_1. Interpret your interval estimate.

25.5 Model (25.3) was employed to study the number of requests for information received by a tourist agency during the past 40 quarters. The data follow, with most of the quarters omitted for conciseness. Here, Y_t denotes thousands of requests received in quarter t, X_{t1} is a linear trend variable, and X_{t2}, X_{t3}, and X_{t4} are indicator variables to represent quarterly seasonal effects.

t:	1	2	3	4	$\cdots$	37	38	39	40
X_{t1}:	1	2	3	4	$\cdots$	37	38	39	40
X_{t2}:	0	1	0	0	$\cdots$	0	1	0	0
X_{t3}:	0	0	1	0	$\cdots$	0	0	1	0
X_{t4}:	0	0	0	1	$\cdots$	0	0	0	1
Y_t:	11.68	6.86	4.32	6.20	$\cdots$	19.99	14.34	11.22	13.28

The estimated regression equation and mean square error are:

$$\hat{Y}_t = 11.68 + .206X_{t1} - 5.25X_{t2} - 8.42X_{t3} - 6.59X_{t4} \qquad MSE = .1775$$

   a. In which of the four quarters does the seasonal component reach its peak? In which does it reach its trough? Explain.

   b. Obtain point estimates of the expected numbers of requests to be received in quarters 41 through 44.

   c. Are random effects in the quarterly numbers of requests received relatively small or large here? Explain.

25.6 The following residuals were obtained in an application of regression model (25.1):

t:	1	2	3	4	5	6	7	8	9	10
e_t:	−5.58	−.60	2.10	6.64	5.28	7.46	7.71	7.57	2.68	2.06

t:	11	12	13	14	15	16	17	18	19	20
e_t:	5.52	2.27	−4.69	−8.21	−5.31	−12.43	−11.17	−10.68	−9.44	−5.98

t:	21	22	23	24	25	26	27	28	29	30
e_t:	−3.55	2.07	5.21	4.18	−.13	6.73	1.44	4.69	3.24	.93

Plot the residuals against time order. Do the errors appear to be serially correlated? Comment.

25.7 Refer to Problem 25.1. The following residuals were obtained in the regression analysis:

t:	1	2	3	4	5	6	7	8
e_t:	−.45	−2.61	6.05	5.39	−3.87	−12.03	4.63	−8.03
t:	9	10	11	12	13	14	15	16
e_t:	8.71	−12.46	8.21	6.54	−4.71	14.12	−6.21	3.12
t:	17	18	19	20	21	22	23	24
e_t:	4.87	3.70	−13.63	−5.30	−4.55	9.28	.95	−1.72

Plot the residuals against time order. Do the errors appear to be serially correlated? Comment.

*25.8 Data for the past 12 years on number of applicants accepted to a master's program at a state university (X_t), number of applicants who actually entered the program (Y_t), and least squares residuals obtained in fitting regression model (19.1) to the observations, follow.

t:	1	2	3	4	5	6
X_t:	330	342	348	355	365	373
Y_t:	140	151	154	161	167	171
e_t:	1.76	1.57	−1.02	−.55	−3.87	−7.33

t:	7	8	9	10	11	12
X_t:	375	367	376	381	375	385
Y_t:	170	176	184	192	185	192
e_t:	−10.19	3.26	2.87	6.21	4.81	2.48

a. Plot the residuals against time order. Do the error terms appear to be positively autocorrelated? Explain.

b. Fit the first-differences model (25.13) and state the estimated regression equation.

25.9 Data for the past 11 years on number of payroll-processing accounts serviced by Empire Data Processing (X_t), annual payroll-processing revenue in thousands of dollars received by Empire (Y_t), and least squares residuals obtained in fitting regression model (19.1) to the observations, follow.

t:	1	2	3	4	5	6
X_t:	20	25	30	40	35	45
Y_t:	402	526	603	803	756	972
e_t:	26.7	22.5	−28.8	−85.4	−4.1	−44.7

t:	7	8	9	10	11	
X_t:	50	50	45	50	55	
Y_t:	1,172	1,172	1,072	1,165	1,258	
e_t:	27.0	27.0	55.3	20.0	−15.3	

a. Make a residual plot to aid in assessing whether the errors are time-dependent. State your findings.

b. Fit the first-differences model (25.13) and state the estimated regression equation.

*25.10 Refer to Problem 25.8. The fitted first-differences model is $\hat{Y}'_t = .66145X'_t$.

a. Calculate MSE for this fitted model. How many degrees of freedom are associated with MSE here?

b. Obtain a 90 percent confidence interval for β_1. Interpret your interval estimate.

c. The number of applicants accepted for year 13 is 380. Predict by means of a point estimate the number of applicants entering the program in year 13.

25.11 Refer to Problem 25.9. The fitted first-differences model is $\hat{Y}'_t = 20.88X'_t$.

a. Obtain the residuals for this fitted model and plot them in time order. Does the pattern appear to be consistent with time-independent errors? Explain.

b. Calculate MSE. How many degrees of freedom are associated with MSE here?

c. Obtain a 95 percent confidence interval for β_1. Interpret the interval estimate.

d. The firm has contracted to service 50 payroll-processing accounts in year 12. Predict by means of a point estimate the total payroll-processing revenue to be received in year 12.

EXERCISES

25.12 Refer to model (25.3).

a. What is the implication if $\beta_1 = 0$?

b. What is the implication if $\beta_4 = 0$? If $\beta_2 = \beta_3 = \beta_4 = 0$?

25.13 An analyst stated: "Quarterly sales of this product are primarily affected by three factors—the size of the current target market, advertising expenditures, and quarter of the year. The current size of the market has an immediate impact. Advertising expenditures in a given quarter affect sales in the following two quarters. Summer-related factors, which tend to boost sales in the second and third quarters, are absent in the first and fourth quarters. There are still other factors that are not as important, but together they tend to have a linear trend effect on sales."

a. Construct a regression model that might be suitable here. What does each regression coefficient denote?

b. Does your model contain any of the following? (1) a trend component, (2) a seasonal component, (3) a distributed time lag. If so, identify the relevant terms.

c. Will all of the independent variables in your model have to be predicted in order to forecast sales? Explain.

25.14 (Based on Optional Topic Section 15.4.) Refer to Problem 25.6. Conduct a lower-tail test for randomness of the error terms, using the runs-up-and-down test in Section 15.4. Control the α risk at .05. State the alternatives, decision rule, and conclusion. Is your conclusion here consistent with the one drawn from a visual examination of the residual plot?

25.15 With the first-differences model (25.13), if X_t is the same in two consecutive time periods, does this imply Y_t will also remain unchanged? Explain.

STUDIES

25.16 (Computer needed.) Refer to the trimester enrollment data in Problem 24.21. Use a computer routine to fit the general regression model (25.1) to the enrollment data, employing a linear trend variable X_1 (with $X_{t1} = 1$ in the first trimester of 1974) and two indicator variables (X_2 and X_3) for differential seasonal effects for trimesters II and III, respectively.

    a. State the estimated regression equation.

    b. Obtain point estimates of the expected enrollment in each trimester of 1982.

    c. Examine the residuals for the aptness of model (25.1). Is there any evidence of significant time dependence of the error terms? Discuss.

25.17 Consider the first-order autoregressive error model: $\epsilon_t = \rho \epsilon_{t-1} + u_t$. In Appendix Table C–9, the first 15 entries in column 8 constitute a random sequence of u_t observations from the $N(0, 1)$ distribution.

    a. Letting $\epsilon_0 = 0$, compute ϵ_t, $t = 1, 2, \ldots, 15$, under the assumption that: (1) $\rho = .8$, (2) $\rho = -.8$, (3) $\rho = 0$.

    b. Plot the three sets of ϵ_t values obtained in **a** in time order in three separate graphs. Contrast the patterns of values in the graphs.

26
Price and Quantity Indexes

Price and quantity indexes are summary measures of relative price and quantity changes over time in a set of items. Thus, the Consumer Price Index summarizes relative price changes in goods and services purchased by urban households; the Producer Price Indexes measure relative changes in prices received in primary markets by producers of commodities in all stages of processing; and the Index of Industrial Production measures relative changes in output in manufacturing, mining, and utilities.

Price and quantity indexes are not only used as summary measures of price and quantity changes, but are also employed in statistical analysis. Regression models, for instance, frequently contain price or quantity indexes as independent variables. Further, price indexes are employed to "adjust" time series in dollar units so that the series can be studied without the effects of price changes. In this chapter, we discuss price and quantity indexes in their several roles. First, we take up the construction of price indexes and consider various uses of them. Then we discuss quantity indexes. Finally, we briefly consider three major published indexes: Consumer Price Index, Producer Price Indexes, and Industrial Production Index.

26.1   PRICE RELATIVES AND LINK RELATIVES

First we take up the measurement of price changes for a single item, such as gasoline, butter, or socks.

Price Relatives

Price relatives are useful for studying the pattern of relative changes in the price of an item over time:

(26.1)   A *price relatives series* expresses the unit price of an item in each period as a percent of the item's unit price in the base period.

☐ Example

The prices of a petroleum-based compound during the period 1977–1980 are shown in Table 26.1, column 1. The corresponding price relatives, calculated on the base period 1977, are shown in column 2. Thus, the price relative for 1978 is $100(22/20) = 110$. This means that the price of the compound in 1978 was 110 percent as great as the price in 1977. The other price relatives are calculated and interpreted in a similar manner. ☐

Any time period that facilitates the comparisons of interest may be used as the base period. The choice of 1977 as the base period in our example facilitates study of the increase in the price of the compound during a period of transition.

Since the price relative for the base period is 100, the base period for a price relatives series is usually identified for reporting purposes in the manner shown in Table 26.1, e.g., $1977 = 100$.

Interpretation of Price Relatives.   Price relatives must be interpreted with care. We discuss briefly four important considerations:

1. The absolute magnitudes of price relatives are affected by the choice of the base period, but their proportional magnitudes are not affected by this choice. We illustrate this by presenting in column 3 of Table 26.1 the price relatives calculated on the base period 1980. The price relative for 1977 now is $100(20/50) = 40$, not 100 as for the 1977 base period series. However, the proportional magnitudes of the price relatives in columns 2 and 3 are the same. Thus, the price relatives for 1978 and 1977, with base period 1977, have the ratio $110/100 = 1.1$. For the price relatives with the 1980 base period, this ratio is the same, i.e., $44/40 = 1.1$. Similarly, for both series of price relatives the ratio of the 1979 price relative to the 1978 price relative is 1.64.

2. A comparison of price relatives for two different variables indicates nothing about the absolute prices unless we know the absolute magnitudes in the base period. Thus, information that the price relative for large grade A eggs in Montreal last month was 120 and that for Toronto was 123 tells us nothing about how the egg prices in the two cities compared unless we know the prices in the base period.

3. In analyzing price relatives, we must distinguish between *percent points of change* and *percent change*. To illustrate the difference, refer to the price rela-

TABLE 26.1   *Price relatives and link relatives series for petroleum-based compound during 1977–1980*

Year	(1) Unit Price	(2) Price Relatives 1977 = 100	(3) Price Relatives 1980 = 100	(4) Link Relative
1977	20	100	40	—
1978	22	110	44	110
1979	36	180	72	164
1980	50	250	100	139

tives with base period 1977 in column 2 of Table 26.1. Between 1979 and 1980, the percent points of change was $250 - 180 = 70$. On the other hand, the percent change was $100[(250 - 180)/180] = 100(70/180) = 38.9$. Thus, percent points of change refers to the absolute change in the price relatives and is dependent on the choice of the base period. Percent change, in contrast, refers to the relative change in the price relatives and, as we noted previously, does not vary with the choice of the base period.

4. A price relative below 100 indicates that the price in the given period is less than the price in the base period.

Link Relatives

Link relatives are useful in studying relative period-to-period price changes instead of changes from a fixed base period. For instance, link relatives enable us to study whether or not the price of an item is increasing at a constant rate, an increasing rate, or a decreasing rate.

(26.2) A *link relatives series* expresses the unit price of an item in each period as a percent of the item's unit price in the immediately preceding period.

☐ Example

The link relatives for the petroleum-based compound example are given in column 4 of Table 26.1. The link relative for 1978 is $100(22/20) = 110$; thus, the price in 1978 was 110 percent as great as that in 1977. The link relative for 1979 is $100(36/22) = 164$; thus, the price in 1979 was 164 percent as great as that in 1978, and so on. Note that the annual percent increase declined at the end of the period (column 4), even though the amount of annual price increase did not decline during the period (column 1).   ☐

Comment

When the values in any time series (e.g., annual sales, monthly production) are expressed relative to the value in the base period, the resulting ratios expressed as percentages are called *percent relatives*. Price relatives are therefore a special type of percent relative, computed from a price series.

Similarly, link relatives may be computed for time series other than price series.

26.2 PRICE INDEXES BY METHOD OF WEIGHTED AGGREGATES

Need for Price Indexes

In some applications, information on price changes for a single item in the form of price relatives is all that is required. For instance, during the mid-1970s energy costs began rising rapidly and many rental agreements in office buildings and apartment houses had clauses allowing rents to increase automatically as fuel costs increased. A typical clause would specify that the annual rent will increase by a given dollar amount for each one percent point increase in the price relative for the relevant fuel.

In many other applications, however, knowledge of individual price relatives is insufficient. In New York City, for instance, many apartment buildings have been "rent stabilized." This involves controlling rents by law at levels prevailing in a base period. Rent increases are allowed only to the degree that operating costs have increased because of price increases. In this situation, the prices of many items, such as electricity, water, and gas, are involved. Price relatives for these individual items will not provide suitable information about the aggregate effect of the individual price changes on the cost of operating an apartment building.

Price indexes are summary measures that combine the price changes for a group of items, using weights to give each item its appropriate importance. The Consumer Price Index is such an index, measuring the combined effect of price changes in many goods and services purchased by urban households. It has been estimated that about half of the population of the United States is affected directly by statutes and agreements tied to movements of the Consumer Price Index.

Two basic methods of calculating price indexes are widely used, the method of weighted aggregates and the method of weighted average of relatives. We shall explain each of these in turn, using the following illustration.

◻ Illustration

Officials of a large school district needed to develop a price index for maintenance supplies for the school buildings in the district. This was part of an effort to develop a variety of indexes, including ones for salaries and wages, instructional materials, and capital items. The set of indexes has been useful in budgeting and cost analysis, and also in negotiations with legislators and state officials for state aid.

In compiling the price index for school maintenance supplies, it was not feasible to include all supply items in the index because a large number of items are used. Instead, a sample of supply items was selected to be representative of all items used. The items selected are shown in Table 26.2, column 1. Also shown in this table, in columns 3, 4, and 5, are the unit prices of these items for 1978, 1979, and 1980, respectively. Finally, column 2 of Table 26.2 presents typical quantities consumed for each of the supply items. ◻

Method of Weighted Aggregates

The method of weighted aggregates for compiling a price index is conceptually very simple. We shall first explain the notation and terminology to be used. The *schedule of items* is the list of items included in the price index. Often, the schedule of items consists of a sample of all items of interest, as in our school maintenance supplies example. Sometimes, it is possible to include all items of interest in the schedule.

The price of the ith item in the schedule in any *given period t* is denoted by P_{it}, and the price of this item in the *base period* by P_{i0}. The typical quantity of the ith item that is consumed is denoted by Q_{ia}, where the subscript a stands for an average or typical period. These quantities are used as weights reflecting the importance of the items.

The method of weighted aggregates simply compares the cost of the typical quantities at period t prices with the cost of the same quantities at base period

TABLE 26.2 *Calculation of price index series for school maintenance supplies example by method of weighted aggregates*

	(1)	(2)	(3)	(4)	(5)	(6)	(7)	(8)
				Unit Prices				
i	Schedule of Items	Quantity Q_{ia}	1978 P_{i0}	1979 P_{i1}	1980 P_{i2}	$P_{i0}Q_{ia}$	$P_{i1}Q_{ia}$	$P_{i2}Q_{ia}$
1	Window glass	15 sheets	10.93	12.57	12.26	163.95	188.55	183.90
2	Fluorescent tube	130 boxes	22.60	25.58	26.98	2,938.00	3,325.40	3,507.40
3	Cleaner	290 cans	14.62	17.14	17.77	4,239.80	4,970.60	5,153.30
4	Floor wax	100 cans	41.71	57.82	64.00	4,171.00	5,782.00	6,400.00
5	Paint	175 cans	7.58	7.97	8.83	1,326.50	1,394.75	1,545.25
6	Mop head	200 pieces	2.84	3.14	3.31	568.00	628.00	662.00
					Total	13,407.25	16,289.30	17,451.85

$$I_{78} = 100\frac{13,407.25}{13,407.25} = 100.0$$

$$I_{79} = 100\frac{16,289.30}{13,407.25} = 121.5$$

$$I_{80} = 100\frac{17,451.85}{13,407.25} = 130.2$$

prices. The cost at period t prices is:

$$\text{Cost at period } t \text{ prices} = \sum_i P_{it}Q_{ia}$$

where the summation is over all the items in the schedule. The cost at base period prices is:

$$\text{Cost at period 0 prices} = \sum_i P_{i0}Q_{ia}$$

The price index for period t by the method of weighted aggregates is simply the ratio of these two costs, expressed as a percent:

(26.3) The price index for period t, denoted by I_t, by the *method of weighted aggregates* is:

$$I_t = 100\frac{\sum_i P_{it}Q_{ia}}{\sum_i P_{i0}Q_{ia}}$$

where: P_{i0} is the unit price of the ith item in period 0 (base period)
P_{it} is the unit price of the ith item in period t (given period)
Q_{ia} is the quantity weight assigned to the ith item

The price index calculated by the method of weighted aggregates has a simple empirical interpretation. It shows how much the typical quantities of the items in the schedule cost at period t prices relative to the cost of the same items and quantities at base period prices.

☐   Example

The price index series for school maintenance supplies for 1978, 1979, and 1980 is calculated by the method of weighted aggregates in Table 26.2. The costs of the typical quantities at the prices of each year are shown in columns 6, 7, and 8. For example, the cost of 15 sheets of window glass at 1978 prices is $15(10.93) = 163.95$ while at 1979 prices it is $15(12.57) = 188.55$. The aggregate costs of the schedule of items at the prices of the three periods are shown at the bottom of columns 6, 7, and 8, respectively, and the index numbers are calculated at the bottom of the table.

Since the same items and quantities are priced in each period, the relative differences in the index numbers are attributable wholly to changes in the prices of the items. Thus, the price index $I_{79} = 121.5$ indicates that the prices of school maintenance supplies increased, in terms of their aggregate effect, by 21.5 percent between 1978 and 1979. Since the index for 1980 ($I_{80} = 130.2$) is above the index for 1979 ($I_{79} = 121.5$), the prices increased between 1979 and 1980 also. We can calculate the percent change by expressing the percent point change, $130.2 - 121.5 = 8.7$, as a percent of the 1979 index, and obtain $100(8.7/121.5) = 7.2$ percent as the relative price increase between 1979 and 1980.   ☐

Comments

1. Just as with price relatives, the proportional magnitudes of the index numbers in Table 26.2 are not affected by which period is chosen as the base period.
2. When the typical quantity weights Q_{ia} are the quantities consumed in the base period, the weighted aggregates price index is called a *Laspeyres* price index:

(26.4)
$$\text{Laspeyres price index} = 100 \frac{\sum_i P_{it} Q_{i0}}{\sum_i P_{i0} Q_{i0}}$$

3. When the quantity weights are the quantities consumed in each given period t, the weighted aggregates price index is called a *Paasche* price index:

(26.5)
$$\text{Paasche price index} = 100 \frac{\sum_i P_{it} Q_{it}}{\sum_i P_{i0} Q_{it}}$$

With a Paasche index series, all comparisons must be made with the base period only. For all other comparisons, the quantity weights are not constant and differences in index numbers therefore reflect both price and quantity changes. If, for example, a Paasche price index series were to be calculated for the school maintenance supplies example with base period 1978, the 1979 price index would utilize quantities consumed in 1979 while the 1980 index would utilize quantities consumed in 1980. Since the quantities consumed in different years generally are not the same, a comparison of the 1979 and 1980 Paasche indexes on a 1978 base would reflect both price and quantity changes.

The Paasche price index is rarely used in practice.

4. The period from which the quantity weights are derived for the weighted aggregates price index is called the *weight period*. The base period for the price index and the weight period need not coincide, as we have seen.

Maintenance of Price Index Series

A key issue in the compilation of price index series involves the handling of items to be included in the index and the weights to be assigned to the items. Should the schedule of items and the weights be changed frequently so they are always up to date, or should they be held fixed over a relatively long sequence of years? If frequent changes are made, changes in the price index series over some years will reflect both changes in prices and changes in the composition of the index. Another disadvantage of frequent revision is that frequent updating of the schedule of items and of the weights may be very expensive. On the other hand, an index can become badly outdated if the schedule of items and the weights are held fixed over too long a period.

The procedure generally followed in practice is to hold the schedule of items and the weights essentially fixed for 5 to 10 years and then to revise them. The schedule and weights need not be held exactly fixed, since procedures are available for making interim adjustments without disturbing the index. For instance, a new item can be substituted for an item that has been taken off the market, and seasonal foods can be included in the schedule in season.

In addition to periodic updating of the schedule of items and weights, revisions often also include upgrading of data collection and tabulation procedures and modernizing of definitions. A new series of index numbers is then initiated with each revision. The new series can be "spliced" into the preceding series, if desired, by procedures to be illustrated shortly.

26.3 PRICE INDEXES BY METHOD OF WEIGHTED AVERAGE OF RELATIVES

A second method of compiling a price index series which is used in practice is the method of weighted average of relatives. This method is also conceptually quite simple. It utilizes the price relatives for each item in the schedule. The index for period t is a weighted average of the price relatives for that period for all items in the schedule.

Unlike the method of weighted aggregates, quantity weights cannot be used here. Recall that a price relative is the ratio of the price of an item in a given period to its price in the base period, expressed as a percent. In the ratio, the units cancel out and the relative is therefore unit-free. If quantity weights were to be used in our school maintenance supplies example and multiplied against unit-free relatives, we would obtain a mixture of units, such as sheets of window glass, boxes of fluorescent tubes, and cans of cleaner. Clearly, it would not be appropriate to add these terms.

Usually, *value weights* are used instead. Value weights are dollar values that reflect the importance of each item in the schedule. For example, the value weights may be obtained from the typical quantities consumed, Q_{ia}, by multiplying these quantities by typical prices. We shall denote the value weight for the ith item by V_{ia}.

Thus, the method of weighted average of relatives simply calls for a weighted average of the price relatives $100(P_{it}/P_{i0})$ for the items in the schedule, using value weights V_{ia}.

(26.6) The price index for period t, denoted by I_t, by the *method of weighted average of relatives* is:

$$I_t = \frac{\sum_i \left(100 \frac{P_{it}}{P_{i0}}\right) V_{ia}}{\sum_i V_{ia}}$$

where: P_{i0} is the unit price of the ith item in period 0 (base period)
P_{it} is the unit price of the ith item in period t (given period)
V_{ia} is the value weight assigned to the ith item

☐ Example

We return to our school maintenance supplies example, and shall calculate a price index series by the method of weighted average of relatives with base period 1978. The value weights will be based on the typical quantities and the unit prices in 1978 from Table 26.2, i.e.:

$$V_{ia} = P_{i0}Q_{ia}$$

Table 26.3 contains the necessary calculations. In column 1 is the schedule of items, and in column 2 are the value weights as obtained from Table 26.2. For example, the value weight for window glass is $V_{1a} = P_{10}Q_{1a} = 10.93(15) = 163.95$. The sum of the value weights is $\Sigma V_{ia} = 13,407.25$.

Columns 3, 4, and 5 contain the price relatives for the three years, based on the prices in Table 26.2. Note that all price relatives are expressed on a 1978 base because the price index is to have 1978 as the base period. For example, the 1979 price relative for window glass is $100(12.57/10.93) = 115.00$.

For each year, we take a weighted average of the price relatives in the appropriate column. The weighting of the price relatives is done in columns 6, 7, and 8, and the indexes are obtained at the bottom of the table. The index of 121.5 for 1979 indicates that prices of school maintenance supplies increased, in terms of their aggregate effect, by 21.5 percent between 1978 and 1979. ☐

Comments

1. The price indexes in Tables 26.2 and 26.3 are identical. This happened because we used base year prices P_{i0} in obtaining the value weights for the price index series by the method of weighted average of relatives. In that case, the method of weighted average of relatives index (26.6) reduces to the method of weighted aggregates index (26.3), i.e.:

$$\frac{\sum_i \left(100 \frac{P_{it}}{P_{i0}}\right) V_{ia}}{\sum_i V_{ia}} = \frac{\sum_i \left(100 \frac{P_{it}}{P_{i0}}\right) P_{i0}Q_{ia}}{\sum_i P_{i0}Q_{ia}} = 100 \frac{\sum_i P_{it}Q_{ia}}{\sum_i P_{i0}Q_{ia}}$$

TABLE 26.3 *Calculation of price index series for school maintenance supplies example by method of weighted average of relatives*

(1)	(2)	(3)	(4)	(5)	(6)	(7)	(8)
		Price Relatives (1978 = 100)			Price Relative × Value Weight		
		1978	1979	1980	1978	1979	1980
Schedule of Items i	Value Weight $V_{ia} = P_{i0}Q_{ia}$	$100\dfrac{P_{i0}}{P_{i0}}$	$100\dfrac{P_{i1}}{P_{i0}}$	$100\dfrac{P_{i2}}{P_{i0}}$	$\left(100\dfrac{P_{i0}}{P_{i0}}\right)V_{ia}$	$\left(100\dfrac{P_{i1}}{P_{i0}}\right)V_{ia}$	$\left(100\dfrac{P_{i2}}{P_{i0}}\right)V_{ia}$
1 Window glass	163.95	100.00	115.00	112.17	16,395	18,854	18,390
2 Fluorescent tube	2,938.00	100.00	113.19	119.38	293,800	332,552	350,738
3 Cleaner	4,239.80	100.00	117.24	121.55	423,980	497,074	515,348
4 Floor wax	4,171.00	100.00	138.62	153.44	417,100	578,184	639,998
5 Paint	1,326.50	100.00	105.15	116.49	132,650	139,481	154,524
6 Mop head	568.00	100.00	110.56	116.55	56,800	62,798	66,200
Total	13,407.25				1,340,725	1,628,943	1,745,198

$$I_{78} = \frac{1,340,725}{13,407.25} = 100.0$$

$$I_{79} = \frac{1,628,943}{13,407.25} = 121.5$$

$$I_{80} = \frac{1,745,198}{13,407.25} = 130.2$$

If prices other than base period prices are used for obtaining the value weights V_{ia}, the two methods will not lead to identical results. Generally, the differences in the indexes by the two methods will not be great.

2. Our earlier discussion about the need for periodic updating of the schedule of items and weights applies also to price index series compiled by the method of weighted average of relatives.

26.4 BASE PERIOD FOR INDEX SERIES

We now take up a number of considerations concerned with the base period of an index series.

Choice of Base Period

Any period within the coverage of the index series can be used as the base period. Of course, when a special purpose index is designed to measure price changes occurring since a particular period, such as since the lifting of price controls, that period would be taken as the base period. Whenever possible, the base period should involve relatively normal or standard conditions because many index users assume that the base period represents such conditions. Sometimes an interval of two or more years is selected as the base period. In this case, the index numbers are calculated so that the indexes for the base period years average to 100. Examples of base periods used for indexes compiled by statistical agencies of the federal governments in the United States and Canada include 1947–49, 1958, 1957–59, 1967, and 1971.

Shifting the Base of an Index Series

Sometimes it is necessary to shift the base period of an index series which is already compiled—as when two index series on different base periods need to be placed on a common base period to facilitate comparison. Usually it is not possible to recalculate a published index on the new base period because needed data are not available. However, a shortcut method can be employed which does not entail recalculation of the index.

Consider the following price index series with 1979 as the base period:

Year:	1976	'77	'78	'79	'80	'81
Price index:	75	88	92	100	110	122

Suppose the base period is to be shifted to 1976. We simply divide each index number by .75, the index number for the new base period in decimal form. The index number for 1976 becomes $75/.75 = 100$, and that for 1977 becomes $88/.75 = 117.3$. The new price index series with base period 1976 is:

Year:	1976	'77	'78	'79	'80	'81
Price index:	100.0	117.3	122.7	133.3	146.7	162.7

Because the proportional magnitudes of the index numbers are not affected by this shift in the base period, the shifted series conveys the same information about year-to-year relative price changes as the original series.

When the index series is calculated by the method of weighted aggregates employing fixed quantity weights, the shortcut procedure yields results identical to those obtained by recalculating the index series on the new base period. With most other formulas, however, the shortcut procedure only approximates such results.

Splicing

We noted earlier that in compiling index numbers on an ongoing basis, the usual procedure is to revise the index periodically and to initiate a new series with each revision. The new series can be joined or spliced to the older series to yield a single continuous series by a procedure similar to that just described for shifting the base period of an index series. Suppose the following two price index series are to be joined:

Year:	1972	'73	'74	'75	'76	'77	'78	'79	'80
Old series:	100	103	106	113	130				
Revised series:					100	120	126	131	134

We simply shift the level of one of the series so that both have a common value for 1976. For instance, to obtain a combined series with 1972 as the base period, we multiply every index in the revised series by 1.30. The spliced series is:

Year:	1972	'73	'74	'75	'76	'77	'78	'79	'80
Spliced series:	100	103	106	113	130	156	164	170	174

26.5 COLLECTION OF PRICE AND QUANTITY DATA

We now take up some important problems pertaining to the collection of price and quantity data for price indexes.

Collection Methods

All of the data collection methods discussed in Chapter 1 are used in collecting data for price indexes. Price and quantity data are obtained by observation, personal interview, and self-enumeration. Often sampling is used, as in the Consumer Price Index where probability samples of retail outlets and of items carried by the selected outlets are employed to obtain price data. Sampling is also used for the Consumer Price Index in periodic consumer expenditures surveys to obtain quantity data for weights.

Specification Pricing

In collecting price data for an item over a period of years, it is important to hold quality and other price-determining characteristics as constant as possible. This typically requires that detailed specifications be developed and adhered to in pricing the item. Here is an example of such a specification:

> Interior latex paint. Professional or commercial grade latex interior house paint, matte or flat finish, off-white, first line or quality.

Note that color, finish, and quality are specified because each of these characteristics can affect the price.

Quality Changes

A different problem in collecting and using price data in ongoing index series is the adjustment for quality changes in the items covered by the index. In principle, a price change that is caused by a quality change should not be reflected in the price index. In practice, minor quality changes are ignored. On the other hand, attempts are made in better price indexes to adjust for important quality changes. However, the conceptual and measurement problems are difficult. If the price of a car increases by $200 because of improved emission control equipment, should this increase be treated as a price increase (and reflected in the index), or as a price adjustment necessitated by a quality improvement (and not reflected in the index)? If an oven door is modified by the manufacturer to exclude the glass window while the list price remains unchanged, should this be treated as a price increase? If so, what should be the magnitude of the imputed increase? Extensive research has been undertaken on how these problems connected with quality changes should be handled.

Weights

In some indexes, the schedule includes all or almost all of the items used in the activity. Thus in a price index for highway construction, where a relatively small number of major items account for most of the total cost, it suffices for practical purposes to limit the schedule to these major items. Here, each weight reflects the importance of the particular item.

Often, however, the schedule must be limited to a sample of items, since thousands of items may be involved in the activity and it is neither feasible nor necessary to include them all. Each item is then selected for the schedule to represent a class of related items. In such cases, the weight normally reflects the importance of the entire class and not just of the item itself. For instance, a particular type of insulation is selected for a price index for residential building construction to represent all types of insulation and vapor barriers. The weight for the particular type of insulation in the schedule represents the importance of the entire class of insulation and vapor barriers.

26.6 USES OF PRICE INDEXES

We noted earlier that price indexes are summary measures of price changes. Indexes such as the Consumer Price Index and the Producer Price Indexes are widely followed as indicators of price changes. In addition, price indexes play important roles in a variety of applications. We now consider several of these.

Measuring Real Earnings

As prices change, the quantities of goods and services that can be purchased by a fixed sum of money change. In economic analysis, it is frequently important to measure changes in *real earnings* (i.e., quantities of goods and services that can be purchased). Price indexes are a basic tool in making this measurement.

☐   Example

Table 26.4 shows, in column 1, spendable average weekly earnings of private nonagricultural workers in the United States with three dependents during the period 1975–1979. These actual earnings data are expressed in *current dollars*. For example, 1975 earnings are expressed in terms of the buying power of a wage earner's dollar in 1975, and earnings in 1976 are expressed in terms of the buying power of the dollar in 1976. Column 1 of Table 26.4 shows that the earnings in current dollars increased steadily during the period.

The extent to which prices of goods and services purchased for daily living by workers and their dependents changed during the period is shown in column 2 by the Consumer Price Index for Urban Wage Earners and Clerical Workers. This index is expressed on a 1975 base period in the table. Note that prices in 1976 stood at 105.8 percent of their 1975 level. Thus, families, on the average, had to spend \$1.058 in 1976 for every \$1 in 1975 in purchasing goods and services for daily living. Consequently, the average money earnings of \$155.87 in 1976, expended at the prices of that year, were equivalent in purchasing power to 155.87/1.058 = \$147.33 at 1975 prices.

The average money earnings in the other years at 1975 prices are obtained by the same procedure: each current earnings figure is divided by the decimal value of the price index for that year. Thus, the money earnings of \$169.93 in 1977 are equivalent to earnings of 169.93/1.126 = \$150.91 at 1975 prices. Column 3 of Table 26.4 contains the average

TABLE 26.4 *Spendable weekly earnings at current and 1975 prices*

Year	(1) Weekly Earnings	(2) Consumer Price Index (1975 = 100)	(3) Weekly Earnings at 1975 Prices	(4) Link Relative	(5) Percent Relative (1975 = 100)
				Relatives for Column 3	
1975	145.65	100.0	145.65	—	100.0
1976	155.87	105.8	147.33	101.2	101.2
1977	169.93	112.6	150.91	102.4	103.6
1978	180.71	121.2	149.10	98.8	102.4
1979	194.82	135.0	144.31	96.8	99.1

SOURCE OF BASIC DATA: *Monthly Labor Review*, June 1980, pp. 82, 85.

weekly earnings data at 1975 prices. These data are said to be in *constant dollars* or *1975 dollars* since they are expressed in terms of the purchasing power of the dollar in 1975.

Relative changes in constant-dollar earnings data indicate the relative changes in purchasing power associated with the money earnings, i.e., the relative changes in *real earnings*. To show these relative changes explicitly, we have expressed the constant-dollars data as link relatives in column 4 of Table 26.4 and as percent relatives in column 5 (1975 = 100). The link relatives show that real earnings increased between 1975 and 1977 and then declined during the remainder of the period. The percent relatives show that all of the gain in real earnings between 1975 and 1977 was lost by 1979. ☐

Comment

We noted earlier that proportional magnitudes in an index series are not affected by the choice of the base period, but that the absolute magnitudes are affected. The same holds for constant-dollars series. Thus, if in our weekly earnings example the Consumer Price Index had been expressed on a 1979 base, the constant-dollars earnings would differ from column 3 of Table 26.4, but the percent relatives and link relatives based on these 1979 constant-dollars data would remain exactly the same.

Measuring Quantity Changes

Many important business and economic series on volume of activity are expressed in dollars, such as total retail sales and GNP. Changes in dollar volume reflect quantity changes, price changes, or both. Often, there is interest in the quantity changes alone. For example, if retail sales this year are five percent higher than last year's sales, is this due to price increases only or did the physical volume of goods sold increase? We can measure relative changes in quantities by use of price indexes. The procedure is similar to that for expressing current-dollars earnings as constant-dollars earnings, as we shall now illustrate.

☐ Example

Table 26.5, column 1, shows for the period 1978–1981 annual dollar sales by a manufacturer of household appliances. A price index for the selling prices of the appliances sold by the manufacturer is shown in column 2 of Table 26.5. The base year of the index is 1978. In column 3, sales are expressed at constant 1978 prices. For instance, 1979 sales of $43,538

TABLE 26.5 *Sales by manufacturer of household appliances at current and 1978 prices*

Year	(1) Annual Sales ($000)	(2) Product Price Index (1978 = 100)	(3) Sales at 1978 Prices ($000)	(4) Relatives for Column 3 Link Relative	(5) Percent Relative (1978 = 100)
1978	38,500	100	38,500	—	100.0
1979	43,538	103	42,270	109.8	109.8
1980	49,050	105	46,714	110.5	121.3
1981	54,950	107	51,355	109.9	133.4

thousand are equivalent to sales of 43,538/1.03 = $42,270 thousand at 1978 prices. Columns 4 and 5, respectively, present link relatives and percent relatives for the constant-dollars sales in column 3. These relatives reflect changes in the quantity of appliances sold, since prices are held constant. The link relatives show that the quantity of appliances sold increased by about 10 percent per year during the period. The percent relatives show, in turn, that by 1981 the quantity sold was about 33 percent higher than in 1978. □

Comment

In converting current-dollars series into constant dollars, the price index must be relevant to the series. Thus, we would not use, say, a food price index to adjust the series on appliances sales, or a Consumer Price Index to adjust the sales of a steel manufacturer.

Escalator Clauses in Contracts

Escalator clauses keyed to price indexes are widely employed. They are used in long-term contracts for the production of manufactured goods and leasing of office space, where sellers' costs can go up or down markedly because of price changes beyond their control. The most familiar use is for cost-of-living adjustments in labor contracts. In a typical escalator clause, wage rates will increase automatically by some fraction of a percent (e.g., two-thirds of a percent) for each one percent increase in an index such as the Consumer Price Index for Urban Wage Earners and Clerical Workers. The clause may or may not provide for wage rate decreases if the index declines.

Escalator clauses are also used to make automatic adjustments in pension and retirement benefits, funding for school programs, poverty threshold specifications, construction contracts, and alimony and child support agreements.

Price Comparisons between Locations

Our discussion up to this point has involved comparisons of prices over time. Price indexes also are used for comparisons between different locations. For example, the U.S. Department of State maintains indexes of living costs abroad which are used in setting up allowances for personnel serving at overseas posts. An index for an overseas post measures the cost of representative goods and services required by a family for a U.S. "pattern of living" at that post, relative to the cost of goods and services for a comparable pattern of living in Washington, D.C.

The indexes are calculated quarterly. For instance, in March 1979 the index value for Paris was 139. This indicates that the cost of the representative goods and services in Paris in March 1979 was about 39 percent greater than the cost of comparable goods and services in Washington, D.C.

26.7 QUANTITY INDEXES

A quantity index is a summary measure of relative changes over time in quantities of a set of items—e.g., of the relative changes since 1975 in the quantities of automobiles, trucks, and other vehicles imported by a country. In such index series, the quantities can change from one period to another, but the prices or other weights remain fixed. The index formulas are analogous to those for price indexes:

(26.7) The quantity index for period t is denoted by I_t. The quantity index by the *method of weighted aggregates* is:

(26.7a)
$$I_t = 100 \frac{\sum_i Q_{it} P_{ia}}{\sum_i Q_{i0} P_{ia}}$$

The quantity index by the *method of weighted average of relatives* is:

(26.7b)
$$I_t = \frac{\sum_i \left(100 \frac{Q_{it}}{Q_{i0}}\right) V_{ia}}{\sum_i V_{ia}}$$

Comment

A special problem in constructing quantity indexes arises in the choice of weights. The usual price or value weights may not be appropriate here. Consider a company that wishes to measure quantity changes in the outputs of different types of electronic circuits assembled from purchased components. Here, price weights or value weights for the different types of circuits would reflect mainly the prices or values of the purchased components, and not the quantities of output of the company per se. Weights derived from the typical number of manhours expended in assembling a unit of each type of circuit, or from the value added to a unit of each type of circuit in the assembly, would be more appropriate. Value added would be the value of the circuit when shipped from the plant less the cost of the purchased components and other purchased goods and services (e.g., fuel, electricity, containers) used in assembling the circuit and packaging it for shipment. The Index of Industrial Production utilizes value-added weights in many of its segments.

26.8 THREE MAJOR INDEXES

In this chapter, we have cited the Consumer Price Index, the Producer Price Indexes, and the Index of Industrial Production. These indexes are major indicators of price and quantity changes in the U.S. economy. Similar indexes are compiled in Canada and other countries for their economies. We now briefly describe the main features of the three U.S. indexes.

Consumer Price Index

Indexes in the Consumer Price Index (CPI) system are published monthly by the Bureau of Labor Statistics, U.S. Department of Labor. Two versions of the CPI are published—the CPI for All Urban Consumers and the CPI for Urban Wage Earners and Clerical Workers. The two index versions are based on schedules, or *market baskets*, of goods and services that are representative of purchases by families in the two respective population groups. Periodic consumer expenditures surveys are taken to determine the purchasing patterns on which the market baskets are based. Since the two population groups have somewhat different purchasing patterns, the respective market baskets differ. Pricing of the items included in the market baskets is done in outlets selected to be representative of places where households in the respective groups shop. The indexes are calculated by a method that gives results equivalent to those of formula (26.3) for the method of weighted aggregates.

For each version of the CPI, the All-Items Index covers all categories of goods and services purchased. Component indexes are also compiled which cover major categories, such as food and beverages and housing. At still lower levels of aggregation, component indexes are compiled for specific types of commodities and services.

Producer Price Indexes

Indexes in the Producer Price Indexes (PPI) system are published monthly by the Bureau of Labor Statistics, U.S. Department of Labor. This system was previously called the Wholesale Price Index. Producer Price Indexes measure average changes in prices received in primary markets of the United States by producers of commodities at all stages of processing. The indexes are based on a sample of about 10,000 respondents for some 2,800 major commodities. The responses are obtained monthly, primarily through mail questionnaires. Price indexes for individual commodities are aggregated into indexes for successively broader commodity groups according to three separate systems of classification: (1) stage of processing (i.e., finished goods, semifinished goods, crude goods), (2) type of commodity, and (3) industry sector. Weights are based on value of shipments and are obtained from periodic industrial censuses and special surveys. The indexes are calculated by the method of weighted average of relatives.

Industrial Production Index

The Industrial Production Index (IPI) is a system of quantity indexes published monthly by the Board of Governors of the Federal Reserve System. The indexes measure changes in the physical volume of output in manufacturing, mining, and utilities in the United States. The system contains 235 individual series at various levels of aggregation. Output data are supplied by industry, and the index is calculated by the method of weighted average of relatives. Value-added or manhour weights are used.

PROBLEMS

*26.1 The prices (in dollars) of two varieties of hedges at local nurseries during the period 1977–1981 follow. The prices are for three- to four-foot shrubs.

Year:	1977	1978	1979	1980	1981
Honeysuckle:	6.00	6.25	7.00	7.50	8.25
Lilac:	8.00	8.50	9.00	10.00	10.75

a. For each of the two price series, calculate: (1) price relatives on a 1977 base, (2) link relatives.
b. Which variety experienced the greater relative price increase between 1977 and 1981? Which experienced the greater relative price increase between 1980 and 1981?

26.2 Prices (in cents) per 20 grams of protein for two food products carried by a supermarket chain during 1977–1980 follow.

Year:	1977	1978	1979	1980
Peanut butter:	16	17	18	20
Bologna:	60	63	67	70

a. For each of the two price series, calculate: (1) price relatives on a 1977 base, (2) link relatives.
b. What is the percent point change in the price relatives between 1979 and 1980 for peanut butter? For bologna? What are the corresponding percent changes?
c. Which price series shows the greater absolute increase between 1977 and 1980? Which shows the greater relative increase? Is any contradiction involved here? Explain.
d. If 1980 instead of 1977 were used as the base period for the two series of price relatives, would the magnitudes of the price relatives differ from those in a? Would the proportional magnitudes of the price relatives in each series differ? Explain.

*26.3 A firm manages a complex of vacation cottages at a large resort. A list of items representative of all types of items that must be replaced periodically in routine maintenance of appliances and plumbing follows, together with quantity weights for the representative items and the unit prices paid in 1978, 1979, and 1980. This information is to be used in compiling a price index for appliances and plumbing supplies.

Item	Quantity	Unit Prices ($) 1978	1979	1980
Ice-making machine	1	687.00	738.00	805.00
Room airconditioner	12	174.00	187.00	207.00
Color television, table model	15	425.00	438.00	458.00
Sink faucet	80	15.00	16.25	18.50

a. Obtain the aggregate cost of the schedule of items in each period. Why does this cost differ from period to period?
b. Calculate the price index series by the method of weighted aggregates. Use 1978 as the base period. Interpret the index number for 1980.

26.4 The following data for part-time personnel of the Marlowe Guard Agency show hourly wage rates and numbers of manhours utilized in 1979, 1980, and 1981 for each job classification:

Job Classification	Wage Rate ($)			Manhours Utilized (*thousands*)		
	1979	1980	1981	1979	1980	1981
Plant protection	6.50	7.15	7.70	50	58	55
Construction-site patrol	5.65	6.25	6.75	40	44	42
Escort service	9.60	10.50	11.40	48	47	48

Calculate an index series of wage rates for the part-time personnel by the method of weighted aggregates. Employ 1979 as the base period and 1979 manhours as weights. Interpret the index number for 1981.

26.5 Refer to Problem 26.4. Do the base period and the weight period of the index coincide? Need this always be the case? Explain.

*26.6 Refer to Problem 26.3.
   a. Calculate a price index series by the method of weighted average of relatives. Use 1978 as the base period and 1978 prices in the value weights.
   b. Must your results in a be the same as those in Problem 26.3b? Explain.
   c. If the base period had been specified as 1979 for the index series, what price relatives would have been utilized?

26.7 Refer to Problem 26.4.
   a. Calculate an index series of wage rates by the method of weighted average of relatives. Use 1979 as the base period and derive the value weights from 1979 wage rates and manhours.
   b. Must your results in a agree with those in Problem 26.4? Explain.

26.8 Why are value weights used when averaging price relatives while it suffices to use quantity weights when averaging unit prices? Explain.

26.9 A research group studied family budgets for police officers and fire fighters in a state and developed a list of goods and services representative of the 1978 standard of living of families of the state's police officers and fire fighters. The total cost of this list was calculated to be $9,890 at 1978 prices. In a subsequent report, the group calculated the total cost of this same list at the prices of each of the next three years:

Year:	1978	1979	1980	1981
Dollar cost:	9,890	11,896	13,330	15,025

   a. Calculate a "consumer price index" series for the families covered by the study, using 1978 as the base year. Explain how the index numbers are interpreted.
   b. Which method—weighted aggregates or weighted average of relatives—did you implicitly use in calculating the index series in a? Explain.
   c. Can we conclude from the research group's data that a police officer's family having a cost of living of $9,890 in 1978 actually spent $15,025 on living costs in 1981? Discuss.

26.10 Refer to Problem 26.3. You have been asked to examine whether the index should be updated. What are some of the questions you would investigate?

26.11 In a large city, rents are stabilized at the levels existing in April 1967. Rent increases are allowed only to the degree that operating costs have increased since April 1967 because of price increases.

a. In compiling an appropriate index to measure price changes in operating costs for rental units in this city, which base period would be best? Why?

b. Give an example where the choice of base period is not clear-cut.

*26.12 A price index of raw materials purchased by the Zarthan Company has been compiled on a 1971 base, and is as follows for the period 1969–1981:

Year:	1969	'70	'71	'72	'73	'74	'75
Index:	96.2	99.1	100.0	101.1	103.7	105.2	107.9
Year:	'76	'77	'78	'79	'80	'81	
Index:	111.4	119.0	129.4	133.0	133.6	139.4	

Shift the index series to a 1979 base. Are the proportional magnitudes of the index numbers maintained by this shift? Explain.

26.13 Refer to Problem 26.3.

a. Confirm that shifting the base of the index series to 1980 by the shortcut method gives the same results as those obtained by recalculating the index series on a 1980 base from the original data.

b. When will the shortcut procedure not yield results identical to those obtained by recalculating the index series on the new base period?

*26.14 Splice the following two price index series into one continuous series with a 1978 base:

Year:	1976	'77	'78	'79	'80	'81
Old series (1976 = 100):	100	111	110			
New series (1980 = 100):			84	92	100	115

26.15 The following two price index series are to be joined into a continuous series:

Year:	1975	'76	'77	'78	'79	'80	'81
Old series (1975 = 100):	100	109	113	120			
New series (1978 = 100):				100	109	119	130

a. Splice the two series into one continuous series with a 1975 base period, and calculate the link relatives for this continuous series.

b. Answer a, using a 1981 base period.

c. How do the link relatives in a and b compare? Is this a coincidence? Explain.

*26.16 Spendable average weekly dollar earnings of U.S. manufacturing workers with three dependents during 1975–1979 were:

Year:	1975	1976	1977	1978	1979
Earnings:	166.29	181.32	200.06	214.87	232.07

SOURCE: *Monthly Labor Review*, June 1980, p. 82.

a. Use the Consumer Price Index in Table 26.4 to calculate each year's earnings in constant 1975 dollars. Interpret the constant-dollars earnings for 1979.

b. For the constant-dollars earnings in a, calculate: (1) percent relatives on a 1975 base, (2) link relatives.

c. Compare your results in **b** with those in Table 26.4 for all private nonagricultural workers with three dependents and state your findings.

d. Would your comparison in **c** be affected if the spendable earnings were expressed in constant 1979 dollars? Explain.

26.17 Data on the value of shipments by a manufacturing firm and an index of selling prices for the firm's products during 1976–1981 follow.

Year:	1976	'77	'78	'79	'80	'81
Value of shipments (*$ million*):	10.45	10.01	10.30	12.75	14.88	18.45
Product price index:	124	108	100	112	119	133

a. Obtain the value of shipments at 1978 prices for each year.

b. From the constant-dollars shipments data in **a**, calculate: (1) percent relatives on a 1976 base, (2) link relatives.

c. Has the increase in the value of the shipments between 1976 and 1981 been due primarily to the increase in the selling prices? Have the relative year-to-year changes in the constant-dollars series been fairly stable during the period? Discuss.

26.18 Annual imports and exports of merchandise by a developing country (in millions of national currency units) during 1976–1981, as well as indexes of prices paid for imported merchandise and prices received for exported merchandise, were as follows:

Year:	1976	'77	'78	'79	'80	'81
Imports:	380	383	435	420	525	706
Exports:	397	410	435	485	545	620
Imports index:	100	106	110	140	233	249
Exports index:	100	102	106	118	132	150

a. Adjust the imports and exports data for price level changes. Compare the unadjusted and adjusted series and state your findings.

b. Is the difference between unadjusted imports and exports for any one year a meaningful measure? Is the difference between adjusted imports and exports for any one year a meaningful measure? Explain.

26.19 Two cost-of-living escalator clauses are under discussion in contract negotiations. One clause calls for a 4¢ change in the wage rate for each 1 percent change in the Consumer Price Index for Urban Wage Earners and Clerical Workers. The other clause calls for a 4¢ change in the wage rate for each 1 percent point of change in the same index. Are the two clauses equivalent? Explain.

26.20 A speaker, noting that state welfare and social service payments are automatically increased to keep pace with increases in the Consumer Price Index, pointed out that television costs are included in the index. He objected to the inclusion of "luxuries" in an index used to determine increases in welfare payments. Does the inclusion of television in the Consumer Price Index automatically cause the index to increase by more than it would otherwise? Explain.

26.21 a. The value of the Dallas–Fort Worth Consumer Price Index for All Urban Consumers was 228.2 in October 1979, while the value of the corresponding index for Houston was 244.2 in the same month. Each index is on the base 1967 = 100. Can any conclusions be reached from these two index values about comparative living costs in the two areas in October 1979? Explain.

b. U.S. Department of State Indexes of Living Costs Abroad had values of 103 and 158 for Panama City and Brussels, respectively, in April–June 1979, the base for each being the comparable cost of living in Washington, D.C. Can any conclusions be drawn from these two index values about comparative living costs in Panama City and Brussels during the period? Explain.

*26.22 A list of representative items of work and materials for highway construction in a state, together with quantities used and typical unit prices, follows.

	Quantity Used (million units)			Typical Unit Price ($)
Item	1979	1980	1981	
Roadway excavation	50 yd³	54	56	1.30
Portland cement	2 bbl	2.7	2.9	13.11
Bituminous surfacing	1 yd²	1.1	1.1	19.84
Structural steel	70 lb	88	100.0	.58
Structural concrete	.14 yd³	.21	.28	165.65

a. Construct a quantity index series for highway construction by the method of weighted aggregates. Use 1979 as the base period.

b. Interpret the index number for 1980.

26.23 Refer to Problem 26.22. Construct a quantity index series for highway construction by the method of weighted average of relatives. Use 1980 as the base period and employ 1980 quantities in the value weights. Interpret the index number for 1981.

EXERCISES

26.24 Refer to Table 26.2. Calculate the price index series based on 1978–1979 = 100. Does the use of a two-year base period affect the year-to-year relative changes in the price index here? Explain.

26.25 In congressional hearings on governmental price statistics, a committee member asked whether the method of weighted average of relatives gives double importance to prices as compared to the method of weighted aggregates, since in the former method prices enter both the price relatives and the value weights. Comment.

26.26 Lead pencils are one of the items to be included in a new price index of school supplies. What are some price-affecting characteristics you would consider in preparing a specification for obtaining price data for this item?

26.27 Several years ago some manufacturers of clothes washers planned to eliminate the warm rinse option because rinsing is a mechanical function and water temperature plays no part in the actual rinsing process. This change would yield an eight percent saving in energy use of clothes washers. However, many users appeared unwilling to accept the concept of cold rinsing. An increase in customer complaints and service calls could be anticipated if the warm rinse option were eliminated. There would be no price changes in the washers.

a. From the point of view of collecting price data for a price index, should this design change be treated as a quality improvement (hence, a price decrease) or as a quality deterioration (hence, a price increase)? Discuss.

b. What problem in the compilation of ongoing price index series is illustrated here?

STUDIES

26.28 Total annual construction costs of restaurants built by a fast-food chain during 1975–1978 follow, together with a construction cost index for restaurants of this type.

Year:	1975	1976	1977	1978
Cost (*$ thousand*):	695	1,501	3,503	938
Construction cost index:	100	108	126	135

Today's value of the construction cost index is 180. What is the approximate replacement cost today of the restaurants constructed during the period 1975–1978—that is, what would be the total construction cost if all of these restaurants were built today?

26.29 Explain how you would construct a daily index of stock prices for common stocks traded on a major exchange. In your explanation, describe: (1) the types of stocks that would appear in your schedule of items, (2) the quantity or value weights to be used, (3) choice of the base period, (4) whether the method of weighted aggregates or the method of weighted average of relatives would be used in computing the index. Also comment on any special factors that would need to be considered in compiling and maintaining the index.

26.30 a. Consult the *Monthly Labor Review* to obtain the index values of the Consumer Price Index for All Urban Consumers for each of the most recent six months reported. Also obtain the index values for the following components: (1) food and beverages, (2) housing, (3) apparel and upkeep, (4) transportation.

b. Calculate the percent changes in the All-Items Index and in each of the four component indexes for the six-month period covered by your data. Which components showed larger relative price changes than the All-Items Index? Which ones showed smaller relative changes?

Appendix A
Mathematical Review

A.1   SUMMATION NOTATION

Single Summation

A number of formulas in this book use the symbol Σ (Greek capital sigma) to designate summation. The expression:

$$\sum_{i=1}^{n} X_i$$

is a shorthand form of the sum of the n values $X_1, X_2, \ldots, X_n$. In other words:

(A.1)
$$\sum_{i=1}^{n} X_i = X_1 + X_2 + \cdots + X_n$$

The subscript i on X is a label or index given to each value of X so that the n values may be distinguished from one another.

Given next are interpretations of several expressions that employ the Σ symbol:

(A.2)
$$\left(\sum_{i=1}^{n} X_i \right)^2 = (X_1 + X_2 + \cdots + X_n)^2$$

(A.3)
$$\sum_{i=1}^{n} X_i^2 = X_1^2 + X_2^2 + \cdots + X_n^2$$

(A.4)
$$\sum_{i=1}^{n} c = c + c + \cdots + c = nc$$

where: c is a constant

(A.5)
$$\sum_{i=1}^{n} cX_i = cX_1 + cX_2 + \cdots + cX_n = c\left(\sum_{i=1}^{n} X_i \right)$$

(A.6) $\displaystyle\sum_{i=1}^{n} (X_i + Y_i) = (X_1 + Y_1) + (X_2 + Y_2) + \cdots + (X_n + Y_n) = \sum_{i=1}^{n} X_i + \sum_{i=1}^{n} Y_i$

☐ Example

Suppose $n = 3$, $c = 11$, $X_1 = 3$, $X_2 = -1$, and $X_3 = 7$. Then:

$$\sum_{i=1}^{3} X_i = 3 - 1 + 7 = 9$$

$$\left(\sum_{i=1}^{3} X_i\right)^2 = (3 - 1 + 7)^2 = 9^2 = 81$$

$$\sum_{i=1}^{3} X_i^2 = 3^2 + (-1)^2 + 7^2 = 59$$

$$\sum_{i=1}^{3} 11 = 11 + 11 + 11 = 3(11) = 33$$

$$\sum_{i=1}^{3} 11X_i = 11 \sum_{i=1}^{3} X_i = 11(9) = 99$$

☐

Double Summation

Some formulas use the double summation expression, such as $\displaystyle\sum_{i=1}^{r} \sum_{j=1}^{c} X_{ij}$. The double subscript ij on X is again a label or index given to each value of X so that the values may be distinguished from one another. In this case, the subscript i ranges from 1 to r and the subscript j ranges from 1 to c. The definition of $\displaystyle\sum_{i=1}^{r} \sum_{j=1}^{c} X_{ij}$ is:

(A.7) $\displaystyle\sum_{i=1}^{r} \sum_{j=1}^{c} X_{ij} = \sum_{i=1}^{r} \left(\sum_{j=1}^{c} X_{ij}\right) = \sum_{j=1}^{c} X_{1j} + \sum_{j=1}^{c} X_{2j} + \cdots + \sum_{j=1}^{c} X_{rj}$

where, for instance:

$$\sum_{j=1}^{c} X_{1j} = X_{11} + X_{12} + \cdots + X_{1c} \qquad \sum_{j=1}^{c} X_{2j} = X_{21} + X_{22} + \cdots + X_{2c}$$

☐ Example

Consider the following table of X_{ij} values:

	$j = 1$	$j = 2$	$j = 3$
$i = 1$	4	-3	1
$i = 2$	5	10	-6

X_{ij} is the value at the intersection of the ith row and jth column in the table; e.g., $X_{11} = 4$, $X_{12} = -3$, $X_{21} = 5$. Note that:

$$\sum_{j=1}^{3} X_{1j} = 4 - 3 + 1 = 2 \qquad \sum_{j=1}^{3} X_{2j} = 5 + 10 - 6 = 9$$

Hence:

$$\sum_{i=1}^{2} \sum_{j=1}^{3} X_{ij} = \sum_{j=1}^{3} X_{1j} + \sum_{j=1}^{3} X_{2j} = (4 - 3 + 1) + (5 + 10 - 6) = 11$$

Observe that the double summation is simply the sum of all six values in the table.

☐

Comment

In all of our uses of double summation, the order of summation can be reversed; i.e.:

(A.8)
$$\sum_{i=1}^{r} \sum_{j=1}^{c} X_{ij} = \sum_{j=1}^{c} \sum_{i=1}^{r} X_{ij}$$

For the previous example, we have:

$$\sum_{j=1}^{3} \sum_{i=1}^{2} X_{ij} = \sum_{i=1}^{2} X_{i1} + \sum_{i=1}^{2} X_{i2} + \sum_{i=1}^{2} X_{i3} = (4 + 5) + (-3 + 10) + (1 - 6) = 11$$

Abbreviated Notation

Where the context makes it clear which values are to be summed, it is common to omit the range of the summation index or even the index itself. In this case, we write:

$$\sum_{i} X_i \quad \text{or} \quad \Sigma X_i \qquad \text{instead of} \qquad \sum_{i=1}^{n} X_i$$

and:

$$\sum_{i} \sum_{j} X_{ij} \quad \text{or} \quad \Sigma \Sigma X_{ij} \qquad \text{instead of} \qquad \sum_{i=1}^{r} \sum_{j=1}^{c} X_{ij}$$

Abbreviated summation notation of this type is used frequently in this Text.

A.2 ALGEBRAIC RULES FOR EXPONENTS AND LOGARITHMS

Exponents

The following is a summary of algebraic rules for exponents. Here a, b, m, and n denote numbers. Each rule is illustrated.

	Rule	Illustration
(A.9)	$a^0 = 1$ if $a \neq 0$	$3^0 = 1$
(A.10)	$a^m a^n = a^{m+n}$	$.6^2(.6^3) = .6^5 = .07776$
(A.11)	$(a^m)^n = a^{mn}$	$(3^2)^4 = 3^8 = 6{,}561$
(A.12)	$a^{-n} = 1/a^n$	$10^{-2} = 1/10^2 = .01$
(A.13)	$a^m/a^n = a^{m-n}$	$5^2/5^4 = 5^{-2} = .04$
(A.14)	$(ab)^n = a^n b^n$	$[5(7)]^2 = 5^2(7^2) = 1{,}225$
(A.15)	(i)$a^{1/n} = \sqrt[n]{a}$ if $a > 0$	$8^{1/3} = \sqrt[3]{8} = 2$
	(ii)$a^{1/2} = a^{.5} = \sqrt[2]{a}$ is understood to	
	mean $\sqrt{a}$, the square root of a	$9^{1/2} = \sqrt[2]{9} = \sqrt{9} = 3$
(A.16)	$a^{m/n} = \sqrt[n]{a^m}$ if $a > 0$	$5^{3/2} = \sqrt[2]{5^3} = 11.180$

Logarithms

Definition. The symbol $\log_b A$, for any positive number A, denotes the exponent to which b must be raised to equal A. In other words, if $\log_b A = C$, then by definition, $b^C = A$. The number b is called the *base* of the logarithm and may be any positive number except 1. Logarithms with base 10 are called *common* logarithms. Logarithms to base e (where $e = 2.71828\ldots$) are called *natural* or *Naperian logarithms*. $C = \log_b A$ is called the *logarithm* or *log* of A to base b, and A is called the *antilogarithm* or *antilog* of C to base b. Frequently, natural logarithms are denoted by *ln* (pronounced "lawn") rather than by $\log_e$. In this Text, and in the following examples in particular, all logarithms are common logarithms (i.e., to base 10) and the base designation is omitted.

☐   Examples

1. $\log 1 = 0$ because $10^0 = 1$
2. $\log 10 = 1$ because $10^1 = 10$
3. $\log 100 = 2$ because $10^2 = 100$
4. $\log .1 = -1$ because $10^{-1} = 1/10 = .1$
5. $\log 3.162 = .5$ because $10^{.5} = \sqrt{10} = 3.162$
6. antilog $2 = 100$ because $10^2 = 100$
7. antilog $.5 = 3.162$ because $10^{.5} = 3.162$ ☐

Rules. The following is a summary of the algebraic rules for logarithms. Here, A and B denote positive numbers while c denotes any number. Each rule is illustrated.

	Rule	Illustration
(A.17)	$\log AB = \log A + \log B$	$\log[(10)(100)] = \log 10 + \log 100 = 1 + 2 = 3$
(A.18)	$\log A/B = \log A - \log B$	$\log(3.162/10) = \log 3.162 - \log 10 = .5 - 1 = -.5$
(A.19)	$\log A^c = c \log A$	$\log(100)^3 = 3 \log 100 = 3(2) = 6$
		$\log \sqrt[3]{10} = \log 10^{1/3} = \dfrac{1}{3}\log 10 = \dfrac{1}{3}(1) = .3333$

Calculator Usage

Most calculators have one or more of the following operator keys for computing exponents and logarithms:

Key	Operation	Illustration
10^x	Computes the antilog of x to base 10	antilog $.5 = 10^{.5} = 3.162$
log x	Computes the log of x to base 10	log $3.162 = .5$
e^x	Computes the antilog of x to base e	$\text{antilog}_e 2.0 = e^2 = (2.71828)^2 = 7.39$
ln x	Computes the log of x to base e	ln $7.39 = \log_e 7.39 = 2.0$
y^x	Raises y to the power x	antilog $.5 = 10^{.5} = 3.162$
		$\text{antilog}_e 2.0 = e^2 = (2.71828)^2 = 7.39$

A.3 SET NOTATION, OPERATIONS, AND ALGEBRAIC RULES

Basic Concepts

A *set* is a collection of distinct objects called *elements* of the set. We shall specify a set by listing its elements within a pair of braces. For example, $\{a, c, i, s, t\}$ is the set of letters in the word *statistics* and $\{0, 1, 2, 3\}$ is the set of number of defective units that can occur in a sample of size three.

If a is an element of set A, then we write $a \in A$. If a is not an element of set A, then we write $a \notin A$. For instance, if $S = \{a, c, i, s, t\}$, then $c \in S$ while $m \notin S$. If all the elements of a set A are also elements of set B, then we say A is a *subset* of B and we write this symbolically as $A \subseteq B$. For instance, if $T = \{a, c, t\}$ and $S = \{a, c, i, s, t\}$, then $T \subseteq S$. Two sets are equal if they contain the same elements.

If I denotes the set of all elements of interest in a particular context, we call I the *universal set* for that context. If A is a subset of I, then A^* denotes the *complement* of A (relative to I) and represents the set of all elements of I that are not elements of A. For instance, if $I = \{1, 2, 3, 4, 5\}$ and $E = \{2, 4\}$, then $E^* = \{1, 3, 5\}$. The set that contains no elements is called the *empty set* and is denoted by $\varnothing$.

Set Operations

The two most common operations involving pairs of sets are set union and set intersection.

Union. For any pair of sets A and B, $A \cup B$ denotes the set of all elements that belong to either A or B or both A and B, and is called their *set union*. For example, if $A = \{1, 11, 21\}$ and $B = \{1, 3, 8, 11\}$, then $A \cup B = \{1, 3, 8, 11, 21\}$.

Intersection. For any pair of sets A and B, $A \cap B$ denotes the set of all elements that belong to both A and B, and is called their *set intersection*. For example, if A and B are the sets in the preceding example, then $A \cap B = \{1, 11\}$.

Algebraic Rules for Set Operations

Name of Rules	Rules
(A.20) *Associative Rules*	$A \cup (B \cup C) = (A \cup B) \cup C$
	$A \cap (B \cap C) = (A \cap B) \cap C$
(A.21) *Commutative Rules*	$A \cup B = B \cup A$
	$A \cap B = B \cap A$
(A.22) *Distributive Rules*	$A \cap (B \cup C) = (A \cap B) \cup (A \cap C)$
	$A \cup (B \cap C) = (A \cup B) \cap (A \cup C)$
(A.23) *Idempotency Rules*	$A \cup A = A$
	$A \cap A = A$
(A.24) *I and Ø Rules*	$A \cup \varnothing = A$
	$A \cup I = I$
	$A \cap \varnothing = \varnothing$
	$A \cap I = A$
(A.25) *Complementation Rules*	$A \cup A^* = I$
	$A \cap A^* = \varnothing$
	$(A \cup B)^* = A^* \cap B^*$
	$(A \cap B)^* = A^* \cup B^*$
(A.26) *Involution Rules*	$(A^*)^* = A$
	$\varnothing^* = I$
	$I^* = \varnothing$

A.4 PERMUTATIONS AND COMBINATIONS

Permutations

If r objects are selected from a set of n different objects and arranged in a particular order, the arrangement is called a *permutation of n objects taken r at a time*. The number of permutations of n different objects taken r at a time is denoted by $_nP_r$ and equals:

(A.27)
$$_nP_r = n(n - 1)(n - 2) \cdots (n - r + 1)$$

The number of permutations of n different objects taken together is:

(A.27a)
$$_nP_n = n! = n(n - 1)(n - 2) \cdots (1)$$

The notation $x!$ (read "x factorial") stands for $x(x - 1)(x - 2) \cdots (1)$. Thus, $5! = 5(4)(3)(2)(1) = 120$. By definition, $0! = 1$.

☐ Examples

1. An experiment requires that two persons be selected from a group of four persons—A, B, C, and D—the first person selected being tested on day 1 and the second person tested on day 2. The number of different permutations of the $n = 4$ persons taken $r = 2$ at a time by (A.27) is $_4P_2 = 4(3) = 12$. The 12 permutations are:

$$\begin{array}{cccccc} AB & AD & BC & CA & CD & DB \\ AC & BA & BD & CB & DA & DC \end{array}$$

Note that the order of persons matters—AB and BA, for instance, are different permutations, because the testing of the two persons follows different sequences.

2. In Example 1, if all $n = 4$ persons are taken for the experiment and scheduled on different days, the total number of different permutations by (A.27a) is $4! = 4(3)(2)(1) = 24$. ☐

Permutations of Like Objects Consider n objects that are of c different types. Suppose there are r_1 objects of type 1, r_2 of type 2, etc., where $r_1 + r_2 + \cdots + r_c = n$. The number of permutations (i.e., distinguishable ordered arrangements) when all n objects are taken together equals:

(A.28)
$$\frac{n!}{r_1! r_2! \cdots r_c!}$$

☐ Example

$n = 5$ bottles come off a filling line, $r_1 = 2$ having acceptable fills (A) and $r_2 = 3$ defective fills (D). One possible sequence or permutation for the five bottles as they are processed is ADDDA. The number of permutations by (A.28) here is:

$$\frac{5!}{2!\ 3!} = \frac{120}{2(6)} = 10$$ ☐

Combinations

A set of r objects selected from n different objects, considered without regard to their order in the set, is called a *combination of n objects taken r at a time*. The number of distinct combinations of n objects taken r at a time is denoted by $_nC_r$ and equals:

(A.29)
$$_nC_r = \frac{n!}{r!(n-r)!}$$

Note that $_nC_r = {_nC_{n-r}}$.

☐ Example

A litter contains five pups—A, B, C, D, and E—four of which are to be sold to a family. The number of different combinations of these $n = 5$ pups taken $r = 4$ at a time by (A.29) is $_5C_4 = 5!/4!1! = 5$. These five combinations are:

$$\begin{array}{ccccc} ABCD & ABCE & ABDE & ACDE & BCDE \end{array}$$

Note that the order of the pups is immaterial here—ABCD and DCBA, for instance, are the same combination. ☐

Equivalence to Binomial Coefficient. $_nC_r$ is equivalent to the number of permutations of like objects defined in (A.28) when there are $c = 2$ groups of like objects. Since $_nC_r$ is the same as the binomial coefficient $\binom{n}{r}$ defined in (6.4), it can be seen that the binomial coefficient equals the number of permutations that can be formed from n elements, of which r are of one type and $n - r$ are of a second type.

Appendix B
Chi-Square, *t*,
and *F* Distributions

In this appendix, we discuss the chi-square, *t*, and *F* distributions, which are all related to the normal distribution and are widely used in statistical analysis. For each distribution, we describe its main characteristics and explain how to use its table of percentiles. In a final section, we consider the theoretical relations between the chi-square, *t*, *F*, and normal distributions.

In Figure B.1 (p. 714), we have set out in summary form, for subsequent reference, several of the important characteristics of the chi-square, *t*, and *F* distributions, as well as of the standard normal distribution.

B.1 CHI-SQUARE DISTRIBUTIONS

Characteristics

The χ^2 (Greek chi-square) distribution is a continuous probability distribution. It has one parameter, ν (Greek nu), which is called its *degrees of freedom* (abbreviated *df*). The parameter ν may be any positive whole number. Each different value of ν corresponds to a different member of the χ^2 family of distributions. The random variable associated with the χ^2 distribution is denoted by $\chi^2(\nu)$ and may take on any nonnegative value, i.e., $0 \leq \chi^2(\nu) < \infty$. Graphs of the χ^2 probability density function for several values of ν are shown in Figure B.1. Figure B.1 also summarizes some of the properties of the χ^2 distribution, which we now discuss.

Shape. From the graphs of the χ^2 distribution in Figure B.1, it is seen that the χ^2 distribution is unimodal and right-skewed, although the skewness becomes smaller as ν increases. In fact, it can be shown that as ν increases, the shape of the χ^2 distribution approaches that of a normal distribution.

Mean and Variance. The mean and variance of the $\chi^2(\nu)$ random variable are:

(B.1)
$$E\{\chi^2(\nu)\} = \nu$$

(B.2)
$$\sigma^2\{\chi^2(\nu)\} = 2\nu$$

For example, $\chi^2(5)$ has a mean of 5 and a variance of $2(5) = 10$. Observe that the expected value is equal to the degrees of freedom ν and that the variance equals twice the degrees of freedom. Hence, both the mean and variance of the χ^2 distribution increase as ν increases.

Determining Percentiles for Chi-Square Distributions

Table of Percentiles. Table C–2 contains selected percentiles of χ^2 distributions for various degrees of freedom from 1 to 100. We shall let $\chi^2(a; \nu)$ denote the $100a$ percentile of a χ^2 random variable with ν degrees of freedom, i.e.:

(B.3) $$P[\chi^2(\nu) \leq \chi^2(a; \nu)] = a$$

The illustration at the top of Table C–2 shows this relationship.

☐ Examples

1. For 5 degrees of freedom, find $\chi^2(.90; 5)$. From Table C–2, we see that the 90th percentile of the χ^2 distribution with 5 degrees of freedom is $\chi^2(.90; 5) = 9.24$. Hence, the probability is .90 that $\chi^2(5)$ is less than or equal to 9.24.
2. For $df = 60$, find the 5th percentile. From Table C–2, we see that $\chi^2(.05; 60) = 43.19$. Hence, the probability is .05 that $\chi^2(60)$ is less than or equal to 43.19. ☐

Normal Approximation. It was noted earlier that the shape of the χ^2 distribution approaches that of a normal distribution as ν increases. This allows one to approximate χ^2 probabilities for large values of ν by normal probabilities. A good approximation of a chi-square percentile $\chi^2(a; \nu)$ in terms of a standard normal percentile $z(a)$ is given by:

(B.4) $$\chi^2(a; \nu) \simeq \frac{1}{2}\left[z(a) + \sqrt{2\nu - 1}\right]^2$$

☐ Examples

1. For 100 degrees of freedom, find the 95th percentile using the normal approximation. We know from Table C–1 that $z(.95) = 1.645$. Hence, we obtain by (B.4):

$$\chi^2(.95; 100) \simeq \frac{1}{2}\left[1.645 + \sqrt{2(100) - 1}\right]^2 = 124.1$$

Incidentally, the exact percentile as shown in Table C–2 is 124.3, so the approximation is quite good.
2. For $df = 250$, find the 25th percentile. Since we know from Table C–1 that $z(.25) = -.67$, we obtain:

$$\chi^2(.25; 250) \simeq \frac{1}{2}\left[-.67 + \sqrt{2(250) - 1}\right]^2 = 234.8$$ ☐

Problems

*B.1 Consider a $\chi^2(2)$ random variable.
    a. Obtain the 90th and 10th percentiles of the probability distribution.
    b. What is the value of the mean? The standard deviation?

      c. Are the percentiles in **a** equally spaced about the mean? What does this fact imply about the symmetry of the probability distribution?

B.2 Refer to Problem B.1. Answer all parts for a $\chi^2(4)$ random variable.

*B.3 **a.** Obtain $\chi^2(.95; 8)$. What does this value represent?

      b. Estimate $\chi^2(.05; 145)$ using the normal approximation.

B.4 **a.** Obtain $\chi^2(.01; 12)$. What does this value represent?

      b. Estimate $\chi^2(.90; 113)$ using the normal approximation.

B.2   *t* DISTRIBUTIONS

Characteristics

The *t* distribution is a continuous probability distribution. Like the χ^2 distribution, it has only one parameter, ν, which is called its *degrees of freedom (df)*. The parameter ν may be any positive whole number. Each different value of ν corresponds to a different member of the family of *t* distributions. The random variable associated with the *t* distribution is denoted by $t(\nu)$ and may take on any value, i.e., $-\infty < t(\nu) < +\infty$.

    Graphs of the *t* probability density function for two values of ν are shown in Figure B.1. Figure B.1 also presents some of the properties of the *t* distribution.

Shape. From the graphs of the *t* distribution in Figure B.1, it is seen that the *t* distribution is symmetrical and almost bell-shaped, and its appearance is like the standard normal distribution when ν is large. In fact, it can be shown mathematically that the *t* distribution approaches the standard normal distribution as ν becomes large.

Mean and Variance. The mean and variance of the $t(\nu)$ random variable are:

(B.5)
$$E\{t(\nu)\} = 0$$

(B.6)
$$\sigma^2\{t(\nu)\} = \frac{\nu}{\nu - 2} \qquad \text{when } \nu > 2$$

The variance is undefined for $\nu \leq 2$.

    For the $t(6)$ random variable, for instance, the mean and variance are 0 and $6/(6 - 2) = 1.50$, respectively. For $t(60)$, the mean still is 0 but the variance now is only 1.03.

    Note that the *t* distribution has a mean of 0 for all ν, and that the variance approaches 1 as ν gets larger. These facts are consistent with the earlier statement that the *t* distribution approaches the standard normal distribution as ν gets larger, since the latter distribution has mean 0 and variance 1.

Determining Percentiles for *t* Distributions

Table of Percentiles. Table C–3 contains selected percentiles of *t* distributions for various degrees of freedom ν. We shall let $t(a; \nu)$ denote the 100*a* percentile of

FIGURE B.1 *Summary of characteristics of standard normal, chi-square, t, and F distributions*

Distribution	Random variable	Sample space	Parameters	Mean	Variance	Skewness	Density functions
Standard normal	Z	$(-\infty, +\infty)$	None	0	1	Symmetrical	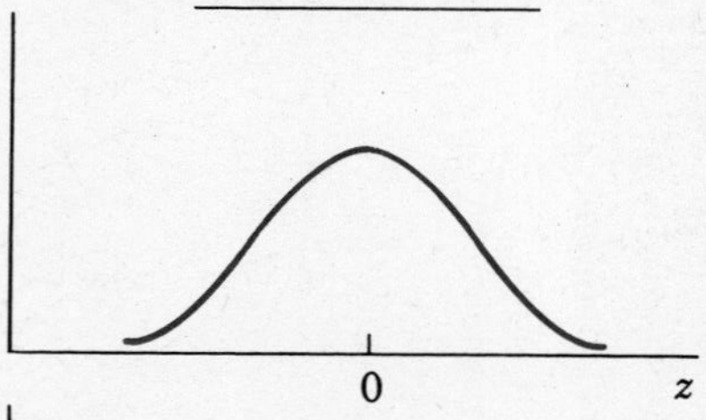
χ^2	$\chi^2(v)$	$(0, \infty)$	v	v	$2v$	Right-skewed	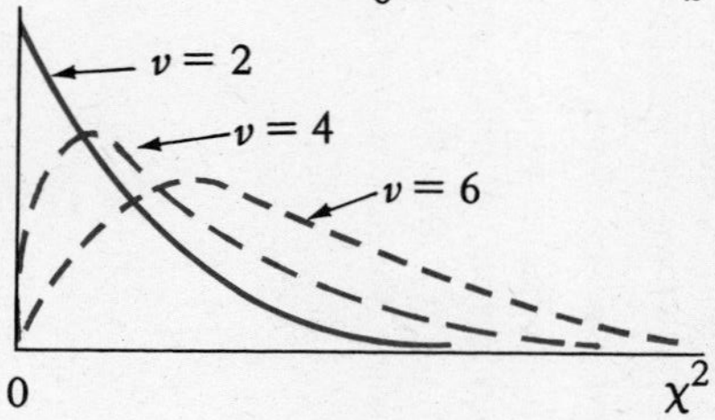
t	$t(v)$	$(-\infty, +\infty)$	v	0	$\dfrac{v}{(v-2)}$ $(v>2)$	Symmetrical	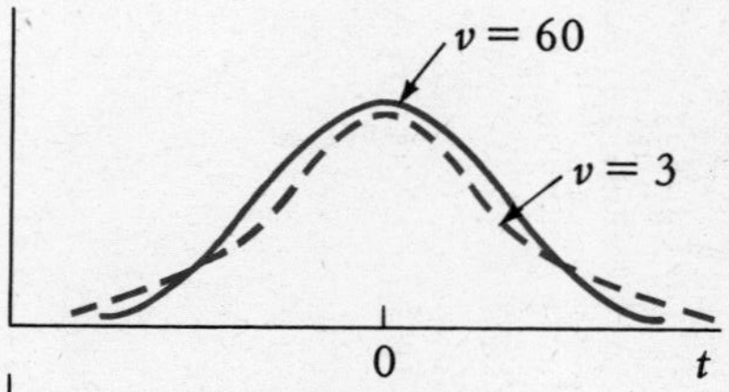
F	$F(v_1, v_2)$	$(0, \infty)$	v_1, v_2	$\dfrac{v_2}{v_2-2}$ $(v_2>2)$	$\dfrac{2v_2^2(v_1+v_2-2)}{v_1(v_2-2)^2(v_2-4)}$ $(v_2>4)$	Right-skewed	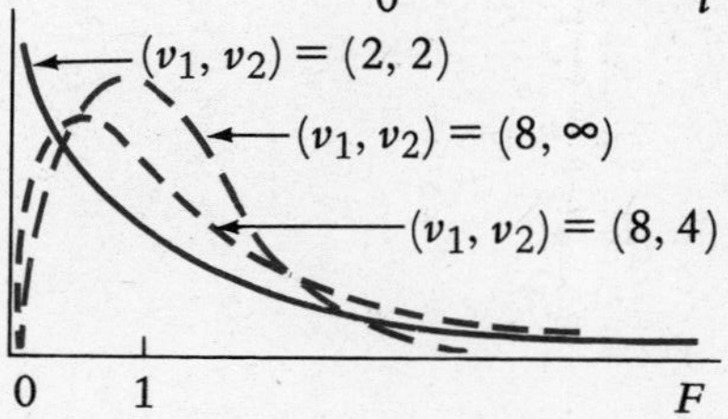

a t random variable with ν degrees of freedom; i.e.:

(B.7)
$$P[t(\nu) \leq t(a; \nu)] = a$$

The illustration at the top of Table C–3 shows this relationship.

☐ Examples

1. For 10 degrees of freedom, find the 95th percentile. From Table C–3, we see that $t(.95; 10) = 1.812$. Hence, the probability is .95 that $t(10)$ is less than or equal to 1.812.
2. For $df = 30$, find the 90th percentile. From Table C–3, we see that $t(.90; 30) = 1.310$.
3. For $df = 10$, find the 5th percentile. Because of the symmetry of the t distribution about its mean 0, the 5th percentile is the negative of the 95th percentile. The same is true for the standard normal distribution, as illustrated in Figure 7.4. Hence, $t(.05; 10) = -t(.95; 10) = -1.812$. Note that the symmetry property has made it unnecessary to show percentiles less than the 50th in Table C–3. ☐

Normal Approximation. Since the t distribution approaches the standard normal distribution as ν increases, for large ν the percentiles of the standard normal distribution may be used to approximate those of the t distribution. In fact, the last row in Table C–3, which corresponds to $\nu = \infty$, has the same percentiles as those for the standard normal distribution. For instance, we know from Table C–1 that $z(.975) = 1.960$. In Table C–3, we see that $t(.975; \infty) = 1.960$. As Table C–3 shows, once ν is larger than about 30, the standard normal distribution provides approximate percentiles for the t distribution which are adequate for most practical purposes. Thus, $t(.975; 217) \simeq 1.960$.

Problems

*B.5 Consider a $t(5)$ random variable.
    a. Obtain the 95th and 5th percentiles of the probability distribution.
    b. What is the value of the mean? The standard deviation?

B.6 Refer to Problem B.5. Answer all parts for a $t(4)$ random variable.

B.7 Plot the cumulative probability function of the $t(5)$ random variable and that of the standard normal random variable on the same graph. Compare and contrast the two functions.

*B.8 **a.** Obtain $t(.90; 15)$ and $t(.05; 21)$. What does the latter value represent?
    b. Estimate $t(.01; 250)$ using the normal approximation.

B.9 **a.** Obtain $t(.995; 11)$ and $t(.10; 18)$. What does the latter value represent?
    b. Estimate $t(.975; 410)$ using the normal approximation.

B.3 *F* DISTRIBUTIONS

Characteristics

The F distribution is a continuous probability distribution. It has two parameters, ν_1 and ν_2, which are positive whole numbers called *degrees of freedom* (*df*). To each pair (ν_1, ν_2) corresponds a different member of the family of F distributions.

The random variable associated with the *F* distribution is denoted by:

$$F(\nu_1, \nu_2)$$

numerator   denominator
df   df

where the parameter in the left position denotes the *numerator degrees of free-dom* and the parameter in the right position denotes the *denominator degrees of freedom.*

Graphs of the *F* probability density function for several pairs of (ν_1, ν_2) values are shown in Figure B.1. Figure B.1 also summarizes some of the properties of the *F* distribution.

Shape.   From the graphs of the *F* distribution in Figure B.1, it is seen that the *F* distribution is unimodal and right-skewed. The mode of the *F* distribution (i.e., the value of *F* for which the density is largest) approaches 1 as both degrees of freedom get large.

Mean and Variance.   The mean and variance of the $F(\nu_1, \nu_2)$ random variable are:

(B.8)
$$E\{F(\nu_1, \nu_2)\} = \frac{\nu_2}{\nu_2 - 2} \qquad \text{when } \nu_2 > 2$$

(B.9)
$$\sigma^2\{F(\nu_1, \nu_2)\} = \frac{2\nu_2^2(\nu_1 + \nu_2 - 2)}{\nu_1(\nu_2 - 2)^2(\nu_2 - 4)} \qquad \text{when } \nu_2 > 4$$

The mean is undefined for $\nu_2 \leq 2$, and the variance is undefined for $\nu_2 \leq 4$.

For the $F(3, 5)$ random variable, for instance, the mean and variance are, respectively:

$$\frac{5}{5 - 2} = 1.67 \qquad \frac{2(5^2)(3 + 5 - 2)}{3(5 - 2)^2(5 - 4)} = 11.11$$

Hence, the standard deviation of $F(3, 5)$ is $\sqrt{11.11} = 3.33$.

Note from (B.8) and (B.9) that the mean approaches 1 as ν_2 gets large, while the variance approaches 0 as both degrees of freedom become large.

Determining Percentiles for *F* Distributions

Because the *F* distribution has two parameters, any extensive tabulation of it requires considerable space. Table C–4 contains the 95th and 99th percentiles of *F* distributions for selected numerator and denominator degrees of freedom. As usual, we shall let $F(a; \nu_1, \nu_2)$ denote the 100*a* percentile of random variable $F(\nu_1, \nu_2)$, i.e.:

(B.10)
$$P[F(\nu_1, \nu_2) \leq F(a; \nu_1, \nu_2)] = a$$

The illustration at the top of Table C–4 shows this relationship.

Percentiles below 50 percent are not presented in Table C–4 because they can be obtained from the percentiles presented by means of the following relationship:

(B.11)
$$F(a; \nu_1, \nu_2) = \frac{1}{F(1 - a; \nu_2, \nu_1)}$$

Note the reversal of the degrees of freedom in the denominator of (B.11).

☐ Examples

1. Find the 95th percentile for $F(3, 5)$. From Table C-4, we find that the 95th percentile of the F distribution with 3 degrees of freedom for the numerator and 5 degrees of freedom for the denominator is $F(.95; 3, 5) = 5.41$. Hence, the probability is .95 that $F(3, 5)$ is less than or equal to 5.41.
2. Find the 99th percentile for $F(10, 6)$. From Table C-4, we find $F(.99; 10, 6) = 7.87$.
3. Find the 5th percentile of $F(3, 5)$. To use the relationship (B.11), we first need to find $F(.95; 5, 3)$. From Table C-4, we find this to be $F(.95; 5, 3) = 9.01$. Hence, by (B.11):

$$F(.05; 3, 5) = \frac{1}{F(.95; 5, 3)} = \frac{1}{9.01} = .111$$

Thus, the probability is .05 that $F(3, 5)$ is less than or equal to .111.

4. Find $F(.01; 6, 10)$. We found in Example 2 that $F(.99; 10, 6) = 7.87$. Hence, $F(.01; 6, 10) = 1/7.87 = .127$. ☐

Problems

*B.10 Consider an $F(2, 5)$ random variable.
   a. Obtain the 99th and 1st percentiles of the probability distribution.
   b. What is the value of the mean? The standard deviation?
   c. Are the percentiles in a equally spaced about the mean? What does this fact imply about the symmetry of the probability distribution?

B.11 Refer to Problem B.10. Answer all parts for an $F(2, 6)$ random variable.

*B.12 a. Obtain $F(.95; 2, 40)$. What does this value represent?
   b. Obtain $F(.01; 8, 3)$.

B.13 a. Obtain $F(.99; 8, 7)$. What does this value represent?
   b. Obtain $F(.05; 4, 4)$.

B.4 RELATIONS BETWEEN Z, t, χ^2, and F

We first describe the relations of χ^2, t, and F to the standard normal random variable Z, and then explain how Z, t, and χ^2 may be viewed as special cases of the random variable F.

Relation of Chi-Square Distribution to Standard Normal Distribution

The χ^2 distribution is related directly to the standard normal distribution. Let Z_1, $Z_2, \ldots, Z_\nu$ be ν independent standard normal variables. Then:

(B.12)
$$\chi^2(\nu) = Z_1^2 + Z_2^2 + \cdots + Z_\nu^2$$

where: Z_i's are independent $N(0, 1)$

In other words, the sum of ν independent squared standard normal variables is a χ^2 random variable with ν degrees of freedom.

Comment

We can find $E\{\chi^2(\nu)\}$ by applying property (5.15a) of the expectation operator (expected value of a sum is the sum of the expected values) to definition (B.12). We obtain:

$$E\{\chi^2(\nu)\} = E\{Z_1^2\} + E\{Z_2^2\} + \cdots + E\{Z_\nu^2\}$$

For the standard normal variable Z, it follows from (5.8a) that $\sigma^2\{Z_i\} = 1 = E\{Z_i^2\}$, because $E\{Z_i\} = 0$. Consequently:

$$E\{\chi^2(\nu)\} = 1 + 1 + \cdots + 1 = \nu$$

which is the result given in (B.1).

Relation of t Distribution to Standard Normal Distribution

We have just shown that the χ^2 distribution is related to the standard normal distribution. The $t(\nu)$ random variable is related to both the $\chi^2(\nu)$ and Z random variables, as follows:

(B.13)
$$t(\nu) = \frac{Z}{\left[\dfrac{\chi^2(\nu)}{\nu}\right]^{1/2}}$$

where: Z and $\chi^2(\nu)$ are independent

In other words, a t random variable with ν degrees of freedom is obtained by: (1) taking a standard normal random variable Z and a χ^2 random variable with ν degrees of freedom which are independent of one another, and (2) forming the ratio of Z and the positive square root of $\chi^2(\nu)$ divided by its degrees of freedom ν.

Comment

We can see intuitively why the t distribution approaches the standard normal distribution when ν is large by considering the random variable $\chi^2(\nu)/\nu$ in the denominator of (B.13). This random variable has expectation:

$$E\left\{\frac{\chi^2(\nu)}{\nu}\right\} = \frac{1}{\nu}E\{\chi^2(\nu)\} = \frac{\nu}{\nu} = 1 \qquad \text{by (5.6b) and (B.1)}$$

while the variance of the random variable is:

$$\sigma^2\left\{\frac{\chi^2(\nu)}{\nu}\right\} = \frac{1}{\nu^2}\sigma^2\{\chi^2(\nu)\} = \frac{2\nu}{\nu^2} = \frac{2}{\nu} \qquad \text{by (5.11b) and (B.2)}$$

Note that the variance approaches 0 as ν gets large. Hence, $\chi^2(\nu)/\nu$ becomes concentrated within an arbitrarily small neighborhood of the mean value 1 as ν approaches ∞. This suggests that we can think of t as approximately $Z/\sqrt{1} = Z$ when ν is large.

Relation of F Distribution to Standard Normal Distribution

It was just shown that the χ^2 distribution is related to the standard normal distribution. The $F(\nu_1, \nu_2)$ random variable in turn is related to two independent $\chi^2(\nu_1)$ and $\chi^2(\nu_2)$ random variables, as follows:

$$(\text{B.14}) \qquad F(\nu_1, \nu_2) = \frac{\dfrac{\chi^2(\nu_1)}{\nu_1}}{\dfrac{\chi^2(\nu_2)}{\nu_2}}$$

where: $\chi^2(\nu_1)$ and $\chi^2(\nu_2)$ are independent

In other words, to obtain an F random variable, one divides two independent χ^2 random variables by their respective degrees of freedom and forms the ratio of these ratios.

Comment

The symmetrical relationship between the numerator and denominator in (B.14) makes it easy to see that:

$$(\text{B.15}) \qquad F(\nu_1, \nu_2) = \frac{1}{F(\nu_2, \nu_1)}$$

This fact makes it possible to compute the $100a$ percentiles of $F(\nu_1, \nu_2)$ from knowledge of the $100(1 - a)$ percentiles of $F(\nu_2, \nu_1)$ by the relationship (B.11).

Z, t, and χ^2 as Special Cases of F

Relations between Random Variables. The random variables Z, t, and χ^2 may all be considered special cases of the random variable F:

$$(\text{B.16}) \qquad Z^2 = F(1, \infty)$$

$$[t(\nu)]^2 = F(1, \nu)$$

$$\frac{\chi^2(\nu)}{\nu} = F(\nu, \infty)$$

The relationship between Z and F follows, because $\nu_2 = \infty$ implies, as we have seen, that $\chi^2(\nu_2)/\nu_2$ is essentially equivalent to 1. Thus:

$$F(1, \infty) = \frac{\dfrac{\chi^2(1)}{1}}{1} = Z^2$$

The other relations follow in the same manner.

Relations between Percentiles. The percentiles of the Z, t, and χ^2 distributions are related to the percentiles of the F distribution, in ways illustrated by the following examples.

☐   Examples

1. Consider $z(.975) = 1.960$. We have $(1.960)^2 = 3.84 = F(.95; 1, \infty)$.
2. Consider $t(.975; 10) = 2.228$. We have $(2.228)^2 = 4.96 = F(.95; 1, 10)$.
3. Consider $\chi^2(.95; 8) = 15.51$. We have $15.51/8 = 1.94 = F(.95; 8, \infty)$.   ☐

Comment

The reason why the corresponding percentiles for the Z and F distributions in Example 1 are 97.5 and 95 percent, respectively, and similarly for the t and F distributions in Example 2, is related to the fact that Z^2 and t^2 equal F. Hence, both negative and positive outcomes of Z and t lead to the same F value.

To show this for Example 1, consider the following probability statement for Z:

$$P[z(.025) \leq Z \leq z(.975)] = .95$$

But we know that $z(.025) = -z(.975)$, so we can write the probability statement as follows:

$$P[-z(.975) \leq Z \leq z(.975)] = .95$$

In turn, this can be expressed as follows:

$$P\{Z^2 \leq [z(.975)]^2\} = .95$$

because an inequality of the form $-a \leq b \leq a$ is mathematically equivalent to $b^2 \leq a^2$.

Since $Z^2 = F(1, \infty)$ by (B.16), it follows that $[z(.975)]^2$ must equal $F(.95; 1, \infty)$.

Exercises

B.14 **a.** Given $\sigma^2\{Z^2\} = 2$ for a standard normal random variable Z, show that $\sigma^2\{\chi^2(\nu)\} = 2\nu$.

b. Show that $\chi^2(\nu_1 + \nu_2) = \chi^2(\nu_1) + \chi^2(\nu_2)$ when $\chi^2(\nu_1)$ and $\chi^2(\nu_2)$ are statistically independent.

B.15 **a.** Explain why $[t(\nu)]^2 = F(1, \nu)$.

b. Find $F(.95; 1, 3)$ from Table C–3.

B.16 **a.** Explain why $\chi^2(\nu)/\nu = F(\nu, \infty)$.

b. Find $F(.99; 8, \infty)$ from Table C–2.

Appendix C
Tables

TABLE C-1   *Cumulative probabilities and percentiles of the standard normal distribution*

(a)
Cumulative Probabilities

Entry is area *a* under the standard normal curve from $-\infty$ to $z(a)$

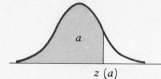

$z(a)$

z	.00	.01	.02	.03	.04	.05	.06	.07	.08	.09
.0	.5000	.5040	.5080	.5120	.5160	.5199	.5239	.5279	.5319	.5359
.1	.5398	.5438	.5478	.5517	.5557	.5596	.5636	.5675	.5714	.5753
.2	.5793	.5832	.5871	.5910	.5948	.5987	.6026	.6064	.6103	.6141
.3	.6179	.6217	.6255	.6293	.6331	.6368	.6406	.6443	.6480	.6517
.4	.6554	.6591	.6628	.6664	.6700	.6736	.6772	.6808	.6844	.6879
.5	.6915	.6950	.6985	.7019	.7054	.7088	.7123	.7157	.7190	.7224
.6	.7257	.7291	.7324	.7357	.7389	.7422	.7454	.7486	.7517	.7549
.7	.7580	.7611	.7642	.7673	.7704	.7734	.7764	.7794	.7823	.7852
.8	.7881	.7910	.7939	.7967	.7995	.8023	.8051	.8078	.8106	.8133
.9	.8159	.8186	.8212	.8238	.8264	.8289	.8315	.8340	.8365	.8389
1.0	.8413	.8438	.8461	.8485	.8508	.8531	.8554	.8577	.8599	.8621
1.1	.8643	.8665	.8686	.8708	.8729	.8749	.8770	.8790	.8810	.8830
1.2	.8849	.8869	.8888	.8907	.8925	.8944	.8962	.8980	.8997	.9015
1.3	.9032	.9049	.9066	.9082	.9099	.9115	.9131	.9147	.9162	.9177
1.4	.9192	.9207	.9222	.9236	.9251	.9265	.9279	.9292	.9306	.9319
1.5	.9332	.9345	.9357	.9370	.9382	.9394	.9406	.9418	.9429	.9441
1.6	.9452	.9463	.9474	.9484	.9495	.9505	.9515	.9525	.9535	.9545
1.7	.9554	.9564	.9573	.9582	.9591	.9599	.9608	.9616	.9625	.9633
1.8	.9641	.9649	.9656	.9664	.9671	.9678	.9686	.9693	.9699	.9706
1.9	.9713	.9719	.9726	.9732	.9738	.9744	.9750	.9756	.9761	.9767
2.0	.9772	.9778	.9783	.9788	.9793	.9798	.9803	.9808	.9812	.9817
2.1	.9821	.9826	.9830	.9834	.9838	.9842	.9846	.9850	.9854	.9857
2.2	.9861	.9864	.9868	.9871	.9875	.9878	.9881	.9884	.9887	.9890
2.3	.9893	.9896	.9898	.9901	.9904	.9906	.9909	.9911	.9913	.9916
2.4	.9918	.9920	.9922	.9925	.9927	.9929	.9931	.9932	.9934	.9936
2.5	.9938	.9940	.9941	.9943	.9945	.9946	.9948	.9949	.9951	.9952
2.6	.9953	.9955	.9956	.9957	.9959	.9960	.9961	.9962	.9963	.9964
2.7	.9965	.9966	.9967	.9968	.9969	.9970	.9971	.9972	.9973	.9974
2.8	.9974	.9975	.9976	.9977	.9977	.9978	.9979	.9979	.9980	.9981
2.9	.9981	.9982	.9982	.9983	.9984	.9984	.9985	.9985	.9986	.9986
3.0	.9987	.9987	.9987	.9988	.9988	.9989	.9989	.9989	.9990	.9990
3.1	.9990	.9991	.9991	.9991	.9992	.9992	.9992	.9992	.9993	.9993
3.2	.9993	.9993	.9994	.9994	.9994	.9994	.9994	.9995	.9995	.9995
3.3	.9995	.9995	.9995	.9996	.9996	.9996	.9996	.9996	.9996	.9997
3.4	.9997	.9997	.9997	.9997	.9997	.9997	.9997	.9997	.9997	.9998

1 Tail

2 Tail

$1 - 99.87$

.0013

2

.0026

TABLE C-1 *Cumulative probabilities and percentiles of the standard normal distribution* (Continued)

(b)
Selected Percentiles

Entry is $z(a)$ where $P[Z \leq z(a)] = a$

a:	.10	.05	.025	.02	.01	.005	.001
$z(a)$:	-1.282	-1.645	-1.960	-2.054	-2.326	-2.576	-3.090
a:	.90	.95	.975	.98	.99	.995	.999
$z(a)$:	1.282	1.645	1.960	2.054	2.326	2.576	3.090

EXAMPLE: $P(Z \leq 1.96) = .9750$ so $z(.9750) = 1.96$.
TEXT REFERENCE: Use of this table is discussed on pp. 169–173.

TABLE C-2 *Percentiles of the chi-square distribution*

Entry is $\chi^2\,(a;\nu)$ where $P[\chi^2(\nu) \le \chi^2(a;\nu)] = a$

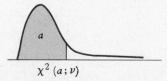

$\chi^2\,(a;\nu)$

df ν	.005	.010	.025	.050	.100	.900	.950	.975	.990	.995
1	0.0⁴393	0.0³157	0.0³982	0.0²393	0.0158	2.71	3.84	5.02	6.63	7.88
2	0.0100	0.0201	0.0506	0.103	0.211	4.61	5.99	7.38	9.21	10.60
3	0.072	0.115	0.216	0.352	0.584	6.25	7.81	9.35	11.34	12.84
4	0.207	0.297	0.484	0.711	1.064	7.78	9.49	11.14	13.28	14.86
5	0.412	0.554	0.831	1.145	1.61	9.24	11.07	12.83	15.09	16.75
6	0.676	0.872	1.24	1.64	2.20	10.64	12.59	14.45	16.81	18.55
7	0.989	1.24	1.69	2.17	2.83	12.02	14.07	16.01	18.48	20.28
8	1.34	1.65	2.18	2.73	3.49	13.36	15.51	17.53	20.09	21.96
9	1.73	2.09	2.70	3.33	4.17	14.68	16.92	19.02	21.67	23.59
10	2.16	2.56	3.25	3.94	4.87	15.99	18.31	20.48	23.21	25.19
11	2.60	3.05	3.82	4.57	5.58	17.28	19.68	21.92	24.73	26.76
12	3.07	3.57	4.40	5.23	6.30	18.55	21.03	23.34	26.22	28.30
13	3.57	4.11	5.01	5.89	7.04	19.81	22.36	24.74	27.69	29.82
14	4.07	4.66	5.63	6.57	7.79	21.06	23.68	26.12	29.14	31.32
15	4.60	5.23	6.26	7.26	8.55	22.31	25.00	27.49	30.58	32.80
16	5.14	5.81	6.91	7.96	9.31	23.54	26.30	28.85	32.00	34.27
17	5.70	6.41	7.56	8.67	10.09	24.77	27.59	30.19	33.41	35.72
18	6.26	7.01	8.23	9.39	10.86	25.99	28.87	31.53	34.81	37.16
19	6.84	7.63	8.91	10.12	11.65	27.20	30.14	32.85	36.19	38.58
20	7.43	8.26	9.59	10.85	12.44	28.41	31.41	34.17	37.57	40.00
21	8.03	8.90	10.28	11.59	13.24	29.62	32.67	35.48	38.93	41.40
22	8.64	9.54	10.98	12.34	14.04	30.81	33.92	36.78	40.29	42.80
23	9.26	10.20	11.69	13.09	14.85	32.01	35.17	38.08	41.64	44.18
24	9.89	10.86	12.40	13.85	15.66	33.20	36.42	39.36	42.98	45.56
25	10.52	11.52	13.12	14.61	16.47	34.38	37.65	40.65	44.31	46.93
26	11.16	12.20	13.84	15.38	17.29	35.56	38.89	41.92	45.64	48.29
27	11.81	12.88	14.57	16.15	18.11	36.74	40.11	43.19	46.96	49.64
28	12.46	13.56	15.31	16.93	18.94	37.92	41.34	44.46	48.28	50.99
29	13.12	14.26	16.05	17.71	19.77	39.09	42.56	45.72	49.59	52.34
30	13.79	14.95	16.79	18.49	20.60	40.26	43.77	46.98	50.89	53.67
40	20.71	22.16	24.43	26.51	29.05	51.81	55.76	59.34	63.69	66.77
50	27.99	29.71	32.36	34.76	37.69	63.17	67.50	71.42	76.15	79.49
60	35.53	37.48	40.48	43.19	46.46	74.40	79.08	83.30	88.38	91.95
70	43.28	45.44	48.76	51.74	55.33	85.53	90.53	95.02	100.4	104.2
80	51.17	53.54	57.15	60.39	64.28	96.58	101.9	106.6	112.3	116.3
90	59.20	61.75	65.65	69.13	73.29	107.6	113.1	118.1	124.1	128.3
100	67.33	70.06	74.22	77.93	82.36	118.5	124.3	129.6	135.8	140.2

SOURCE: Tabulated values adapted by permission from C. M. Thompson, "Table of Percentage Points of the Chi-Square Distribution," *Biometrika*, Vol. 32 (1941), pp. 188–189.

EXAMPLE: $\chi^2(.900;\ 4) = 7.78$ so $P[\chi^2(4) \le 7.78] = .900$.

TEXT REFERENCE: Use of this table is discussed on p. 712.

TABLE C-3 *Percentiles of the t distribution*

Entry is $t(a; \nu)$ where $P[t(\nu) \le t(a; \nu)] = a$

a

$t(a; \nu)$

Watch
1 Tail
2 Tail

df ν	.90	.95	.975	.99	.995	.9995
1	3.078	6.314	12.706	31.821	63.657	636.619
2	1.886	2.920	4.303	6.965	9.925	31.599
3	1.638	2.353	3.182	4.541	5.841	12.924
4	1.533	2.132	2.776	3.747	4.604	8.610
5	1.476	2.015	2.571	3.365	4.032	6.869
6	1.440	1.943	2.447	3.143	3.707	5.959
7	1.415	1.895	2.365	2.998	3.499	5.408
8	1.397	1.860	2.306	2.896	3.355	5.041
9	1.383	1.833	2.262	2.821	3.250	4.781
10	1.372	1.812	2.228	2.764	3.169	4.587
11	1.363	1.796	2.201	2.718	3.106	4.437
12	1.356	1.782	2.179	2.681	3.055	4.318
13	1.350	1.771	2.160	2.650	3.012	4.221
14	1.345	1.761	2.145	2.624	2.977	4.140
15	1.341	1.753	2.131	2.602	2.947	4.073
16	1.337	1.746	2.120	2.583	2.921	4.015
17	1.333	1.740	2.110	2.567	2.898	3.965
18	1.330	1.734	2.101	2.552	2.878	3.922
19	1.328	1.729	2.093	2.539	2.861	3.883
20	1.325	1.725	2.086	2.528	2.845	3.850
21	1.323	1.721	2.080	2.518	2.831	3.819
22	1.321	1.717	2.074	2.508	2.819	3.792
23	1.319	1.714	2.069	2.500	2.807	3.768
24	1.318	1.711	2.064	2.492	2.797	3.745
25	1.316	1.708	2.060	2.485	2.787	3.725
26	1.315	1.706	2.056	2.479	2.779	3.707
27	1.314	1.703	2.052	2.473	2.771	3.690
28	1.313	1.701	2.048	2.467	2.763	3.674
29	1.311	1.699	2.045	2.462	2.756	3.659
30	1.310	1.697	2.042	2.457	2.750	3.646
40	1.303	1.684	2.021	2.423	2.704	3.551
60	1.296	1.671	2.000	2.390	2.660	3.460
120	1.289	1.658	1.980	2.358	2.617	3.373
∞	1.282	1.645	1.960	2.326	2.576	3.291

EXAMPLE: $t(.95; 10) = 1.812$ so $P[t(10) \le 1.812] = .95$.

TEXT REFERENCE: Use of this table is discussed on pp. 713–715.

TABLE C-4 *Percentiles of the F distribution*

Entry is $F(a; \nu_1, \nu_2)$ where $P[F(\nu_1, \nu_2) \leq F(a; \nu_1, \nu_2)] = a$

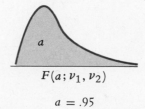

$$F(a; \nu_1, \nu_2)$$

$$a = .95$$

denominator df	numerator df								
	1	2	3	4	5	6	7	8	9
1	161.4	199.5	215.7	224.6	230.2	234.0	236.8	238.9	240.5
2	18.51	19.00	19.16	19.25	19.30	19.33	19.35	19.37	19.38
3	10.13	9.55	9.28	9.12	9.01	8.94	8.89	8.85	8.81
4	7.71	6.94	6.59	6.39	6.26	6.16	6.09	6.04	6.00
5	6.61	5.79	5.41	5.19	5.05	4.95	4.88	4.82	4.77
6	5.99	5.14	4.76	4.53	4.39	4.28	4.21	4.15	4.10
7	5.59	4.74	4.35	4.12	3.97	3.87	3.79	3.73	3.68
8	5.32	4.46	4.07	3.84	3.69	3.58	3.50	3.44	3.39
9	5.12	4.26	3.86	3.63	3.48	3.37	3.29	3.23	3.18
10	4.96	4.10	3.71	3.48	3.33	3.22	3.14	3.07	3.02
11	4.84	3.98	3.59	3.36	3.20	3.09	3.01	2.95	2.90
12	4.75	3.89	3.49	3.26	3.11	3.00	2.91	2.85	2.80
13	4.67	3.81	3.41	3.18	3.03	2.92	2.83	2.77	2.71
14	4.60	3.74	3.34	3.11	2.96	2.85	2.76	2.70	2.65
15	4.54	3.68	3.29	3.06	2.90	2.79	2.71	2.64	2.59
16	4.49	3.63	3.24	3.01	2.85	2.74	2.66	2.59	2.54
17	4.45	3.59	3.20	2.96	2.81	2.70	2.61	2.55	2.49
18	4.41	3.55	3.16	2.93	2.77	2.66	2.58	2.51	2.46
19	4.38	3.52	3.13	2.90	2.74	2.63	2.54	2.48	2.42
20	4.35	3.49	3.10	2.87	2.71	2.60	2.51	2.45	2.39
21	4.32	3.47	3.07	2.84	2.68	2.57	2.49	2.42	2.37
22	4.30	3.44	3.05	2.82	2.66	2.55	2.46	2.40	2.34
23	4.28	3.42	3.03	2.80	2.64	2.53	2.44	2.37	2.32
24	4.26	3.40	3.01	2.78	2.62	2.51	2.42	2.36	2.30
25	4.24	3.39	2.99	2.76	2.60	2.49	2.40	2.34	2.28
26	4.23	3.37	2.98	2.74	2.59	2.47	2.39	2.32	2.27
27	4.21	3.35	2.96	2.73	2.57	2.46	2.37	2.31	2.25
28	4.20	3.34	2.95	2.71	2.56	2.45	2.36	2.29	2.24
29	4.18	3.33	2.93	2.70	2.55	2.43	2.35	2.28	2.22
30	4.17	3.32	2.92	2.69	2.53	2.42	2.33	2.27	2.21
40	4.08	3.23	2.84	2.61	2.45	2.34	2.25	2.18	2.12
60	4.00	3.15	2.76	2.53	2.37	2.25	2.17	2.10	2.04
120	3.92	3.07	2.68	2.45	2.29	2.17	2.09	2.02	1.96
∞	3.84	3.00	2.60	2.37	2.21	2.10	2.01	1.94	1.88

TABLE C-4 *Percentiles of the F distribution* (Continued)

$a = .95$

numerator df										denominator df
10	12	15	20	24	30	40	60	120	∞	
241.9	243.9	245.9	248.0	249.1	250.1	251.1	252.2	253.3	254.3	1
19.40	19.41	19.43	19.45	19.45	19.46	19.47	19.48	19.49	19.50	2
8.79	8.74	8.70	8.66	8.64	8.62	8.59	8.57	8.55	8.53	3
5.96	5.91	5.86	5.80	5.77	5.75	5.72	5.69	5.66	5.63	4
4.74	4.68	4.62	4.56	4.53	4.50	4.46	4.43	4.40	4.36	5
4.06	4.00	3.94	3.87	3.84	3.81	3.77	3.74	3.70	3.67	6
3.64	3.57	3.51	3.44	3.41	3.38	3.34	3.30	3.27	3.23	7
3.35	3.28	3.22	3.15	3.12	3.08	3.04	3.01	2.97	2.93	8
3.14	3.07	3.01	2.94	2.90	2.86	2.83	2.79	2.75	2.71	9
2.98	2.91	2.85	2.77	2.74	2.70	2.66	2.62	2.58	2.54	10
2.85	2.79	2.72	2.65	2.61	2.57	2.53	2.49	2.45	2.40	11
2.75	2.69	2.62	2.54	2.51	2.47	2.43	2.38	2.34	2.30	12
2.67	2.60	2.53	2.46	2.42	2.38	2.34	2.30	2.25	2.21	13
2.60	2.53	2.46	2.39	2.35	2.31	2.27	2.22	2.18	2.13	14
2.54	2.48	2.40	2.33	2.29	2.25	2.20	2.16	2.11	2.07	15
2.49	2.42	2.35	2.28	2.24	2.19	2.15	2.11	2.06	2.01	16
2.45	2.38	2.31	2.23	2.19	2.15	2.10	2.06	2.01	1.96	17
2.41	2.34	2.27	2.19	2.15	2.11	2.06	2.02	1.97	1.92	18
2.38	2.31	2.23	2.16	2.11	2.07	2.03	1.98	1.93	1.88	19
2.35	2.28	2.20	2.12	2.08	2.04	1.99	1.95	1.90	1.84	20
2.32	2.25	2.18	2.10	2.05	2.01	1.96	1.92	1.87	1.81	21
2.30	2.23	2.15	2.07	2.03	1.98	1.94	1.89	1.84	1.78	22
2.27	2.20	2.13	2.05	2.01	1.96	1.91	1.86	1.81	1.76	23
2.25	2.18	2.11	2.03	1.98	1.94	1.89	1.84	1.79	1.73	24
2.24	2.16	2.09	2.01	1.96	1.92	1.87	1.82	1.77	1.71	25
2.22	2.15	2.07	1.99	1.95	1.90	1.85	1.80	1.75	1.69	26
2.20	2.13	2.06	1.97	1.93	1.88	1.84	1.79	1.73	1.67	27
2.19	2.12	2.04	1.96	1.91	1.87	1.82	1.77	1.71	1.65	28
2.18	2.10	2.03	1.94	1.90	1.85	1.81	1.75	1.70	1.64	29
2.16	2.09	2.01	1.93	1.89	1.84	1.79	1.74	1.68	1.62	30
2.08	2.00	1.92	1.84	1.79	1.74	1.69	1.64	1.58	1.51	40
1.99	1.92	1.84	1.75	1.70	1.65	1.59	1.53	1.47	1.39	60
1.91	1.83	1.75	1.66	1.61	1.55	1.50	1.43	1.35	1.25	120
1.83	1.75	1.67	1.57	1.52	1.46	1.39	1.32	1.22	1.00	∞

TABLE C-4 *Percentiles of the F distribution* (Continued)

$$a = .99$$

denominator df	\multicolumn{9}{c}{numerator df}
	1	2	3	4	5	6	7	8	9
1	4052	4999.5	5403	5625	5764	5859	5928	5981	6022
2	98.50	99.00	99.17	99.25	99.30	99.33	99.36	99.37	99.39
3	34.12	30.82	29.46	28.71	28.24	27.91	27.67	27.49	27.35
4	21.20	18.00	16.69	15.98	15.52	15.21	14.98	14.80	14.66
5	16.26	13.27	12.06	11.39	10.97	10.67	10.46	10.29	10.16
6	13.75	10.92	9.78	9.15	8.75	8.47	8.26	8.10	7.98
7	12.25	9.55	8.45	7.85	7.46	7.19	6.99	6.84	6.72
8	11.26	8.65	7.59	7.01	6.63	6.37	6.18	6.03	5.91
9	10.56	8.02	6.99	6.42	6.06	5.80	5.61	5.47	5.35
10	10.04	7.56	6.55	5.99	5.64	5.39	5.20	5.06	4.94
11	9.65	7.21	6.22	5.67	5.32	5.07	4.89	4.74	4.63
12	9.33	6.93	5.95	5.41	5.06	4.82	4.64	4.50	4.39
13	9.07	6.70	5.74	5.21	4.86	4.62	4.44	4.30	4.19
14	8.86	6.51	5.56	5.04	4.69	4.46	4.28	4.14	4.03
15	8.68	6.36	5.42	4.89	4.56	4.32	4.14	4.00	3.89
16	8.53	6.23	5.29	4.77	4.44	4.20	4.03	3.89	3.78
17	8.40	6.11	5.18	4.67	4.34	4.10	3.93	3.79	3.68
18	8.29	6.01	5.09	4.58	4.25	4.01	3.84	3.71	3.60
19	8.18	5.93	5.01	4.50	4.17	3.94	3.77	3.63	3.52
20	8.10	5.85	4.94	4.43	4.10	3.87	3.70	3.56	3.46
21	8.02	5.78	4.87	4.37	4.04	3.81	3.64	3.51	3.40
22	7.95	5.72	4.82	4.31	3.99	3.76	3.59	3.45	3.35
23	7.88	5.66	4.76	4.26	3.94	3.71	3.54	3.41	3.30
24	7.82	5.61	4.72	4.22	3.90	3.67	3.50	3.36	3.26
25	7.77	5.57	4.68	4.18	3.85	3.63	3.46	3.32	3.22
26	7.72	5.53	4.64	4.14	3.82	3.59	3.42	3.29	3.18
27	7.68	5.49	4.60	4.11	3.78	3.56	3.39	3.26	3.15
28	7.64	5.45	4.57	4.07	3.75	3.53	3.36	3.23	3.12
29	7.60	5.42	4.54	4.04	3.73	3.50	3.33	3.20	3.09
30	7.56	5.39	4.51	4.02	3.70	3.47	3.30	3.17	3.07
40	7.31	5.18	4.31	3.83	3.51	3.29	3.12	2.99	2.89
60	7.08	4.98	4.13	3.65	3.34	3.12	2.95	2.82	2.72
120	6.85	4.79	3.95	3.48	3.17	2.96	2.79	2.66	2.56
∞	6.63	4.61	3.78	3.32	3.02	2.80	2.64	2.51	2.41

TABLE C-4 *Percentiles of the F distribution* (Continued)

$$a = .99$$

numerator *df*										denominator *df*
10	12	15	20	24	30	40	60	120	∞	
6056	6106	6157	6209	6235	6261	6287	6313	6339	6366	1
99.40	99.42	99.43	99.45	99.46	99.47	99.47	99.48	99.49	99.50	2
27.23	27.05	26.87	26.69	26.60	26.50	26.41	26.32	26.22	26.13	3
14.55	14.37	14.20	14.02	13.93	13.84	13.75	13.65	13.56	13.46	4
10.05	9.89	9.72	9.55	9.47	9.38	9.29	9.20	9.11	9.02	5
7.87	7.72	7.56	7.40	7.31	7.23	7.14	7.06	6.97	6.88	6
6.62	6.47	6.31	6.16	6.07	5.99	5.91	5.82	5.74	5.65	7
5.81	5.67	5.52	5.36	5.28	5.20	5.12	5.03	4.95	4.86	8
5.26	5.11	4.96	4.81	4.73	4.65	4.57	4.48	4.40	4.31	9
4.85	4.71	4.56	4.41	4.33	4.25	4.17	4.08	4.00	3.91	10
4.54	4.40	4.25	4.10	4.02	3.94	3.86	3.78	3.69	3.60	11
4.30	4.16	4.01	3.86	3.78	3.70	3.62	3.54	3.45	3.36	12
4.10	3.96	3.82	3.66	3.59	3.51	3.43	3.34	3.25	3.17	13
3.94	3.80	3.66	3.51	3.43	3.35	3.27	3.18	3.09	3.00	14
3.80	3.67	3.52	3.37	3.29	3.21	3.13	3.05	2.96	2.87	15
3.69	3.55	3.41	3.26	3.18	3.10	3.02	2.93	2.84	2.75	16
3.59	3.46	3.31	3.16	3.08	3.00	2.92	2.83	2.75	2.65	17
3.51	3.37	3.23	3.08	3.00	2.92	2.84	2.75	2.66	2.57	18
3.43	3.30	3.15	3.00	2.92	2.84	2.76	2.67	2.58	2.49	19
3.37	3.23	3.09	2.94	2.86	2.78	2.69	2.61	2.52	2.42	20
3.31	3.17	3.03	2.88	2.80	2.72	2.64	2.55	2.46	2.36	21
3.26	3.12	2.98	2.83	2.75	2.67	2.58	2.50	2.40	2.31	22
3.21	3.07	2.93	2.78	2.70	2.62	2.54	2.45	2.35	2.26	23
3.17	3.03	2.89	2.74	2.66	2.58	2.49	2.40	2.31	2.21	24
3.13	2.99	2.85	2.70	2.62	2.54	2.45	2.36	2.27	2.17	25
3.09	2.96	2.81	2.66	2.58	2.50	2.42	2.33	2.23	2.13	26
3.06	2.93	2.78	2.63	2.55	2.47	2.38	2.29	2.20	2.10	27
3.03	2.90	2.75	2.60	2.52	2.44	2.35	2.26	2.17	2.06	28
3.00	2.87	2.73	2.57	2.49	2.41	2.33	2.23	2.14	2.03	29
2.98	2.84	2.70	2.55	2.47	2.39	2.30	2.21	2.11	2.01	30
2.80	2.66	2.52	2.37	2.29	2.20	2.11	2.02	1.92	1.80	40
2.63	2.50	2.35	2.20	2.12	2.03	1.94	1.84	1.73	1.60	60
2.47	2.34	2.19	2.03	1.95	1.86	1.76	1.66	1.53	1.38	120
2.32	2.18	2.04	1.88	1.79	1.70	1.59	1.47	1.32	1.00	∞

SOURCE: Tabulated values adapted from Table 5 of Pearson and Hartley, *Biometrika Tables for Statisticians*, Volume 2, 1972, published by the Cambridge University Press for the Biometrika Trustees, with the permission of the authors and publishers.

EXAMPLE: $F(.99; 8, 24) = 3.36$ so $P[F(8, 24) \leq 3.36] = .99$.

TEXT REFERENCE: Use of this table is discussed on pp. 716–717.

TABLE C-5 Binomial probabilities

Entry is probability $P(X = x) = \binom{n}{x} p^x (1-p)^{n-x}$

n	x	.01	.02	.03	.04	.05	.06	.07	.08	.09		
						p						
2	0	0.9801	0.9604	0.9409	0.9216	0.9025	0.8836	0.8649	0.8464	0.8281	2	
	1	0.0198	0.0392	0.0582	0.0768	0.0950	0.1128	0.1302	0.1472	0.1638	1	
	2	0.0001	0.0004	0.0009	0.0016	0.0025	0.0036	0.0049	0.0064	0.0081	0	2
3	0	0.9703	0.9412	0.9127	0.8847	0.8574	0.8306	0.8044	0.7787	0.7536	3	
	1	0.0294	0.0576	0.0847	0.1106	0.1354	0.1590	0.1816	0.2031	0.2236	2	
	2	0.0003	0.0012	0.0026	0.0046	0.0071	0.0102	0.0137	0.0177	0.0221	1	
	3	0.0000	0.0000	0.0000	0.0001	0.0001	0.0002	0.0003	0.0005	0.0007	0	3
4	0	0.9606	0.9224	0.8853	0.8493	0.8145	0.7807	0.7481	0.7164	0.6857	4	
	1	0.0388	0.0753	0.1095	0.1416	0.1715	0.1993	0.2252	0.2492	0.2713	3	
	2	0.0006	0.0023	0.0051	0.0088	0.0135	0.0191	0.0254	0.0325	0.0402	2	
	3	0.0000	0.0000	0.0001	0.0002	0.0005	0.0008	0.0013	0.0019	0.0027	1	
	4	0.0000	0.0000	0.0000	0.0000	0.0000	0.0000	0.0000	0.0000	0.0001	0	4
5	0	0.9510	0.9039	0.8587	0.8154	0.7738	0.7339	0.6957	0.6591	0.6240	5	
	1	0.0480	0.0922	0.1328	0.1699	0.2036	0.2342	0.2618	0.2866	0.3086	4	
	2	0.0010	0.0038	0.0082	0.0142	0.0214	0.0299	0.0394	0.0498	0.0610	3	
	3	0.0000	0.0001	0.0003	0.0006	0.0011	0.0019	0.0030	0.0043	0.0060	2	
	4	0.0000	0.0000	0.0000	0.0000	0.0000	0.0001	0.0001	0.0002	0.0003	1	
	5	0.0000	0.0000	0.0000	0.0000	0.0000	0.0000	0.0000	0.0000	0.0000	0	5
6	0	0.9415	0.8858	0.8330	0.7828	0.7351	0.6899	0.6470	0.6064	0.5679	6	
	1	0.0571	0.1085	0.1546	0.1957	0.2321	0.2642	0.2922	0.3164	0.3370	5	
	2	0.0014	0.0055	0.0120	0.0204	0.0305	0.0422	0.0550	0.0688	0.0833	4	
	3	0.0000	0.0002	0.0005	0.0011	0.0021	0.0036	0.0055	0.0080	0.0110	3	
	4	0.0000	0.0000	0.0000	0.0000	0.0001	0.0002	0.0003	0.0005	0.0008	2	
	5	0.0000	0.0000	0.0000	0.0000	0.0000	0.0000	0.0000	0.0000	0.0000	1	
	6	0.0000	0.0000	0.0000	0.0000	0.0000	0.0000	0.0000	0.0000	0.0000	0	6
7	0	0.9321	0.8681	0.8080	0.7514	0.6983	0.6485	0.6017	0.5578	0.5168	7	
	1	0.0659	0.1240	0.1749	0.2192	0.2573	0.2897	0.3170	0.3396	0.3578	6	
	2	0.0020	0.0076	0.0162	0.0274	0.0406	0.0555	0.0716	0.0886	0.1061	5	
	3	0.0000	0.0003	0.0008	0.0019	0.0036	0.0059	0.0090	0.0128	0.0175	4	
	4	0.0000	0.0000	0.0000	0.0001	0.0002	0.0004	0.0007	0.0011	0.0017	3	
	5	0.0000	0.0000	0.0000	0.0000	0.0000	0.0000	0.0000	0.0001	0.0001	2	
	6	0.0000	0.0000	0.0000	0.0000	0.0000	0.0000	0.0000	0.0000	0.0000	1	
	7	0.0000	0.0000	0.0000	0.0000	0.0000	0.0000	0.0000	0.0000	0.0000	0	7
8	0	0.9227	0.8508	0.7837	0.7214	0.6634	0.6096	0.5596	0.5132	0.4703	8	
	1	0.0746	0.1389	0.1939	0.2405	0.2793	0.3113	0.3370	0.3570	0.3721	7	
	2	0.0026	0.0099	0.0210	0.0351	0.0515	0.0695	0.0888	0.1087	0.1288	6	
	3	0.0001	0.0004	0.0013	0.0029	0.0054	0.0089	0.0134	0.0189	0.0255	5	
	4	0.0000	0.0000	0.0001	0.0002	0.0004	0.0007	0.0013	0.0021	0.0031	4	
	5	0.0000	0.0000	0.0000	0.0000	0.0000	0.0000	0.0001	0.0001	0.0002	3	
	6	0.0000	0.0000	0.0000	0.0000	0.0000	0.0000	0.0000	0.0000	0.0000	2	
	7	0.0000	0.0000	0.0000	0.0000	0.0000	0.0000	0.0000	0.0000	0.0000	1	
	8	0.0000	0.0000	0.0000	0.0000	0.0000	0.0000	0.0000	0.0000	0.0000	0	8
9	0	0.9135	0.8337	0.7602	0.6925	0.6302	0.5730	0.5204	0.4722	0.4279	9	
	1	0.0830	0.1531	0.2116	0.2597	0.2985	0.3292	0.3525	0.3695	0.3809	8	
	2	0.0034	0.0125	0.0262	0.0433	0.0629	0.0840	0.1061	0.1285	0.1507	7	
	3	0.0001	0.0006	0.0019	0.0042	0.0077	0.0125	0.0186	0.0261	0.0348	6	
	4	0.0000	0.0000	0.0001	0.0003	0.0006	0.0012	0.0021	0.0034	0.0052	5	
	5	0.0000	0.0000	0.0000	0.0000	0.0000	0.0001	0.0002	0.0003	0.0005	4	
	6	0.0000	0.0000	0.0000	0.0000	0.0000	0.0000	0.0000	0.0000	0.0000	3	
	7	0.0000	0.0000	0.0000	0.0000	0.0000	0.0000	0.0000	0.0000	0.0000	2	
	8	0.0000	0.0000	0.0000	0.0000	0.0000	0.0000	0.0000	0.0000	0.0000	1	
	9	0.0000	0.0000	0.0000	0.0000	0.0000	0.0000	0.0000	0.0000	0.0000	0	9
		.99	.98	.97	.96	.95	.94	.93	.92	.91	x	n
						p						

TABLE C-5 *Binomial probabilities* (Continued)

n	x	.01	.02	.03	.04	.05	.06	.07	.08	.09		
						p							
10	0	0.9044	0.8171	0.7374	0.6648	0.5987	0.5386	0.4840	0.4344	0.3894	10	
	1	0.0914	0.1667	0.2281	0.2770	0.3151	0.3438	0.3643	0.3777	0.3851	9	
	2	0.0042	0.0153	0.0317	0.0519	0.0746	0.0988	0.1234	0.1478	0.1714	8	
	3	0.0001	0.0008	0.0026	0.0058	0.0105	0.0168	0.0248	0.0343	0.0452	7	
	4	0.0000	0.0000	0.0001	0.0004	0.0010	0.0019	0.0033	0.0052	0.0078	6	
	5	0.0000	0.0000	0.0000	0.0000	0.0001	0.0001	0.0003	0.0005	0.0009	5	
	6	0.0000	0.0000	0.0000	0.0000	0.0000	0.0000	0.0000	0.0000	0.0001	4	
	7	0.0000	0.0000	0.0000	0.0000	0.0000	0.0000	0.0000	0.0000	0.0000	3	
	8	0.0000	0.0000	0.0000	0.0000	0.0000	0.0000	0.0000	0.0000	0.0000	2	
	9	0.0000	0.0000	0.0000	0.0000	0.0000	0.0000	0.0000	0.0000	0.0000	1	
	10	0.0000	0.0000	0.0000	0.0000	0.0000	0.0000	0.0000	0.0000	0.0000	0	10
12	0	0.8864	0.7847	0.6938	0.6127	0.5404	0.4759	0.4186	0.3677	0.3225	12	
	1	0.1074	0.1922	0.2575	0.3064	0.3413	0.3645	0.3781	0.3837	0.3827	11	
	2	0.0060	0.0216	0.0438	0.0702	0.0988	0.1280	0.1565	0.1835	0.2082	10	
	3	0.0002	0.0015	0.0045	0.0098	0.0173	0.0272	0.0393	0.0532	0.0686	9	
	4	0.0000	0.0001	0.0003	0.0009	0.0021	0.0039	0.0067	0.0104	0.0153	8	
	5	0.0000	0.0000	0.0000	0.0001	0.0002	0.0004	0.0008	0.0014	0.0024	7	
	6	0.0000	0.0000	0.0000	0.0000	0.0000	0.0000	0.0001	0.0001	0.0003	6	
	7	0.0000	0.0000	0.0000	0.0000	0.0000	0.0000	0.0000	0.0000	0.0000	5	
	8	0.0000	0.0000	0.0000	0.0000	0.0000	0.0000	0.0000	0.0000	0.0000	4	
	9	0.0000	0.0000	0.0000	0.0000	0.0000	0.0000	0.0000	0.0000	0.0000	3	
	10	0.0000	0.0000	0.0000	0.0000	0.0000	0.0000	0.0000	0.0000	0.0000	2	
	11	0.0000	0.0000	0.0000	0.0000	0.0000	0.0000	0.0000	0.0000	0.0000	1	
	12	0.0000	0.0000	0.0000	0.0000	0.0000	0.0000	0.0000	0.0000	0.0000	0	12
15	0	0.8601	0.7386	0.6333	0.5421	0.4633	0.3953	0.3367	0.2863	0.2430	15	
	1	0.1303	0.2261	0.2938	0.3388	0.3658	0.3785	0.3801	0.3734	0.3605	14	
	2	0.0092	0.0323	0.0636	0.0988	0.1348	0.1691	0.2003	0.2273	0.2496	13	
	3	0.0004	0.0029	0.0085	0.0178	0.0307	0.0468	0.0653	0.0857	0.1070	12	
	4	0.0000	0.0002	0.0008	0.0022	0.0049	0.0090	0.0148	0.0223	0.0317	11	
	5	0.0000	0.0000	0.0001	0.0002	0.0006	0.0013	0.0024	0.0043	0.0069	10	
	6	0.0000	0.0000	0.0000	0.0000	0.0000	0.0001	0.0003	0.0006	0.0011	9	
	7	0.0000	0.0000	0.0000	0.0000	0.0000	0.0000	0.0000	0.0001	0.0001	8	
	8	0.0000	0.0000	0.0000	0.0000	0.0000	0.0000	0.0000	0.0000	0.0000	7	
	9	0.0000	0.0000	0.0000	0.0000	0.0000	0.0000	0.0000	0.0000	0.0000	6	
	10	0.0000	0.0000	0.0000	0.0000	0.0000	0.0000	0.0000	0.0000	0.0000	5	
	11	0.0000	0.0000	0.0000	0.0000	0.0000	0.0000	0.0000	0.0000	0.0000	4	
	12	0.0000	0.0000	0.0000	0.0000	0.0000	0.0000	0.0000	0.0000	0.0000	3	
	13	0.0000	0.0000	0.0000	0.0000	0.0000	0.0000	0.0000	0.0000	0.0000	2	
	14	0.0000	0.0000	0.0000	0.0000	0.0000	0.0000	0.0000	0.0000	0.0000	1	
	15	0.0000	0.0000	0.0000	0.0000	0.0000	0.0000	0.0000	0.0000	0.0000	0	15
20	0	0.8179	0.6676	0.5438	0.4420	0.3585	0.2901	0.2342	0.1887	0.1516	20	
	1	0.1652	0.2725	0.3364	0.3683	0.3774	0.3703	0.3526	0.3282	0.3000	19	
	2	0.0159	0.0528	0.0988	0.1458	0.1887	0.2246	0.2521	0.2711	0.2818	18	
	3	0.0010	0.0065	0.0183	0.0364	0.0596	0.0860	0.1139	0.1414	0.1672	17	
	4	0.0000	0.0006	0.0024	0.0065	0.0133	0.0233	0.0364	0.0523	0.0703	16	
	5	0.0000	0.0000	0.0002	0.0009	0.0022	0.0048	0.0088	0.0145	0.0222	15	
	6	0.0000	0.0000	0.0000	0.0001	0.0003	0.0008	0.0017	0.0032	0.0055	14	
	7	0.0000	0.0000	0.0000	0.0000	0.0000	0.0001	0.0002	0.0005	0.0011	13	
	8	0.0000	0.0000	0.0000	0.0000	0.0000	0.0000	0.0000	0.0001	0.0002	12	
	9	0.0000	0.0000	0.0000	0.0000	0.0000	0.0000	0.0000	0.0000	0.0000	11	
	10	0.0000	0.0000	0.0000	0.0000	0.0000	0.0000	0.0000	0.0000	0.0000	10	
	11	0.0000	0.0000	0.0000	0.0000	0.0000	0.0000	0.0000	0.0000	0.0000	9	
	12	0.0000	0.0000	0.0000	0.0000	0.0000	0.0000	0.0000	0.0000	0.0000	8	
	13	0.0000	0.0000	0.0000	0.0000	0.0000	0.0000	0.0000	0.0000	0.0000	7	
	14	0.0000	0.0000	0.0000	0.0000	0.0000	0.0000	0.0000	0.0000	0.0000	6	
	15	0.0000	0.0000	0.0000	0.0000	0.0000	0.0000	0.0000	0.0000	0.0000	5	
	16	0.0000	0.0000	0.0000	0.0000	0.0000	0.0000	0.0000	0.0000	0.0000	4	
	17	0.0000	0.0000	0.0000	0.0000	0.0000	0.0000	0.0000	0.0000	0.0000	3	
	18	0.0000	0.0000	0.0000	0.0000	0.0000	0.0000	0.0000	0.0000	0.0000	2	
	19	0.0000	0.0000	0.0000	0.0000	0.0000	0.0000	0.0000	0.0000	0.0000	1	
	20	0.0000	0.0000	0.0000	0.0000	0.0000	0.0000	0.0000	0.0000	0.0000	0	20
		.99	.98	.97	.96	.95	.94	.93	.92	.91	x	n
						p							

X=r
IN
H is terms

TABLE C-5 Binomial probabilities (Continued)

n	x	.10	.15	.20	.25	.30	.35	.40	.45	.50		
2	0	0.8100	0.7225	0.6400	0.5625	0.4900	0.4225	0.3600	0.3025	0.2500	2	
	1	0.1800	0.2550	0.3200	0.3750	0.4200	0.4550	0.4800	0.4950	0.5000	1	
	2	0.0100	0.0225	0.0400	0.0625	0.0900	0.1225	0.1600	0.2025	0.2500	0	2
3	0	0.7290	0.6141	0.5120	0.4219	0.3430	0.2746	0.2160	0.1664	0.1250	3	
	1	0.2430	0.3251	0.3840	0.4219	0.4410	0.4436	0.4320	0.4084	0.3750	2	
	2	0.0270	0.0574	0.0960	0.1406	0.1890	0.2389	0.2880	0.3341	0.3750	1	
	3	0.0010	0.0034	0.0080	0.0156	0.0270	0.0429	0.0640	0.0911	0.1250	0	3
4	0	0.6561	0.5220	0.4096	0.3164	0.2401	0.1785	0.1296	0.0915	0.0625	4	
	1	0.2916	0.3685	0.4096	0.4219	0.4116	0.3845	0.3456	0.2995	0.2500	3	
	2	0.0486	0.0975	0.1536	0.2109	0.2646	0.3105	0.3456	0.3675	0.3750	2	
	3	0.0036	0.0115	0.0256	0.0469	0.0756	0.1115	0.1536	0.2005	0.2500	1	
	4	0.0001	0.0005	0.0016	0.0039	0.0081	0.0150	0.0256	0.0410	0.0625	0	4
5	0	0.5905	0.4437	0.3277	0.2373	0.1681	0.1160	0.0778	0.0503	0.0312	5	
	1	0.3280	0.3915	0.4096	0.3955	0.3601	0.3124	0.2592	0.2059	0.1562	4	
	2	0.0729	0.1382	0.2048	0.2637	0.3087	0.3364	0.3456	0.3369	0.3125	3	
	3	0.0081	0.0244	0.0512	0.0879	0.1323	0.1811	0.2304	0.2757	0.3125	2	
	4	0.0004	0.0022	0.0064	0.0146	0.0283	0.0488	0.0768	0.1128	0.1562	1	
	5	0.0000	0.0001	0.0003	0.0010	0.0024	0.0053	0.0102	0.0185	0.0312	0	5
6	0	0.5314	0.3771	0.2621	0.1780	0.1176	0.0754	0.0467	0.0277	0.0156	6	
	1	0.3543	0.3993	0.3932	0.3560	0.3025	0.2437	0.1866	0.1359	0.0938	5	
	2	0.0984	0.1762	0.2458	0.2966	0.3241	0.3280	0.3110	0.2780	0.2344	4	
	3	0.0146	0.0415	0.0819	0.1318	0.1852	0.2355	0.2765	0.3032	0.3125	3	
	4	0.0012	0.0055	0.0154	0.0330	0.0595	0.0951	0.1382	0.1861	0.2344	2	
	5	0.0001	0.0004	0.0015	0.0044	0.0102	0.0205	0.0369	0.0609	0.0938	1	
	6	0.0000	0.0000	0.0001	0.0002	0.0007	0.0018	0.0041	0.0083	0.0156	0	6
7	0	0.4783	0.3206	0.2097	0.1335	0.0824	0.0490	0.0280	0.0152	0.0078	7	
	1	0.3720	0.3960	0.3670	0.3115	0.2471	0.1848	0.1306	0.0872	0.0547	6	
	2	0.1240	0.2097	0.2753	0.3115	0.3177	0.2985	0.2613	0.2140	0.1641	5	
	3	0.0230	0.0617	0.1147	0.1730	0.2269	0.2679	0.2903	0.2918	0.2734	4	
	4	0.0026	0.0109	0.0287	0.0577	0.0972	0.1442	0.1935	0.2388	0.2734	3	
	5	0.0002	0.0012	0.0043	0.0115	0.0250	0.0466	0.0774	0.1172	0.1641	2	
	6	0.0000	0.0001	0.0004	0.0013	0.0036	0.0084	0.0172	0.0320	0.0547	1	
	7	0.0000	0.0000	0.0000	0.0001	0.0002	0.0006	0.0016	0.0037	0.0078	0	7
8	0	0.4305	0.2725	0.1678	0.1001	0.0576	0.0319	0.0168	0.0084	0.0039	8	
	1	0.3826	0.3847	0.3355	0.2670	0.1977	0.1373	0.0896	0.0548	0.0312	7	
	2	0.1488	0.2376	0.2936	0.3115	0.2965	0.2587	0.2090	0.1569	0.1094	6	
	3	0.0331	0.0839	0.1468	0.2076	0.2541	0.2786	0.2787	0.2568	0.2188	5	
	4	0.0046	0.0185	0.0459	0.0865	0.1361	0.1875	0.2322	0.2627	0.2734	4	
	5	0.0004	0.0026	0.0092	0.0231	0.0467	0.0808	0.1239	0.1719	0.2188	3	
	6	0.0000	0.0002	0.0011	0.0038	0.0100	0.0217	0.0413	0.0703	0.1094	2	
	7	0.0000	0.0000	0.0001	0.0004	0.0012	0.0033	0.0079	0.0164	0.0312	1	
	8	0.0000	0.0000	0.0000	0.0000	0.0001	0.0002	0.0007	0.0017	0.0039	0	8
9	0	0.3874	0.2316	0.1342	0.0751	0.0404	0.0207	0.0101	0.0046	0.0020	9	
	1	0.3874	0.3679	0.3020	0.2253	0.1556	0.1004	0.0605	0.0339	0.0176	8	
	2	0.1722	0.2597	0.3020	0.3003	0.2668	0.2162	0.1612	0.1110	0.0703	7	
	3	0.0446	0.1069	0.1762	0.2336	0.2668	0.2716	0.2508	0.2119	0.1641	6	
	4	0.0074	0.0283	0.0661	0.1168	0.1715	0.2194	0.2508	0.2600	0.2461	5	
	5	0.0008	0.0050	0.0165	0.0389	0.0735	0.1181	0.1672	0.2128	0.2461	4	
	6	0.0001	0.0006	0.0028	0.0087	0.0210	0.0424	0.0743	0.1160	0.1641	3	
	7	0.0000	0.0000	0.0003	0.0012	0.0039	0.0098	0.0212	0.0407	0.0703	2	
	8	0.0000	0.0000	0.0000	0.0001	0.0004	0.0013	0.0035	0.0083	0.0176	1	
	9	0.0000	0.0000	0.0000	0.0000	0.0000	0.0001	0.0003	0.0008	0.0020	0	9
		.90	.85	.80	.75	.70	.65	.60	.55	.50	x	n

p

TABLE C-5 *Binomial probabilities* (Continued)

n	x	.10	.15	.20	.25	.30	.35	.40	.45	.50		
						p						
10	0	0.3487	0.1969	0.1074	0.0563	0.0282	0.0135	0.0060	0.0025	0.0010	10	
	1	0.3874	0.3474	0.2684	0.1877	0.1211	0.0725	0.0403	0.0207	0.0098	9	
	2	0.1937	0.2759	0.3020	0.2816	0.2335	0.1757	0.1209	0.0763	0.0439	8	
	3	0.0574	0.1298	0.2013	0.2503	0.2668	0.2522	0.2150	0.1665	0.1172	7	
	4	0.0112	0.0401	0.0881	0.1460	0.2001	0.2377	0.2508	0.2384	0.2051	6	
	5	0.0015	0.0085	0.0264	0.0584	0.1029	0.1536	0.2007	0.2340	0.2461	5	
	6	0.0001	0.0012	0.0055	0.0162	0.0368	0.0689	0.1115	0.1596	0.2051	4	
	7	0.0000	0.0001	0.0008	0.0031	0.0090	0.0212	0.0425	0.0746	0.1172	3	
	8	0.0000	0.0000	0.0001	0.0004	0.0014	0.0043	0.0106	0.0229	0.0439	2	
	9	0.0000	0.0000	0.0000	0.0000	0.0001	0.0005	0.0016	0.0042	0.0098	1	
	10	0.0000	0.0000	0.0000	0.0000	0.0000	0.0000	0.0001	0.0003	0.0010	0	10
12	0	0.2824	0.1422	0.0687	0.0317	0.0138	0.0057	0.0022	0.0008	0.0002	12	
	1	0.3766	0.3012	0.2062	0.1267	0.0712	0.0368	0.0174	0.0075	0.0029	11	
	2	0.2301	0.2924	0.2835	0.2323	0.1678	0.1088	0.0639	0.0339	0.0161	10	
	3	0.0852	0.1720	0.2362	0.2581	0.2397	0.1954	0.1419	0.0923	0.0537	9	
	4	0.0213	0.0683	0.1329	0.1936	0.2311	0.2367	0.2128	0.1700	0.1208	8	
	5	0.0038	0.0193	0.0532	0.1032	0.1585	0.2039	0.2270	0.2225	0.1934	7	
	6	0.0005	0.0040	0.0155	0.0401	0.0792	0.1281	0.1766	0.2124	0.2256	6	
	7	0.0000	0.0006	0.0033	0.0115	0.0291	0.0591	0.1009	0.1489	0.1934	5	
	8	0.0000	0.0001	0.0005	0.0024	0.0078	0.0199	0.0420	0.0762	0.1208	4	
	9	0.0000	0.0000	0.0001	0.0004	0.0015	0.0048	0.0125	0.0277	0.0537	3	
	10	0.0000	0.0000	0.0000	0.0000	0.0002	0.0008	0.0025	0.0068	0.0161	2	
	11	0.0000	0.0000	0.0000	0.0000	0.0000	0.0001	0.0003	0.0010	0.0029	1	
	12	0.0000	0.0000	0.0000	0.0000	0.0000	0.0000	0.0000	0.0001	0.0002	0	12
15	0	0.2059	0.0874	0.0352	0.0134	0.0047	0.0016	0.0005	0.0001	0.0000	15	
	1	0.3432	0.2312	0.1319	0.0668	0.0305	0.0126	0.0047	0.0016	0.0005	14	
	2	0.2669	0.2856	0.2309	0.1559	0.0916	0.0476	0.0219	0.0090	0.0032	13	
	3	0.1285	0.2184	0.2501	0.2252	0.1700	0.1110	0.0634	0.0318	0.0139	12	
	4	0.0428	0.1156	0.1876	0.2252	0.2186	0.1792	0.1268	0.0780	0.0417	11	
	5	0.0105	0.0449	0.1032	0.1651	0.2061	0.2123	0.1859	0.1404	0.0916	10	
	6	0.0019	0.0132	0.0430	0.0917	0.1472	0.1906	0.2066	0.1914	0.1527	9	
	7	0.0003	0.0030	0.0138	0.0393	0.0811	0.1319	0.1771	0.2013	0.1964	8	
	8	0.0000	0.0005	0.0035	0.0131	0.0348	0.0710	0.1181	0.1647	0.1964	7	
	9	0.0000	0.0001	0.0007	0.0034	0.0116	0.0298	0.0612	0.1048	0.1527	6	
	10	0.0000	0.0000	0.0001	0.0007	0.0030	0.0096	0.0245	0.0515	0.0916	5	
	11	0.0000	0.0000	0.0000	0.0001	0.0006	0.0024	0.0074	0.0191	0.0417	4	
	12	0.0000	0.0000	0.0000	0.0000	0.0001	0.0004	0.0016	0.0052	0.0139	3	
	13	0.0000	0.0000	0.0000	0.0000	0.0000	0.0001	0.0003	0.0010	0.0032	2	
	14	0.0000	0.0000	0.0000	0.0000	0.0000	0.0000	0.0000	0.0001	0.0005	1	
	15	0.0000	0.0000	0.0000	0.0000	0.0000	0.0000	0.0000	0.0000	0.0000	0	15
20	0	0.1216	0.0388	0.0115	0.0032	0.0008	0.0002	0.0000	0.0000	0.0000	20	
	1	0.2702	0.1368	0.0576	0.0211	0.0068	0.0020	0.0005	0.0001	0.0000	19	
	2	0.2852	0.2293	0.1369	0.0669	0.0278	0.0100	0.0031	0.0008	0.0002	18	
	3	0.1901	0.2428	0.2054	0.1339	0.0716	0.0323	0.0123	0.0040	0.0011	17	
	4	0.0898	0.1821	0.2182	0.1897	0.1304	0.0738	0.0350	0.0139	0.0046	16	
	5	0.0319	0.1028	0.1746	0.2023	0.1789	0.1272	0.0746	0.0365	0.0148	15	
	6	0.0089	0.0454	0.1091	0.1686	0.1916	0.1712	0.1244	0.0746	0.0370	14	
	7	0.0020	0.0160	0.0545	0.1124	0.1643	0.1844	0.1659	0.1221	0.0739	13	
	8	0.0004	0.0046	0.0222	0.0609	0.1144	0.1614	0.1797	0.1623	0.1201	12	
	9	0.0001	0.0011	0.0074	0.0271	0.0654	0.1158	0.1597	0.1771	0.1602	11	
	10	0.0000	0.0002	0.0020	0.0099	0.0308	0.0686	0.1171	0.1593	0.1762	10	
	11	0.0000	0.0000	0.0005	0.0030	0.0120	0.0336	0.0710	0.1185	0.1602	9	
	12	0.0000	0.0000	0.0001	0.0008	0.0039	0.0136	0.0355	0.0727	0.1201	8	
	13	0.0000	0.0000	0.0000	0.0002	0.0010	0.0045	0.0146	0.0366	0.0739	7	
	14	0.0000	0.0000	0.0000	0.0000	0.0002	0.0012	0.0049	0.0150	0.0370	6	
	15	0.0000	0.0000	0.0000	0.0000	0.0000	0.0003	0.0013	0.0049	0.0148	5	
	16	0.0000	0.0000	0.0000	0.0000	0.0000	0.0000	0.0003	0.0013	0.0046	4	
	17	0.0000	0.0000	0.0000	0.0000	0.0000	0.0000	0.0000	0.0002	0.0011	3	
	18	0.0000	0.0000	0.0000	0.0000	0.0000	0.0000	0.0000	0.0000	0.0002	2	
	19	0.0000	0.0000	0.0000	0.0000	0.0000	0.0000	0.0000	0.0000	0.0000	1	
	20	0.0000	0.0000	0.0000	0.0000	0.0000	0.0000	0.0000	0.0000	0.0000	0	20
		.90	.85	.80	.75	.70	.65	.60	.55	.50	x	n
						p						

EXAMPLE: For $n = 12$, $p = .25$, and $x = 3$, $P(X = 3) = .2581$. For $n = 15$, $p = .55$, and $x = 10$, $P(X = 10) = .1404$.

TEXT REFERENCE: Use of this table is discussed on p. 149.

Individual Density Function

TABLE C-6 *Poisson probabilities*

Entry is probability $P(X = x) = \dfrac{\lambda^x \exp(-\lambda)}{x!}$

x	λ								
	.1	.2	.3	.4	.5	.6	.7	.8	.9
0	0.9048	0.8187	0.7408	0.6703	0.6065	0.5488	0.4966	0.4493	0.4066
1	0.0905	0.1637	0.2222	0.2681	0.3033	0.3293	0.3476	0.3595	0.3659
2	0.0045	0.0164	0.0333	0.0536	0.0758	0.0988	0.1217	0.1438	0.1647
3	0.0002	0.0011	0.0033	0.0072	0.0126	0.0198	0.0284	0.0383	0.0494
4	0.0000	0.0001	0.0003	0.0007	0.0016	0.0030	0.0050	0.0077	0.0111
5	0.0000	0.0000	0.0000	0.0001	0.0002	0.0004	0.0007	0.0012	0.0020
6	0.0000	0.0000	0.0000	0.0000	0.0000	0.0000	0.0001	0.0002	0.0003

x	λ								
	1.0	1.5	2.0	2.5	3.0	3.5	4.0	4.5	5.0
0	0.3679	0.2231	0.1353	0.0821	0.0498	0.0302	0.0183	0.0111	0.0067
1	0.3679	0.3347	0.2707	0.2052	0.1494	0.1057	0.0733	0.0500	0.0337
2	0.1839	0.2510	0.2707	0.2565	0.2240	0.1850	0.1465	0.1125	0.0842
3	0.0613	0.1255	0.1804	0.2138	0.2240	0.2158	0.1954	0.1687	0.1404
4	0.0153	0.0471	0.0902	0.1336	0.1680	0.1888	0.1954	0.1898	0.1755
5	0.0031	0.0141	0.0361	0.0668	0.1008	0.1322	0.1563	0.1708	0.1755
6	0.0005	0.0035	0.0120	0.0278	0.0504	0.0771	0.1042	0.1281	0.1462
7	0.0001	0.0008	0.0034	0.0099	0.0216	0.0385	0.0595	0.0824	0.1044
8	0.0000	0.0001	0.0009	0.0031	0.0081	0.0169	0.0298	0.0463	0.0653
9	0.0000	0.0000	0.0002	0.0009	0.0027	0.0066	0.0132	0.0232	0.0363
10	0.0000	0.0000	0.0000	0.0002	0.0008	0.0023	0.0053	0.0104	0.0181
11	0.0000	0.0000	0.0000	0.0000	0.0002	0.0007	0.0019	0.0043	0.0082
12	0.0000	0.0000	0.0000	0.0000	0.0001	0.0002	0.0006	0.0016	0.0034
13	0.0000	0.0000	0.0000	0.0000	0.0000	0.0001	0.0002	0.0006	0.0013
14	0.0000	0.0000	0.0000	0.0000	0.0000	0.0000	0.0001	0.0002	0.0005
15	0.0000	0.0000	0.0000	0.0000	0.0000	0.0000	0.0000	0.0001	0.0002

x	λ								
	5.5	6.0	6.5	7.0	7.5	8.0	9.0	10.0	11.0
0	0.0041	0.0025	0.0015	0.0009	0.0006	0.0003	0.0001	0.0000	0.0000
1	0.0225	0.0149	0.0098	0.0064	0.0041	0.0027	0.0011	0.0005	0.0002
2	0.0618	0.0446	0.0318	0.0223	0.0156	0.0107	0.0050	0.0023	0.0010
3	0.1133	0.0892	0.0688	0.0521	0.0389	0.0286	0.0150	0.0076	0.0037
4	0.1558	0.1339	0.1118	0.0912	0.0729	0.0573	0.0337	0.0189	0.0102
5	0.1714	0.1606	0.1454	0.1277	0.1094	0.0916	0.0607	0.0378	0.0224
6	0.1571	0.1606	0.1575	0.1490	0.1367	0.1221	0.0911	0.0631	0.0411
7	0.1234	0.1377	0.1462	0.1490	0.1465	0.1396	0.1171	0.0901	0.0646
8	0.0849	0.1033	0.1188	0.1304	0.1373	0.1396	0.1318	0.1126	0.0888
9	0.0519	0.0688	0.0858	0.1014	0.1144	0.1241	0.1318	0.1251	0.1085
10	0.0285	0.0413	0.0558	0.0710	0.0858	0.0993	0.1186	0.1251	0.1194
11	0.0143	0.0225	0.0330	0.0452	0.0585	0.0722	0.0970	0.1137	0.1194
12	0.0065	0.0113	0.0179	0.0263	0.0366	0.0481	0.0728	0.0948	0.1094
13	0.0028	0.0052	0.0089	0.0142	0.0211	0.0296	0.0504	0.0729	0.0926
14	0.0011	0.0022	0.0041	0.0071	0.0113	0.0169	0.0324	0.0521	0.0728
15	0.0004	0.0009	0.0018	0.0033	0.0057	0.0090	0.0194	0.0347	0.0534
16	0.0001	0.0003	0.0007	0.0014	0.0026	0.0045	0.0109	0.0217	0.0367
17	0.0000	0.0001	0.0003	0.0006	0.0012	0.0021	0.0058	0.0128	0.0237
18	0.0000	0.0000	0.0001	0.0002	0.0005	0.0009	0.0029	0.0071	0.0145
19	0.0000	0.0000	0.0000	0.0001	0.0002	0.0004	0.0014	0.0037	0.0084
20	0.0000	0.0000	0.0000	0.0000	0.0001	0.0002	0.0006	0.0019	0.0046
21	0.0000	0.0000	0.0000	0.0000	0.0000	0.0001	0.0003	0.0009	0.0024
22	0.0000	0.0000	0.0000	0.0000	0.0000	0.0000	0.0001	0.0004	0.0012
23	0.0000	0.0000	0.0000	0.0000	0.0000	0.0000	0.0000	0.0002	0.0006
24	0.0000	0.0000	0.0000	0.0000	0.0000	0.0000	0.0000	0.0001	0.0003
25	0.0000	0.0000	0.0000	0.0000	0.0000	0.0000	0.0000	0.0000	0.0001

TABLE C-6 *Poisson probabilities* (Continued)

	λ								
x	12	13	14	15	16	17	18	19	20
0	0.0000	0.0000	0.0000	0.0000	0.0000	0.0000	0.0000	0.0000	0.0000
1	0.0001	0.0000	0.0000	0.0000	0.0000	0.0000	0.0000	0.0000	0.0000
2	0.0004	0.0002	0.0001	0.0000	0.0000	0.0000	0.0000	0.0000	0.0000
3	0.0018	0.0008	0.0004	0.0002	0.0001	0.0000	0.0000	0.0000	0.0000
4	0.0053	0.0027	0.0013	0.0006	0.0003	0.0001	0.0001	0.0000	0.0000
5	0.0127	0.0070	0.0037	0.0019	0.0010	0.0005	0.0002	0.0001	0.0001
6	0.0255	0.0152	0.0087	0.0048	0.0026	0.0014	0.0007	0.0004	0.0002
7	0.0437	0.0281	0.0174	0.0104	0.0060	0.0034	0.0019	0.0010	0.0005
8	0.0655	0.0457	0.0304	0.0194	0.0120	0.0072	0.0042	0.0024	0.0013
9	0.0874	0.0661	0.0473	0.0324	0.0213	0.0135	0.0083	0.0050	0.0029
10	0.1048	0.0859	0.0663	0.0486	0.0341	0.0230	0.0150	0.0095	0.0058
11	0.1144	0.1015	0.0844	0.0663	0.0496	0.0355	0.0245	0.0164	0.0106
12	0.1144	0.1099	0.0984	0.0829	0.0661	0.0504	0.0368	0.0259	0.0176
13	0.1056	0.1099	0.1060	0.0956	0.0814	0.0658	0.0509	0.0378	0.0271
14	0.0905	0.1021	0.1060	0.1024	0.0930	0.0800	0.0655	0.0514	0.0387
15	0.0724	0.0885	0.0989	0.1024	0.0992	0.0906	0.0786	0.0650	0.0516
16	0.0543	0.0719	0.0866	0.0960	0.0992	0.0963	0.0884	0.0772	0.0646
17	0.0383	0.0550	0.0713	0.0847	0.0934	0.0963	0.0936	0.0863	0.0760
18	0.0255	0.0397	0.0554	0.0706	0.0830	0.0909	0.0936	0.0911	0.0844
19	0.0161	0.0272	0.0409	0.0557	0.0699	0.0814	0.0887	0.0911	0.0888
20	0.0097	0.0177	0.0286	0.0418	0.0559	0.0692	0.0798	0.0866	0.0888
21	0.0055	0.0109	0.0191	0.0299	0.0426	0.0560	0.0684	0.0783	0.0846
22	0.0030	0.0065	0.0121	0.0204	0.0310	0.0433	0.0560	0.0676	0.0769
23	0.0016	0.0037	0.0074	0.0133	0.0216	0.0320	0.0438	0.0559	0.0669
24	0.0008	0.0020	0.0043	0.0083	0.0144	0.0226	0.0328	0.0442	0.0557
25	0.0004	0.0010	0.0024	0.0050	0.0092	0.0154	0.0237	0.0336	0.0446
26	0.0002	0.0005	0.0013	0.0029	0.0057	0.0101	0.0164	0.0246	0.0343
27	0.0001	0.0002	0.0007	0.0016	0.0034	0.0063	0.0109	0.0173	0.0254
28	0.0000	0.0001	0.0003	0.0009	0.0019	0.0038	0.0070	0.0117	0.0181
29	0.0000	0.0001	0.0002	0.0004	0.0011	0.0023	0.0044	0.0077	0.0125
30	0.0000	0.0000	0.0001	0.0002	0.0006	0.0013	0.0026	0.0049	0.0083
31	0.0000	0.0000	0.0000	0.0001	0.0003	0.0007	0.0015	0.0030	0.0054
32	0.0000	0.0000	0.0000	0.0001	0.0001	0.0004	0.0009	0.0018	0.0034
33	0.0000	0.0000	0.0000	0.0000	0.0001	0.0002	0.0005	0.0010	0.0020
34	0.0000	0.0000	0.0000	0.0000	0.0000	0.0001	0.0002	0.0006	0.0012
35	0.0000	0.0000	0.0000	0.0000	0.0000	0.0000	0.0001	0.0003	0.0007
36	0.0000	0.0000	0.0000	0.0000	0.0000	0.0000	0.0001	0.0002	0.0004
37	0.0000	0.0000	0.0000	0.0000	0.0000	0.0000	0.0000	0.0001	0.0002
38	0.0000	0.0000	0.0000	0.0000	0.0000	0.0000	0.0000	0.0000	0.0001
39	0.0000	0.0000	0.0000	0.0000	0.0000	0.0000	0.0000	0.0000	0.0001

EXAMPLE: For $\lambda = 14$ and $x = 8$, $P(X = 8) = .0304$.

TEXT REFERENCE: Use of this table is discussed on p. 152.

TABLE C-7 *Cumulative probabilities of the exponential distribution*

Entry is area *a* under the exponential curve from 0 to *x*(*a*)

λx	.00	.01	.02	.03	.04	.05	.06	.07	.08	.09
0.0	0.0000	0.0100	0.0198	0.0296	0.0392	0.0488	0.0582	0.0676	0.0769	0.0861
0.1	0.0952	0.1042	0.1131	0.1219	0.1306	0.1393	0.1479	0.1563	0.1647	0.1730
0.2	0.1813	0.1894	0.1975	0.2055	0.2134	0.2212	0.2289	0.2366	0.2442	0.2517
0.3	0.2592	0.2666	0.2739	0.2811	0.2882	0.2953	0.3023	0.3093	0.3161	0.3229
0.4	0.3297	0.3363	0.3430	0.3495	0.3560	0.3624	0.3687	0.3750	0.3812	0.3874
0.5	0.3935	0.3995	0.4055	0.4114	0.4173	0.4231	0.4288	0.4345	0.4401	0.4457
0.6	0.4512	0.4566	0.4621	0.4674	0.4727	0.4780	0.4831	0.4883	0.4934	0.4984
0.7	0.5034	0.5084	0.5132	0.5181	0.5229	0.5276	0.5323	0.5370	0.5416	0.5462
0.8	0.5507	0.5551	0.5596	0.5640	0.5683	0.5726	0.5768	0.5810	0.5852	0.5893
0.9	0.5934	0.5975	0.6015	0.6054	0.6094	0.6133	0.6171	0.6209	0.6247	0.6284
1.0	0.6321	0.6358	0.6394	0.6430	0.6465	0.6501	0.6535	0.6570	0.6604	0.6638
1.1	0.6671	0.6704	0.6737	0.6770	0.6802	0.6834	0.6865	0.6896	0.6927	0.6958
1.2	0.6988	0.7018	0.7048	0.7077	0.7106	0.7135	0.7163	0.7192	0.7220	0.7247
1.3	0.7275	0.7302	0.7329	0.7355	0.7382	0.7408	0.7433	0.7459	0.7484	0.7509
1.4	0.7534	0.7559	0.7583	0.7607	0.7631	0.7654	0.7678	0.7701	0.7724	0.7746
1.5	0.7769	0.7791	0.7813	0.7835	0.7856	0.7878	0.7899	0.7920	0.7940	0.7961
1.6	0.7981	0.8001	0.8021	0.8041	0.8060	0.8080	0.8099	0.8118	0.8136	0.8155
1.7	0.8173	0.8191	0.8209	0.8227	0.8245	0.8262	0.8280	0.8297	0.8314	0.8330
1.8	0.8347	0.8363	0.8380	0.8396	0.8412	0.8428	0.8443	0.8459	0.8474	0.8489
1.9	0.8504	0.8519	0.8534	0.8549	0.8563	0.8577	0.8591	0.8605	0.8619	0.8633
2.0	0.8647	0.8660	0.8673	0.8687	0.8700	0.8713	0.8725	0.8738	0.8751	0.8763
2.1	0.8775	0.8788	0.8800	0.8812	0.8823	0.8835	0.8847	0.8858	0.8870	0.8881
2.2	0.8892	0.8903	0.8914	0.8925	0.8935	0.8946	0.8956	0.8967	0.8977	0.8987
2.3	0.8997	0.9007	0.9017	0.9027	0.9037	0.9046	0.9056	0.9065	0.9074	0.9084
2.4	0.9093	0.9102	0.9111	0.9120	0.9128	0.9137	0.9146	0.9154	0.9163	0.9171
2.5	0.9179	0.9187	0.9195	0.9203	0.9211	0.9219	0.9227	0.9235	0.9242	0.9250
2.6	0.9257	0.9265	0.9272	0.9279	0.9286	0.9293	0.9301	0.9307	0.9314	0.9321
2.7	0.9328	0.9335	0.9341	0.9348	0.9354	0.9361	0.9367	0.9373	0.9380	0.9386
2.8	0.9392	0.9398	0.9404	0.9410	0.9416	0.9422	0.9427	0.9433	0.9439	0.9444
2.9	0.9450	0.9455	0.9461	0.9466	0.9471	0.9477	0.9482	0.9487	0.9492	0.9497
3.0	0.9502	0.9507	0.9512	0.9517	0.9522	0.9526	0.9531	0.9536	0.9540	0.9545
3.1	0.9550	0.9554	0.9558	0.9563	0.9567	0.9571	0.9576	0.9580	0.9584	0.9588
3.2	0.9592	0.9596	0.9600	0.9604	0.9608	0.9612	0.9616	0.9620	0.9624	0.9627
3.3	0.9631	0.9635	0.9638	0.9642	0.9646	0.9649	0.9653	0.9656	0.9660	0.9663
3.4	0.9666	0.9670	0.9673	0.9676	0.9679	0.9683	0.9686	0.9689	0.9692	0.9695
3.5	0.9698	0.9701	0.9704	0.9707	0.9710	0.9713	0.9716	0.9718	0.9721	0.9724
3.6	0.9727	0.9729	0.9732	0.9735	0.9737	0.9740	0.9743	0.9745	0.9748	0.9750
3.7	0.9753	0.9755	0.9758	0.9760	0.9762	0.9765	0.9767	0.9769	0.9772	0.9774
3.8	0.9776	0.9779	0.9781	0.9783	0.9785	0.9787	0.9789	0.9791	0.9793	0.9796
3.9	0.9798	0.9800	0.9802	0.9804	0.9806	0.9807	0.9809	0.9811	0.9813	0.9815

λx	0.0	0.1	0.2	0.3	0.4	0.5	0.6	0.7	0.8	0.9
4.0	0.9817	0.9834	0.9850	0.9864	0.9877	0.9889	0.9899	0.9909	0.9918	0.9926
5.0	0.9933	0.9939	0.9945	0.9950	0.9955	0.9959	0.9963	0.9967	0.9970	0.9973
6.0	0.9975	0.9978	0.9980	0.9982	0.9983	0.9985	0.9986	0.9988	0.9989	0.9990
7.0	0.9991	0.9992	0.9993	0.9993	0.9994	0.9994	0.9995	0.9995	0.9996	0.9996
8.0	0.9997	0.9997	0.9997	0.9998	0.9998	0.9998	0.9998	0.9998	0.9998	0.9999
9.0	0.9999	0.9999	0.9999	0.9999	0.9999	0.9999	0.9999	0.9999	0.9999	0.9999

EXAMPLE: For $\lambda = .6$ and $x = 1.3$, $\lambda x = .6(1.3) = .78$ so $P(X \leq 1.3; \lambda = .6) = .5416$.

TEXT REFERENCE: Use of this table is discussed on p. 177.

TABLE C-8 *Table of random digits*

Row	1–5	6–10	11–15	16–20	21–25	26–30	31–35
				Column			
1	13284	16834	74151	92027	24670	36665	00770
2	21224	00370	30420	03883	94648	89428	41583
3	99052	47887	81085	64933	66279	80432	65793
4	00199	50993	98603	38452	87890	94624	69721
5	60578	06483	28733	37867	07936	98710	98539
6	91240	18312	17441	01929	18163	69201	31211
7	97458	14229	12063	59611	32249	90466	33216
8	35249	38646	34475	72417	60514	69257	12489
9	38980	46600	11759	11900	46743	27860	77940
10	10750	52745	38749	87365	58959	53731	89295
11	36247	27850	73958	20673	37800	63835	71051
12	70994	66986	99744	72438	01174	42159	11392
13	99638	94702	11463	18148	81386	80431	90628
14	72055	15774	43857	99805	10419	76939	25993
15	24038	65541	85788	55835	38835	59399	13790
16	74976	14631	35908	28221	39470	91548	12854
17	35553	71628	70189	26436	63407	91178	90348
18	35676	12797	51434	82976	42010	26344	92920
19	74815	67523	72985	23183	02446	63594	98924
20	45246	88048	65173	50989	91060	89894	36036
21	76509	47069	86378	41797	11910	49672	88575
22	19689	90332	04315	21358	97248	11188	39062
23	42751	35318	97513	61537	54955	08159	00337
24	11946	22681	45045	13964	57517	59419	58045
25	96518	48688	20996	11090	48396	57177	83867
26	35726	58643	76869	84622	39098	36083	72505
27	39737	42750	48968	70536	84864	64952	38404
28	97025	66492	56177	04049	80312	48028	26408
29	62814	08075	09788	56350	76787	51591	54509
30	25578	22950	15227	83291	41737	59599	96191
31	68763	69576	88991	49662	46704	63362	56625
32	17900	00813	64361	60725	88974	61005	99709
33	71944	60227	63551	71109	05624	43836	58254
34	54684	93691	85132	64399	29182	44324	14491
35	25946	27623	11258	65204	52832	50880	22273
36	01353	39318	44961	44972	91766	90262	56073
37	99083	88191	27662	99113	57174	35571	99884
38	52021	45406	37945	75234	24327	86978	22644
39	78755	47744	43776	83098	03225	14281	83637
40	25282	69106	59180	16257	22810	43609	12224
41	11959	94202	02743	86847	79725	51811	12998
42	11644	13792	98190	01424	30078	28197	55583
43	06307	97912	68110	59812	95448	43244	31262
44	76285	75714	89585	99296	52640	46518	55486
45	55322	07598	39600	60866	63007	20007	66819
46	78017	90928	90220	92503	83375	26986	74399
47	44768	43342	20696	26331	43140	69744	82928
48	25100	19336	14605	86603	51680	97678	24261
49	83612	46623	62876	85197	07824	91392	58317
50	41347	81666	82961	60413	71020	83658	02415

SOURCE: Excerpt from *Table of 105,000 Random Decimal Digits.* Interstate Commerce Commission, Bureau of Transport Economics and Statistics, May 1949.

TEXT REFERENCE: This table is discussed on pp. 193–194.

TABLE C-9 *Table of random standard normal numbers*

Row	\multicolumn{10}{c}{Column}									
	1	2	3	4	5	6	7	8	9	10
1	0.464	0.137	2.455	−0.323	−0.068	0.296	−0.288	1.298	0.241	−0.957
2	0.060	−2.526	−0.531	−0.194	0.543	−1.558	0.187	−1.190	0.022	0.525
3	1.486	−0.354	−0.634	0.697	0.926	1.375	0.785	−0.963	−0.853	−1.865
4	1.022	−0.472	1.279	3.521	0.571	−1.851	0.194	1.192	−0.501	−0.273
5	1.394	−0.555	0.046	0.321	2.945	1.974	−0.258	0.412	0.439	−0.035
6	0.906	−0.513	−0.525	0.595	0.881	−0.934	1.579	0.161	−1.885	0.371
7	1.179	−1.055	0.007	0.769	0.971	0.712	1.090	−0.631	−0.255	−0.702
8	−1.501	−0.488	−0.162	−0.136	1.033	0.203	0.448	0.748	−0.423	−0.432
9	−0.690	0.756	−1.618	−0.345	−0.511	−2.051	−0.457	−0.218	0.857	−0.465
10	1.372	0.225	0.378	0.761	0.181	−0.736	0.960	−1.530	−0.260	0.120
11	−0.482	1.678	−0.057	−1.229	−0.486	0.856	−0.491	−1.983	−2.830	−0.238
12	−1.376	−0.150	1.356	−0.561	−0.256	−0.212	0.219	0.779	0.953	−0.869
13	−1.010	0.598	−0.918	1.598	0.065	0.415	−0.169	0.313	−0.973	−1.016
14	−0.005	−0.899	0.012	−0.725	1.147	−0.121	1.096	0.481	−1.691	0.417
15	1.393	−1.163	−0.911	1.231	−0.199	−0.246	1.239	−2.574	−0.558	0.056
16	−1.787	−0.261	1.237	1.046	−0.508	−1.630	−0.146	−0.392	−0.627	0.561
17	−0.105	−0.357	−1.384	0.360	−0.992	−0.116	−1.698	−2.832	−1.108	−2.357
18	−1.339	1.827	−0.959	0.424	0.969	−1.141	−1.041	0.362	−1.726	1.956
19	1.041	0.535	0.731	1.377	0.983	−1.330	1.620	−1.040	0.524	−0.281
20	0.279	−2.056	0.717	−0.873	−1.096	−1.396	1.047	0.089	−0.573	0.932
21	−1.805	−2.008	−1.633	0.542	0.250	−0.166	0.032	0.079	0.471	−1.029
22	−1.186	1.180	1.114	0.882	1.265	−0.202	0.151	−0.376	−0.310	0.479
23	0.658	−1.141	1.151	−1.210	−0.927	0.425	0.290	−0.902	0.610	2.709
24	−0.439	0.358	−1.939	0.891	−0.227	0.602	0.873	−0.437	−0.220	−0.057
25	−1.399	−0.230	0.385	−0.649	−0.577	0.237	−0.289	0.513	0.738	−0.300
26	0.199	0.208	−1.083	−0.219	−0.291	1.221	1.119	0.004	−2.015	−0.594
27	0.159	0.272	−0.313	0.084	−2.828	−0.439	−0.792	−1.275	−0.623	−1.047
28	2.273	0.606	0.606	−0.747	0.247	1.291	0.063	−1.793	−0.699	−1.347
29	0.041	−0.307	0.121	0.790	−0.584	0.541	0.484	−0.986	0.481	0.996
30	−1.132	−2.098	0.921	0.145	0.446	−1.661	1.045	−1.363	−0.586	−1.023
31	0.768	0.079	−1.473	0.034	−2.127	0.665	0.084	−0.880	−0.579	0.551
32	0.375	−1.658	−0.851	0.234	−0.656	0.340	−0.086	−0.158	−0.120	0.418
33	−0.513	−0.344	0.210	−0.735	1.041	0.008	0.427	−0.831	0.191	0.074
34	0.292	−0.521	1.266	−1.206	−0.899	0.110	−0.528	−0.813	0.071	0.524
35	1.026	2.990	−0.574	−0.491	−1.114	1.297	−1.433	−1.345	−3.001	0.479
36	−1.334	1.278	−0.568	−0.109	−0.515	−0.566	2.923	0.500	0.359	0.326
37	−0.287	−0.144	−0.254	0.574	−0.451	−1.181	−1.190	−0.318	−0.094	1.114
38	0.161	−0.886	−0.921	−0.509	1.410	−0.518	0.192	−0.432	1.501	1.068
39	−1.346	0.193	−1.202	0.394	−1.045	0.843	0.942	1.045	0.031	0.772
40	1.250	−0.199	−0.288	1.810	1.378	0.584	1.216	0.733	0.402	0.226
41	0.630	−0.537	0.782	0.060	0.499	−0.431	1.705	1.164	0.884	−0.298
42	0.375	−1.941	0.247	−0.491	−0.665	−0.135	−0.145	−0.498	0.457	1.064
43	−1.420	0.489	−1.711	−1.186	0.754	−0.732	−0.066	1.006	−0.798	0.162
44	−0.151	−0.243	−0.430	−0.762	0.298	1.049	1.810	2.885	−0.768	−0.129
45	−0.309	0.531	0.416	−1.541	1.456	2.040	−0.124	0.196	0.023	−1.204
46	0.424	−0.444	0.593	0.993	−0.106	0.116	0.484	−1.272	1.066	1.097
47	0.593	0.658	−1.127	−1.407	−1.579	−1.616	1.458	1.262	0.736	−0.916
48	0.862	−0.885	−0.142	−0.504	0.532	1.381	0.022	−0.281	−0.342	1.222
49	0.235	−0.628	−0.023	−0.463	−0.899	−0.394	−0.538	1.707	−0.188	−1.153
50	−0.853	0.402	0.777	0.833	0.410	−0.349	−1.094	0.580	1.395	1.298

SOURCE: Reprinted from Beyer, W. H., Ed., *Handbook of Tables for Probability and Statistics*, 2nd ed., copyright The Chemical Rubber Company, 1968, p. 484, with the permission of CRC Press, Inc.

TABLE C-10   *Confidence level associated with interval* $L_r \leq \eta \leq U_r$

Entry is confidence level for given n and r

Largest/Smallest

n	r						
	1	2	3	4	5	6	7
2	.500						
3	.750						
4	.875	.375					
5	.938	.625					
6	.969	.781	.312				
7	.984	.875	.547				
8	.992	.930	.711	.273			
9	.996	.961	.820	.492			
10	.998	.979	.891	.656	.246		
11	.999	.988	.935	.773	.451		
12	1.000	.994	.961	.854	.612	.226	
13	1.000	.997	.978	.908	.733	.419	
14	1.000	.998	.987	.943	.820	.576	.209
15	1.000	.999	.993	.965	.882	.698	.393

EXAMPLE: For $n = 7$ and $r = 1$, confidence interval $L_1 \leq \eta \leq U_1$ has confidence coefficient .984.
TEXT REFERENCE: Use of this table is discussed on pp. 374–375.

TABLE C-11   *Percentiles of the* $D(n)$ *distribution*

Entry is $D(a; n)$ where $P[D(n) \leq D(a; n)] = a$

n	a			n	a		
	.90	.95	.99		.90	.95	.99
5	.51	.56	.67	30	.22	.24	.29
10	.37	.41	.49	35	.20	.22	.27
15	.30	.34	.40	40	.19	.21	.25
20	.26	.29	.35	45	.18	.20	.24
25	.24	.26	.32	50	.17	.19	.23
				$n > 50$	$\dfrac{1.22}{\sqrt{n}}$	$\dfrac{1.36}{\sqrt{n}}$	$\dfrac{1.63}{\sqrt{n}}$

SOURCE: Tabulated values adapted by permission from Table 1 of L. H. Miller, "Table of Percentage Points of Kolmogorov Statistics," *Journal of the American Statistical Association*, Vol. 51 (1956), pp. 111–121.
EXAMPLE: $D(.95; 25) = .26$ so $P[D(25) \leq .26] = .95$.
TEXT REFERENCE: Use of this table is discussed on p. 419.

Appendix D
Data Set

This data set contains financial information for 383 firms classified into eight industry groups. Each line contains the data for one firm. The financial information presented is as follows:

Column	Variable
1	1980 net assets
2	1980 net income
3	1980 net sales
4	1981 net assets
5	1981 net income
6	1981 net sales

NOTE: All figures in millions of dollars.

	(1)	(2)	(3)	(4)	(5)	(6)

INDUSTRY - CRUDE OIL PRODUCERS

	(1)	(2)	(3)	(4)	(5)	(6)
1	33.037	2.281	8.498	35.197	2.908	11.537
2	320.156	17.661	370.733	355.848	24.378	384.491
3	307.332	23.718	31.049	321.408	26.656	49.118
4	41.395	2.489	10.226	45.792	3.641	11.559
5	144.090	10.934	84.864	180.534	12.407	101.029
6	46.611	2.585	12.297	60.950	4.389	16.143
7	9.767	0.923	4.705	12.350	0.350	6.124
8	3.252	-0.174	1.098	3.141	-0.155	1.017
9	98.651	2.564	17.522	114.403	3.825	26.788
10	19.605	0.730	6.761	23.518	1.027	6.827
11	21.404	0.572	4.720	20.982	0.809	4.474
12	24.046	0.508	3.730	26.714	0.614	4.001
13	173.007	13.208	54.328	198.311	8.936	60.160
14	98.261	0.563	13.442	93.740	-0.792	13.269
15	6.958	-0.359	2.479	5.708	-0.558	1.521

	(1)	(2)	(3)	(4)	(5)	(6)
16	23.682	-0.907	4.312	23.996	0.502	5.940
17	273.146	13.774	56.038	308.475	14.895	71.194
18	24.983	1.990	11.279	27.505	2.723	13.755
19	43.478	3.444	17.569	52.257	4.490	18.148
20	8.996	-0.176	1.107	14.643	0.270	1.953
21	307.403	11.521	68.764	322.025	20.836	67.442
22	365.377	7.070	40.995	365.423	10.577	48.238
23	17.987	1.064	5.173	27.188	1.631	6.310
24	446.506	31.259	128.664	480.794	37.461	146.655
25	67.860	4.824	17.067	94.029	6.101	20.074
26	321.744	80.528	182.449	547.854	85.031	202.052
27	255.357	13.991	127.806	312.386	17.638	149.302
28	157.964	5.631	41.197	237.078	7.459	45.764
29	18.203	0.099	4.679	23.921	0.190	5.594
30	141.333	17.173	118.166	255.155	20.555	123.335
31	40.345	2.464	5.326	42.460	2.740	5.509
32	20.308	2.028	4.097	23.382	2.966	4.814
33	3483.039	-64.824	3557.532	3458.764	26.561	3673.144
34	32.736	2.368	15.155	41.457	3.339	17.123
35	2085.829	63.693	986.517	2340.733	79.184	1093.102
36	20.862	1.355	4.783	22.203	1.957	11.781
37	31.069	1.420	2.716	33.268	1.808	3.532
38	89.658	1.351	11.331	90.445	0.394	11.260
39	27.925	2.133	9.669	48.831	2.581	11.885
40	730.017	5.802	182.627	772.251	6.900	194.079
41	69.692	1.696	17.690	87.668	2.742	24.122
42	14.672	1.679	5.006	23.909	0.883	4.906
43	32.733	-4.766	7.270	24.743	-0.419	5.346
44	53.121	2.079	9.596	63.963	2.708	10.996
45	3.517	0.020	2.294	3.318	-0.169	2.045
46	17.077	1.264	2.849	24.944	1.354	4.289
47	17.203	-0.026	4.929	22.230	1.580	5.334
48	52.296	-1.500	5.165	62.005	-2.318	6.365
49	59.748	1.933	14.828	60.552	3.020	16.871
50	14.660	0.408	2.529	14.579	0.687	3.282
51	36.516	2.921	9.442	38.493	1.322	9.608
52	316.345	-2.618	34.545	374.205	1.041	40.330
53	4.915	0.248	0.949	3.895	0.579	1.309
54	43.039	-1.887	0.909	47.474	-4.212	1.291
55	44.989	-4.253	34.668	34.052	-0.277	24.866
56	7.587	0.375	2.267	8.077	0.158	2.279
57	67.094	3.470	20.651	65.312	3.742	20.831
58	17.042	0.169	6.398	16.720	0.801	5.216
59	16.315	0.678	6.141	17.071	0.689	8.510
60	49.406	-5.277	22.226	49.457	0.655	21.437

INDUSTRY — TEXTILE PRODUCTS

	(1)	(2)	(3)	(4)	(5)	(6)
61	98.291	7.814	23.778	105.288	9.729	26.808
62	120.361	7.722	194.511	129.735	7.054	206.686
63	86.612	3.178	97.496	80.549	2.732	95.772
64	78.705	3.819	128.598	88.090	3.939	125.086
65	1876.768	54.190	2331.511	1945.958	66.969	2451.761

	(1)	(2)	(3)	(4)	(5)	(6)
66	103.526	3.730	170.813	111.185	5.577	234.989
67	242.555	23.159	393.291	278.809	21.080	441.342
68	271.921	9.486	430.817	283.952	11.119	447.119
69	2227.966	118.818	2208.021	2630.037	188.623	2601.829
70	177.614	6.144	81.555	153.874	6.087	84.332
71	378.621	-3.456	421.747	372.959	5.374	494.932
72	168.253	7.424	190.932	177.291	2.470	184.217
73	35.087	3.366	43.570	45.125	2.701	48.507
74	33.296	3.984	52.414	35.537	1.828	46.413
75	3.325	0.467	5.114	3.484	0.616	6.287
76	17.075	2.749	32.083	21.426	2.294	30.235
77	216.766	10.673	306.821	228.358	10.130	329.509
78	34.459	-0.941	53.361	26.707	0.576	44.423
79	8.806	1.569	10.966	12.289	1.176	15.003
80	37.476	1.667	29.588	50.316	2.095	40.524
81	119.556	5.018	182.525	125.491	6.896	218.244
82	73.948	5.983	61.690	81.329	7.090	69.895
83	40.027	2.716	68.498	40.460	0.274	55.874
84	25.240	0.212	29.230	18.237	-1.393	24.659
85	421.222	12.058	599.505	466.554	10.720	634.442
86	63.540	0.477	76.270	70.280	2.561	94.361
87	48.743	2.615	103.068	54.090	3.059	111.544
88	42.856	1.044	50.015	44.796	1.072	54.776
89	146.614	7.079	236.523	153.471	7.162	250.703
90	180.981	3.816	231.001	181.323	6.205	273.785
91	70.844	2.534	81.564	82.104	3.800	101.269
92	468.790	11.371	439.613	482.852	19.201	538.514
93	55.277	2.275	71.302	54.014	1.929	77.232
94	22.546	3.213	44.455	32.187	2.238	49.048
95	833.932	-0.867	1162.439	869.886	20.994	1279.287
96	171.237	15.247	221.647	201.282	9.659	241.229
97	1100.631	22.140	996.824	1173.978	20.669	1062.485
98	30.260	5.281	40.137	31.820	3.974	38.814
99	7.016	0.828	10.774	11.883	0.441	11.043
100	345.825	8.413	479.578	367.921	13.765	551.020
101	38.295	3.827	61.619	38.497	1.457	56.399
102	19.868	-0.421	20.733	16.748	0.134	22.599
103	145.220	-7.868	152.852	153.824	0.540	156.510
104	19.741	2.003	34.386	23.351	1.675	32.993
105	359.381	22.725	433.403	369.360	22.073	477.104
106	55.337	4.389	89.018	58.428	5.198	101.382
107	12.768	1.098	17.998	15.499	1.962	26.777
108	7.871	1.127	14.630	8.037	1.000	16.386
109	9.796	0.995	26.989	12.007	0.740	23.657
110	23.693	-0.497	39.135	28.103	1.181	50.421
111	23.205	-0.959	20.879	21.246	-0.312	23.698
112	24.434	2.514	41.562	25.793	2.797	41.965

INDUSTRY - TEXTILE APPAREL MANUFACTURERS

	(1)	(2)	(3)	(4)	(5)	(6)
113	52.288	2.218	73.175	67.090	0.680	84.460
114	48.435	1.196	48.429	57.810	1.528	59.755
115	74.096	9.403	88.318	104.981	2.502	78.569

	(1)	(2)	(3)	(4)	(5)	(6)
116	27.379	-1.439	29.088	24.694	-1.898	27.760
117	55.835	4.922	87.476	66.343	5.515	106.024
118	242.812	17.905	396.723	303.727	20.268	465.007
119	148.376	4.791	236.810	155.930	4.385	268.119
120	415.396	15.732	674.024	427.906	18.221	738.672
121	21.657	1.188	39.635	26.793	1.544	57.129
122	18.498	1.903	46.088	21.041	2.076	49.336
123	65.791	0.668	102.888	81.270	0.510	116.853
124	28.219	0.775	35.752	33.504	1.717	44.686
125	149.359	8.139	222.170	136.670	-11.507	210.068
126	30.541	3.245	66.321	32.527	4.520	76.073
127	22.911	0.852	43.243	24.662	1.563	46.969
128	832.097	34.706	1764.240	832.483	17.298	1883.795
129	30.256	2.086	32.864	31.564	2.824	36.671
130	27.655	2.129	50.676	32.011	1.925	56.214
131	216.942	4.570	237.709	229.044	11.120	330.275
132	337.690	13.973	502.230	362.123	19.156	571.204
133	15.891	2.666	30.633	19.193	2.622	32.883
134	10.361	0.864	17.812	17.169	1.458	27.200
135	29.480	0.671	44.515	38.321	0.899	51.431
136	63.427	3.087	102.076	66.513	3.761	112.988
137	305.329	22.288	407.090	331.933	24.886	448.536
138	256.412	8.012	381.683	284.656	9.007	414.912
139	81.581	2.854	124.636	112.385	3.025	157.575
140	81.716	4.925	139.016	97.902	6.113	169.551
141	327.182	26.162	546.893	414.532	33.781	680.541
142	12.766	0.543	13.666	13.032	0.748	15.683
143	16.697	0.895	48.217	24.152	1.705	61.023
144	132.795	5.630	245.480	156.450	6.822	290.168
145	54.738	0.612	105.897	61.331	1.035	118.966
146	55.727	0.258	70.245	50.776	0.618	81.732
147	23.634	1.305	43.990	27.333	1.301	46.182
148	70.405	5.416	108.545	77.150	6.133	123.516
149	8.087	1.841	11.514	10.656	2.612	14.926
150	8.512	1.034	18.546	10.629	0.085	21.148
151	13.585	0.999	22.880	16.143	1.754	25.259
152	5.678	0.190	15.668	6.433	0.679	19.428
153	131.981	7.638	256.439	159.208	10.760	274.423
154	56.938	0.397	78.640	66.077	2.029	102.272
155	227.935	9.580	342.660	232.482	11.943	402.085
156	12.685	-2.021	32.874	19.903	2.144	36.647
157	47.874	-1.069	38.075	52.057	1.156	49.028
158	95.637	6.440	179.816	116.532	7.490	192.254
159	82.075	9.783	154.724	85.915	10.965	172.012
160	88.359	4.057	154.888	100.690	5.634	196.680
161	28.130	2.763	30.108	38.817	3.841	40.127
162	60.762	0.771	85.118	58.273	1.477	87.549
163	12.898	2.438	26.583	13.940	2.300	24.775
164	13.769	1.040	21.777	14.909	0.425	21.395
165	108.794	5.114	85.822	139.760	5.939	84.229
166	19.011	1.143	25.318	25.038	1.457	29.369
167	183.176	19.873	372.702	219.225	23.540	401.209
168	29.655	3.310	38.295	39.998	4.444	50.903
169	212.932	9.343	352.894	233.429	12.458	381.388
170	56.958	2.520	81.279	54.895	2.263	79.415

	(1)	(2)	(3)	(4)	(5)	(6)
171	23.844	2.036	42.379	26.703	2.331	48.497
172	10.981	1.426	21.230	10.976	1.094	21.998
173	9.692	0.535	26.622	13.096	1.230	31.720
174	9.468	2.792	30.912	13.334	2.979	38.223
175	17.057	1.546	42.827	22.557	1.886	55.133
176	25.695	2.485	69.533	25.526	2.137	57.854
177	6.090	0.244	15.926	6.808	0.348	19.719
178	33.182	1.696	55.619	38.151	2.128	63.864
179	3.479	0.733	8.636	4.374	0.822	10.179
180	11.818	0.922	34.672	11.987	0.626	33.195
181	12.220	0.689	27.842	16.128	1.066	36.828
182	14.957	0.377	31.217	15.548	0.306	30.129
183	10.427	0.738	19.093	12.913	1.345	25.784
184	8.240	1.894	26.476	10.938	2.580	31.406
185	41.907	3.740	55.481	42.941	5.017	69.854
186	14.525	2.531	19.825	17.563	2.954	22.887
187	23.706	1.462	51.307	30.623	3.676	62.905

INDUSTRY — PAPER

	(1)	(2)	(3)	(4)	(5)	(6)
188	43.060	2.734	36.353	47.135	3.687	43.569
189	994.853	11.333	879.184	1562.930	32.293	1887.666
190	1391.506	42.559	1330.982	1471.317	59.544	1501.588
191	688.492	14.175	697.108	697.367	23.582	757.052
192	106.703	1.665	60.862	108.282	2.905	64.626
193	595.544	19.236	479.868	622.023	24.566	553.250
194	483.549	4.340	499.886	486.714	6.282	532.888
195	2751.196	93.590	2658.893	2802.900	138.694	2825.996
196	1266.933	42.779	1266.331	1300.139	75.060	1364.198
197	1173.023	31.466	1425.591	1169.280	35.165	1523.867
198	443.204	14.236	477.943	482.408	22.390	509.535
199	1292.187	30.853	1226.328	1387.853	55.782	1367.681
200	1159.126	35.612	1007.854	1171.518	52.115	1098.618
201	30.312	-1.411	36.129	31.779	0.009	41.804
202	29.018	0.855	46.539	31.579	0.894	52.793
203	774.129	35.259	699.699	812.662	52.402	812.175
204	715.082	6.776	581.754	736.833	17.685	637.162
205	71.040	0.905	68.168	67.924	0.049	77.653
206	34.637	0.529	43.365	61.263	1.253	58.621
207	188.390	6.534	188.456	200.197	9.129	204.339
208	140.033	1.328	121.527	138.510	2.707	128.775
209	32.912	1.220	43.876	32.955	1.523	45.754
210	4.529	0.510	7.367	20.428	0.865	34.267
211	196.256	15.668	107.785	221.863	16.562	120.837

INDUSTRY — ELECTRONIC COMPUTER EQUIPMENT

	(1)	(2)	(3)	(4)	(5)	(6)
212	25.632	1.949	32.442	43.231	1.473	42.881
213	23.537	1.002	36.713	25.615	1.787	43.357
214	1.561	-1.052	0.556	7.428	1.161	9.076
215	32.403	2.252	20.462	46.818	5.261	40.937

	(1)	(2)	(3)	(4)	(5)	(6)
216	100.240	0.583	68.700	89.427	1.971	80.715
217	202.692	14.310	198.246	259.761	20.655	253.197
218	67.605	2.660	101.272	73.362	1.558	103.194
219	2947.183	88.722	2627.272	3025.063	103.402	2869.352
220	429.474	-18.076	148.771	427.247	1.611	196.320
221	245.226	-0.635	161.723	318.560	-0.417	193.342
222	18.563	-1.916	13.323	20.268	-5.549	18.988
223	19.358	2.641	28.844	20.193	2.664	30.895
224	2232.655	82.053	2462.315	2484.858	121.577	3009.492
225	22.095	-5.515	4.946	72.295	3.164	35.520
226	29.426	-11.480	17.244	28.844	1.085	21.221
227	127.468	4.134	98.294	183.368	-18.051	90.477
228	4.948	-2.367	2.993	12.141	0.432	17.326
229	6.774	0.207	8.756	6.965	-0.128	7.960
230	15.032	0.166	5.643	18.720	0.458	7.942
231	4.103	-2.654	3.421	6.225	0.398	5.210
232	1.675	-0.170	2.916	5.018	0.425	6.581
233	6.719	-0.089	11.108	8.278	0.150	8.751
234	38.596	-9.167	11.995	54.448	-2.866	39.982
235	25.140	-6.001	5.243	45.206	-7.097	17.661
236	10.176	0.015	15.275	9.651	0.158	17.472
237	10.504	-5.063	4.001	13.862	-2.997	7.304
238	2.608	-2.577	0.244	13.271	-3.406	4.585
239	6.305	-0.721	6.704	5.696	0.363	6.029
240	21.044	-3.411	18.021	29.722	0.397	29.925
241	12.481	-0.004	14.348	19.111	1.239	21.639
242	21.854	-6.966	5.434	39.536	0.649	29.502
243	24.700	1.396	10.855	46.787	2.331	20.521
244	2.627	-1.922	3.133	4.724	0.485	8.428
245	84.004	-0.594	52.551	81.315	0.551	58.084
246	22.095	-5.515	4.946	72.295	3.164	35.520
247	8.431	-1.910	11.278	13.859	1.554	21.128
248	11.413	0.952	25.596	17.487	1.200	29.834

INDUSTRY — ELECTRONICS

	(1)	(2)	(3)	(4)	(5)	(6)
249	53.546	1.040	64.782	51.957	-0.262	63.373
250	184.036	4.031	181.903	187.940	5.932	177.285
251	277.861	32.481	323.525	329.313	44.808	407.816
252	53.838	2.600	66.057	62.830	3.218	78.744
253	107.916	0.377	96.629	74.338	0.486	88.244
254	37.926	0.432	49.600	40.133	0.363	52.207
255	45.367	0.136	54.903	50.225	1.021	53.225
256	204.968	-10.585	260.669	256.404	10.430	302.260
257	401.301	7.347	372.515	444.267	12.485	430.150
258	75.383	0.124	114.122	78.563	1.412	124.160
259	59.889	-1.107	44.851	70.428	0.921	81.252
260	34.603	-0.269	24.867	40.125	0.405	37.557
261	28.712	0.251	43.508	30.841	0.824	46.930
262	42.916	2.754	80.719	72.028	5.021	133.688
263	826.922	47.500	1765.533	852.024	55.584	1977.791
264	134.881	-33.251	197.293	145.097	2.596	200.795
265	6.286	0.510	8.934	7.568	0.447	9.742

	(1)	(2)	(3)	(4)	(5)	(6)
266	30.609	0.833	41.858	33.084	2.480	61.578
267	24.926	-1.374	42.449	27.066	1.567	48.981
268	2.946	0.131	6.263	17.348	1.007	15.119
269	5.177	0.358	6.352	7.451	0.394	8.192

INDUSTRY - ELECTRONIC COMPONENTS

270	491.453	-121.041	383.297	359.448	1.530	346.415
271	9.783	2.141	13.716	13.555	3.042	19.243
272	258.823	16.783	399.929	308.337	21.618	482.695
273	103.572	5.219	119.848	116.888	6.386	133.649
274	76.125	8.100	114.130	91.839	13.427	145.126
275	20.064	0.304	25.379	21.133	0.436	28.065
276	3.056	¬0.524	4.917	3.395	0.135	6.094
277	5.797	0.250	4.926	5.563	0.263	5.584
278	35.080	1.527	33.776	41.402	1.955	37.660
279	9.651	¬0.023	11.004	11.212	0.662	14.495
280	1.773	-0.381	1.743	1.530	-0.347	1.746
281	8.006	¬0.151	5.378	7.277	-0.450	5.463
282	9.466	¬0.576	11.754	10.025	0.259	16.598
283	39.955	¬0.473	46.724	44.133	2.816	58.257
284	17.887	¬1.012	19.080	20.594	0.119	24.358
285	9.280	0.576	14.646	10.831	0.883	18.355
286	47.550	-3.671	46.603	47.781	0.736	59.519
287	516.657	8.370	691.172	492.824	15.976	751.541
288	143.136	7.641	215.780	168.244	9.998	256.349
289	32.473	2.124	41.907	36.204	2.799	54.490
290	5.827	0.209	8.509	5.909	0.447	8.334
291	76.873	1.912	112.952	82.094	3.776	131.262
292	1066.853	41.017	853.079	1061.011	40.784	945.749
293	14.997	0.076	14.318	15.819	0.107	16.286
294	25.923	-0.298	24.091	20.142	-0.078	25.411
295	6.323	0.713	8.705	7.938	1.332	11.734
296	14.228	0.544	20.119	15.215	0.913	22.640
297	104.990	5.530	132.893	120.159	7.270	163.156
298	72.548	0.036	27.895	70.039	1.215	33.645
299	173.356	-10.831	159.120	179.659	0.900	197.973
300	7.470	0.278	8.725	8.586	-0.063	6.373
301	782.829	45.526	1031.749	856.462	64.841	1273.987
302	79.974	1.783	95.707	70.361	2.803	106.215
303	60.445	3.497	41.531	64.317	4.081	46.513
304	115.657	-2.241	89.658	97.123	-1.172	86.283
305	20.165	1.678	18.487	22.599	2.363	26.527
306	46.189	0.869	40.148	45.390	1.763	47.731
307	19.156	0.544	29.685	18.650	1.220	36.424
308	14.436	-0.649	15.104	15.593	0.398	21.765
309	29.106	0.463	42.485	28.966	-2.460	31.948
310	41.592	0.344	35.806	42.399	0.117	38.969
311	11.767	0.872	16.624	13.177	1.269	21.109
312	38.790	2.979	52.419	41.032	4.119	66.423
313	5.176	-4.242	0.942	8.205	-4.928	1.434
314	47.106	-1.933	63.322	48.975	1.654	74.446
315	8.145	0.633	10.328	12.899	0.998	12.196

	(1)	(2)	(3)	(4)	(5)	(6)
316	19.962	-0.556	12.374	29.624	2.673	31.011
317	12.108	-1.935	6.030	10.303	-2.020	7.509
318	13.831	0.041	21.518	15.193	0.867	28.862
319	5.531	0.166	4.906	22.394	3.043	23.941
320	7.756	0.703	8.489	10.240	1.195	13.842
321	12.112	-0.138	12.853	13.072	0.173	13.025
322	14.881	0.616	13.302	16.331	0.790	16.332
323	3.147	-1.080	2.326	8.936	0.554	6.132
324	29.650	0.996	26.993	30.896	1.823	30.811
325	4.872	0.054	5.705	5.319	0.327	6.101
326	4.698	0.189	5.544	5.503	0.493	6.942

INDUSTRY — AUTO PARTS AND ACCESSORIES

	(1)	(2)	(3)	(4)	(5)	(6)
327	26.590	1.233	42.840	24.798	1.520	47.912
328	135.421	4.886	166.394	151.732	5.196	201.787
329	28.351	2.479	55.229	36.032	3.740	78.438
330	163.767	7.584	245.157	177.817	10.103	277.402
331	55.625	6.309	125.118	64.745	6.515	137.080
332	1619.015	56.835	2159.730	1667.520	79.920	2380.186
333	1300.048	63.967	1550.054	1378.921	80.020	1732.303
334	550.678	7.669	729.176	554.082	19.981	905.594
335	19.386	0.339	23.566	21.273	1.813	36.507
336	370.446	47.122	439.957	407.549	53.730	495.959
337	665.819	38.073	858.155	757.080	60.519	1109.792
338	1129.579	77.004	1398.144	1279.626	95.029	1651.144
339	45.738	1.922	52.758	59.208	3.275	68.441
340	271.663	17.948	363.968	318.553	19.668	391.314
341	13.335	1.434	37.071	20.886	1.843	49.414
342	74.743	5.457	82.798	99.174	7.077	93.439
343	408.271	19.320	461.924	540.327	26.291	636.240
344	20.415	0.936	30.617	21.497	1.314	32.713
345	76.976	3.370	109.582	90.557	8.109	154.571
346	195.265	11.321	271.581	213.925	17.285	334.040
347	68.306	3.506	98.653	76.343	5.053	118.558
348	586.732	66.840	716.704	613.967	71.024	802.459
349	179.264	6.973	325.448	218.990	15.256	365.481
350	71.284	5.912	146.322	85.826	5.816	149.899
351	29.886	0.198	31.795	25.449	0.467	32.767
352	132.485	24.014	153.326	166.323	27.626	172.155
353	32.521	0.586	36.569	31.236	0.899	36.162
354	134.451	12.932	240.858	160.434	15.551	295.538
355	293.471	15.574	386.584	329.299	19.229	452.600
356	136.326	3.669	204.375	147.780	6.360	205.760
357	88.918	5.212	145.113	131.481	7.707	210.109
358	5.366	-1.079	8.114	7.156	0.016	11.256
359	44.238	4.294	83.835	49.625	4.618	89.282
360	93.654	3.217	94.851	94.370	2.408	99.816
361	75.159	5.798	99.889	100.290	9.242	133.685
362	166.925	7.289	275.438	184.460	8.555	309.597
363	382.097	17.291	616.742	407.795	13.424	665.240
364	37.384	3.980	74.313	45.877	3.784	83.476
365	40.821	3.195	88.393	54.390	5.249	106.061

	(1)	(2)	(3)	(4)	(5)	(6)
366	44.951	1.839	54.470	48.009	2.600	63.508
367	12.157	1.281	18.538	14.886	1.700	22.333
368	1506.522	90.891	2088.407	1668.217	102.747	2278.139
369	62.097	2.229	146.312	66.840	3.008	175.388
370	567.894	51.458	554.328	594.967	57.116	635.523
371	167.801	10.939	289.663	191.273	13.426	319.323
372	69.428	3.241	111.572	75.824	4.955	126.815
373	10.842	1.933	26.641	15.981	2.643	31.759
374	25.402	0.707	32.731	27.937	1.717	34.171
375	23.277	1.408	43.447	25.307	2.403	47.335
376	46.671	3.430	59.015	53.533	4.732	70.728
377	49.480	4.429	103.741	52.981	4.807	117.489
378	13.272	0.292	32.507	14.407	−0.028	35.084
379	24.986	0.324	30.752	25.325	0.088	31.196
380	64.899	4.386	108.813	73.919	7.440	129.973
381	83.894	9.303	90.550	108.963	14.074	96.532
382	28.883	3.166	59.360	36.202	4.031	72.837
383	8.668	0.205	18.535	8.514	0.108	21.288

Answers to Selected Problems

CHAPTER 2

2.3 a.

Industry	Number of Companies
Crude oil producers	60
Textile products	52
Textile apparel manufacturers	75
Paper	24
Electronic computer equipment	37
Electronics	21
Electronic components	57
Auto parts and accessories	57
Total	383

2.5 a. 15, 25, 25, 30, 50, 50, 50, 50, 50, 60, 65, 70, 75, 75, 80, 80, 90, 90, 100, 100, 100, 100, 100, 125, 150 **b.** (1) 15, 150 (2) 28 percent (3) 50 and 100

2.7 a.

Amount of Donation ($)	Number of Donations
12.5–under 37.5	4
37.5–under 62.5	6
62.5–under 87.5	6
87.5–under 112.5	7
112.5–under 137.5	1
137.5–under 162.5	1
Total	25

2.12 a.

Response Time (*minutes*)	Percent of Alarms
2.5–under 7.5	10
7.5–under 12.5	36
12.5–under 17.5	30
17.5–under 22.5	18
22.5–under 27.5	6
Total	100

2.18 **a.**

Less Than This Time (*minutes*)	Cumulative Percent of Alarms
2.5	0
7.5	10
12.5	46
17.5	76
22.5	94
27.5	100

b. 85 percent

2.22 **a.** (1) 14.3 percent (2) 20.4 percent (3) 37.4 percent (4) retardation

b.

Admit	Number of Patients
No	4,904
Yes	818
Total	5,722

2.25 **a.**

	Sex of Officer		
Age	Male	Female	Total
Under 30	1	2	3
30–39	2	0	2
40 and over	1	0	1
Total	4	2	6

2.28 **a.** Average Weekly Wages and Salaries ($)

16	5	9										
17	2	5	5	6	8	9						
18	2	2	3	5	6	6	7	9	9			
19	1	2	5	7	8	8	9					
20	0	0	1	1	2	3	4	5	5	7	8	9	9	9
21	0	0	0	0	5	5	8					
22	1	1	5	5	7	7						
23	1	1	5	7								
24	5	6										
25	0	3	5									

The class intervals would be: 160–169, 170–179, . . . , 250–259.

b. 19 areas

2.30 **a.** 7.5

b.

Packer:	1	2	3	4	5	6	7	8
Residual:	−1.5	−2.5	+6.5	−2.5	−2.5	−3.5	+6.5	−.5

CHAPTER 3

3.2 **a.** 33, 2.75   **3.7** 8.127

3.9 13.7   **3.12** **a.** 2.67

3.15 3   **3.19** 13.17

3.29 **a.** Array: 15.7, 20.5, 20.8, 20.9, 21.5, 21.8, 22.7, 24.2, 28.4, 29.2 **b.** 20.8, 24.2

3.31 **a.** 9.58, 17.33 **b.** 21.39 **3.34** **a.** 7, 3

3.36 7.75 **3.38** 4.386, 2.09

3.41 28.378, 5.33

3.45 **a.** At most .25. **b.** At least .9375.

3.47 38.9 percent **3.51** A: +1.29, B: +1.78

3.54 **a.** 1.20, 1.40, 1.25 **b.** 1.28059

3.56 342.86

CHAPTER 4

4.3 **a.** $T_1 = \{o_2, o_3, o_4\}$, $T_1^* = \{o_1, o_5\}$ **b.** $T_2 = \{o_1, o_2, o_3, o_4\}$; neither mutually exclusive nor complementary.

4.7 (1) $\{o_1, o_5\}$ (2) $\{o_2, o_3, o_4, o_6, o_7, o_8\}$ (3) $\{o_1, o_3, o_4, o_5, o_6, o_7, o_8\}$ (4) $\{o_5\}$ (5) $\{o_1, o_3, o_4, o_5\}$ (6) $\emptyset$

4.13 **a.** (1) $P(A_1)$ (2) $P(A_1 \cap B_1)$ (3) $P(B_1 | A_1)$ (4) $P(A_1 | B_1)$ **b.** (1) .2 (2) .2 (3) 1 (4) .67

4.14 **b.**

B_i	$P(B_i)$
B_1	.3
B_2	.7
Total	1.0

c.

B_i	$P(B_i | A_2)$
B_1	.125
B_2	.875
Total	1.000

4.18 (1) .3 (2) .1 (3) .2

4.21 **b.** The probability tree takes the following form:

$$P(A_1) = .2 \quad \begin{array}{l} P(B_1 | A_1) = 1 \\ P(B_2 | A_1) = 0 \end{array} \quad \begin{array}{l} P(A_1 \cap B_1) = .20 \\ P(A_1 \cap B_2) = 0 \end{array}$$

$$P(A_2) = .8 \quad \begin{array}{l} P(B_1 | A_2) = .125 \\ P(B_2 | A_2) = .875 \end{array} \quad \begin{array}{l} P(A_2 \cap B_1) = .10 \\ P(A_2 \cap B_2) = .70 \end{array}$$

4.24 **a.** .08, .50, .167

b.

	B_1	B_2	B_3	Total
A_1	.10	.30	.08	.48
A_2	.40	.10	.02	.52
Total	.50	.40	.10	1.00

4.28

		Week 2		
		A	B	C	Total
	A	.09	.18	.03	.30
Week 1	B	.18	.36	.06	.60
	C	.03	.06	.01	.10
	Total	.30	.60	.10	1.00

4.31 .009

4.34 **a.** $P(A_1) = .2$, $P(A_2) = .8$, $P(B_1 | A_1) = 1$, $P(B_1 | A_2) = .125$, $P(A_1 | B_1) = .67$, $P(A_2 | B_1) = .33$

CHAPTER 5

5.2 **b.** (1) .60 (2) .10 (3) .80 (4) .35

5.5 **b.**

x:	0	1	2	3	4
$P(X \leq x)$:	.60	.80	.90	.95	1.00

$P(X \leq 3) = .95$

5.8 (1) .30 (2) .03 (3) .93 (4) .03 (5) .405

5.10 **a.**

x	P(x)
0	.57
1	.36
2	.07
Total	1.00

b.

x	$P(X = x \mid Y = 0)$
0	.541
1	.405
2	.054
Total	1.000

5.12 .75

5.16 **a.** $Y = 28X$ **b.** 81.20

c.

y:	28	56	84	112
$P(y)$:	.1	.3	.2	.4

5.19 1.288, 1.135 **5.22** **a.** .889

5.24 **a.** 854.5, 29.2

5.27 **a.**

y:	−.661	.220	1.101	1.982	2.863
$P(y)$:	.60	.20	.10	.05	.05

b. 0, 1

5.29 **a.** 5.80, 2.18

b.

t:	2	3	4	5	6	7	8
$P(t)$:	.01	.06	.13	.20	.28	.16	.16

c. 0, 1.48

5.32 14.5, 5.45 **5.34** **a.** (1) .917 (2) .083

5.37 **a.** 6.8, 35 **b.** −4.0 **c.** −2.0

5.39 **a.** −.41

5.41 **a.** 1.10, 0, 1.62, .60 **b.** 1.10, 0, 1.22, 1.22

CHAPTER 6

6.3 **a.** $X = 0, 1, 2, 3, 4$ **b.** .3164, .2109, .9492

6.5 (1) .3851 (2) .2355 (3) .2985 (4) .7442

6.7 **b.** 1.0, .750, .866

6.9 **a.** .9321, 1.0

6.13 **a.** .0302, .2158 **b.** 3.5, 3.5 **6.16** **a.** .1954, .5665 **c.** 4, 2

6.20 **a.** (1) .0714 (2) .2143 (3) .3571 **b.** 6.5, 4.03

6.23 **a.** (1) .30 (2) .67 **b.** 1.8, .748

CHAPTER 7

7.1 **a.** $f(x) = 1/50$ **b.** 275, 208.33   **c.** .40, .50
7.3 **b.** (1) $(X - 200)/10$ (2) $X/10$ (3) $X - 5$
7.5 **a.** (1) .5000   (2) .6628   (3) .8554   (4) .9938   **b.** (1) .5000   (2) .0668
(3) .1056   **c.** (1) 0   (2) 1.32   (3) −1.32   **d.** (1) 2.326   (2) −1.645
(3) 1.960
7.7 **a.** (1) .8413 (2) .0918 (3) .8904   **b.** 2,900.50   **c.** 2,699.50 to 2,900.50
7.12 **a.** .005   **b.** (1) .3935 (2) .7135 (3) .1353   **c.** 200
7.15 **a.** Exponential with $\lambda = 3$.   **b.** .333 hour, .5276

CHAPTER 8

8.2 37, 6.95   **8.4** **b.** .34, .624
8.16 **a.** *AB AC AD BC BD CD*, six   **b.** 1/6

CHAPTER 9

9.1 **a.** 1.74, 1.74   **b.** .00949, .00424
9.9 .003
9.11 **a.** 30.303, 2.145, approximately normal   **b.** .9372   **c.** 24.78 to 35.83, 22.49
to 38.12
9.15 **a.** Normal, .9726   **b.** 68.05 to 69.95   **c.** 68.23 to 69.77
9.17 **a.** .40, .663

b.

$\bar{X}$:	0	.5	1	1.5	2
$P(\bar{X})$:	.49	.28	.18	.04	.01

c. .40, .469

CHAPTER 10

10.5 **a.** $s\{\bar{X}\} = .125$ **b.** 3.22 and 3.72, 95.4 percent   **c.** $3.264 \leq \mu \leq 3.676$
d. .90 or 90 percent
10.10 152   **10.13** $107.7 \leq \mu \leq 117.9$
10.16 \$6.53 million $\leq \tau \leq$ \$7.35 million
10.19 **a.** $\mu \geq 3.264$
10.22 **a.** $L(\lambda) = \lambda^3 \exp(-2\lambda)/6$

b.

λ:	0	1	1.5	2	3
$L(\lambda)$:	.0	.0226	.0280	.0244	.0112

CHAPTER 11

11.3 **a.** H_0: $\mu \leq 2.0$, H_1: $\mu > 2.0$ **b.**

μ	$P(H_1; \mu)$	$P(\text{error})$
2.00	.0301	.0301
2.15	.1736	.8264
2.30	.5000	.5000
2.45	.8264	.1736
2.60	.9699	.0301

     c. Maximum $\alpha = .0301$, maximum $\beta = .9699$. **d.** $P(H_0; \mu = 2.5) = .1056$

11.7 **a.** If $\bar{X} \leq 12.735$, conclude H_0; if $\bar{X} > 12.735$, conclude H_1. Conclude H_1.
     b. $P(H_0; \mu = 13.00) = .3156$

11.9 161

11.12 **a.** If $\bar{X} \geq -4.635$, conclude H_0; if $\bar{X} < -4.635$, conclude H_1. Conclude H_1.
     b. $P(H_0; \mu = -5.00) = .4052$

11.15 **a.** H_0: $\mu = 16.50$, H_1: $\mu \neq 16.50$ **b.**

μ	$P(H_1; \mu)$	$P(\text{error})$
13.50	.8849	.1151
15.00	.5082	.4918
16.50	.2302	.2302
18.00	.5082	.4918
19.50	.8849	.1151

     d. (1) $P(H_1; \mu = 16.50) = .2302$ (2) $P(H_0; \mu = 18.00) = .4918$, $P(H_0; \mu = 13.50) = .1151$

11.18 If $13.27 \leq \bar{X} \leq 19.73$, conclude H_0; if $\bar{X} < 13.27$ or $\bar{X} > 19.73$, conclude H_1. Conclude H_0.

11.21 **a.** 2,425 **b.** If $15.60 \leq \bar{X} \leq 17.40$, conclude H_0; if $\bar{X} < 15.60$ or $\bar{X} > 17.40$, conclude H_1. Conclude H_0.

11.23 H_0: $\mu = 10$, H_1: $\mu \neq 10$. If $8.88 \leq \bar{X} \leq 11.12$, conclude H_0; if $\bar{X} < 8.88$ or $\bar{X} > 11.12$, conclude H_1. Conclude H_1.

11.26 **a.** If $z^* \leq 2.326$, conclude H_0; if $z^* > 2.326$, conclude H_1. **b.** $z^* = 2.16$. Conclude H_0.

11.29 **a.** .0154

CHAPTER 12

12.1 **b.** Binomial **c.**

$\bar{p}$	$P(\bar{p})$
0	.0000
1/9	.0000
2/9	.0000
3/9	.0001
4/9	.0008
5/9	.0074
6/9	.0446
7/9	.1722
8/9	.3874
1	.3874

$E\{\bar{p}\} = .90$, $\sigma\{\bar{p}\} = .10$

     e. (1) .3874 (2) .3874 (3) .9470

12.5 **a.** $np = 10$, $n(1 - p) = 40$ **b.** 10, 2.8284 **c.** (1) .0262 (2) .0262 (3) .1428 **d.** (1) .0262 (2) .0262 (3) .1428

12.7 **a.** .40, .03098 **b.** (1) .0537 (2) .8926 **c.** .9988 instead of .8926.

12.10 **a.** .80, .0267 **b.** $.748 \leq p \leq .852$

12.14 246

12.19 **a.** H_0: $p \geq .70$, H_1: $p < .70$ **b.** $\sigma\{\bar{p}\} = .04583$. If $\bar{p} \geq .593$, conclude H_0; if $\bar{p} < .593$, conclude H_1. **c.** $\bar{p} = .65$. Conclude H_0.

12.23 **a.** 124 **b.** If $\bar{p} \geq .604$, conclude H_0; if $\bar{p} < .604$, conclude H_1.

12.28 **a.** $.00067 \leq p \leq .368$ **b.** $p \leq .368$

CHAPTER 13

13.1 **a.** $-14.0 \leq \mu_2 - \mu_1 \leq 54.0$ **13.3** **a.** $33.0 \leq \mu_2 - \mu_1 \leq 123.0$

13.5 **a.** H_0: $\mu_2 - \mu_1 = 0$, H_1: $\mu_2 - \mu_1 \neq 0$. If $-34.0 \leq \bar{Y} - \bar{X} \leq 34.0$, conclude H_0; if $\bar{Y} - \bar{X} < -34.0$ or $\bar{Y} - \bar{X} > 34.0$, conclude H_1. $\bar{Y} - \bar{X} = 20$. Conclude H_0.

13.7 H_0: $\mu_2 - \mu_1 = 0$, H_1: $\mu_2 - \mu_1 \neq 0$. If $-1,035 \leq \bar{Y} - \bar{X} \leq 1,035$, conclude H_0; if $\bar{Y} - \bar{X} < -1,035$ or $\bar{Y} - \bar{X} > 1,035$, conclude H_1. $\bar{Y} - \bar{X} = 1,962$. Conclude H_1.

13.9 $1.82 \leq \mu_2 - \mu_1 \leq 2.12$ **13.11** $1,373 \leq \mu_2 - \mu_1 \leq 2,551$

13.13 H_0: $\mu_2 - \mu_1 \leq 0$, H_1: $\mu_2 - \mu_1 > 0$. If $\bar{D} \leq .149$, conclude H_0; if $\bar{D} > .149$, conclude H_1. $\bar{D} = 1.970$. Conclude H_1.

13.15 H_0: $\mu_2 - \mu_1 \geq 0$, H_1: $\mu_2 - \mu_1 < 0$. If $\bar{D} \geq -.0847$, conclude H_0; if $\bar{D} < -.0847$, conclude H_1. $\bar{D} = -.18$. Conclude H_1.

13.16 **a.** $-.024 \leq p_2 - p_1 \leq .046$

13.18 H_0: $p_2 - p_1 = 0$, H_1: $p_2 - p_1 \neq 0$. If $-.035 \leq \bar{p}_2 - \bar{p}_1 \leq .035$, conclude H_0; if $\bar{p}_2 - \bar{p}_1 < -.035$ or $\bar{p}_2 - \bar{p}_1 > .035$, conclude H_1. $\bar{p}_2 - \bar{p}_1 = .011$. Conclude H_0.

13.20 **a.** $.801 \leq \sigma^2 \leq 1.945$ **b.** $.895 \leq \sigma \leq 1.395$

13.23 H_0: $\sigma^2 = 16.32$, H_1: $\sigma^2 \neq 16.32$. If $9.89 \leq 24s^2/16.32 \leq 45.56$, conclude H_0; if $24s^2/16.32 < 9.89$ or $24s^2/16.32 > 45.56$, conclude H_1. $24(18.9225)/16.32 = 27.8$. Conclude H_0.

13.26 **a.** $1.109 \leq \sigma_2^2/\sigma_1^2 \leq 4.680$ **b.** $1.05 \leq \sigma_2/\sigma_1 \leq 2.16$

13.28 H_0: $\sigma_2^2/\sigma_1^2 \leq 1$, H_1: $\sigma_2^2/\sigma_1^2 > 1$. If $s_2^2/s_1^2 \leq 2.20$, conclude H_0; if $s_2^2/s_1^2 > 2.20$, conclude H_1. $s_2^2/s_1^2 = 1.23$. Conclude H_0.

CHAPTER 14

14.1 $1/\sqrt{1 - (n/N)} = 1.015$. $s\{\bar{X}\}$ is 1.5 percent larger when the finite population correction is omitted.

14.3 $196.1 \leq \mu \leq 203.9$ **14.6** 379

14.9 $.455 \leq p \leq .645$ **14.11** 470

14.13 $452.05 \leq \mu \leq 486.51$

14.17 **a.** 300, 200, 100 **b.** 389, 162, 49

14.23 25, 240

14.25 **a.** Reject the lot. **b.** Accept the lot.

14.27 **a.** UCL = 356, LCL = 314

CHAPTER 15

15.1   **a.** 1   **b.** .273

15.3   **a.** $Md = 6.2$ **b.** $2.1 \le \eta \le 12.5$

15.5   12

15.7 $r = 7$, $67.9 \le \eta \le 1{,}169.3$

15.10   Conclude H_1: $\eta \ne 13$. $\alpha = .006$.

15.13   **a.** H_0: $\eta = 62.50$, H_1: $\eta \ne 62.50$ or, equivalently, H_0: $p = P(X_i > 62.50) = .5$, H_1: $p = P(X_i > 62.50) \ne .5$. If $.448 \le \bar{p} \le .552$, conclude H_0; if $\bar{p} < .448$ or $\bar{p} > .552$, conclude H_1. $\bar{p} = .560$. Conclude H_1.   **b.** Decision rule based on $n = 248$ same to three decimal places. $\bar{p} = .565$. Conclude H_1.

15.15 H_0: $\eta = 13$, H_1: $\eta \ne 13$ or, equivalently, H_0: $p = P(X_i > 13) = .5$, H_1: $p = P(X_i > 13) \ne .5$. If $2 \le B \le 10$, conclude H_0; if $B < 2$ or $B > 10$, conclude H_1. $B = 1$. Conclude H_1.

15.18 H_0: $\delta = 0$, H_1: $\delta \ne 0$. If $152.79 \le S_2 \le 224.21$, conclude H_0; if $S_2 < 152.79$ or $S_2 > 224.21$, conclude H_1. $S_2 = 221$. Conclude H_0.

15.22 $.3 \le \eta_D \le 1.8$

15.23 H_0: $\eta_D = 0$, H_1: $\eta_D \ne 0$ or, equivalently, H_0: $p = P(D_i > 0) = .5$, H_1: $p = P(D_i > 0) \ne .5$. If $4 \le B \le 11$, conclude H_0; if $B < 4$ or $B > 11$, conclude H_1. $B = 12$. Conclude H_1.

15.25 H_0: $\eta_D = 0$, H_1: $\eta_D \ne 0$. If $-69.02 \le T \le 69.02$, conclude H_0; if $T < -69.02$ or $T > 69.02$, conclude H_1. $T = 93$. Conclude H_1.

15.28   **b.** H_0: Sequence generated by a random process, H_1: Sequence generated by a process containing persistence. If $R \ge 25.002$, conclude H_0; if $R < 25.002$, conclude H_1. $R = 23$. Conclude H_1.

15.31   **b.**

i:	4	5
X_i:	6	4
s_{-i}:	3.050	2.915
J_i:	1.148	1.823

$.43 \le \sigma \le 5.38$

CHAPTER 16

16.1   **a.** (1) H_1: The distribution is not Poisson.   (2) 7   (3) 18.48   **b.** (1) H_1: The distribution is not uniform with $a = 2$ and $b = 10$.   (2) 5   (3) 9.24   **c.** (1) H_1: The distribution is not normal.   (2) 9   (3) 16.92

16.3 H_0: The number of calls is Poisson-distributed with $\lambda = 1.5$, H_1: The number of calls is not Poisson-distributed with $\lambda = 1.5$. If $X^2 \le 7.81$, conclude H_0; if $X^2 > 7.81$, conclude H_1. $X^2 = 23.92$. Conclude H_1.

16.6   **a.** H_0: The number of no-shows is Poisson-distributed, H_1: The number of no-shows is not Poisson-distributed. If $X^2 \le 6.25$, conclude H_0; if $X^2 > 6.25$, conclude H_1. $X^2 = 1.64$. Conclude H_0.

16.8 H_0: Service life is normally distributed, H_1: Service life is not normally distributed. If $X^2 \le 4.61$, conclude H_0; if $X^2 > 4.61$, conclude H_1. $X^2 = .67$. Conclude H_0.

16.11   **a, b.** $D(.90; 10) = .37$. Selected values for the cumulative sample function and the confidence band are:

x:	-1	3	6	9	12	15	18	21	25	35
$S(x)$:	.1	.2	.3	.4	.5	.6	.7	.8	.9	1.0
$L(x)$:	0	0	0	.03	.13	.23	.33	.43	.53	.63
$U(x)$:	.47	.57	.67	.77	.87	.97	1.0	1.0	1.0	1.0

16.13 H_0: The distribution for the generating process is $N(0, 1)$, H_1: The distribution for the generating process is not $N(0, 1)$. $D(.90; 10) = .37$. Selected values for the confidence band and the hypothesized cumulative probability function are:

x:	-1.029	$-.957$	$-.869$	$-.298$	$-.238$	.418	.479	.525	.551	1.064
$F(x)$:	.15	.17	.19	.38	.41	.66	.68	.70	.71	.86
$L(x)$:	0	0	0	.03	.13	.23	.33	.43	.53	.63
$U(x)$:	.47	.57	.67	.77	.87	.97	1.0	1.0	1.0	1.0

Decision rule: If $F(x)$ falls entirely within the confidence band, conclude H_0; otherwise conclude H_1. In this case, conclude H_0.

CHAPTER 17

17.1   **a.**   (1) .0450   (2) .0027   **b.**

f_1, f_2, f_3, f_4	$P(f_1, f_2, f_3, f_4)$
1, 1, 0, 0	.30
1, 0, 1, 0	.10
1, 0, 0, 1	.10
0, 1, 1, 0	.06
0, 1, 0, 1	.06
0, 0, 1, 1	.02
2, 0, 0, 0	.25
0, 2, 0, 0	.09
0, 0, 2, 0	.01
0, 0, 0, 2	.01

Outcome $(1, 1, 0, 0)$ is most probable.

17.5   **b.** $.322 \leq p_1 \leq .512$

17.7   **a.** H_0: $p_1 = .35$, $p_2 = .45$, $p_3 = .20$, H_1: The probabilities are not those stated in H_0. If $X^2 \leq 9.21$, conclude H_0; if $X^2 > 9.21$, conclude H_1. $X^2 = 17.54$. Conclude H_1.

     b.

f_i	F_i	$f_i - F_i$
75	63.0	12.0
54	81.0	-27.0
51	36.0	15.0

17.10   **a.** H_0: Price movements in the two weeks are statistically independent, H_1: Price movements in the two weeks are not statistically independent. If $X^2 \leq 7.78$, conclude H_0; if $X^2 > 7.78$, conclude H_1. $X^2 = 77.35$. Conclude H_1.

     b.   Residuals $f_i - F_i$:

		Movement in Second Week		
		Increase	No Change	Decrease
Movement	Increase	15.40	-9.40	-6.00
in First	No Change	-9.12	13.52	-4.40
Week	Decrease	-6.28	-4.12	10.40

17.13 **a.** H_0: The preference patterns in the two study populations are identical, H_1: The preference patterns in the two study populations are not identical. If $X^2 \leq 4.61$, conclude H_0; if $X^2 > 4.61$, conclude H_1. $X^2 = 1.95$. Conclude H_0.
b. $-.025 \leq p_{12} - p_{11} \leq .152$
17.16 $.979 \leq p_1/p_2 \leq 1.971$

CHAPTER 18

18.5 -4
18.7 **a.** .7888 **b.** 17.5 **c.** (1) 1.8 (2) .5793
18.10 **b.** $\hat{Y} = 72.776 - 1.7726X$ **18.11** 160, 25.283
18.13 **b.** 39.1
18.18 **a.**

i:	1	2	3	4	5	6	7	8
e_i:	10.575	-3.875	-4.975	-4.425	$-.725$	.875	$-.025$	2.575

18.20 **a.**

Source	SS	df	MS
Regression	134.717	1	134.717
Error	25.283	6	4.214
Total	160.000	7	

b. σ^2, σ **d.** .8420, $-.918$

CHAPTER 19

19.1 **a.** 32.006, .8069 **b.** $30.03 \leq E\{Y_h\} \leq 33.98$
19.6 **a.** $26.61 \leq Y_h \leq 37.40$ **19.9** $-2.540 \leq \beta_1 \leq -1.005$
19.12 $H_0: \beta_1 = 0, H_1: \beta_1 \neq 0$. If $-.7671 \leq b_1 \leq .7671$, conclude H_0; if $b_1 < -.7671$ or $b_1 > .7671$, conclude H_1. $b_1 = -1.7726$. Conclude H_1. Alternatively, if $|t^*| \leq 2.447$, conclude H_0; if $|t^*| > 2.447$, conclude H_1. $t^* = -5.654$. Conclude H_1.
19.16 $-8.225 \leq \beta_0 \leq 16.875$ **19.18** -4.104
19.20 $H_0: \beta_1 = 0, H_1: \beta_1 \neq 0$. If $F^* \leq 5.99$, conclude H_0; if $F^* > 5.99$, conclude H_1. $F^* = 31.97$. Conclude H_1.
19.24 See residuals in answer to Problem 18.18a.

CHAPTER 20

20.1 **a.** $E\{Y\} = -5 + .15X_1 - .08X_2, 6$
20.6 **a.** $\hat{Y} = -1.707 + .01681X_1 - 3.507X_2$ **b.** 16.010

c.

Source	SS	df	MS
Regression	6,444.79	2	3,222.4
Error	55.14	29	1.901
Total	6,499.93	31	

$R^2 = .9915$
d. 1.901

20.9 **a.** (1) .900, .700 (2) .900, .894 **b.** No

20.11 .7130, −.844

20.15 H_0: $\beta_1 = \beta_2 = 0$, H_1: Not both $\beta_k = 0$ $(k = 1, 2)$. If $F^* \leq 5.42$, conclude H_0; if $F^* > 5.42$, conclude H_1. $F^* = 1{,}695$. Conclude H_1.

20.18 **a.** H_0: $\beta_1 = 0$, H_1: $\beta_1 \neq 0$. If $|t^*| \leq 2.878$, conclude H_0; if $|t^*| > 2.878$, conclude H_1. $t^* = 4.479$. Conclude H_1. **b.** $.718 \leq \beta_2 \leq 1.440$

20.20 **a.** $15.50 \leq E\{Y_h\} \leq 16.52$ **b.** $13.14 \leq Y_h \leq 18.88$

20.24 **a.**

i:	1	2	3	4	5	6	7	8	9	10	11	12
e_i:	−.8	−.7	−.7	.3	.8	−.3	1.7	.2	.3	−.8	−1.0	1.0

b. $\widehat{Y} = 4.084 + 5.8478X − .10736X^2$. A fragment of the data setup for computer input follows.

i	X_i	X_i^2	Y_i
1	5	25	30
2	10	100	51
...	...	...	...
12	6	36	36

c. H_0: $\beta_2 = 0$, H_1: $\beta_2 \neq 0$. If $|t^*| \leq 2.262$, conclude H_0; if $|t^*| > 2.262$, conclude H_1. $t^* = −2.803$. Conclude H_1.

20.27 **a.** (1) 42.26 (2) 38.55 **b.** $2.14 \leq \beta_2 \leq 5.30$

CHAPTER 21

21.5 **a.** (1) 3 (2) 24 (3) 27 **b.** (1) 3 (2) 76 (3) 79 **c.** (1) 4 (2) 11 (3) 15

21.7 **a.** (1) 15 (2) 5 (3) 78 (4) 78 (5) 81 **b.** Programmed text.

c.

Source	SS	df	MS
Treatments	130	2	65.00
Error	460	12	38.33
Total	590	14	

21.9 **a.** H_0: $\mu_1 = \mu_2 = \mu_3$, H_1: Not all μ_j $(j = 1, 2, 3)$ are equal. $F(2, 12)$. **b.** If $F^* \leq 3.89$, conclude H_0; if $F^* > 3.89$, conclude H_1. $F^* = 1.696$. Conclude H_0.

21.11 **a.** $398 \leq \mu_2 \leq 422$ **b.** $153 \leq \mu_3 − \mu_2 \leq 187$

21.14 $−.77 \leq \mu_2 − \mu_4 \leq .17$, $.03 \leq \mu_1 − \mu_4 \leq .97$

21.17 **a.**

i	$j = 1$	$j = 2$	$j = 3$
1	1	2	8
2	−3	−9	0
3	9	0	−5
4	−8	−1	4
5	1	8	−7

21.19 **a.** $Y_i = \beta_0 + \beta_1 X_{i1} + \beta_2 X_{i2} + \epsilon_i$, $i = 1, 2, \ldots, 15$

b. A fragment of the data setup for computer input follows.

X_1	X_2	Y
0	0	86
...	...	...
0	0	86
1	0	80
...	...	...
1	0	86
0	1	88
...	...	...
0	1	73

c. H_0: $\beta_1 = \beta_2 = 0$, H_1: Not both $\beta_k = 0$ ($k = 1, 2$). If $F^* \leq 3.89$, conclude H_0; if $F^* > 3.89$, conclude H_1. $F^* = 1.696$. Conclude H_0. d. $79.0 \leq \beta_0 \leq 91.0$
e. $-15.5 \leq \beta_1 \leq 1.5$

CHAPTER 22

22.2

			A_j	
S_i	$P(S_i)$	0	1	2
0	.4	0	−15	−30
1	.5	−110	−90	−105
2	.1	−220	−200	−180

22.7   a. A_1, 200

b.

		A_j		
S_i	1,000	1,500	2,000	2,500
Above average	300	200	100	0
Average	100	0	50	100
Below average	0	50	100	150

A_3 is the minimax regret act.

22.9   a. A_2, 270   b.   Yes, no

22.12   15.6, 2.8

22.17   b.   Optimal strategy is to undertake research and development and, if this alternative proves unsuccessful, to negotiate a merger arrangement. 80.6.

22.19   a.   Risk-averse   b.   (1) $EU = .4688$ (2) $EU = .5100$. Prefers $30 thousand for certain.

CHAPTER 23

23.1   a. A_1, 1,300   b.   1,450, 150

23.3   a.

$(6, K)$:	$(6, 0)$	$(6, 3)$	$(6, 6)$	$(6, -1)$
$EP(6, K)$:	1,302	1,325	1,300	500

b. $(6, 1)$, 1,371   c. A_1

23.6   15

23.8 a.

n:	0	100	200	500	1,000
$BNEP(n)$:	0	56	78	87	83

$n = 500$ leads to the largest expected payoff net of sampling costs.

b.

n:	0	100	200	500	1,000
$EVSI(n)$:	0	63	86	98	99
$ENGS(n)$:	0	56	78	87	83

23.12 a.

S_i:	0	.1	.2	.3	.4
$P(S_i \mid X_0)$:	.6261	.2695	.0525	.0361	.0158

b. A_1, 1,872

CHAPTER 24

24.1 a.

Year:	1962	'63	'64	'65	'66	'67	'68	'69	'70	'71
M.A.:		3.63	3.70	3.43	3.17	3.17	3.40	3.43	3.43	3.57

Year:	'72	'73	'74	'75	'76	'77	'78	'79	'80	'81
M.A.:	4.03	4.23	4.50	4.53	4.60	4.50	4.27	4.03	3.80	

24.3 3.42, 3.43
24.7 a. $T_t = 3.29 + .0514286X_t$, $X_t = 1$ at 1962, X in one-year units. **c.** 51,429 metric tons **d.** 4.6786, 5.380

24.8

Year:	1962	'63	'64	'65	'66	'67	'68	'69	'70	'71
$C \cdot I$:	98.8	109.1	113.2	100.1	81.8	86.1	95.9	97.3	85.3	92.0

Year:	'72	'73	'74	'75	'76	'77	'78	'79	'80	'81
$C \cdot I$:	103.7	117.7	103.6	119.7	115.7	104.6	108.1	94.9	84.4	88.0

24.11 a. $T'_t = 3.305272 + .01569859X_t$, $X_t = 1$ at 1967, X in one-year units.
24.12 a. 3.7 percent **b.** 4,808.8, 20,416.5
24.15 a.

Year:	1962	'63	'64	'65	'66	'67	'68	'69	'70	'71
C:			100.6	98.1	95.4	92.2	89.3	91.3	94.8	99.2
I:			112.5	102.1	85.7	93.3	107.4	106.5	89.9	92.7

Year:	'72	'73	'74	'75	'76	'77	'78	'79	'80	'81
C:	100.5	107.3	112.1	112.3	110.3	108.6	101.5	96.0		
I:	103.2	109.7	92.4	106.6	104.9	96.3	106.5	98.9		

24.20 **b.**

Quarter	1976	1977	1978	1979	1980	1981
1		73.6	79.1	71.6	76.9	73.6
2		95.7	93.8	104.8	101.8	103.3
3	133.4	135.6	134.6	127.4	127.7	
4	96.2	91.1	95.6	94.1	93.3	

c.

Quarter:	1	2	3	4
Index:	73.1	101.1	132.4	93.4

24.24 **a.**

Quarter:	1	2	3	4
Deseasonalized sales:	4.10	4.01	4.13	3.81

b. 3.16

24.26 **a.**

Quarter:	1	2	3	4
Forecast:	170	235	308	217

b. 229.8, 919.3

24.28 **a.**

t:	2	3	4	5	6	7	8
F_t:	636.2	634.5	643.9	647.5	650.5	650.3	653.0

b.

t:	2	3	4	5	6	7	8
F_t:	635.3	635.0	637.3	638.8	640.5	641.4	643.0

CHAPTER 25

25.1 **a.** 305.63, −13.63 **b.** (1) 131.29, .211 (2) 87.44, −.105 **c.** 232.14
 d. .421, yes

25.3 **a.** $215.63 \leq Y_{26} \leq 248.65$ **b.** H_0: $\beta_4 = 0$, H_1: $\beta_4 \neq 0$. If $|t^*| \leq 2.093$, conclude H_0; if $|t^*| > 2.093$, conclude H_1. $t^* = 9.101$. Conclude H_1.

25.8 **b.** $\hat{Y}'_t = .66145X'_t$

25.10 **a.** 18.643, 10 **b.** $.366 \leq \beta_1 \leq .957$ **c.** 188.7

CHAPTER 26

26.1 **a.**

Year:	1977	1978	1979	1980	1981
(1) Honeysuckle:	100.0	104.2	116.7	125.0	137.5
Lilac:	100.0	106.3	112.5	125.0	134.4
(2) Honeysuckle:		104.2	112.0	107.1	110.0
Lilac:		106.3	105.9	111.1	107.5

26.3 **a, b.**

Year	Aggregate Cost	Index
1978	10,350	100.0
1979	10,852	104.9
1980	11,639	112.5

26.6 a.

Year	Index
1978	100.0
1979	104.9
1980	112.5

26.12

Year:	1969	'70	'71	'72	'73	'74	'75
Index:	72.3	74.5	75.2	76.0	78.0	79.1	81.1

Year:	'76	'77	'78	'79	'80	'81
Index:	83.8	89.5	97.3	100.0	100.5	104.8

26.14

Year:	1976	'77	'78	'79	'80	'81
Index:	90.9	100.9	100.0	109.5	119.0	136.9

26.16 a, b.

Year:	1975	1976	1977	1978	1979
Constant-dollars earnings:	166.29	171.38	177.67	177.29	171.90
Percent relatives:	100.0	103.1	106.8	106.6	103.4
Link relatives:		103.1	103.7	99.8	97.0

26.22 a.

Year	Index
1979	100.0
1980	122.0
1981	135.6

APPENDIX B

B.1 a. 4.61, .211
　　b. 2, 2
　　c. No
B.3 a. 15.51
　　b. 117.9
B.5 a. 2.015, −2.015
　　b. 0, 1.29
B.8 a. 1.341, −1.721
　　b. −2.326
B.10 a. 13.27, .0101
　　b. 1.67, 3.73
　　c. No
B.12 a. 3.23
　　b. .132

Index